ASTRONOMY:
THE COSMIC JOURNEY

A cliff on the planet Mars. This cliff forms one side of gorge that was cut by flowing water, probably more than two billion years ago. The cliff, not unlike parts of our own Grand Canyon, shows layers of blocky rock (top), a slope of debris, and the canyon floor with deposits of sand dunes (bottom). This photo symbolizes several themes of cosmic exploration. First, it shows that we can find recognizable Earth-like features on at least some other planets. Second, it exemplifies how astronomy interfaces with other disciplines, such as planetary geology, in unraveling the history of the universe. Third, it reveals the type of location where we may some day be able to look for evidence of life on other planets. Layers in such Martian cliffs may include rocks that were created billions of years ago when Mars had a more habitable, water-rich environment. Do such rock layers contain fossils of ancient alien organisms? (Mars Global Surveyor photo MO7-01689: NASA, Jet Propulsion Lab, and Malin Space Science Systems).

Astronomy: The Cosmic Journey

SIXTH EDITION

WILLIAM K. HARTMANN
Planetary Science Institute
(Science Applications International Corporation)
Affiliate Faculty, University of Arizona
and University of Hawaii at Hilo

CHRIS IMPEY
Associate Professor, Steward Observatory,
University of Arizona

Australia • Canada • Mexico • Singapore • Spain • United Kingdom • United States

Sponsoring Editor: *Keith Dodson*
Marketing Team: *Kelly McAllister and Laura Hubrich*
Editorial Assistant: *Faith Riley*
Production Editor: *Tom Novack*
Production Service: *Matrix Productions*
Manuscript Editor: *Pat Lewis*
Permissions Editor: *Sue Ewing*
Technology Project Manager: *Sam Subity*
Project Development Editor: *Marie Carigma-Sambilay*
Indexer: *Jeanne Busemeyer*
Cover Design: *Ray R. Neuhaus*
Cover Illustration: *William K. Hartmann*
Interior Illustration: *Precision Graphics*
Photo Researcher: *Linda R. Rill*
Print Buyer: *Nancy Ponziera*
Typesetting: *Thompson Type*
Printing and Binding: *Von Hoffman Press*

For more information about this or any other Brooks/Cole product, contact:
BROOKS/COLE
511 Forest Lodge Road
Pacific Grove, CA 93950 USA
www.brookscole.com
1-800-423-0563 (Thomson Learning Academic Resource Center)

Printed in the United States of America

10 9 8 7 6 5 4 3 2 1

Library of Congress Cataloging-in-Publication Data
Hartmann, William K.
Astronomy : the cosmic journey/ William K. Hartmann, Chris Impey.—6th ed.
p. cm.
Includes bibliographical references and index.
ISBN 0-534-38249-5
1. Astronomy. I. Impey, Chris. II. Title
QB45.2.H37 2002
520—dc21 2001052654

ABOUT THE AUTHORS

William K. Hartmann is a senior scientist at the Planetary Science Institute in Tucson. He was the first winner of the Carl Sagan Medal from the American Astronomical Society, for communicating planetary science to the public. He is known for research on the formation and evolution of planetary systems, discovering the giant multiring lunar impact basins, coauthoring the most widely accepted theory of lunar origin, and codiscovering "asteroid" Chiron's cometary outburst in 1988. He has participated in NASA's Mariner 9 and Mars Global Surveyor missions to Mars. His paintings of planetary landscapes and other astronomical subjects have been published, collected internationally, and commissioned by the NASA Fine Arts Program. In addition to *Astronomy: The Cosmic Journey,* he is the author of *Moons and Planets* (Wadsworth) and co-author of *The Universe Revealed* (Brooks/Cole). Hartmann has collaborated on five popular, illustrated books on astronomy: *The Grand Tour, Out of the Cradle, Cycles of Fire, The History of Earth,* and a Russian-American astronomical art collection, *In the Stream of Stars.* He has also published a book of his text and photos on the Sonoran Desert, *Desert Heart,* and a novel of Mars exploration, *Mars Underground.* Asteroid 3341 was named after him in 1987.

Chris Impey is a University Distinguished Professor and Deputy Department Head at Steward Observatory of the University of Arizona. He is an experienced teacher of undergraduate and graduate astronomy courses, and has won seven teaching awards at the college and university level. He has active research programs on quasars and ultrafaint galaxies, which have been supported by numerous grants from the National Science Foundation and NASA. He has published more than 120 research articles in the astronomical literature and has edited two conference proceedings. Impey has used most of the world's largest telescopes in Hawaii, California, Arizona, Russia, Chile, and Australia. His research has also involved satellite observations from X-ray to far infrared wavelengths. He has had fifteen projects approved and funded with the Hubble Space Telescope. For five years, Impey was Associate Director of the NASA Arizona Space Grant Consortium, which supports a variety of educational and outreach programs. He has also codirected a successful Masters program for high school science teachers at the University of Arizona. Impey is lead author of the Brooks/Cole textbook *The Universe Revealed.* He is heavily involved in the development of new teaching technologies for astronomy, including interactive web-based tools, experts systems, and methods for delivering content to wireless devices.

Brief Contents

Contents

Preface

Ever since the publication of the first edition in 1978, *Astronomy: The Cosmic Journey* has provided hundreds of thousands of students an innovative introduction to the universe. It emphasized the long-term march of science, and helped introduce the comparative planetology approach for studying the solar system and placing our planetary system in the context of stellar evolution. It also pioneered a comprehensive art program that mixed professional photos, amateur photos, and scientifically accurate paintings of astronomical subjects. In this new edition, we take yet another radical step forward, based on three trends that have occurred since the first edition.

First, the color photography in texts has dramatically increased, driving up production costs. Ours was the first astronomy text to present an integrated package of professional photos, amateur photos, paintings, and diagrams to illustrate the nature of the cosmos. But competition among publishers to pack texts with color has gotten out of hand, in terms of costs to students.

Second, at the same time, the explosion of astronomical knowledge has expanded successive editions of most texts into encyclopedic dinosaurs, leading critics to protest correctly that they are written as if the goal were to train all students to be professional astronomers. This expansion of detailed content has also driven up the price of most recent editions to realms as astronomical as the content. A few attempts to remedy this with short texts have led to skimpy treatments of important subjects.

Third, the World Wide Web has entered the picture. Everyone talks about it, but no one is sure how to integrate it into the world of textbooks and teaching. Not all students have easy access yet, and even with easy access, there is a need for a basic print document that can be carried and studied in any environment.

A TWENTY-FIRST-CENTURY TEXT

The philosophy of our response to these factors is simple: We let paper do what paper does best, which is to provide a transportable summary of the basic material along with simple black and white reference illustrations; at the same time, we let the Web do what it does best, which is to provide the low-cost medium for color, animated, and interactive teaching materials. The result is a twenty-first-century teaching package: a basic black-and-white textbook with access to a rich, Web-based teaching package. This cuts the cost to students well below that of alternative "dinosaur" editions, while actually producing more and richer teaching material.

We present this book with the belief that cosmic awareness plays an interesting role in the world today. With human footprints on the Moon, telescopes in orbit, announcements of planets around other stars, radio telescopes listening for messages from alien creatures (who may or may not exist), technicians looking for celestial and planetary sources of energy to support our civilization, and rapid evolution of the socio-political structure of our planet, an astronomy book published today enters a world different from the one that greeted books a generation ago. Astronomy has broadened to involve our basic circumstances and our future in the universe. Astronomy offers adventure for all people—an outward exploratory thrust that may one day be seen as an alternative to mindless consumerism, ideological bickering, and wars whose roots lie in attempts to control dwindling resources on a closed, finite Earth.

Today's astronomy students not only seek an up-to-date summary of astronomical facts; they ask, as people have asked for ages, about our basic relationships to the rest of the universe. They may study astronomy partly to seek points of contact between science and other human endeavors: philosophy, religion, history, politics, environmental action, and even the arts. Science fiction writers and the special effects artists on recent films help today's students realize that the unseen worlds of space are real places—not abstract concepts. Today's students are citizens of a more real, more vast cosmos than that conceptualized by students of a decade ago.

THE STRUCTURE OF OUR BOOK

We've retained the basic structure of the earlier editions. We begin with the viewpoint of ancient people on Earth and work outward across the universe. This method of organization automatically (if loosely) reflects the order of humanity's discoveries about astronomy and provides a unifying theme of increasing distance and scale.

The subtitle thus refers to three separate cosmic journeys that we undertake simultaneously. First, we travel forward through historical time, where we see how humans slowly and sometimes painfully evolved our present picture of the universe. Second, we journey outward through space, where we see how our expanding frontiers have revealed the geography of the universe. Beginning in the Earth-Moon system, we move outward to our local system of planets, the nearby stars, our own enormous galaxy, and finally the encompassing universe of other galaxies. Finally, we travel back through vast reaches of cosmic time. Familiar features of the Earth are typically only a few hundred million years old. The solar system is about 4.6 billion years old. Our galaxy is roughly 12 to 15 billion years old. The universe itself began (or began to reach its present form) an estimated 15 billion years ago.

Because astronomy touches many areas of life and philosophy, we allow the text to encompass a wide range of relevant topics, including space exploration, financing of science, cosmic sources of energy, the checkout counter's barrage of astrology and other pseudoscience, and the possibility of life on other worlds, as well as the conventional "hard science" of Martian rocks, stellar energy sources, and galactic structure. This variety of topics shows how basic scientific research touches all areas of life, and lets readers ponder the relation between science and priorities in our society.

USING OUR BOOK: TIPS FOR TEACHERS

The arrangement of text material into eight parts and 28 chapters should give instructors some flexibility in tailoring a course according to their interest. For example, those who are not interested in the early development of astronomical ideas could use Part A only as optional reading. It would also be possible to omit some of the interesting but challenging material on the early universe contained in Chapter 27.

Each part gives some historical background, describes recent discoveries and theories, and then discusses advances that might occur if society continues to support research. This more or less chronological approach has several purposes. Since there is often a certain logic to the order in which discoveries were made, the historical emphasis helps students remember not only the facts, but also the evidence for the facts. Second, a historical discussion allows us to introduce basic concepts in a more interesting way than simply reciting definitions. Our lives are richer if we realize that some seemingly modern concepts descend from historical struggles, or may have thousand-year-old roots. Third, there is a widespread fallacy that the only progress worth mentioning is that of the last few decades. Astronomy, of all subjects, shows clearly that (to paraphrase Newton) we see as far as we do because we stand on the shoulders of past generations. Exploration of the universe is a continuing human enterprise. The Copernican Revolution is a 500-year struggle to see ourselves not as the center of the universe, but as a part of something larger than ourselves—a struggle still playing itself out in our world. As we try to maintain and improve our civilization, that is an important lesson for a science course to teach.

Another principle we have followed is to treat astronomical objects in an *evolutionary* way, to show the sequence of development of matter in the universe. Stars, pulsars, black holes, and other celestial bodies are linked in evolutionary discussion, rather than listed as different types of objects detected by different observational techniques. We have also not hesitated to mention nonscientific approaches to cosmology and evolution, such as "creationist" and fundamentalist concepts encouraged in various American states' school systems by their state legislatures but later thrown out after courtroom battles.

MERGING THE TEXT WITH THE WEB

As noted above, this edition of *Astronomy: The Cosmic Journey* is more than just a paper-based textbook. Recognizing that the Web is an ideal platform for supplying

- the latest in dazzling four-color photos from telescopes and satellites,
- illuminative animations to help explain inherently dynamical astronomical and physical phenomena, and
- study materials that allow students to instantly receive feedback on their understanding of important concepts,

we have included a 12-month access code to Web Tutor© with each copy of the book. Powered by either Web CT© or Blackboard©, the leading course management systems in education today, the *Astronomy: The Cosmic Journey* Web Tutor provides a complete version of the text, including full-color graphics and numerous related animations; a wealth of quizzing and problem material; and the ability for instructors to customize their course and keep the course current by drawing on the richness of the Web.

★ This symbol is the link to materials on the Web Tutor Web site. When this symbol appears in the text, it means there is related material on the site. This may be more specific detail on the subject of the adjacent paragraph, or passages of broader enrichment text, or more pictorial material, usually in color. Often, we put one key black-and-white illustration in the text, but added a number of color illustrations on the same topic on the Web site. The ability to understand the colors and browse among various examples is of great pedagogic value. For a deeper level of study, or perhaps to find material for term papers, students can examine the extra text passages.

MATH TREATMENT

In the main part of the text, mathematics is almost nonexistent. The book can thus be used for a descriptive course. Ten basic equations are distributed through the book in optional boxes. The general content of each box is included qualitatively in the text, but the boxes introduce a higher level of physics and math, allowing a more quantitative course to be taught. Sample calculations using these equations now appear in every box, and the *Advanced Problems* at the ends of chapters use the optional equations.

The ten optional basic equations are introduced in the text as needed. Teachers offering a more quantitative course can integrate them into the course work; teachers offering more descriptive courses may skip over them. The ten boxes discuss:

I. *The Small-Angle Equation,* useful for calculating apparent sizes of objects at known distances.

II. *Newton's Universal Law of Gravitation,* illustrating the simplicity of gravitational attraction between bodies throughout the universe.

III. *Calculating Circular and Escape Velocities,* useful for deriving speeds or masses in co-orbiting systems (planetary, binary star, galactic).

IV. *Measuring Temperatures of Astronomical Bodies: Wien's Law,* which shows how radiation measurements can reveal the temperatures of distant objects.

V. *The Definition of Mean Density,* a simple concept for gaining information about the nature of material inside planets and stars.

VI. *Typical Velocities of Atoms and Molecules in a Gas,* by which we characterize temperature, as well as collision energies when the atoms or molecules smash into each other. These energies, in turn, control the types of chemical or nuclear reactions that can occur.

VII. *The Doppler Effect: Approach and Recession Velocities,* which shows how spectral measures can reveal radial velocities of distant objects.

VIII. *The Stefan-Boltzmann Law: Rate of Energy Radiation,* which shows how temperature and luminosity measurements can reveal sizes of radiating sources.

IX. *The Hubble Law and the Age of the Universe,* which shows how the relationship between the distance and recession velocity of galaxies can lead to an estimate of the age of the universe.

X. *The Relativistic Redshift,* a modification of equation VII, which explains phenomena that occur at high velocities close to the velocity of light.

ILLUSTRATIONS AND OTHER FEATURES

1. In addition to the classic large telescope photos and recent NASA photos, we have included three other categories of illustrations:

a. Photos from recently published research papers, kindly provided by various authors and institutions.

b. Photos by amateurs with small and intermediate instruments, often used to show sky locations of well-known objects in the large-telescope photos. These can help readers to visualize and locate these objects in the sky, a difficult task if based on classic large-telescope photos alone. Photographic data provided with many of these pictures may be used in setting up student projects in sky photography.

c. Scientifically realistic paintings that show how various objects might look firsthand to observers in space. Discussion of features shown in the paintings illustrates a synthesis of scientific data from various sources.

2. Key concepts are shown in **boldface** type. These are repeated in *Concepts* lists at the end of chapters and defined in a *Glossary* at the end of the book.

3. Limited numbers of references to technical and non-technical sources appear in the text. They are there partly to help students and teachers find more material for projects and partly to help instructors emphasize that statements should be verifiable. Non-technical sources are included in the *References* section; others are in the Instructor's Manual.

END-OF-CHAPTER MATERIALS

1. *Chapter Summaries* review basic ideas of the chapter and sometimes synthesize material from several preceding chapters.

2. *Concepts* lists include the important concepts appearing in **boldface** in the text. Reviewing the Concepts is a good way for the student to review the content of each chapter.

3. *Problems* are aimed at students with nonmathematical backgrounds.

4. *Advanced Problems* usually involve simple arithmetic or algebra and are often applications of the ten basic equations. These can be omitted in nonmathematical courses.

5. *Projects* are intended for class use where modest observatory or planetarium facilities are available. The intent is to get students to do astronomical observing or experimenting.

SUPPLEMENTARY MATERIAL

1. *Appendices* include *Powers of 10* and *Units of Measurement. Supplemental Aids in Studying Astronomy* are listed in the Instructors' Manual.

2. The *Glossary* defines all terms included in the *Concepts* lists, as well as other key terms.

3. The *References* section includes nontechnical references useful for student papers; widely available journals and magazines are emphasized in this group, including most astronomy articles appearing in *Scientific American* in recent years. Technical references are listed in the Instructor's Manual.

4. The *Index* includes names and terms.

5. *Star Maps* for the seasons are found after the index. Since more detailed, larger maps are usually available in classrooms and laboratories, these star maps have been simplified, emphasizing the plane of the solar system and the plane of the galaxy, and indicating major constellations mentioned or illustrated in the text.

ACKNOWLEDGMENTS

Our thanks go to many people who helped produce this book. We have tried to incorporate as many of their suggested corrections and improvements as possible; final responsibility for weaknesses remains ours.

For the Fifth Edition they were, from the University of Arizona: John Bieging, John Black, Adam Burrows, Rob Kennikutt, Jim Liebert, Don McCarthy, Ed Olszewski, Marcia Rieke, Gary Schmidt, and Ray White; Catherine O'Sullivan, Tucson, AZ; Virginia Trimble, University of Maryland; Gary Mechler, Pima Community College; and Arthur W. Wiggins, Oakland Community College. We would particularly like to thank Mike Shull, University of Colorado, for an incisive review of the second half of this book. We also thank a wide circle of colleagues, especially in Tucson and at the University of Hawaii, for helpful discussions.

We thank the additional reviewers for this Sixth Edition: Jeff Brown, Washington State University; Susan Hartley, University of Minnesota, Duluth; T. J. Keefe, Community College of Rhode Island; Brad Peterson, Ohio State University; Eileen Ryan, New Mexico Highlands University; and John Safko, University of South Carolina.

Thanks also to the staffs of Steward Observatory (University of Arizona) and the Planetary Science Institute for making our work easier. We also acknowledge the help and inspiration of our students at the University of Arizona and the University of Hawaii.

In locating and providing photographs, Dale Cruikshank, Walter Feibelman, Steven Larson, Alfred McEwen, Alan Stockton, Don Strittmatter, Alan Toomre, and Dave Webb were helpful. Don Bane and Jurrie Van der Woude (Jet Propulsion Laboratory), David Moore and Nigel Sharp (National Optical Astronomy Observatory), Margaret Weems and Pat Smiley (National Radio Astronomy Observatory), and David Malin (Anglo-Australian Observatory) gave kind assistance with their institutional files. We especially thank Chesley Bonestell, Ron Miller, Pamela Lee, Andrei Sokolov, and many other members of the International Association of Astronomical

Artists for stimulating discussions about paintings of astronomical objects. Floyd Herbert and Mike Morrow helped in making several wide-angle photos with clock-driven mounts.

The staff at Wadsworth–Brooks/Cole has worked hard to create a useful and beautiful product, and made the job a pleasure at the same time. We thank Sam Subity, and especially Keith Dodson, for working with us on the design of the new Web site materials. Thanks to Vernon Boes, Marie Carigma-Sambilay, Sue Ewing, Roy Neuhaus, Tom Novack, Nancy Panziera, and Faith Riley, as well as Merrill Peterson and Jaye Caldwell at Matrix Productions, for the exceptional work on production and design, cheerful dispositions, hospitality, and care of the subject, not to mention care for the harried authors.

William K. Hartmann
Chris Impey
Tucson

Invitation to the Cosmic Journey

If you awoke one day to discover that you had been put on a strange island, your first project after getting food and water would be to try to find out where you were. This is just what has happened to all of us. We have all been born on an island in space: the Earth. We are all passengers on a cosmic journey. Astronomy is the process of finding out where we are—our place in space and our position in the unfolding history of the universe.

This definition of astronomy would not always have satisfied people. At one time we thought we knew where we were: at the center of the universe, with the Sun, stars, planets, and satellites all revolving around us. Modern astronomy contradicts that old theory. The definition is not quite complete, though, because it doesn't really explain what astronomers do.

OUR DEFINITION OF ASTRONOMY

Oddly enough, as science progresses, astronomy becomes harder to define. When astronomy was restricted to observations by earthbound viewers, it was easily defined as "the science of objects in the sky." But now that we have actually begun to explore space, other disciplines have become involved. Is the astronaut who chips a rock sample off the Moon practicing astronomy? What about the researcher who studies the sample once it reaches the Earth? What of the nuclear physicist who measures properties of nuclear reactions going on at the center of stars? What about the biologist interested in life on Mars? Their work should be included in our definition of astronomy.

There is a popular misconception that as various scientists pursue new research, knowledge proliferates into an endless complex of specialized disciplines. Indeed, students are usually taught this way, with advanced courses probing into narrower and narrower specialties. According to this idea, research is like climbing a tree, starting near the trunk and proceeding to ever more specialized branches of knowledge.

But science can also be compared to working our way *down* the tree of knowledge. At first we see a bewildering variety of seemingly unrelated phenomena—twigs of knowledge. After analyzing these twigs, we see that they meet in a branch. As workers map details on the branch, someone with vision may discover that it is joined to another branch. As we work our way toward the trunk, more and more branches join. By grasping the relations between the major branches close to the trunk, we can better understand the higher branches and twigs. This is why different specialties are connecting with astronomical research; astronomy is becoming more general.

A historical example illustrates this process. Around 1600, Galileo and others conducted many experiments to understand seemingly unrelated types of motion, such as the motion of falling objects and the motion of the Moon. The wide variety of motions and accelerations seemed impossible to explain in a simple way. By about 1680, however, Isaac Newton realized that they were connected by simple relationships between mass, acceleration, and gravitational attraction. As a result of Newton's insight, today's college students can learn to understand motions better than the Galileos of the past!

In the same way, students can grasp the relationship between the Earth and the cosmos better than the explorers who mapped astronomical frontiers in the past. Better than any generation in human history, we can begin to sense where we are, at least in relation to other physical objects in the universe.

This book treats astronomy not as a narrow set of academic observational results or abstruse physical laws, but as a voyage of exploration with practical effects on humanity. A broad view of the major discoveries of the last years, the last decades, and even the last centuries is at least as important as the latest factual detail, because the broad view helps us understand how scientific research will continue to affect us and perhaps offer us new options for living. In an era of energy crises, food shortages, and environmental threats, when badly applied technology has created such horrors as biological weapons and hydrogen bombs, educated people are called on to make judgments about their own actions—the kinds of jobs they have, the materials they consume, and the actions of organizations they work for. In short, we are called to judge what kind of civilization should continue. Thus we all need to know how our world is affected by its cosmic surroundings and to understand the scientific and social procedures available for obtaining and extending that knowledge.

Therefore, this book is organized according to the following definition of **astronomy:**

Astronomy is the study of all matter and energy in the universe, emphasizing the concentration of this matter and energy in evolving bodies such as planets, stars, and galaxies. It recognizes that we observers—humanity—are part of the universe, and that our home, the Earth, is only one of the many places in the universe, but also the special point from which our voyage of exploration has started.

The reason for emphasizing humanity is that astronomy influences how we think about ourselves and our role in the universe. The growing perception of environmental problems—for example, the thinning of the ozone layer—has shown that we must look at the Earth as a single astronomical body in order to survive. As the poet Archibald MacLeish said after one of the Apollo flights, we have now seen the Earth as "small and blue and beautiful" and ourselves as riders on the Earth together." Thus this book regards astronomy as including space flights and studies of rocks from our neighbor planets, just as much as telescopic observations of remote galaxies.

A SURVEY OF THE UNIVERSE

The ensuing chapters start with subjects on Earth and then move outward through space. Before beginning, however, we give a brief preview of the universe, starting at the largest scale.

The **universe** is everything that exists. To the best of our knowledge, the universe consists of untold thousands of clusters of galaxies (Figure I-1). **Galaxies** are swarms of billions of stars. Galaxies differ in form. Some are football shaped, some irregular. Our galaxy, like many others, is a disk with about a hundred billion stars arrayed in curving spiral arms. Most galaxies, including ours, are surrounded by a halo of **globular star clusters,** each a spheroidal mass of hundreds of thousands of stars. If we made a model of our galaxy big enough to cover North America, Earth would not be as big as a basketball, or even a BB. It would be about the size of a large molecule.

Scattered inside our galaxy, mostly in the spiral arms, are groupings of stars called **open star clusters** and clouds of dust and gas called **nebulae.** Individual stars are scattered at random. Each **star** is an enormous ball of gas, mostly hydrogen. In the center of a typical star, atoms are jammed together so closely that their nuclei interact to create the heat and light of the star.

The **Sun** is 3 hundred million billion kilometers (200,000,000,000,000,000 mi) from the center of our galaxy and 40 thousand billion kilometers from the nearest star. A hydrogen sphere more than a million kilometers in diameter, the Sun is the center of the **solar system**—the small system made up of the Sun and its family of orbiting rocky and icy **planets.** Circling around most planets are smaller bodies called **satellites.** Between planets are various other small bodies called **asteroids** and **comets.**

About 150 million kilometers (km) from the Sun is the **Earth.** The **Moon,** Earth's satellite, is only 384,000 km away. The Earth itself is nearly 13,000 km across, a tiny speck among myriads of larger and smaller specks in the universe.

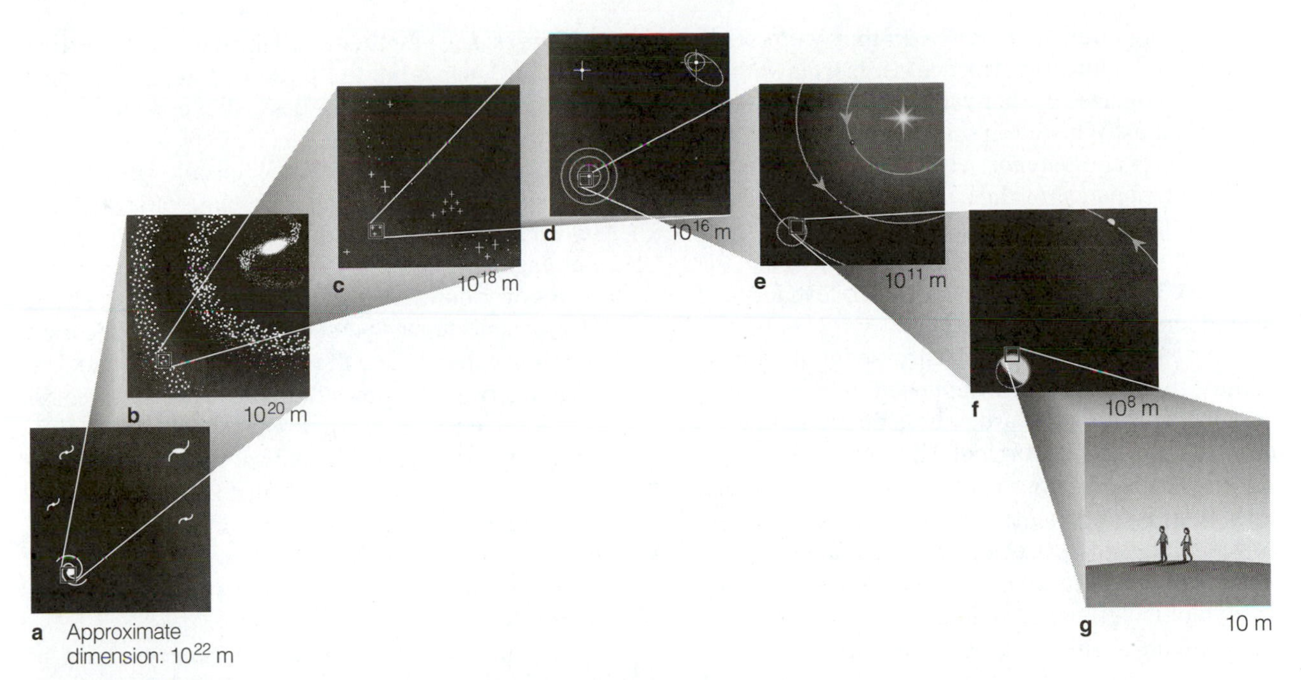

Figure I-1 A journey through the universe from large-scale structure of galaxies to the human scale. The change in scale from one picture to another ranges from a factor of 100 to a factor of 100,000. In cartoon form, **a** shows galaxies; **b** and **c,** parts of our galaxy; **d,** our solar system and the nearest star; **e,** part of our solar system; **f,** the Earth-Moon system; and **g,** human observers. Scale in meters is given in terms of powers of 10.

A WORD ABOUT MATHEMATICS

Our preceding survey of the universe shows how hard it is to express huge astronomical quantities in ordinary units. For reason the system of expressing numbers as **powers of 10** is very useful. For example, instead of being laboriously written as the cumbersome number 300,000,000,000,000,000 km, the distance from the Sun to the galactic center becomes simply 3×10^{17} km; the distance from the Earth to the Sun is 1.5×10^{8} km. A reader uncertain about this system should consult Appendix 1 at the end of the book. We will use this system occasionally, as convenience dictates.[1]

When using any equation to calculate physical quantities, you must remember to use a consistent set of units. For example, the English system of inches, ounces, and so on must not mix in the same equation with the metric system. The metric system, long used by scientists, is becoming universal, and we will use it here, often giving English equivalents. The metric system is *much* easier to use than the English system because its units come in multiples of 10. One centimeter = 10 millimeters, 1 meter = 100 centimeters, and so on. Gone are complicated conversions like 1 mile = 5280 feet. Specifically, in this book we will give preference to the **SI metric system**—the units adopted in the Systeme International (International System), such as meters, kilograms, and so on. In some cases, astronomical units such as *parsecs* will be explained and adopted. These units of measurement are summarized in Appendix 2.

A NOTE ABOUT NAMES OF PEOPLE

One thing that distinguishes science from the pseudoscience now so popular is that scientific assertions are backed by evidence. Because much of this evidence has already been published elsewhere, it is not always discussed in detail, but is merely referenced. This is usually done by listing the name of the author and the date of publication (for example, "Smith, 1988"); the reference is then listed in the back of the book so that a reader who wants more information can locate the source. With so many wildly distorted

[1]We also use the term *billion* here to mean 1,000,000,000, or 10^{9}, which is the usual American meaning. The British use *billion* to mean 1,000,000,000,000, or 10^{12}.

claims being published today as if they were scientific, this referencing of evidence is important in helping you recognize whether you have reliable material in your hands. Whether in science, business, or politics, readers and listeners should learn to demand documentation of assertions and, if in doubt, look up the documents to judge their quality. Aside from promoting this intellectual tradition, we believe that the extensive bibliography will be useful to teachers and to students preparing term papers.

Students are sometimes distressed to find these names, thinking they are supposed to learn names and dates. The names are only to document assertions, not to be memorized. The emphasis in this book is on the exciting nature of the universe that astronomy is revealing, not on the astronomers themselves. Many of the Greek and Renaissance astronomers mentioned in Chapters 2 and 3 are major figures in our Western intellectual heritage, however, and students should learn about them.

A HINT ON USING THIS BOOK

When reading a text, many students underline everything that seems important. We and our publishers have taken care of part of that job by putting terms that represent important concepts in boldface and then listing them in order at the end of the chapter. A useful study procedure is to read a chapter through and then go over the concept list. Try to define each term and use it in a sentence. If you have trouble with a term, go back and check it in the chapter or look it up in the glossary. If you are comfortable with all the terms in each concept list, you will have gone a long way toward mastering the material and will be able to enjoy many articles about astronomy in newspapers and magazines.

FACE TO FACE WITH THE UNIVERSE

The universe is immense; we are perhaps audacious even to attempt studying it. Yet something noble in the human spirit urges us to seek knowledge of the universe and our place in it.

Throughout history, scientific discoveries in the heavens have sometimes illuminated and sometimes conflicted with religions, metaphysical, and philosophical conceptions. Persecution, legal action, even wars have resulted from the challenge to old beliefs by new knowledge. In this book we present scientific theories and the *evidence* for them with the understanding that our knowledge is limited and slowly improving. New evidence is what usually distinguishes new knowledge from old superstition.

Astronomy gives us a new cosmic perspective from which many conflicts of the past, often founded on amazingly few facts, now seem inconsequential. For example, we have given up the medieval arguments over how many angels can dance on the head of a pin, and stopped persecuting those who advocate the theory that the Earth moves around the Sun. Unfortunately, world news demonstrates that we still seem unable to give up warring over ideologies, whether Christian vs. Muslim vs. Jew, Protestant vs. Catholic, or capitalist vs. socialist. Astronomy, however, unifies humanity by showing us that we are all in this together, facing the unknown—a word that aptly describes much of the universe, its past, and its future.

The universe is no abstract concept; it consists of real, physical places. This book offers descriptions of those places, often with photos or scientifically realistic paintings to show how we think they look.

Suppose that we were able to explore these places and that in all these vast reaches, among all the stars and galaxies, we found no one else—no living creatures anywhere except ourselves. Or suppose that we discovered nonhuman intelligent life—alien civilizations, perhaps advanced and incomprehensible or perhaps very simple. Our contacts with them could alter our history far more profoundly than the first contacts between Native Americans and Europeans in 1492.

These are two wildly different possibilities. One is likely to be true. Either boggles the mind. We have just barely entered the century in which we may be able, if research continues, to decide which scenario is more realistic.

CONCEPTS

astronomy
universe
galaxy
globular star cluster
open star cluster
nebula
star
Sun
solar system
planet
satellite
asteroid
comet
Earth
Moon
powers of 10
SI metric system

PROBLEM

1. Scientific discoveries are sometimes described as conflicting with religious or philosophical beliefs. Before continuing this book, examine your present beliefs or expectations and comment on which of the following views best match your own:

a. There are concepts in astronomy that conflict with my own religious or philosophical views.

b. Astronomical concepts are consistent with my religious or philosophical views; there is no problem.

c. If there is a conflict, scientific hypotheses will evolve and eventually become consistent with my religious or philosophical views.

d. Science measures the reality of the physical world only; there is no need for this to be consistent with an inner psychological or spiritual reality.

e. Science tries to reflect how all nature behaves; thus science and a valid religious philosophy of daily life should be consistent with each other if they are to contribute to a well-integrated personality.

PART 1

The Early Discoveries

Prehistoric "standing stones" erected around 2000 B.C. in what is now the English village of Avebury symbolize early humanity's attempts to interact with the cosmos. Similar constructions nearby at Stonehenge are aligned on the Sun's position at sunrise on the date of the summer solstice. (Photo by WKH)

CHAPTER 1

Prehistoric Astronomy: Origins of Science and Superstition

WHAT THE READER SHOULD WATCH FOR IN THIS CHAPTER

Astronomy arose from prehistoric humanity's practical needs to measure or predict various phenomena associated with the sky, such as the time of day, the date, the phases of the Moon, and the beginnings of the seasons. A number of prehistoric ruins, such as Stonehenge from around 2400 B.C., show the sophistication of ancient techniques for observing the seasons. At the same time, there was interest in predicting unusual events such as eclipses or the unusual prominence of specific planets. Hand in hand with these practical efforts, superstitions arose that various patterns in nature could control human events. The superstition that patterns in the sky control human events is called astrology. Astrology along with astronomical knowledge arose in many ancient societies. Today we know that eclipses occur when the Earth and Moon cast shadows on each other as they move around each other. ■

In 1906 the American astronomer Percival Lowell wrote:

Smoke from multiplying factories . . . has joined with electric lighting to help put out the stars. These concomitants of an advancing civilization have succeeded above the dreams of the most earth-centered in shutting off sight of the beyond, so that today few city-bred children have any conception of the glories of the heavens which made of the Chaldean shepherds astronomers in spite of themselves.

This observation seems even more apt today. It is difficult to realize how important the sky once was in daily affairs. In addition, we have forgotten how much we depend on ideas that, although they seem obvious today, were developed only through centuries of struggle to understand nature.

There are several reasons to review the ancient discoveries. First, they were often very basic. We forget how hard they were to recognize. For example, we think it obvious that the Earth moves round the Sun, but early scholars were severely criticized for saying so. It's important to know how these arguments unfolded and how they were won.

Second, discoveries were often made just when society reached a point where they *could* be made, because the needed technical devices or theoretical concepts had just been invented. Therefore, a historical approach helps place important concepts in a logical order.

Third, it's interesting to discover that many common ideas today come from astronomical traditions of the past. For example, although we favor a decimal numerical system, we have 24 hours in a day instead of 10, and 360 degrees in a circle instead of 100. As we will see in this chapter, the reason comes from ancient astronomy and mathematics.

THE EARLIEST ASTRONOMY: MOTIVES AND ARTIFACTS (c. 30,000 B.C.)

Archaeology and anthropological studies of present-day primitive tribes[1] shed light on the earliest astronomical practices. Imagine yourself an intelligent hunter of 30,000 y (years) ago. Agriculture is not yet practiced. You have to depend on hunting and gathering for food. You've been taught that certain celestial cycles are important. You want to know when berries on the mountain will ripen or when certain migratory birds will arrive at a nearby lake, and such knowledge requires knowledge of the seasons, best derived from sky phenomena. When you travel far, you want to be able to find your way home, and celestial objects are your only reliable guides in unfamiliar landscapes.

If you are a woman, you also want to know in what season the child you may carry will be born and when your next menstrual period will occur. Living mostly outdoors, women undoubtedly used the $29\frac{1}{2}$-d (days) cycle of the Moon's phases (for example, from one full moon to the next) to keep track of their 27- to 30-d menstrual cycles. Some experiments indicate that menstrual cycles of women sleeping in the light of an artificial moon become synchronized to its phases, and some scientists have even speculated that the human menstrual cycle evolved to match the period of the Moon's phases (Luce, 1975).

For such reasons, early people began to keep records of events in the sky. There may be actual evidence of the earliest such records. Curious notations—groupings of scratches or notches—found on thousands of artifacts, scattered over Europe, Africa, and Asia and dating as far back as 30,000 B.C., may be the beginnings of calendar systems—perhaps counts of the number of days between phases of the Moon. The makers of these tools were not the brutish cave dwellers of some stereotypes; their craftsmanship is demonstrated by the magnificent paintings of their prey—mammoths, boars, reindeer—that they left on European cave walls. Many early cultures did not use our familiar calendar, but rather a lunar calendar, which kept time primarily in months. Each month started from the first sighting of the new moon—the thin crescent moon in the evening sky—so named because it began each monthly cycle (Figure 1-1).

[1]Time grows short for these studies. The last few pockets of primitive culture are being discovered and integrated into surrounding cultures by scientific expeditions and even guided tours.

Figure 1-1 Ancient peoples used phases of the Moon as one way of keeping time. The configuration shown here, called "the old moon in the new Moon's arms," is seen in the evening sky a few days after a new moon, when the portion not lit by the Sun is dimly lit by light reflected off the Earth. Many primitive calendars used this phase to define the beginning of each month. (Photo by WKH)

This interpretation is supported by contemporary calendar sticks made by modern aborigines. For example, a lunar calendar stick carved by Indian Ocean islanders shows a series of grooves almost identical to the Paleolithic examples of 30,000 y earlier.

The next step in achieving knowledge of the sky, and also in achieving cultural identity, was the keeping of long-term records. This began even before societies had written languages. North American Indian calendar sticks, for example, recorded historical and natural events by sequences of carvings. The calendar stick in Figure 1-2, for example, records an earthquake and other events. Since the Indians who made it had no written language, it can be "read" only by the carver. It is dramatic evidence of how records of astronomical events may first have accumulated, thus allowing the discovery of long-term astronomical cycles.

CALENDAR REFINEMENTS (10,000–3000 B.C.)

Around 10,000 B.C., the domestication of animals and the cultivation of crops began. This agricultural revolution brought two new incentives for understanding the sky. First, to know when to plant, people had to be able to predict seasonal changes days in advance.

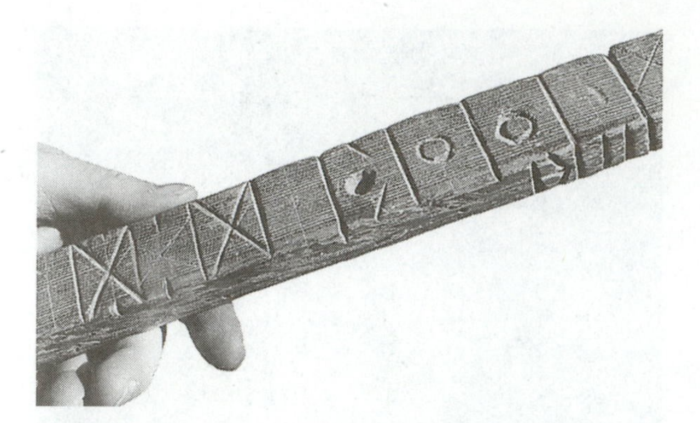

Figure 1-2 Portion of a calendar stick carved by Tohono O'odham Indians of Arizona. The stick, begun in 1841 and passed on to the carver's son, carries a record of social and natural events, including an earthquake, for the years 1841 to 1939. (Calendar stick from collection of Arizona State Museum, University of Arizona. Photo by WKH)

Second, in the stable villages made possible by agriculture, people kept better records; the mere existence of records gave thinkers the opportunity to discover seasonal and annual cycles of celestial events.

This stage of development is shown in historical records of Native Americans in the American Southwest. The desert-dwelling O'odham people of Arizona, for example, named the 12 lunar cycles (months) that made up the year—among them a "Green Moon" (March, plants leafing) and a "Hungry Moon" (May, no wild foods). This was obviously the beginning of an annual calendar system. But since each lunar cycle is about $29\frac{1}{2}$d, 12 such cycles would be only 354 d, not a full year. Thus the lunar-based calendar must be readjusted each year.

People began to get a better sense of the yearly cycle by counting days between certain dramatic annual events. In Egypt, for instance, the Nile flooded at a certain time each year. At high northern latitudes people looked eagerly for the Sun to return to the northern sky after the cold, short days of winter. In some early societies, day counts of this cycle gave estimates of about 360 d, later refined to 365 and then $365\frac{1}{4}$.

Accumulated observational experience revealed the correlation between this annual cycle of seasons and the apparent movements of the Sun. The Sun does not follow the same path across the sky every day. In the summer at northern latitudes (for example, Europe and the United States), the Sun is high at midday; in the winter, it is low in the southern sky at midday. Thus careful observations of the Sun's motion relative to other features of the sky would have helped determine the exact number of days in the year and the exact dates of each observation.

Organized observations of the daily position of the Sun were being made throughout the world thousands of years ago so that people could keep track of the date in order to plant crops and carry out ceremonies at the correct time. Mesopotamian calendars were being designed by 4000 to 3000 B.C. In the nineteenth and twentieth centuries, anthropologists found modern pretechnological peoples using similar means to create sophisticated systems of time measurement.

OTHER EARLY DISCOVERIES

Before describing ways in which these principles were utilized and extended, we pause to describe other aspects of the sky discovered or defined by early people. These aspects are important aids in describing many astronomical phenomena.

North Celestial Pole Camp out for a night and you will discover that the stars slowly wheel around a single point in the sky. In the Northern Hemisphere, that single point is called the north celestial pole; in the Southern Hemisphere, the south celestial pole. The wheeling of the stars is due to Earth's rotation on its axis, and therefore one complete circuit takes about 24 h (hours).

The two **celestial poles** can be thought of as the projections of the Earth's axis onto the sky, or the spots that would be targeted by vertical searchlights at the North and South Poles. You can visualize the sky as a giant dome overhead, as seen in Figure 1-3. Recognizing the celestial pole helped in navigation, because it indicates the north or south direction and can be used to measure an observer's latitude (a point to which we will return shortly).

North Star A bright star, **Polaris** (shown in Figure 1-4), happens to lie near the north celestial pole. Thus it stands almost still all night above the northern horizon as other stars wheel around it. It therefore serves as an excellent reference beacon.[2] For nearly 1000 y, Earth's north polar axis has pointed

[2]Note that a magnetic compass points only approximately north, because Earth's magnetic field is not aligned with true north. Because the field varies, the compass needle also varies in direction by small amounts from year to year. In any case, the compass was not available to the ancients. It was apparently first invented by the Olmecs in Mexico around 1400–1000 B.C. Magnetic materials were later found by the Greeks around 600 B.C., and compasses were recorded in China around 300 B.C. to A.D. 100 (Carlson, 1975).

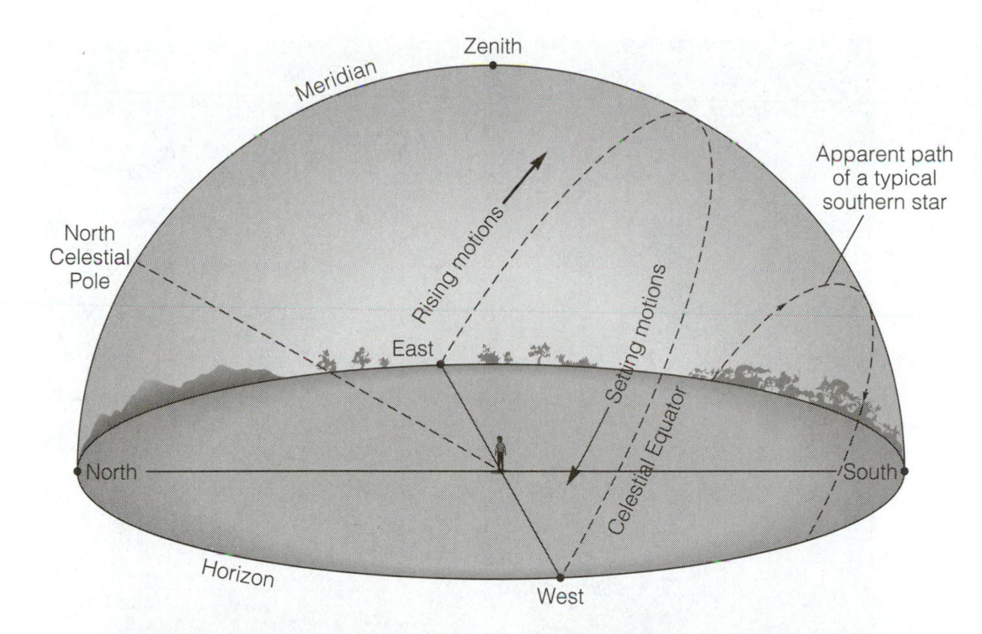

Figure 1-3 General properties of the sky as defined by the Earth's daily rotation. The sky is shown as seen by observers at midnorthern latitudes, such as the United States.

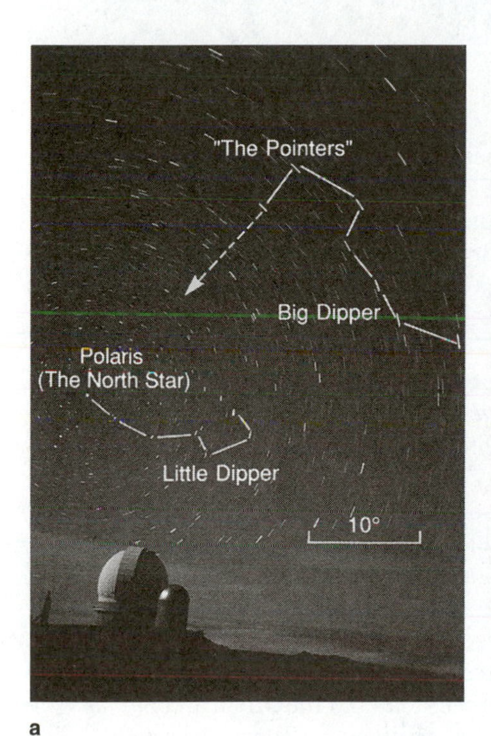

Figure 1-4 A view of the northern sky showing the Big Dipper and Little Dipper and the pointers at the end of the Big Dipper's bowl, which help viewers find the North Star. **a** Ten-minute time exposure with a stationary camera on a tripod shows star trails as Earth rotates through about $2\frac{1}{2}$°. **b** Four-minute exposure with a small motor driving the camera to track the stars represents approximately the naked-eye view. Notice that the observatory dome is slightly blurred in this exposure, as the camera itself has rotated through 1° to track the stars. (Both photos by WKH at Mauna Kea Observatory, Hawaii, with 24-mm wide-angle lens, f2.8, on Ektachrome ISO 1600 film)

within a few degrees of Polaris, and thus Polaris has long been known as the **North Star.**

Zenith and Meridian Wherever an observer stands, the point directly overhead is called the **zenith.** The **meridian** is an imaginary north-south arc from horizon to horizon through the celestial pole and the zenith. The zenith and meridian are shown in Figure 1-3. This line marks the highest point that each star reaches above the horizon on any night; it is the dividing line between the rising stars and setting stars. For this reason it is especially useful in timekeeping. In the daytime, the moment when the Sun crosses the meridian is called noon; in very early times, therefore, people used this astronomical concept to divide the day. Before the Sun crosses the meridian, the time is *ante*-meridian, or A.M.; after the Sun crosses the meridian, it is *post*-meridian, or P.M.

Celestial Equator As shown in Figure 1-3, the imaginary circle lying 90° from both the north and the south celestial poles is the **celestial equator.** It

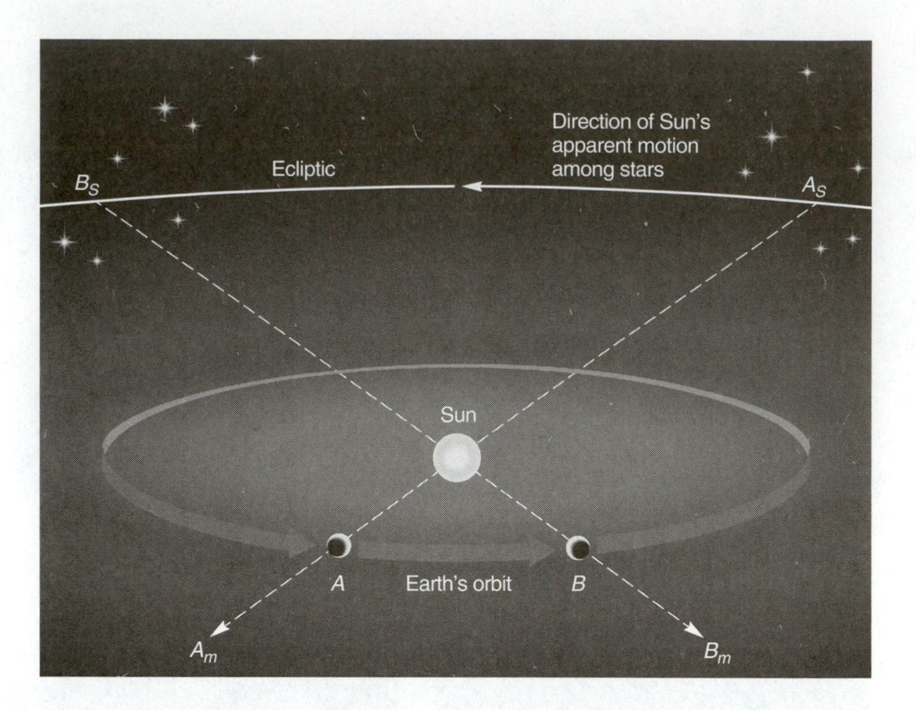

Figure 1-5 Explanation of observations shown in Figure 1-6. As Earth moves in a few weeks from point *A* to point *B* in its orbit, the Sun first appears among background stars A_s and then against stars B_s. Similarly, the midnight sky, opposite the Sun, first contains stars in directions A_m and later those in direction B_m.

is the projection of the Earth's equator onto the sky and is parallel to the daily east-to-west paths of the stars as they wheel around the celestial poles. The celestial equator helps divide the sky into easily recognizable portions.

The 360° Circle Our division of the circle into 360 units, or degrees, and the hour into 60 minutes, and the minute into 60 seconds, follows a Babylonian practice of about 3000 y ago. It probably comes from the ancient estimate of 360 d for one year and from the fact that 360 is easily divisible into 2, 3, 4, 5, or 6 equal parts. We think the decimal (10-based) system of counting and writing numbers is easier, but to early people the ability to divide a unit (for instance, a bag of wheat) into 2, 3, 4, 5, or 6 parts was important for merchants and administrators. Later societies replaced the sexagesimal (60-based) system in most uses *except* the expression of angles and time. It's interesting that our watches carry this 3000-y-old remnant of ancient times with us every day.

The Ecliptic By observing the relative positions of the Sun and stars at dawn or dusk, one can establish that the Sun appears to shift nearly a degree to the east each day relative to the stars, as shown in Figures 1-5 and 1-6. (This can be confirmed by noting that the Sun "moves" 360° in 365 d.) The cause of this shift is Earth's motion around the Sun, also seen in Figure 1-5. Ancient observers found that the Sun traces the *same path* through the stars year after year. The path differs from the celestial equator (being tipped toward it by $23\frac{1}{2}°$ and crossing it at two points). This path of the Sun among the stars is a circle extending all the way around the sky and is called the **ecliptic.** We now know that the ecliptic is the same as the plane of Earth's orbit around the Sun, as can be visualized in Figure 1-5.

The Planets Ancient observers—priest-astronomers and wakeful shepherds alike—found that five of the brighter starlike objects were not fixed like the rest. They came to be known in the Western world as **planets,** from the Greek word for "wanderers."[3] Not until after the telescope was invented, in about 1610, was there any proof that these planets were globes like our Earth or that three faint ones had gone undiscovered. For this reason the ancients spoke of five planets (or sometimes seven, because they sometimes counted the Moon and Sun as wandering bodies, or planets). Today, we speak of nine planets, including our own.

[3] Many of our technical terms have Greek roots because modern Western traditions are directly linked to Greek civilization through the Roman and medieval worlds. However, many of the concepts discussed here were discovered in pre-Greek times and later renamed by the Greeks. Roman and medieval writers, often unaware of the earlier work, revered the Greek thinkers and kept their terms.

a

b

Figure 1-6 Observation of heliacal risings. **a** Shortly before sunrise some stars are already well up in the sky. Others, just above the horizon, are barely visible in the orange glow of dawn. The latter stars have just risen; in a few moments the sky will be too bright to see them. On the previous day at sunrise, they were lost in the glow. Therefore, this is the first day of the year on which these stars are visible; it is called the date of heliacal rising. (Photo by WKH at Mauna Kea Observatory, Hawaii; fixed 35-mm camera; wide-angle lens; 1-min exposure at f2.8 on commercially available 3M ASA 1000 film.) **b** The same principle in diagrammatic form. On date A, the star closest to the Sun is too close to be visible at sunrise. A day or so later, on date B, it is far enough from the Sun to be visible for a few moments at sunrise—the date of heliacal rising.

The Zodiac Ancient observers tracking the paths of the planets from night to night among the stars discovered that the planets never stray out of a zone about 18° wide centered on the ecliptic. Many star patterns, or constellations, along this zone were said to resemble animals, so the zone came to be called the **zodiac** (after Greek for "animals"—the same root as *zoo*). The constellations located within the 18° zone came to be known as the signs of the zodiac. They include such familiar examples as Scorpius (the scorpion) and Leo (the lion)—see the star maps on the last pages of the book.

Heliacal Risings and Settings Any given star rises and sets slightly earlier each night. The **heliacal** (he-LIE-ah-cal) **rising** of a star *occurs on the first day each year when the star can be seen just before dawn.* (*Heliacal* means "near the sun," from the Greek *helios,* "sun.") One day earlier, the star rises a few minutes *after* the sky gets too bright for the star to be seen. On the day of heliacal rising, the star glimmers near the horizon just a minute or so before the sky gets too bright, as shown in Figure 1-6. On the day after, the star rises slightly earlier and can be seen for several minutes before the sky lightens. Similarly, **heliacal setting** occurs on the last day each year when the star can be seen at dusk. On the next day, by the time darkness falls, the star has already set.

Why would anyone living in, say, 4000 B.C. care about the heliacal rising or setting of a certain star?

Figure 1-7 View of the summer constellation Scorpius, rising in the east. Field of view top to bottom is about 80° in this wide-angle photo. Scorpius is one of the few constellations that closely resemble their namesakes, in this case a scorpion with two curving claws at top and curved tail at bottom. A double star marks the stinger at the end of the tail. In the heart of the scorpion, the brightest star, Antares, is distinctly redder than adjacent stars—a difference that can be seen with the naked eye. This time exposure also shows the softly glowing starfields and dark dust clouds of the Milky Way, which are visible to the naked eye on a very dark night far from city lights. (Photo by WKH, Mauna Kea Observatory, 25-min guided time exposure, 24-mm lens, f2.8, Ektachrome ISO 1600 film)

The answer is that each heliacal rising or setting occurs on the same date every year. Depending on the star, that date might be the first day of spring or the day when planting should start. Without clocks, calendars, and morning newspapers on the doorstep, heliacal risings and settings would tell the date with an accuracy of a day or two. It is hardly surprising that the ancient Egyptian calendar began with the heliacal rising of Sirius, which marked the beginning of the Nile's annual flooding.

Similarly, agriculture in Java was once scheduled according to the heliacal rising of Orion's belt. Australian aborigines begin their spring when the Pleiades, a prominent star cluster, rise in the evening sky. The Pima peoples of Arizona also divided their agricultural year according to the Pleiades (Castetter and Bell, 1942):

Pleiades rising in summer, start planting;
At the zenith at dawn, too late to plant more;
Past the zenith, time for corn harvest;
One quarter down from the zenith, time for
deer hunting;
Setting, time for harvest feast.

ORIGIN OF THE CONSTELLATIONS

Early observers named groups of stars for their resemblance to familiar animals, objects, and mythical characters—such as Scorpius (which really resembles a scorpion, see Figure 1-7), Cygnus the swan, or Orion (the hunter), and many others. These groups of stars are **constellations.** The constellations served as memory aids in learning to recognize the patterns in the sky. People in different cultures saw different patterns, often derived from their mythologies; for example, the ancient Chinese constellations differ from Western constellations. Oddly enough, the star group called Ursa Major, or the Great Bear, was seen as a bear in Europe, Asia, North America, and even ancient Egypt, where there are no bears; for this reason, Gingerich (1984) suggests the bear identification may go all the way back to ice-age Euro-Asia and spread from there all the way to America.

Most of our own familiar constellations came out of the Near East. As early as 3300 B.C., Leo (the lion) and Taurus (the bull) often appear in combat on Mesopotamian artifacts. Star symbols sometimes shown on Leo's shoulders indicate that the artisans had the constellation in mind rather than a real lion. The struggle motif may refer to Leo chasing Taurus out of the sky: As Leo rises, Taurus sets. These images originated when the risings and settings of Leo and Taurus coincided with equinox and solstice dates, which were critical in early calendars. Later, they gained a mythical importance of their own, and the lion-bull struggle persisted in art until A.D. 1200 (Hartner, 1965).

★ To learn about astronomical and archaeological sleuthing about how the constellations were invented, read the section in our web site on this subject.

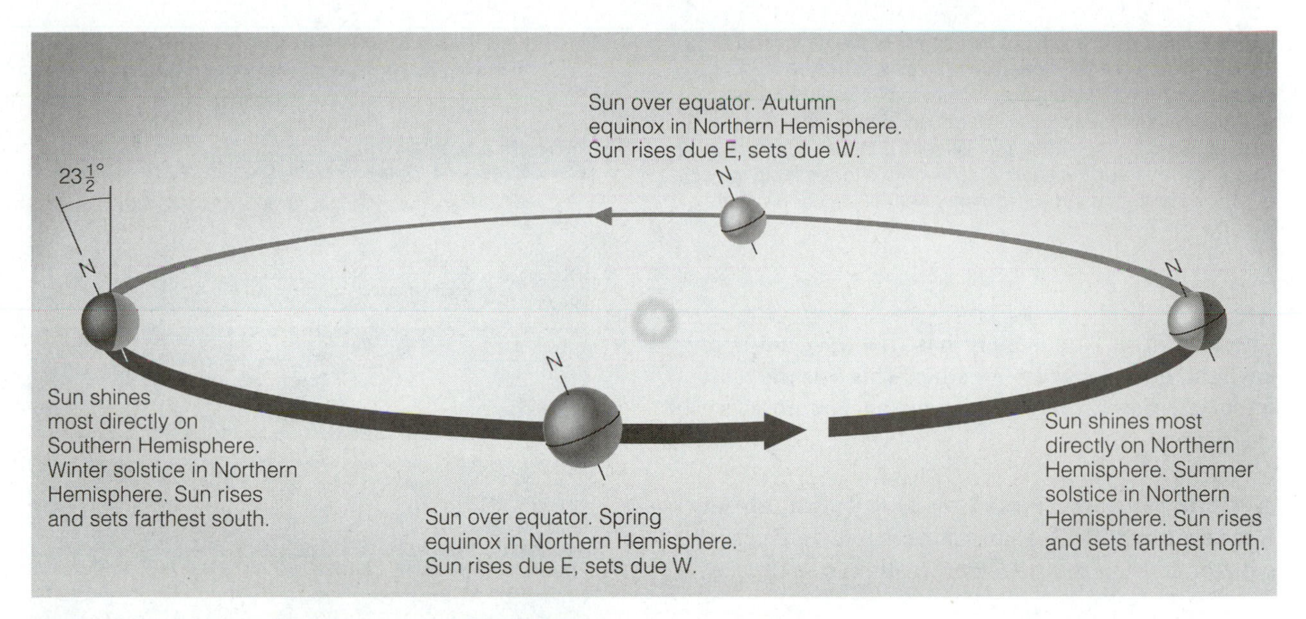

Figure 1-8 The cause of the seasons. Earth's axis is tipped at $23\frac{1}{2}°$ to its orbit plane in a constant direction as it moves around the Sun. Therefore the Sun shines more on the Northern Hemisphere during one season (northern summer) and more on the Southern 6 mo later (northern winter). See text for further discussion.

As shown in the four star maps at the end of this book, different constellations are visible in the evening sky during different seasons. Today they no longer have mythical significance, but it is pleasant to greet them each year as old friends who signal the arrival of a new season.

THE SEASONS: SOLSTICES, EQUINOXES, AND THEIR APPLICATIONS

Most people think the Sun rises in the east and sets in the west. This is roughly true, but it is only an approximation. As the seasons progress, the Sun rises and sets at different points on the horizon. Consider an observer in the United States or elsewhere in the Northern Hemisphere. In summer, the Sun rises in the *north*east, passes high overhead at noon, and sets in the *north*west. In winter, the Sun rises in the *south*east, passes low in the southern sky at noon, and sets in the *south*west. Only on two dates, the beginning of spring and the beginning of autumn, does the Sun rise exactly due east and set exactly due west (see Figure 1-3, where such a path is shown).

The reasons for this have to do with the **cause of the seasons:** the tip of the Earth's equator by $23\frac{1}{2}°$ to the plane of the ecliptic, as seen in Figure 1-8. As Earth moves around the Sun during a given year, the axis always stays pointed in the same direction—always toward the North Star (or, to be more precise, toward the north celestial pole). Thus, as seen in the figure, the Northern Hemisphere is tipped toward the Sun for part of the year and away from the Sun during the other part of the year. When the Northern Hemisphere is tipped toward the Sun, the Sun rises and sets farther around toward the northern part of the horizon; the Sun also rises higher in the sky and shines more directly down on us; the weather grows warmer, and we call the seasons spring and summer.

Thus there are four special days each year, called **equinoxes** and **solstices:**

Spring Equinox (First Day of Spring, about March 21) The Sun crosses the celestial equator moving north.[4] It rises due east and sets due west. This was an important day to ancients in the Northern Hemisphere because it marked the return of the Sun to "their" sky, bringing warmth. The spring equinox is also called the vernal equinox.

Summer Solstice (First Day of Summer, about June 22) The Sun reaches a point farthest north of the celestial equator. It rises and sets farthest north.

[4]As a memory aid, recall that the Sun crosses the *equ*ator on the *equ*inox.

Figure 1-9 Annual cycle of movement of the sunset position along the horizon, as viewed from a fixed position. Photos taken about the 22nd of each month, facing west. Sun sets due west of the equinoxes in March and September. It sets farthest north on the summer solstice in June, and farthest south on the winter solstice in December. (Wide-angle photos by Gayle Hartmann, Amy Hartmann, Kelly Rehm, and William Hartmann)

The Northern Hemisphere has the most hours of daylight, ensuring warm weather. This was the most important day of the year in many ancient northern calendars.

Autumn Equinox (First Day of Autumn, about September 23) The Sun again crosses the celestial equator, moving south. Colder weather is on the way.

Winter Solstice (First Day of Winter, about December 22) The Sun is farthest south of the celestial equator. It rises and sets farthest south. There are the fewest hours of daylight in the Northern Hemisphere.

Part of this sequence can be seen in Figure 1-9, which is a series of photos looking west at sunset and taken one month apart. In the first photos, around winter solstice, the Sun sets in the southwest (left), but as we move toward summer, the sunset position slides to the north (right). We might call this cyclical shift in the sunset and sunrise positions the **solstice principle.**

Ancient Applications of the Solstice Principle

These concepts were very important to primitive people because they had no dated newspapers arriving on their doorsteps and no calendars to hang on their walls. Observations of solstices and equinoxes offered a way to reset the calendar each year. For example, if you lived in an agricultural community, and tradition dictated that the best time for planting was mid-April, how would you know when mid-April had arrived? Priest-astronomers might have begun counting days at the time of spring equinox and might know that the optimum planting time was 24 d after equinox.

Several lines of evidence indicate that such ideas were practiced. For example, Spanish explorers recorded that Peruvian Inca Indians determined the date of solstice in this way and conducted ceremonies at that time, sending runners through the Inca empire to spread the word of the beginning of a new year. Similarly, Pueblo Indians in the Southwest observed sunrise and sunset positions to determine the solstice or equinox dates.

But one of the most extraordinary lines of evidence reaches back much earlier in time. Around 2800–2200 B.C. at **Stonehenge,** in England, prehistoric builders constructed a large circular embankment with a broad avenue leading outward about $\frac{1}{2}$ km ($\frac{1}{4}$ mi) directly toward the horizon position of summer solstice sunrise, as shown in Figure 1-10. Subsequently, as late as 2100 to 1500 B.C., additional huge stones were brought in to construct a monument in the center of the ring. These included 30- to 50-ton stone blocks brought as far as 30 km (20 mi) and some 5-ton stones hauled about 380 km (240 mi). These stones can be seen in Figures 1-10 and 1-11. A large stone called the heelstone (Fig. 1-11) was placed in the avenue in such a way that an observer at the center would see the Sun rise over that stone *on the day of summer solstice* as shown in Figure 1-12. Archaeologists have found additional post holes near this stone, suggesting that a grid of wooden posts might have allowed "fine-tuning" of the observations to determine on which day the Sun rose farthest north. As shown in the map in Figure 1-10, other astronomical alignments have also been found in Stonehenge, suggesting that the builders may have been much more sophisticated, keeping track of motions of the Moon as well as the Sun and possibly predicting eclipses, though this idea remains controversial.

Of course, we don't know whether Stonehenge was used as a practical observatory to *measure* dates of astronomical events like summer solstice or whether it was more of a ceremonial temple where, perhaps, solstice-related rituals were held. Many other prehistoric standing-stone monuments exist in Great Britain, but astronomical orientations are clear only for some of them (cf. Figure 1-13). There is some evidence that the earliest builders at Stonehenge had precise observing in mind and that builders in later centuries were more interested in elaborate ritual, because the huge stones that were among the latest additions seem to show less clear astronomical alignment. Perhaps they were added just to enhance a sense of mystic magnificence to what had become a quasi-religious temple. In any case, more than 4000 y ago the builders of Stonehenge had a strong awareness of the solstices.

The fact that Stonehenge is designed around a solstice alignment was pointed out as long ago as 1740 by English scholar Dr. W. Stukely. Yet in the 1880s and 1890s, when American astronomer Samuel Langley and English astronomer J. N. Lockyer spoke of Stonehenge as an astronomical observatory, most scholars thought ancient people could not have been sophisticated enough to orient large structures this way. Today we have much more evidence that early civilizations did make sophisticated use of the solstice principle and other astronomical knowledge.

For example, one of the largest temples ever built, the Temple of Amon-Re in Karnak, Egypt, built around 1400 B.C., covered twice the area of St. Peter's in Rome. Its central hallway, about 370 m (1200 ft) long, was aligned within $\frac{1}{2}°$ of the summer solstice sunset position (Figure 1-14). Other Egyptian temples had similar orientations, and still others were oriented toward heliacal risings and settings of certain stars. Because of precession (the wobble of the Earth on its axis over time), the rising and setting points of stars shift slowly over the centuries. Some of these temples show evidence of rebuilding as if to correct for such shifts, even though the rebuilding often ruined the symmetry of the temple by changing the angle of the central hall. The Egyptians are known to have made astronomical observations: Chicago's Oriental Institute has part of a device for charting star positions that was made by the young Pharaoh Tutankhamen (the famous "King Tut") around 1350 B.C.

★ Further information on precession, plus a number of additional examples of ancient ruins with possible astronomical significance are discussed on our web site.

New Year's Day and the Solstice Principle in Ancient Europe

Why does our New Year start on January 1? Does anything special happen on that date in terms of Earth's motion around the Sun? No. Ancient Rome and certain other cultures began the New Year in mid to late March, the time of the spring equinox, when the Sun moved into the northern half of the sky and warmer weather arrived. But many northern European cultures celebrated the New Year on about December 22, winter solstice, because that's when the days began to get longer and the long, cold, gloomy northern nights began to shorten. In 152 B.C. the Roman Senate reportedly moved New Year's Day from March 15 to January 1 to allow a military campaign scheduled for the new year to begin early. Most authorities agree that the early Christian church, uncertain of the historical birthdate of Jesus, chose December 25 for the Christ-mass celebration to co-opt major pagan celebrations of the equinox and New Year season.

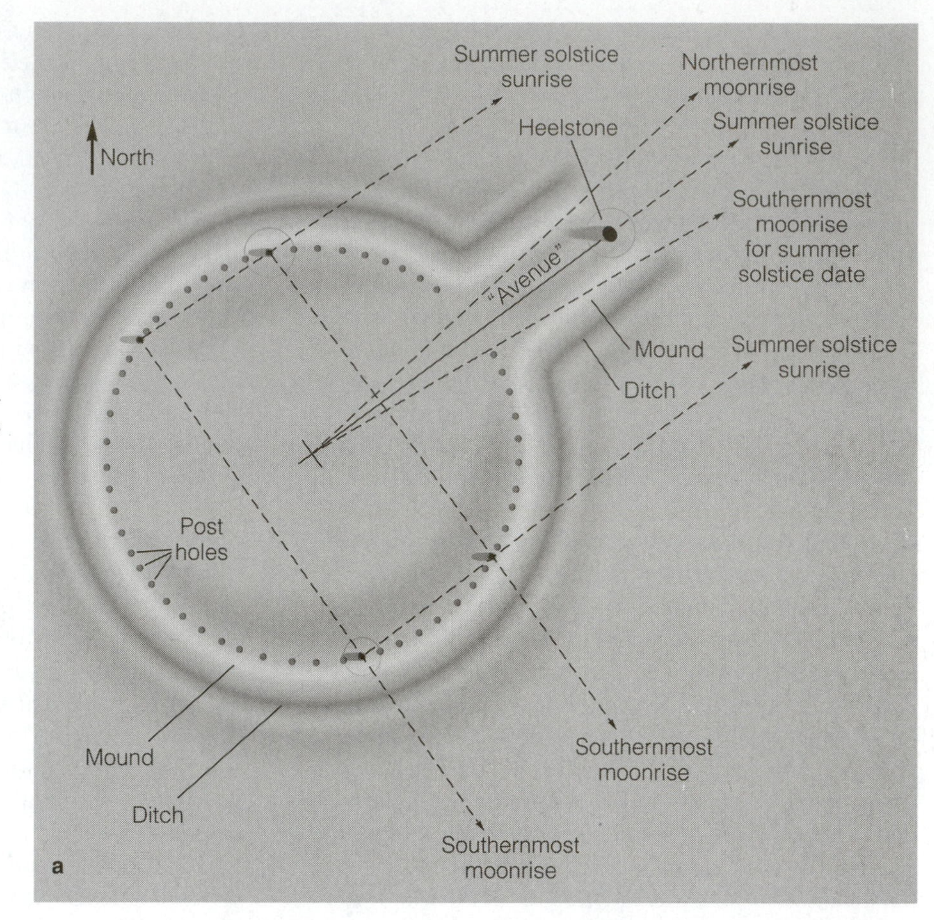

Figure 1-10 **a** Aerial plan of the original Stonehenge construction around 2500 B.C. An outer ditch, mound, and ring of posts or stones were cut by an avenue leading toward summer solstice sunrise. As seen from the center, the summer solstice Sun rose over the heelstone, partway down the avenue. Selected moonrises could also be observed and timed. **b** Aerial photo shows modern Stonehenge. Central ring of giant stones was added around 2000 B.C. White path is a twentieth-century addition. The avenue leads out of the right of the picture, and the heelstone is near the road. (Georg Gerster, Photo Researchers, Inc.)

b

Figure 1-11 Stonehenge (rear) and its heelstone (foreground), which marks the position of sunrise on the date of summer solstice. (Photo by WKH)

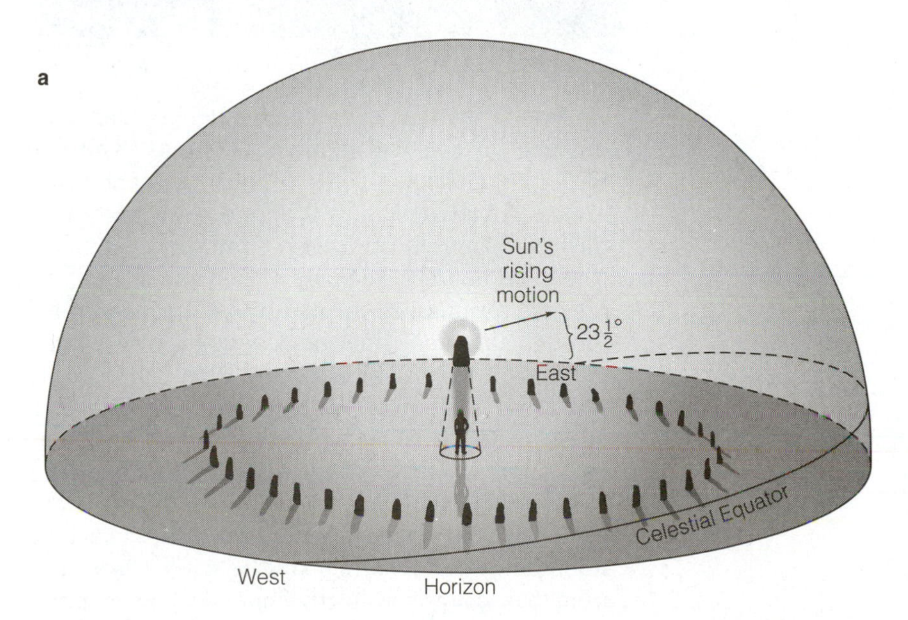

Figure 1-12 The Stonhenge principle. **a** The Sun's position at solstice is $23\frac{1}{2}°$ from the celestial equator. The rising (or setting) Sun on this date is thus as far as possible from due east or west, respectively. This position on the horizon is marked by a pillar or other prominent marker. Additional markers in a ring could allow observations of other selected risings or settings during the year. **b** Sunrise on the morning of summer solstice from the center of Stonehenge, showing the Sun's position over the top of the distant heelstone. (Georg Gerster, Photo Researchers, Inc.)

Figure 1-13 Circles of standing stones are dotted throughout Great Britain and northern Europe. Astronomical alignments at most sites are less clear than at Stonehenge and are still being studied. While some may have been for calendrical functions, others may have had nonastronomical purposes. These giant stones were erected in prehistoric times as part of a vast double-ring complex some miles north of Stonehenge at Avebury. The Avebury complex is so large that the rings of stones enclose part of a village. The purpose of the complex is uncertain. (Photo by WKH)

Figure 1-14 The ruins of the long hall in the Egyptian temple of Amon-Re, built around 1400 B.C. Nineteenth-century astronomer-archaeologist Norman Lockyer used this illustration when he pointed out in 1894 that this main hall is aligned toward the summer solstice sunset position.

Interestingly, vestiges of the ancient March solstice New Year's Day persisted in historic times. Some older European communities used to start the New Year on March 25. The early Christian church had a feast day on this date to commemorate the conception of Jesus—possibly another co-opting of the pagan calendar. In *Tess of the D'Urbervilles,* novelist Thomas Hardy records that farm labor contracts in nineteenth-century England were still calculated from that date.

Even Christian cathedrals inherited some of the ancient solstice/equinox traditions. The old (330–1506) and new (1506–) cathedrals of St. Peter in Rome were reportedly oriented so accurately to the east that on the morning of the spring equinox "the great doors were thrown open and as the Sun rose, its rays passed through the outer door, then through the inner door, and penetrating straight through the nave, illuminated the High Altar" (Lockyer, 1894). Certain other Christian cathedrals and missions were oriented toward sunrise on the day of the saint to whom they were dedicated. This sounds suspiciously like temple orientation traditions in ancient Egypt and elsewhere. Probably these ancient traditions were absorbed by the early Christians from their cultures. Ghostly fingers of ancient practices reach into the present to touch us!

Modern Applications of the Solstice Principle

A totally different application of the solstice principle is appearing in modern times. As fossil energy supplies dwindle, we are more concerned about utilizing solar energy in our buildings. The cheapest way is by clever building orientation and roof overhang to let sunlight in during winter and keep it out during summer.[5] In a sunny climate, most of the heat input through windows and walls comes in afternoon and morning when the Sun is low and strikes these sur-

[5]This is called passive solar utilization because it requires no further machinery or energy consumption. Active solar utilization implies further machines, such as fans to distribute warm air once it is heated by the Sun.

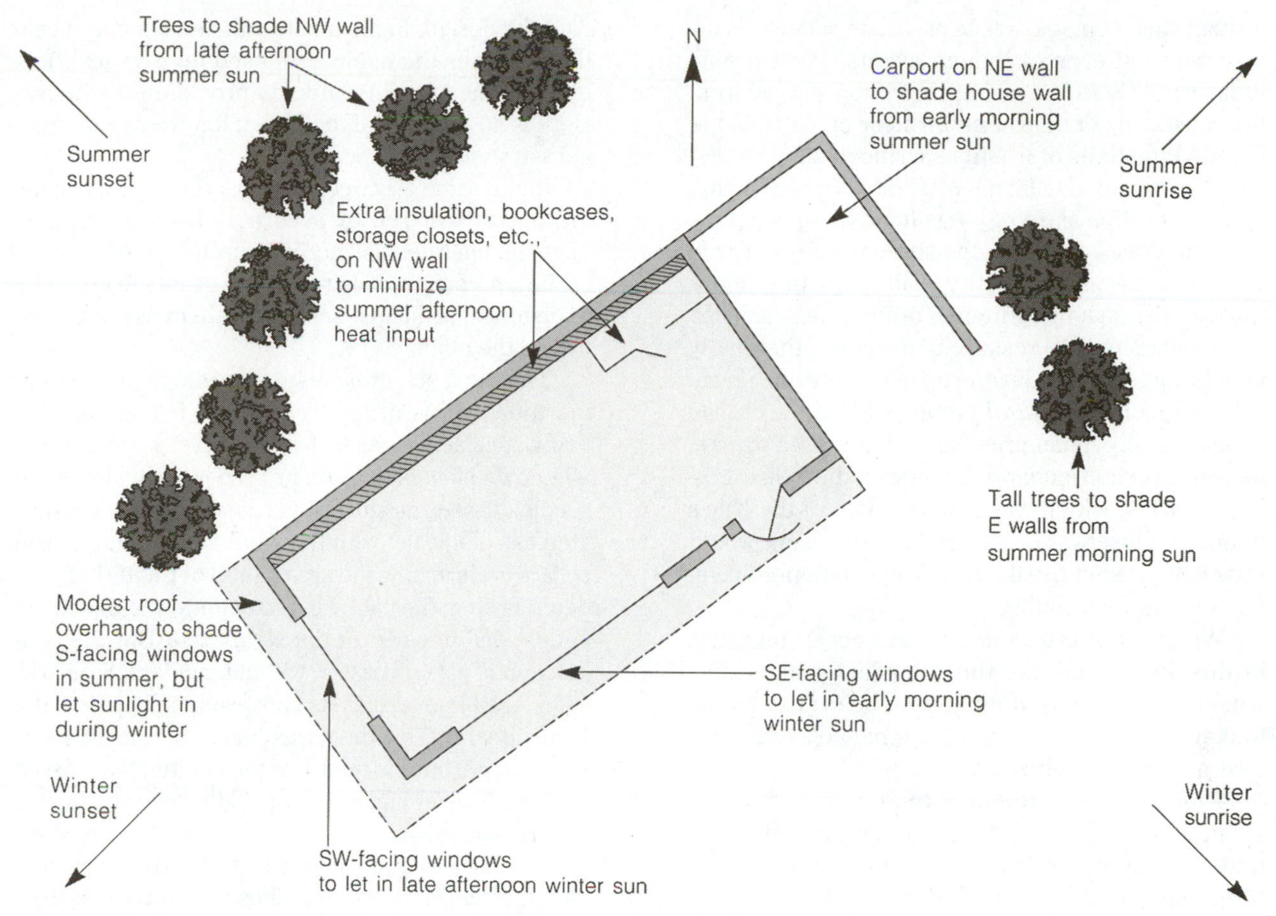

Figure 1-15 The solstice principle as applied to environmental architecture and landscaping. Northeast and northwest walls are shaded to reduce heat from the summer morning and afternoon sun. Southern windows allow sunshine to enter and heat the house in winter.

faces full on. (Midday heat input through roofs must be minimized by thick insulation.) As shown in Figure 1-15, house design can utilize the solstice principle. Southeast- and southwest-facing windows, for example, can be unshaded to provide "free" warmth around winter solstice, and northwest-facing windows can be covered with external shields in summer to prevent late afternoon sunlight from pouring unwanted heat into the house.

ASTROLOGY: ANCIENT ORIGINS OF A CONTINUING SUPERSTITION

If you lived in a society where the calendar was determined by observations of the stars, where priest-astronomers counted days from the summer solstice and studied heliacal risings in order to announce when to plant crops, and where the death of a king was remembered as happening in "the year the sun went dark and the stars came out in the daytime," you could see how easy it might be to make a serious philosophical error. Instead of realizing that the stars offer a practical way to date and coordinate human affairs, you might come to believe that the stars *control* or at least influence human affairs. This superstition is called **astrology.** We call it a superstition for the same reason that we say it is superstitious to believe in invisible elves in your garden: No experimental evidence supports the belief, and predictions based on the belief have no better than random accuracy.

An occupational hazard of being an astronomer is to be introduced at parties as an astrologer. This mistake is annoying because astrology is not part of modern astronomy but a pseudoscience associated with astronomy as it was practiced 3000 y ago.

The pseudoscientific nature of astrology can be understood best by exploring astrology's roots. Astrology can be traced back at least 3000 y, when it flourished with other ancient magical beliefs. A common form of ancient magic was to associate patterns in nature with patterns of human events. Most people today would scoff at having their futures read from flight patterns of migrating birds or patterns in the bloody intestines of freshly sacrificed animals. Yet these were popular forms of divination in Babylon and Rome when astrology was flourishing, and they have the *same basic logic.* In astrology, the pattern in nature is that made by the planets as they move through the imaginary figures of the constellations.[6]

Archaeological research suggests that early people correctly used patterns in the sky in a practical way to foretell *natural* events, such as the change in seasons. Egyptian priest-astronomers, as we saw, designed their calendar to begin with the heliacal rising of Sirius, which was also the date of the Nile's flooding. Observation of the heliacal rising would thus help predict the flood, an important prediction for agricultural planning.

When priest-astronomers were observing signs in the sky to calibrate the calendar or determine when to plant crops, it must have been easy for onlookers to conclude mistakenly that celestial signs forecast *human* events.

Human minds are quick to assume that if one event follows another, the second is caused by the first. Sometimes this is true and sometimes not. Centuries ago, logicians identified and named this error with the Latin phrase *post hoc, ergo propter hoc* (following this, therefore *because* of this). Yet this type of error is still common today. Historical evidence suggests that astrology grew out of this error. For example, if a king died a month after an eclipse, some early observers reasoned that an eclipse causes or foretells the death of a king.

There is some evidence to suggest that early astronomical knowledge actually degenerated into myth and pseudoscience in some societies. For example, the best astronomical alignments at Stonehenge were probably those constructed first, around 2500 B.C. Later construction, around 1800 B.C., muddled the design. In the nineteenth and twentieth centuries we see the monument used for mystical rites by costumed Druids and self-proclaimed witches. Here is an example of the loss of knowledge of original astronomical purpose.

In the same way, records of astronomical events, originally gathered by accurate observation, may have degenerated as later interpreters added mystical interpretations of astronomical events to accounts of historical incidents, describing them as if one controlled the other.

The most scientific test of astrology is to examine the success of its predictions. It has failed in many spectacular cases. For example, a grouping of all known planets in Libra in 1186 led astrologers to predict disastrous storms, because Libra was associated astrologically with the wind. People dug storm cellars in Germany; the archbishop of Canterbury ordered fasting; the palace in Byzantium was walled up; people fled to caves in the Near East. But the conjunction of planets passed without incident. Similarly, in 1524, all known planets clustered in Aquarius, the Water Bearer. This time astrologers predicted a second Deluge, but the month of the conjunction passed without disaster and was reportedly drier than usual (Ashbrook, 1973).

Similarly, pseudoscientists and astrologers predicted cataclysmic earthquakes and other disasters due to the so-called Jupiter effect—increased gravitational stresses caused by a rough alignment of Jupiter, Earth, and other planets on the same side of the Sun in March 1982. The month and year came and went without the predicted cataclysms.

★ A further discussion of the historical development of astrology, and historical instances where it failed to give good predictions, are discussed on our web site.

Problems with Astrology

A test of astrology is whether it is consistent with observations. The basic practice is to forecast the influences for a particular day by casting a *horoscope.* The horoscope is usually presented as a circular chart, showing the positions of the planets, Sun, and Moon with respect to the constellation signs at the moment being studied. For example, since Venus was the goddess of love, prominence of Venus might indicate a loving influence. The prominence of the war god Mars might suggest an aggressive influence. This technique follows the rules laid down by astrologers centuries ago.

★ But application of the ancient rules creates an embarrassing problem for astrologers today. As-

[6]An interesting example of this all-too-easy mental connection—between changing patterns in the sky and mystic patterns of human relationships—was given by French diarist Anaïs Nin during a 1935 visit to a planetarium. A practicing psychoanalyst, she had been thinking of the "human tangles" of her patients. Her diary records: "The planetarium . . . restored to me a sense of space and I could detach myself from the haunting patients. . . . I saw, instead of stars, relationships moving like constellations, moving away and towards each other . . . turning, moving, according to an invisible design, according to influences we have not yet been able to measure, analyze, contain."

trologers nearly 4000 y ago designated 12 "signs," or portions of the zodiac, corresponding to the 12 zodiacal constellations. Like the planetary gods and goddesses, each constellation and its sign were said to have a certain personality trait. At the time of astrology's beginnings, the point where the Sun crossed the celestial equator on March 21 (spring equinox) and entered the Northern Hemisphere was in the constellation Aries and was called the "first point of Aries." Thus a person born between March 21 and April 19 was said to be an "Aries," since the Sun lay in this constellation and thus in this sign. The original astrologers of that era thought this pattern was fixed. But around 130 B.C., Hipparchus discovered precession, which shifts the first point of Aries relative to the constellations (see our web site). Thus the signs no longer correspond to their respective constellations!

Astrologers during Ptolemy's time tried to patch up the scheme. They defined the signs at 30° intervals measured from the March equinox point and claimed that the signs were more important than the actual constellations. Astrologers still claim that a person born between March 21 and April 19 is an Aries, but during most of this period the Sun is really in Pisces! Thus astrology has become removed from the realities of the sky.

Another embarrassment is that we now know that the stars move independently; thousands of years from now the constellation patterns themselves will have changed. Although one could patch up astrology further by inventing new constellations and associated psychological characteristics to cover this problem, the discrepancy shows that astrology as it is now practiced is not consistent with observed reality.

Another example will arise in the near future. When astronauts land on Mars, they will see no war god Mars in the sky. They will be standing on him! Instead, they will see a new planet, Earth, crossing the zodiac. Will astrologers invent a new personality influence associated with Earth's position in the Martian sky when the first baby is born on Mars? The old astrological rules will hardly apply!

What Should We Do?

Because its logic is identical to that of ancient magic, because it is inconsistent with the twentieth-century sky, and because it fails to predict accurately, we conclude that astrology does not warrant our belief or our attention in planning our lives.

We might just smile at astrology as another amusing human foible if it weren't for the nagging suspicion that it threatens healthy civilization. For example, a 1975 Gallup poll indicated that about 12% of all Americans take astrology quite seriously. In an age of space exploration and serious nuclear threats, do we really want ancient magic to be an influence in human affairs? It is sad to realize that millions of people seeking advice about real problems in their lives might spend their (sometimes meager) money on the products of astrologers and pious con artists. As astronomer Bart Bok pointed out in 1975: "The astrologer can refine his interpretations to any desired extent—the end product becoming increasingly more expensive as further items are added." Nowadays, computers are used to calculate planetary positions while casting horoscopes, giving a scientific gloss that misleads those who confuse scientific tools with scientific method. Thus naive members of the public are fooled into thinking that astrology has some scientific basis.

Philosophically, the continuing practice of astrology helps remind us that the human mind has an amazing capacity to come up with ideas—many of which are wrong—and then to develop intricate systems of thought based on those ideas without ever establishing that the ideas are correct.[7]

★ An epilog on our web site describes further examples of pseudoscience that are merchandised as scientific truth in books and tabloids at supermarket checkout stands. This epilog also discusses in more detail how to distinguish science from pseudoscience—not through the hypotheses themselves, but rather through the way *evidence* is used to confirm or refute the hypothesis.

If we agree that astrology is little more than ancient superstition that wastes people's money and mental energy, should we try to suppress it? Newspapers routinely reject suggestions that they drop astrology columns, which they claim are harmless

[7] *Comment by WKH:* I once met an industrious young man who had amassed a library of books on arcane details of various authors' astrological systems. His goal was to make more accurate astrological forecasts to guide his own life and to perfect a gambling system he could apply in Las Vegas. He had once had a string of successful bets in a football pool and was convinced that a fatal bet in which he lost all his winnings was merely the result of an erroneous reading of the horoscope.

As a student I rented a room from an ill, elderly woman of low income. I was aghast to discover that she had accumulated a substantial library of books and tracts purchased from a radio and TV evangelist. The books included drawings of various horned and tailed "demons," which, according to the books, were responsible for various illnesses. It was depressing to see the energy and money she needed for life being squandered in this way.

Figure 1-16 The March 1970 total eclipse as seen from the village of Atatlan, Mexico. The sky has darkened dramatically as the Moon's shadow falls across the scene. (Photo by WKH)

entertainment that promotes sales. Of course, most advice in astrology columns is innocuous. It might be said of astrology, like fairy stories, that if you believe it hard enough, it will work. For example, if you read morning advice to avoid frivolous expenses today, you may indeed be more prudent than usual. But one would hope we could find better sources of inspiration than ancient magic that is pseudoscience at best and fraudulent waste at worst. In a free country the answer is not suppression. In a free country with a free press, we hope that educated, pragmatic citizens will pick and enjoy the best from among competing published materials and look for evidence of accuracy before committing themselves to a system of belief.

ECLIPSES: OCCASIONS FOR AWE

So far we have treated the Moon and the Sun as independent bodies that can be tracked for practical purposes. Cycles of the Moon divide the year into about 12 equal months; solstices and equinoxes, defined by sunrise positions, divide the year into four seasons. Ancient people who discovered these facts may have gained a sense of well-being: At least *some* things in their environment could be counted on from year to year. Calendars could be made and agriculture regulated. The predictable, friendly Sun could be worshiped as a deity, always providing light and warmth.

What, then, could be made of **eclipse,** the sporadic occasions when the Sun or, more commonly, the Moon disappears while above the horizon? In the few minutes of a total solar eclipse, the sky turns dark, stars can be seen in daytime, and an uncanny chill and gloom settle over the land, as indicated in Figure 1-16. During the hours of a total lunar eclipse, the Moon turns blood-red. Small wonder that the Greek root of eclipse, *ekleipsis,* means abandonment or that the ancient Chinese pictured a solar eclipse as a dragon trying to devour the Sun. Eclipses must have seemed an unpredictable menace to the scheme of things.

We know eclipses awed early people. In Greece, the poet Archilochus, observing the solar eclipse of April 6, 648 B.C., was moved to write: "Nothing can be sworn impossible . . . since Zeus, father of the Olympians, made night from midday, hiding the light of the shining Sun, and sore fear came upon men." The Greek historian Herodotus reported that a war between the Lydians and the Medes stopped when an eclipse surprised the competing armies during a battle in the 580s B.C. A hasty peace was cemented by a double marriage of couples from the opposing camps.

Knowledge of what causes eclipses allayed people's fears of them. And if people knew how to predict eclipses, they could become very powerful. For instance, in February 1504, Christopher Columbus was having trouble convincing Jamaicans that they were obliged to feed him and his crew. Seeing that his almanac foretold an eclipse of the Moon on February 24, Columbus warned the Indians that the Christian god would punish them by turning the Moon to blood. The scoffing natives changed their attitude when the eclipse began on schedule.

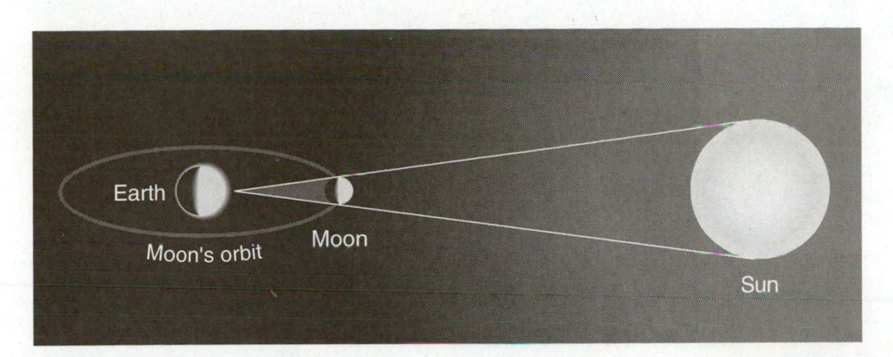

Figure 1-17 Geometry of an eclipse of the Sun (not to scale). The shadow of the Moon falls on Earth.

This event may have inspired Mark Twain, whose Connecticut Yankee established his authority in King Arthur's court by predicting a total solar eclipse and then "commanding" the eclipse to end. This imaginative scene suggests the political prestige gained by some of the first priest-astronomers who discovered how to predict eclipses.[8]

In addition to providing us with interesting historical reading, the ancient astronomers who recorded eclipses centuries ago provided another valuable service: Their records are still used today to help analyze a slight increase in the length of the day, at a variable rate of about 0.002 s (second) per century (Stephenson, 1982).

Cause and Prediction of Eclipses

Whatever their emotions and motives, some early people learned how to predict eclipses—and this was an early step toward recognizing that we live on a world among worlds orbiting in space.

To understand how eclipses can be predicted, we should first understand the **causes of eclipses.** Two important things to remember about eclipses are that (1) they happen when the shadow of one celestial body falls on another, and (2) what you see during an eclipse depends on your position with respect to the shadow. If you see an eclipse at all, either you are in the shadow (a celestial body has come between you and the source of light), or you are looking at the shadow from a distance as it falls on some surface.

[8]Even animals are affected by eclipses. I have seen roosters crowing and birds flocking to trees at midday because of the gloom during a solar eclipse, causing the creatures to believe evening has fallen. There is a story about an eclipse in Colorado during which an astronomer arrived too late to get a prime observing site. He hurriedly set up his equipment in an empty chicken coop to protect his instruments from the wind, and then spent most of the eclipse trying to shoo away the chickens, who dutifully reported to the roost when darkness fell.

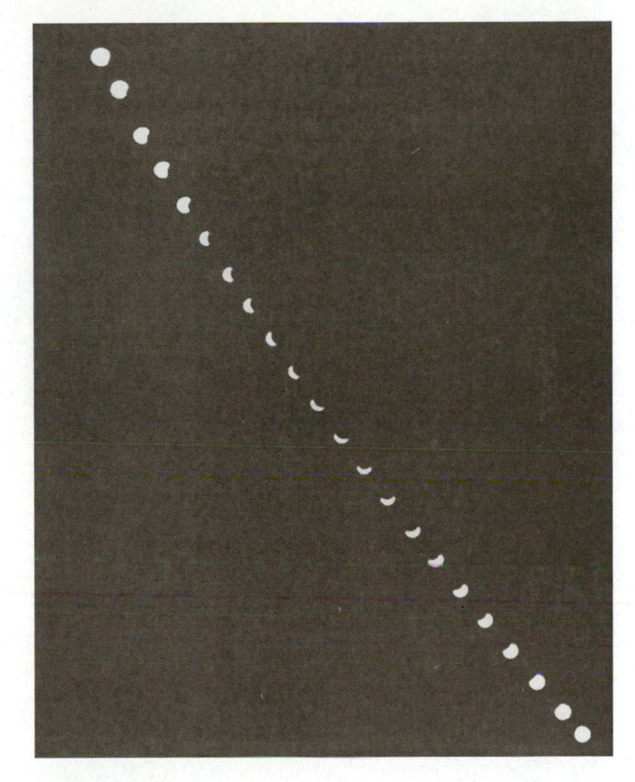

Figure 1-18 Multiple exposure showing the Sun's movement across the sky during a partial solar eclipse on July 10, 1972. At maximum, 72% of the Sun was covered by the Moon. Photos were made 6 min apart with a fixed 35-mm camera and a dark filter. (NASA photo by A. K. Stober)

Earthbound observers see two types of eclipses: solar and lunar. **Solar eclipses** occur when the Moon, on its $29\frac{1}{2}$-d journey around the Earth, happens to pass between Earth and the Sun as illustrated in Figure 1-17. A **total solar eclipse** occurs if the Moon completely covers the Sun as seen by an earthbound observer. A **partial solar eclipse,** shown in Figure 1-18, occurs if the Moon is "off center" and covers only part of the Sun. By coincidence, the Moon happens to have the same angular size as the Sun—about $\frac{1}{2}°$, or a little less than the angular size of

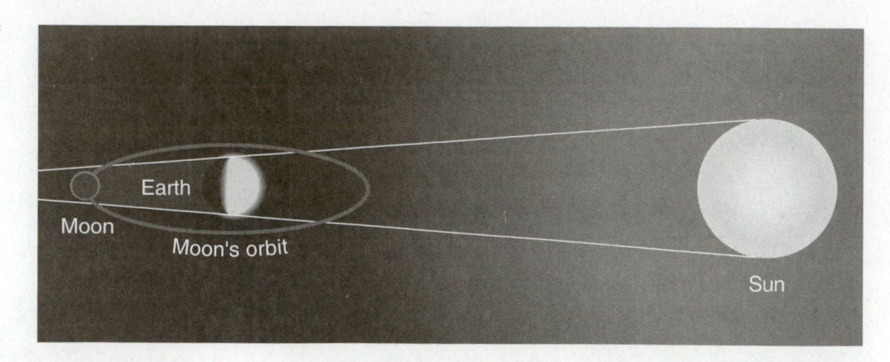

Figure 1-19 Geometry of an eclipse of the Moon (not to scale). The Moon passes through Earth's shadow.

Figure 1-20 A nearly total lunar eclipse. Sunset-colored reddish light refracted through Earth's atmosphere colors the dim, umbral part of the shadow (right). The silvery crescent on the left is in the penumbral part of the shadow, partly lit by the Sun. This photo gives a good impression of the visual appearance of a lunar eclipse. (Photo by Stephen M. Larson)

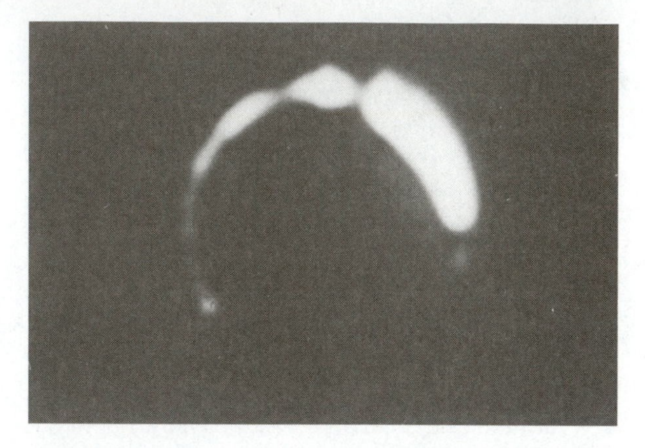

Figure 1-21 An eclipse of the Sun by Earth, photographed from the surface of the Moon. This television image, made by Surveyor III in 1967, shows a ring of light that is Earth's atmosphere, back-lit by the Sun, which lies behind the black disk of Earth. The brightest light is coming through the atmosphere over eastern Asia; local clouds affect the amount of light transmitted. (NASA)

a little fingernail at arm's length.[9] Therefore, the Moon usually just covers the Sun during a solar eclipse. But if the Moon happens to be at the farthest point in its orbit, it has a smaller angular size than usual and doesn't quite cover the Sun. This causes an **annular solar eclipse,** in which the observer sees a ring (Latin *annulus*) of light, which is the rim of the Sun, surrounding the Moon.

Lunar eclipse occurs when the Moon, on its journey around Earth, passes through a point exactly on the opposite side of Earth from the Sun. This point lies in the shadow cast by Earth (Figure 1-19). It takes the Moon a few hours to pass through Earth's shadow, during which time the Moon usually turns an astonishing copper-red because the only sunlight reaching the Moon passes through Earth's atmosphere, as seen in Figure 1-20. In effect the Moon is illuminated only by the colored light of a sunset.

Eclipses can also happen in other parts of the solar system. For example, Figure 1-21 is a photo of a solar eclipse as seen from the Moon's surface, and Figure 1-22 shows an eclipse on Mars.

Umbral and Penumbral Shadows

In solar and lunar eclipses, the shadows of Earth and the Moon are not sharply defined. Because the Sun as seen from Earth has an angular size of $\frac{1}{2}°$, sunlight is slightly diffuse; its rays come from slightly different directions.

Thus shadows—including eclipse shadows—cast by sunlit objects near Earth are not sharply defined. The inner, core area of a shadow—the part that receives no light at all—is named from the Latin word for shadow, **umbra.** The outer, fuzzy boundary is called the **penumbra.** An observer in the umbra

[9] If this concept is unclear, see the discussion of the small-angle equation in Optional Basic Equation I in Chapter 2.

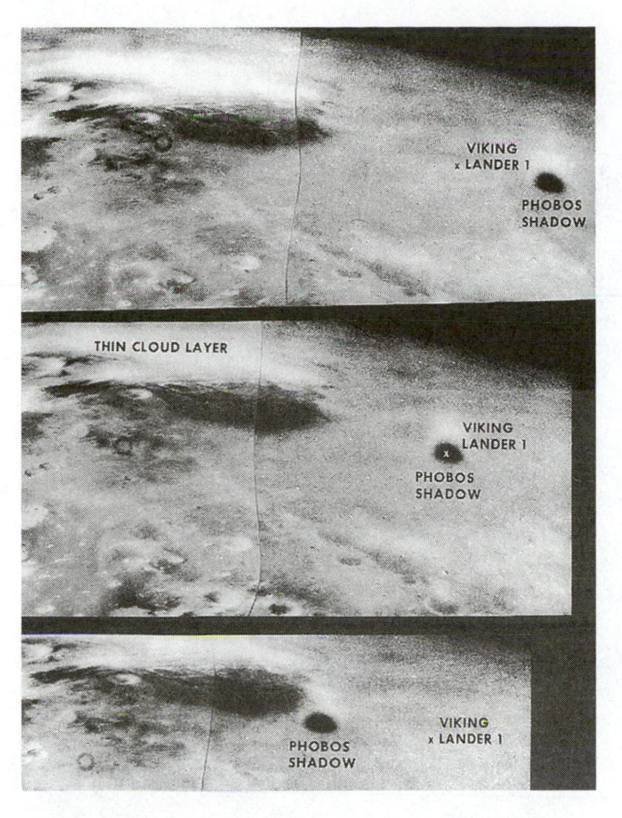

Figure 1-22 Oblique orbital views of an eclipse happening on the surface of Mars. The penumbral shadow of the small Martian moon Phobos moves from right to left. The shadow, about 90 km wide, passed over the site of the Viking I lander, helping scientists locate the lander's position. (Contrast-enhanced NASA Viking I orbiter photo)

sees the Sun entirely obscured; an observer in the penumbra sees the Sun only partly obscured. Figure 1-23 shows some varieties of solar eclipses, illustrating these principles. If you hold your hand at eye level above smooth ground in sunlight, its shadow will have a central, dark umbra and a penumbra about a centimeter wide. If you hold your hand high and spread your fingers, their shadows will be indistinct. An ant in the umbra would see a total solar eclipse; an ant in the penumbra would see a partial solar eclipse.

The relative sizes of umbra and penumbra depend on the distance between the shadow-casting body and the surface on which the shadow appears. During a lunar eclipse, Earth's shadow on the Moon has an umbra several times the Moon's diameter and a considerably larger penumbra. Total lunar eclipses, with the Moon in the umbra, can last up to $1\frac{3}{4}$ h.

On the other hand, the Moon's umbral shadow on Earth does not exceed a 267-km (166-mi) diameter. Due to the motion of this shadow across Earth, total solar eclipses cannot last more than $7\frac{1}{2}$ min. In an annular eclipse, the Moon is too far away to produce any umbral shadow, and the eclipse has no true total phase, as seen in Figure 1-23 (case C).

Frequency of Eclipses

To be able to predict eclipses, we must understand the intervals between them. Discovery of these intervals several thousand years ago marks the beginning of our ability to predict seemingly mysterious celestial events. To understand the technique, imagine a viewpoint in space. If the Moon's orbit around Earth lay exactly on the plane of Earth's orbit around the Sun, the Moon would pass exactly between Earth and the Sun every time around, as can be visualized from Figures 1-17 and 1-19. It would also enter the Earth's shadow on every pass, and a solar and a lunar eclipse would occur every month.

But the Moon's orbit is tipped by an angle of 5° out of Earth's orbital plane, as sketched in Figure 1-24. Therefore the Moon is likely to pass "above" or "below" the Earth-Sun line and Earth's shadow. Since the two orbital planes must, by simple geometry, intersect in a line, there is a line (NN′ in Figure 1-24) that contains the **node,** the only two points where the Moon passes through Earth's orbital plane, or ecliptic. Thus we now can understand the origin of the ancient word *ecliptic:* An eclipse can happen only when the Moon is passing through the ecliptic. That the ancients gave this name to the Sun's annual path shows their interest in eclipses. The line between the two nodes, NN′, is called the **line of nodes.** To produce an eclipse, the Moon must be near one of the nodes, and even then *an eclipse will occur only if the line of nodes is pointing at the Sun.*

When is the line of nodes pointed at the Sun? If no external forces acted on the Moon's orbit, the orbital plane would stay fixed with respect to the stars, and the line of nodes would line up with the Sun twice a year, as can be seen in Figure 1-25. Eclipses *could* occur at these two possible times each year, but *only* if the Moon moved through points N and N′ at these moments. Predicting eclipses would require alertness during only two intervals per year.

But there is one last complication. The Moon's orbit is not fixed with respect to the stars. It is disturbed by various forces. As a result of these forces, the Moon's orbit shows what is called a **regression of nodes:** The orbital plane swings slowly around, always keeping its 5° tilt to Earth's orbital plane. This regression of nodes is analogous to the precession of

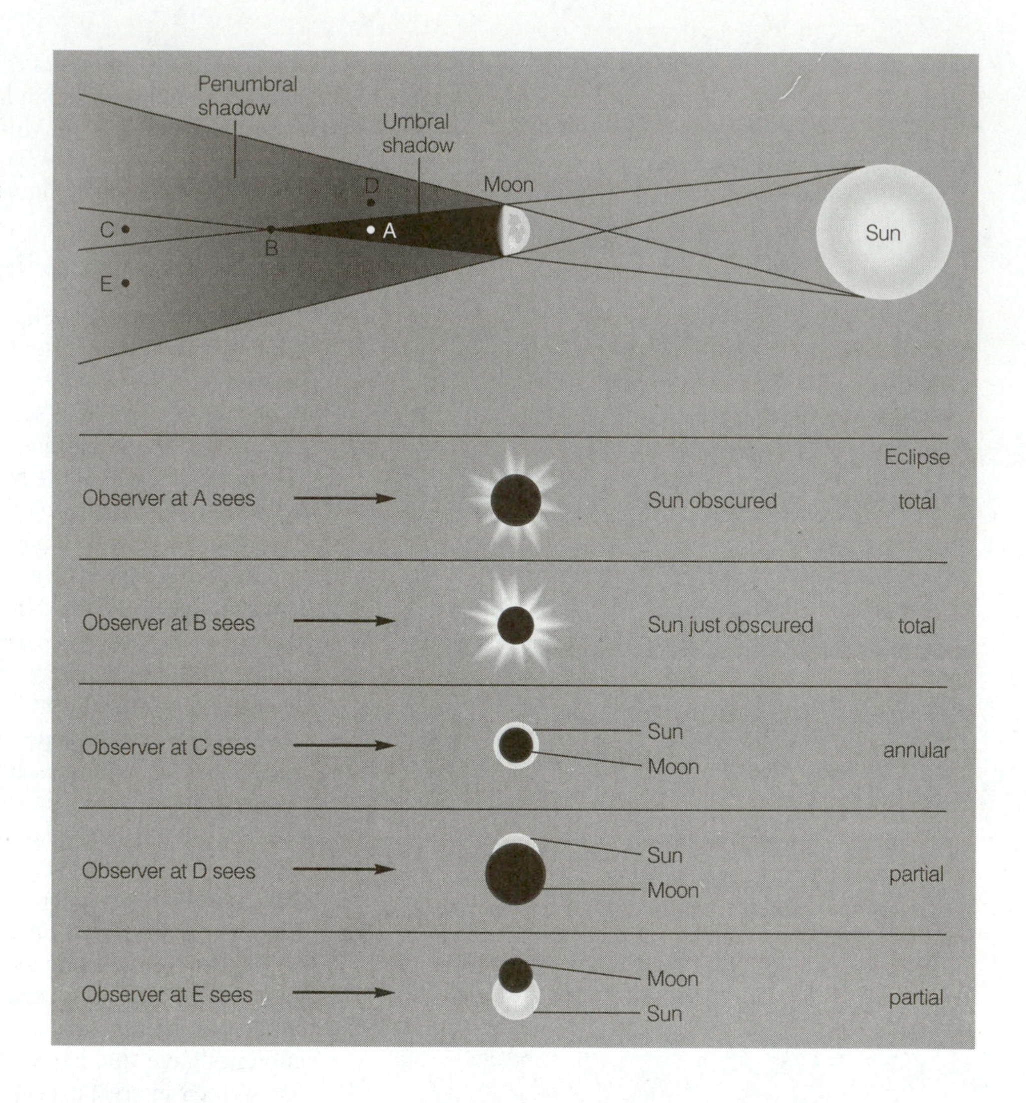

Figure 1-23 Geometry and configurations of solar eclipses by the Moon. Top diagram shows umbral and penumbral shadows of the Moon. Observers at different points see different kinds of eclipses, as shown here. By chance, the tip of the umbra (B) lies very close to Earth's surface.

the Earth's axis. The line of nodes NN′ rotates slowly with respect to the stars, taking 18.61 y to complete one rotation. Therefore the line of nodes aligns itself with the Sun twice in a period somewhat less than a year. This period, called the *eclipse year,* turns out to be 346.6 d.

Thus several simultaneous cycles must be in phase to produce an eclipse: the 29.5-d cycle of lunar revolution with respect to the Sun; the 1-y cycle of Earth's revolution; and the 18.6-y lunar regressional cycle. These overlapping cycles cause subtle periodicities in the occurrence of eclipses. These periodicities helped ancient astronomers discover how to predict eclipses, as we will see in the next sections.

Discovery of the Saros Cycle (c. 1000 B.C.?)

From the information just given, we can understand one periodicity among eclipses. The Moon passes nearest the Sun's direction every 29.5 d (month), and the node N lines up with the Sun every 346.6 d (eclipse year). A lunar or solar eclipse occurs if the two cycles coincide. How often do they coincide? Using figures more exact than those just given, we find that 223 lunar months equal 6585.321 d, while 19 eclipse years equal 6585.781 d. Thus the two cycles come almost exactly into phase (to within only 0. 46 d) every 6585 d, or 18 y, 11 d.

This interval is called the **saros cycle** (a Greek name from an earlier Assyrian-Babylonian word). Early astronomers discovered that if, in a given year, a particular sequence of eclipses occurred, a similar sequence would probably occur after one saros. (Shorter periodicities, such as 41 and 47 months, also exist but are less accurate. Various periodicities were discovered and used by early astronomers to predict eclipses.)

Because the Moon's umbral shadow on Earth is so small, a fixed observer has only a small chance of

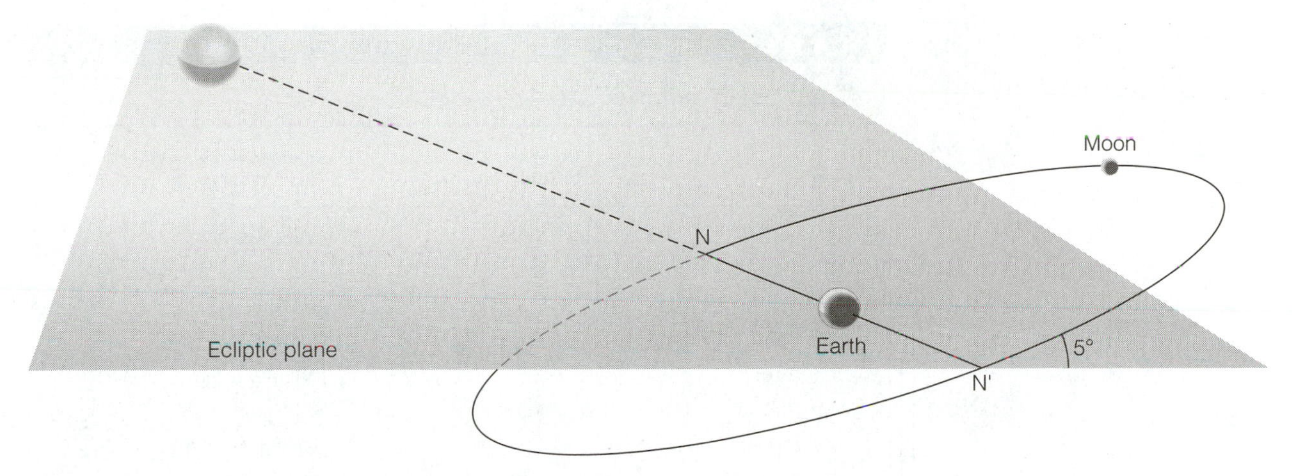

Figure 1-24 The Moon's orbit lies out of the ecliptic plane. Thus only at points N and N′ can Earth, Moon, and Sun be aligned to produce an eclipse.

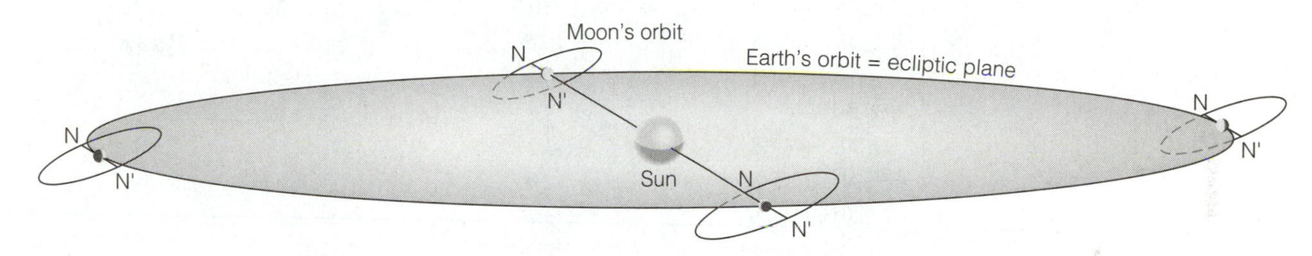

Figure 1-25 As Earth moves around the Sun, there are only two periods each year when the nodal line NN′ of the Moon's orbit aligns with the Sun so that eclipses can occur.

seeing any given solar eclipse. But because Earth's umbra is large and the Moon can be seen from the whole night hemisphere of Earth, half the Earth will see every lunar eclipse (barring cloud cover). Total lunar eclipses are therefore relatively common for any observer, and they were thus easier than solar eclipses for ancient astronomers to predict using the saros and other cycles. Modern astronomers use computers and orbital theory, not cycles, to predict all eclipses accurately.

On the average a given location may witness a lunar eclipse nearly every year and a partial solar eclipse nearly every other year, but a total solar eclipse only about once every four centuries. Many of the dates are related by cycles.

Let us now return to the viewpoint of ancient earthbound peoples. Once records were made and kept, astronomers could benefit from their community's past observations of eclipses. If they discovered periodicities, they could make rough predictions. We know this happened by 700 B.C. in the Mediterranean area and probably by about A.D.500 in Central America, because records found in Assyrian libraries and Mayan cities discuss the prediction of eclipses.

A Famous Eclipse in History and the Discovery of Eclipse Cycles

Thales (c. 580 B.C.) Legend states that one of the first known astronomers, Thales of Miletus, predicted the eclipse that stopped the battle between warring Greek factions in the 580s B.C. Thales may have made this prediction by knowing of the saros cycle from Mesopotamian records, or he may have discovered a useful 3-saros periodicity of 669 lunar months, which would have been prominent in the eclipse records of his region for the preceding 125 y. You can repeat the discovery of the saros cycle by studying Table 1-1, which lists total solar eclipses for 1994 to 2025.

★ Our web site discusses additional examples of eclipses in history, including the Mayan discovery of eclipse cycles (on which they based their calendar!) and an interesting attempt by historians to use eclipses to date the crucifixion of Jesus.

TABLE 1-1

Total Solar Eclipses, A.D. 1994–2025

Date	Duration of Totality (min)	Selected Regions Where Totality Visible
1994 Nov. 3	4.6	Chile, Brazil
1995 Oct. 24	2.4	Iran, India, Vietnam
1997 Mar. 9	2.8	NE Asia
1998 Feb. 26	4.4	Central America
1999 Aug. 11	2.6	Europe, India
2001 Jun. 21	4.9	S. Africa
2002 Dec. 4	2.1	S. Africa, Australia
2003 Nov. 23	2.0	Antarctica
2005 Apr. 8	0.7	S. Pacific
2006 Mar. 29	4.1	Africa, Russia
2008 Aug. 1	2.4	Siberia, China
2009 Jul. 22	6.6	India, China, S. Pacific
2010 Jul. 11	5.3	S. Pacific
2012 Nov 13	4.0	Australia, S. Pacific
2013 Nov. 3	1.7	Africa
2015 Mar. 20	2.8	N. Atlantic, Arctic
2016 Mar. 9	4.2	SE Asia, Pacific
2017 Aug. 21	2.7	United States
2019 Jul. 2	4.5	Pacific, S. America
2020 Dec. 14	2.1	S. Pacific, S. America
2021 Dec. 4	1.9	Antarctica
2023 Apr. 20	1.3	Indian and Pacific Oceans
2024 Apr. 8	4.5	Mexico and central-NE U.S.

Note: Brackets show some pairs of eclipses separated by one saros cycle. You can find other examples in the table. Sequences of eclipses separated by one saros cycle are often visible, at least in partial form, from the same site.

SUMMARY

Unaided by telescopes, unknown geniuses of prehistoric times made many of the most basic astronomical discoveries. They (1) recognized and tracked five planets; (2) discovered the celestial poles and the four directional coordinates defined by daily star motions; (3) learned to use heliacal and solstitial risings and settings to formulate calendars; (4) designed constellations as memory aids for learning the sky; (5) recognized the ecliptic and the zodiac as the paths of the Sun and planets, respectively; (6) may have recognized some effects of precession, which causes stars to shift their positions relative to the celestial poles and the celestial equator; and (7) discovered eclipse-related cycles.

Stable societies encouraged astronomy, and vice versa. Astronomy probably contributed to civilization by creating calendars and encouraging record keeping. The discovery of eclipse cycles in particular required record keeping over many years.

Around 2600 B.C. there may have been a golden age in the Mediterranean world, when modern constellations were sighted and named on or near Crete, major pyramids were carefully oriented on north-south axes in Egypt, and the Stonehenge solstitial observatory copied some of these ideas in England. By around 1400 B.C., building temples with astronomical alignments was a well-developed art in Egypt.

While the recognition of solstices, eclipse cycles, and so on did have applications in ancient societies, it did not lead to visualization of Earth's movement through space. As we will see in the next chapter, the first glimmers of a Sun-centered system came in Greek times, around 200 B.C. These concepts were not fully clarified until the most recent "moments" in the long story of human civilization. Of the 150 generations that have lived since astronomy emerged around 3000 B.C., only the last dozen generations (since about A.D. 1600) have well understood the concept of Earth as a ball moving around the Sun.

CONCEPTS

celestial poles
Polaris
North Star
zenith
meridian
celestial equator
ecliptic
planet
zodiac
heliacal rising
heliacal setting
constellation
circumpolar zone (web site)
equatorial zone (web site)
precession (web site)
cause of the seasons
equinox
solstice
solstice principle
Stonehenge
astrology
eclipse
causes of eclipses
solar eclipse
total solar eclipse
partial solar eclipse
annular solar eclipse
lunar eclipse
umbra
penumbra
node
line of nodes of Moon's orbit
regression of nodes of Moon's orbit
saros cycle

PROBLEMS

1. Suppose you are standing facing north at night. Describe, as a consequence of the Earth's rotation, the apparent direction of motion of each of the following:

a. A star just above the north celestial pole

b. A star just below the north celestial pole

c. A star to the left of the north celestial pole

d. A star on the northern horizon

2. Answer the parts of Problem 1 but for an observer in the Southern Hemisphere facing south and looking near the south celestial pole.

3. What is your latitude? What is the approximate elevation of Polaris above your horizon?

4. What is the radius of the north circumpolar zone of constellations as seen from your latitude?

5. How many degrees from the south celestial pole is the southernmost constellation that you can see from your latitude?

6. Solar eclipses are slightly more common than lunar eclipses, but many more people have observed lunar eclipses than solar eclipses. Why?

7. Does the Moon cast an umbral shadow on Earth during an annular solar eclipse? Why or why not?

8. How would a lunar eclipse look if Earth had no atmosphere? Compare the Earth's appearance from the Moon during such an imaginary eclipse with its actual appearance from the Moon.

9. Why would lunar and solar eclipses each occur once per month if the Moon's orbit lay exactly in the ecliptic plane?

10. Compare ancient European and American cultures of around A.D. 200 to 1000. (For more information on this question check our web site.)

a. How much did each know about eclipses?

b. How many years apart did they achieve a comparable ability to predict eclipses?

c. When, if at all, did they begin to understand the causes of eclipses?

d. During this period, were the two cultures' advances in technology similar to their astronomical advances?

e. Do you think that either of these two types of advances (or the two combined) offers a valid measure of cultural achievement?

ADVANCED PROBLEM

11. Using the following diagram, prove that the angle of elevation λ of the north celestial pole above the horizon equals the latitude of the observer. Why does the angle from the celestial equator to the zenith equal the latitude?

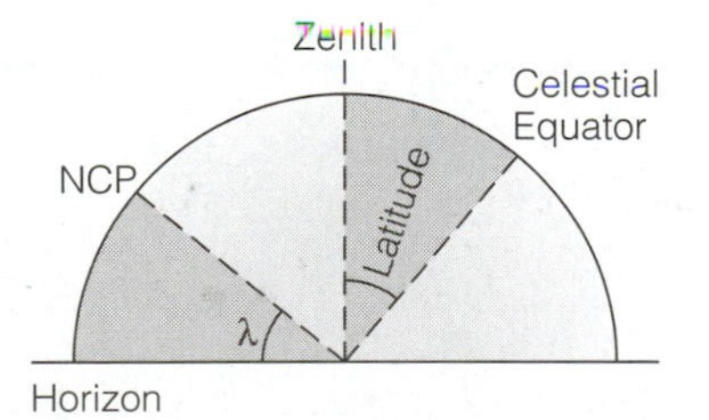

PROJECTS

1. Starting on the date of the new moon (shown on many calendars), observe the sky at dusk and record whether the Moon is visible. Repeat each evening for several weeks and record the Moon's appearance. Repeat at the next new moon. How many days is the Moon visible between the new moon and the first quarter? Between the first quarter and the full moon? How might these counts relate to clusters of grooves, such as 3, 6, 4, 8, . . . , found on prehistoric tools? Can you prove or only speculate that the prehistoric records are lunar calendars?

2. From the preceding project, determine how much later the Moon sets or rises each night.

3. If a planetarium is available, arrange for a demonstration showing the following:

- **a.** The position of the north celestial pole
- **b.** The position of the celestial equator
- **c.** The daily motion of the stars
- **d.** The prominent constellations
- **e.** The daily motion of the Sun with respect to the stars
- **f.** The position of the ecliptic
- **g.** Planetary motions

4. Measure the angle of elevation λ of the North Star above the horizon. A protractor can be used as shown to make the measurement. Compare the result with your latitude.

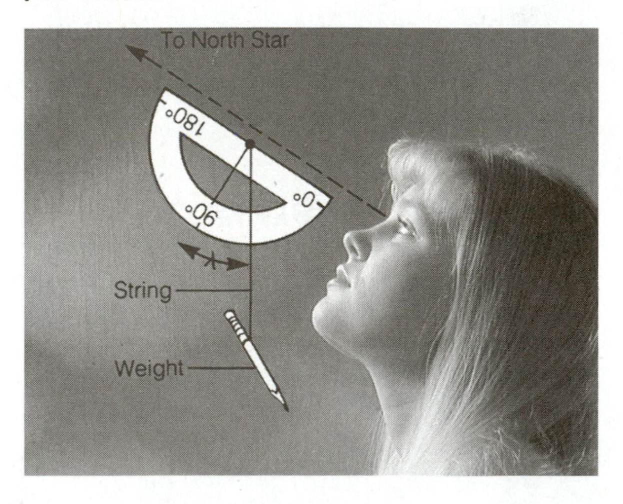

5. Identify a bright planet and draw its position among nearby stars each night for several weeks. (A sketch covering about $10° \times 10°$ should suffice.) Does the planet move with respect to the stars? How many degrees per day? (The latter result will differ from one planet to another and from one week to another.)

6. Find a viewing area with a clear western horizon and determine the date of heliacal setting for some bright star or star group. If several students work independently on the same star, compare results. With how many days' uncertainty can the heliacal setting date be identified?

7. From a viewing area with a clear western horizon, chart sunset positions with respect to distant hills, trees, or buildings for several days and demonstrate the motion of the sunset point from day to day. Do the same for a few days around winter or summer solstice and demonstrate the reversed drift of the sunset position. How accurately can you measure the solstice date in this way? Does the Sun approach the horizon vertically or at an angle?

8. Using a light bulb across the room for the Sun, a small ball for the Moon, and yourself as a terrestrial observer, simulate total, partial, and annular solar eclipses.

9. Using the same props, demonstrate a total eclipse of the Moon. Show why a lunar eclipse occurs only during a full moon.

10. If an eclipse of the Moon occurs while you are taking this course (see Table 1-2), observe it. Compare the visibility of surface features on the Moon before the eclipse, in the penumbra, and in the umbra. Confirm that the curved shadow of Earth defines a disk bigger than the Moon. What color is the Moon and why? Would the Moon be lighter or darker during eclipse if there were an unusually large number of storm clouds around the "rim" of the Earth, as seen from the Moon?

11. Make star trail photos like that of Figure 1-4 or as on the web site by setting your camera at night on some interesting foreground, opening the lens wide open, focusing on infinity, and exposing on a "fast" film (ISO number = 400 or more), for 1 minute, 5 minutes, 10 minutes, 30 minutes, and an hour. Compare the star trails and the sky brightness on the different exposures.

CHAPTER 2

Historic Advances: Worlds in the Sky

WHAT THE READER SHOULD WATCH FOR IN THIS CHAPTER

Historical written records show that the Mesopotamians, Greeks, and Chinese were among the first to develop written catalogs of astronomical phenomena along with attempts to explain what they saw. The Greeks, in particular, around 500 B.C. were the first to realize that geometric principles, such as relationships among angles and triangles, applied in space as well as on Earth. Using these ideas, a number of Greek scholars by around 200 B.C. recognized that Earth is round, measured its approximate diameter, proved that the Sun is further away than the Moon, and crudely estimated the relative distances of the Moon and the Sun. Much of this knowledge was lost when Mediterranean civilization collapsed and many old books were destroyed around A.D. 400–600. Some of it was preserved by Arabic and Islamic scholars and reintroduced into Europe after about A.D. 1000. Mayan astronomers in Middle America were repeating some of the early steps, such as recording eclipses and movements of planets, but most records of their work have been lost. ■

In their epic *Gilgamesh,* the Mesopotamians forecast that their own works would vanish as the wind. As they predicted, their names were lost and many of their discoveries degenerated into myth. But some of their knowledge of nature, along with other ideas from ancient Egypt and the regions of present-day Syria, Iran, Iraq, and neighboring countries, reached the Greeks.

The Greeks, with their blend of practicality and imagination, expanded this knowledge rapidly. We know these first scientists of 2500 y ago as human beings! Historical records give their names and sometimes their personalities and biographical information. Partly because of their work, we now accept without question the Earth's roundness, the nearness of the spherical Moon, and other commonplace but subtle ideas. If you doubt the sophistication of these early scientists, try to measure the diameter of Earth and the relative distances between Earth, the Moon, and the Sun without any telescopes or electronic devices. *They* did.

Because our modern discoveries and our whole approach to science depend on the underlying conceptual or philosophical framework from the past, we use this chapter to look at the attitudes toward nature that the Greeks inherited and developed. We finish with some notes on astronomy in non-Western cultures.

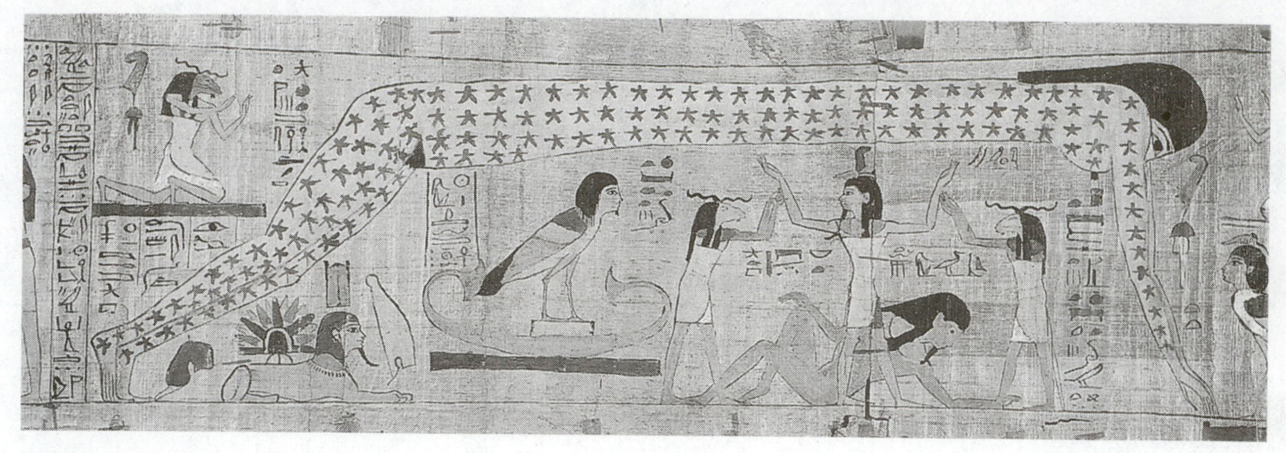

Figure 2-1 An ancient Egyptian conception of the universe. Stars are distributed over the body of the sky goddess, who arches over Earth. Some Egyptian drawings showed several goddesses, arching one over another—an idea that may have carried over to Greek and medieval times, when the planets and stars were visualized as distributed in concentric shells. (Giraudon/Art Resource)

EARLY COSMOLOGIES AND ABSTRACT THINKING (2500–100 B.C.)

Neither Greek scholars nor their systematic observations burst upon the Mediterranean scene from nowhere. The *idea* of thinking about abstract physical concepts can be traced back to early **cosmologies,** or theories about the origin and nature of the universe. For example, certain Egyptian cosmologies assigned different roles to different godlike personages who interacted with the real world. The Memphite theology (c. 2500 B.C.) spoke of an intelligence that organized the "divine order" of the universe (see Figure 2-1). According to Wilson (1951), this theory's "insistence that there was a creative and controlling intelligence, which fashioned the phenomena of nature and which provided, from the beginning, rule and rationale, was a high peak of pre-Greek thinking." This kind of thinking represented a step toward astronomical science, because it assigned different gods, intelligences, or forces to different natural events and sought relations among them. Early naturalists expressed these relations in myths; later naturalists expressed them in *laws,* or generalizations derived from repeated observations.

It is interesting to trace how cosmological thought evolved. Many ideas seem to have carried over directly from the world of 3000 y ago to us today. For instance, the dominant god among all Egyptian gods came to be Amon-Re, to whom the solstitial temple at Karnak was dedicated (see Chapter 1). Around 1350 B.C. the revolutionary Akhenaten, husband of the famous Queen Nefertiti, created a new religion based on the idea of a single god, Aton, the solar disk. Akhenaten's heresy was short-lived because he was overthrown by priests of the old religion. They installed a new young pharaoh, Tutankhamen, whose treasure-filled tomb astonished the world when it was discovered in 1922.

Out of this world, around 1300 B.C., escaped a tribe of nomads—the Hebrews—whose book of monotheistic religious thought (influenced by Akhenaten?) is the core of much religious thought in the Western world today. Some scholars believe their Psalm 104 may be a direct transcription of a hymn to the Sun written by Pharaoh Akhenaten himself and still preserved (Pritchard, 1955). The pharaoh praised the Sun in almost biblical terms: "How manifold it is, what you have made! . . . You created the world according to your desire. . . ." Psalm 104 echoes these ideas: "O Lord, how manifold are thy works! In wisdom hast thou made them all: the Earth is full of thy riches!"

The Hebrews developed a specific cosmology : "In the beginning God created the heavens and the earth. . . ." This cosmological theory asserted that Earth was created in six stages, or days: (1) light, day, night; (2) sky; (3) dry land, ocean, plants; (4) Sun, Moon, stars; (5) sea creatures, birds; and (6) humans. This cosmology dominated much Western art and science (Figure 2-2).

Cosmological theories of this type, which were especially common in the Middle East, stimulated new questions about relations between phenomena

Figure 2-2 English poet/engraver William Blake's version of the Western creation myth reflects the idea that the universe was created according to rational principles. This idea contributed to the birth of modern science, for it allowed philosophers to study the evolution of the universe without considering sudden changes in natural laws that might be caused by a pantheon of capricious gods. (Photograph courtesy of CORBIS-Bettmann)

in the universe. Out of this thinking came an important new idea: Nature could be investigated and not just contemplated. *Independent of questions about gods,* facts could be learned about our world by systematic observations and experiments. Repeatable results could be obtained. A naturalist in Alexandria could get the same results as a naturalist in Athens. **Science** (from Latin "to know") is simply the process of learning about nature by applying this technique: Questions are formulated that can be answered by observations or experiments, which are then carried out. To paraphrase a definition by Edward Wilson in a book about the unification of knowledge (1998): *Science is the systematic enterprise that gathers knowledge about the world and condenses it into testable laws and principles.* In fact, this approach helped teach humans about our relation to the universe and thus helped to refine philosophic/religious ideas and make them more consistent with the realities of the natural world.

This system of scientific observation and recording of nature was developed to the greatest extent in Greece. In addition to richly provocative Egyptian and Hebrew philosophies, the Greek world received a legacy of astronomical concepts, such as the ecliptic, the zodiac, solstices, and the saros cycle, as well as generations of astronomical observations of eclipses and planet motions from Mesopotamian (and perhaps European) sources. This inheritance, plus the Greeks' inclination to philosophize about natural phenomena, led to a Greek renaissance around 500 B.C.

THE SYSTEM OF ANGULAR MEASUREMENT

One of the most important inheritances that the Greeks received from the earlier world was the sexagesimal (60-based) system of measuring angles and time, which was mentioned in Chapter 1. Just as there are 60 min of time in one hour and 60 s in a minute, the system of angular measurement uses the following definitions and symbols:

1 degree	=	1°	=	$\frac{1}{360}$ of a circle
60 minutes of arc	=	$60'$	=	1°
60 seconds of arc	=	$60''$	=	$1'$

To give a better idea of 1 second of arc: It is the angular size of a tennis ball seen at a distance of about 8 mi.

Applying this numerical system for measuring angles, the Greeks developed not only rules of geometry (such as Euclid's geometry and Pythagoras' famous theorem about the hypotenuse of a right triangle), but also ways of measuring phenomena in the sky. With sighting devices, Greeks and other early observers measured the positions of planets relative to the fixed pattern of background stars and the number of degrees of the Sun above the southern horizon as it crossed the meridian at noon in different seasons. These measures first revealed the *detailed* systematics of the movements of celestial bodies.

An important concept is the difference between linear measure and angular measure. **Linear measure** gives the actual length of something in linear units such as inches, meters, or miles. **Angular measure** gives the angle covered by an object (or, alternatively, the apparent separation between two objects) at a given distance from the observer, in angular units such as degrees. The verb **subtend** refers to the angle covered by such an object: For example, we might say a distant object subtends 1°. A useful rule of thumb is that your thumbnail at arm's length subtends about 1°. The disks of the Sun and Moon always subtend $\frac{1}{2}°$. (There is no special reason for this, it just works out by chance.) The pointers in the Big Dipper (see Figure 1-4) subtend about 5°. As the Greeks knew, when you see a distant object (such as a ship at sea or the Moon), you can directly measure not its linear size or distance, but only its angular size. We unconsciously *infer* linear distance of many objects by recognizing the object (such as a ship) and knowing roughly how big it is; we similarly *infer* the linear size by estimating an object's distance and judging it must be as big as a house, a dog, and so on. The treachery of such inferences shows up, for example, when people report unfamiliar aerial objects such as a bright meteor, called a fireball. People commonly report that a fireball "looked as big as a dinner plate," but this statement is literally meaningless; they really perceive only angular size, not linear size. To specify angular size correctly, they would have to say, "It looked as big as a dinner plate at a distance of 50 feet" or "It looked twice the angular size of the Moon." Similarly, the common report that a fireball "must have landed just over the hill" is almost always wrong. The speaker misjudges the distance because he *assumes* a certain linear size after he

OPTIONAL BASIC EQUATION I

The Small-Angle Equation

Angles and linear measures can be combined in an extremely useful and simple equation called the **small-angle equation,** which involves the angular size of an object, its linear size, and its distance. If any two of these quantities are known, the third can be calculated. Let us call the angular size α, expressed in seconds of arc. Let the diameter of the object be d and its distance D. Then the small-angle equation is

$$\frac{\alpha}{206{,}265} = \frac{d}{D}$$

The number 206,265 is called a *constant of proportionality;* it stays the same in all applications of the equation.*

Consider an example. Suppose a friend who is 2.0 m tall is standing across a field, where he subtends an angle of $\frac{1}{2}°$, or 1800″, as shown in Figure 2-3. How far away is he? We want to solve the equation for D. Rearranging the equation, we have

$$D = 206{,}265d/\alpha$$

Using SI metric units (see Appendix 2), we would write $d = 2.0$m. Thus, expressing the equation in SI metric units and powers of 10, we would have

$$D = \frac{206{,}265d}{\alpha}$$

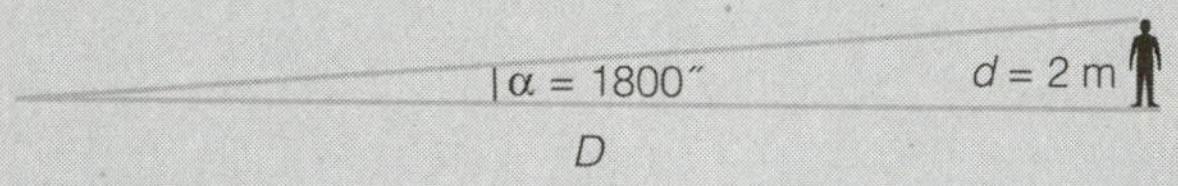

*The number 206,265 is actually the number of seconds of arc in 1 radian (rad). A radian is an angle of about 57°.3, defined as the angle subtended at the center of a circle by one radius of the circle laid along the circumference. The radian has many applications in geometry.

Figure 2-3 An application of the small-angle equation. If your friend is 2 m tall and subtends an angle of $\frac{1}{2}°$ (or 1800″), his distance is D is 230 m.

$$D = \frac{2.06\ (10^5)\ 2}{1.8\ (10^3)} = 2.3 \times 10^2 \text{ m} = 230 \text{ m}$$

Your friend is about one-sixth mi away.

As the Greeks realized, exactly the same geometry can be used to investigate astronomical distances. The Greeks could not get good measurements of the Moon's diameter, but only its angular size α, which is roughly $\frac{1}{2}°$, or 1800″. If we use the modern knowledge that the Moon is about 3500 km in diameter, we can estimate its distance just as we did for the friend's distance above (Figure 2-4). In SI metric units, d would be 3.5×10^6 m. The equation would read:

$$D = \frac{206{,}265d}{\alpha} \simeq \frac{2.06(10^5)3.5(10^6)}{1.8(10^3)}$$

$$\simeq 4 \times 10^8 \text{ m} \simeq 4 \times 10^5 \text{ km}$$

or about 400,000 km.

Several mathematical notes should be observed. First, the symbol $\simeq$ means "approximately equal to"; it is useful whenever approximate values (such as $\frac{1}{2}°$) are involved.

observes only the angular size. Fireballs are typically in the upper atmosphere 60 mi or even 100 mi from the observer!

As we will see in a moment, Greek thinkers carefully separated angular measures from linear measures and used angular measures together with clever logic to estimate linear sizes and distances of the Sun, Moon, and other objects.

The word **resolution** refers to the smallest angular sizes that can be discriminated with optical systems. The human eye, for example, can resolve an angle of about 2′ thus we can see details covering about $\frac{1}{15}$ of the lunar disk. The largest planetary disks subtend only about 1′ and are thus too small for the eye to resolve.

These principles help us analyze photographs and other images. For example, as shown in Figure 2-5, a typical snapshot subtends an angle of about 40°, about the portion of a scene that the eye concentrates on. With modern 35-mm film-based cameras or good-quality digital cameras, the typical 40°-wide snapshot can resolve 2′ details and thus presents a view comparable to what the eye sees. However, 40° views made with smaller or cheaper cameras (such as instamatic and disk cameras) are limited by the inherent graininess of film and often do not resolve details as small as 2′. Thus they look grainy and do not resolve as much as the eye can see from the same spot. American TV images suffer a similar limitation, resolving less detail than the eye can see in the typical

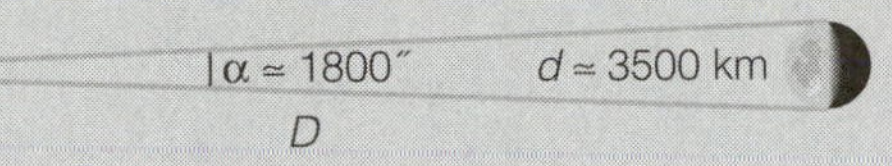

Figure 2-4 The same geometry as in Figure 2-3 can be applied to measure interplanetary distances, such as the distance to the Moon.

Second, this calculation shows the economy of writing the numbers as powers of 10—for example, 3.5×10^6 m instead of 3,500,000 m.

Third, the answer is given only to an accuracy of one significant figure. **Significant figures** are the number of digits known for certain in a quantity. For example, π to one significant figure is 3; to three significant figures, 3.14. Generally, an answer to a calculation should have no more significant figures than does the least accurate number in the equation.† In this case that number was the angular size of the Moon, $\frac{1}{2}$°, or 0.5°, with only one significant figure.

A fourth note is that, following our rule given in the prologue, we convert data to SI metric units. (In our small-angle equation this is unnecessary, since the dimensions of d and D are the same and cancel out; but it is a useful habit.)

†This statement is especially important to users of electronic hand calculators, which blindly print answers with seven or eight significant figures, even when the accuracy of the input is only two significant figures.

The small-angle equation has many applications in astronomy. It lets us calculate the size of distant objects, once we know their distance.

Sample Problem 1 How big are the smallest craters we can see on the Moon with a backyard telescope? *Solution:* Many backyard telescopes can resolve angular detail as small as one second of arc, or 1″. The Moon is 384,000 km away. So we are asking: How big is an object that subtends 1″ at that distance? The student should use the small-angle equation to solve for d and confirm that such a telescope can show craters as small as $1\frac{1}{4}$ km across (roughly a mile across) on the Moon.

Sample Problem 2 The Sun has a diameter of about 1.4 million km and is 150 million km away. Prove that it subtends and angle of roughly $\frac{1}{2}$°. *Solution:* We need to solve the equation for the angle α. This gives

$$\alpha = \frac{d}{D}\, 206{,}265$$

Substituting the values for diameter d and distance D, we get 1925″. since there are 3600″ in one degree, this would be equivalent to 0.5348°. Because the d and D values were given to only two significant figures, the third and fourth digits in the answers cited above are meaningless, and we must round off to two significant figures: 0.53°, or about $\frac{1}{2}$°.

Figure 2-5 These views toward the northeast from the center of Stonehenge (compare map, Figure 1-10) illustrate angular measurements applied to photography with different lenses. **a** This view using an ultra-wide-angle fisheye lens subtends a horizontal angle of 120°. **b** This view with a standard wide-angle lens subtends 65°. **c** This view with a normal lens subtends 40°; this is about the field of view of most snapshots, postcard views, paintings, and so on. **d** This telephoto view subtends only 15°. (Photos by WKH with a 35-mm camera. Lens focal length of *a–d:* 15 mm, 24 mm, 50 mm, and 135 mm, respectively. Curvature of pillars at edges of *a* is distortion common with ultra-wide-angle lenses.)

40° view, which explains the move toward higher definition TV, used in many other countries. Conversely, modern motion pictures photographed and projected with 70-mm film have a dramatically realistic presence because, like life, they can present the eye with a 40° view containing more detail than the eye can resolve.

The ancient system of angular measure, using degrees, minutes of arc, and seconds of arc, is still used today by surveyors, engineers, navigators, astronomers, and others. Modern large telescopes can often resolve details as small as 0.5′, but rarely can do better than this because of the shimmering quality of heat waves in the atmosphere. Satellite telescopes that have been launched into orbit have been able to resolve still smaller angles. Later in this book we will encounter more applications of angular measurement.

EARLY GREEK ASTRONOMY (C. 600 B.C. TO A.D. 150)

Around 600 B.C. the Greeks began vigorously applying logic and observation to learn about the universe. They talked more of tangible physical "elements" and less of metaphysical relations. One of their big advances was to realize that they could apply the same geometric principles to measure cosmic distances that they used to measure farmyards or work out theorems about triangles.

One of the first known Greek thinkers was Thales of Miletus (a Greek-dominated town in present-day Turkey). Living about 636–546 B.C., Thales was a noted statesman, geometer, and astronomer. He is best known for predicting the peacemaking solar eclipse mentioned in Chapter 1. He probably

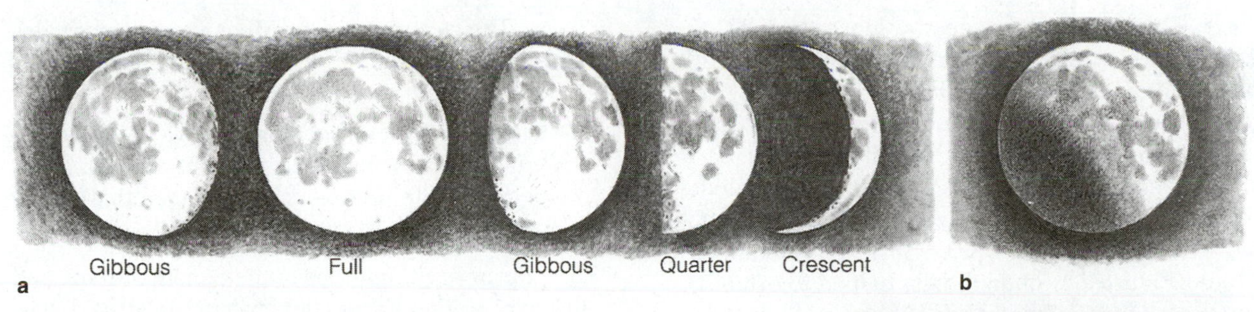

Figure 2-6 Evidence of the Moon's true nature. **a** The phases correctly suggested to some Greeks that the Moon is not a disk but a sphere illuminated by the Sun. **b** The Earth's curved shadow on the Moon during every lunar eclipse suggested that Earth too is spherical. **c** Contrary to popular conception, the Moon is visible in the day during part of the month, as well as at night. Thus its phase can be studied in relation to the Sun, showing that the phases match those of a sphere illuminated by the Sun. (Photo by WKH, Sonora, Mexico)

knew some Mesopotamian astronomical concepts—perhaps the saros cycle, the lengths of seasons, and the daily changing position of the Sun among the constellations of the ecliptic. Thales also reportedly speculated that the Sun and stars were not gods, as was then usually thought, but balls of fire. Of course, Thales could not prove his idea, but he got other Greeks thinking in terms of tractable, physical ideas.

Thales' school produced several notable thinkers. Anaximander (611–547 B.C.) made astronomical and geographical maps; speculated on the relative distances of the Sun, Moon, and planets from our Earth; and argued that the matter from which things are made is an eternal substance. Heraclitus (535?–475? B.C.) made this remarkable comment:

This ordered cosmos, which is the same for all, was not created by any one of the gods or by mankind, but it was ever and shall be ever-living Fire, kindled in measure and quenched in measure. . . . The fairest universe is but a dust-heap piled up at random.

Obviously, human beings were getting serious about trying to figure out the nature of the universe around us!

The Pythagoreans: A Spherical, Moving Earth (c. 500 B.C.)

Pythagoras (flourished 540–510 B.C.), famous for his theorem on right triangles, was also one of the first experimental scientists. Pythagoras proposed the unusual idea that Earth is spherical. He may have gotten this idea by studying the phases of the Moon. The line separating the lit side of the Moon (or any planetary body) from the unlit side (the **terminator**) changes its curvature as the Moon's phases progress, thus revealing that the Moon is spherical rather than flat, as shown in Figure 2-6. By analogy, then, the Earth and other bodies would also be spherical.

In southern Italy, Pythagoras founded a school that had wide influence around 450 B.C. We are uncertain, however, which thinkers in this school should be credited with which ideas. Pythagoras himself put the Earth at the center of the universe, but later Pythagoreans proposed that it moves, like the Moon and the planets, around a distant center. The universe, they said, is spherical with a central "fire" containing a force that controls all motion. Around it, in order outward from the center, move Earth, the Moon, the Sun, the five planets, and the stars. This system presages by more than 2000 y Copernicus' revolutionary, correct proposal that the planets move around the Sun (see Chapter 3). The idea of a spherical Earth persisted among some Greeks, though it was not universally accepted.

★ Anaxagoras (500?–428 B.C.) is credited with deducing the true cause of eclipses. Thereafter, the observed roundness of the Earth's shadow on the Moon (Figure 2-4) undoubtedly helped to establish the theory that Earth itself is a spherical body. After residing in Athens for 30 y, Anaxagoras was charged with impiety and banished for saying that the Sun was an incandescent stone even larger than Greece.

Aristotle: The Earth Back at the Center (c. 350 B.C.)

The most influential Greek scientist-philosopher was Aristotle (384–322 B.C.). His views were built on earlier knowledge but were biased in favor of absolute symmetry, simplicity, and an abstract idea of perfection. Aristotle's universe was spherical and finite, with the Earth at the center. Planets and other bodies moved in a multitude of spherical shells centered on the Earth. The shells were supposed to turn with varying rates, which explained the observed changeable motions of the planets.

Aristotle is credited with founding modern scientific investigation. His school at Lyceum (a grove near Athens) contained a library, a zoo, and lavish physical and biological research equipment paid for by his onetime pupil Alexander the Great, then ruler of Greece. In the Middle Ages, when research lapsed, Aristotle came to be regarded as the final authority, and so his rejection of the Pythagorean idea and his placement of Earth at the center of the solar system turned out to delay progress in astronomy. However, there was little reason for him to choose a Sun-centered over an Earth-centered system, since both views were consistent with observations known in his time.

Aristotle was right about several important astronomical ideas, however:

1. He thought the Moon is spherical.

2. He argued that the Sun is farther away than the Moon because:

 a. The Moon's crescent phase shows that it passes between the Earth and the Sun.

 b. The sun appears to move more slowly in the sky than the Moon. (This second argument is not rigorous, but the first is.)

3. He thought the Earth is spherical because:

 a. The curvature of the Moon's terminator rules out its being a disk, and the Earth seemed likely to be like the Moon in this respect.

 b. As a traveler goes north, more of the northern sky is exposed while the southern stars sink below the horizon—a circumstance that would not arise on a flat Earth.

The apparent motions of the Sun, the Moon, and the stars around Earth could be explained, said Aristotle, either by their actually moving around us or by Earth moving. But Aristotle concluded that Earth is stationary and gave a very powerful argument. If Earth were moving, we ought to be able to see changes in the relative configurations of the various stars, just as when you walk down a path, you see changes in the relative positions of nearby and distant trees. If you line up a tree in the middle distance with a very distant tree and then step to one side, the nearby tree will seem to shift to the side of the distant one. Such a shift in position due to motion is called **parallax,** or a *parallactic shift.* If Earth were moving in a straight line, we would see a continuous parallactic shift of the nearer stars with respect to more distant stars; and if Earth were moving around some distant center, we would see a periodic parallactic shift back and forth among the stars. But a visual survey of the stars and the constellations over time showed no evidence of such a shift. So, reasoned Aristotle, Earth must not move.

The reasoning was sound, but the stars are too far away to produce noticeable parallactic shifts for the unaided eye during a human lifetime. In the same way, distant mountains show little parallactic shift from a car speeding down an interstate highway, even though nearby trees whiz by in seconds. Stellar parallaxes were sought for years and not discovered until 1838, confirming that the Earth really does move with respect to all the stars.

Aristotle died shortly after being forced to leave Athens for allegedly teaching that prayer and sacrifices to the Greek gods were useless.

Aristarchus: Relative Distances and Sizes of the Moon and the Sun (250 B.C.)

Aristarchus (310?–230? B.C.) of Samos (an island off present-day Turkey) must have been a brilliant man. He dramatically extended the Greek methods of seeking quantitative data. Unfortunately, little is known of him, because only one of his original works survived: "On the Sizes and Distances of the Sun and

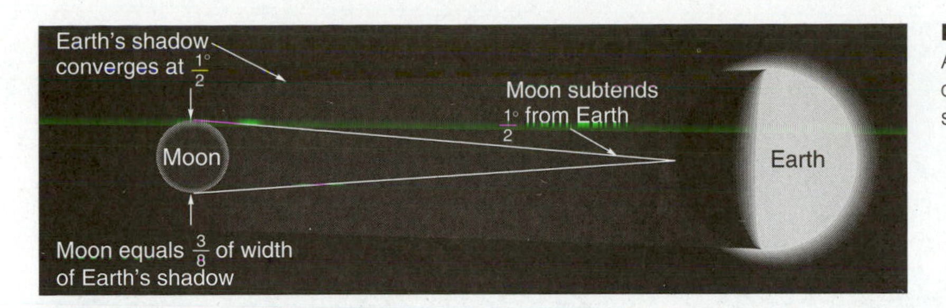

Figure 2-7 The geometry by which Aristarchus estimated the relative sizes of the Moon and Earth (not drawn to true scale).

Moon." However, his other astronomical works are quoted by other Greek authors and show astonishing advances.

Aristarchus devised a way to measure the relative distances of the Sun and Moon from Earth, based on the geometry of the Moon's orbit and phases. From this he correctly inferred that the Sun is much farther away than the Moon.

Aristarchus also formulated a way to measure the relative sizes of Earth and Moon. We can reconstruct his calculations, which used what he knew about eclipses. As shown in Figure 2-7, Aristarchus imagined a total eclipse of the Moon in progress. He drew a circle to represent the Earth. Because he knew the Sun's angular diameter is $\frac{1}{2}°$, he could represent Earth's umbral shadow by lines leading away from Earth and converging at an angle of $\frac{1}{2}°$. Aristarchus also knew that the Moon looks $\frac{3}{8}$ as big as Earth's shadow, so wherever he placed the Moon in the Earth's shadow, its diameter would have to be $\frac{3}{8}$ the distance across the shadow at that point. But where to place the Moon's circle? Aristarchus knew that the Moon, like the Sun, has an angular diameter of $\frac{1}{2}°$ as seen from Earth. These two criteria, when combined, specified the position and size of the Moon, as trial and error with a diagram like Figure 2-7 will show.

What about the Sun? Aristarchus estimated that the Sun is 18 to 20 times as far away as the Moon, based on observations of the moment when the Moon's disk was half illuminated. He then applied the observation that the Sun's angular diameter is $\frac{1}{2}°$, like the Moon's, thus fixing its relative size. Using these methods, Aristarchus concluded that the Moon is one-third as big as Earth and that the Sun is about seven times as big as Earth. The correct figures are closer to one-fourth and 100, but Aristarchus was on the right track. More importantly, he showed how direct logic and geometry could reveal amazing facts about our situation in space—facts that other cultures had never dreamed of.

Aristarchus made still another contribution. Because he thought the Sun is much bigger than Earth, he guessed (without many supporting observations) that the Sun, not Earth, must be the central body in the system. For this an outraged critic declared he should be indicted for impiety.

Although Aristarchus made some quantitative errors, he was nonetheless far ahead of later scholars who thought the Earth was flat. Aristarchus correctly visualized the Moon in orbit around a spherical Earth and Earth in orbit around the Sun, and he developed a method of measuring interplanetary distances. These ideas were not confirmed for another 2000 y!

Eratosthenes: Earth's Size (200 B.C.)

As Greece declined and Rome prospered, Greek scholars became resident intellectuals in many parts of the Mediterranean world. Eratosthenes (276?–192? B.C.) was a researcher and librarian at the great **Alexandrian library** in Egypt. By the time of Eratosthenes, this institution was the leading intellectual community and research site in the world. Eratosthenes reportedly completed a catalog of the 675 brightest stars and measured the $23\frac{1}{2}°$ inclination of the Earth's polar axis to the ecliptic pole, as shown in Figure 2-8. As described in Figure 1-8, this is the tilt that causes our seasons.

Eratosthenes is most famous for using angular geometric relations to measure Earth's size. Told that at summer solstice the Sun shone directly down a well near Aswan, he noted that the Sun's direction was off vertical by $\frac{1}{50}$ of a circle on the same date at Alexandria (Figure 2-9). He realized this difference had to be due to the curvature of the Earth and concluded that Earth's circumference was 50 times the distance from Alexandria to the site of the well. Measuring that distance, he multiplied by 50 and got an estimate of the Earth's circumference. His estimate was probably within 20% of the right answer. This Greek master clearly understood the shape and approximate size of Earth 1700 y before Columbus!

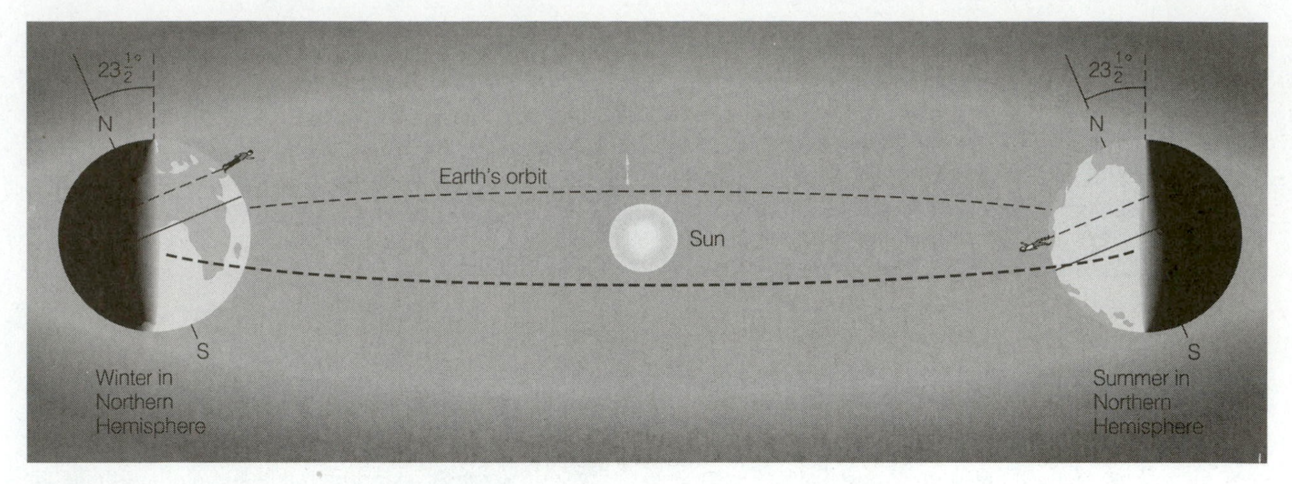

Figure 2-8 The seasons are caused by the $23\frac{1}{2}°$ angle of tilt between Earth's North Pole and the north pole of the plane of Earth's orbit around the Sun. As shown by the human figure, an observer at a fixed northern latitude finds the noontime Sun more nearly overhead in summer than in winter. This difference, which would not arise if the tilt were zero, makes summer days hotter and longer. Eratosthenes measured the difference between the summer and winter noontime Sun's elevation and was able to use it to deduce the $23\frac{1}{2}°$ tilt angle. (The angular difference between the summer and winter noontime Sun elevations twice the tilt angle—a statement the student may try to verify.)

Figure 2-9 The geometry of Eratosthenes' measurement of the size of Earth. When the Sun was directly over a certain well, Eratosthenes measured the Sun's angle θ from the zenith at Alexandria, a known distance D away. He found D was $\frac{1}{50}$ of the way around Earth. He could thus find Earth's circumference and diameter.

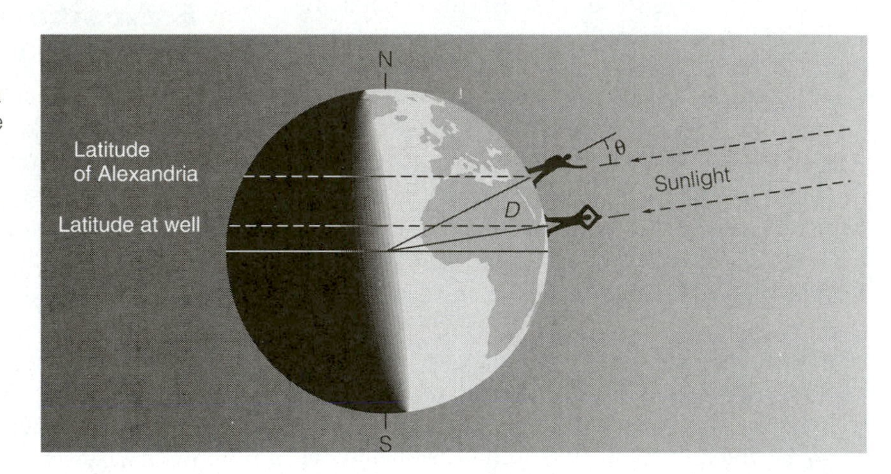

Hipparchus: Star Maps and Precession (c. 130 B.C.)

From his observatory on the island of Rhodes, Hipparchus (160?–125? B.C.) observed the positions of astronomical bodies as accurately as possible and compiled a catalog of some 850 stars. His exhaustive observations—all done, of course, without a telescope—along with material he inherited from Babylon, enabled him to predict with reasonable accuracy the position of the Sun and Moon for any date. Hipparchus has been called antiquity's greatest astronomer.

The most important discovery attributed to Hipparchus is precession (though astronomers of earlier centuries may have been aware of its effects; see the material on the web site for Chapter 1). Comparing his own measurements of star positions with materials handed down to him from centuries before, Hipparchus found that, with respect to the background stars, there had been curious shifts in the positions of the north celestial pole, the vernal and autumnal equinoxes, and other coordinates. The whole celestial equator was oriented somewhat differently with respect to the stars! Could the old maps be wrong? Hipparchus concluded instead that the whole coordinate system of the celestial equator and the poles was drifting slowly with respect to the distant stars. This drift came to be known as **precession** or *precession of the equinoxes.*

In modern terms, *precession is the result of a wobble of the spinning Earth due to forces produced by the Sun and the Moon.* Just as a spinning top describes a conical wobble when it is pulled downward by the force of the Earth's gravity, the spinning Earth's polar axes describe a conical wobble with respect to the fixed stars; hence each celestial pole describes a circle among the stars, shown in Figure 2-10b. The circle, $23\frac{1}{2}°$ in angular radius, is centered on the ecliptic poles. Thus, as mentioned in the web site material for the last chapter, in different millennia different stars become the North Star; a complete cycle takes about 26,000 y. As seen in Figure 2-10b, the star Alpha Draconis was the North Star around 2500 B.C., when Stonehenge was being built. All star coordinates, which are measured with respect to the celestial equator, therefore change slightly each year.

Hipparchus also contributed to the description of solar and planetary motions. Although a few of Hipparchus' contemporaries accepted Aristarchus' idea that the Earth moves around the Sun, Hipparchus and most other astronomers thought this model unnecessarily complex. His own observations showed that the Sun's apparent motion with respect to the stars is not uniform from day to day; he therefore concluded (incorrectly) that the Sun moves around the Earth in a circular orbit whose center is slightly offset from the Earth.

Like so many ancient works, most of Hipparchus' writings have disappeared except in others' reports—according to which Hipparchus also studied the relative distance from Earth to the Moon and the Sun. He calculated that the Moon is $29\frac{1}{2}$ Earth diameters away, close to the correct value of about 30. Hipparchus apparently realized that the Sun must be much farther away than Aristarchus' estimate of 18 to 20 times the lunar distance.

Ptolemy: Planetary Motions (A.D. 150)

Claudius Ptolemy (flourished c. A.D. 140) was another scholar associated with the Alexandrian library. His fame as an astronomer is based on a 13-volume work, *The Mathematical Collection.* Passed on to the Arabs after the destruction of the library, the work became known as *al-Megiste* (The Greatest). European translations were called the Almagest, and the book was famous for more than a thousand years.

Ptolemy extended Hipparchus' star catalog to 1022 entries, correcting older reported star positions to compensate for precession.[1] But his best-known contribution was a method for predicting the positions of the Sun, Moon, and planets, called the **epicycle theory,** or *Ptolemaic theory.* Following Hipparchus, Ptolemy incorrectly assumed that the Earth is near the center of the planetary system, as shown in Figure 2-11. In order outward from the Earth, he placed in *circular* orbits the Moon, Mercury, Venus, the Sun, Mars, Jupiter, and Saturn. To explain why planets apparently do not move at uniform rates, Ptolemy devised combinations of circular motions. Each planet, he said, moves in a circle called the *deferent,* whose center is offset from Earth. The planet does not move strictly on this circle but in a smaller circle called an *epicycle,* whose center moves along the deferent just as the Moon moves around Earth while the Earth moves around the Sun (Figure 2-12). The epicycle theory was simply a case of adding "wheels within wheels" until there were enough wheels to explain the observed irregular movements of the planets. Astronomers who applied this theory and its later changes during the Middle Ages could predict the actual positions of planets within a few degrees. At that time, such predictions were used more often for casting astrological horoscopes than for astronomical studies of the universe!

In spite of its usefulness in ancient times, Ptolemy's theory was incorrect. Today we know that planets move in elliptical orbits around the Sun, not in circles around the Earth. Ptolemy has been criticized for abandoning Aristarchus' Sun-centered system and for biasing his theory toward a supposed perfection of the circle, thus delaying the introduction of the true elliptical orbits. But the system fitted the observations available in Ptolemy's time. Aristotle's old argument still seemed true: The Earth could not move around a distant center, because no one had observed parallax. Certainly, no one had yet realized they were observing elliptical motions. And Ptolemy's system worked fairly well. The trouble with Ptolemy's choice of Aristotle over Aristarchus lies not in its being a mistake, but in its historical effects: The *Almagest* became the Bible of ancient astronomy, and the erroneous Earth-centered system held sway for 1400 y.

[1] Correction for precession continues today. For precise setting of a large, modern telescope, a star's coordinates published for "epoch 2000" or any other year must be corrected for the current date, or the star could be missed. The correction is generally done by a computer operating as part of the telescope.

a

Figure 2-10 **a** This photo shows stars wheeling around the North Star in a 10-min time exposure. Constellation patterns of the Little Dipper and Draco the Dragon are shown. (Wide-angle photo by WKH, 24-mm lens, f2.8, Ektachrome ISO 1600 film.) **b** (*Opposite*) Map of the same region identifies constellations and shows changing position of the north celestial pole in other centuries. Because of precession, the north celestial pole around 2500 B.C. lay near Alpha Draconis. In A.D. 8000 it will lie near the star Alderamin, and in A.D. 15,000, near the very bright star Vega.

The Loss of Greek Thought (c. A.D. 500)

Alexandria, where Cleopatra first fascinated Julius Caesar and Mark Antony around 47 B.C., was the world's intellectual center by A.D. 250. With Rome's fall and the world in disorder in 410, maintaining the great library of ancient discoveries became more and more difficult.

★ Among the last guardians of the old knowledge in Alexandria was one of the first known woman astronomers, Hypatia (c. A.D. 375–415). Widely admired for her intelligence and beauty, she wrote a commentary on Ptolemy's work and invented astronomical navigation devices, but was murdered by a mob during one of the riots that plagued Alexandria during its decline. In A.D. 640, after a 14-mo siege by the Arabs, Alexandria fell. The library buildings were burned, and the best collection of Greek knowledge was lost. Because there was no printing, there were few other reference books. For a description of the library and more about its disastrous fall, see Krisciunas (1988).

Table 2-1 summarizes the dramatic Greek advances.

ANCIENT ASTRONOMY BEYOND THE MEDITERRANEAN

With the fall of Alexandria in A.D. 640, the rate of new discoveries declined in the West, and Europe slipped into the Dark Ages. But intellectual progress occurred in other cultures.

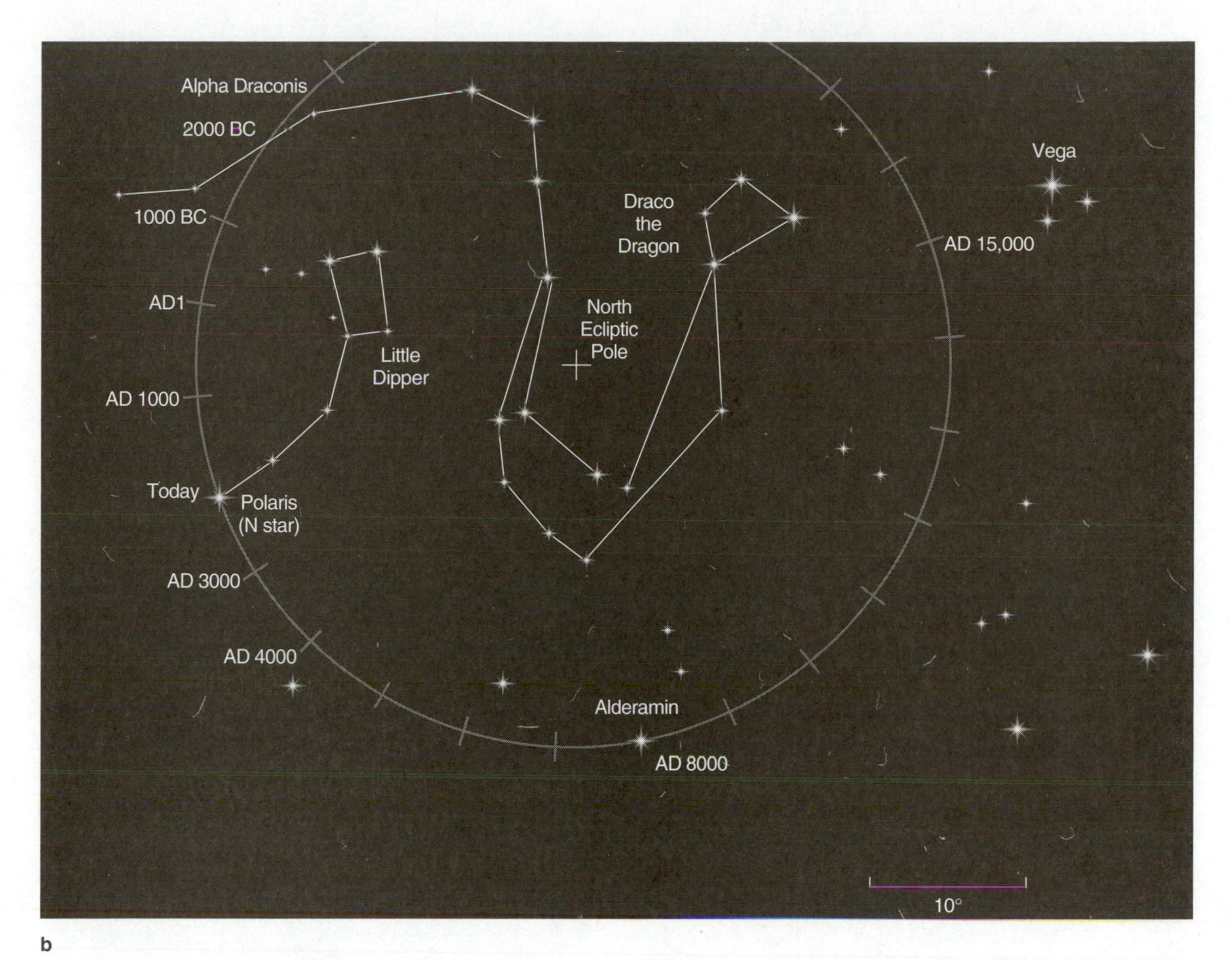

b

Islamic Astronomy

The fall of Alexandria occurred at about the time of Muhammad, the founder of Islam. Much of the Greek and Alexandrian knowledge passed into the hands of the Arabs and Islamic societies. The motivation to understand astronomy remained strong in the world of that time. For example, the time of Passover and Easter was calculated not according to what we regard as the "ordinary" 365-d solar calendar, but rather according to the phases of the Moon (this is still true). Similarly, Calif Umar I, around A.D. 640 interpreted Muhammad's writing in the Koran as requiring a lunar calendar, which is still used in Islamic countries. (See Gingerich, 1986, for a good review of Islamic astronomy.) Thus early Islamic scholars were interested in predicting lunar phases and other astronomical phenomena.

Around A.D. 760, about a century after Muhammad, Islamic leaders in the new capital of Baghdad began to sponsor translations of old Greek texts, spurring interest in new astronomical observations. The next known measurement of the Earth's circumference was made near Baghdad in A.D. 820 and was only 4% too large. Similarly, Arab astronomer Muhammad al-Battani (c. 850–929; known later in Europe as Albategnius) made only a 4% error in his measurement of the eccentricity, or noncircularity, of the Earth's orbit. (He would have called it the Sun's orbit around the Earth.) By A.D. 1000 the Islamic empire had spread to Spain, and astronomical tables were published with the 0° reference longitude in Córdoba (rather than in Greenwich, England, as in the modern longitude system, introduced when Britannia ruled the waves).

Although Mesopotamian-Greek-Arab astronomy was the most direct contributor to modern Western conceptions, other astronomical centers, which initially arose independently, were also important in human development.

Figure 2-11 A medieval conception of the universe from 1537 shows the persistence of Aristotle and Ptolemy's ideas. The central sphere is labeled "Terra immobilis," or immovable Earth. Around it are shells of water, air, and fire and then shells carrying the Moon, Sun, planets, and stars. The term *seventh heaven* comes from the idea that the seventh (outer) sphere was the finest and purest. (The Granger Collection, New York)

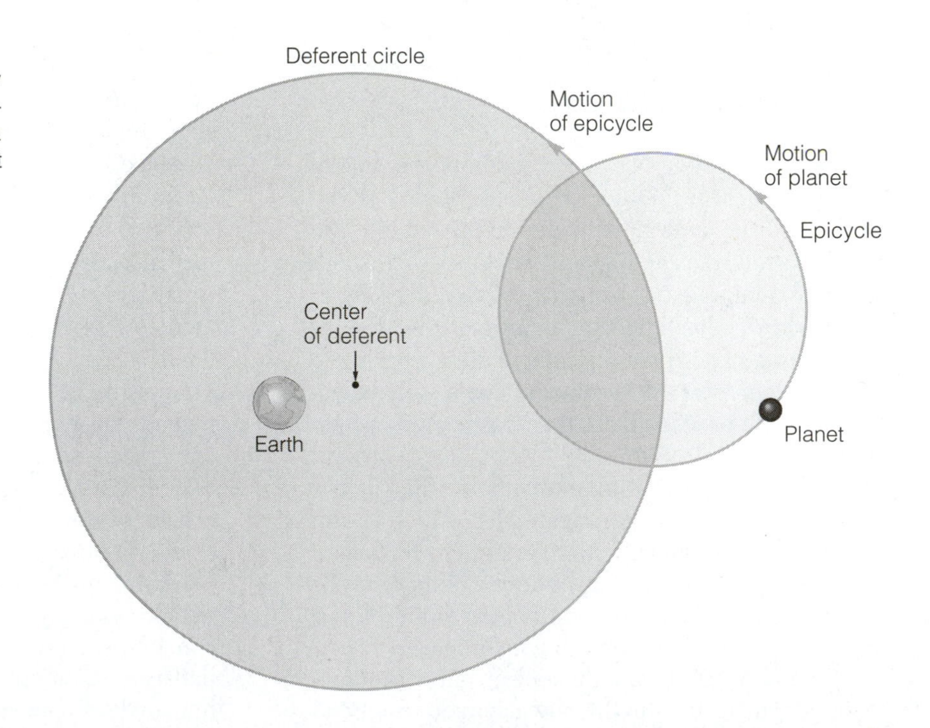

Figure 2-12 Ptolemy's system for explaining planetary motions. Ptolemy thought each planet moved in a circular epicycle, whose center moved in a circular orbit (deferent) around a point near the Earth.

TABLE 2-1

Astronomical Discoveries of the Greeks

Observation	Inference	Observer Commonly Quoted
Curved lunar terminator	Moon round	Pythagoreans
Round shadows during lunar eclipses	Earth round	Pythagoreans
Crescent phases of Moon	Moon between Earth and Sun	Aristotle
Different stars at zenith at different latitudes	Earth round	Aristotle
No evident stellar parallax observed by naked eye	Distance Earth moves is small compared with distances to stars	Aristotle
Relative sizes and angles of Moon and Earth's shadow	Moon smaller than Earth; Sun bigger than Earth	Aristarchus
Angle from first quarter to last quarter moon slightly less than 180°	Sun tens of times farther away from Earth than Moon	Aristarchus
Relation between angular shift of zenith distance of Sun and linear distance traveled on Earth	Calculable circumference of Earth	Eratosthenes
North celestial pole's shift with respect to constellations	Precession	Hipparchus

Astronomy in India: A Hidden Influence

Astronomical practices in India date back to about 1500 B.C. The first known astronomy text, describing planetary motions and eclipses and dividing the ecliptic into 27 or 28 sections, appeared around 600 B.C. By this time India had come into contact with the Mesopotamian and Greek worlds, and influences probably traveled both ways.

Texts dating from around A.D. 450 use Greek computational methods and refer to the longitudes of both Alexandria and Benares, a major Indian astronomical center. Arabs who later visited India wrote of Brahmagupta (588–660?) as one of the greatest Indian astronomers, who reportedly helped introduce Greco-Indian astronomy to the Arabs.

Unfortunately, most records of this fertile early period of Indian astronomy were destroyed during invasions in the 1100s. The great center at Benares was destroyed in 1194, and various university libraries of Buddhist and other ancient literature were burned in religious wars. A massive observatory—one of the world's five major observatories by the 1700s—was reestablished at Benares in later centuries (and damaged again by invading religious fanatics).

Astronomy in China: An Independent Worldview

Chinese astronomy began very early. According to legend, Chinese astronomers were predicting eclipses before 2000 B.C., but more realistic historians place this accomplishment closer to 1000 B.C. Chinese astronomy flourished at the same time as Greek astronomy. Both were probably influenced by mideastern cultures. For many centuries after 100 B.C., ancient Chinese observers kept the world's best records of the mysterious "guest stars" (exploding stars that flare up occasionally—we will encounter them in Chapter 21). The Chinese also made some of the best ancient observations of positions of Halley's comet and other comets.

Much of this was motivated by astrological superstition, but at the same time Chinese astronomers developed strikingly modern conceptions of the universe (Needham, 1959). According to one Chinese view around 120 B.C., "The Earth is constantly in motion, never stopping, but men do not know it; they are like people sitting in a huge boat with the windows closed; the boat moves but those inside feel nothing." This was well ahead of Aristotle's view of the same era, in which a stationary Earth occupied the universe's center. Another text of this period stated, "All time that has passed from antiquity until now is called *chou;* all space in every direction . . . is called *yü.*" The Chinese thus adopted the term *yü-chou* for the universe, prefiguring the modern concept of space-time, the four-dimensional continuum encompassing all that has existed.

A Chinese astronomer reportedly rediscovered precession around A.D. 336, when European science was entering the Dark Ages. Not all Chinese discoveries were independent; ideas flowed in a weak and hidden current all the way from Europe to China, but during some centuries, the Chinese had the world's most sophisticated astronomy.

Astronomy in Japan: A Borrowed Science

★ In view of the technological leadership of Japan in today's world, it is surprising that Japan is not known for its ancient science. Around A.D. 500 and 600, the Japanese court imported scholars from Korea and China who brought calendar systems and other astronomical information, mixed with astrology. Japanese Buddhists held an Earth-centered idea of the universe. From around 1000 to the 1600s, Japan isolated itself from outside ideas, and science stagnated. After the ports were reopened in the 1700s, Chinese and Western ideas spread in Japan.

Native American Astronomy: Science Cut Short

Many people underestimate the sophistication of some of the Native American societies. The high point of their protoscience developed between A.D. 100 and 1200 in a few localized areas, especially in Central America. Here, as noted in Chapter 1, Native Americans built astronomically aligned observatories, tracked the planets, and devised calendars based on eclipse cycles. Their influence reached as far north as the "frontier" of New Mexico-Arizona and the Mississippi Valley. European explorers found Native Americans in some regions maintaining sophisticated calendars by observing solstice dates, planetary positions, and so on.

Particularly interesting is the astronomy that was flourishing in the American Southwest when the Spanish arrived, because it has been preserved and practiced until modern times in the New Mexico pueblos and Arizona Hopi villages. Ethnographers of the 1800s and early 1900s recorded many of the ancient traditions passed down by Indian priests. With interest in archaeoastronomy surging since the 1970s, many of these records are being restudied in order to interpret possible astronomical functions of ancient Indian ruins. Zeilik (1985) gives a good survey of Pueblo astronomy. Priests were charged with studying sunrise and sunset positions in order to *predict* dates such as solstices for religious ceremonies and to set times for planting crops. Using the principles of Stonehenge, they determined dates either by observing the Sun's position on the horizon from a certain spot relative to distant mountains or by measuring the Sun's rising or setting position by observing the pattern of light and shadow cast through windows onto special markers placed in opposing walls. The latter method supports the theory that specially shaped windows found in the ruins of Casa Grande, built in central Arizona around A.D. 1350, were solstice observing sites, as shown dramatically in Figure 2-13.

Closer to the equator, the Mayans created a strange calendar that was still in use when the Europeans arrived. As noted in Chapter 1, it was designed partly to be useful for predicting eclipses. Mayans celebrated the beginning of the new year on July 26! What could have led them to this choice? Once again, it was astronomical observation, but in a non-European tradition. In the tropics ($23\frac{1}{2}°$ north latitude to $23\frac{1}{2}°$ south), but not in Europe, the Sun passes directly overhead at noon on two dates during the year. Perhaps because their horizons were obscured by dense jungle, the Mayans paid as much attention to *zenith* observations of the Sun as to observations of the Sun's position on the horizon. Near the latitude of a major Mayan site, Edzna, in Yucatan, the Sun passes through the zenith, on its way to more southern latitudes, at noon on July 26. Recent archaeological studies show that Edzna was a major city of some 20,000 people in the first centuries A.D. In the courtyard in front of the main five-story pyramid, a cleverly designed stone pedestal (Figure 2-14) allowed priests to measure the important "New Year's Day" when the Sun passed through the zenith (Thomsen, 1984; Malmstrom, 1987, private communication). Probably it was in or near this prehistoric city that the early Mayan astronomers first codified and then ceremonialized July 26 as their New Year's Day.

a

b

Figure 2-13 A prehistoric American Indian observatory, Casa Grande, Arizona, dating from c. A.D.1350. Windows, originally on an upper floor in this four-story adobe structure, were cut in different shapes, apparently to facilitate astronomical observation. **a** A late afternoon view of a "solstice window." **b** A view through the same window at sunset on the summer solstice shows it was built so that the sunset position on the horizon was revealed by a diagonal view through the cylindrical shape. This orientation allowed determination of the day of the solstice each year. Similar methods were used by Pueblo Indians in historic times to calibrate their calendars. (Photos by WKH)

a

b

Figure 2-14 The Mayan site of Edzna lies almost exactly at latitude where the Sun passes overhead on the date of the beginning of the Mayan new year, July 26. In the main temple courtyard **a,** the Mayans erected a stone device to measure this event. At the precise moment when the Sun shines straight down from the zenith, the shadow of the top of the stone covers the entire shaft. At other dates and times (as in photo **b**), part of the shaft is in sunlight. (Photos courtesy V. H. Malmstrom, Dartmouth College)

The Mayans' "alien" astronomy is perhaps the most fascinating example of incipient Native American science. Although it survived until historic times and produced written records of complex planetary observations, astronomical conferences, eclipse predictions, and calendars, all of this was stamped out by the zealous Europeans who were destroying non-Christian practices, and many of the written records were destroyed.

SUMMARY

We have seen how various cultures, preliterate and technological alike, moved toward certain concepts about Earth as a world among other worlds in space. Some of these concepts were purely practical; some, abstract. These movements came in fits and starts and were scattered throughout the world. Progress toward knowledge has not been continuous. Cultures have advanced and regressed, depending on their stability and vigor; scientific knowledge has been lost and regained.

Many of the discoveries reviewed here dealt with the relation of Earth, the Moon, and the planets. Table 2-1 furnishes a good review of the Greeks' key observations. Their advances, among all those of antiquity, were most important in influencing Western scientific thought. Some of the Greek researchers had realized that Earth is a globe and had measured its diameter, as well as the relative sizes and distances of the Moon and Sun. This knowledge was lost to the European world with the destruction of the Alexandrian library, but was partially saved and resurrected in the Islamic world and reintroduced some centuries later into Europe by Islamic scholars. Although conceptual models of the universe differed from culture to culture, all cultures moved toward discovering astronomical relationships.

CONCEPTS

cosmology	significant figures (from Optional Basic Equation)
science	subtend
degree	resolution
minute of arc	terminator
second of arc	parallax
linear measure	Alexandrian library
angular measure	precession
small-angle equation (from Optional Basic Equation)	epicycle theory

PROBLEMS

1. How critical can we be of early theorists who believed the Earth is at the center of the universe? Explain, considering the following questions:

a. Did they have any basis for not putting the Earth at the center?

b. Did either possibility fit the available observation?

c. Did any of the Greeks *prove* that any celestial bodies do or do not revolve around a central Earth?

2. As the Moon goes through its phases:

a. Why is its terminator usually curved?

b. At what lunar phase is the terminator straight?

c. At what lunar phase is the terminator not seen?

3. Contrast the types of astronomical observations available to the Greeks with those available today.

4. Do you agree with the comment by Plato (cited on the web site) implying that astronomical events may have influenced the beginnings of philosophizing about the universe? Could astronomical phenomena have influenced religious concepts?

5. Do you think Aristotle's faith in symmetry and "perfection" helped or hindered his investigations of the universe?

6. Why is the Moon's umbral shadow on the Earth much smaller than the Moon itself? (*Hint:* See Figure 2-7.)

7. How does Hipparchus' discovery of precession prove the existence of earlier, careful astronomical records of stars' positions?

8. Do you think destructive events such as the pillaging of the Alexandrian library or the sacking of the observatory and library at Benares are significant or insignificant in world history? (The answer, of course, requires that you define *significant.*)

9. Does the Sun pass through the zenith every day on the equator? If not, on what dates does it do so?

ADVANCED PROBLEMS

10. At latitude 40° N will the Sun ever pass through the zenith? At latitude $23\frac{1}{2}°$ N? If so, on what date(s)?

11. The angular diameter of the Sun is roughly $\frac{1}{2}°$, and its distance from Earth is about 150 million km.

a. Use the small-angle equation to estimate the Sun's diameter.

b. Could Aristarchus determine the ratio of the Sun's diameter to its distance?

c. Why could the Greeks, such as Aristarchus, not determine the Sun's linear diameter or distance?

12. American TV pictures have about 435 resolution elements (sometimes called picture elements, or *pixels*) along the width of the picture.

a. Assuming a TV picture is photographed with a lens giving the standard snapshot field of view of 40°, calculate the angular size of the smallest details resolved, and compare this with what the eye could see from the same viewpoint.

b. Most European television systems have 20 to 50% more scanning lines across the screen and more resolution elements than American TVs (except in England, where they have about 23% less). Comment on the resolution and sharpness of these TV images.

13. A backyard telescope with aperture greater than about 15 cm (6 in) can reveal features with an angular diameter 1″. The Moon is about 400,000 km away. What size lunar crater could you see with such a telescope?

14. If you landed on a small satellite and found that walking in a 1-m straight line caused the stars in front of you to rise 1° farther above the horizon (while stars overhead also shifted by 1°), what would be the satellite's circumference? What would be its diameter?

15. The photo in Figure 2-6 was taken in the Northern Hemisphere. What time of day was it? (*Hint:* Consider what time of day it would have been if the Moon were in a crescent phase with the horns of the crescent pointing down to the left.)

PROJECTS

1. Using a distant, strong light source such as the Sun or a light bulb, a small ball to represent the Moon, and your eye to represent a terrestrial observer show that crescent phases of the Moon prove that it passes between the Earth and the Sun.

2. If you travel far enough during vacation to change your latitude significantly, compare measurements of the elevation of the North Star (or the Sun during daytime, taking care not to stare directly at it) made from various points in your trip. A sighting device like that described for the problems in Chapter 1 can be used. (Point the device at the Sun by watching its shadow; don't look directly at the Sun.) How accurately can latitude be determined in this way? Measure the number of kilometers or miles corresponding to your change of latitude, and estimate the circumference of the Earth. (This method is similar to Eratosthenes'.) The project can be done as a class effort, with different people's reports of elevation angles plotted against their latitude to give a curve showing how elevation angle changes with latitude. Coordinate with your instructor.

3. With a camera and fairly fast black-and-white film, such as Tri-X, make time exposures of the night sky. Try different exposures, such as 1 min, 5 min, and 1 h. Make one series including the North Star, one toward the eastern or western horizon, and one toward the southern horizon. Explain the patterns made by the trails.

4. During a camping trip or late-evening outing, pick an equatorial constellation in the sky and follow its motion. Make a series of sketches at different hours, showing its position with respect to the horizon. Do the same for a circumpolar constellation and contrast the results.

5. Using star maps, trace out the position of the celestial equator in the sky. Compare this with the position of the Sun's path, the ecliptic.

CHAPTER

3

Discovering the Layout of the Solar System

WHAT THE READER SHOULD WATCH FOR IN THIS CHAPTER

There are specific observational clues to the arrangement of planets in the solar system, such as the motions of Mercury and Venus versus the motions of the other planets. Many Greek and nearly all medieval astronomers thought the Earth was in the center of the planetary system and that all other worlds moved around our world. This reinforced the medieval view that we were a unique and special creation—the center of the whole universe. In the 1500s and 1600s, a profound revolution in science and philosophy reversed this view and proved that the Sun was the center of our planetary system and that Earth was just one of several planets moving around the Sun. The student should be particularly aware of the contributions of Copernicus, Tycho, Kepler, Galileo, and Newton to this revolution. Copernicus hypothesized that the Sun, not Earth, was the center of the system. Tycho observed planetary positions. Kepler took the observations and proved the planets move around the Sun in elliptical orbits. Galileo pioneered the use of the telescope and discovered mountains on the Moon and moons moving around Jupiter, confirming that Earth was not the unique center of the universe. Newton's law of universal gravitation, and his laws of motion, explained the elliptical orbits and revealed that all bodies in the universe attract other bodies by a force called gravity. Because of the work of these five naturalists, humanity learned for the first time the true nature of the planetary system. ■

Astronomy waxed and waned in various parts of the world for a thousand years between A.D. 500 and 1500. But throughout that time, no one realized that the dots of light drifting among the stars from night to night were worlds like Earth, spherical globes pursuing their own orbits around the Sun. Most people knew nothing about Aristarchus' estimates of the distances of the Moon and the Sun. The planets were just dots of light, supernatural orbs, or gods in the sky, moving around Earth, as Ptolemy had specified. The revolutionary change in this view was destined to come from Europe. As a result, we now conceive of the **solar system** as a system of bodies with the Sun at the center and all the planets orbiting around it, along with innumerable, small interplanetary bodies, called asteroids and comets, also orbiting around the Sun.

The change came from an explosion in European knowledge during the Renaissance—the two or three centuries of intense inquiry and exploration culminating in the 1500s.[1] Already, in 1492, Columbus's findings had led to the realization

[1] It is intriguing to speculate how history might have differed if a different culture had made these breakthroughs first—say, the Mayans or Aztecs!

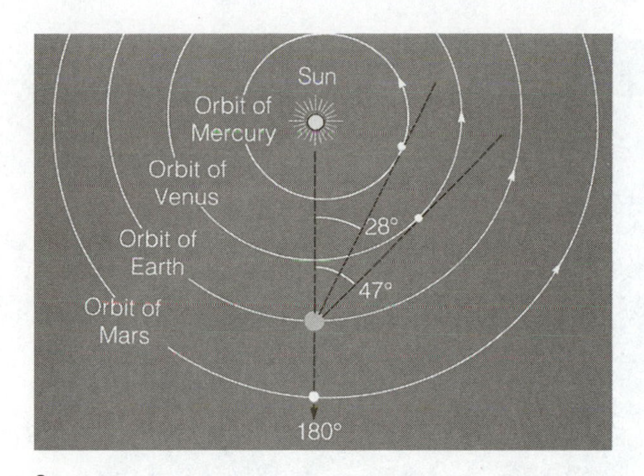

a b

Figure 3-1 Mercury and Venus always appear close to the Sun and therefore can be seen well only in the dawn or evening sky. **a** As seen from Earth, Mercury can never be more than 28° from the Sun, and Venus never more than 47°. Sometimes they can transit in front of the Sun, as seen from Earth. Exterior planets, however, can appear at opposition, 180° from the Sun. **b** Typical appearance of Venus in the dawn or dusk sky. (Photo by WKH, 24-mm-wide-angle lens, f2.8, 25 s on 3M ISO 1000 film)

that the world was not the bottom half of a Heaven-above-Earth-below universe, but rather a sphere. By 1522, Magellan's crew had sailed around it. How did Renaissance naturalists figure out what lay in the sky beyond?

CLUES TO THE SOLAR SYSTEM'S CONFIGURATION

Centuries of accumulated observations of planets' positions in the sky (gathered mostly to refine the calculations of astrologers) led to new ideas about the arrangements of the planets, Earth, the Moon, and the Sun.

We will examine, first, some clues that were discovered even before the telescope was invented. Then we will discuss how the true nature of the solar system was discovered. One important clue is that, to an earthly observer, the planet Mercury never strays more than 28° from the Sun and Venus never strays more than 47°. This means that Mercury and Venus are always fairly close to the direction of the Sun. They are usually in the daytime sky, although, of course, the sky is too bright for them to be seen. (Bright Venus can sometimes be seen in the daytime sky if you know exactly where to look.) All other planets can appear at any angular distance from the Sun along the zodiac. This observation indicates that Mercury and Venus lie closer to the Sun than Earth does, whereas the other six planets are outside Earth's orbit, as demonstrated in Figure 3-1a.

★ Final proof of this arrangement may have come when ancient astronomers observed actual **transit,** or passages of Venus directly between the Earth and the Sun. Only a planet closer to the Sun than Earth could pass between Earth and the Sun. During a transit of Venus, a sharp observer looking through fog or a smoked glass can see Venus as a tiny black spot moving across the Sun, but Mercury is too small to see without a telescope. Transits of Mercury are more common. The next transits of Mercury are May 7, 2003 and November 8 and 9, 2006, and the next transits of Venus are June 8, 2004 and June 6, 2012. Transits of Mercury are more common than those of Venus. More transit information is on our web site.

Retrograde Motion

Planets further from the Sun than we are slowly shift position from night to night, usually west to east relative to the background stars. Of these planets, Mars most dramatically exhibits a motion that puzzled all early solar system analysts. Observers who plotted its position from night to night found that as Mars approaches a point opposite the Sun in the midnight sky, it slows down, reverses itself, and drifts westward in so-called **retrograde motion** for some days before resuming its normal eastward (prograde) motion.

We now know that Mars does not really reverse its motion in its orbit around the Sun. The appearance

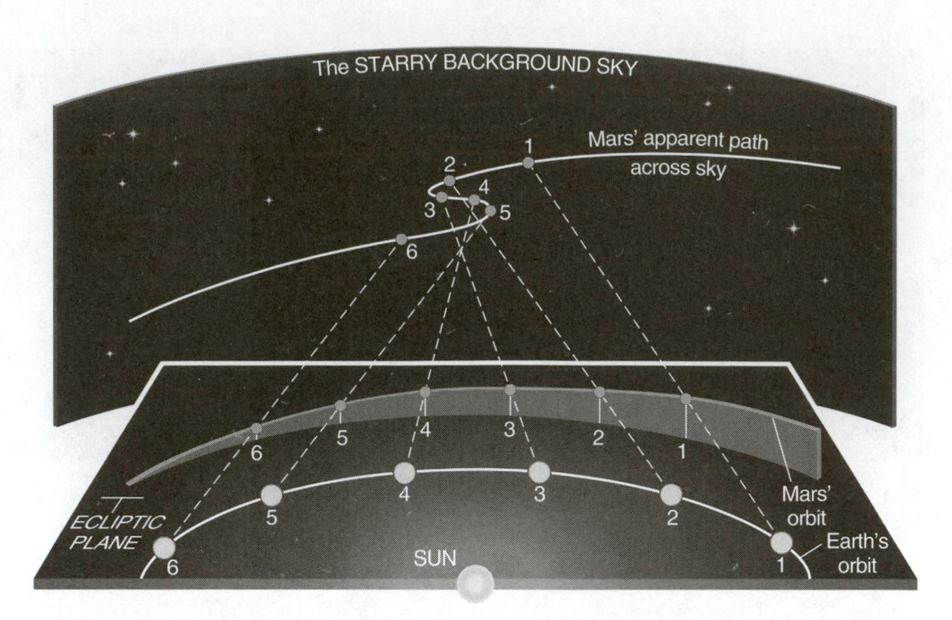

Figure 3-2 A schematic model showing why Mars appears to have a complicated path across the sky. Positions of Earth and Mars are shown for six dates. Note that Mars moves slower than Earth and at a slight angle to Earth's orbit. As Earth overtakes Mars, Mars' apparent path relative to the background stars changes speed and direction. This phenomenon revealed major problems with the Ptolemaic theory of the solar system and led to Kepler's discovery of elliptical planetary orbits around the Sun.

is an illusion caused by the motion of Mars *relative to earthly observers,* as shown in Figure 3-2. The Earth moves around the Sun faster than Mars does, following an "inside track," closer to the Sun. Mars seems to go backward whenever we overtake and pass it.

PROBLEMS WITH THE PTOLEMAIC MODEL

In Chapter 2, we showed how ancient scientists arrived at a solar system model with planets moving in deferent circles and epicycles around Earth, as shown in Figure 2-12. But the motions of the various planets, especially Mars, were destined to cause trouble. Mercury's and Venus' motions were explained by putting their orbits between Earth and the Sun and adjusting their deferent and epicycle rates of motion so they would never get more than 47° and 28°, respectively, from the Sun. Similarly, the Ptolemaic astronomers tried to explain and predict Mars' motions by adjusting its epicycle motion so it would sometimes appear to be moving east to west.

The main goal of this model was to predict the positions of planets for astrological applications. The models attempted to break down the seemingly irregular motions of the planets into combinations of circular motions or uniform cycles. Eudoxus (c. 360 B.C.) reportedly represented planetary motions by a combination of motions of 27 rotating concentric spheres carrying the planets; Aristotle (c. 360 B.C.) reportedly used 55 spheres; Apollonius of Perga (c. 220 B.C.) reportedly increased the number of possible combinations of cycles by introducing epicycles; and Ptolemy perfected the **epicycle model** around A.D. 140. Ptolemy's version of the epicycle model is known as the **Ptolemaic model** of the solar system and is seen in Figure 3-3.

In the *Almagest,* Ptolemy showed how a properly chosen set of epicycles could account for all observations of planetary positions accumulated by his time, and he made relatively accurate predictions of planetary positions. Arab astronomers introduced this scheme into Europe in the early Middle Ages. By the 1200s and 1300s, however, astronomers in Damascus realized that further adjustments were needed and discussed adding small epicycles to the main epicycles.

In 1252, King Alfonso X of Castile, acting as a medieval version of the National Science Foundation, supported a 10-y project conducted by Arab and Jewish astronomers to calculate the extensive Alfonsine Tables, an almanac of predicted planetary positions based on the Ptolemaic model. These tables became the basis of planetary predictions for the next three centuries. However, the solar system now seemed very complicated because of the many overlapping cycles. (On seeing the complexity of the Ptolemaic epicycles, Alfonso is said to have remarked that had he been present at the creation, he could have suggested a simpler arrangement!) New epicycles had to be added to make positions agree with observations, and the calculations became even more complicated.

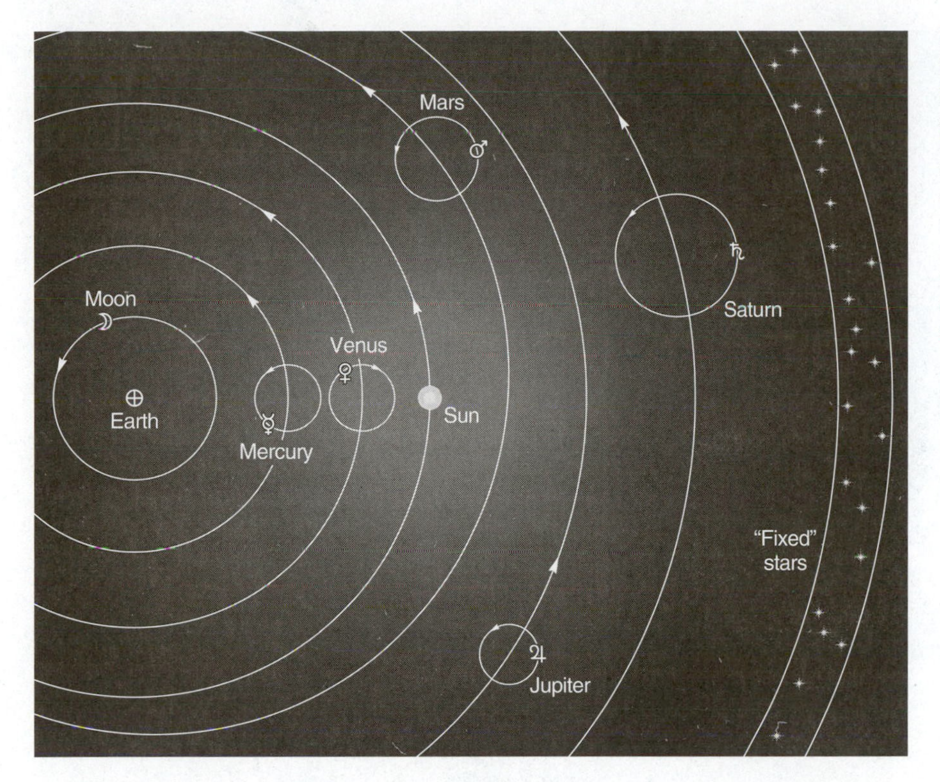

Figure 3-3 The solar system as it might have been conceived by a Ptolemaic astronomer between A.D. 100 and 1500. The diagram is actually simplified, since the Ptolemaic astronomers imagined even more complex hierarchies of multiple epicycles and off-center motions to explain planetary movements. The symbols on the chart are ancient astronomical (and astrological) signs for the planets.

Around 1340 the English scholar William of Occam enunciated his famous principle, called **Occam's razor,** applicable to all branches of science and philosophy. It was called a "razor" because it helped scientists in any given field cut through a thicket of competing theories to find the best theory. Paraphrased in brief, it said:

> **Among competing theories, the best theory is the simplest theory—that is, the one with the fewest assumptions.**

In other words: *The best theory is the one requiring the fewest complications in order to fit the available observations.* This idea may have contributed to suspicion of the tacked-on epicycles. Some European scholars, concerned about how the planets moved and remained suspended, regressed to Aristotle's more simplistic idea of spherical shells (see Figure 2-11). Purbach (c. 1460) thought the spheres were made of transparent crystal with special hollows for the epicycles; Fracastoro (c. 1550) proposed a scheme with 79 spheres!

According to astronomical historian Owen Gingerich (1973a, 1973b), the growing dissatisfaction with the Ptolemaic system was aesthetic (a result of its increasing unwieldiness) as much as intellectual (a dissatisfaction with its results). In addition, obvious errors of a few degrees had accumulated in the predicted positions. One naturalist proposed that perhaps some sort of new model might be developed that would be more "pleasing to the mind." This observer was Nicolaus Copernicus.

THE COPERNICAN REVOLUTION

The **Copernican revolution** was a profound intellectual revolution that abolished the old theory of an Earth-centered universe with the discovery that the Sun is at the center of the solar system, with Earth moving around it. The Copernican revolution took about a century and a half, from roughly 1540 to 1690. It involved five very famous scientists: Copernicus, Tycho, Kepler, Galileo, and Newton.[2]

Copernicus' Theory

Nicolaus Copernicus (Figure 3-4) was born February 12, 1473, the son of a Polish merchant. During his university education in Italy, he became excited by the surging scientific thought of that country. He associated with several astronomers and mathematicians

[2]For historical reasons, Tycho and Galileo are commonly referred to by their first names.

Figure 3-4 Nicolaus Copernicus (1473–1543). Copernicus is holding a model with a central Sun circled by Earth, and Earth by the Moon. (CORBIS-Bettmann)

and made his first astronomical observations at age 24. A few years later, a cathedral post gave him the economic security to continue his observations. At age 31 he observed a rare conjunction that brought all five known planets as well as the Moon into the constellation of Cancer. He found that their positions departed by several degrees from predictions in his set of Alfonsine Tables.

Familiar with several classical alternatives to Ptolemy's system, Copernicus analyzed planetary motions by various methods, including use of small "epicyclets." He soon realized that the solar system would be simpler and the prediction of planetary positions easier if the Sun were placed at the center and Earth placed as one of the Sun's orbiting planets. This was in keeping with the idea of Occam's razor: a simpler theory, getting rid of the epicyclets that were being tacked on to satisfy each new observation, might be more likely to be correct. In 1512, he circulated a short comment (*Commentariolus*) containing the essence of his new thesis: The Sun is the center of the solar system, the planets move around it, and the stars are immeasurably more distant. This comment was not widely distributed, however, and few of Copernicus' acquaintances realized that the work he was pursuing would scandalize and revolutionize the medieval world.

He continued his studies but, fearing controversy, delayed publication for many years. Finally, encouraged by visiting colleagues, including some in the clergy, he allowed the *Commentariolus* to be more widely circulated. News of Copernicus' work spread rapidly. Late in life, Copernicus prepared a synthesis of all his work, *De Revolutionibus* (*On Revolutions,* 1543). In this book he laid out and explained the evidence about the solar system's arrangement:

Venus and Mercury revolve around the Sun and cannot go farther away from it than the circles of their orbits permit [for example, the 47° figure mentioned earlier for Venus]. . . . *According to this theory, then, Mercury's orbit should be included inside the orbit of Venus. . . . If, acting upon this supposition, we connect Saturn, Jupiter, and Mars with the same center, keeping in mind the greater extent of their orbits . . . we cannot fail to see the explanation of the regular order of their motions. This proves sufficiently that their center belongs to the Sun.*

Having thus laid out the correct nature of the solar system, Copernicus tackled the solar system's relation to the stars:

The extent of the universe . . . is so great that, whereas the distance of the Earth from the Sun is considerable in comparison with the other planetary orbits, it disappears when compared to the sphere of the fixed stars. I hold this to be more easily comprehensible than when the mind is confused by an almost endless number of circles, which is necessarily the case with those who keep the Earth in the middle of the universe. Although this may appear incomprehensible and contrary to the opinion of many, I shall, if God wills, make it clearer than the Sun, at least to those who are not ignorant of mathematics.

Now the stage was set for turmoil. Church officials and most intellectuals held that the Earth was at the center. The printer of *De Revolutionibus,* a Lutheran minister, had tried to defuse the situation by extending its title to *On the Revolutions of Celestial Orbs,* as if to imply that the Earth was not necessarily included. He had also inserted a preface stating that the new theory need not be accepted as physical reality but could be seen merely as a convenient model for calculating planetary positions. This gambit did not deter medieval critics. Already Copernicus himself had come under fire from Protestant fundamentalists: In 1539, Martin Luther had called him "that fool [who would] reverse the entire art of astronomy. . . . Joshua bade *the Sun* and not the Earth to stand still!"

In a world of strong dogmas, tampering with established ideas was dangerous. The Reformation era of ideological clashes was no exception. In the 1530s, Michael Servetus had been criticized for certain writings on astrology and astronomy; in 1553, he was burned at the stake as a heretic for professing a mysterious theology that offended both Protestants and Catholics.[3]

Copernicus himself missed the height of the violent debate. He was ill in his last year, and the first copies of his book were reportedly delivered to him on the day of his death, in 1543, at age 70. But the Copernican revolution was under way.

About 1584, a 36-y-old Italian theologian and naturalist, Giordano Bruno, became known for tracts that combined science, mysticism, and theology to vigorously defend the Copernican view as a new vision of the universe that no longer saw Earth as unique, but rather saw the universe as a unity in which all stars and planets were as important as Earth. He believed the warring Protestant and Catholic factions in Europe could be brought together by a better appreciation of the new scientific view of Earth's and humanity's place in the universe (Lerner and Gosselin, 1973, 1986).

Bruno went beyond the Copernican cosmology. The stars, he said, were all worlds like the Sun. Many planets might orbit around them, offering abodes for other races. Bruno traveled throughout Europe, lecturing on the theological and political implications of astronomy and science, often with remarkable imagery that was at once poetic and scientific. For example, he discussed optics using "light" both in the physical and in the theological sense as "divine light." When he returned to Italy in 1592, he was arrested by the Inquisition, a church court established to detect and punish heresy. In 1600, after eight years of investigation of his philosophical and political views, he was burned at the stake.

In 1575, a correspondent wrote to the astronomer Tycho Brahe: "No attack on Christianity is more dangerous than the infinite size and depth of the heavens." Was the Earth to be taken as merely a minor province of the universe? If so, where was heaven? In 1616, the Catholic church banned reading of *De Revolutionibus* "until corrected." It was corrected in 1620 by removing nine sentences asserting that it contained actual fact, not just theory.

Figure 3-5 Tycho Brahe (1546–1601), as shown in an old print. Silver plate on his nose covers a dueling scar.

These sad incidents illustrate the continuing problem of reconciling differences. Historian Will Durant wrote:

The heliocentric astronomy compelled men to reconceive God in less provincial, less anthropomorphic terms; it gave theology the strongest challenge in the history of religion. Hence the Copernican revolution was far profounder than the Reformation; it made the differences between Catholic and Protestant dogmas seem trivial.

Tycho Brahe's Sky Castle

Tycho Brahe (Figure 3-5) was a flamboyant naturalist who wore a silver nose to cover a dueling mutilation. With funds from the king of Denmark, he built the first modern European observatory, named Uraniborg (Sky Castle), at his island home near Copenhagen. From his observations, all made with the naked eye (the telescope had not yet been invented), he made catalogs of star and planet positions. By demonstrating that stars and other bodies show no angular shift in position as our position shifts with rotation of Earth, Tycho proved that stars and

[3] Both Protestants and Catholics were involved in outrageous suppression. John Calvin himself masterminded Servetus' execution, although, in a fit of moderation, he recommended beheading instead of burning. Servetus, a man of wide learning and varied interests, had improved geographic data on the Holy Land and discovered blood circulation in the lungs.

planets were many times farther away than the Moon, for which he *could* detect a shift (called *parallax*).

At age 16 Tycho noticed errors in predicted planetary positions in the same Alfonsine Tables that Copernicus had used. At age 25, in 1572, Tycho saw a temporarily bright exploding star (further discussion in Chapter 19). By demonstrating that it had no parallactic shift, he disproved the popular belief that it was an object in the Earth's atmosphere or near the Earth-Moon system. In 1577, he observed a bright comet and showed that it too was a remote object, far beyond the Moon.

These discoveries were critical in overturning pre-Copernican theories. They meant that new objects could appear in the supposedly unchangeable heavens and that planets could not be attached to crystalline spheres, because such spheres would be smashed by the comets.

These observations inspired Tycho to catalog the precise positions of the stars and planets, which he did between 1576 and 1596. Unable to convince himself that the Earth could move, Tycho invented a compromise solar system in which the Earth was central and stationary, but other planets were placed in the correct sequence from the Sun.

After the king of Denmark withdrew his pension, Tycho moved to Prague in 1599, where he was joined in 1600 by a 30-year-old assistant named Johannes Kepler. When Tycho died in 1601, Kepler inherited the great compendium of Tycho's observations, with all its potential for fruitful analysis.

Figure 3-6 Johannes Kepler (1571–1630). (CORBIS-Bettmann)

Kepler's Laws

Devoutly religious and a believer in astrology, **Johannes Kepler** (Figure 3-6) was sure that planetary motions must be governed by hidden regularities—"the harmony of the spheres." With Tycho's material, Kepler first went to work on the orbit of Mars, which presented the most notable case of retrograde motion. This movement had plagued astronomical theorists since Ptolemy. Kepler found something astonishing: After all the centuries of debate over the arrangement of circular orbits, the orbit that fitted Mars' motion best was not a circle at all, but an ellipse. **Ellipses** are roughly egg-shaped figures that can range from nearly circular to highly flattened, elongated loops. Each ellipse is symmetric around two inner points called **foci** (singular **focus**). Kepler found that Mars' orbit is an ellipse that is almost circular and that the Sun lay exactly at one focus. Eventually, this was found to be true of every planet's orbit. We will describe these elliptical, or so-called Keplerian, orbits in more detail in the next chapter.

Kepler went on to discover two other related principles, and these *three laws of planetary motion* were published in two books, *New Astronomy* (1609) and *The Harmony of the Worlds* (1619). **Kepler's laws** describe how the planets move (without attributing this motion to any more general physical laws), show that the Sun is the central body, and allow accurate prediction of planetary positions:

> **1. Each planet moves in an ellipse with the Sun at one focus.**
>
> **2. The line between the Sun and the planet sweeps over equal areas in equal time intervals.**
>
> **3. The ratio of the cube of the semimajor axis to the square of the period (of revolution) is the same for each planet. This is sometimes called the harmonic law.)[4]**

[4]The major axis of an ellipse is its longest diameter; the semimajor axis is half that. Since most planets in the solar system have nearly circular orbits, the semimajor axes of their orbits are essentially the orbital radii.

Although the planetary orbits are ellipses, they are only slightly elliptical—that is, they are nearly circular. This is why the Ptolemaic system of circles worked as well as it did.

In the case of the Earth, the semimajor axis, or average distance from the Sun, is 1 **astronomical unit** (AU), and the other planets' distances are measured in multiples of this unit. Working in astronomical units and years, we can easily confirm the third law numerically, using a as the semimajor *axis* and P as the period. For the Earth,

$$\frac{a^3}{p^2} = \frac{(1\text{ AU})^3}{(1\text{ y})^2} = \frac{1}{1} = 1.00$$

The same formulation works for other planets.

Kepler's laws explained the apparent retrograde motion of Mars: One consequence of the laws, taken together, is that any planet moves faster than any other planet farther away from the Sun. Thus Earth moves faster in its orbit than Mars (see Figure 3-2). Therefore, Earth catches up to Mars like the faster driver on the inside track at a race. As explained earlier in this chapter, this creates the illusion of retrograde motion.

Galileo's Observations

Kepler's laws might not have gone so far in establishing the Copernican model of the solar system had it not been for the contemporary invention of the telescope and for extensive observations by an Italian scientist, **Galileo Galilei** (Figure 3-7). Unlike Kepler, Galileo had a superbly practical turn of mind. For example, after reportedly watching the regular swing of a lamp in the Pisa cathedral, he applied the periodic motion of the pendulum to regulate clocks. As early as 1597, Galileo wrote to Kepler: "Like you, I accepted the Copernican position several years ago. . . . I have not dared until now to bring [my writings on this] into the open."

Galileo perfected the telescope and began astronomical observations with it in late 1609. By 1610 he had made some of the most important observations ever. For example, he found four satellites revolving around Jupiter—proving at last that some bodies do not revolve around the Earth. Also, he found that the planet Venus undergoes a variety of phases, from crescent to nearly full. The full phase implies a situation with Venus on the far side of the Sun. Note in Figure 3-3 that in the Ptolemaic theory, Venus' epicycle was between the Sun and Earth, implying that only crescent phases could exist. *Here, then, was proof that the Ptolemaic model of Venus' orbit was wrong.* The Copernican model, however, fit the observation. As a final example, Galileo discovered mountains on the Moon and spots on the Sun, showing that these were not polished celestial orbs, as the ancients surmised. In particular, Galileo emphasized that the Moon was a *world,* with geological features, like Earth. These discoveries electrified European intellectuals. Other early telescope users, such as Thomas Harriot in England, duplicated Galileo's observations of Jupiter's moons, Venus' phases, lunar mountains, and sunspots by the end of 1610, but were not as widely known as Galileo.

Figure 3-7 Galileo Galilei (1564–1642). (CORBIS-Bettmann)

Because Galileo wrote in Italian rather than Latin, he built a popular following outside the universities. Reactionary academics and churchmen saw Galileo as a threat and soon began criticizing him. His invitations to them to look through his telescope and see for themselves led nowhere. Some looked and said they saw nothing; some refused to look; some said that if the telescope had been worth anything, the Greeks would have invented it.

From 1613 to 1633, Galileo was in frequent contact with church authorities, even in Rome. In 1616, a cardinal read Galileo an order that he must not "hold or defend" Copernican theory, though he could discuss it as a "mathematical supposition." In 1632, Galileo's great book *Dialogue of the Two Chief World Systems* appeared. It featured a fictionalized debate between Copernican and Ptolemaic advocates. In 1633, 69-y-old Galileo was ordered to Rome to stand trial before the Inquisition, where a curious episode occurred. The court produced a purported copy of the cardinal's order of 1616 telling Galileo not to "hold or defend" Copernicanism. The Inquisition's copy, still in the files today, also ordered him not to "teach" or "discuss" it in speech or in writing (contrary to the cardinal's 1616 order). The document in the Inquisition files is not the original and lacks the names of the alleged witnesses. Historians now suspect this document was a fraud created to frame Galileo.

The Inquisition jurors were inclined to be lenient only if Galileo repudiated his work. The elderly Galileo saw no point in getting himself killed; his book was already published, and he had faith that intelligent people could see the truth through telescopes or in print. So he recited a prepared recantation and was sentenced to prison, a sentence commuted by the pope to house arrest on Galileo's own estate, where he died in 1642.[5]

Newton's Synthesis

In spite of the Inquisition, the evidence for a Sun-centered solar system accumulated so rapidly that the Copernican revolution was almost complete. The main element still lacking was an overall theoretical scheme that would draw together Kepler's empirical laws of planetary motion into a concise physical explanation of the behavior of the solar system. To be intellectually satisfying, this theory needed to start with a few universal principles and show that the Keplerian orbits and the Galilean satellite motions *had* to exist as a consequence of these principles. The man who achieved this synthesis—the man usually deemed the greatest physicist who ever lived—was **Isaac Newton** (Figure 3-8).

Isaac Newton was a father of physics and astronomy. Between the ages of 23 and 25, while attending Cambridge, he almost single-handedly developed calculus, discovered the principle of gravitational attraction and certain properties of light, and invented the reflecting telescope (in which a curved mirror replaces a lens). Newton once said that he made his discoveries "by always thinking about them," a trait that no doubt contributed to his reputation for absentmindedness.

Figure 3-8 Isaac Newton (1642–1727). (The Granger Collection, New York)

At age 41, Newton began writing his famous *Principia,* a revolutionary compendium of physics, and published it three years later in 1687. He became president of the Royal Society at 60, died at 84 in 1727, and was buried in Westminster Abbey. Of Newton, Alexander Pope wrote:

Nature and Nature's laws lay hid in night:
God said, "Let Newton be!" and all was light.

In 1726, an acquaintance gave this account of Newton's early thoughts on gravity, based on conversations with Newton (Ball, 1972):

The first thoughts [were those] he had when he retired from Cambridge in 1666 on account of the plague. As he sat alone in a garden, he fell into a speculation on the power of gravity: that as this power is not found sensibly diminished at the remotest [height] to which we can rise, neither at the tops of the loftiest buildings, nor even on the summits of the highest mountains, it appeared to him reasonable to conclude that this power must extend much farther than was usually thought; why not as high as the Moon, said he to himself?

[5]There are several postscripts. In 1757, Galileo's book was removed from a list of books banned by the Catholic church. In 1983, Pope John Paul II took up Galileo's case, and in 1992 he formally proclaimed that the Catholic church had erred in condemning Galileo. He also stated that science is a valid realm of knowledge, "which reason can discover by its own power."

The Moon must be attracted to Earth by some force, Newton thought, because it does not travel in a straight line, as it would if no force were pulling on it. Reasoning in this way, Newton was able to deduce one of the most important discoveries in the history of science. It is called **Newton's law of universal gravitation:**

> **Every particle in the universe attracts every other particle with a force proportional to the product of their masses and inversely proportional to the square of the distance between them.**

We will explore some of the ramifications of this law in the next chapter. Here we will simply note that the attraction of every particle for every other particle gave a quantitative explanation at last of why the planets follow orbits around the Sun instead of flying off into interstellar space in a straight line: The massive Sun, in the center of our solar system, *attracts* the planets. If the Sun suddenly vanished, the planets would indeed fly away!

Once Newton discovered that masses attract each other gravitationally, he concluded that gravity is the *only* force involved in keeping the planets moving around the Sun. But this solution led to another riddle for natural philosophers of the day: How could the Sun influence the planets in their Keplerian orbits if it never touched them and always stayed at such a great distance from them? And how could the planets stay in the sky without spheres to hold them up?

Newton answered all these questions of "action at a distance" with three simple laws of motion and his law of gravity, the basis of most modern physics except for the corrections made necessary by work on relativity during the present century. These laws were enunciated in Newton's book *Principia* in 1687. *They are quite unlike Kepler's three laws.* They are not merely empirical rules based on observation, but *fundamental postulates* from which Kepler's laws and many other phenomena can be predicted. **Newton's laws of motion** are:

> **1. A body at rest stays at rest, and a body in motion moves at constant speed in a straight line unless a net force acts on it.**
>
> **2. For every force acting on a body, there is a corresponding acceleration proportional to and in the direction of the force and inversely proportional to the mass of the body. In other words, force = mass × acceleration.**
>
> **3. For every force (sometimes called action) on one body, there is an equal and opposite force (called reaction) acting on another body.**

The elliptical orbits discovered by Kepler follow from Newton's laws. An important exercise in advanced astronomy courses is to derive all three of Kepler's laws from Newton's laws. This exercise shows that if Newton's laws are true, the Copernican theory and Kepler's laws also have to be true. This was a fantastic advance! A mathematician, sitting in a room and starting with Newton's simple law about gravitational attraction and the assumption that the Sun is a massive body, could predict that planets must go around the Sun in the kinds of orbits that it took observers 2000 y to discover by direct observation.

Newton's laws thus tidied up the miscellaneous observations of preceding centuries and completed the Copernican revolution. If the force of gravitation had a *different* form than given by Newton's universal law, then orbits might be nonelliptical. The law of gravitation, when combined with the first and second laws of motion, shows why planets do not move in straight lines but are always deflected toward the Sun. The third law explains why rockets work and ultimately led to artificial satellites and spacecraft orbiting moons and planets in Keplerian orbits, as we will see in the next chapter.

By the time of Newton's death, at age 84 in 1727, the solar system was conceived essentially as we see it today, lacking only the discovery of the three outer planets, Uranus, Neptune, and Pluto. Subsequent astronomical observations have shown that Newton's laws also apply in all other parts of the universe that we can see. They correctly predict properties of certain pairs of stars that orbit around each other, far beyond our own solar system.

The Solar System at the End of the Copernican Revolution

We can hardly overemphasize what an important philosophical and scientific advance was wrought by Copernicus, Tycho, Kepler, Galileo, and Newton, as summarized in Table 3-1. To see its effect, compare Figure 3-9 with Figure 3-3. Figure 3-9 contrasts the Copernican solar system, as perceived after Kepler and Galileo, with the Ptolemaic system in Figure 3-3. Earth is no longer at the center but is relegated to an orbit like any other planet. Mercury and Venus are no longer trapped on deferents between Earth and Sun. None of the planets follow epicycles. Jupiter and

TABLE 3-1

Five Key Figures in the Copernican Revolution

Name	Dates	Contribution
Nicolaus Copernicus	1473–1543	Proposed circular motions of planets around Sun
Tycho Brahe	1546–1601	Recorded planets' positions
Johannes Kepler	1571–1630	Analyzed Tycho's records; deduced elliptical orbits and laws of planetary motion
Galileo Galilei	1564–1642	Made telescopic discoveries supporting Copernican model
Isaac Newton	1642–1727	Formulated laws of gravity and used them to explain elliptical planetary orbits

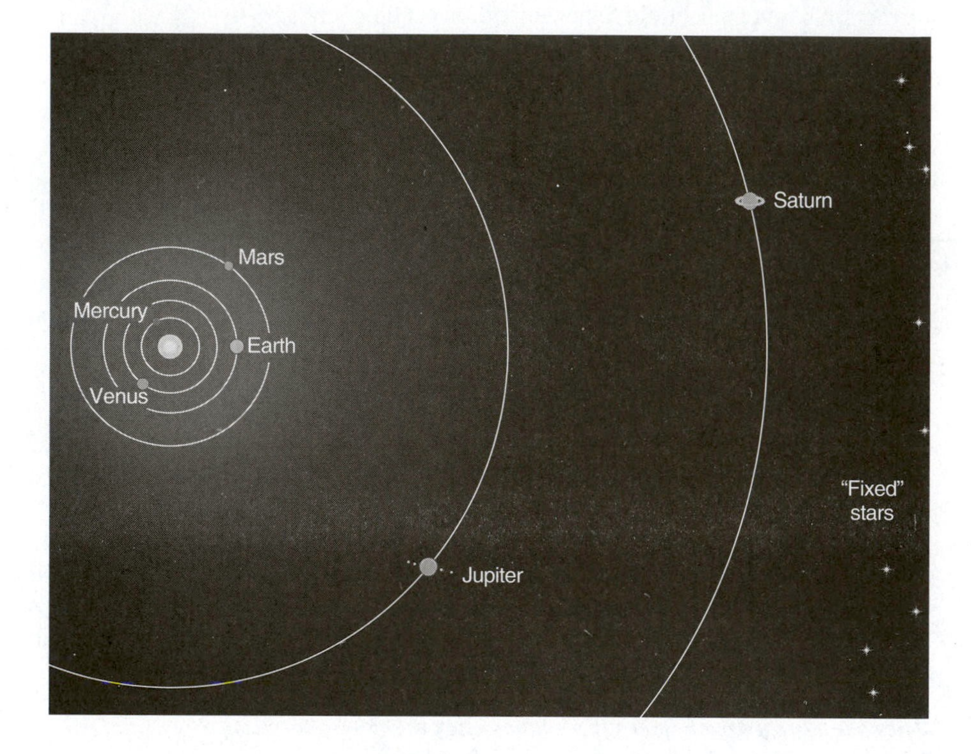

Figure 3-9 The solar system as it might have been conceived by an astronomer around 1700, at the end of the Copernican revolution. This diagram shows the true orbits of the then-known planets to true scale. The view is essentially the correct, modern view, except that the outermost planets (Uranus, Neptune, and Pluto) and the asteroids had not yet been discovered. Compare with Figure 3-3 to see the change from the Ptolemaic view.

several other planets are centers of their own systems of satellites.

BODE'S RULE

A curious relationship discovered by the German astronomer Titius and popularized by his colleague Johann Bode in 1772 is helpful in memorizing the distances of the planets from the Sun. **Bode's rule** is: Write down a row of 4s, one for each planet, and add the sequence, 0, 3, 6, 12, 24, and so on, doubling each time, as shown in Table 3-2. By dividing the sums by 10, you get the number of astronomical units (AUs) between each planet and the Sun. (In memorizing this rule, be sure to start with a 0 when you add the numbers; check it by being sure that Earth comes out at 1.0 AU.)

Because Bode's rule, unlike Kepler's laws, does not necessarily follow from Newton's laws, it is considered more descriptive than explanatory, and not a law of physics. It is merely a handy way to remember planetary positions. Nonetheless, it indicates that the planets formed in such a way that each

TABLE 3-2

Bode's Rule: Distances of Planets from the Sun

	Mercury	Venus	Earth	Mars	Asteroids	Jupiter	Saturn	Uranus	Neptune	Pluto
	4	4	4	4	4	4	4	4	4	4
	0	3	6	12	24	48	96	192	—	384
Predicted Distance	0.4	0.7	1.0	1.6	2.8	5.2	10.0	19.6	—	38.8
Actual Distance	0.4	0.7	1.0	1.5	2.8	5.2	9.5	19.2	30.0	39.4

Note: All distances are expressed in astronomical units (1 AU = average distance of Earth from the Sun).

planet was nearly twice as far from the Sun as the next inner planet. This was because the planets had to be far enough apart that the gravitational forces of one did not disturb the formative process of its neighbor.

Bode's rule has played an interesting role in the discoveries about the solar system. In 1781, Bode's rule was strengthened with the discovery of Uranus at its predicted position, about twice as far from the Sun as Saturn. Astronomers then noted that the rule predicted a planet between Mars and Jupiter. German observers, nicknamed "celestial police," set out to find the missing planet, but an Italian observer beat them to it. In 1801, the Italian discovered the first and largest asteroid, Ceres, at just the right distance! Ceres might have become known as the smallest planet, except that within a few years the "celestial police" found three more asteroids at about the same distance. Today we know of thousands of asteroids between Mars and Jupiter. The asteroids more or less confirmed Bode's rule, though in this case the planet-forming process did not go to completion.

In 1846, the discovery of Neptune somewhat reduced the credibility of Bode's rule by putting a planet outside a predicted position, though the 1930 discovery of Pluto did put a small planet at roughly the next predicted position.

THE SOLAR SYSTEM AS WE KNOW IT TODAY

Figure 3-10 shows the solar system as we know it today. It is more complex and interesting than the simple system of Sun, Earth, and five other planets known to the ancients. First, we see the orbits of eight major planets (solid curves) spaced evenly and more or less obeying Bode's rule. Four small planets, including Earth, are close to the Sun; four much larger "giant" planets orbit further away from the Sun. We will study the properties of all these planets in later chapters.

Second, we note a number of interplanetary bodies. Many of these are asteroids—rocky bodies only a fraction the size of the smallest planets and located mostly in a belt between Mars and Jupiter. Others are comets. A comet's orbit, similar to that of Halley's comet, is shown at top left.

Third, we note that the so-called ninth planet, Pluto, is different from most planets. It has a considerably more elliptical orbit than other planets. Moreover, Pluto crosses inside Neptune's orbit and is the only planet that crosses another's orbit. Pluto is also much smaller than the other planets—even smaller than our Moon and only about 60% as big as Mercury. Its orbit is also unusually inclined, running well "above" the plane of the solar system, where the other planets' orbits are concentrated. Pluto is thus unusual among the planets. In 1977, astronomers discovered an object (now called Chiron) one-tenth as big as Pluto on an eccentric orbit between Uranus and Saturn. In the 1990s, astronomers began discovering many more faint bodies in the region of Pluto. For these reasons, if Pluto had been discovered today, it probably would not have been listed as a full-fledged planet, but rather as one of the largest of these interplanetary bodies. More Pluto-sized bodies may eventually be discovered in Pluto's region. Pluto is about three times as big as the largest known asteroid, a 1000-km-diameter object named Ceres, in the asteroid belt. We can go on calling Pluto the ninth planet for now, but let us remember that we have not yet learned everything there is to know about the solar system!

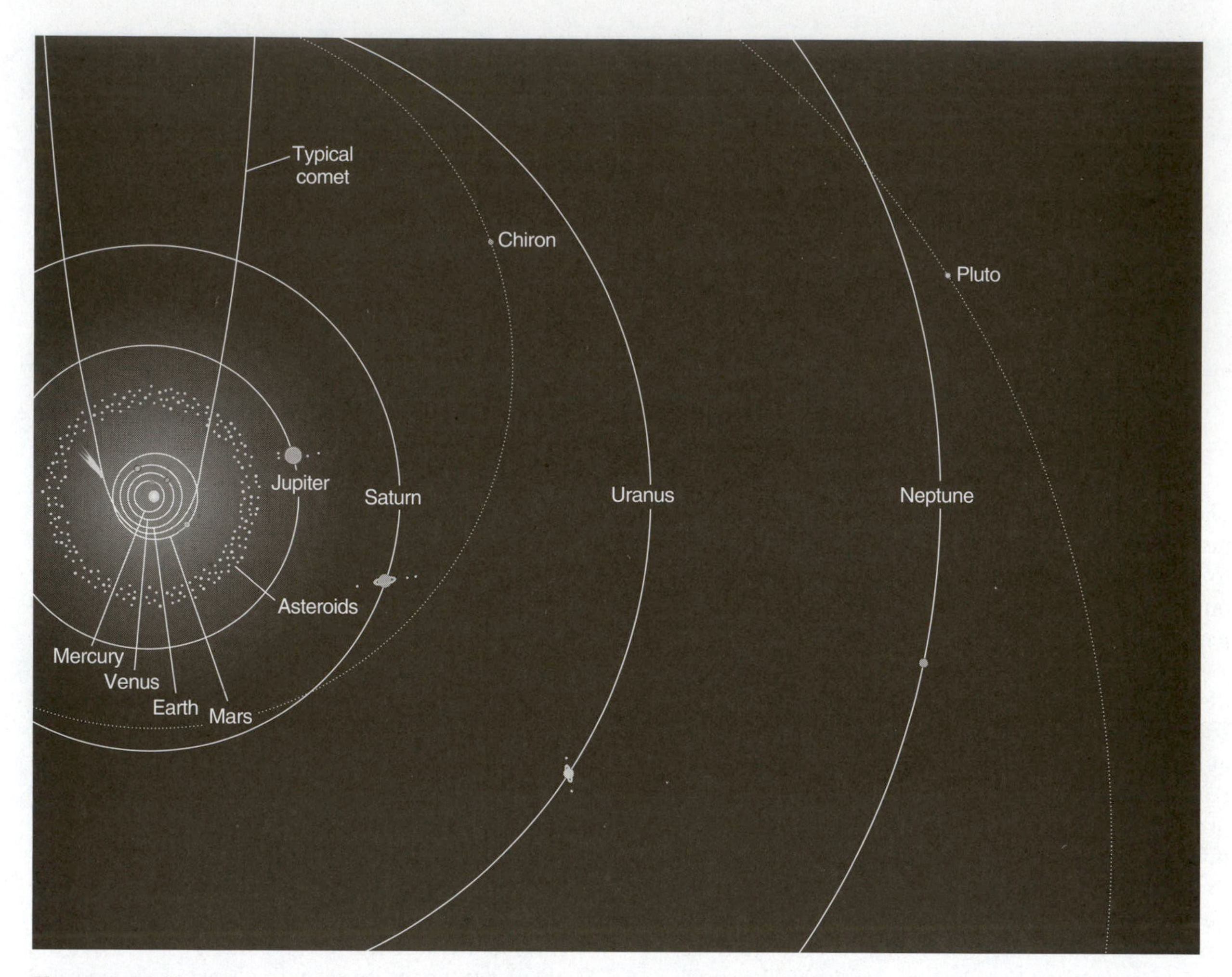

Figure 3-10 The arrangement of the solar system as it is now known, shown approximately to scale. Orbits of the eight main planets and a typical comet are shown, plus positions of typical asteroids. The orbits of Pluto and the unusual asteroidlike object Chiron are also shown as dotted. Note eccentricities of the orbits of Pluto, Mercury, and Mars.

SUMMARY

The planets are tiny specks circling the Sun. If you backed off far enough to see the system as a whole, the outer giants would hardly be noticeable, and the inner planets would be lost in the glare of the Sun. This conception of the solar system was accepted only after one of the major intellectual upheavals in human history took place about four centuries ago. The key to this Copernican revolution was the work of the five scientists listed in Table 3-1, which you should review. Of special importance were Kepler's three laws, which described how planets moved, and Newton's laws of motion and gravity, which revealed the underlying forces that explain Kepler's laws.

CONCEPTS

solar system
transit
retrograde motion
epicycle model
Ptolemaic model
Occam's razor
Copernican revolution
Tycho Brahe
Johannes Keper
ellipse
focus
Kepler's laws
astronomical unit
Galileo Galilei
Isaac Newton
Newton's law of universal gravitation
Newton's laws of motion
Bode's rule

PROBLEMS

1. To an observer north of the plane of the solar system, do the planets appear to revolve around the Sun clockwise or counterclockwise? Which way to an observer south of the plane? (*Hint:* All planets revolve in the same direction as the Earth rotates, from west to east.)

2. Which planet moves the largest number of degrees per day in its orbit around the Sun? Which the least?

3. Can the full moon ever occult Venus (pass between Earth and Venus)? Draw a sketch to show why or why not.

4. One of Galileo's telescopic discoveries was that Venus, like the Moon, goes through a complete cycle of phases, from narrow crescent to full. How did this disprove the Ptolemaic model, which restricts Venus to positions between the Sun and the Earth?

5. Which planet can come closer to Jupiter: Earth or Uranus? (*Hint:* Use Bode's rule.)

6. Does Mars' actual orbital motion around the Sun change in any special way during the time it exhibits retrograde motion?

7. If Venus, Earth, Mars, and Jupiter are in a straight line on the same side of the Sun, what phenomenon does an observer on Earth see? What would an observer on Mars see? An observer on (or near) Jupiter?

8. Do you think Copernicus and other major figures in the Copernican revolution would have viewed themselves as revolutionaries? Why? Contrast the causes of scientific revolutions and political revolutions. What roles do factual discoveries, strong personalities, controversy, and publicity play in each? Which are more dangerous to human life? Which have the more lasting effects?

ADVANCED PROBLEMS

9. A small telescope will show the disk of Mars and Jupiter when they are closest to Earth. Assume that Mars approaches within 50 million kilometers and Jupiter within 630 million kilometers. Mars has a diameter of 6787 km, and Jupiter's diameter is 142,800 km. Use the small-angle equation to calculate the annular sizes of the two planets in seconds of arc. Which appears larger in angular size in the telescope?

10. Suppose we have a large telescope that, on a good night, resolves details as small as $\frac{1}{2}$ second of arc. Based on Problem 9, what are the smallest features it could reveal on Mars?

PROJECTS

1. On a piece of notebook paper, try to make a scale drawing of the orbits of the planets, based on orbital radii listed in Table 8-1 ("Distance from Primary" heading). Which orbits are hard to show clearly? What size dots could represent Earth and Jupiter at this scale?

2. Make a large wall chart showing the orbits of the planets out to Saturn in scale. Mark the motion of each planet in one-day or one-week intervals, as appropriate. Using the *Astronomical Almanac,* an astronomy magazine, or a similar source, locate each of the planets' relative positions for the current date. Update the chart during the semester and watch for alignments that represent conjunctions, elongations, and so on. Confirm these in the night sky.

3. If a planetarium is nearby, arrange for a demonstration of planetary motions from night to night. Demonstrate the apparent retrograde motion of Mars.

4. With a telescope of at least 2-in. aperture, examine Jupiter and confirm that it is attended by four prominent satellites, as discovered by Galileo. (On any given night, one or more satellites may be obscured by Jupiter or its shadow.) By following Jupiter from night to night, confirm that the satellites move around Jupiter. This proves the Copernican dictum that not all celestial objects move in Earth-centered orbits.

PART

2

Two Methods for Exploring Space: Understanding Gravity and Understanding Light

NASA's Infrared Telescope Facility is only one of numerous state-of-the-art observatories at 14,000-ft elevation on the summit of the extinct volcano Mauna Kea, in Hawaii. Sun and wind carve grotesque shapes in winter ice deposits. (Photo by WKH)

CHAPTER

4

Gravity and the Conquest of Space

WHAT THE READER SHOULD WATCH FOR IN THIS CHAPTER

The basic principles of space flight follow from Newton's laws of gravity and motion, and Newton himself realized that satellites could be launched into orbit around Earth. Newton's law of gravity states that each object or mass in the universe attracts every other mass. The force of attraction between two objects is equal to the product of their two masses, divided by the square of the distance between them. This type of law, where an effect decreases according to the square of distance, is called an inverse square law. Using Newton's laws, it is easy to derive the speed that an object needs in order to go into a circular orbit around Earth (circular velocity) and the still greater speed needed to escape into interplanetary space. These laws apply to all spacecraft and orbital motion in the universe, even stars moving around other stars. Some of the major events in humanity's exploration of space, such as the first artificial satellite (Sputnik I, 1957), the first spaceflight around the Earth (Yuri Gagarin, 1961), and the first landings on the Moon (Apollo program, 1969–72), are described. ■

Part 1 of this book showed how humans arrived at a conception of the Earth as one of several planets moving around the Sun, with the stars far beyond. But how can we proceed from there to learn more about the physical nature of the astronomical bodies around us? To explore our space environment, we need to apply a variety of tools: both figurative tools, like physical theories of gravity and light waves, and literal tools, like telescopes and spaceships. Note that by *space environment* we mean not empty space but the rich diversity of planets, moons, gas, dust, energy fields, and stars that stretch in all directions outward from Earth.

★ There are two ways to explore our space environment. We can actually go there, or we can interpret light signals coming from there. In this chapter we will concentrate on the first method: how humans learned to understand, and then overcome, the bonds of gravity, so that we can send instruments and people to our "neighborhood," the planets, moons, asteroids, and comets of the solar system. Note that the stars are too far away to visit (at least for now), and the next chapter will show how we use light waves—nature's messages from space—to gain information about them.

NEWTON'S LAW OF GRAVITATIONAL FORCE

The last chapter described the life of the great scientist Isaac Newton. One of his greatest accomplishments was to recognize some simple principles that describe how gravitational attraction works. This allowed him to calculate the force that one body exerts on another. This breakthrough allows all humans to master their environment in many ways: The astronomer can calculate the orbital motion of a moon around a planet; the rocket scientist can calculate the power needed to lift a cargo into orbit; the civil engineer can calculate the stresses in a bridge spanning a river.

Whether or not Newton was inspired by watching an apple fall in his garden, he started thinking about why an object falls toward Earth. He concluded that the material in Earth exerts an attraction on any material nearby, whether an apple, a stone, or a human. When he thought about the Moon, he realized that it, too, must be attracted toward Earth. Remember that **Newton's first law of motion** (Chapter 3) says:

> **A body remains stationary or moves in a straight line unless a force acts on it.**

Since the Moon moves in a curve around Earth, Newton reasoned, a force must be acting on it, and that force must be coming from Earth to keep the Moon circling around Earth. Newton realized that he could calculate from the Moon's known orbital motion how fast it "fell away" from a hypothetical straight line. He then compared this acceleration rate with that of a falling body at the Earth's surface. After gathering accurate data, Newton showed that whereas the Moon is 60 times farther than the Earth's surface from the Earth's center, its gravitational acceleration is about 1/3600, or $1/60^2$, of the acceleration experienced at the Earth's surface. Newton thus discovered that gravitational force diminishes as the *inverse square of the distance*. Gravitational force is thus said to follow an **inverse square law.**

This result is not surprising. If a force or a substance spreads out from a point in straight lines in all directions, it must become less concentrated as it gets farther from that point. Light, radio waves, and water spray from a fast-rotating sprinkler are examples. Newton, in a leap of imagination, concluded that gravity works the same way. Earth (or any individual atom of it) acts like a source of gravity, but the farther away you go from the source, the weaker the force.

The inverse square relation is illustrated by Figure 4-1. Imagine light from a candle shining into a pyramid. If the pyramid is cut at the point where its base has an area of 1 cm^2, then all the candlelight entering the tip of the pyramid passes through this square centimeter. Twice as far from the light, the base of the pyramid is twice as wide and so has four times the area, but it receives the same amount of light. The original radiation is now dispersed over 4 cm^2. Thus the base receives one-fourth as much light per unit area at twice any given distance. Similarly, three times as far from the light, the light is dispersed over nine times the area, and the base receives one-ninth as much light per unit area. Hence light radiation follows the inverse square law.

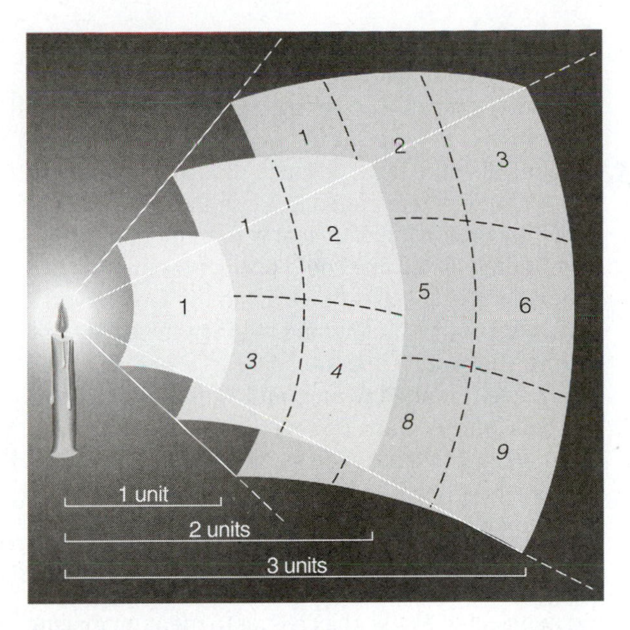

Figure 4-1 The inverse square law. At twice the distance from the source, the light is spread over four times the area. At three times the distance, the light is spread over nine times the area.

Newton thus knew how distance affected gravity, but he did not know what else affected it. He eventually showed that the gravitational attraction between two bodies is proportional to the amount of material in each body—that is, to its **mass.** The more mass, the more attraction. Thus if either mass doubles, the force between them doubles; if the distance between their centers doubles, the force drops by a factor of 4. The relationship is discussed further as Optional Basic Equation II.

Mass should not be confused with *weight.* The weight of an object is merely the gravitational force with which the Earth pulls the object against the Earth's surface. *Weight* is merely a convenient name for gravitational force. The weight of an object would be different on different planets because the gravity of different planets differs. The mass, or amount of material, of a body remains the same regardless of the body's location.

Note an aspect of the scientific method that is illustrated here. Neither Newton nor we claim to know any "ultimate underlying cause" of gravity. We treat gravity only as a property manifested by matter, a property we can confirm by experiment and by observations of nature.

★ Note also the huge advance that Newton and his generation gave to humanity. Until that time, no one really understood how to deal with motion, force, acceleration, etc. No one could predict the motion of a falling object. The great theorist of the Greek world, Aristotle, tried to understand motions by saying that among the "four elements" of earth, air, fire, and water, earth tended to sink and fire tended to rise, and this is what caused movement in the universe. What an amazing advance when Newton showed us not only how to conceptualize, but also how to calculate numerically, the rate of motion of a wagon rolling down a hill or the forces needed to support a large building, or the amount of heat transferred from a warm object to a cold object—not to mention making the leap of realizing that the same forces and the same mathematics used to predict the motion of a falling apple can also predict the motions of the moons and planets! For more on the relation of Greek science to modern ideas, see the discussion at this point in the text on our web site.

Circular Velocity: How to Launch a Satellite

Newton realized that an object could be launched into orbit around the Earth. His *Principia* contains a diagram of an Earth satellite fired from a cannon on a mountaintop, with the barrel pointed parallel to the ground (Figure 4-2). He merely applied his first law of motion. In the case of the cannonball, the force of gravity pulls the cannonball toward the ground. If the launch speed is too slow, the cannonball falls to the ground near the cannon (curve *A* in the figure). At a higher speed, it travels farther (curve *B*). At a high enough speed, it curves toward the ground, but the

OPTIONAL BASIC EQUATION II

Newton's Universal Law of Gravitation

Once Newton realized that the gravitational force between two objects is proportional to their masses and inversely proportional to the distance between them, he could express this more simply in math than in words:

$$F = G\,\frac{Mm}{R^2}$$

where

F = force of gravitational attraction between two bodies

G = gravitational constant, measured to be $6.67 \times 10^{-11}\ \mathrm{N \cdot m^2/kg^2}$, in SI units

M = mass of larger body

m = mass of smaller body

R = distance between centers of M and m

The gravitational constant, *G*, may require some explanation. It is called a constant of proportionality. To take another example, in the familiar equation for the area of a circle, $A = \pi R^2$, the Greek letter pi is used to stand for the number 3.1416, which is the constant of proportionality between the variables A (area) and R^2 (square of radius). Similarly, using SI units of kilograms for m and meters for R, scientists measure the numerical value of $G = 6.67 \times 10^{-11}$ for all known conditions.

A striking fact is that this equation is true throughout the known universe, according to spacecraft and telescopic measurements. The equation and the value for G are true whether measuring the gravitational force between two large lead balls in the lab, the Earth and your body, the Earth and the Moon, the Sun and Mars, Saturn and its moons, or two stars orbiting around each other. Hence the equation is known as the *universal law* of gravitation. It seemed remarkable to Newton and his contemporaries in 1687 that the reasoning of a single human could reveal the elegant simplicity, $F = GMm/R^2$, behind such a profound phenomenon of nature.

Note that since we use SI units in this book, we express mass in kilograms and distance in meters, and we get the SI unit for force, called the newton, which may be unfamiliar. Europeans and other users of the metric system commonly express amounts of material in terms of mass (kilograms); newtons usually appear only in physics and engineering problems. (Another slight complication is that Americans and English express amount of material in terms of weight or force units [pounds], instead of mass units. This will not affect problem solving in this book. You should remember the approximate rule of thumb that 1 kg has a weight of

surface of Earth, being round, curves away at the same rate. Thus the projectile never reaches the ground, but travels all the way around Earth and returns to the mountaintop on circular orbit *D*.[1] This speed—the speed at which an object must move parallel to the surface of a body in order to stay in circular orbit around it—is called the **circular velocity.**

The farther from Earth or another central body, the less the force of gravity that must be overcome, and therefore the lower the circular velocity. At Earth's surface, 6378 km from the Earth's center, the circular velocity is 8 km/s, or nearly 18,000 mph. The Moon, 384,000 km from that same center, moves at only about 1 km/s in its circular orbit.

In short, launching a satellite into Earth orbit is a seventeenth-century idea! One must simply get an object above the atmosphere (so that air resistance isn't a problem), point it in a direction parallel to the ground, and accelerate it to 8 km/s.

Escape Velocity

An orbiting body's closest point to the Earth is called its **perigee;** its farthest point is its **apogee.** As a body is launched at higher and higher speeds, the apogee point is farther and farther away. Each orbit is an **ellipse**—a type of curve describing the closed orbit of one body around a second body. A high enough launch speed would send a body from perigee near the Earth to an apogee far beyond the Moon. At a slightly higher speed, the body would travel infinitely far from the Earth (neglecting the

[1]Newton's mountaintop cannon neglects two realities: In the length of a cannon barrel, no shell could be accelerated to the necessary speed (8 km/s) without shattering; and the Earth's atmosphere would retard the satellite, making it fall back to the ground. Thus rockets must be used, and the projectile must be launched above the atmosphere.

2.2 lb on the Earth's surface. In any problem that starts with an amount of material expressed in pounds, convert to kilograms before proceeding with calculations using the SI units of this book.)

Sample Problem 1 Suppose you are flying your 10-ton (10,000-kg) spaceship at a distance of 10,000 km past a planet, and you detect a gravitational force of 40,000 newtons deflecting toward the planet. Determine the mass of the planet. *Solution:* We want to solve for *M*, rather than *F*. Rearranging the equation, we have

$$M = \frac{F R^2}{Gm}$$

Substituting in values, using units of newtons, kilograms, and meters, we have

$$M = \frac{40{,}000\ (10^7)^2}{6.67\ (10^{-11})\ (10^4)} = 6.00\ (10^{24})\ \text{kg}$$

Note from Table 8-1 that this planet is almost exactly the mass of Earth. This problem is instructive because it shows how scientists can measure the mass of planets, moons, or stars by measuring forces on objects moving near them, such as space vehicles or natural satellites.

Sample Problem 2 If the Moon were moved twice as far away, how much weaker would the Earth's gravitational pull on it be? *Solution:* We simply note that force is inversely proportional to the square of the distance. Thus if *R* increases by a factor 2, *F* decreases by a factor 4. The force would be one-fourth as much as for the present Moon.

Sample Problem 3 The Earth is about 3.7 times as big as the Moon and has about 81 times as much mass. Prove that an astronaut on the Moon weighs about one-sixth as much as on Earth. *Solution:* In this problem we don't have to worry about absolute weights, pounds, or kilograms of mass, because we are asked only to get the *ratio* of weights, or forces of attraction, on Earth and Moon. To get the ratio we can simply divide the force experienced on the Moon by the force experienced on Earth. The student will find

$$\frac{f_m}{f_e} = \frac{M_m}{M_e}\left(\frac{R_e}{R_m}\right)^2$$

Note that the beauty of doing the problem this way is that the gravitational constant *G*, as well as the actual Earth weight of the astronaut, *m*, cancel out and we don't have to be bothered with these numbers. Substituting 1/81 for the mass ratio and 3.7 for the size ratio, we find that the astronaut weighs 1/5.9 as much on the Moon as on Earth, or about one-sixth as much.

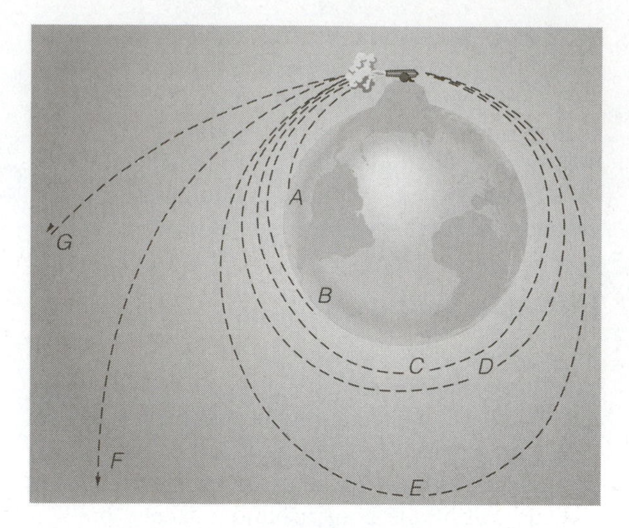

Figure 4-2 Newton realized that a vehicle launched parallel to the surface of Earth from a mountaintop could travel varying distances depending on its launch velocity. At low speed, it would drop along curve *A*, hitting nearby. At a higher speed, it could travel halfway around Earth on curve *B*. A slightly higher speed would put it in an elliptical orbit with a low point (or perigee) at *C*. A still higher speed, called circular velocity, would put it in a special type of elliptical orbit, the circular orbit *D*. A slightly higher speed would create an elliptical orbit with the farthest point (or apogee) at *E*. Escape velocity would create a parabolic orbit *F* that never returns to the Earth. A still higher speed creates a hyperbolic orbit *G*, which also never returns.

gravitational influence of the Sun), following a curve the shape of a **parabola** (curve *F* in Figure 4-2), and it would never come back. This unique speed, which allows the object to escape the Earth forever, is called the **escape velocity,** or *parabolic velocity.* At a point near the surface of the Earth, the escape velocity is about 11 km/s; it is less at more distant points. Launched at a still higher speed, a body travels a similar curve called a **hyperbola** (such as curve *G*) and does not return. Thus a launch speed exceeding escape velocity is called a *hyperbolic velocity.*

In addition to the Earth, each other body in the universe, such as a planet, moon, or star, has its own gravitational field. Hence a unique escape velocity applies at the surface of each body.

ROCKETS AND SPACESHIPS

How can a body be propelled to circular velocity or escape velocity? Jules Verne, in *From the Earth to the Moon,* imagined using a 900-ft-long cannon and 400,000 lb of explosive. But as just noted, a cannon is impractical.

A more realistic technology has since been devised. A spacecraft must carry its own means of propulsion and operate in the vacuum of space. The second point rules out propeller or jet aircraft. Around the beginning of the twentieth century, several experimenters and visionaries realized that rockets were ideal. Their use was first recorded in China and Europe in the 1200s. Rockets work essentially by **Newton's third law of motion:**

> **For every action, there is an equal and opposite reaction.**

For example, if you sit on a wagon and throw a large mass (like a cinder block) out the back, the wagon coasts forward; the force needed to expel the mass causes an opposite force on the vehicle. In the same way, the force used to expel high-velocity gases out the back of a rocket nozzle pushes the rocket forward with equal force. This force is called **thrust.**

The Russian experimenter Konstantin Tsiolkovsky, beginning in 1898, and the American Robert Goddard, in the 1920s, studied and fired rockets. Both were mavericks, however, and their work was almost ignored by their contemporaries. In the 1920s in Germany, Hermann Oberth, who remarked that Verne's book was an inspiration, published several books on rocket-powered space travel. Oberth's work attracted a group of enthusiasts, including Wernher von Braun, whose astronautical experiments were converted into the V-2 guided missile program under the Nazis. At the end of World War II, about 125 German rocket experts, including von Braun, moved to the United States and continued the chain of space-travel development that stretches back to Lucian, Newton, and Verne. A remarkable footnote to our times is Oberth's survival through war and political vicissitudes to attend as a NASA guest the launch of the first successful Moon flight in 1969.

The First Satellites

After several secret postwar studies of satellites had been made, President Eisenhower announced in 1955 that the United States would launch a satellite during the International Geophysical Year (1957–58). This was to be a civilian program using a nonmilitary rocket called Vanguard. Within days, Soviet scientists announced their plan to launch satellites larger than the American one. This plan was not taken seriously in the West: Americans viewed the Soviets as the Soviets portrayed themselves in their poster art—as unsophisticated, shirtsleeved tractor drivers.

On October 4, 1957, the Soviet Union astonished the world by launching the first artificial satellite, the 83-kg (184-lb) instrumented sphere shown in Figure 4-4. It was named Sputnik I (Russian for "satellite"). In November the half-ton Sputnik II went up, carrying a dog as a biological test. Because it was easily visible to the naked eye when it passed overhead at dusk, Sputnik II electrified the Western citizenry. In December, under hasty orders, American technicians tried to launch a small satellite in one of the Vanguard test rockets, but as millions watched on live television, the rocket blew up on the launch pad.

These three months produced a crisis in Western confidence and soul searching in American education (illustrated in the charming and historically realistic 1999 film, *October Sky*). After years of chafing at the bit, the Army team under von Braun was

OPTIONAL BASIC EQUATION III

Calculating Circular and Escape Velocities

One of Newton's achievements in systematizing physics was that his results, once derived for a specific case such as the Earth, could be generalized. The basic physical laws, such as Newton's law of gravitation, apply to bodies orbiting around other planets or around stars. From the law of gravitation, we can *derive* other equations of interest. For example, for any small satellite orbiting an object of much greater mass M, the circular velocity is

$$V_{\text{circ}} = \sqrt{\frac{GM}{R}}$$

where

G = Newton's gravitational constant = 6.67×10^{-11} $\text{N} \cdot \text{m}^2/\text{kg}^2$, in SI units

M = mass of central body

R = distance of orbiter from center of central body

See Figure 4-3. This is the third of 11 simple but important equations that are presented in this text. This simple equation shows that the greater the mass M, the greater the circular velocity; and the greater the distance R, the less the circular velocity.

A useful fact to memorize is that the escape velocity of a small satellite is *always* $\sqrt{2}$ times the circular velocity at that distance from the planet or primary body. Thus, for example, the circular velocity of the Earth just above the atmosphere is about 8 km/s, and the escape velocity of that location is about 11 km/s.

Sample Problem 1 Show that the circular velocity not far above the Earth's surface is about 8 km/s. *Solution:* Expressing units in the SI system, we start with the mass and radius of the Earth:

$$M = 5.98 \times 10^{24}\ \text{kg}$$
$$R = 6.38 \times 10^{6}\ \text{m}$$

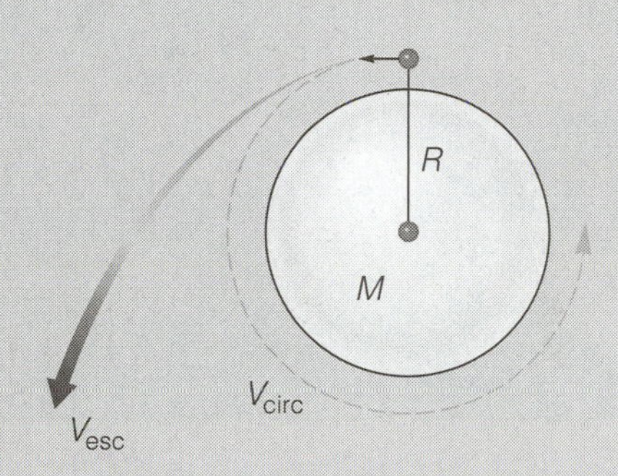

Figure 4-3 The equation for circular velocity gives the velocity V_{circ}, required to place a small object in circular orbit (dashed line) around any large mass M from an initial position at any distance R. The velocity for escape V_{esc} from this same position is always $\sqrt{2}\ V_{\text{circ}}$.

Substituting these values into the equation, we find that a vehicle must reach a speed of 7.91×10^3 m/s, or about 8 km/s, to stay in circular orbit, confirming the value quoted above.

Sample Problem 2 The Moon is 384,000 km from the Earth. Suppose a spaceship is following the Moon's circular orbit at a great distance from the Moon (so that it is affected primarily by the Earth's gravity, not the Moon's). By how much would it have to increase its velocity to escape from Earth altogether? *Solution:* Using the same reasoning as before, but changing the value of R to 3.84×10^8 m, we find a circular velocity of only 1.02 km/s. Multiplying by $\sqrt{2}$, we find a circular velocity of 1.45 km/s. Therefore the spaceship has to increase its speed by only 0.43 km/s, or about 430 m/s. For present-day spaceships this is a relatively small change in speed!

Figure 4-4 The first artificial satellite. Russian artist A. Sokolov and cosmonaut A. Leonov collaborated on this painting of Sputnik I.

given the go-ahead in November; 84 days later, on January 31, 1958, the first American satellite, Explorer I, was orbited. Vanguard I, a $1\frac{1}{2}$-kg (3-lb) sphere, went into orbit in March. Sputnik III went up in May. At $1\frac{1}{2}$ tons,[2] Sputnik III was 56 times as massive as the three American satellites combined and intensified American anxiety during that spring.

The first satellites were designed primarily to probe the nearby environment of space. Among their discoveries were the **Van Allen belts** (doughnut-shaped zones of energetic atomic particles surrounding Earth) and Earth's slight bulge in the Southern Hemisphere, which gives our planet a slight pear shape.[3]

The First Manned Space Flights

American engineers concentrated on miniaturizing precision instruments, while Soviet engineers, lacking the technology to miniaturize, concentrated on rocket power. The Soviets' large, powerful rockets gave them an edge during early space missions. After putting the first probe on the Moon and photographing the Moon's far side for the first time in 1959, the Russians began to test the biological possibilities of space flight with dogs, some of which they recovered from orbit. On April 12, 1961, in a 5-ton craft, a 27-y-old Russian, Yuri Gagarin, became the first person to orbit the Earth, which he did in 108 min (see Figure 4-5.)

The first American manned rocket flight was Alan Shepard's 15-min suborbital flight on May 5, 1961. On August 7, Russian cosmonaut Gherman Titov made 17 orbits in a full-day flight. America's first single-orbit flight came 6 mo later on February 20, 1962, when John Glenn piloted a Mercury capsule. One scientific result of these flights was to allay fears that the Van Allen radiation belts or meteoroids might prevent manned space flight.

THE DECISION TO EXPLORE THE MOON: SCIENCE AND NATIONAL POLICY

Space exploration obviously has not been a purely scientific effort divorced from politics. The pacing of the whole enterprise has been determined by social judgments about national prestige and by funding decisions. The technical resources needed to attack a major scientific problem have become so complex that they demand not a single genius in a hand-built laboratory but a coordinated, well-managed program with heavy financial, political, and industrial support. The financial and political commitment must last throughout the program. Even a wealthy Isaac Newton could hardly be expected to manufacture the heavy castings, special glass, and solid-state electronics necessary for sophisticated astronomical observations today, not to mention a fueled, 111-m (364-ft) Saturn rocket! Newton's work might have been lost in a society that did not favor inquiry, disseminate opinions, and preserve results. Similarly, the end of

[2]One metric ton approximately equals the familiar English ton. Specifically, 1 metric ton = 1000 kg = 2200 lb.

[3]Ironically, it is now known that Sputnik II first detected the Van Allen radiation belts. But because the Russians did not have a global tracking network and did not tell other countries how to decode the radio signals, the Russians themselves did not get enough tracking data from Sputnik II to recognize the belts. NASA researcher A. J. Dessler (1984) has noted: "Because of their perceived need for secrecy, the Russians missed making one of the most dramatic discoveries in space science." Instead of being named the Van Allen belts (after an Iowa scientist who built the detectors), they would have been called the Vernov radiation belts (for the corresponding Russian scientist)!

Figure 4-5 The launch of Vostok I and Yuri Gagarin, the first human to circle the Earth, in April 1961.

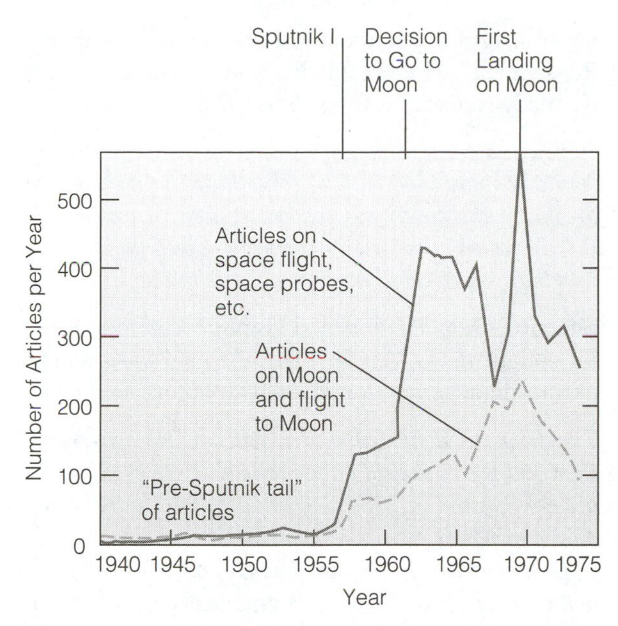

Figure 4-6 The sudden dawn of the space age. The evolution of public awareness of space exploration is reflected in the number of related articles published each year in major periodicals. The coming potential of space flight caught the media and the public unprepared when the first satellite was launched in 1957; prior to that time, the media tended to ignore the possibility or treat it as a "Buck Rogers" fantasy.

theoretical dreaming and the beginning of a practical effort to overcome gravity and explore space involved its own set of social conditions.

Figure 4-6 is one way of showing these social conditions. Articles on space flight and lunar exploration had a long but sparse pre-Sputnik history. Sputnik took the public by surprise. Soon an avalanche of articles on space followed, creating a strong public awareness of space by the beginning of the Apollo program.

Verne imagined a voyage to the Moon undertaken by a group of shrewd Yankees and funded by an international subscription. In the late 1930s, Robert Heinlein imagined it as a commercial venture taking place in 1978, funded by an industrial tycoon. In reality, it was to be a $20 billion government-inspired enterprise, conducted by an agency whose greatest problem was not the technology but the coordination of widespread resources necessary for the undertaking.

Between 1957 and 1961, planners had roughed out technical requirements and timetables for a lunar voyage. Though President Kennedy, who took office in January 1961, sought national goals that would spur creative effort, he was at first skeptical about a lunar program, particularly because of the existing Russian lead. The president's science adviser recalls Kennedy remarking (Logsdon, 1970):

"If you had a scientific spectacular on this earth that would be more useful . . . or something that is just as dramatic and convincing as space, then we would do it." We talked about a lot of things . . . and the answer was that you couldn't make another choice.

Shepard's May 5 suborbital flight gave some reason for optimism. On May 25, 1961, the goal was set in an extraordinary presidential speech before Congress:

The dramatic achievements in space which occurred in recent weeks should have made clear to us all, as did the Sputnik in 1957, the impact of this adventure on the minds of men everywhere. . . .

I believe that this nation should commit itself to achieving the goal, before this decade is out, of landing a man on the Moon and returning him safely to earth.

After Kennedy's assassination in November 1963, the NASA program, especially the Apollo Moon-landing program, became almost a memorial to the president. This prevented funding cutbacks until the program was completed. After the sixth successful landing, in 1972, a few proposed additional landings were canceled, and funding for planetary exploration began a long-term decline.

Figure 4-7 The Russian manned spaceship Soyuz (Union, lower left) approaches the early 1980s Soviet space Station Salyut 7 (top). "Wings" are solar panels that generate electrical power. (Painting by leading Russian space artist Andrei Sokolov)

AFTER APOLLO

The last remaining Apollo spacecraft was used in a short-lived effort at global cooperation in space exploration. In 1975, this spacecraft, carrying three American astronauts, linked in orbit with a Russian Soyuz spacecraft carrying two cosmonauts. The explorers shared good-humored handshakes, conducted mutual experiments, and televised pictures of each others' countries to audiences below in a gesture of goodwill. The Apollo-Soyuz project was remarkable; the challenges included an initial political agreement between Presidents Nixon and Kosygin in 1972, design of a mating tunnel to dock the two vehicles, travel of engineers and astronauts between the two countries, and learning rudiments of each other's languages on the part of astronauts and technicians. The program proved that if political leaders are willing to set challenging, cooperative goals, technical communities in the involved countries can respond with enthusiasm—a more hopeful situation than having these communities engage primarily in bomb building!

Little real follow-up occurred in terms of international collaboration in space exploration after the Apollo-Soyuz flight. The Soviet Union and America resumed a competitive stance. In the 1970s and 1980s, both countries experimented with temporary space stations occupied intermittently by astronauts (Figure 4-7), to prove that humans could live and work in space. Both countries also launched successful probes to other planets during this period. All planets except Pluto were studied at close range with these robotic probes. In the cases of Venus and Mars, robotic probes have also landed on the surfaces, sending back photos and environmental data.

Meanwhile, American engineers developed a fleet of four Space Shuttles intended to meet most of the nation's launch needs (Figure 4-8). The first 24 flights, from 1981 through 1985, were highly successful. Milestones included launches of space probes from the shuttle bay, flights of the first American female and black astronauts, flight of the European Space Agency's Spaceman (a research lab manned

Figure 4-8 The Space Shuttle gave humans new capability to launch, construct, and repair payloads in space. This vertical view, looking down onto the Earth, shows the shuttle Challenger against a background of clouds during a 1983 flight. (NASA)

by a joint European-American crew), and demonstration of an efficient drug-manufacturing technique utilizing special physical and chemical processes during weightless conditions. In his 1984 State of the Union address, President Reagan called for a permanently inhabited U.S. space station that would be built by shuttle crews and would be bigger than the Russian stations.

In 1986, however, during the 25th shuttle launch, disaster struck. An explosion destroyed the shuttle Challenger, killing the racially mixed crew of two women and five men. The American space science program was delayed for two years, until shuttles resumed flying in 1988. Subsequent missions included the 1989 launch of the Magellan probe, which has mapped the surface of Venus; the 1989 launch of the Galileo probe, which has made the first close-up photos of an asteroid and is continuing on toward Jupiter (a flight marred by a stuck antenna that will limit data); and the 1991 launch of the largest astronomical telescope in orbit, the *Hubble Space Telescope* (an instrument marred by an embarrassing optical flaw—see the next chapter). In 1992, a replacement shuttle brought the U.S. fleet back to four spaceships.

During this period, the Soviet program made striking progress, especially during the period of Gorbachev's *glasnost* program and the final years of the Soviet regime (see the *Scientific American* review by Banks and Ride, 1989). For example, in 1986 they launched the Mir space station, which was expanded to a very large size in the following years by the addition of several more modules. Russian cosmonauts lived in Mir for more than a year at a time—a period long enough to fly to Mars. In 1987, they docked a space observatory with the Mir station, which then observed a supernova (a type of exploding star) that appeared that year. In 1987, they launched the world's largest rocket booster, Energia, and also a shuttle of their own. In 1989, they carried out a partially successful robotic mission to Mars' moon, Phobos.

Military space programs expanded in both countries during the cold war. By the early 1990s, the U.S. Department of Defense had more funding for space operations than all of NASA; most of it was secret and not aimed at scientific research. At that time, communism in the Soviet Union collapsed and the Soviet Union itself disbanded, ending the "space race" between socialist and capitalist countries and leading to some debate about the purpose of science research funding in the United States. Some argued that the nation needed science and engineering Ph.D.'s mainly to fight the cold war, and the administrator of NASA publicly raised the issue of how many Ph.D.'s were really needed to explore the universe. The 1992 congressional budget debate canceled the in-progress development of an advanced American probe to a comet. The administration also proposed turning off a successful probe in orbit around Venus before its useful life was over, in order to save money. By 2001, cash-strapped Russia could not maintain the Mir space station and let it crash into the Pacific Ocean.

In the 1990s, the U.S. space program regrouped around smaller, cheaper missions, funded at lower levels than the earlier grand gestures, such as the Apollo flights. This approach has had mixed success, ranging from a successful probe that landed on Mars in 1997 to two crashes of probes onto Mars in 1999, including the embarrassing loss of what was supposed to be an orbiter, when the cash-strapped engineering staff forgot to convert English units into metric units. The latter fiasco emphasized the additional embarrassment that the United States is one of the last three nations on Earth to give up cumbersome English units, which even England itself has abandoned! Much of the funding of the U.S. program is aimed at expanding the new international space station, with joined modules constructed by the United States, Russia, and European countries. This is a long-term joint investment to develop a human capability in space.

Europe, Japan, and China are becoming more active in space exploration. The European Space Agency has planned future space projects, including the Cassini probe to Saturn in partnership with the United States. In 1989, Japan placed a small probe in orbit around the Moon, a harbinger of a projected Japanese lunar research program. In 1999, the Chinese orbited a capsule capable of carrying an astronaut, apparently in preparation for a program of human space exploration.

SPACE EXPLORATION AND SCIENCE: COSTS AND RESULTS

Space exploration and modern science in general are expensive. The results will be clearer to our grandchildren than to us, but we can at least compare some costs and benefits.

Costs

During the 10 y while we geared up to land on the Moon, 1959–69, NASA's total expenditures (including aircraft development as well as space research) were $35 billion, or about 2.5% of the total budgeted U.S. expenditures. During the deficit spending to build up the military and compete with the Soviet Union under President Reagan, NASA's budget dropped to only about 1% of the total budget, where it remained through much of the early 1990s. In the 2000 budget all general science research and technology support is only 1% of the budget, and space exploration is only part of that.

★ Some critics assert that we should cut space spending in order to solve our fiscal and social problems, but these figures show that cutting all civilian science plus the entire space program could save only 1% of the budget and thus make little impact. On the other hand, to put things in perspective, if we could cut military programs by 10%, we could double the size of NASA, double general science and energy technology (thus expanding the programs we will need in the present century), and still have billions left over to spend on social programs and reducing the enormous debt incurred in the 1980s.

Another approach to research funding is to share costs through international joint programs. Examples to date include the probe built by several European countries to explore Halley's comet and the Canada-France-Hawaii telescope project, which built a large observatory in Hawaii. This approach not only saves each country money on specific projects, but teaches all of us, as humans, how to conduct international technical programs, which will be more and more important as we enter the era when we have to manage the resources of the whole planet and solve planet-scale problems such as ozone layer damage or climate change due to carbon dioxide release.

Practical Results of Space Exploration

Seven centuries elapsed between the first use of rockets (probably in a Mongol battle in 1232) and their first scientific application (an atmospheric research flight conducted by Goddard in 1929). Constructive benefits have accumulated only in recent years. For example, the first weather satellite, launched in 1960, ushered in an era of continuous weather monitoring, saving lives and dollars by predicting the approach of hurricanes and other weather disturbances (Figure 4-9). It is hard to realize that only two generations ago hurricanes were totally unpredictable, hence known as "acts of God," in the parlance of insurance contracts.

Another benefit of space exploration is the communications satellite. Intercontinental TV, radio, and telephone links are made possible by a group of satellites orbiting far above the equator at 42,000 km (26,000 mi) from Earth's center. Because satellites at this altitude revolve around Earth with an orbital period of 24 hours, they stay above the same spot on Earth. This is why backyard satellite-TV dish antennas can be pointed at a fixed spot in the sky: They aim at the position of one of these distant satellites.[4]

★ Although some of the satellite phone links are being replaced with optical cables, such satellites provide critical communication networks, making

[4]This system was first proposed in the late 1940s by science fiction writer Arthur C. Clarke. In practice today, it causes curious pauses in transcontinental phone calls. If your call goes by normal ground links from the East Coast to the West Coast, the gap between your question and your friend's answer is about 0.2 s, the normal response time for the brain to frame an answer. However, if your call is beamed up to a satellite about 36,000 km above the surface, then these radio waves must complete two round trips (144,000 km) at the speed of light (300,000 km/s) between question and answer. This takes roughly 0.5 s. Added to the 0.2 s mentioned above, it gives a characteristic awkward pause of 0.7 s between verbal exchanges.

Figure 4-9 Use of space photography in weather forecasting has become familiar on the evening news. This photo made from the Applications Technology Satellite (ATS) on September 27, 1971, shows U.S. cloud cover and two tropical storms: Ginger (right) and Olivia (lower left). Advance storm warnings reduce property damage in the United States and other countries. (NASA)

ours the first generation to have live coverage of distant wars as well as global cultural events, the Olympic Games, and worldwide technical conferences. Many writers have predicted that this improved communication will strengthen our sense of human community on the planet, producing a "global village"—just as the bickering American colonies eventually came to accept a common identity.

Intangible Results

There are three important intangible results of space exploration. First, it creates a cosmic perspective. Much of the world listened on Christmas Eve 1968 when astronauts made the first flight around the Moon, and one of the astronauts described Earth on a live broadcast as "the only color in the universe—very fragile. . . . It reminded me of a Christmas tree ornament." By no coincidence, the ecological movement blossomed just at the time of those first lunar flights, when astronauts brought back the first pictures of Earth seen from a large distance as a globe hanging in empty space (Figure 4-10), forcing a dramatic realization that we share the same finite globe.

Second, space exploration provides a frontier and a sense of adventure that is important to human well-being (Figure 4-11). After several decades when frontiers seemed to be closing, we now see a new frontier opening above us. Princeton physicist Freeman Dyson (1969) noted at the time of the first lunar landings:

We are historically attuned to living in small exclusive groups, and we carry in us a stubborn disinclination to treat all men as brothers. On the other hand, we live on a shrinking and vulnerable planet which our lack of foresight is rapidly turning into a slum. Never again on this planet will there be unoccupied land . . . freedom from bureaucracy, freedom for people to get lost and be on their own. Never again on this planet. But what about somewhere else?

Finally, spaceflight offers the theoretical prospect to escape from natural or human-made disasters that might befall the Earth. Rocket pioneer Wernher von Braun put it more dramatically after the first lunar landing (Lewis, 1969): "The ability for man to walk and actually live on other worlds has virtually assured mankind of immortality."

LOOKING TO THE FUTURE

Numerous projects by various nations promise exciting results in planetary exploration and general astronomy in the early twenty-first century. The Galileo and Mars Global Surveyor missions of the

Figure 4-10 The image of our Earth hanging in space, as seen from partway to the Moon, provided humanity with a new perception of our planet as a fragile and finite globe, helping to raise environmental consciousness. In this view, much of North America is under winter cloud cover, but the Southwest and Baja California are prominent. (NASA photo by Apollo 13 astronauts)

1990s continue to make close-up observations of Jupiter, Jupiter's moons, and Mars. The American/ European Cassini mission is expected to study Saturn and Saturn's moons, including parachuting a probe onto the mysterious, cloud-shrouded surface of the moon Titan (see Chapter 11). The international space station, though controversial, will allow us to develop human capability in space for the benefit of future generations (Barnsley, 1996). Numerous other robotic missions to asteroids, comets, Mars, and other bodies are planned for the next decade. Various space telescopes are in the works to complement the *Hubble Space Telescope* at X-ray and infrared wavelengths inaccessible to that telescope.

Looking even further into the future, the U.S. presidential National Commission on Space presented a bold vision of the twenty-first century—a vision including interplanetary flights to support research colonies on the Moon and Mars and exploration for space resources. Its report (National Commission on Space, 1986) compared this enterprise to the opening of the American West by government survey expeditions and government-subsidized railroads in the 1800s.

Such ideas promise the adventure of learning how to carry out activities in a new environment (Figure 4-11). But they go beyond adventure. Space exploration has taught us that Earth is a Hawaii in a universe of Siberias and that we need to preserve it. Space operations may ultimately help solve energy shortages, raw material shortages, and pollution problems on Earth. For example, large satellites could collect pollution-free solar energy and beam it down to large (10-km) "antenna farms" that would replace power-generating plants. This system could probably provide energy with less environmental damage than coal-fired or nuclear generating plants, which pollute the biosphere. Although the cost of such a solar energy space program would be high (the Apollo program cost about $20 billion over 10 y), we must balance it against the roughly $50 billion that the United States spends annually in OPEC countries for oil. Once burned, the oil is gone for good, whereas each investment in solar energy gives us new capabilities for the future as space scientist James Arnold pointed out as early as 1980.

We now know from astronomical studies that many asteroids contain large amounts of pure metal, and it may become economical to retrieve these resources and process them in space, rather than continuing to process low-grade ores on Earth with fossil fuels. To be visionary about it, we can imagine that in the very long-term future, transferring much of our heavy industry into space may allow our own planet to begin to heal itself and revert to its more natural state, without sacrificing the standard of living associated with our technological society. In that way, humans would have truly expanded their horizons to live in their cosmic environment.

Figure 4-11 A 1984 test of a free-flying "manned maneuvering unit" allowed astronauts to fly away from the Space Shuttle to pursue construction and repair activities. During such a flight, the astronaut is an independent satellite of the Earth controlled by small jets of compressed gas. (NASA)

SUMMARY

The idea of carrying people or instruments into space existed long before the technological possibility. Newton's theory of gravitational attraction made it possible to calculate how fast objects would have to go in order to orbit around Earth or escape from Earth. The development of rocket technology early in the twentieth century provided the means to reach these speeds.

Political and social factors, including strong public support for space exploration, allowed the implementation of these means. The technology of space travel has been used in four areas: (1) improving communication, weather prediction, and manufacturing processes; (2) helping us search for new nonrenewable resources, while helping us realize that Earth and its supply of resources are finite; (3) exploring other bodies in the solar system; and (4) improving telescopic observations of stars and galaxies far beyond the solar system. Space flight has provided a sense of adventure and exploration unmatched since the Renaissance voyages to the New World.

CONCEPTS

Newton's first law of motion
inverse square law
mass
circular velocity
perigee
apogee
ellipse
parabola
escape velocity
hyperbola
Newton's third law of motion
thrust
Van Allen belts

PROBLEMS

1. If an astronaut is flying in circular orbit just above the atmosphere and wants to escape from Earth, how much additional velocity does he or she need?

2. Rocket engineers typically speak of the difference between one orbit and another in terms of velocity difference, for example, in meters per second. Why is this? (Consult Problem 1 if necessary.)

3. Because Earth rotates, the equatorial regions move at about 1600 km/h (1000 mph). Why is it easiest to launch a satellite in a west-to-east orbit over the equator?

4. You have just rowed a rowboat to a point at rest next to a dock. You step off the rowboat toward the dock.

a. Why does the rowboat move away from the dock?

b. In this instance, how are you analogous to the exhaust from a rocket?

5. Do you think a proposal to spend $20 billion to land humans on the Moon for the first time would be endorsed by Congress or the public if it had happened this year instead of 1961? Explain your answer, accounting for similarities or differences in public attitudes during these two periods.

6. Do you believe that manned flights to other planets, satellites, or asteroids will be common by the end of the twenty-first century? Do you think this would have positive or negative effects on social progress, intellectual stimulation, the economy, the environment, the availability of energy and materials, and other characteristics of our civilization?

ADVANCED PROBLEMS

7. The lowest point on earth's land surface is 392 m below sea level, at the Dead Sea in Israel and Jordan. The highest point is 8847 m above sea level, on Mt. Everest. Using a mean sea level radius for Earth of 6371 km, and assuming your average weight is measured at sea level, calculate your weight near the Dead Sea and on Mt. Everest. Use SI units as needed, but convert your answer back into pounds. Does Earth's surface make much difference in gravity from place to place?

8. Suppose you landed on a planet with a mass of 0.11 Earth-masses and a radius 0.53 times that of Earth.

a. Compare your weight on that planet with your weight on Earth.

b. Using data in Table 8-1 (pp. 158–159), compare this situation with that of an astronaut on Mars.

9. Suppose an astronaut is orbiting Earth at the same distance as the Moon in a circular orbit, but not near the Moon. How much faster would he or she have to move to escape Earth altogether?

10. a. Prove that, as stated in the text, a satellite at about 42,000 km from the Earth's center above the equator would stay above the same point on Earth, thus being useful as a "fixed" communications or weather satellite. (*Hint:*Use the circular velocity equation to get speed and then calculate the time required to traverse the circumference of the circular orbit.)

b. Why would the "fixed" position over one point on Earth apply only in such an orbit above the equator? (*Hint:* Consider a circular orbit of the same size, but passing over the North and South Poles.)

CHAPTER 5

Light and the Spectrum: Messages from Space

WHAT THE READER SHOULD WATCH FOR IN THIS CHAPTER

Light emitted from the Sun, stars, and all solid bodies consists of a mixture of colors called the spectrum. Each different color corresponds to a different wavelength of light. Some colors can't be seen by the naked eye. Thus X-rays, ultraviolet light, visible light, infrared light, and radio waves are merely different forms of light, which is also called electromagnetic radiation. The hotter the radiating object, the bluer the overall color being emitted. Thus an astronomer can measure the overall color and get information on temperature. The process of emission of light can be described at the level of atoms, and especially electrons (which are one of the kinds of particles making up atoms). When an electron in an atom of a gas is disturbed, radiation of a certain color, that is, a certain wavelength, is emitted. Each type of element radiates its own set of colors. Thus an astronomer observing the spectrum of the emitted light from the gas can measure the specific wavelengths and determine the composition of the atoms emitting the light. This chapter gives several additional related examples of ways astronomers can measure properties of the spectrum of light emitted by planets and stars and infer the properties of those objects, such as their temperature, composition, atmospheric pressure, and motions. The important overall lesson is to realize how much information we can gain about conditions on alien planets and stars without actually visiting them. ■

If we send astronauts to the surface of the Moon or a robot to look at Pluto, we are reaching out to "touch" other parts of the universe. But most parts of the universe are too far away to reach in person or even with our space probes. To get information about these regions we have to rely on nature's messages from them reaching us in the form of light. In this chapter we will study light and other forms of electromagnetic radiation.[1] We will then apply this information in this and the next few chapters to see how information can be gained about planets. In later chapters we will expand on this information and apply it to stars and galaxies.

[1] *Note to teachers and students:* Basic concepts useful in the next chapters are introduced here. The student will be able to develop familiarity with them as we discuss the planets. More concepts, such as Kirchhoff's laws, the Doppler effect, and the Stefan-Boltzmann law are introduced and applied to the Sun and stars in Chapters 15 and 16. This appears a useful way to introduce these physical concepts a little at a time, as needed.

THE NATURE OF LIGHT: WAVES VS. PARTICLES

Suppose you stand by a quiet swimming pool where a cork is floating. You disturb the cork by jiggling it. You will notice that this disturbance causes a set of waves to move out across the water. The waves have a certain spacing from one crest to the next, called the **wavelength.** They move at only one fixed speed. As they move by a given point on the water's surface, that point moves up and down; the number of these up-and-down pulsations per second at any one spot is called the **frequency** of the wave.[2]

★ These waves provide a useful analogy, but not a perfect description, of some **wavelike properties of light.** For instance, just as a water wave expands from its source, light spreads out in all directions from its source. **Visible light** has a tiny wavelength, around 400 to 700 nm (0.0000004–0.0000007 m—a nanometer, abbreviated nm, is a billionth of a meter; see Appendix 1). Radiation with still shorter wavelengths exists, but it is too deep violet for our eyes to perceive; it is called **ultraviolet radiation.** Light with wavelengths longer than red light is called **infrared radiation.** Extremely long-wavelength infrared waves are called *radio waves;* radio waves are just another form of radiation that we cannot see. The term **electromagnetic radiation,** or just *radiation,* is used for light of any wavelength.[3] Just as the speed of water waves is constant, the **speed of light** through empty space is constant, about 300,000 km/s (186,000 mi/s). At this speed, it takes about 8 min for light emitted by the Sun to reach Earth. (If the Sun suddenly stopped emitting light, we wouldn't know it until 8 min later.) The speed of all electromagnetic waves, including those of much longer and shorter wavelengths than light, has the same value. It has been measured, for example, by the time interval necessary to communicate with a distant spacecraft.

Now suppose you shoot a bullet at something. The bullet has a certain energy. Unlike a wave, which takes a while to pass a fixed object and then die out, the bullet delivers all its energy to its target at the moment it hits. The bullet is an analog for some peculiar **particlelike properties of light.** For instance, light's energy is concentrated in individual units, instead of being spread out along the wave. The energy-containing unit of light is called the **photon.** It can be visualized roughly as a microscopic, bulletlike particle that moves at the speed of light, yet has a certain wavelength associated with it. An important concept is that each color of light corresponds to a photon of different wavelength and energy. *The bluer the light, the shorter the wavelength and the more energetic the photon.*

From the 1600s to the 1800s, scientists argued about whether light is "really" a wave or a particle. For example, Isaac Newton argued for a particle theory of light, while the Dutch astronomer-optician Christian Huygens (1625–95) argued for a wave theory.

Both were right. For example, light is wavelike because both light waves and water waves can bend slightly around corners. If a swimming pool has an inward-protruding corner or wall, you can observe that a wave bends slightly as it goes past the corner, so that some energy (wave motion) reaches a target slightly behind the wall as seen from the wave's source. Light does the same thing, indicating a wavelike property. A bullet, however, would move in a straight line past the wall and could not hit such a target. This process of light bending its path slightly as it passes an edge is called **diffraction;** it limits the sharpness that can be achieved in the image formed by a telescope.

On the other hand, Albert Einstein described one of the important particlelike properties of light in 1905.[4] This is the so-called photoelectric effect—an effect in which light rays can knock electrons out of metal surfaces. The effect can be explained only if the light energy arrives in individual "bullets," or photons, rather than being "smeared out" along the wave.

In summary, light spreads out through space something like a wave in which energy is carried by

[2] *Note to advanced students:* The velocity of a wave is equal to the wavelength (the distance between two crests) times the frequency (1 divided by the time interval between crests as they pass a fixed point). Check the dimensions: Wavelength × frequency has dimensions of distance/time, that is, dimensions of velocity. Note that a light wave and its color can be specified by either its wavelength or its frequency. Some scientists tend to characterize waves by their frequency (as in radio parlance where stations are listed by cycles/s), but astronomers tend to refer to waves by their wavelength—a practice we will follow in this book.

[3] However, the word *radiation* must be read with caution, since physicists and engineers sometimes speak of streams of atomic particles, such as protons or electrons, as radiation. For example, charged particles escaping from radioactive material are sometimes called *radiation,* or better, *particle radiation.* These atomic particles are not electromagnetic radiation. Other examples are *radiation belts* and *cosmic rays,* both of which are particles, not electromagnetic radiation.

[4] Interestingly, Einstein eventually got his Nobel Prize primarily for this work, not directly for his better-known work on relativity.

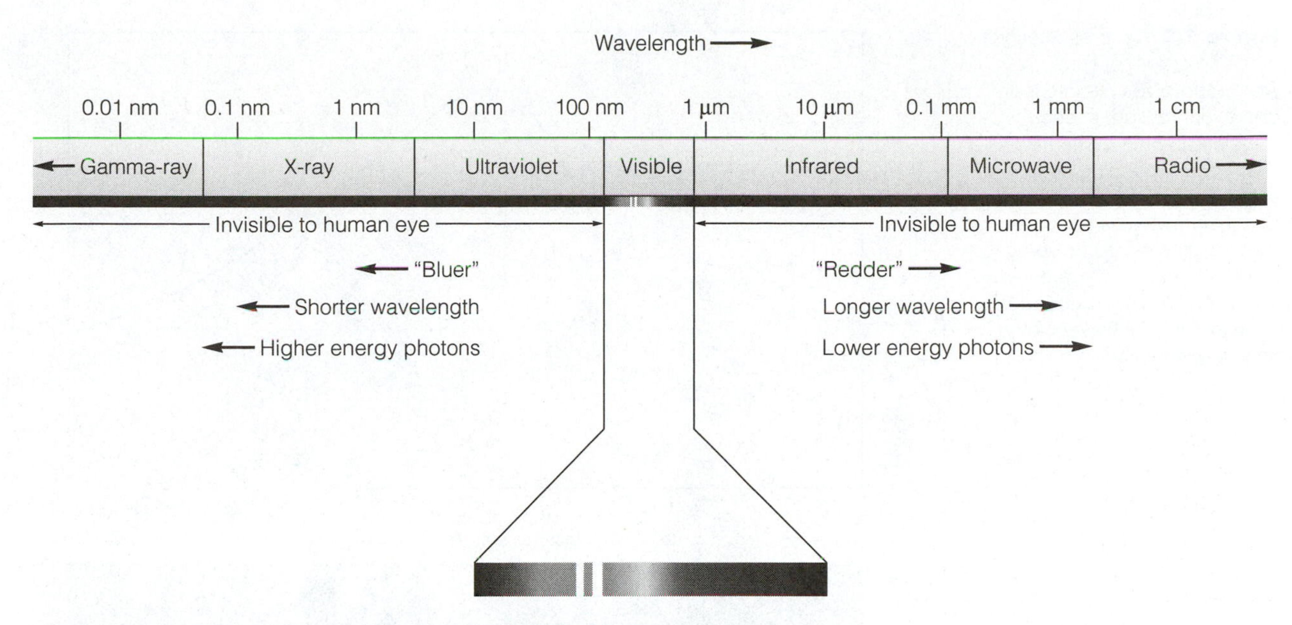

Figure 5-1 The spectrum. The chart shows a wide range of wavelengths with names attached to different regions.

tiny, particlelike photons. Which is light: a wave or a particle? Consider a platypus. It has some ducklike and some beaverlike properties, but it is neither. Similarly, light has some wavelike and some particlelike properties, but it is neither a pure wave nor a pure particle. Considered in microscopic detail, light is a phenomenon not wholly familiar in terms of analogs in our everyday world. We use the analogy of a wave or a particle to give us words to describe its properties, but the analogy and the words are not quite good enough for our purpose.

THE SPECTRUM

★ An arrangement of all colors, in order of wavelength (or in order of photon energy), is called the **spectrum** (plural: *spectra*). Newton discovered he could see the spectrum of visible light by passing sunlight through a glass prism. When a narrow, pencil-like beam of light goes through the prism, it spreads out into a band of various colors. Newton closed the shutters in a room, let in a narrow beam of sunlight, passed it through a prism, and cast the band of colors on the wall. Check our web site to see a color figure showing how Newton did this.

★ Water droplets in a rainstorm act like little prisms and allow us to see the spectrum—the same arrangement of colors from violet to red—when we look at a rainbow. Newton's discovery proved that "white" light from the Sun is really made up of all the different colors. The spectrum extends to far shorter and longer wavelengths than those to which the eye is sensitive. Figure 5-1 shows the names attached to different parts of the spectrum. As this figure indicates, the radio waves we receive on our radios and TVs are simply long-wavelength versions of the light we see with our eyes. (The figures of the color spectrum can be studied better in color on our web site.)

Figure 5-2 shows an important way astronomers present this information. The bar along the bottom labels the colors. This bar, like the band in Figure 5-1, shows the kind of picture we could get if we simply photographed a projected spectrum. But instead of photographing, suppose we used a little photocell, like a photographer's exposure meter, to scan along the spectrum, left to right, to measure the amount of energy (that is, light intensity) at each wavelength. Then we could plot the results as the graph in the upper part of Figure 5-2. This allows us to see the amount of light of each color. For example, Figure 5-2 shows that sunlight's strongest intensity is reached at greenish-yellow wavelengths. Astronomers often present the spectrum in this way.

Astronomers' study of the properties of the spectrum of different objects is called **spectroscopy.** *Spectroscopy is the most important method for learning about remote planets, stars, and galaxies.*

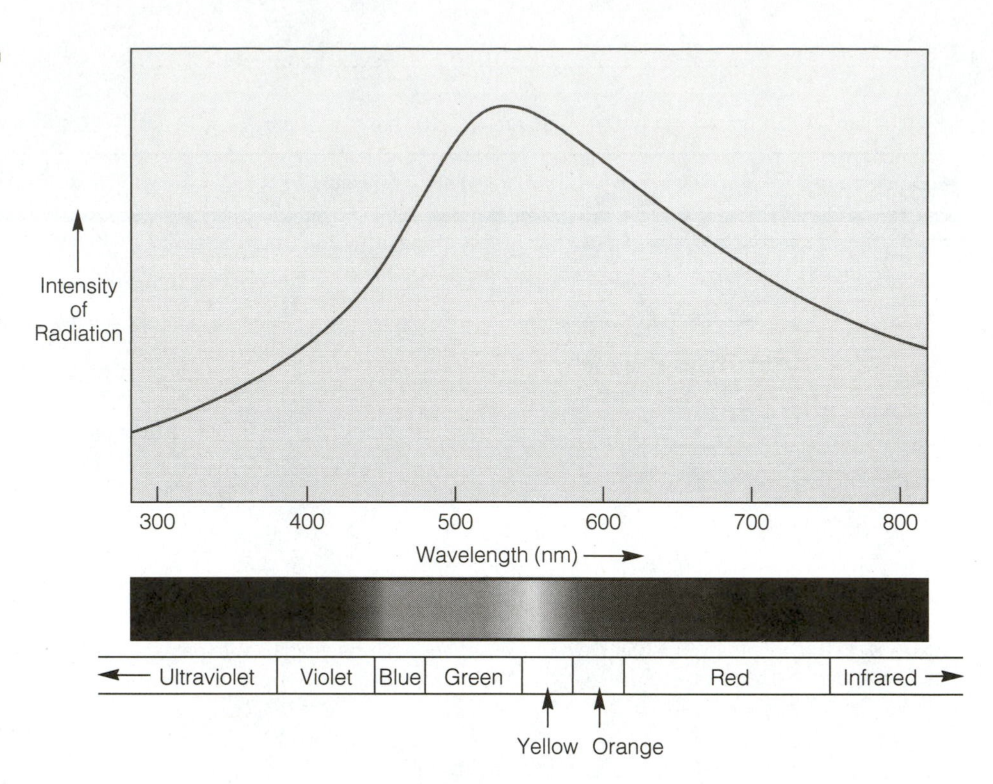

Figure 5-2 A representation of the spectrum of sunlight. The graph gives the intensity, or amount, of light at each wavelength and the names of the corresponding colors along the bottom. The Sun's dominant radiation occurs at wavelengths we perceive as greenish-yellow; we perceive the blend of *all* the Sun's radiation at different wavelengths as yellowish-white.

ORIGINS OF LIGHT: ELECTROMAGNETIC DISTURBANCES

Consider again the cork in the swimming pool. We had to disturb it to get a wave. If you put a second cork in the swimming pool and watch as you make a wave, you would see it bob up and down as the wave passed by. The cork would be a *detector* of the passing wave. In the same way, if you had a tiny enough electrically charged particle, or a tiny enough compass, you could detect light waves passing by. As the wave passes, the electric charge would vibrate with the frequency of the wave, and the compass needle would oscillate rapidly with this same frequency. For this reason, physicists say that light is a disturbance of electric and magnetic **fields** in space. A field is an important concept in much of physics. A field is said to exist in a certain volume of space if some physical effect can be measured and assigned a numerical value *at every point* throughout that volume. For example, at any point near Earth, Earth attracts a given "test particle" with a measurable force; therefore a gravitational field is said to exist. Similarly, where a compass measures a certain response at any point, a magnetic field is said to exist. Also, when an electrically charged particle experiences a force at any point, an electric field exists.

The charged particle and the compass just mentioned are actual tools with which we can detect electric and magnetic fields in space or in your room. They prove that light involves pulsations in these electromagnetic fields (coexisting electric and magnetic fields). This is why we say that all light, of whatever wavelength—gamma-ray, X-ray, ultraviolet, visible, infrared, microwave, or radio—is **electromagnetic radiation.**

If the second floating cork were hit by a bullet or by another drifting cork, this would disturb its motion and cause it to radiate a new wave. In the same way, a drifting atomic particle, such as an electron, can be disturbed in such a way as to release electromagnetic radiation. Each electron has a certain amount of energy associated with it by virtue of its motion, spin, and other properties. If an electron is disturbed in such a way as to reduce its total energy, it can release energy in the form of radiation—a new photon that speeds away from the electron. (At what speed? The speed of light.) This photon has an energy equal to the energy the electron lost. Thus the wavelength and color of the emitted radiation depend entirely on the amount of energy lost by the electron. If the electron is unattached to an atom, its energy can change by any amount, and hence it can radiate photons of any wavelength. A group of free electrons, if disturbed (for example by heating), thus radiate an array of photons of all wavelengths. This produces a spectrum

TABLE 5-1

Temperature Comparisons: Fahrenheit, Centigrade, and Absolute

	Fahrenheit	Centigrade	Absolute
Absolute zero	−459°F	−273°C	0 K
Freezing of water	−32°F	0°C	273 K
Room temperature	68°F	20°C	293 K
Boiling of water	212°F	100°C	373 K
Surface of Venus	891°F	477°C	750 K
Molten lava (typical)	2102°F	1150°C	1423 K
Cool star's surface			3000 K
Sun's surface			5700 K
Hot star's surface			10,000 K

like that in Figure 5-1—a continuous band of colors called a **continuum,** or **continuous spectrum.**

Temperature, Thermal Radiation, and Wien's Law

Continuum radiation is constantly emitted all around us. Electrons (and atoms and molecules) in all gases, liquids, and solids are in constant motion, jostling each other. **Temperature** is simply a measure of the average speed of these motions, which are thus called **thermal motions** (from the prefix *thermo-*, for heat). This is an important concept—that the temperature of a substance simply measures the thermal motions of the atoms in the substance.

To measure temperature, scientists use the **absolute temperature scale,** in which 0 refers to the state of no thermal motion—the coldest temperature possible. Each degree on the absolute scale is the same as a centigrade degree. The units are called kelvins (after a famous physicist), abbreviated K. Thus one kelvin equals 1°C. Table 5-1 compares the Fahrenheit, centigrade, and absolute temperatures of some relevant objects. Note that the centigrade scale is *defined* with freezing water at 0°C and boiling water at 100°C. Listings in °F and °C are not given for stars because temperatures this high are generally discussed only in K units.

Because thermal motions constantly disturb electrons, *all objects constantly radiate a continuum spectrum of radiation.* The higher the temperature, the faster the motions and the more radiation is emitted. Radiation that depends on the temperature of the radiating material in this way is called **thermal radiation.**

It is important to realize that the light you see from a rock or a planet, or from this book, is *not* thermal radiation, but **reflected radiation.** It is light from the Sun or a lamp, bouncing off the object.

The reason we can't see thermal radiation given off by objects around us is found in a law that is crucial for all of astronomy. It was discovered in 1898 by the German physicist Wilhelm Wien (pronounced VEEN) and is called **Wien's law:**

> **The hotter an object, the bluer the radiation it emits, and the more total energy is emitted.**

This effect is familiar in everyday life. Ordinarily, an electric stove burner emits no visible light. As it heats up, it radiates dull red light, then a brighter orange-red glow. At a still higher temperature, it would be white-hot, emitting a dazzling yellowish-white light. Thus the hotter the stove, the more the color moves from red toward blue. If the burner could be heated enough, its radiation would become distinctly bluish. This effect is shown in terms of spectra in Figure 5-3, where we see the dominant light change from red to blue and increase in intensity as an object is heated. Note that *bluer* always means toward shorter wavelength. Still higher temperatures would lead to dominantly ultraviolet radiation!

Now you can understand why you can't see the thermal radiation from ordinary objects or planets: They are too cold, and hence the radiation is too red and too faint; they give off infrared radiation, as shown by the objects at 300 K, 500 K, and even 1000 K, in Figure 5-3. Objects at these temperatures, such as rocks and planets, do radiate, but the radiation has

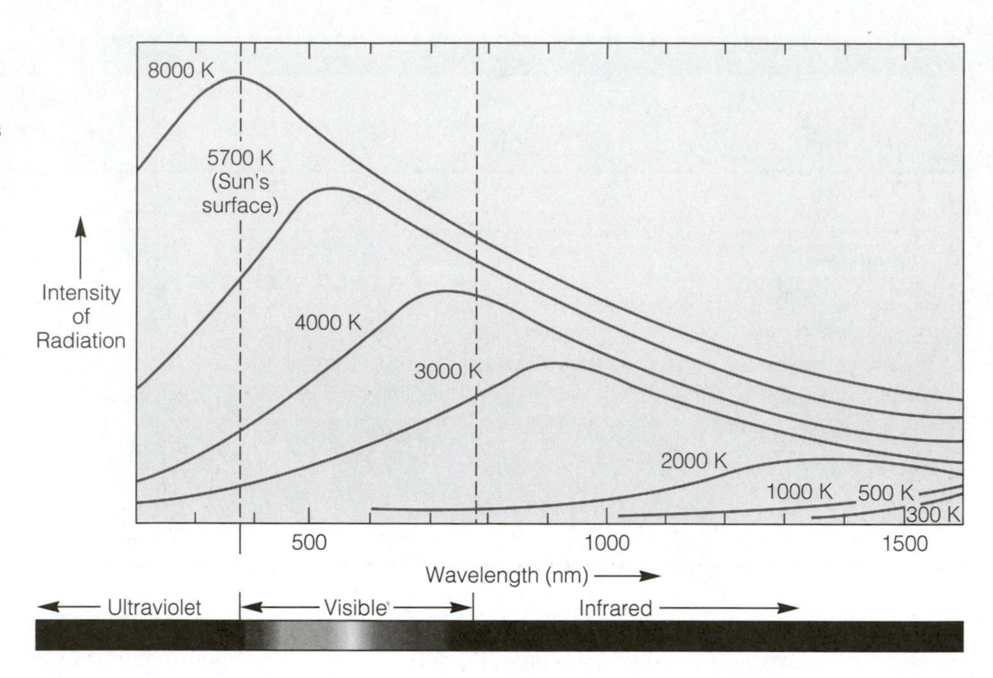

Figure 5-3 Schematic diagrams of the spectrum of a single object (a nail, lava, or a planet) as its temperature is changed. At room temperatures (~300 K), the object glows only in the infrared, with radiation invisible to the eye. As predicted by Wien's law, the dominant color becomes more blue as the temperature increases. As it heats, the object glows first deep red, then orange, and so on. At 5700 K, the material radiates light like the Sun's (cf. Figure 5-2). The total energy radiated by the object (which is proportional to the area under the curve) also increases as the temperature grows.

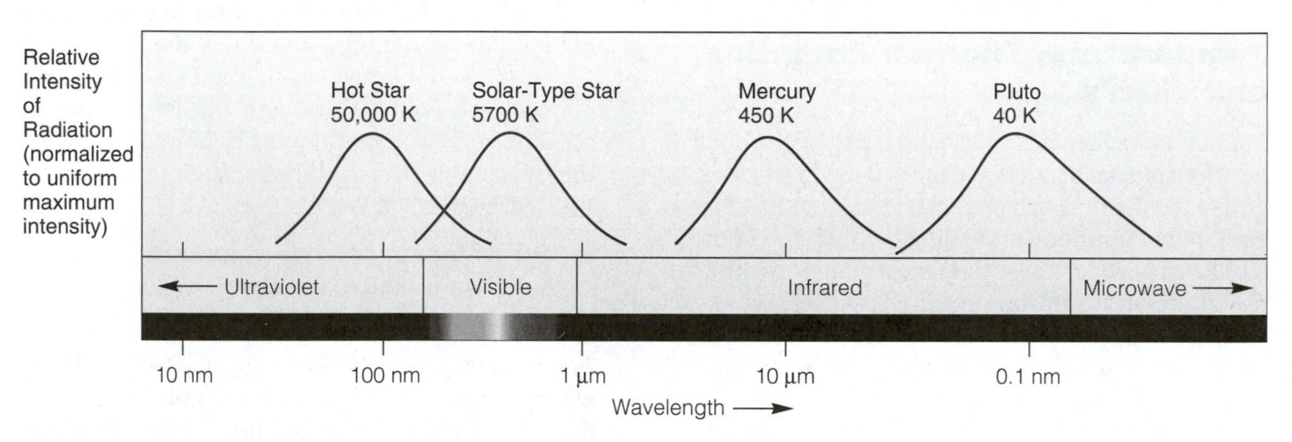

Figure 5-4 Schematic spectra of thermal radiation from four different celestial objects, adjusted to the same intensity for clarity. As predicted by Wien's law, hotter objects are bluer. The temperature can be determined by the wavelength of the dominant radiation. Planets are so cold they radiate their thermal radiation only in the infrared part of the spectrum; infrared detectors are needed to measure them. Similarly, ultraviolet detectors help study the hottest stars.

such long wavelength that our eyes are not sensitive to it. However, we can build instruments that detect infrared or ultraviolet radiation.

Measuring the Temperature of a Planet (or a Star)

Now we can see one of the ways that light carries messages from space. Instruments to detect infrared radiation of planets and stars were first built around World War II and are still being improved. With such infrared detectors, we can measure the temperature of a planet by measuring the color (wavelength) at which its thermal radiation peaks. As shown in Figure 5-4, an object with peak radiation at a certain infrared wavelength has one temperature, while a star with peak radiation at visible wavelengths is hotter. The peak thermal radiation lies in the infrared parts of the spectrum for all planets and moons, from Mercury to Pluto. They are all too cold to emit appreciable visible light; we see them not by their *own* light but by the sunlight they reflect.

EMISSION LINES AND BANDS

We've been describing ways to produce light as electrons are disturbed. This process, in which a particle emits a photon, is called **emission.** We started off picturing a freely drifting electron (like a cork floating in a pool), not attached to an atom. Now consider an electron orbiting around the nucleus of an atom. As soon as we consider electrons attached to atoms, extraordinary modifications occur to the emission process, and these are critical to all of astronomy.

Atoms and Emission Lines

Figure 5-5 shows the schematic structure of an atom, with its central nucleus and some orbital paths available to a specific electron. This electron may be only one of many, depending on the element: hydrogen, helium, and so on. In some ways, the atom is like a tiny solar system—that is, a relatively massive central object and tiny orbiting particles. But there is an extraordinary difference. In the solar system we can put a rocket into any orbit we choose; each orbit would have its own velocity and hence energy. But in an atom, electrons can occupy only *certain* orbits. These are called **energy levels,** since each orbit has one specific energy. Physicists describe this by saying the atom has quantized energy levels, because only certain quantities of orbital energy are possible. This was discovered in 1913 by Danish physicist Niels Bohr. In Figure 5-5, the solid circle represents the orbit occupied by the electron, and the dashed circles are other allowed energy levels. At the top are a few of the infinite number of energy levels outside the atom, where energy levels are not discrete.

Now let us disturb the electron as before (hitting it with a photon or a neighboring atom) so that it drops to a lower energy level. (In fact, this process can also happen spontaneously without outside disturbance.) As it drops, it emits a photon whose energy equals the difference in energy between the two levels. But note that being inside an atom, it can experience only certain, fixed changes in energy level, as shown in Figure 5-6. For instance, if it starts in level 3, it can drop only to level 2 or 1. Thus it can emit only certain amounts of energy corresponding

OPTIONAL BASIC EQUATION IV

Measuring Temperatures of Astronomical Bodies: Wien's Law

Wien's law tells which wavelength W corresponds to the maximum amount of radiation, given the temperature T. Using SI metric units, the wavelength is given in meters and the temperature in kelvins, abbreviated K. (These units are sometimes called *degrees kelvin,* since they define a temperature scale in degrees; but the correct SI terminology is *kelvins.*)

The law is

$$W = \frac{0.00290}{T}$$

The number 0.00290 is a constant of proportionality and remains the same in all applications of the law. Thus, as T increases, W decreases, giving shorter wavelengths and hence bluer light.

Sample Problem 1 The Sun has a surface temperature of about 5700 K. Calculate the wavelength of the strongest solar radiation. *Solution:* Using SI metric units and powers of 10, we get

$$W = \frac{2.9 \times 10^{-3}}{5.7 \times 10^{3}} = 5.1 \times 10^{-7}\ \text{m}$$

This, of course, is exactly in the middle of the range to which the eye is sensitive. (Otherwise we could not see sunlight!) Light of this wavelength corresponds to a greenish-yellow color. We see the blend of all solar wavelengths as yellowish-white.

Sample Problem 2 A certain planet (not unlike Earth) has a mean temperature of 290 K (around room temperature). Characterize its thermal radiation. *Solution:* Inserting $T = 290$ K in the equation, we find a wavelength $W = 10^{-5}$ m, or 10 μm. This is well out into the infrared—invisible to the eye but detectable by instruments.

Sample Problem 3 A certain star, cooler than the Sun, has a temperature of 2900 K. Characterize the color of its radiation. *Solution:* Since it is cooler than the Sun, it must be at least somewhat redder. But how much? Is its peak radiation orange, red, or what? Inserting $T = 2900$ K in the equation, we find that $W = 10^{-6}$ m, or 1 μm. This means the peak radiation is in the infrared—just redder than the eye can see. But much radiation will spill into the nearby red and orange parts of the visible spectrum (see curves in Figure 5-3), so the star will look orangish-red.

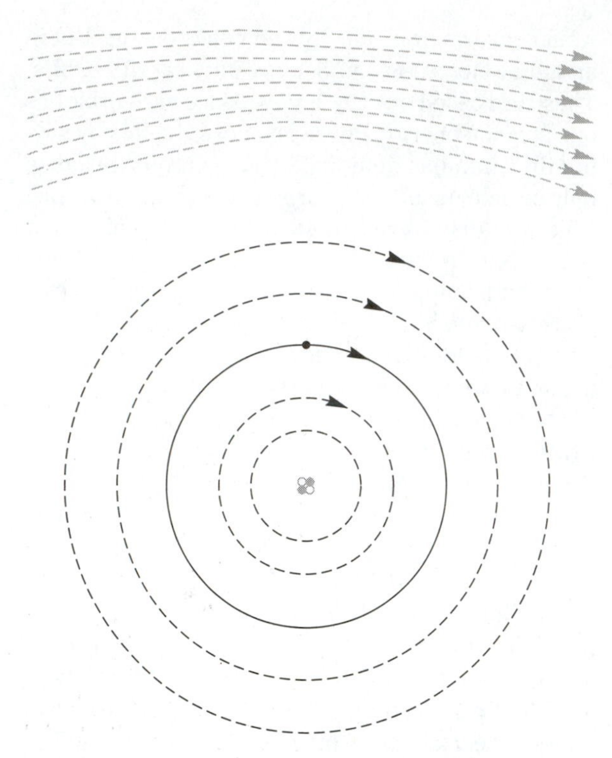

Figure 5-5 Schematic view of an atom's structure. If an electron (dot) finds itself in an orbit (solid line) around the nucleus of an atom, there are only certain other orbits, or *energy levels,* that it can occupy in the atom (dashed circles). However, at large distances from the nucleus (top) virtually any orbits, or energy levels, are possible.

to these differences in energy level; it can emit only photons of certain wavelength or color. As a result, *each element can emit only certain wavelengths of light.* Figure 5-7 shows, for example, the emissions produced by a sample of hydrogen gas containing neutral (that is, uncharged) atoms, with their electrons in different energy levels. These emissions are called **emission lines** because they appear in a projected spectrum as *lines* or narrow bars of color, as seen in Figure 5-7. You can help your understanding of the process by looking at these figures in color on our web site, which shows the relationship of the different emission lines to different colors. Hydrogen, for example, has a prominent emission line of red color that shows up in various places in the universe.

Here, then, is an astounding and useful fact! If you see a certain set of emission lines, you can match it with a certain element and infer that atoms of *that* element are present and glowing in the distant object. You can tell a lot about the object's composition, even without having a sample!

Note that if the electrons in an atom are in their lowest possible energy level, that atom cannot produce an emission line. This is because the electrons cannot drop to any lower energy level. An atom in which all electrons are in the lowest possible energy level is said to be in its **ground state.** An atom in which one or more electrons are in energy levels higher than the lowest available ones is said to be in an **excited state.**[5] Excited states usually last only a short time before the electrons *decay* to the lowest available energy level, producing the ground state. Atoms generally need to be disturbed to produce and maintain excited states. For this reason, hot gases are more apt to produce emission lines than cold gases, because atoms in hot gases collide faster and more often.

Molecules, Crystals, and Emission Bands

The electron structure in a molecule is more complex than in an atom. The electron's path may take it around two or more nuclei. Thus the emission line structure from a molecule is not as simple as that from an atom. For example, in a gas containing water molecules (H_2O), we get more complex emission lines than in a gas containing single H and O atoms. The molecule has various ways of responding to a disturbance in addition to having its electrons change energy levels—for example, it may vibrate like two balls linked with a spring. As a result, the energy levels are vastly more numerous, and the resulting emission lines blend together. Instead of sharp emission lines, we get blended clusters of barely separated emission lines over a range of wavelengths, as shown in Figure 5-8. The resulting broader emission feature from a molecule is called an **emission band.** The rest of the story is the same, though: A given molecule (such as H_2O) can produce only certain emission bands. Thus we can identify the molecules glowing in a given remote source just by detecting their emission bands.

In order for atoms and molecules to produce clear emission lines and bands, they must be detached from one another, as in gases. If the atoms were linked together, they would form molecules, and if the molecules were linked, they would form a solid or liquid. In most solid or liquid substances, the electron structure is so complex that emissions are not confined to one wavelength, but are smeared out. Therefore, emission features of solids and liquids are

[5] Each energy level can be occupied only by a certain number of electrons. For example, the lowest level can take only two.

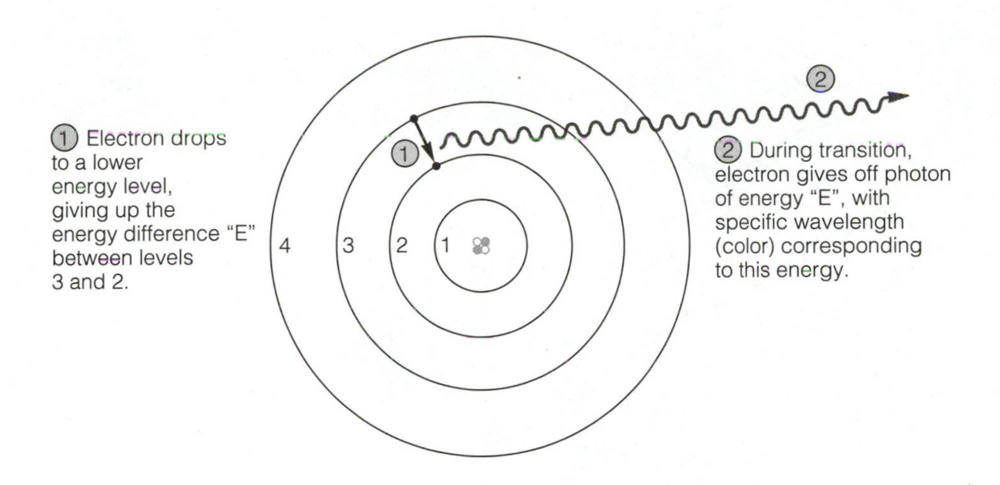

Figure 5-6 Schematic view of emission of a photon of light as an electron drops from a higher to a lower energy level in an atom.

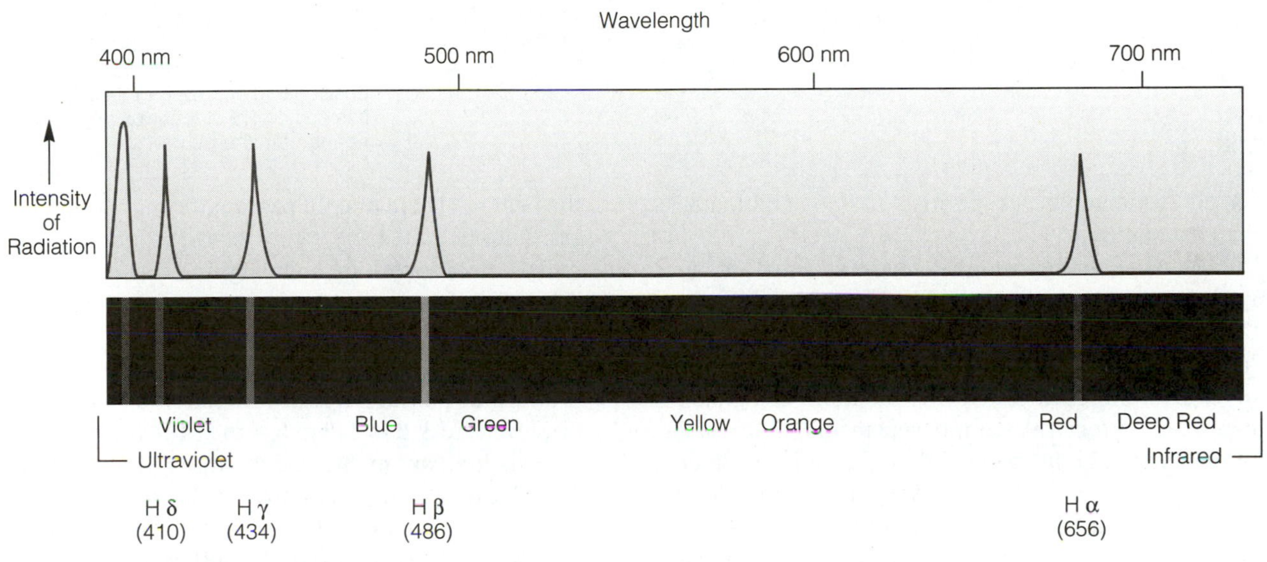

Figure 5-7 A visible-light spectrum consisting only of the emission lines produced by glowing, neutral hydrogen gas. Wavelength scale is given at the top. The emission line at the red end, called the hydrogen alpha line, is especially prominent in many astronomical examples of glowing gas, causing many astronomical objects to glow with reddish light.

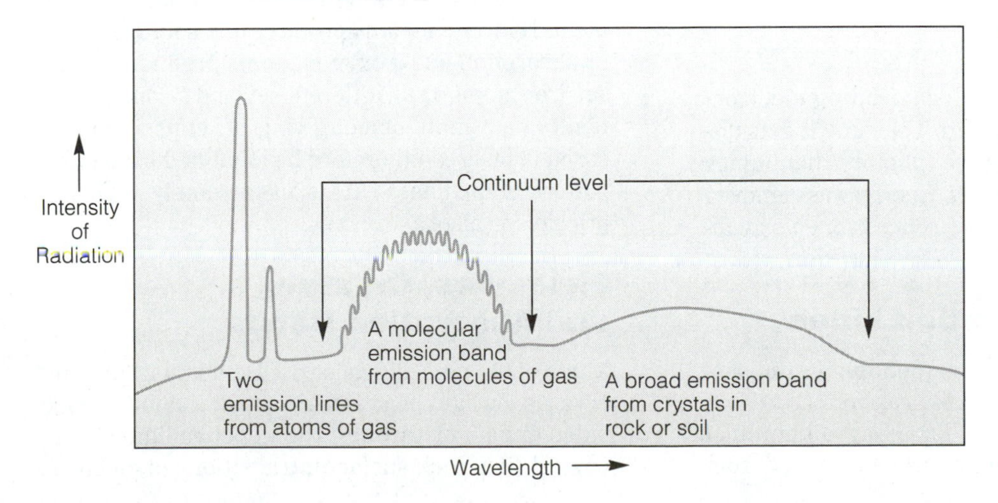

Figure 5-8 Schematic examples of emission lines and bands as they might appear in a spectrum. The molecular emission band shows a "fine structure" associated with the structure of the molecule.

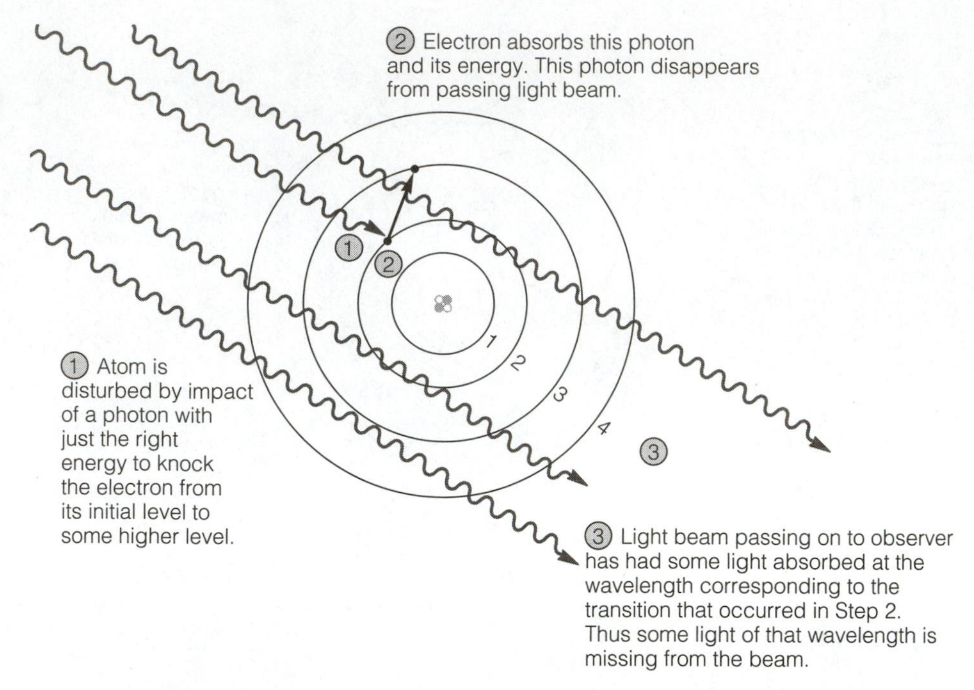

Figure 5-9 Schematic view of absorption of light as an electron is knocked from a lower to a higher energy level in an atom.

barely discernible. Most emission lines and bands arise from gases.

However, an important exception occurs among rock-forming minerals, which make up most planetary surfaces and interstellar dust grains. Most rock-forming minerals are crystals, which can be thought of as giant molecules. In a crystal, atoms are joined in a fixed pattern that simply repeats as the crystal grows bigger. Light can penetrate through the outer millimeter or so of some crystals and then reflect back out toward the viewer. Very broad, faint emission bands can result, as shown in Figure 5-8. Warm grains of silicate material near other stars have been identified from their emission bands, for example.

ABSORPTION LINES AND BANDS

Let us return to an atom with an orbiting electron. Figure 5-9 shows such an atom. This time it gets disturbed by a passing light wave (photon) that bumps it up into a higher energy level. Energy was removed from the beam to do this. This process of energy removal from a light beam is called **absorption.**

Atoms and Absorption Lines

★ Only the specific energy corresponding to the specific transition (level 2 to 3 in the case of Figure 5-9) can be absorbed in an atom. Thus only a photon of specific energy or wavelength can be removed from the beam. A beam of light passing through a cloud of such atoms will have many of these photons removed, thus absorbing some light of that color. As a result, the light in a narrow interval of the spectrum is lost, and this missing interval is called an **absorption line.** Various absorption lines can result from the various possible upward transitions—for example, level 2 to 3, 2 to 4, 2 to 5, 1 to 2, 1 to 3, and so on. The visible part of the spectrum, with absorption lines due to hydrogen, is shown in Figure 5-10. Check out the color version of this figure on our web site to see how the lines correspond to different colors.

Since the *intervals* between energy levels in every atom *of any specific element* are the same, regardless of whether absorption or emission is occurring, the pattern of emission and absorption lines for a given element is always the same for any given element. Hydrogen atoms produce one set of emission or absorption lines; oxygen atoms produce another set; iron atoms a still different set; and so on. Any element in a remote object giving off emission or absorption lines can therefore be identified from either pattern of the lines. This is an extremely important principle in all of astronomy.

Molecules, Crystals, and Absorption Bands

Much of the preceding also applies to molecules and crystals. As light penetrates through a cloud of molecules (a gas) or through the upper millimeter of a crystal (in a rock surface), transitions of electrons

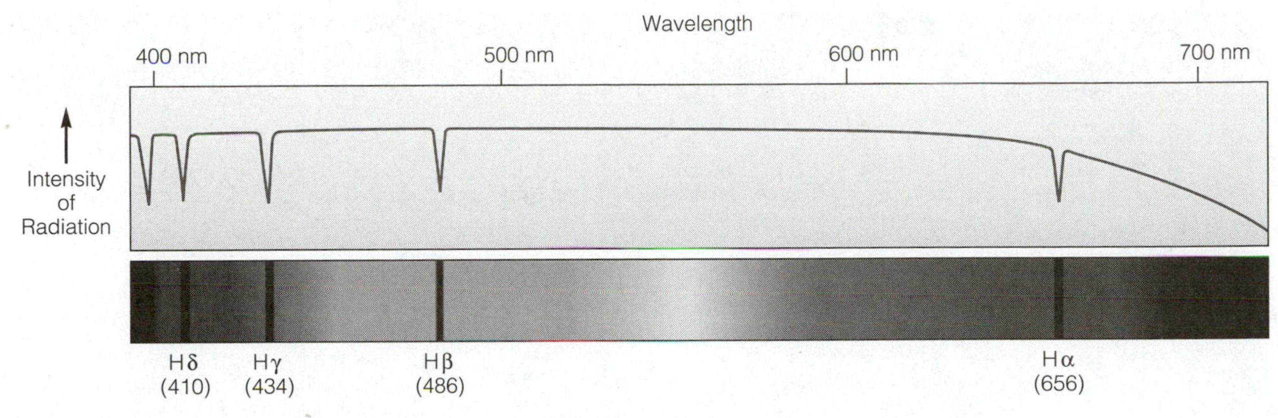

Figure 5-10 Visible portion of the spectrum showing absorption lines due to hydrogen. Such a spectrum would be seen if hydrogen gas lay between the observer and a light source with a continuous spectrum; the hydrogen absorbs only the specific "missing" colors. These absorption lines have the same positions as the emission lines in Figure 5-7.

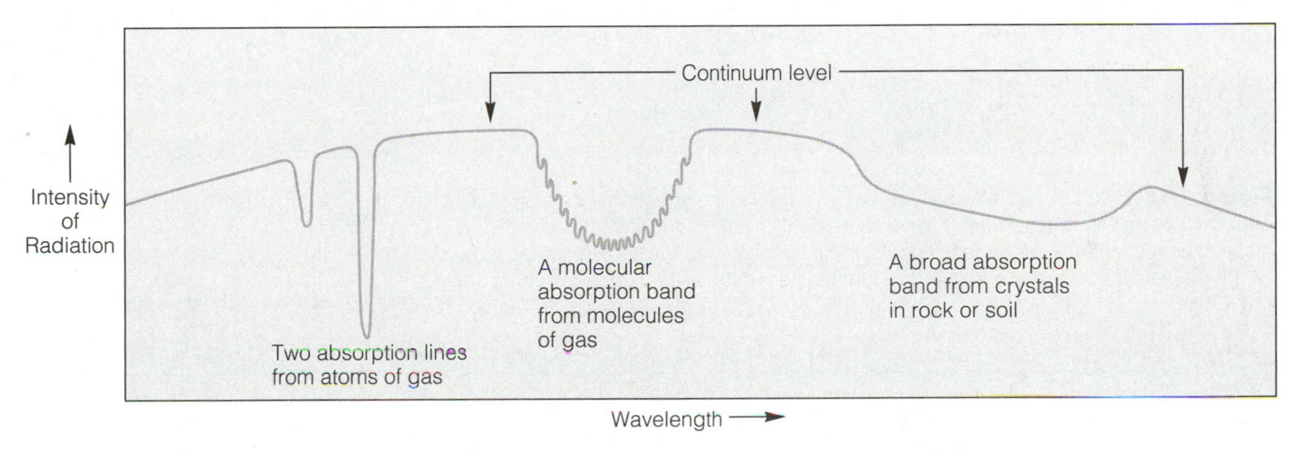

Figure 5-11 Schematic examples of absorption lines and bands as they might appear in a spectrum. The molecular absorption band shows "fine structure" associated with the structure of the molecule.

among its numerous, closely spaced energy levels produce **absorption bands.** The composition of the gas or the crystal can be identified if the absorption bands themselves can be detected and identified. Figure 5-11 shows schematic examples of absorption lines and absorption bands. The absorption bands of a substance have the same wavelength intervals as its emission bands.

Measuring the Composition of an Atmosphere of a Planet (or a Star)

Now we see how light carries even more messages—namely, a tabulation of the elements composing the light-emitting material! In practice, this statement is a bit optimistic because some elements or compounds have less prominent emission or absorption features than others, so identifying *all* the elements or compounds of a distant object is not necessarily easy.

To take an example, suppose we look at a planet with an atmosphere. Light passes into the atmosphere and back out, as shown in Figure 5-12. As the light passes through the gas, it acquires absorption lines and bands that allow us to identify some of the atoms and molecules in the gas. Some, such as carbon dioxide (CO_2), are easy to detect, and others, such as nitrogen (N_2), are hard to detect from their spectra. Thus, while the carbon dioxide of Venus and Mars was discovered by Earth-based spectroscopy in the first half of the 1900s, the abundant

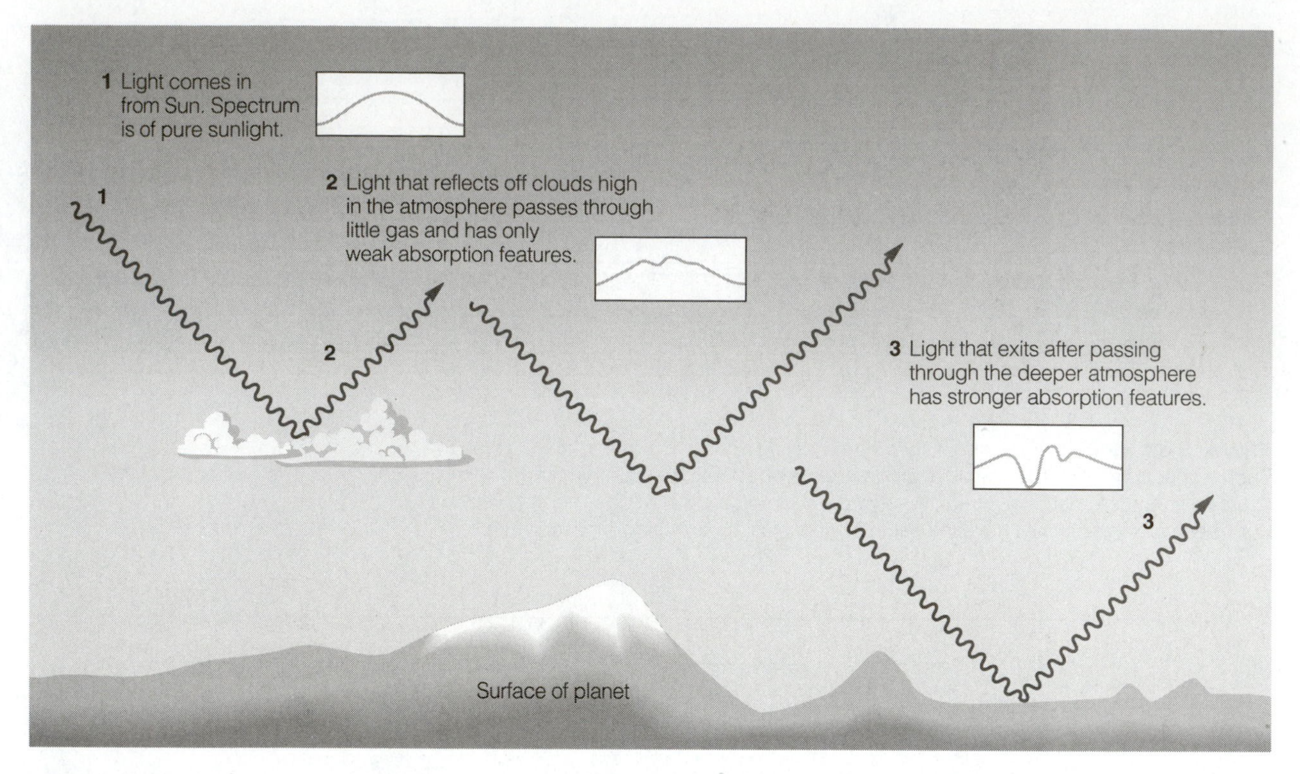

Figure 5-12 Production of absorption lines or bands in the spectrum of a planet as light passes through the atmosphere of the planet. Identification of the lines or bands allows identification of at least some constituents in the atmosphere. The same process occurs as light radiates from a star outward through the star's surrounding gaseous atmosphere.

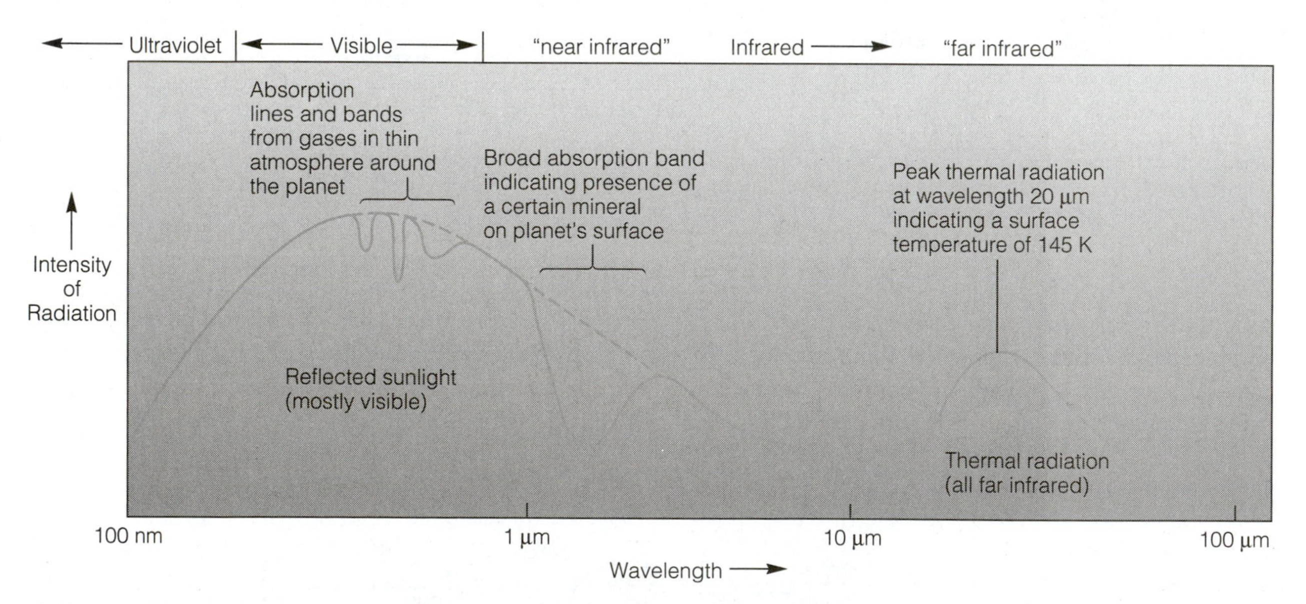

Figure 5-13 Simplified spectrum of an imaginary planet from the ultraviolet to infrared wavelengths, showing the reflected sunlight and the thermal radiation from the planet itself. The dashed curves show the shape of the reflected solar spectrum that we would see if there were no absorption lines and bands. A thin atmosphere creates some absorption lines and bands, but lets enough light through to the surface that we see a broad absorption band (at about 1.5 μm) due to minerals on the planet's surface. (See text for further description.)

nitrogen of Saturn's moon, Titan, was not discovered until a Voyager spacecraft made a close pass by Titan in 1980.

ANALYZING SPECTRA

How to Measure Properties of a Planet's Surface from a Distance

As mentioned earlier, the absorption bands caused by mineral crystals are broad, often shallow, and hard to detect. For this reason, spectroscopy has, we might say, only a 60% success record in identifying planetary surface materials. There are two approaches to using the spectrum to measure planets' materials: measuring colors and detecting actual absorption bands. Measuring the color of an object is essentially measuring the relative intensities of the broad wavelength ranges in different parts of the spectrum. A red object looks red because it reflects more light at red wavelengths than at bluer wavelengths. A measurement of color often restricts the possible materials on a surface. For instance, most satellites of Saturn have a bluish-white color similar to ice, but inconsistent with the red rocks and rock-forming minerals on Mars. Indeed, after the bluish-white colors of these satellites were measured, later measurements of specific absorption bands confirmed that they are surfaced with various kinds of ice, such as frozen water and frozen methane.

Figure 5-13 shows some results that might be obtained if we could scan a planet's spectrum all the way from the ultraviolet to the infrared. (This diagram is somewhat idealized, since different instruments are needed in different regions. One astronomer may obtain the spectrum only in one region. To construct Figure 5-13 might require the combined work of many astronomers using different instruments at different observatories.) In the visible region, we see reflected sunlight. Superimposed on it are some absorption lines and bands that tell us the planet has a gaseous atmosphere containing certain elements and compounds; these absorption features arose as shown in Figure 5-12. In the near infrared, just beyond 1 μm (micrometer = 0.001 mm) wavelength, we see a broad absorption feature characteristic of a rock-forming mineral. This gives us some indication of the rock type on the surface, indicating at the same time that clouds are not thick enough to block most of the light from reaching the surface. Finally, in the far infrared at 20 μm wavelength, we see the thermal emission from the planet. Noting that the dominant wavelength is at 20 μm, we can use Wien's law to estimate the surface temperature on the planet. In this case, it is a cold 145 K (–198°F).

The Spectrum and Our Atmosphere: Seeing into Space from the Earth

Figure 5-14 illustrates several important points about the relationship between astronomy and the Earth's atmosphere. First, clouds block much of our view of space. One reason for putting observatories on top of mountains is to get above the clouds. Another reason is to get above the dense, shimmery air that makes telescopic star images dance and twinkle when seen from sea level.

But a more important reason can be seen by comparing Figure 5-14a made in normal visible light, with Figure 5-14b, made in infrared radiation—that is, light that has somewhat longer wavelengths than the reddest light our eyes can detect. Many atmospheric molecules, such as water molecules, strongly absorb infrared light just beyond the reddest wavelengths we can see. Figure 5-14b is an image of Earth made with light of a particular infrared wavelength that is absorbed by water molecules. Thus, this picture is especially effective at showing the distribution of water-vapor haze and clouds. You can see that water-vapor haze and clouds blanket most of Earth's atmosphere. Thus a sea-level astronomer who wants to look for water-vapor absorptions on, say, Mars, has a problem: The infrared radiation that carries this message from Mars is blocked by water vapor in our own atmosphere before it ever reaches a telescope on the ground, as can be seen in Figure 5-15. Even the thin, high atmosphere blocks many wavelengths. For example, the **ozone (O_3) layer,** about 20 to 50 km up, absorbs nearly all ultraviolet radiation (thus protecting us from severe sunburn).

These absorptions of certain wavelengths by our atmosphere are a main reason for putting modern observatories on high mountains. High observatories are above most of our atmosphere's water vapor, which is concentrated in the bottom 3 km (9000 ft) of the atmosphere and absorbs much of the infrared light from planets and stars. At the same time, high-altitude observatories escape the clouds and the shimmering, hazy air of low elevations. Thus, for example, astronomers at 14,000-ft Mauna Kea Observatory in Hawaii can measure most of the infrared spectrum almost every night, because they

a b

Figure 5-14 Views of Earth based on two different wavelengths of light. **a** View in visible light, showing clouds and continents. Africa dominates this image. **b** View of the same part of Earth at a wavelength of reflected solar infrared light. Much of this light is absorbed by water-vapor gas in the air. Due to absorption, water-vapor haze obscures most of Earth's surface. (European Meteorological Satellite images, courtesy of C. R. Chapman)

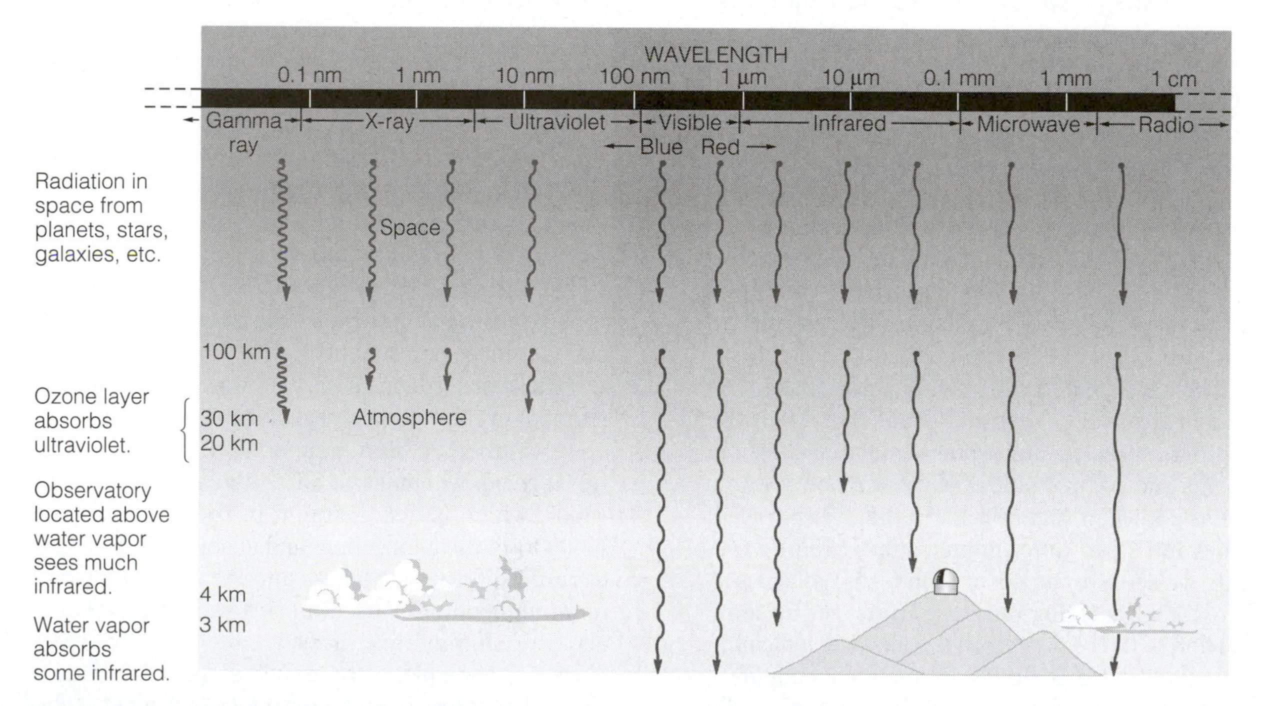

Figure 5-15 Electromagnetic spectrum is shown at top. All these types of radiation from various astronomical sources can be detected by a telescope in space (upper middle), but many of them are absorbed by various gases in our atmosphere (bottom).

are above most of the water vapor that absorbs infrared light.

Figure 5-14 and 5-15 illustrate one of the advantages of access to a wide range of wavelengths. Photos using light of different wavelengths penetrate gaseous atmospheres of planets or stars to different depths (because of the different rate of absorption at different wavelengths). Therefore they show different features at different depths in the atmosphere. In this way, comparisons of images at different wavelengths allows us to tell something about the vertical structure of the atmosphere we are observing.

THE THREE FUNCTIONS OF TELESCOPES

A **telescope** has three functions. First and most obvious, it magnifies the image of an object to a larger angular size than we perceive with our naked eye. The term **magnification** applies primarily to a telescope designed for visual observation; the magnification is the apparent angular size of a distant object seen through the eyepiece, relative to its apparent size seen by the naked eye. If a telescope makes something look 10 times larger, we say it has a magnification of 10×.

The second function is **resolution**—the ability to discriminate fine detail. Whereas the eye can resolve angular details only a few minutes of arc across, a telescope might show details a hundred times smaller, only a second of arc across. The resolution of a telescope is given by the ratio of the diameter of the telescope aperture to the wavelength of the light being measured. Larger telescopes can resolve smaller angles and are said to have *higher resolution. In principle,* a telescope with a large aperture of 4 m has a resolution of 0.02 second of arc. However, *in practice,* the blurring of incoming light due to the Earth's atmosphere reduces the resolution to about 1 second of arc.

The telescope's third function is **light-gathering power**—the ability to collect light and reveal fainter details than the naked eye can see. When light from a distant planet or star reaches Earth, a certain number of photons strike each square centimeter of Earth each second. The pupil of the eye has a diameter less than a centimeter and can receive only a limited number of photons per second. But a telescope collects *all* the photons striking a lens or mirror many centimeters across. This makes a much brighter image.

Early telescopes were designed entirely for observers to look through. In modern professional telescopes, the human eye is replaced by instruments that can make more precise measurements. An example can be seen in Figure 5-16. Thus modern astronomers rarely look through their giant telescopes! Nevertheless, the three functions still apply. A large telescope used with film to take pictures, for example, can be compared to an ordinary camera instead of to the eye: It obtains a more magnified image, a more clearly resolved image, and a brighter image than the camera.

Two Designs for Optical Telescopes

Two basic designs have been used for telescopes. The first to be built, the **refractor,** uses a lens to bend, or refract, light rays to a focus, as in Figure 5-17a. Galileo first used this type astronomically in 1609. The second type, the **reflector,** uses a curved mirror to reflect light rays to a focus, as in Figure 5-17b. Isaac Newton built the first reflector in 1668. Several reflector designs have since been constructed, but the simplest is Newton's, sometimes called the Newtonian reflector. In recent years new designs of backyard telescopes for amateur astronomy have combined lenses and mirrors. Often called **compound telescopes,** these designs can be very compact and portable.

The first large telescopes were refractors, but all major research telescopes now use the reflector design for three reasons. First, each little segment of a lens acts like a prism, splitting white light into its component colors. Since red and blue wavelengths are bent by different amounts, in a refracting telescope there is no single place where light of *all* colors is in focus. Reflecting telescopes avoid this problem because a mirror reflects *all* wavelengths to a single focus. Second, large refractors have large, heavy lenses, which must be supported around the edge, and they may sag slightly in the middle, distorting the image. Reflectors use a curved mirror, which can be supported across its back surface. The largest refractor ever built had an aperture of 1 m, whereas astronomical mirrors 8 m across have been constructed. Finally, large refractors are long and thin, so they suffer from flexure. The standard design for large reflectors is similar to Figure 5-17b, but the secondary mirror is used to bounce the light back down through a hole in the primary mirror. This results in a compact configuration with the heavy mirror and instrumentation at the bottom of the telescope (see Figure 5-16).

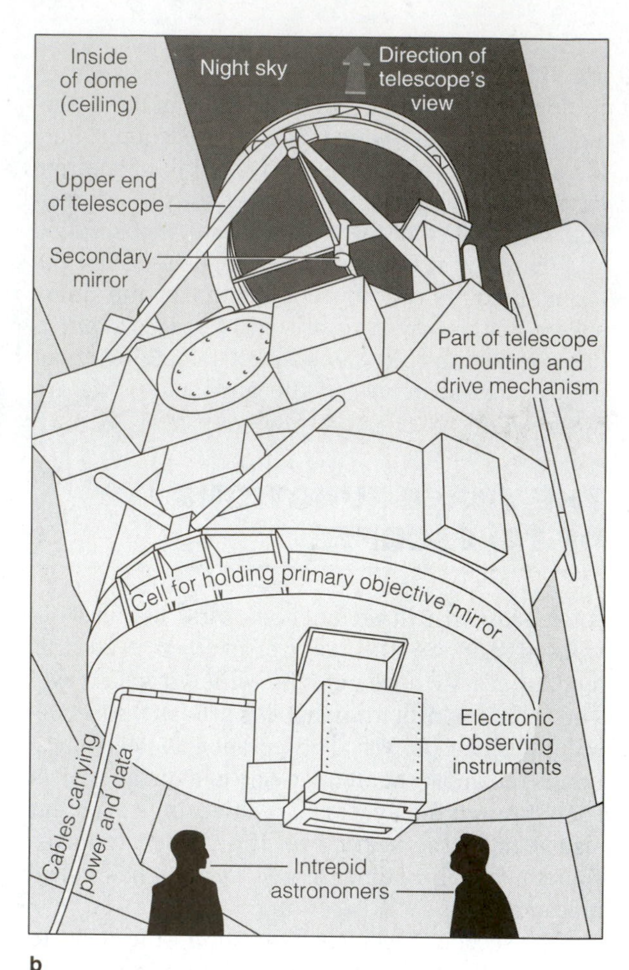

a b

Figure 5-16 a A view of a modern telescope, at Mauna Kea Observatory, Hawaii. In this telescope design, light is focused through a hole in the center of the main mirror, where the astronomer mounts electronic instruments to record data. Data are fed through cables to computer equipment, which performs initial analysis and stores data for further analysis. The dome (top) is open to night air in order to give the clearest observing conditions; temperatures inside the dome often drop below freezing during observing sessions. (Photo by WKH) **b** Schematic diagram of *a*.

Focal Length, Aperture, and the Telescope's Functions

In a telescope, the main lens or mirror is called the **objective,** as shown in Figure 5-17a. The distance from the objective to the place where the image is focused is called the **focal length** of the objective (marked *F* in Figure 5-17a). The diameter of the objective is called the **aperture.**

The three functions of a telescope are controlled by the focal length and the aperture. The longer the focal length, the greater the magnification and the bigger the image. This principle also applies to camera lenses. Lenses with long focal length give big images and are called telephoto lenses. Camera lenses with short focal length are called wide-angle lenses (see Figure 2-5). The wider the aperture, the better the resolution. Note that magnification is not the same as resolution. Magnification alone produces only a big image, not necessarily a sharp image. By increasing focal length, we get a bigger image, but it may be blurry. If we increase the aperture (an expensive process, since we have to buy a bigger lens or mirror), we can get a sharper image. The third function, light-gathering power, is connected with aperture also: The larger the aperture, the more light collected.

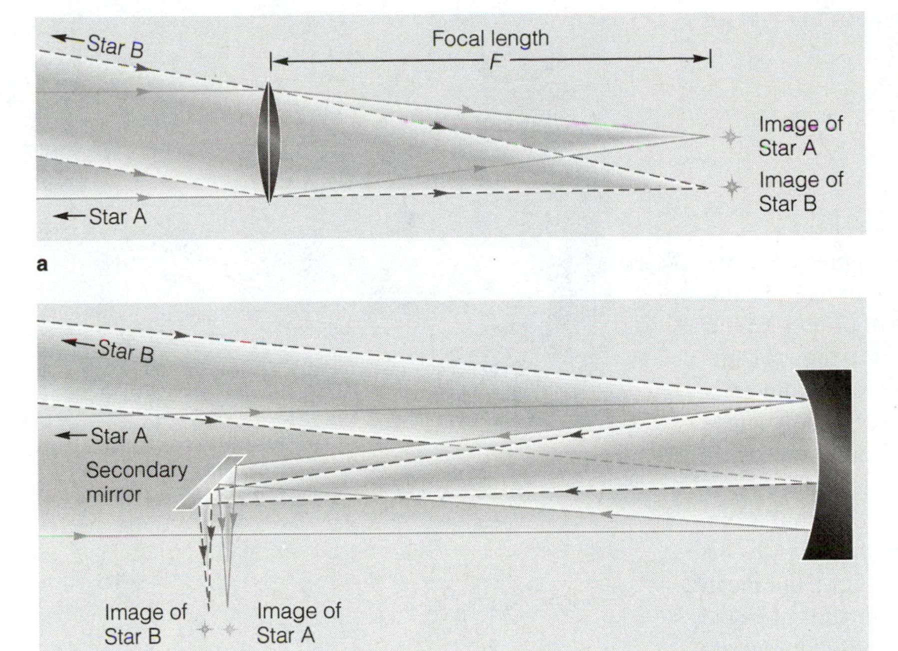

Figure 5-17 **a** Cross section through a lens, showing image formation in a refracting telescope (and most cameras). Light rays from two stars are focused into two images. **b** Cross section through a mirror system, showing image formation in a reflecting telescope. Light rays strike curved mirror (right) and are reflected back toward focus. The focus would normally lie in an inconvenient position in front of the mirror, but a secondary mirror is used to beam light rays to one side for easier access to the image.

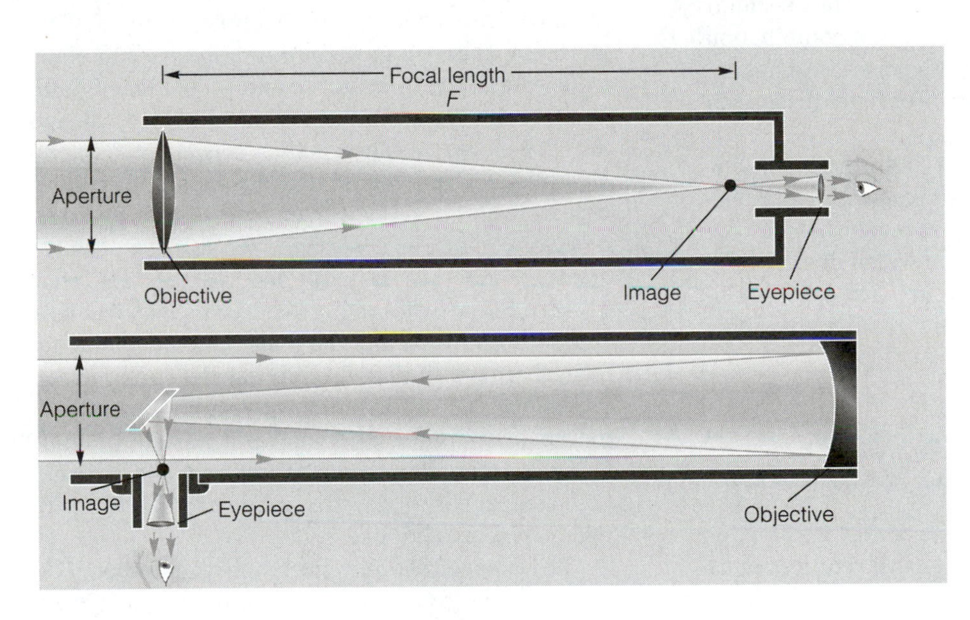

Figure 5-18 Cross section showing how the systems of Figure 5-17 are converted to visual telescopes by the addition of an eyepiece. The eyepiece is a small magnifying lens (or several lenses mounted together) used to examine the image. Tubing from the objective to the eyepiece helps cut out stray light.

USING VISUAL TELESCOPES

An optical telescope designed to be looked through, as opposed to a radio telescope or a camera, may be called a *visual telescope.* The optical systems of Figure 5-17 can be converted into visual telescopes simply by adding an eyepiece (and usually tubing to keep out stray light), as shown in Figure 5-18. The eyepiece is simply a lens or system of lenses designed to magnify the image still more and allow the eye to examine it.

★ One of the first questions asked of backyard telescope enthusiasts is: "How far can you see with that thing?" This is not the right question to ask, because no telescope is limited by distance. Every optical system can see as far as there is an object large enough or bright enough to detect. The naked eye as well as the 5-m (200-in.) Palomar telescope can see the Andromeda galaxy 19 billion billion kilometers away—but the telescope shows more detail and fainter regions because it gathers more light and gives a bigger image. Further information about using visual telescopes is given at this point in the text on our web site.

PHOTOGRAPHY WITH TELESCOPES

★ Instead of viewing an image with an eyepiece, astronomers are more likely to record the image permanently by projecting it onto a piece of photographic film or a digital imaging device, as in a digital camera. Modern imagery with digital devices, from both ground-based and orbital telescopes, is vastly superior to anything that the eye can perceive, even with the largest telescopes. In the case of faint objects, like distant galaxies, this is because the image can accumulate photons of light over many minutes or hours, whereas the eye cannot "store" light, but must operate in "real time." In the case of bright objects, like the Moon or planets, it is because digital recording techniques are so sensitive that they can record images during the instants (e.g., 1/100 of a second) of best seeing, whereas the eye/brain combination averages more than about 1/30 of a second and includes moments of bad seeing. (For more discussion of this, see our web site.)

Many photos of star fields and nebulae in our book and on our web site were taken with small telescopes or ordinary cameras. The reader should consult the picture captions, most of which describe the equipment and exposure used. In many cases, readers can duplicate or improve the results with their own equipment.

PHOTOMETRY

Pictures of astronomical objects are interesting, of course, but much astronomical work requires measuring an object's brightness—the amount of light coming from it at all wavelengths or at selected ranges of wavelengths (such as blue to green). This is called **photometry.** By giving a precise measure of the amount of light at various wavelengths, photometry allows astronomers to measure temperature, composition, and other properties of a remote object.

The most important type of detector used for this purpose is a digitized detector called a charge-coupled device, commonly known as a **CCD.** CCDs are extremely light-sensitive detectors, made possible by microelectronic technology (CCDs are found in most camcorders, for example). The device works by converting incoming photons into electrons, which are then stored and accumulated until being read out and converted into an electrical signal. The electrical signal is then converted to intensity units. CCDs consist of thousands of tiny light detectors in an array, for example, 2000 by 2000 detectors on a side, in a device the size of a postage stamp. Each tiny detector is called a *pixel* (for PICture ELement). Each pixel creates its own measure of brightness, and the many pixels combine to form an image similarly to the way dots of different sizes combine to form a photographic image in a newspaper. Figure 5-19 shows a mosaic of four CCDs, combined to view a larger area of sky. Each time an image is made with this device, there are 16 million pixels to be read out!

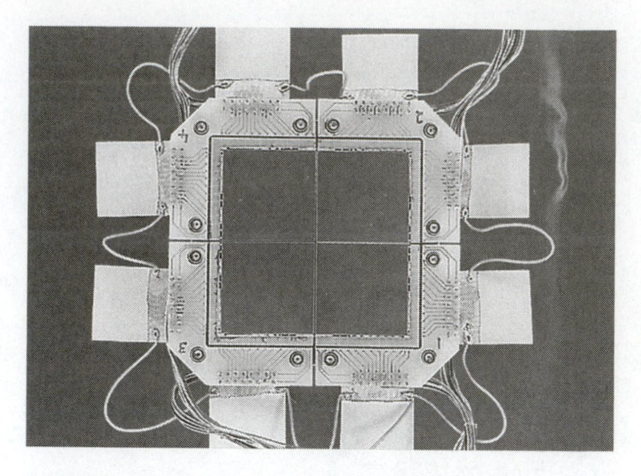

Figure 5-19 An assembled mosaic of four 2048 × 2048 pixel CCDs. The actual size of the array is approximately 70 mm on an edge. The signals are brought out from the two sides of each CCD on gold wires and then out of the vacuum system in which the CCDs are kept cold with liquid nitrogen. (NOAO)

CCDs have revolutionized optical astronomy. A CCD on a large telescope can produce images of light sources that are roughly a billion (10^9) times fainter than the eye can see! Where does this fantastic gain come from? CCDs are 10 times as efficient as the eye in detecting incoming photons, and the aperture of a large reflector is 1000 times the aperture of the pupil of the eye. An additional factor of 100,000 in sensitivity comes from the fact that the brain "reads out" the eye's image every 1/30 of a second (allowing us to see continuous motion), whereas a CCD can collect data on a single target for an hour or more. Astronomers mount CCDs in sealed cameras cooled with liquid nitrogen, which reduces the background signal due to thermal noise. CCDs are nearly perfect detectors, converting almost every incoming photon into an electrical signal. For this reason, the only way astronomers can see even fainter objects is to build larger telescopes and so collect more light. Photographs are not obsolete; they are still the favored medium for wide-angle imaging. In this book, most of the images that cover more than 10 minutes of arc are photographic, and most of the images that cover smaller bits of the sky in detail are electronic.

IMAGE PROCESSING

In the last decade, extraordinary advances have occurred in our ability to process images in order to gain maximum information from them. There are three overlapping subjects of interest: photographic images, digitized images, and false color images.

Photographic Images

Photographic images are composed of microscopic dots arranged in a random pattern; each dot represents a grain in the chemicals composing the film. The more light hitting a given grain, the larger and darker it appears in the film image, which is thus a reversal, or "negative," of the brightnesses in the scene. The negative image is then processed to make a positive print or transparency. Color films, such as Kodachrome, have several layers of light-sensitive chemicals with different color dyes to produce a color image from a single exposure. However, newer techniques allow the image processor to combine separate black-and-white images made with different-colored filters—such as three images made with red, yellow, and blue filters in a spacecraft—to create a single color photo. A major problem in all photography is that film can reproduce only a limited range of contrast. On a clear day under a bright Sun, film usually cannot quite record the full range of brightness from sunlit white surfaces to dark shadows. In the same way, if the bright core of a galaxy is 10,000 times brighter than its faint spiral arms, an exposure long enough to record the arms overexposes the bright central region, giving the false impression of a bright blob rather than an intense starlike nucleus surrounded by a diffuse glow. Modern photographic techniques, such as using variable-density "masks" that screen light from the brightest regions, allow improved images of high-contrast astronomical subjects.

Digitized Images

As described earlier under photometry, some modern sensors such as CCDs have a *grid* of dots, called pixels, in which the brightness is measured for each dot. Imagine a blowup of a newspaper photo. You can see each individual pixel as a dot. Now imagine that each dot is replaced by a number from zero (blackest tones) to 100 (brightest), representing the brightness of the light at that point in the image. Such an image may come not only from a CCD but also from an ordinary photo that has been scanned and measured and converted into such pixels, which is called *digitizing the image.* The position of each dot and its brightness are stored in a computer as numbers. Then the image can be processed in many ways. Suppose we are interested in some subtle contrast features that appear only in the light gray areas between 70 and 90 on the brightness scale. Then we could ask the computer to make an image in which the darkest tones of the print correspond to 70 in the original image and the brightest correspond to 90. Every part of the image that was fainter than 70 now appears as black; every part that was brighter than 90 now appears as white. We now have an extreme-contrast version of all the detail between 70 and 90—which might reveal some interesting new features that we missed before. This is called "stretching" the image.

False Color Images

Another technique that helps to emphasize subtle details of images is to assign different colors to each brightness level. For instance, we could use violet for the dark tones of 1 to 10, blue for 10 to 20, green for 20 to 30, and so on. (Or a different gray tone could be used for each interval.) Or we could divide the brightness scale even finer, with yellow for 50 to 52, orange for 52 to 54, and so on. In this way, we could produce an image in which the colors have nothing to do with the original colors but are used merely as a code to allow us to separate features of slightly different brightnesses. An example is seen in Figure 5-20. Such images are called *false color images.* In the 1980s and 1990s, this technique went too far because false color became a fad among magazine art directors who were attracted by the flashy pictorial quality of the colors. It became counterproductive, because viewers didn't understand what they were seeing when false color images were published without adequate explanation. TV journalists were heard to exclaim over the strange, "fried-egg" appearance of Halley's comet, suggested by Figure 5-20. Such pictures had nothing to do with the actual appearance

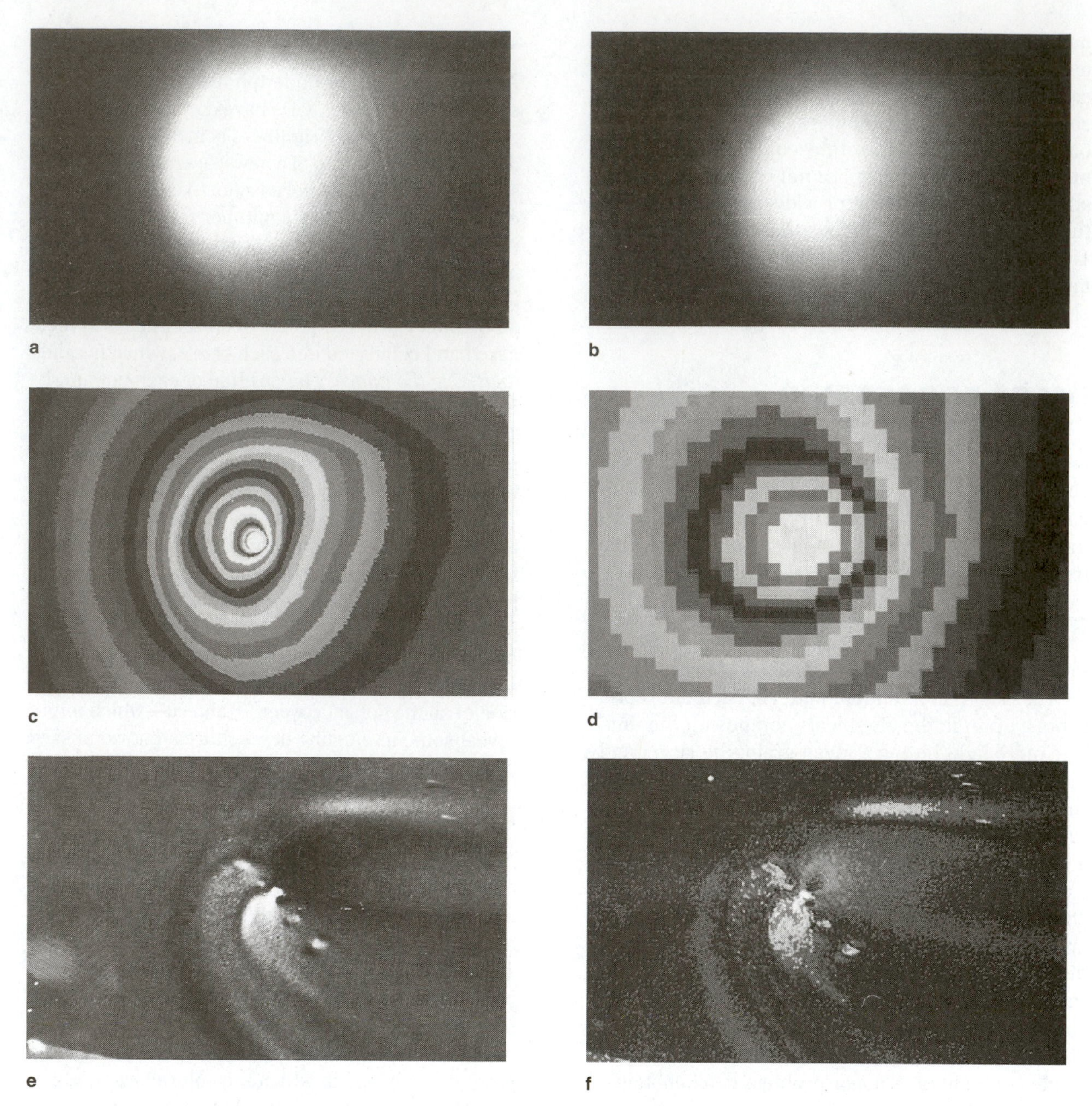

Figure 5-20 New image processing techniques are shown by these images of the head of Halley's comet. **a** A direct image of the comet's head taken by a CCD on a large telescope resembles an ordinary photo and shows an overexposed bright central condensation surrounded by glowing gas. The image has some defects, such as the streak at right, due to internal properties of the CCD. **b** Addition of several images and correction of flaws makes a cleaner image. The central region is still overexposed. **c** The same image in false colors, with different-colored bands used to represent different brightness levels. This image, though much less realistic in terms of visual appearance, reveals the smooth increase of brightness toward the center, even within the regions overexposed on print *a*. **d** An enlargement of the central part of *c* shows the individual pixels composing the image. **e** More sophisticated image enhancement has been used to exaggerate localized intensity boundaries within the seemingly uniform contours. This enhanced image reveals faint spiral jets, present in the earlier images but swamped by the comet's glow. Bright blobs are images of star the comet passed during the several different exposures. The star is barely visible in *b* and *c*, but here it is enhanced by the processing. **f** A false color version of *e* emphasizes the spiral jet structure and the increasing brightness toward the center, but it is even further removed from the appearance the comet would have to the naked eye, since the colors are arbitrary and the overall glow around the comet's head is suppressed to reveal the curved streaks. (All images by Stephen M. Larson, Lunar and Planetary Lab, University of Arizona, from observations made at Boyden Observatory, South Africa, on the day of Giotto probe encounter; see Giotto image in Chapter 13)

of the comet. They were merely false color images with different rings of color representing different brightnesses!

For such reasons, "true color" images are used as much as possible on our web site. "True color" is in quotation marks because at some level of detail it is difficult to reproduce all the nuances of natural color, especially in faint objects. But the colors in most images we've selected do give some impression of the actual colors of the objects. A number of the CCD images also display "true color." As with photography, this is done by combining two or three CCD exposures through different color filters into one final color image.

SPECTROPHOTOMETRY

As we stressed in the first part of this chapter, it is important to measure not only the total amount of light coming from an astronomical object but also the intensity at each wavelength. This process is called **spectrophotometry.** It is done by passing the light through a spectroscopic device that breaks the light into different colors before it enters the electronic detector. Then the detector can measure the amount in each color (that is, wavelength) range. Similarly, photos or TV images can be produced showing the object as seen in different colors (that is, different wavelengths).

LIGHT POLLUTION: A THREAT TO ASTRONOMY

Figure 5-21 shows the pattern of urban lights across the United States as seen on a cloud-free night from space. Many observatories are near rapidly sprawling cities. The glow from these cities lights up the sky, reducing the contrast between faint stars and the dark sky background. This is why it is so hard to get a good view of the sky from urban areas. Moreover, many types of fluorescent streetlights emit emission lines that interfere with astronomical work. These problems have rendered many famous observatory sites, such as Mt. Wilson in southern California, nearly worthless for the most sensitive modern instruments. A classic solution to this problem has been for astronomers to move even farther toward the frontier—as when Percival Lowell established his observatory in Flagstaff, Arizona, in the 1890s or when modern astronomers send telescopes into space. As Lowell himself pointed out, ancient city dwellers before the advent of electricity were much more familiar with the sky and the stars than modern urbanites. Many cities have acted to protect the investment in nearby observatories by adopting only streetlights, such as sodium vapor lights, that emit minimally disruptive wavelengths of light and by installing hoods that block lights from shining up into the sky.

Figure 5-21 A beautiful pattern that threatens modern astronomy. Compositing of satellite photos had produced a montage that shows North American urban lights as seen from space with no cloud cover. Skies over the Eastern megalopolis, southern California, and other regions are "polluted" with light so strongly that neither the eye nor the telescope can see the faintest stars that would otherwise be visible from these regions. (Photo by W. T. Sullivan III, courtesy National Optical Astronomy Observatories)

DETECTING NATURE'S MESSAGES FROM SPACE

One goal of astronomy is to detect and understand nature's messages from space. However, as we have seen, Earth's atmosphere is opaque to most wavelengths of electromagnetic radiation (Figure 5-15). Only a thin band of optical and near infrared wavelengths and a wider band of radio wavelengths penetrate to the ground. For most of the history of astronomy, our knowledge of the universe has been confined to the messages coming through the narrow window of visible light. In the past 30 y, this situation has been transformed by new technologies and

TABLE 5-2

Optical Telescopes Larger than 3 m

Telescope	Observatory and Location	Mirror Diameter (meters)	Date in Operation
Hale	Palomar Observatory, Mt. Palomar, California	5.0	1948
Shane	Lick Observatory, Mt. Hamilton, California	3.0	1959
Mayall	Kitt Peak National Observatory, Kitt Peak, Arizona	4.0	1974
Cerro Tololo	Cerro Tololo Inter-American Observatory, Cerro Tololo, Chile	4.0	1975
Anglo-Australian	Anglo-Australian Observatory, Siding Spring, Australia	3.9	1975
Max Planck Institut	Calar Alto Observatory, Calar Alto, Spain	3.5	1976
European Southern Observatory	European Southern Observatory, Cerro La Silla, Chile	3.6	1976
Bolshoi	Special Astrophysical Observatory, Zelenchukskaya, Russia	6.0	1976
UK Infrared	Mauna Kea Observatory, Mauna Kea, Hawaii	3.8	1979
IR Telescope Facility	Mauna Kea Observatory, Mauna Kea, Hawaii	3.0	1979
Canada-France-Hawaii	Mauna Kea Observatory, Mauna Kea, Hawaii	3.6	1980
Multiple Mirror	Multiple Mirror Telescope Observatory, Mt. Hopkins, Arizona	4.5	1980
Herschel	La Palma Observatory, Canary Islands, Spain	4.2	1987
New Technology	European Southern Observatory, Cerro La Silla, Chile	3.6	1989
Astrophysical Research Consortium	Astrophysical Research Consortium, Apache Point, New Mexico	3.5	1991
Keck	Mauna Kea Observatory, Mauna Kea, Hawaii	10.0	1991
WIYN Consortium	Kitt Peak National Observatory, Kitt Peak, Arizona	3.5	1993
MMT	Multiple Mirror Telescope Observatory, Mt. Hopkins, Arizona	6.5	1994
Very Large Telescope	European Southern Observatory, Parranal, Chile	8.0	1995
Keck-2	Mauna Kea Observatory, Mauna Kea, Hawaii	10.0	1996
Subaru	Japan National Observatory, Mauna Kea, Hawaii	8.3	1997
Hobby-Eberly	McDonald Observatory, Mt. Locke, Texas	10.0	1997
Very Large Telescope-2, 3, 4	European Southern Observatory, Parranal, Chile	3 × 8.0	1999
Gemini North	Mauna Kea Observatory	8.0	1999
Magellan	Las Campanas Observatory, Las Campanas, Chile	8.0	2000
Gemini South	Cerro Tololo Observatory, Cerro Panchen, Chile	8.0	2001
Magellan 2	Las Campances Observatory, Las Campances, Chile	8.0	2003
Large Binocular Telescope	Mt. Graham International Observatory, Mt. Graham, Arizona	2 × 8.4	2004

Notes: The MMT and Keck telescopes are made of multiple mirrors; the Large Binocular Telescope has two 8-m mirrors in a single mount.

the NASA space program. It is as if the blinkers have been removed, and we are now free to explore the entire electromagnetic spectrum.

Ground-Based Telescopes

★ Driven by the need for more light-gathering power, astronomers have built larger and larger optical telescopes. Table 5-2 lists some telescopes of 3-m aperture or larger. There are two main approaches to building large telescopes. The first is to construct large, single mirrors, trying to make them as big, light, and accurate as possible. One of the most ingenious facilities sits under the football stadium at the University of Arizona. Glass chunks are placed in an enormous rotating oven. As the oven is heated, the glass flows freely, and as the oven spins, the mirror takes a roughly parabolic shape. As the oven cools, the mirror solidifies in its final shape, ready for polishing and aluminizing.

Alternatively, a large collecting area can be built up out of a number of smaller mirrors. This approach was first used at the Multiple Mirror Telescope in 1980 in southern Arizona (Figure 5-22). The largest optical telescopes in the world are at the Keck Observatory on Mauna Kea in Hawaii (Figure 5-23).

Figure 5-22 The Multiple Mirror Telescope (MMT) on Mt. Hopkins, Arizona. The original six mirrors have recently been replaced with a single 6.5-m mirror. The small and open design of the building contributes to excellent astronomical seeing. (Multiple Mirror Telescope Observatory)

Figure 5-23 A view of the Keck telescope on Mauna Kea, Hawaii. The Keck telescope has 36 hexagonal mirrors that give it an effective aperture of 10 m, making it the largest optical telescope in the world. A second, similar telescope is under construction on a nearby site. (California Association for Research in Astronomy)

This observatory was funded by a bequest from the world's largest privately owned oil company. Two Keck telescopes are located at the observatory; each has 36 hexagonal segments, each with a diameter of 1.8 m. The effective diameter of each telescope is 10 m. The individual mirrors are controlled by an active system that maintains the mirror's shape to a high degree of accuracy.

★ A **radio telescope** is a device that gathers and concentrates radio waves, analogous to the way an optical telescope gathers and concentrates light waves. The first radio telescope was built by the pioneering electronics engineer and radio ham, Grote Reber, in 1936. For a decade, Reber was the world's only radio astronomer. Radio telescopes serve the same functions as optical telescopes. Radio waves have much larger wavelengths than visible light waves, and they carry much less energy per photon, so radio telescopes have larger surfaces to collect sufficient energy to give a strong signal. On the other hand, since the wavelength is larger, the surface need not be as accurately shaped as the mirror of an optical telescope. Radio telescopes have a curved surface (or *"dish"*) of metal or wire mesh, which reflects the radio waves to a focus, where they are converted into an electrical signal. The largest radio telescope ever constructed is the 305-m Arecibo telescope, built into a natural topographic depression in Puerto Rico. It is not steerable, but can sample a variety of objects in the universe at high resolution, across a narrow strip of the sky.

Telescopes in Space

Large portions of the electromagnetic spectrum are absorbed by Earth's atmosphere, as discussed earlier, and therefore cannot be observed by ground-based telescopes. Hence, they can be observed only from space. This is true of most of the infrared and microwave parts of the spectrum, and of all ultraviolet radiation, X-rays, and gamma-rays. For this

TABLE 5-3

Major Telescopes in Space—Recent, Current, and Planned

Telescope	Countries Involved	Wavelength Region	Mirror Diameter (meters)	Date Launched (or proposed for launch)
Copernicus	US	Ultraviolet	0.9	1972
High Energy Astrophysical Observatory (Einstein)	US	X-ray	0.56	1978
International Ultraviolet Explorer (IUE)	US-Europe	Ultraviolet	0.41	1978
European X-ray Orbiting Satellite (EXOSAT)	Europe	X-ray		1982
Infrared Astronomical Satellite (IRAS)	Holland-UK-US	Infrared	0.57	1983
Hipparcos	Europe	Visible	0.29	1989
Cosmic Background Explorer (COBE)	US	Microwave, Radio		1989
Hubble Space Telescope (HST)	US-Europe	Visible, Ultraviolet	2.4	1990
Roentgen Satellite (ROSAT)	Germany-US	X-ray		1990
Gamma Ray Observatory (GRO)	US	Gamma ray		1991
Extreme Ultraviolet Explorer (EUVE)	US	Ultraviolet	0.6	1991
ASTRO-D	Japan-US	X-ray		1993
Infrared Space Observatory (ISO)	Europe	Infrared	0.6	1995
X-ray Timing Explorer (XTE)	US	X-ray		1996
X-ray Multiple Mirror Mission (XMM)	Europe	X-ray		1998
Far-Ultraviolet Spectroscopic Explorer (FUSE)	US-Europe-Canada	Ultraviolet	0.6	1999
Chandra X-Ray Observatory	US	X-ray		1999
Stratospheric Observatory for Infrared Astronomy (SOFIA)	US	Infrared	2.5	2002
International VLBI Satellite (IVS)	Europe-US-Japan-Russia	Radio	25.0	2002
Space Infrared Telescope Facility (SIRTF)	US	Infrared	0.85	2005
Large Deployable Reflector (LDR)	US	Radio, Microwave	10.0	2006
Space Interferometry Mission (SIM)	US	Visible		2006
Next Generation Space Telescope (NGST)	US-Europe	Visible, Ultraviolet	6.0	2009
Terrestrial Planet Finder (TPF)	US	Visible		2010

Notes: Copernicus, Einstein, EXOSAT, and IRAS are no longer operating. In some cases the funding is in doubt. SOFIA is a telescope in a converted Boeing 747, designed to fly and observe at 40,000 ft.

reason, telescopes of various sizes have been launched into low Earth orbit. Table 5-3 lists the major space observatories that are currently in operation, along with some that are planned for the near future. Because **space astronomy** is expensive, many of these missions are joint ventures between two or more countries. Government agencies from the various countries cooperate on the funding, and scientists from the countries collaborate on interpreting the data.

The dramatic gains possible from space astronomy are best illustrated in the infrared part of the electromagnetic spectrum. As we discussed earlier, water vapor in the atmosphere absorbs most infrared wavelengths so effectively that even a high mountain observatory like Mauna Kea cannot observe all of the infrared spectrum (review Figures 5-14 and 5-15). Furthermore, the Earth and its atmosphere are at a temperature of about 300 K, so it is emitting thermal radiation across the infrared band (see Figures 5-3 and 5-4). All astronomical sources are therefore seen against a bright background of thermal infrared radiation. Across most of the spectrum, the background sky is a *million times darker* in space. In infrared astronomy, the difference between working on the ground and working in space is equivalent to the difference between observing in the day and at night in optical astronomy!

To summarize, space astronomy offers three distinct advantages over Earth-based astronomy. It opens up spectral regions that cannot be observed from the ground. It offers an environment where the optical and infrared backgrounds are very low. It allows telescopes to operate outside the blurring effects of the Earth's atmosphere, where in principle they can achieve their resolution limit. Against these advantages is set the much greater cost and limited lifetime of an observatory in space.

★ The centerpiece of the American space astronomy program is a set of four missions called the Great Observatories. The first of these, the *Hubble Space Telescope* (HST), was deployed by the Space Shuttle in 1990. Euphoria turned to dismay when it was realized that the telescope's 2.4-m primary mirror had been machined to slightly the wrong shape because of an error in the calibration process. This problem was corrected within a few years when astronauts made a special repair mission to the telescope and inserted some correction optics that essentially solved the problem. The HST has produced many breakthrough results as well as stunning images of celestial objects, which have found their way into many media, from textbooks to popular posters, screen savers, and CD covers.

Other space telescopes have been cheaper and highly productive. The *Cosmic Background Explorer* (COBE), launched in 1989, has been spectacularly successful in detecting the thermal radiation from the early hot phase of the universe (Figure 5-24, see also Chapter 27). The *International Ultraviolet Explorer* (IUE) has been working successfully since 1978, producing ultraviolet spectra of a variety of (mostly stellar) sources. The *Infrared Astronomical Satellite* (IRAS), a joint venture of the United States, Britain, and the Netherlands, was launched in 1983. It has produced an important survey of the sky from 10 μm to 100 μm in its one year of operation. The next step in infrared astronomy from space will require larger, cooled telescopes that can make extensive pointed observations. The *European Infrared Space Observatory* (ISO) produced some exciting science, but it should easily be surpassed by NASA's *Space Infrared Telescope Facility* (SIRTF). SIRTF is due for launch in 2002 as the last of NASA's "Great Observatories." SIRTF is expected to advance our understanding of star formation and the evolution of galaxies.

At higher energies and shorter wavelengths, the field of X-ray astronomy came to maturity with the launch of HEAO-2, the *Einstein Observatory,* in 1978. This facility had 1000 times the sensitivity of any X-ray mission that preceded it. Recently, the versatile *Roentgen Satellite* (ROSAT), funded jointly by Germany and the United States, made a survey of the X-ray sky. Further major advances in X-ray sensitivity are expected from the *Advanced X-ray Astrophysics Facility* (AXAF), another NASA mission recently launched. Detecting the highest energy electromagnetic waves is a great technical challenge. In 1990, NASA launched the *Gamma Ray Observatory* (GRO), which has been successful in detecting the high-energy emissions of compact stars and active galaxies.

Figure 5-24 The *Cosmic Background Explorer* (COBE), launched in 1989. The COBE satellite has mapped the microwave sky with unprecedented accuracy and has detected fluctuations that are relics of physical processes in the very early universe. (NASA)

Notice that almost all of the current telescopes in Table 5-3 are smaller than the ground-based telescopes in Table 5-2. It is difficult to launch large telescopes into space (the size of the HST mirror was set by the size of the Space Shuttle cargo bay). Space telescopes offer unique capabilities, but ground-based telescopes cannot be surpassed for sheer light-gathering power and low cost per photon collected.

★ For an additional technique that allows very high resolution studies without going into space, see the section on interferometry, at this point in the text, on our web site.

SUMMARY

Most of our knowledge of the universe comes from deciphering messages carried from extraterrestrial bodies by electromagnetic radiation. To do this, we need to understand the nature of electromagnetic radiation and how it is generated inside atoms and atomic particles. Electromagnetic radiation is released in the form of photons, each of which has its own specific wavelength and energy. The light coming from an object, arranged in order of wavelength, is called the spectrum of the object. Each wavelength corresponds to a "color," ranging from wavelengths that are too short (too blue) to see, through visible colors such as blue and red, to wavelengths too long (too red) to see. Photons of blue light have more energy than photons of red light. So-called white light from the Sun is actually a mixture of all colors of light, as Newton discovered with a prism.

Atoms and molecules each produce absorption and emission lines and bands in the spectrum as photons interact with them. These derive from the energy level structure of their electronic orbits. Each element and compound has its own lines and bands. Materials in a distant object can be identified by obtaining the object's spectrum and identifying emission or absorption lines or bands, if present. Many gases in planetary atmospheres or stars can be identified in this way, but solid surfaces of planets can be only roughly characterized, since solid materials have only weak, poorly defined absorption bands.

Telescopes are devices that not only magnify the angular size of distant objects but also, just as importantly, enable us to gather much more light than is gathered by the eye. Thus we can see objects much fainter than those visible to the unaided eye. The web version of this chapter contains advice on viewing different astronomical objects with small telescopes. Astronomers bolt various kinds of instruments on large telescopes to analyze the light from planets, stars, and so on. One of the most important instruments is the spectrophotometer, which measures the amount of light at each wavelength over a range of wavelengths, allowing precise measurement of the spectrum of the object being observed. We are just entering an era when large telescopes are being placed in orbit above the atmosphere in order to give access to unblurred images throughout all wavelengths of the spectrum. These space telescopes and other types of new instruments are revolutionizing modern astronomy during our generation.

CONCEPTS

wavelength
frequency
wavelike properties of light
visible light
ultraviolet radiation
infared radiation
electromagnetic radiation
speed of light
particlelike properties of light
emission line
ground state
excited state
emission band
absorption
absorption line
absorption band
ozone layer
telescope
magnification
photon
diffraction
spectrum
spectroscopy
fields
continuum
continuous spectrum
temperature
thermal motion
absolute temperature scale
thermal radiation
reflected radiation
Wien's law
emission
energy level
resolution
light-gathering power
refractor
reflector
compound telescope
objective
focal length
aperture
seeing (web site)
photometry
CCD
spectrophotometry
radio telescope
space astronomy
interferometry (web site)

PROBLEMS

1. Photographic films are usually more efficient when measuring more energetic photons. Which would you expect to be easier to photograph: a faint blue, red, or infrared star?

2. The ground cools at night by radiating thermal infrared radiation into space. Explain why the air stays warmer on a moist, cloudy night than on a very dry, clear night.

3. A spacecraft passes close to a hitherto unknown planet. A camera system on the spacecraft photographs surface details quite clearly in all wavelengths except the absorption bands of carbon dioxide. At these wavelengths, the image is featureless. Interpret these results.

4. Why has there been more emphasis on getting ultraviolet and infrared telescopes into orbit than on getting radio telescopes into orbit?

5. You are considering buying a pair of binoculars. Two available pairs each have lenses with 50-mm aperture, but one has a magnifying power of 7× and the other, 20×. Which pair:

a. Gives the larger apparent image size?

b. Gives a greater apparent brightness in the observed image?

c. Is easier to hold still enough to observe? (Remember that the unsteadiness of your hands is magnified as much as the image.)

6. For your birthday, your parents offer you a choice of lenses for your 35-mm camera. One has a focal length of 24 mm, and the other has a focal length of 90 mm. (Focal length numbers are marked on the inner rim of all camera lenses.)

a. Which one would give a bigger image but cover a narrower field of view in terms of degrees?

b. Which one would give a smaller image but cover a wider angular field of view?

c. Which one would be called a telephoto lens?

7. a. Why can an atom in the ground state not produce emission lines?

b. Use this fact to explain why the colorful emission line glows of certain nebulae occur near hot stars, but not in the cold gas of interstellar space.

ADVANCED PROBLEMS

8. Confirm that the thermal infrared radiation in Figure 5.13 could come from a planetary surface at a temperature of 145 K.

9. A planetary astronomer has an instrument that scans the spectrum from 5 to 10 μm wavelength. Looking at a certain moon, he detects thermal infrared radiation that increases from 5 to 10 μm but does not peak in this range. Apparently, it peaks at a longer wavelength than 10 μm. What can you say about this moon's surface temperature?

10. What wavelength of thermal radiation is being emitted with greatest intensity by this book?

11. Your friend's Newtonian reflector telescope has a focal length of 70 in., and he has an eyepiece with a focal length of $\frac{1}{2}$ in.

a. What magnification does the telescope give?

b. With the preceding information, discuss what can be said about the brightness of the image given by this telescope.

c. If you look through the telescope at the Moon, what angular size would the Moon appear to be?

d. Another friend appears and asks how big a lens the telescope has. How would you answer?

12. Two telescopes each have a 4-in. aperture, but one has a short focal length and gives 30× magnification and the other has a long focal length and gives 150× magnification.

a. Which one would give a better view of the planet Mars, which subtends an angle of about 20 seconds of arc?

b. Which one would give a better view of the faint, wispy Andromeda galaxy, which subtends as much as 3°?

PROJECTS

1. Compare views through binoculars of different sizes. (Binoculars carry a designation such as 7 × 35, where the first number is the magnifying power and the second is the aperture.) Confirm that at fixed power, larger apertures give brighter images.

2. Observe and sketch a rainbow. Confirm that the colors appear in order of wavelength as seen in Figure 5-a on the web site.

3. Study a room light equipped with a dimmer switch. Turn the switch from bright to the faintest visible setting. Note the redder color of the light bulb at the dimmest position. Relate this effect to Wien's law.

4. Experiment with two lenses having different focal lengths, such as 10 cm and 1 cm. Measure their focal lengths by focusing sunlight on a surface. Then (repeating experiments that must have been done first by European lensmakers around 1600) hold or mount the two lenses to make a simple refracting telescope like that in Figure 5-18. Tubing is not essential. Estimate your telescope's magnification by comparing the magnified image with a naked-eye view. You may then compare this estimate with the magnification calculated with the formula in footnote a on the web site. REMINDER: DO NOT LOOK DIRECTLY AT THE SUN WITH ANY TELESCOPE BECAUSE THE INTENSE LIGHT WILL BURN YOUR RETINA.

PART

3

Exploring the Earth-Moon System

This remarkable photo showing the Moon in orbit about the Earth was made in 1992 by the Galileo spacecraft from a distance of 4 million miles. The moon's far side is seen; background stars are too faint to be recorded in space photos. (NASA)

CHAPTER 6

Earth as a Planet

WHAT THE READER SHOULD WATCH FOR IN THIS CHAPTER

Earth's age—and the age of the rest of the solar system as well—is 4.6 billion years. The measurement of this number was the subject of a long quest for the last few centuries and led to profound revisions of our conceptions of our planet, especially the early views that it was only a few thousand years old. The measurement is made by utilizing known rates of decay among various radioactive elements. All systems of radioactive elements give the same answer. Earth has a crust of light rocks, a mantle of heavier, partly melted rocks, and a core of nickel-iron metal. Earth's crust and atmosphere have evolved and changed greatly during their 4.6-billion-year history. Movements of crustal "plates" have changed the shape of continents. Emissions of gases from volcanoes and from plants, along with interactions between the gases and the oceans, have changed Earth's atmospheric composition. Volcanism is caused by heating of the interior and partial melting of the outer mantle. Bombardment by asteroids is an ongoing process; the few largest hits, roughly a hundred million years apart, may have had major effects on the evolution of life, including the extinction of dinosaurs and the rise of mammals. ■

Some readers might wonder, "Why discuss Earth in an astronomy book? Why not relegate it to a course in geology?" The answer is dramatized by Figure 6-1. Earth is a world, like many other worlds. The soil in your backyard is a sample of planetary material just as much as soil from the Moon. People have taken a long time to realize the implications of the astronomical discovery that we are spinning through space on a small, finite ball.

When Renaissance astronomers realized that the Earth and other planets were worlds, they assumed that all planets resembled Earth. This led to the idea of "the plurality of worlds"—the assumption that other planets had climates and life-forms like Earth's. Now we realize this is only partially true. Our spaceships have brought us photos of other worlds, showing lonely lava flows, rugged craters, dusty winds, and scudding clouds. We realize now that Earth is just one natural laboratory in which one set of "input" conditions has acted to produce one kind of environment and a certain set of life-forms. If we can understand how Earth's conditions have evolved, we will be better equipped to understand how other input conditions have produced different environments on other planets.

Conversely, astronomical exploration of the planets has helped us evolve a new picture of Earth. We now know that a planet's surface is not a static platform but a dynamically evolving crust. It may be pummeled by meteorites and blanketed with lava flows and, in Earth's case, crumpled by forces arising in the interior and eroded by flowing water, as exemplified in Figure 6-2. Its climate may be affected by subtle variations in solar radiation and, in Earth's case, by pollutants.

Figure 6-1 Earth as a planet. The crescent Earth was photographed by Apollo 12 astronauts on their way to the Moon in 1969. They exclaimed on the beauty of the scene as the Sun emerged from behind Earth. (NASA)

By studying the geological history of Earth, we can discover how these processes work and learn how Earth resembles, and differs from, other worlds.

EARTH'S AGE

A variety of astronomical and geologic evidence indicates that the Earth was formed about 4.6 billion (4.6×10^9) y ago from particles orbiting the Sun. Chapter 14 will detail the process that formed Earth and all the other planets; here we emphasize how we learned Earth's age and how we unraveled the evolutionary processes that will help us understand the conditions on all planets.

Debates about Earth's Age

Our view of Earth as an evolving planet of great age has not come easily. Until a few centuries ago, Earth's age was hardly even considered, except for vague ideas that it was very old. In 1646, the English scholar Sir Thomas Brown wrote that determining Earth's age "without inspiration . . . is impossible and beyond the Arithmetick of God himself." The first step in trying to estimate the Earth's age was the simple but profound realization that there might be ways to answer this question! Around 1650, several scholars in Christian countries hypothesized that Earth's age and history could be deciphered simply by adding up the number of generations recorded in biblical scriptures since the times recorded in the Book of Genesis. Archbishop James Ussher used this method and calculated that the whole cosmos was formed on Sunday, October 23, 4004 B.C., and that humanity was created on Friday, October 28. This method seemed reasonable enough, and it was a good start. As we can now see, however, it dated not the planet Earth, but merely the historical records themselves. Roughly 4000 B.C. is not a bad estimate of the beginnings of Western historical records, oral traditions, and mythological metaphors!

Figure 6-2 Earth's mountain ranges are a unique feature of our planet. Crumpling of the crust creates uplifted areas, and subsequent erosion by water and wind creates types of sharp canyons and peaks not found on other worlds. Water, especially, carves valleys (right center) and deposits the eroded sediments at the valley mouth (light triangular deposit). (French Alps above Val d'Isere; photo by WKH)

★ The second step came in the 1600s to early 1800s, when naturalists combined direct observation with logic. To take one of their best arguments, they realized that at the rate sediments are deposited along various European rivers, many millennia would have been required to accumulate the sediment deposits actually observed. For example, if they measured that a castle at the mouth of a river was being silted in at a rate of $\frac{1}{2}$ m per century, and the river had deposited 500 m of sediments, then it must have taken 1000 centuries, or 100,000 y, to do the job. Thus Earth must be at least 100,000 y old, not 6000 y old. Other naturalists made the interesting observation that if the Earth had formed in a molten state (as some concluded from ancient volcanic lavas), it would have taken thousands of years to cool to today's pleasant temperatures. A striking example of this line of thought came in the 1700s when the Count Buffon, in France, made red-hot iron spheres of various sizes, measured their cooling times, and then made a (gross) extrapolation to the size of Earth, proving that it would require hundreds of thousands of years to cool. Of course, if Earth had its own internal heat sources (which turned out to be true, namely, radioactivity), then the time scale could be even longer. Calculations using laws of heat transfer, developed by Isaac Newton and others, gave the same result. The direct observations overpowered the old hypothesis. Our web site shows how the new date changed many old conceptions of Earth—even the way we view the landscape! By the 1800s, it was clear to all educated people that Earth had to be at least hundreds of thousands or even millions of years old.

Even while evidence for Earth's great age accumulated, a related debate raged over how Earth's landscape came to be formed. In the 1500s, Leonardo da Vinci observed fossil seashells high in the Alps. Echoing the observations of the ancient Greek Xenophanes, who around 490 B.C. remarked on fossil fish found on dry land, Leonardo correctly said those findings proved that the mountaintops had once been under water. This caused a problem for the geologists of the 1700s. If Earth was only 6000 y old, whole mountain ranges must have been thrown up from the sea floor in relatively brief catastrophic episodes, perhaps giant earthquakes. This idea was called **catastrophism.** The new evidence of a longer time scale, plus the careful observation of sand-grain-by-sand-grain erosion, by means of rain, wind, rivers, and floods, led geologists in the 1800s to realize that most Earth landscapes were not shaped by sudden, sporadic catastrophes, but by long, slow, gradual processes of geologic evolution. Whole mountain ranges had been created, worn down, and replaced by seas—sometimes to be replaced by new mountain ranges. This idea was called **uniformitarianism,** because the geologic processes worked slowly and relatively uniformly over time. Around 1820, the uniformitarians won over the catastrophists. The observations convinced researchers across Europe that in spite of earthquakes and disasters, *most* geological change happened in a slow, evolutionary way.

Geologists began to realize that while 100,000 or a million years might be the time required to erode a mountain canyon or fill in a lake, there might have been many cycles of such change, allowing for a still older Earth. Meanwhile, the new concept of changing geologic features fascinated Victorian thinkers. With a sense of wonder, Tennyson wrote, "Where the long street roars hath been the stillness of the central sea . . . the solid lands, like clouds they shape themselves and go."

Study Figure 6-2 in this context. You can see several aspects of the long argument that helped establish the idea of long, slow, geologic evolution. First, such a smooth fold could not have resulted from a sudden catastrophic compression. Rocks shatter in such an event. Second, at measured deformation and erosion rates, Earth must be much more than a few thousand years old to allow time to produce such folds and then to allow erosion to expose them. Third, enormous forces must be at work to compress and crumple regions the size of mountain chains.

The third step in the realization of Earth's true age came in the twentieth century when scientists discovered how to use radioactivity to measure Earth's true age—not 100,000 or a few million years as had been thought, but 4,600,000,000 or 4.6 billion years! This method is so important to all modern conceptions of the universe that we will study it in some detail.

Measuring Rock Ages by Radioactivity

All questions of the Earth's age and evolution rates were clarified by the discovery of a reliable technique to determine the ages of rocks. This technique utilized radioactivity, discovered in 1896. In that year, the French physicist Antoine Becquerel accidentally left some photographic plates in a drawer with some uranium-bearing minerals. Later he opened the drawer and found the plates fogged, as if they had been exposed to light. But they had been in the drawer! Being a good physicist, he did not dismiss the event but investigated. He found that the uranium emitted *rays,* which, like Roentgen's X-rays of 1895, could pass through cardboard. The rays turned out to be energetic particles emitted by unstable atoms. Radioactivity had been discovered.

A **radioactive atom** is an unstable atom that spontaneously changes into a stable form by emitting a particle from its nucleus. The original atom thus

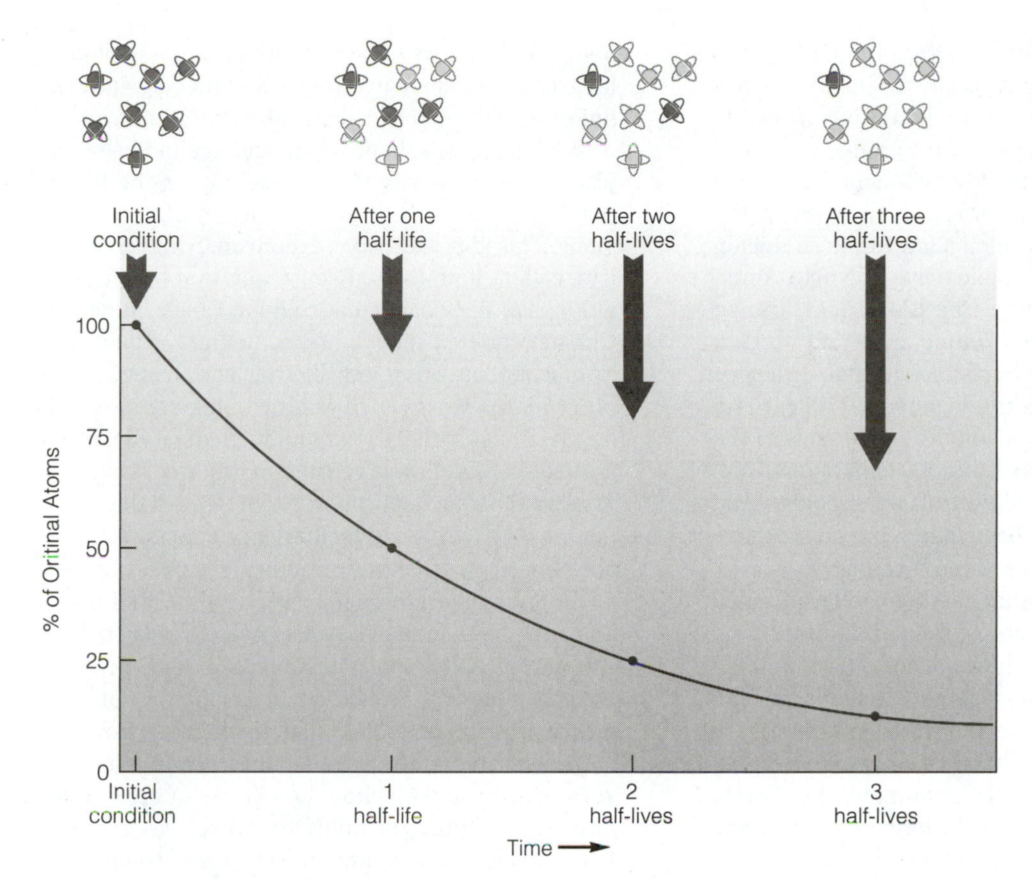

Figure 6-3 Schematic diagram of radioactive decay. Over a period of time, radioactive atoms (dark) will decay into ordinary atoms (light). The period of time for half the atoms of any given isotope to decay is called the half-life for that isotope. The graph shows the decreasing number of radioactive atoms as time passes. Measurement of the percentage of radioactive atoms can thus give the age of the sample.

becomes either a new *element* (change in the number of protons in the nucleus) or a new *isotope* of the same element (no change in the number of protons but a change in the number of neutrons). The original atom is called the **parent isotope,** and the new atom is called the **daughter isotope.**

The time required for half of the original parent isotopes to disintegrate into daughter isotopes is called the **half-life** of the radioactive element. If a billion atoms of a parent isotope are present in a certain mineral grain in a rock, a half billion will be left after one half-life, a quarter billion after the second half-life, and so on. This process is shown schematically in Figure 6-3. As examples, half the rubidium-87 atoms in any given sample change into strontium-87 in 50 billion years; uranium-238 changes (after a chain of successive decays) into lead-206 with an effective half-life of 4.51 billion years; potassium-40 changes into argon-40 with a half-life of 1.30 billion years; and carbon-14 decays into nitrogen-14 with a half-life of only 5570 y.

Early in the twentieth century, physicists realized that here was a way to determine the ages of rocks—the date when a given rock formed. Suppose we could determine the *original* number of parent and daughter isotope atoms in a rock. In part, this can be done by measuring numbers of stable isotope atoms, which usually occur in certain proportions to the unstable isotope atoms in a given fresh mineral. Then, if we simply count the *present* numbers of parent and daughter isotope atoms in the rock, we can tell how many parent atoms have decayed into daughter atoms and hence tell how old the rock is. If half the parent atoms have decayed, the age of the rock equals one half-life of the radioactive parent element being studied. This technique of dating rocks is called **radioisotopic dating,** since radioactive atoms that decay are called radioisotopes.

Note that the quantity being determined in radioisotopic dating of rocks is the time since the rock began to retain the daughter element. In most cases being discussed here, this is the time since the rock solidified from an earlier molten material, such as lava. Once the rock solidifies, any daughter isotope atoms are trapped in the solid mineral structure.

Measuring Earth's Age

To measure Earth's age, you might expect that we would merely search for the oldest known rock on

Earth and assume that it dates from the Earth's creation. In reality, the process is not so simple. Earth is geologically so active that rocks dating from the planet's formation have long since eroded away. For instance, many rocks in the Rocky Mountains formed about 60 million years ago or less; many rocks in the eastern United States formed a few hundred million years ago. Older, more stable parts of North America exist in Canada, where **Earth's oldest known rocks** have been found, dating from 3.96 billion years[1] ago (York, 1993). Such ancient, stable regions are called **continental shields,** after their flat, circular shapes. Several shields in different parts of the world have yielded rocks 3.5 to 3.9 billion years old. In the Australian shield, a few isolated crystals as old as 4.3 billion years have been found, but no complete rocks; they have been destroyed by erosion.

These results mean that Earth must have formed more than 4.3 billion years ago. Another, more complex technique that combines different minerals to treat the Earth as one large rock gives an age around 4.6 billion years for the whole system. As we will see in the next few chapters, dramatic confirming information comes from outside the Earth. Lunar rocks and meteorites show that *all the planets formed within a relatively short interval (about 50 to 90 million years) about 4.6 billion years ago* (Pepin and Phinney, 1976). The **age of Earth** is therefore conclusively known to be 4.6 billion years. The result has been repeated by different scientists in different labs in different countries using different radioactive parent elements in different rocks from different worlds!

Continuing Social Controversies

★ Strangely enough, the exciting and straightforward 400-y quest that produced this conclusive result has mutated into a political/philosophical/religious controversy that in some countries has pitted the naturalists against the religious or traditional authorities. As evidence of Earth's great age accumulated in the 1700s and 1800s, the idea of geological evolution was as controversial as the later concept of biological evolution. Even though the theory of catastrophism was being replaced by observations of long, slow geologic changes, some naturalists and theologians continued to insist that all landforms were created intact, with all the fossils, bent strata, and age indicators in place—perhaps even in one week, as in the literal reading of Genesis or ancient books of other religions. This idea is called creationism. Some creationists, called fundamentalists, argue that the ancient writings of their particular religion (Islam, Judaism, Christianity, etc.) must be taken not just as guides to right living, but absolutely literally and thus be given precedence over science or direct observational evidence. Thus Christian fundamentalists of the 1800s argued that Earth was created in one week and that fossils and strata deposits were "devices of the devil" put in rocks to mislead foolhardy scientists. Remarkably, this line of thought from the 1800s has not died out, but remains a force to be reckoned with. For example, in 1999, fundamentalists gained a majority on the Kansas state Board of Education and removed not only biological evolution but also theories of long-term astronomical evolution in the universe from the subjects tested in the Kansas curriculum. A Kansas veterinarian on the school board was quoted in the international press claiming that there were "legitimate doubts about whether the universe is more than several thousand years old" (Glanz, 1999). This policy was reversed after a subsequent election but the argument goes on in Kansas and other state legislatures. Fortunately, in the United States, all citizens are free to seek the best information and believe what they want, but for scientists, it is sad to see arguments that were defeated in the 1700s and 1800s be resurrected in a new millennium! And it is even sadder to see students carefully shielded from the last 400 y of discovery and exploration. For more on this topic, see the expansion of this section on our web site.

EARTH'S INTERNAL STRUCTURE

What is the inside of the Earth like? In the 1800s, Jules Verne wrote a novel in which intrepid explorers followed a system of caves all the way to the Earth's center. But, in reality, we can't explore the inside directly. The world's deepest drill hole was created in the former Soviet Union, 250 km north of the Arctic Circle (Kozlovsky, 1984). It reached a depth of 12.5 km in 1989. It has yielded interesting data on deep rocks, but is not deep enough to reveal Earth's

[1]Expressing geological ages is a continuing editorial problem. This book uses the most common American convention: 1 billion years = 10^9 y = 1,000,000,000 (abbreviated *1 b. y.*). This would be fine except that the English use 1 billion to mean 10^{12}! Therefore some authors have adopted 10^9 y = 1 aeon (or eon). Other authors have adopted the international prefix *giga-*, which stands for 10^9, and they use 10^9 y = 1 Gy (see Appendix 1).

deep internal structure. The deepest known rocks, brought to the surface by ancient volcanic processes and collected from deposits in South Africa, are believed to have originated at depths of 300 to 400 km in the upper mantle (Haggerty and Sautter, 1990)—but even this is only 5–6% of the way to the Earth's center. Our best clues about the interior come, instead, from earthquake waves that pass through Earth's material. If you make a small splash in a calm swimming pool, waves radiate across the surface from the disturbance. Observers at other edges of the pool can gain information about the disturbance by observing the waves—for example, they can locate the disturbance by comparing the directions from which the waves come.

In the same way, waves are generated in Earth by earthquakes. Waves traveling through the Earth (or other planets) are called **seismic waves.** Some travel along the surface, and others penetrate through the Earth. The velocity and characteristics of the waves depend on the type of rock or molten material they traverse. For example, as shown in Figure 6-4, waves that travel through Earth from an earthquake on the far side traverse deep interior regions. An earthquake produces several different types of waves, and some of these are unable to pass through liquids. Since the outer part of the core is molten, these waves are blocked by the core. As Figure 6-4 indicates, the observer might detect such waves from an earthquake at A, but not from an adjacent earthquake at B. The presence of the Earth's core was discovered in this way.

Such studies have revealed two important types of layering in the Earth: chemical and physical. Chemical layering refers to layers of different composition. Physical layering refers to layers of different mechanical properties, such as rigid layers vs. fluid layers.

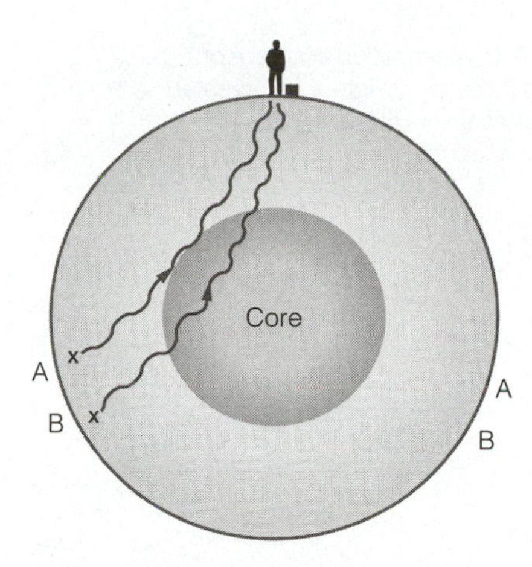

Figure 6-4 Earth's core was discovered in 1906 when observers found that they received modified seismic waves from earthquakes on the far side of the Earth (beyond B from the observer) compared to waves of nearer earthquakes (closer than A). Waves from B and beyond are altered by passing through the core, and certain types of waves from B and beyond do not even reach the observer (see text). The wave paths are curved because rock layers at different depths have different properties.

Chemical Layering of Earth: Core, Mantle, Crust

Chemical layering was the first type of layering recognized. Seismic and other data indicate that the Earth contains a central **core** of nickel-iron metal. The core's radius is about 3500 km, just over half the Earth's radius, as shown in Figure 6-5.

The core is surrounded most of the way out to the surface by a layer of dense rock, called the **mantle.** Near the surface, the densities of the rocks are typically lower. The **crust** is a thin outer layer of lower-density rock about 5 km thick under the oceans and about 30 km thick under the continents.

The core-mantle-crust structure gives us important clues about the history of the Earth and other planets. First, it shows the importance of **differentiation** processes—processes that separate materials of different composition from one another. Most geologists believe that the key differentiation process in the Earth was the melting of much of the inner rock material while the Earth was forming. Apparently, the iron core formed very early, in the first tens of millions of years. The source of the heat was probably that delivered by the explosive impacts of large meteorites that struck the primordial Earth as it was forming (Newsom and Sims, 1991). These were the building blocks of the Earth itself, as we will see later. In the molten or semimolten state of early Earth, heavy materials like metals were able to flow downward toward the center, while lighter, low-density minerals floated toward the surface, where they eventually solidified into a crust of low-density rock. In addition, radioactive minerals released heat as their radioactive atoms decayed. The interior of the Earth was so well insulated by overlying rock that the heat could not escape. This process has maintained a high temperature inside Earth.

As a more detailed example of this theory, we note that one of the lowest-density and most common minerals to form in cooling, molten rock is called *feldspar.* If the theory is right, feldspar should have formed and floated toward the surface. Indeed, we

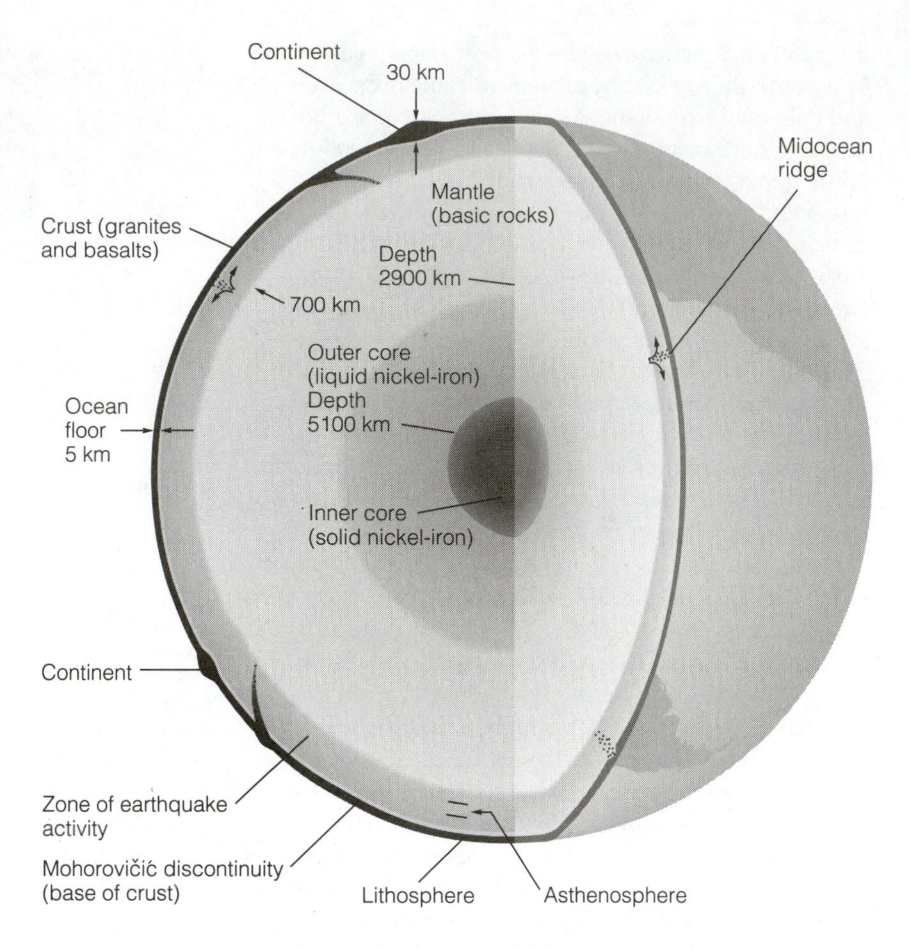

Figure 6-5 A simplified diagram of the Earth's interior structure as revealed by modern seismic studies. Dots indicate typical earthquake positions.

find that surface rocks of both the Earth and the Moon are extremely rich in feldspar. One of the rock types formed from mixtures of minerals rich in feldspar is called **basalt.** Basalt is the common lava that erupts from volcanoes that tap the crust and upper mantle, and basalt is also common on the Moon's surface. Basalt is also the main rock in Earth's oceanic crust.

A still lower-density type of rock is **granite,** the light-colored rock commonly formed from molten materials in continents. Most of the continental crust is granitic. In some ways, continents seem to be a low-density granitic "scum" floating on the denser rocks of the basaltic lower crust and upper mantle.

A simple example of differentiation that can produce such layering occurs during the smelting of metal ores. When the ore is melted, the metal sinks to the bottom of the vat (core) while the bulk of the rock fills the upper part of the vat (mantle). On the surface floats a thin scum of slag, or low-density rock (crust). Note that certain amounts of metals are chemically attracted to minerals in the crust and hence tended to stay with those minerals as they formed. Thus we can have iron and other dense ores in the crustal rocks, even though much of Earth's iron is in the core.

Support for these ideas comes from meteorites, the stone and metal fragments that fall onto the surface of Earth from space. These will be discussed further in Chapter 13, but we note here that they are believed to be fragments from the surfaces and interiors of small interplanetary bodies broken during collisions. Interestingly, they show many of the rock types we have discussed. Nickel-iron meteorites are probably metal fragments from cores of these small worlds, similar to Earth's core. Basaltlike meteorites are probably fragments of mantles and crusts. Thus we see that many worlds may have gone through a differentiation process that led to structures like the Earth's.

Physical Layering of Earth: Lithosphere and Asthenosphere

The second type of layering involves layers of different rigidity. This type of layering is extremely important in determining the types of features we see in the landscapes around us on Earth, as well as on other planets.

Layering by rigidity was created during the cooling of Earth. As you can imagine from the example of the smelter vat, the interior stays molten for a long time because it is hard for the heat to get out, but the surface cools rapidly because it is exposed to surrounding space, and the heat can easily radiate away. Thus, if a planet were melted and left to cool, a solid layer would form on the surface, underlain by a still-molten liquid layer. This picture is complicated by the fact that the layers at different depths have different compositions (due to the differentiation discussed above) and are also at different pressures. They may thus have different solidifying temperatures, and they may actually form alternating layers of solid material, molten material, or partly molten slushy material. A partly molten layer can be visualized as a layer in which some low-melting-point minerals are melted, but high-melting-point minerals are not. (For example, if you heated a mixture of ice and sand, you would get a mixture of solid sand grains with water between them.)

The solid layer at the surface of Earth or any other such planet is called the **lithosphere** (from Greek roots meaning "rock shell"), and it is underlain by a partly melted layer that is much less rigid and less strong. As shown in Figure 6-5, the underlying partly melted layer is called the **asthenosphere** (from Greek roots for "weak shell").

The reason the lithosphere and asthenosphere are important is that their interplay determines the surface features of Earth, such as mountains, seafloors, and continents. To understand this, we need to review some principles of heat flow—to describe how heat gets out of the hot interior of Earth or any other planet, once the planet is heated by radioactivity. Heat always flows from hot to cooler regions, and it can flow by three methods. The first is radiation; an example is the radiant heat that reaches us through space from the Sun. The second is conduction; an example is the flow of heat through a metal cooking pot, whose handle might grow too hot to touch. The third is convection, which is heat flow by movement of the heat-carrying medium; an example is the buoyant ascent of warm air in a thundercloud or the rise of a hot-air balloon. If you put an inch-deep layer of cooking oil in a pan and heat it from below on a stove, you may be able to see patterns of currents, called convection cells, set up in the oil as it convects.

These ideas relate to a planet as follows. As heat is produced inside a planet, its insides try to cool by all three processes. Heat radiates from the surface into space; heat is conducted outward from the hot center through rock; and convection may occur in some of the more fluid layers where currents can flow, like the convection cells in the cooking oil. As the planet cools over hundreds of millions of years, the surface layers solidify, forming a relatively rigid surface layer of rock—the lithosphere. In the Earth, the relatively rigid lithosphere is a layer about 100 km deep, and the relatively fluid asthenosphere is the underlying layer from about 100 to 350 km deep. Even though the asthenosphere is not totally molten, it seems to be plastic enough for sluggish convection currents to flow. Modern geophysicists have likened these currents to the movement of blobs in "lava lamps" (Kellogg and others, 1999; van der Hilst and Karason, 1999). Blobs of hot mantle material rise, creating "hot spots" in the Earth's crust where volcanoes are likely. The currents also drag on the underside of the lithosphere, setting up stresses in it and tending to crack it into separate, large-scale pieces.

Like arctic explorers camping on ice floes floating on the ocean, we live on a rigid layer that "floats" on a more fluid base. Of course, we are not in quite so much danger of a cracking floor as the arctic explorers, but cracks do occur. Figure 6-6 shows a result. As the asthenosphere shifts, it can stretch the lithosphere only so far before the brittle lithosphere cracks. We perceive this as an earthquake.

LITHOSPHERES AND PLATE TECTONICS: AN EXPLANATION OF PLANETARY LANDSCAPES

Tectonics is the study of movements in a planetary lithosphere, such as the movements that cause earthquakes, mountain building, and so on. For more than a century, geologists studied the Earth's tectonics without recognizing the underlying nature of the forces that cause these movements, as just outlined. Only since the 1960s have geologists pieced together a real understanding of how these principles affect the Earth and how they apply to other planets. This new understanding is called the theory of **plate tectonics.**

As the asthenosphere drags on the more brittle lithosphere, it cracks the lithosphere into large, continent-scale pieces called *plates.* Further asthenosphere movements tend to drag and jostle the floating plates, sometimes pulling them apart from each other and sometimes pushing them into each other. Cracks along the margins of plates are usually the

Figure 6-6 Occasionally, nature reminds us that we are living on a thin, brittle crustal planetary lithosphere that is often broken by motions of underlying semifluid layers. This devastation was caused in 1906 in San Francisco by an earthquake and resultant fire. The quake involved movement on the San Andreas fault, which passes under the ocean not far from the downtown area. (Photo courtesy S. M. Larson)

sites of volcanoes and earthquakes, since molten magma from below can squeeze up to the surface through the cracks, and since plate collisions cause stresses as plates rub together, eventually leading to rock fracturing in the form of earthquakes. This can be seen in a map of earthquake locations, such as Figure 6-7. The string of shallow earthquakes down the mid-Atlantic (marked also by volcanoes in Iceland and the Azores) marks a plate boundary where new lava is rising from below. On the ocean floor, it erupts and piles into a feature mapped by oceanographers in the 1960s—the Mid-Atlantic Ridge. Eruption of new lavas at these sites pushes apart the neighboring plates, causing the Americas to drift westward and Europe eastward relative to the plate boundary.

On the west coast of the Americas, the American continental plate collides with the Pacific plates, crumpling and riding up over them as the margins of the Pacific plates slide under the American plates. This crumpling is the explanation of major western American mountain chains like the Sierra Nevada and the Andes. Contorted structures at depths down to 15 km where plate boundaries collide have been mapped with new techniques. Such zones are common sites of earthquakes, as California residents can attest. Fractures caused by earthquakes are called **faults.** Colliding plate regions are laced with faults like the famous San Andreas fault, in California. Many earthquakes are caused by movements along these faults. As shown by Figure 6-6, the devastation can be extreme. Although earthquake prediction is only in its infancy, the U.S. Geological Survey in 1987 predicted a large earthquake in the Riverside–Palm Springs area of southern California by about 2020.

Other scenes of plate tectonic activity can also be glimpsed in Figure 6-7. For instance, years ago the plate containing the subcontinent of India drifted northward, as shown in Figure 6-8. It collided with the Asian plate and caused the massive crumpling that created the Himalayas, marked by shallow earthquakes.

Much geological evidence supports the theory of plate tectonics. Matches of geological provinces of equal date and rock type in eastern South America and western Africa show that they were once part of the same landmass, which has been called Pangaea, as seen in Figure 6-8. Traces of glaciation and other evidence in southern Africa indicate that this landmass drifted north from a position once much closer to the icebound South Pole. The theory of plate tectonics also explains why rock units on Earth older than 1 or 2 billion years are so rare. Most older surfaces have been crumpled beyond recognition or driven downward under other plates, only to be

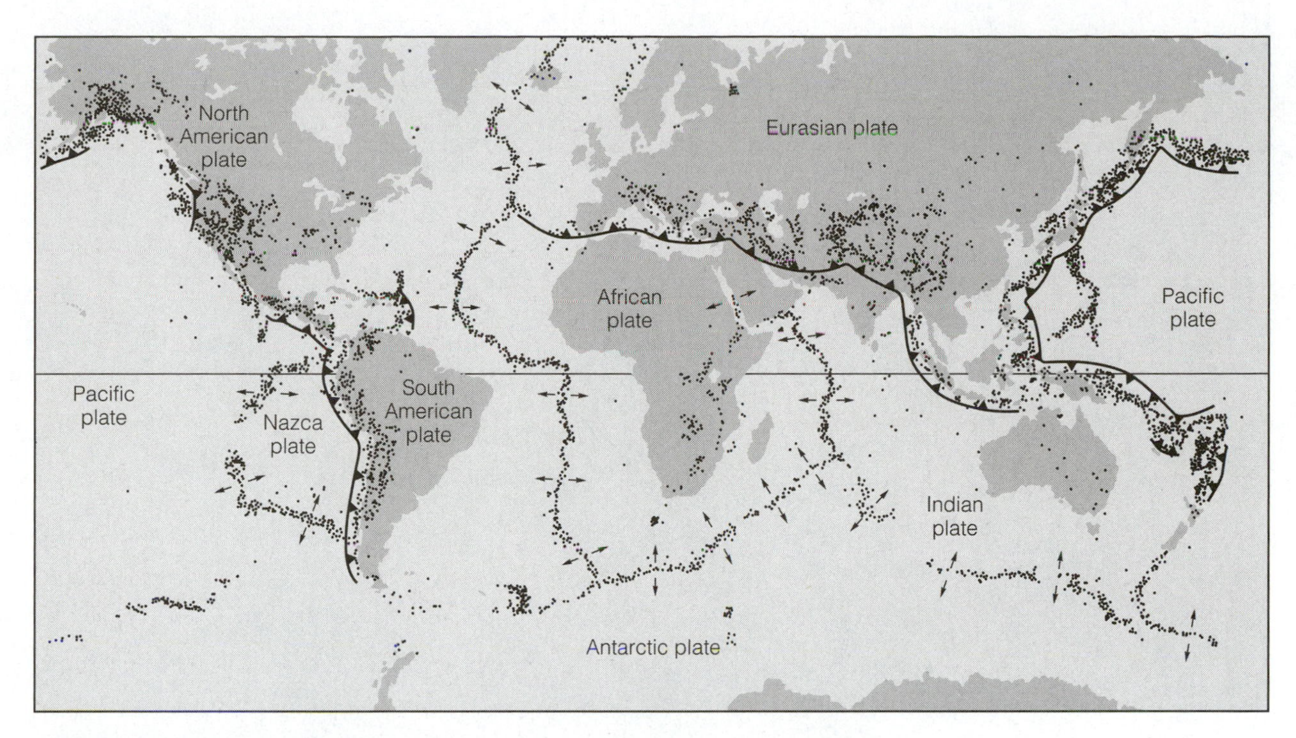

Figure 6-7 A "snapshot" of tectonic activity in Earth's present-day lithosphere. Each dot represents a recorded earthquake. Earthquakes cluster along margins of plates, where tectonic forces cause the most activity. There are two kinds of margins. The lines with arrows on each side are spreading centers, usually midocean ridges, where new magma wells up from the mantle. It pushes the crustal surface in the direction of the arrows across the plate and causes continental drift, whose rates are actually measured by geophysicists. At the far margin, shown as lines with teeth, the material plunges back down into the mantle at an angle under the margin of the adjacent plate. Teeth are on the descending side of the line. Most volcanoes are clustered in the same regions as the earthquakes, because earthquake fractures allow lava to gain access to the surface.

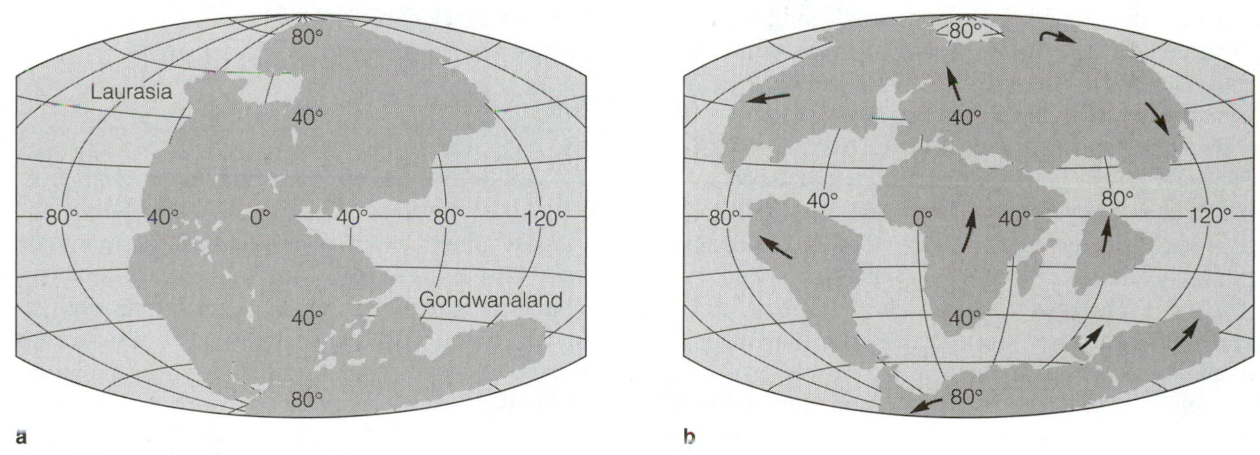

Figure 6-8 Two "frames" in a time-lapse movie of Earth's surface, showing the dramatic movement of continental landmasses due to plate tectonic motions in the last 4% of Earth's history. Geologists are not certain of the motions in earlier times, though there is evidence of previous collisions and splittings. **a** 170 million years ago, Pangaea, composed of two major regions called Laurasia and Gondwanaland, was just beginning to split apart. **b** 70 million years ago, the Atlantic was widening and present landmasses were beginning to take shape. India had not yet collided with Asia to make the Himalayas. (After Ingmanson and Wallace, *Oceanography,* Wadsworth Publishing, 1989)

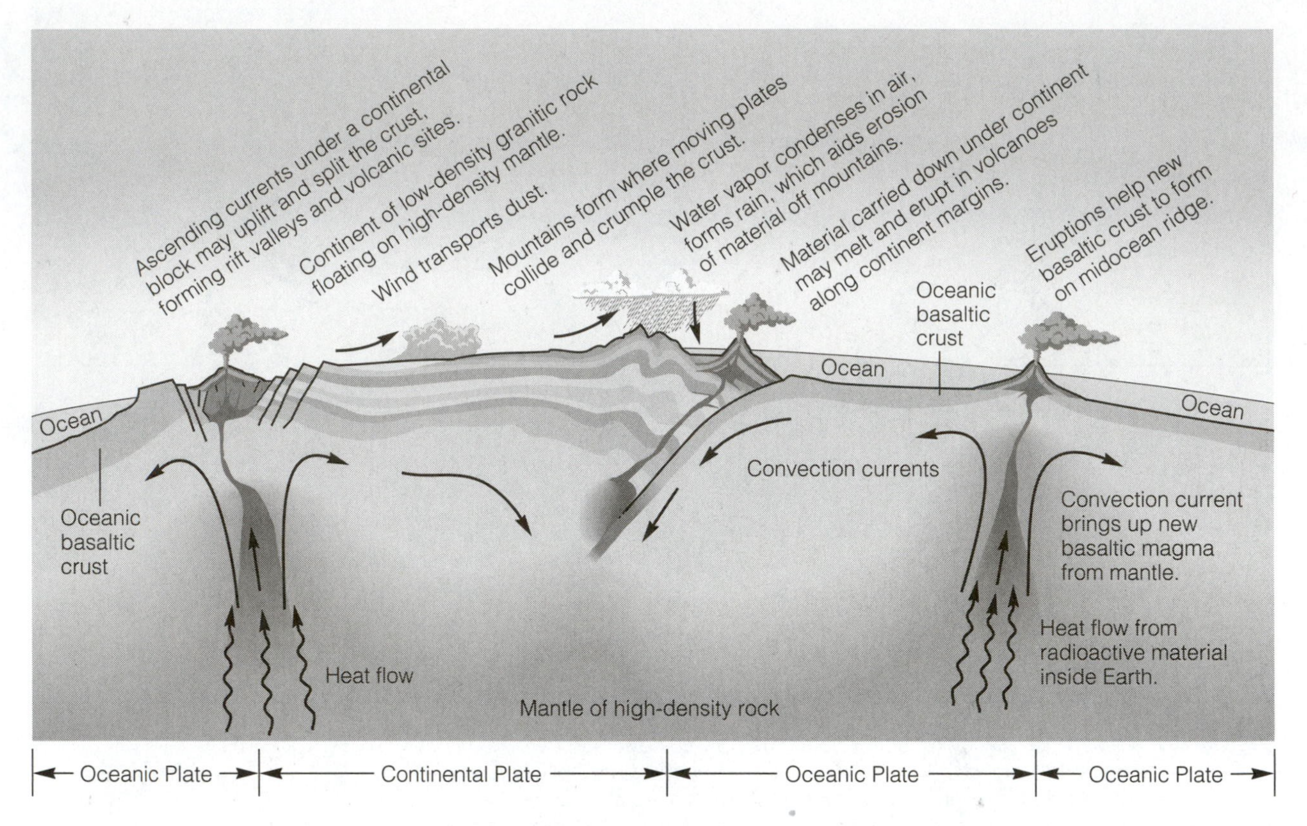

Figure 6-9 Schematic diagram of some of the processes making the Earth more geologically active than most other worlds. Convective heat flow from the interior drives convection currents in the asthenosphere. These in turn crack the more rigid surface layers and move plates. Plate movements cause earthquakes, faulting, the formation of rift valleys (left), and crumpled mountain chains (right). In addition, atmospheric processes and water flow cause rapid erosion. Vertical scale is exaggerated for clarity.

remelted, mixed with mantle material, and perhaps reerupted as new lavas. Some of these complex processes of Earth's active geology can be seen in cross section in Figure 6-9.

Now we can see a connection with the landscapes on other planets, which will be discussed further in later chapters. Smaller worlds, like Mars and the Moon, do not have well-developed crumpled mountain ranges or plate boundaries as shown in Figure 6-9. The reason goes back to the basic principles of heat flow and lithosphere formation. Smaller worlds cool faster than big worlds, so their lithospheres get thicker in the same amount of time. Thus their surfaces are more stable and more protected against asthenosphere currents far below. Lava does not so frequently gain access to the surface. Convection can't so easily drive plates apart or cause them to drift into each other. Therefore the surfaces of Mars and the Moon have much more ancient structures (such as ancient impact craters) than Earth does.

OTHER IMPORTANT PROCESSES IN EARTH'S EVOLUTION

Volcanism is the eruption of molten materials from a planet's interior onto its surface. On Earth, the asthenosphere contains pockets of partly melted materials, as indicated by seismic wave analysis. This underground molten rock, called **magma,** is under pressure, often charged with gas such as steam, less dense than surrounding rock, and highly corrosive. Therefore it tends to work its way to the surface, especially in regions where fractures provide access. When it reaches the surface, it erupts and is then called **lava.** It may shoot in a foamlike form into the air (due to pressure from dissolved gas, like the spray from an agitated can of pop when it is opened) or ooze out and flow for many kilometers across the ground. If enough lava is erupted, it may accumulate into volcanic mountains. During inter-

Figure 6-10 Volcanic activity on Earth builds varied landforms. This crater formed as the ground collapsed around an erupting vent; its light-colored floor is filled with windblown dust. It is on the flank of a shield volcano, a type found on several planets and named for its profile, as seen on the horizon. The rest of the landscape is covered by overlapping lavas and cinders and is dotted by volcanic cones. (Pinacate volcanic field, Mexico; photo by WKH)

Figure 6-11 During Earth's history, eruptions of lava have created new landmasses, pushing back the sea even as the sea eroded coastlines of old landmasses. (1988 Kalapana lava flow, Hawaii; photo by WKH)

vals ranging from years to millions of years, volcanism thus creates landforms ranging from boiling mudpots and flat lava flows to craters and volcanic peaks, as seen in Figures 6-10 and 6-11. Many of these features can be visited in the national parks and monuments of the western United States, such as Craters of the Moon National Monument in Idaho, Sunset Crater National Monument in Arizona, and Yellowstone National Park in Wyoming.

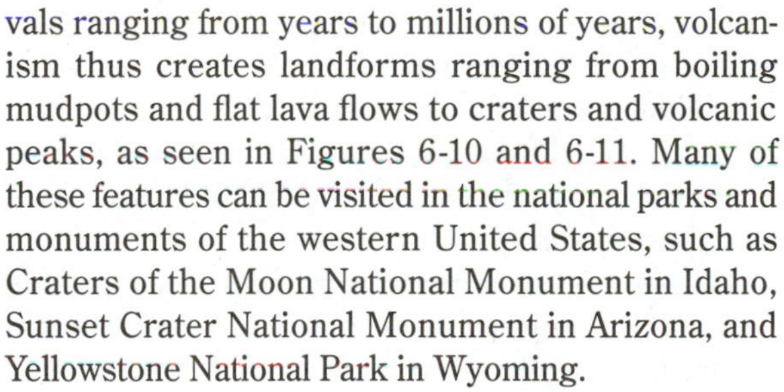

Space exploration has shown that volcanism is one of the most important processes forming landscapes on other planets. Some planets have huge lava flows and volcanoes. Study of the Earth's volcanoes helps us understand these alien landscapes. Conversely, study of the other planets' volcanic features helps us understand relations we see among terrestrial volcanoes.

Among all known planets, the Earth undergoes the most active processes of landform destruction. Largely, this activity is due to the Earth's thick atmosphere and flowing water, which other planets lack. **Erosion** includes all processes by which rock materials are broken down and transported across a planet's surface; such processes include water flow, chemical weathering, and windblown transport of dust. **Deposition** includes all processes by which the materials are deposited and accumulated; such processes include deposition of sediments in lake and ocean bottoms and dropping of windblown dust in dune deposits.

On the Earth, erosion and deposition, especially by flowing water, are the most important landscape-forming processes, but this is not true on all other planets. Because of the extreme amount of flowing water on Earth, erosion and deposition wear away or cover up the most ancient rocks and wore away ancient impact craters so rapidly that they were not even recognized on Earth until the last few decades. Similarly, most familiar mountain ranges do not display their initial forms, which may have been caused by volcanism, fracturing, and folding. Instead, most mountain ranges are merely resistant rock "cores" left after erosion of more massive units of crustal strata.

EARTH'S MAGNETIC FIELD

Physicists use the term **field** to describe any property that can be measured throughout a volume of space. For instance, we know that Earth's gravity field extends far out beyond the Moon because we can measure its effects on the motions of other planets. Similarly, if we move a compass throughout the region around Earth, we discover that there is a **magnetic field,** indicated by the fact that the compass is deflected in a direction roughly toward the North Pole. The compass does not point exactly at the true North Pole (defined by the Earth's spin) but at a point called the magnetic North Pole, which moves slightly from year to year and is currently located among arctic islands off northern Canada.

What causes Earth's magnetism? The interior of Earth cannot be magnetized like a giant bar magnet, because the interior regions are too hot to maintain magnetism. Rather, geophysicists believe that liquid iron in the outer core is slowly convecting and that these motions set up electric currents in the metal. A magnetic field will exist around any electric current, and Earth's field is thus attributed to currents in the liquid metal core. The presence or absence of a magnetic field thus becomes an important indicator of the presence of an iron core inside a planet.

EARTH'S ATMOSPHERE AND OCEANS

Not only have Earth's interior and surface evolved dramatically, but so have the atmosphere and oceans. It is strange to realize that if you had visited the Earth halfway back in its past, it would have seemed an alien planet, with a different atmospheric composition and different continents—not to mention the lack of visible life-forms on the barren landscape! The **early dense atmosphere** is believed to have come from volcanic outgassing of Earth's hot interior as the planet formed. Volcanoes emit primarily steam (H_2O, 58% by weight) and carbon dioxide (CO_2, 24%). Scientists thus believe the early outgassed atmosphere was a hot, dense mixture of water vapor and CO_2 (Kasting, 1993). Some scientists think the earliest atmosphere also acquired hydrogen-rich material from the surrounding interplanetary gas, which was hydrogen-rich, but this is uncertain. As the early dense atmosphere cooled, water vapor condensed into liquid water and formed oceans.

Figure 6-12 The most common "landscape" on Earth is unique in the solar system: An open ocean of liquid water covers about three-fourths of our planet. (Pacific Ocean; photo by WKH)

An important principle is that Earth was at just the right distance from the Sun to allow this to happen. Had Earth been a little further, it would have been too cold, and the water would have frozen. Had it been a little closer, it would have been too warm, and the water would have remained as steam. But (in a kind of "Goldilocks principle"), Earth's temperature was just right, allowing the water to create a vista unique in the solar system—expanses of exposed liquid water oceans (Figure 6-12.)

The condensation of the steam into liquid oceans left a thick atmosphere of CO_2, much denser than today's atmosphere. Thus early Earth was an alien place, with a CO_2 atmosphere! In support of this seemingly strange idea, we note that Earth's neighbor planets, Venus and Mars, both have atmospheres of CO_2 (as we will discuss in more detail in Chapters 9 and 10).

Today's atmosphere, however, is composed of 76% nitrogen molecules (N_2) by weight and 23% oxygen molecules (O_2). How did it get that way? One important process was that the CO_2 of Earth's early atmosphere dissolved in Earth's unique ocean water, where it formed a weak carbonic acid solution. (Most of the CO_2 emitted by modern industrial processes does the same thing, but some stays in the air.) The carbonic acid reacted with minerals of the seafloor, eventually forming carbonate rocks. This tied up most of the early CO_2 in carbonate rocks and sediments, leaving a thinner CO_2 atmosphere. Once plant life evolved, plants began to consume most of the rest of the CO_2 and release O_2 molecules (as they still do today). Thus, amazingly enough, Earth's present atmosphere was created in part by Earth's own life-forms! Plants explain the transition from a CO_2-dominated atmosphere to an atmosphere with 23% O_2 around 2.5 to 2 billion years ago.

Evidence for this startling transition is that sediments deposited more than about 2.5 billion years ago are less oxidized than modern sediments, indicating less oxygen in the air. Also, the earliest fossil life-forms, dating from about 2.5 to 3.5 billion years ago, are types of algae found today only in oxygen-poor environments, such as salt marshes along seacoasts. (For a popular review of early Earth, see Hartmann and Miller, 1991.)

Note that oxygen is a very strange gas to find as a major component of a planet's atmosphere. As you learned in high school chemistry, O_2 is very reactive. It forms rust, burns with hydrogen to make water, and reacts with many other compounds to make oxides. If there were no plants to maintain the atmosphere's oxygen, it would disappear "quickly" in millions or hundreds of millions of years. Thus, if you were flying around the universe in a spaceship and spotted a planet with an oxygen-rich atmosphere, you could be fairly sure of finding life there! The presence of nitrogen, which is the most abundant gas in our atmosphere today, is explained by other processes. First, remember that today's atmosphere is only a small remnant of the thick early atmosphere. Thus, although nitrogen was only the fourth most abundant volcanic gas and only a fraction of the early atmosphere, it ended up dominating the remainder once the CO_2 and water vapor disappeared. Furthermore, extra nitrogen may have been added to Earth late in its formative process by impacts of comets, which carry N atoms in the form of frozen ammonia (NH_3) ice.

Dynamics of Earth's Atmosphere

Why do winds blow? Why don't they just blow themselves out, leaving a calm atmosphere? Any turbulent motions require an energy source to keep them going. The motions of our atmosphere turn out to be another astronomical application of the same physics discussed in the case of motions of materials inside Earth: Heat is transferred through the atmosphere and sets up convective motions. In this case the energy source is sunlight. Suppose we magically stopped all winds. Sunlight would beat down on the noontime side of the Earth, warming the ground and warming the air. The air near the ground would eventually become warm enough to expand and rise. New air would flow in laterally along the ground to take its place; thus winds would arise. If the temperature gradient in the air is steep enough (that is, very warm air near the ground and considerably cooler air aloft), strong convection currents will be set up, lifting the air to form towering thunderhead clouds.

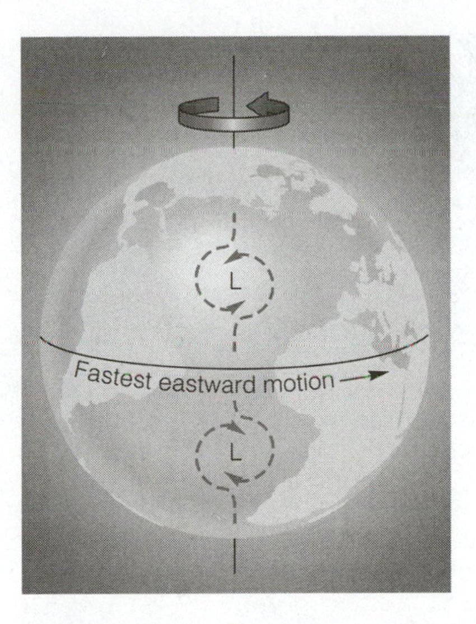

Figure 6-13 Due to the Coriolis drift, airmasses moving toward or away from the equator are deflected east or west, respectively, relative to the ground. Air moving into low-pressure areas (L) produces cyclonic storm patterns; the sense of rotation is opposite in the two hemispheres and is proof of Earth's rotation.

Coriolis Drift

The Earth's rotation is also involved in determining winds and weather patterns. Consider an airmass moving northward off the equator toward a low-pressure region in the atmosphere, as seen in the northern half of Figure 6-13. Since the air has to flow from higher-pressure to lower-pressure regions, it tries to reach the center of the low-pressure region, marked L in the figure. Because Earth is turning, the equatorial airmass would make one trip around the circumference in 24 h. This means an equatorial airmass is being carried eastward at nearly 1700 km/h, whereas material near the pole has no such speed. Therefore, as the equatorial mass moves north toward the low-pressure region, it also has eastward momentum and is moving eastward faster than the ground or local air at that latitude. It is thus deflected east, as shown in Figure 6-13. Meanwhile, a polar

a

b

Figure 6-14 Meteor Crater, in Arizona, testifies to the effects of interplanetary debris hitting Earth. A meteorite's impact formed this 1.2-km crater about 20,000 y ago. **a** Road and museum buildings (left) give scale in the aerial view. **b** View from rim at sunset shows strata distorted by the explosion and boulders thrown out onto the rim. (Photos by WKH)

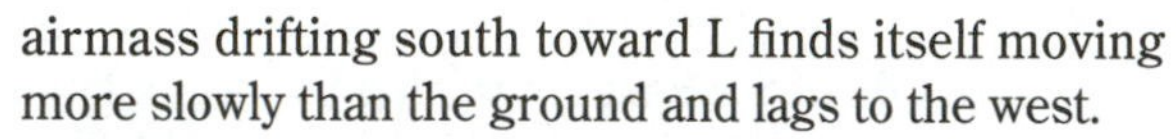

airmass drifting south toward L finds itself moving more slowly than the ground and lags to the west.

Such eastward and westward deflections of winds, due to a planet's rotation, are called **Coriolis drift** (after a French mathematician of the 1800s). Coriolis drift sets up a counterclockwise spiral motion in the air moving into a low-pressure storm system in the Northern Hemisphere, as shown in Figure 6-13. Conversely, in the Southern Hemisphere there is clockwise motion in a spiral, as also seen in the figure. Coriolis drift thus explains the spiral cloud patterns of cyclones and hurricanes, so prominent in space photos of Earth. On nonrotating planets, no Coriolis drifts or spiral patterns would occur.

THE COSMIC CONNECTION

Most people think of the Earth as isolated from astronomical influences, having evolved independently as a result of its own internal forces. But recent research indicates strong *external* astronomical influences on Earth's evolution. For example, the recurrent **ice ages** are caused at least partly by small periodic changes in the configuration of Earth's orbit, producing cooler and warmer eras over periods of tens of thousands of years. The last ice age was about 14,000 y ago, when North America was glaciated as far south as Chicago and Pittsburgh. Another example of the astronomical influence may be subtle changes in solar radiation that affect climate.

A more dramatic effect comes from **impacts of interplanetary debris,** called meteorites, on the Earth throughout geologic history. The interplanetary debris that hit Earth are pieces of rocky material that range in size up to tens of miles across. We will discuss them in more detail in Chapter 13. For now, suffice it to say that many brick-sized meteorites hit Earth every year. The largest ones are much rarer, however, hitting Earth only every hundred million years or so, according to astronomers' calculations based on the observed orbits.

Meteorite impact craters are circular explosion pits created by the larger impacts. Some are fresh-looking and make spectacular landmarks, as shown by a famous example in Arizona (Figure 6-14). Most, however, including examples up to 180 km (110 mi) across, are older and are barely visible because they are heavily eroded. As a result, the existence of terrestrial meteorite impact craters was widely recognized only in the mid-twentieth century.

Until the 1980s, geologists thought that impacts were completely unimportant in the "big picture" of Earth's evolution. That changed with a dramatic discovery in the 1980s. To understand this, study Table 6-1. Note that geologists have divided Earth's history into periods, based on characteristic fossil life-forms and on radioisotopic dates. This table is called the **geological time scale.** It chronicles slow,

TABLE 6-1

Geological Time Scale

Era	Age (millions of years)	Period	Life-Forms	Events
	0			
		Quarternary	Humanity	Technological environmental modification; extraterrestrial travel; ice ages
	3			
Cenozioc		Tertiary		
			Mammals	Building of Rocky Mountains
	65			
		Rapid extinction of ~75% of species, including dinosaurs		
		Cretaceous		Large meteorite impact (65 million years ago); continents taking present shape
	130			
Mesozoic	180	Jurassic	Dinosaurs	
		Triassic	Reptiles	
	240			
		Rapid extinction of ~90% of species		
		Permian	Conifers	Building of Appalachian Mountains
	280			
	310	Pennsylvanian	Ferns	Pangaea breaking apart
Paleozoic	340	Mississippian		
	405	Devonian	Early land plants	
	450	Silurian	Fishes	
	500	Ordovician		
		Cambrian	Trilobites	Earliest abundant fossils (trilobites, etc.)
	570			
		Ediacarian	Small soft forms	
	640			
	1000		First macroscopic life-forms; sexually reproducing life-forms	Growth of protocontinents; scattered fossils
Proterozoic				
	2000			
			Oxygen-producing microbes	Oxygen increasing in atmosphere
	2600			
Archeozoic	3000			Crustal and atmospheral evolution
			Microscopic life	Earliest fossils (algae)
	3600			
				Oldest rocks; crustal formation? Heavy meteoritic cratering
	4500			
Formative	4600			Formation of Sun and Planets (see Chap. 14)
Presolar	12,000?			Formation of Milky Way galaxy (see Chap. 23)
	16,000?			Origin of universe (see Chap. 27)

Source: Data in part from Morris (1987).

long-term shifts in Earth's biology and environment and also a few abrupt changes.

As we discussed earlier in the chapter, geologists from about 1820 onward recognized that the dominant geological changes that form Earth's landscapes are slow. Similarly, Charles Darwin, around 1860, developed the now-accepted theory that animals and plants also evolved slowly, generally from simpler to more complex species as mutations developed. This is the **theory of natural selection,** sometimes called the *survival of the fittest.* These ideas explained the gradual, long-term shifts in environment and species, but they do not explain the more abrupt changes, such as the one that ended the Paleozoic Era (meaning era of ancient life). During that episode, roughly 90% of species died out in only a few million years! The second most drastic change in the fossil record was the end of the Mesozoic Era (era of middle life), when about 75% of species died out in no more than a few million years.

For years no one knew what caused these sudden **mass extinctions** of species. Then, around 1980, everything changed. Scientists discovered that a worldwide 65-million-year-old soil layer, the stratum marking the end of the Mesozoic era, contained concentrations of elements common in many meteorites. From the total amount of meteoritic elements, scientists concluded that a very large meteorite 10 km (6 mi) across—that is, a fragment of an asteroid—crashed from space into Earth 65 million years ago, as shown in Figure 6-15 (Alvarez and others, 1980). Since this soil layer falls exactly at the boundary between the Cretaceous period and the Tertiary period, it is called the Cretaceous-Tertiary boundary layer, and the impact is usually called the **Cretaceous-Tertiary impact** (or K-T impact using traditional geological abbreviations). Most scientists now believe that this impact caused disastrous climate changes that killed off the dinosaurs and other species.

Among the lines of evidence for this revolutionary new idea are the following discoveries in the Cretaceous-Tertiary boundary layer of soil:

- The meteorite element concentrations
- Worldwide soot of a quantity representing the burning of most Cretaceous forests
- Grains of quartz that have microfractures caused by the shock of an intense explosion, larger than known volcanic explosions
- Chaotic deposits in coastal areas, caused by giant tsunamis (popularly but incorrectly called tidal waves)
- Glassy spherules, 65 million years old, that are interpreted as solidified droplets of once-melted rock
- A large, 65-million-year-old eroded impact crater, buried under sediments on the coast of Yucatan; this discovery was the "smoking gun" that proved the hypothesis of a giant impact 65 million years ago.

The interpretation of this evidence is that a large meteorite hit just offshore of Yucatan in the ancient Caribbean. Initiating a titanic explosion, it blasted out a crater estimated at 180 to 300 km across, throwing molten material and dust containing shocked quartz into space and into the high atmosphere. The impact created a giant tsunami that swept over many coastal areas. The debris raining into the atmosphere caused a strong radiant heat pulse for some hours, as the sky lit up with brilliant meteors. This ignited forest fires in many areas. The fires alone probably killed off some species at once, although organisms in regions of severe snowstorms or rainstorms survived. However, as if these events weren't enough to damage the environment, the dust thrown into the atmosphere blocked sunlight, probably for some months (Toon and others, 1994). Thus many aquatic plants and plankton, which depend on sunlight, died off more slowly, disrupting the food chain. This explains extinctions that continued over many millennia.

Scientists have sought evidence of a similar impact during the even greater mass extinction that ended the Paleozoic era, 250 million years ago. So far, none has been found. However, a few smaller mass extinctions at other times probably do correlate with impacts. For example, an extinction at the end of the Triassic period 210 million years ago seem to have coincided with the formation of the Manicouagan crater in Canada.

All these data show that Earth's history is more linked to cosmic phenomena than was suspected a few decades ago. The discovery that at least some mass extinctions were caused by asteroid impacts created a revolution in geology and paleontology, because it showed that evolution and climate change have been driven by influences from outside, as well as by terrestrial causes within Earth's biosphere. It even reintroduced an element of catastrophism into geology and biology: while the uniformitarians were correct that long-term changes, such as continental movements, mountain building, and deposition of sedimentary layers, are gradual, nonetheless occasional catastrophic events are significant. Because of

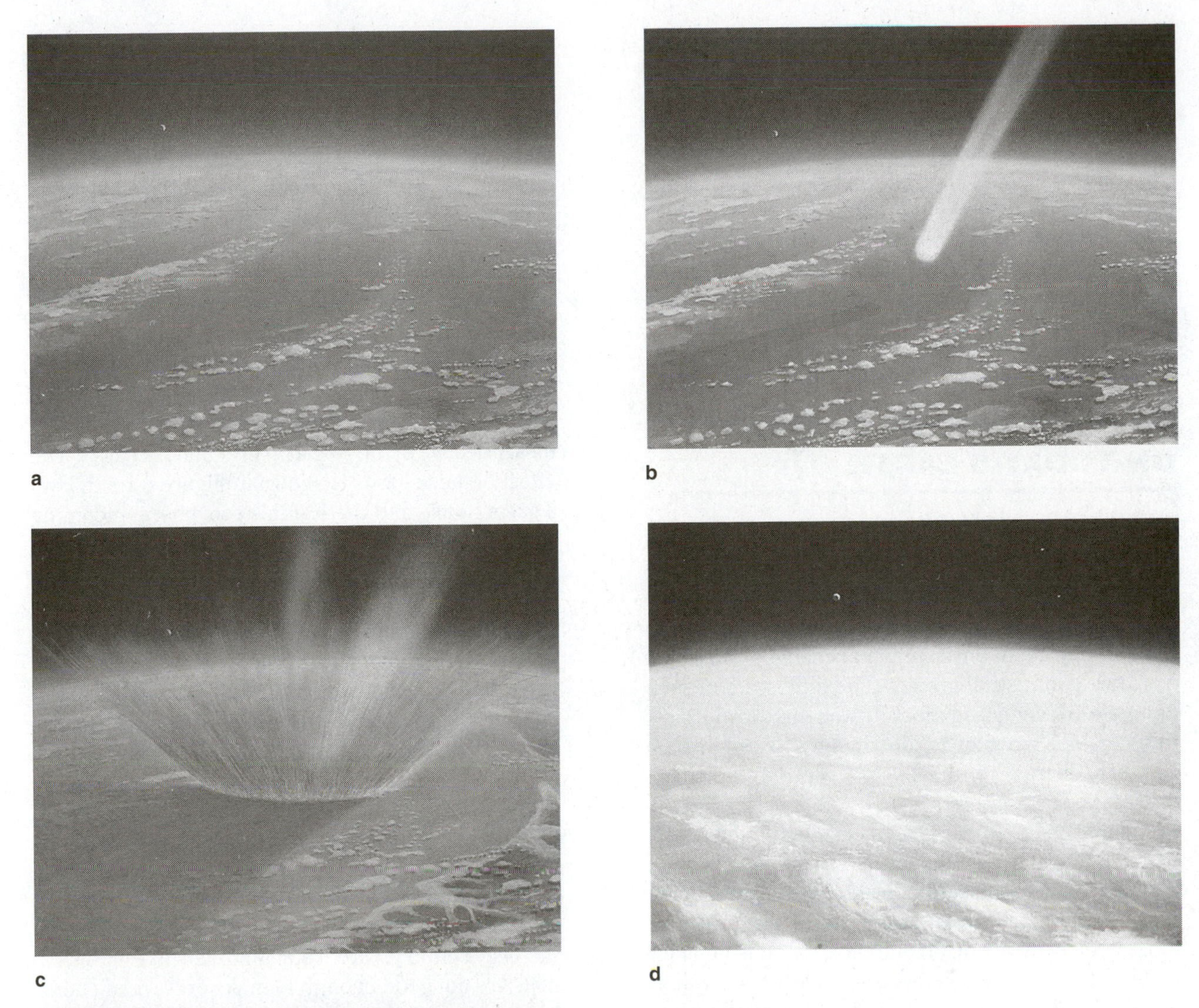

a

b

c

d

e

Figure 6-15 Impact of an interplanetary body about 10 km wide is believed to explain the striking climatic events and extinctions of species at the end of the Mesozoic era 65 million years ago. These views show the scene from an altitude of 100 km (60 mi). Views **a, b,** and **c** show the scene 10 s before impact, 1 s before impact, and 60 s after impact. Massive amounts of dust debris in view **c** fall back into the atmosphere and create a dense pall of dust a month later, as seen in view **d.** View **e** shows the scene 1000 y later, long after the dust has cleared. A 200-km-diameter crater exists, and many species have already become extinct on the surface below. (Paintings by WKH)

the modern controversies over the teaching of science, it is important to note that this has nothing to do with creationism, which involves hypotheses such as the idea that everything was created at once, that species did not evolve, and that Earth is younger than shown by the geophysical evidence. Rather, the modern view is merely the realization that we live in a cosmic shooting gallery, and that although most geological evolution of the earth came bit by bit, in many small changes, occasional big changes have played a secondary role in shaping our environment.

ENVIRONMENTAL CHANGES ON TODAY'S EARTH

As can be seen from this discussion, too-rapid changes to our planet's biosphere can be devastating to life. Evolution cannot keep pace with changes that are too rapid. That is why there is such concern in our century about environmental changes caused by worldwide industrial growth. The time scale of such changes (decades) is much faster than evolution's time scale. Two planetwide changes are of special concern: changes to the ozone layer and to the content of the atmosphere.

The ozone problem is seen clearly in Figure 6-15. In the left part of that diagram, ultraviolet rays from the Sun are being absorbed by the ozone layer of the high atmosphere, 20 to 50 km high. The fact that most of these rays don't reach the ground is good for us because they damage organic molecules. Suntans and skin cancers are two examples of their effects on organic molecules in our skin. However, as is now well known, certain chemicals called chlorofluorocarbons (or CFCs), widely used in air conditioners and other devices in the mid-twentieth century, entered the air and eventually broke down the ozone molecules (O_3). This destroys the ozone layer, lets the ultraviolet rays through, and increases the risk of skin cancer for everyone. This discovery is a good example of astronomy's practical effects, because it was made in part by planetary astronomers studying chemical processes in the atmospheres of Mars and Venus. The ozone-layer breakdown is especially strong at high latitudes. It was directly detected in winter seasons over Antarctica in the 1980s and 1990s, and careful measurements in New Zealand in 1998–99 revealed that 12% more ultraviolet radiation was reaching the ground than at the beginning of the decade. (Other shortwave radiation at wavelengths not affected by ozone did not increase, proving that the effect was due to the depletion of the ozone layer.) In the early 1990s, satellite data suggested that winter breakdown of the ozone layer has started at high northern latitudes as well. However, international treaties in the late twentieth century led to the phasing out of CFC use, and it is believed that the ozone layer will reestablish itself. The incident is proof that humans have achieved the capability of damaging the entire planet.

The second problem is the increase in carbon dioxide due to forest burning and industry. A CO_2 increase of more than 10% has been clearly observed in data taken since 1860. The problem is that CO_2, though a minor constituent of the atmosphere, is one of the most effective of the so-called greenhouse gases. The effect of these gases is shown in Figure 6-16. In both a greenhouse and the Earth's ecosphere, incoming sunlight warms the ground, and the ground reradiates thermal infrared radiation at a wavelength around 10 μm. The **greenhouse effect** is a warming of the air that comes about because the infrared radiation can't get back out of the system easily. In the greenhouse, glass blocks it; in the Earth's atmosphere, CO_2 molecules absorb it, heating the air itself. Certain other gases, including H_2O, also absorb some of the thermal infrared radiation, adding to the effect. (H_2O's effect explains why a cloudy, humid night does not cool down nearly as fast as a clear, dry night.)

Most climatologists expect that the increasing CO_2 will cause an average warming of the climate by the next century. Theoretical models of Earth's climate are not good enough to make exact predictions, but most models suggest warming by 0.5 or 1° C over a few decades and/or greater extremes in temperatures and storm activities. Even small changes in average temperature have big effects. The ice ages involved average temperature drops of only about 5° C. Many little clues suggest that global climate alteration has probably already begun. According to various studies, most of the warmest years of the twentieth century (in terms of planetwide average temperatures) occurred after 1980. Glaciers are retreating in many areas, and during the iceberg season in early 1999, for the first time in 85 y no iceberg warning bulletins were issued by the International Ice Patrol (Wuethrich, 1999). Of course, statistical flukes and other climate effects are hard to rule out (see also Hartmann and Miller, 1991). If such a warming does continue, it could dramatically change the world's economy in the twenty-first century, expanding low-latitude deserts and shifting the best agricultural regions farther toward the poles.

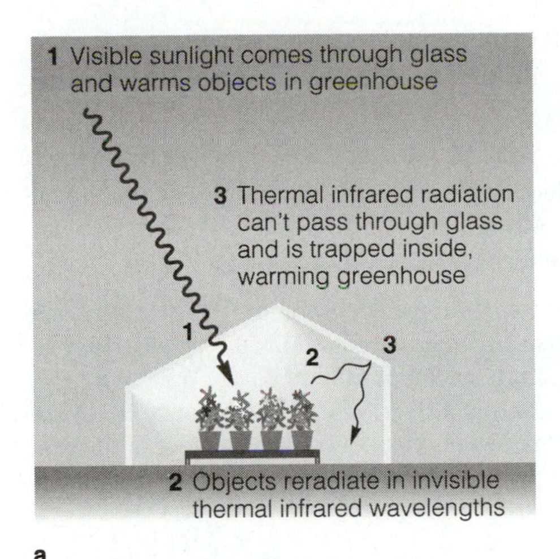

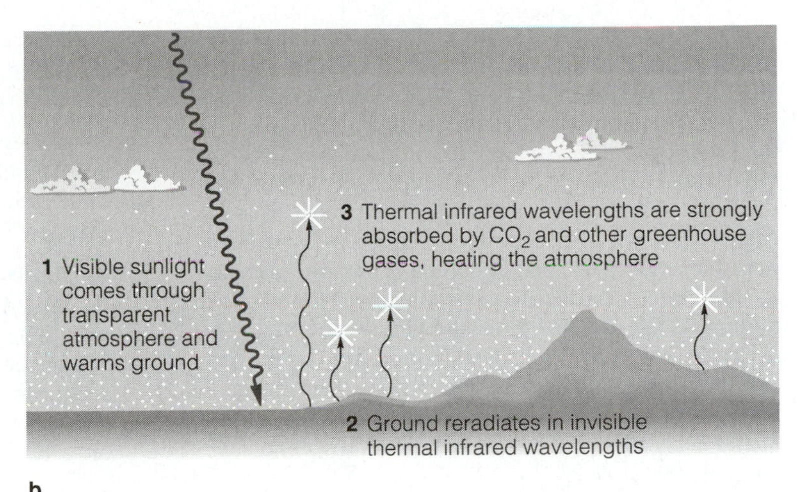

Figure 6-16 Explanation of the greenhouse effect. The greenhouse effect warms both **a** the inside of a greenhouse and **b** the Earth's atmosphere because sunlight can come into the system, but reradiated thermal infrared wavelengths (around 10 μm) can't get out of the system easily.

Industrial and developing nations argue about what steps to take to retard CO_2 production. It is a difficult problem because high CO_2 production (fossil fuel consumption, etc.) is associated with maintaining the pleasant lifestyles in developed countries. In the interests of short-term economic prosperity, several such countries (including the United States in the 1980s and early 1990s) called for more study before action is taken. While it seems plausible that planetwide environmental damage has begun, it is still a few years too early to *prove* any of the anticipated effects. Meanwhile, the waters have been muddied as some political groups and religious fundamentalists have claimed that issues like the ozone and the CO_2 problems are frauds invented by environmentalists and scientists for their own purposes. However, it is unbiased scientific observations, such as monitoring of global temperatures from space as well as on the ground that will help to clarify the rate of change, its causes, and what we can do about it.

SUMMARY

Earth formed 4.6 billion years ago and has gone through many evolutionary changes since then, in terms of atmosphere, continent positions, landscapes, and biology. Geological studies of strata and fossils have established a chronology of events in the history of the planet Earth, and radioisotopic dating of rocks has established the actual ages of these events, as summarized in Table 6-1. Only the last 14% of the Earth's history has yielded enough rock evidence, such as fossils and dates, to provide finer divisions, which geologists called *periods.*

Drawing from material presented in this chapter, we can now construct a thumbnail sketch of Earth's history. In the first few hundred million years, giant meteorites struck much more often than today, scarring the primal landscape with great impact craters. The surface was lifeless. Life probably originated after a few hundred million years of Earth's history. The earliest microscopic fossils date back about 3.5 billion years.

The heating of Earth's interior during its formation led to internal melting. This, in turn, caused differentiation—a draining of metal toward the center to form a nickel-iron core and the floating of lighter minerals to form a low-density crust overlying a dense rock mantle. As Earth cooled and solidified, the surface layer formed a rigid lithosphere overlying a more plastic asthenosphere. Currents in the asthenosphere broke the relatively thin lithosphere into *plates.* Continents repeatedly split and rejoined as rifts broke the lithosphere and plate motions caused crustal masses to collide. Major mountain belts are formed when the continental plates are crumpled at the sites of these collisions.

The evolution of the atmosphere, interior, and life are intimately related. The early atmosphere was probably water vapor and carbon dioxide, but the water vapor condensed to form oceans, and the CO_2 dissolved in the ocean water and ended up being deposited in carbonate rocks. This left nitrogen as the major atmospheric constituent. Oxygen was added in large amounts, beginning around 2 billion years ago, by the action of plants.

Extraordinary events punctuated Earth's gradual evolution. For example, the impact of a 10-km meteorite 65 million years ago caused the extinction of the dinosaurs and many other species. Some other episodes of mass extinction may also have been caused by impacts of interplanetary debris, while others may have involved other causes within the biosphere.

Let us conclude our summary of Earth's history by compressing events into a single day. Life evolved sometime in the early morning, but the fossil-producing trilobites that begin the traditional geological time scale in the Cambrian period did not appear until about 9:30 in the evening. By 10 P.M. there were fishes in the sea, and by 11 P.M. dinosaurs on the land. Mammals did not appear until about 11:40. Human beings, who have been here (depending on your definition of *human*) perhaps 2 million years, made their appearance only 30 s before midnight. The last few thousand years—civilization—occurred in a tenth of a second, represented by the pop of a single flashbulb at midnight. In view of the potential for nuclear conflict and environmental damage, the question is: What will be here a tenth of a second after midnight?

CONCEPTS

catastrophism
uniformitarianism
radioactive atom
parent isotope
daughter isotope
half-life
radioisotopic dating
Earth's oldest known rocks
continental shield
age of the Earth
seismic waves
core
mantle
crust
differentiation
basalt
granite
lithosphere
asthenosphere
plate tectonics
fault
volcanism
magna
lava
erosion
deposition
field
magnetic field
early dense atmosphere
today's atmosphere
Coriolis drift
ice ages
impacts of interplanetary debris
meteorite impact craters
geological time scale
theory of natural selection
mass extinctions
Cretaceous-Tertiary impact
greenhouse effect

PROBLEMS

1. When were the last major earthquakes in your region? Is your region seismically active or inactive? How is it located with respect to the boundaries of tectonic plates?

2. If a rock sample can be shown to contain one-eighth of its original amount of radioactive uranium-238, how old is it?

3. In view of the preservation of ancient craters and the lack of folded mountain ranges on the Moon, would you predict the Moon to have more or less seismic activity than the Earth? Discuss your reasoning.

4. Fossils of apelike predecessors of the genus *Homo* (such as *Australopithecus*), found in Africa, are believed to date back at least 2 to 3 million years. What percentage of the Earth's age is this? Can you accept, philosophically, that events happened during most of the history of the Earth before anyone was around to see, hear, or record them?

5. If the Earth is 4.6 billion years old, about how much more radioactive uranium-238 did it have when it formed? Would the heat-production rate from radioactivity when the Earth formed have been more, less, or the same as it is now?

6. The composition of the Earth's atmosphere is probably much changed from what it first was.

a. What happened to the abundant hydrogen atoms initially present or produced by the breakup of molecules such as methane (CH_4)?

b. Given that much ammonia (NH_3) was initially present and that ammonia molecules break apart when struck by solar photons in the atmosphere, account for one source of the Earth's now abundant nitrogen.

c. What two gases were added abundantly by volcanoes, and where did these two gases finally end up?

7. Describe how you would expect conditions on the Earth to be if Earth were so close to the Sun that the mean surface temperature was above 373 K (100°C). What if the Earth were far enough from the Sun that the mean surface temperature was below 273 K (0°C)? Explain why a view like Figure 6.12 is probably unique to the Earth.

ADVANCED PROBLEMS

8. The average near-surface temperature gradient in the Earth is 20°C/km.

a. Assuming the surface temperature is about 20°C, how many kilometers would one have to drill to reach a depth where water would boil? (The temperature of boiling water is 100°C.) Compare this depth with that of the deepest mines, roughly 3.5 km. (The actual boiling temperature at depth would be somewhat higher than 100°C because of the increased air pressure, an effect we ignore in this problem.)

b. If such depths could be reached economically, dual pipes could be lowered, with water pumped down one pipe, converted to steam, and blown up the other pipe. Steam-powered plants could thus tap the planetary energy source in any part of the world. Describe the possible economic and political consequences of such a project.

9. As seen from Mars, the Sun subtends an angle of about $\frac{1}{3}°$. Suppose that the Earth passes exactly between Mars and the Sun.

a. Is the Earth big enough to cause a total eclipse of the Sun as seen from Mars?

b. Assuming that the human eye can resolve a disk as small as 2 minutes of arc (120 seconds of arc), could the Earth be detected by the unaided eye as it crossed the Sun, as seen from Mars? (*Hint:* Use the small-angle equation. The Earth's diameter is 12,756 km, and its distance from Mars is 60,000,000 km.)

PROJECT

1. Place cooking oil an inch or two deep in a flat pan over low heat and under a single strong light. Because the oil does not readily boil, heat is soon transmitted to the surface in visible convection currents. Note how the currents divide the surface into cells, or regions of ascent, lateral flow, and descent. These cells are analogous to tectonic plates in the Earth's surface layers. Sprinkle a slight skin of flour on the oil's surface to simulate floating continental rocks and watch for examples of continental drift and plate collisions. Experiment with different depths of oil and different temperature gradients (by changing the heat setting), and record the results. Be careful not to turn up the heat too high (especially on a gas stove) to avoid having the cooking oil catch fire!

CHAPTER 7

The Moon

WHAT THE READER SHOULD WATCH FOR IN THIS CHAPTER

The phases of the Moon, such as full moon and first quarter moon, involve the Moon's orbital motion around Earth once each month. Tides are caused mostly by the force of the Moon acting on Earth; tidal forces cause a surprising variety of effects. Apollo astronauts made six flights to the Moon from 1969 to 1972. Data collected by the Apollo astronauts, combined with earlier Earth-based data, show that the Moon formed 4.5 billion years ago, apparently when a giant asteroidlike body hit the primordial Earth and blew material into orbit around Earth; the Moon aggregated from this material. The Moon cooled more quickly than Earth, because it is one-fourth our size. Therefore it does not have the same kind of plate tectonic activity, continental drift, or mountain-building forces. It has essentially no atmosphere or water. It has been heavily cratered by impacts with asteroid fragments and comets ever since it formed, and these craters are preserved because of the lack of erosion. The craters can be seen with a small telescope. Volcanic eruptions, especially from about 4 to 3 billion years ago, formed dark-colored lava plains, which make up the features of the "man in the Moon," which can be seen by the naked eye from Earth. ■

The same Moon that we see today shone down on the breakup of the continents, gleamed in the eyes of the last dinosaurs, and illuminated the antics of the first protohumans. The same Moon was seen by all the historical figures we have mentioned in preceding chapters: Stonehenge builders, Mayan eclipse observers, Aristarchus, al-Battani, Isaac Newton. They all asked what it is and where it came from. Today the Moon is a little different: It has footprints on it and we know some of the answers.

The Moon is the Earth's only natural **satellite**—a body that orbits around a larger body. (The term *moon* is also used generically to mean a satellite.) As a world, it is respectable: It has a diameter of 3476 km, about one-fourth the diameter of the Earth. It is a rocky world splotched with dark gray flows of ancient lavas and dotted with craters formed by explosions when meteorites hit it in the ancient past. The generation living in the 1960s and 1970s was the first to have seen a few of the Moon's starkly beautiful landscapes. We know, not just from instruments but from the *experience* of our astronauts, that the Moon has no air, no water, no weather, no blue sky, no clouds, no life.

In spite of the Moon's familiarity, many people are still confused about even its simple phases. Can you recall which way the horns of the crescent point in the evening sky? Cartoonists often draw the horns pointing down toward the horizon. Not

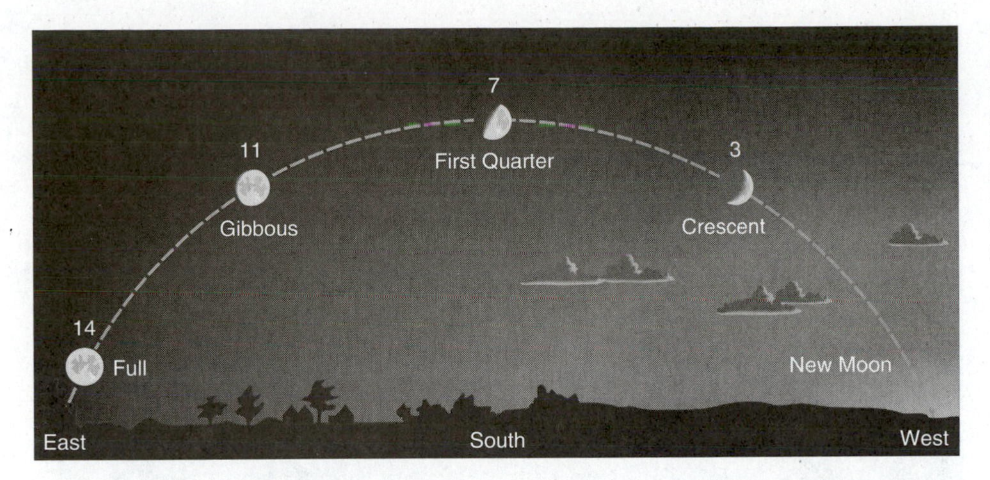

Figure 7-1 Wide-angle view of the sky, looking south, showing some of the phases of the Moon as seen by a Northern Hemisphere observer at sunset on the indicated days of the monthly lunar cycle, counting from the day of new moon.

so. Since the fully illuminated edge of the crescent must face the Sun, which has just set below the horizon, the horns must point upward, away from the horizon. Scoff, too, at the novelist who describes the full moon rising at midnight! To be fully illuminated, the Moon must be opposite the Sun and hence must rise as the Sun sets.

These relations can be seen by studying Figures 7-1 and 7-2, showing the Moon's movements along its orbit. The whole subject of its motions, though seemingly mundane, provides interesting clues about the ancient history of the Earth-Moon pair. We will first consider these motions, then examine the Moon's surface and the astronauts' discoveries, and finally explore the puzzling problem of the Moon's origin.

THE MOON'S PHASES AND ROTATION

As can be seen by comparison of Figures 7-1 and 7-2, the Moon's **phases,** or shapes on different days, are directly caused by the Moon's motion in its monthly orbit around Earth. Let us say that on day 0 the Moon crosses the line between the Earth and the Sun. Here the Moon is nearly in front of the Sun, as seen from the Earth, and is lost in the Sun's glare. On this day the Moon generally cannot be glimpsed. This is called the date of **new moon.** A couple of days later, the Moon has moved far enough from the Sun to be glimpsed in the early evening sky; it is backlit by the Sun, giving it a crescent shape. On day 7, it is 90° from the Sun, a phase called **first quarter** because the Moon is a quarter of the way around its orbit. It looks half-illuminated at this time. For the next week, it is more than half-illuminated, a phase called **gibbous** (hard *g*, as in *give*). On day 14 or 15, it is opposite the Sun and fully illuminated—a phase called **full moon.** This is the day on which the Moon rises at sunset and fills the night sky with its brightest possible light. For the next 2 weeks, the Moon rises after midnight and is visible primarily in the early morning sky. On day 22, the Moon is three-quarters of the way around its orbit and now in a half-lit phase called **third quarter.** On day 29, the Moon is back to the new moon phase.

The term *waxing moon* refers to the first 2-week period, when the Moon is growing more illuminated each day; *waning moon* refers to the second 2-week period, when the Moon is growing slimmer.

The Moon takes 27.3 d to complete one revolution around Earth relative to the stars—an interval called the Moon's **sidereal period**. During this period, Earth moves roughly 27° around the Sun, so the Moon has to move through this additional angle to complete its cycle of phases relative to the Sun. Therefore the cycle of lunar phases takes 29.5 d.

Whenever the Moon is visible, no matter what its phase, we can always distinguish at least some of the dark lava plains that make up the features of the so-called man in the Moon. This is because of a curious characteristic of the Moon's motion: It always keeps the same side facing Earth, as shown by the stylized mountain in Figure 7-2. In 1680, the French astronomer G. D. Cassini explained this characteristic in a statement that is sometimes hard to grasp:

> **The Moon rotates on its axis with a period equal to its orbital revolution period around Earth, so that the same side keeps facing Earth at all times.**

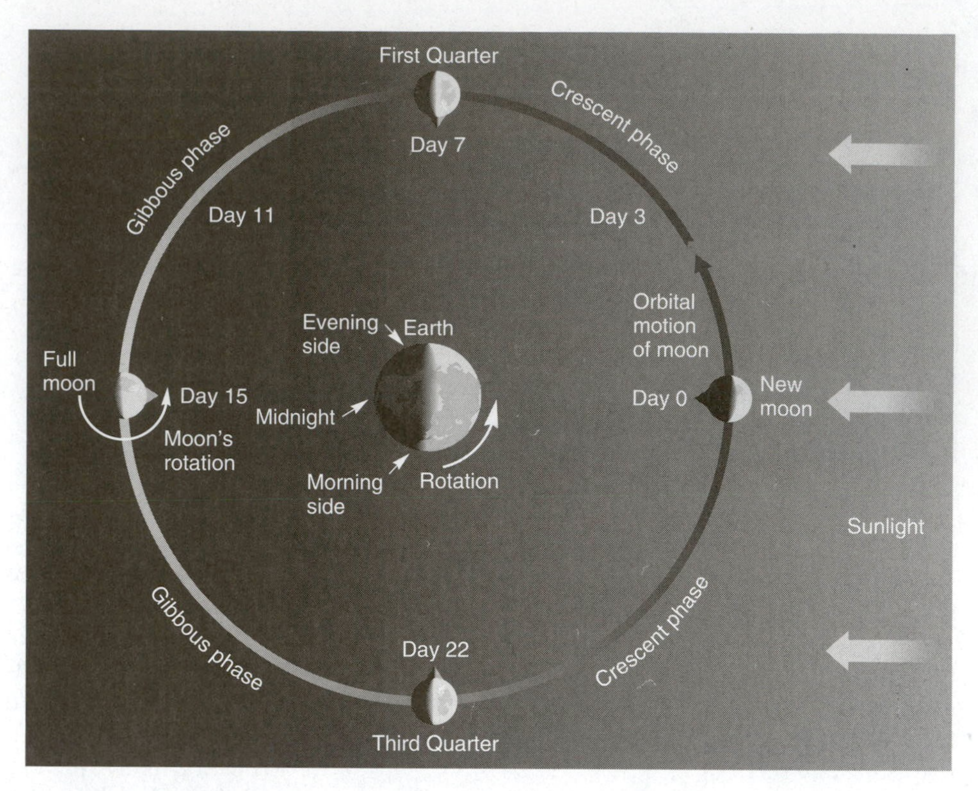

Figure 7-2 The Moon's motion around Earth. Synchronous rotation of the Moon is shown by a mountain (indicated by a triangle) that always faces Earth. The Moon completes one rotation during one complete revolution around Earth. Phases seen by an earthbound observer are indicated at different points in the orbit. First quarter is seen in the evening sky, third quarter in the early morning sky.

The rotation of any satellite in this way is called **synchronous rotation** because it is synchronized with the satellite's own revolution, as shown in Figure 7-2. How can the Moon rotate at all, you might ask, if it always keeps the same side toward the Earth? This question is best answered by an example. Put a chair in the middle of a room. The chair is the Earth; the walls, the distant stars. To represent lunar orbital motion, walk around the chair. If you walk around the chair always facing it, so that an Earthly observer in the chair never sees your back, you will find that you have faced all sides of the room during one circuit. In other words, you have rotated once on your axis at the same time that you made one revolution around the chair. (If you put a strong light in one corner of the room and hold up a ball as you walk around the chair, the observer in the chair can see the cycle of phases on the ball.)

Even writers who should know better sometimes speak of the Moon's "eternally dark side." This mistake comes from a popular belief that the side eternally hidden from us must always be dark. But the far side, just like the near side, has day and night—periods of sunlight and periods of darkness. This can be seen in Figure 7-2. Each period lasts about 2 weeks, since the Moon takes about 4 weeks to make a complete rotation. There is always a dark side, but it isn't necessarily the *far* side.

The Moon's synchronous rotation has another consequence. For an astronaut at any spot on the near side of the Moon, the Earth hangs forever in the same spot in the sky. (Imagine living on the lunar mountaintop in Figure 7-2; Earth would always be overhead.) Nonetheless, Earth goes through a complete cycle of phases every 4 weeks, as seen by a lunar astronaut. Figure 7-3 shows Earth's crescent and gibbous phases as seen from the Moon.

Is there a reason why the Moon keeps one side toward Earth? When asking for a reason, a scientist normally means to ask: "Could the observation be explained by some more fundamental properties of nature so that it becomes a special case of a more general phenomenon?" Here the answer is yes. In the 1780s and 1790s, Joseph Louis Lagrange and Pierre Simon de Laplace used Newton's law of gravity (Chapter 4) to show that if the Moon were slightly egg-shaped or football-shaped, gravitational forces would make the longest axis point toward Earth at all times. Confirming this, space vehicles in the last few decades have shown that one axis of the Moon *is* indeed about 2 or 3 km longer than the others and points steadily toward Earth.

If the Moon always kept *exactly* the same side toward Earth, earthbound observers could never see more than exactly 50% of it. Careful mapping, however, has shown that the Moon wobbles and that dur-

a

b

Figure 7-3 Seen from the Moon, Earth goes through phases. **a** The crescent earth rises, as seen from orbit over the cratered lunar highlands. **b** The gibbous earth seen on another occasion from a similar position. (NASA photos by Apollo 17 and 8 astronauts, respectively, orbiting over the Moon)

ing a period of years 59% of the Moon can be seen, with first one side and then another being turned slightly farther from Earth than its average position.

TIDAL EVOLUTION OF THE EARTH-MOON SYSTEM

Gravitational forces in the Earth-Moon-Sun system cause **tides,** or bulges in the shape of Earth and the Moon. The closer two objects are, the stronger the gravitational force each exerts on the other. Thus the side of the Moon facing the Earth has a stronger force on it than the far side, because the facing side is closer. This is shown by the small arrows in Figure 7-4. As a result, the Moon stretches slightly along this line, limited by the small elasticity of its rock interior. This stretching is called a **body tide.** Similar forces acting on the Earth produce not only a body tide but also an **ocean tide** in the fluid layer of water on the Earth's surface. These tides take the form of bulges on the front and back side of both the Earth and the Moon, since each body comes into equilibrium with the gravitational forces by stretching along the Earth-Moon line.

The Earth's body tides are hard to detect, but its ocean tides are obvious to anyone who visits a beach for more than an hour. They range in height from about $\frac{1}{2}$ m (2 ft) to over 15 m (50 ft). One might suppose that high tide always occurs when the Moon is overhead, with the highest tides at the new moon or full moon, because the tidal forces act along the Earth-Moon line (or, in the strongest case, along the

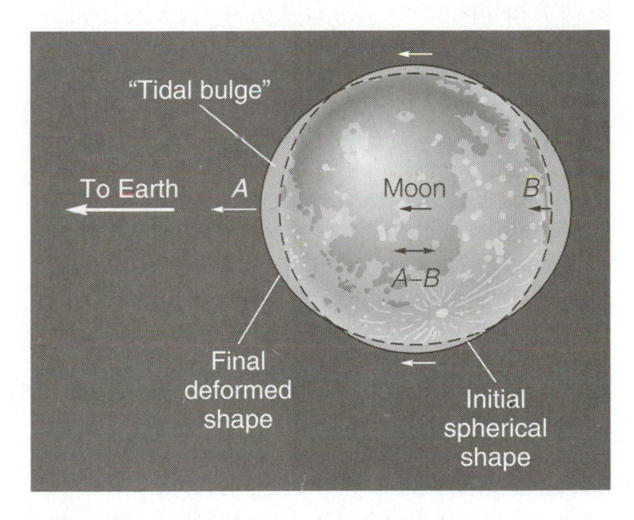

Figure 7-4 The gravitational attraction of the Earth on the Moon is greater on the near side *(A)* than on the far side *(B)*. The difference *(A–B)* acts as a stretching force deforming the Moon from its unstressed spherical shape to a flattened shape with tidal bulges on the near and far sides.

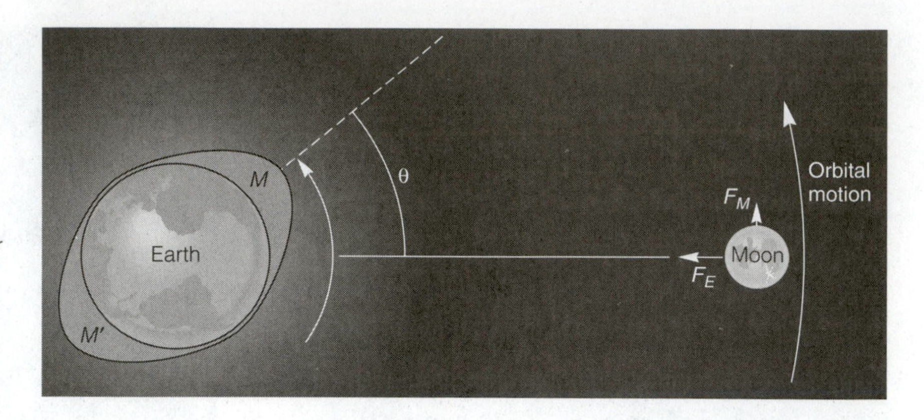

Figure 7-5 The cause of the Moon's tidal recession. Because of Earth's relatively rapid rotation, Earth's tidal bulge *(MM′)* gets dragged off the Earth-Moon line at an angle θ. Thus, in addition to the normal gravitational force of Earth *(F_E)*, there is a net forward force *(F_M)* caused by the difference in attractions between the nearer bulge *(M)* and the farther bulge *(M′)*. This force *(F_M)* pushes the Moon ahead in its orbit, causing it to spiral slowly outward.

Earth-Moon-Sun line). However, this is only a rough tendency; complications result from motions of water around the irregularly shaped oceans and seas and because of the Earth's rotation. Coastline geometry in some places can produce remarkable wave effects associated with tides, but so-called tidal waves are related not to tides but to earthquakes or volcanic activity at sea; they are properly called by their Japanese name, **tsunami**.

Tidal bulges raised on the Earth and the Moon have four major effects, first described in 1898 by George Darwin, son of the famous naturalist. These effects are described in the following four subsections.

Tidal Forces Cause the Moon's Synchronous Rotation

The effect of tides explains why the Moon keeps one side facing Earth. As mentioned earlier, any elongation in the Moon's shape would tend to make one side always face Earth. Tides guarantee such an elongation, since they create bulges. If the Moon were initially spherical and rotating at a nonsynchronous rate, gravitational forces acting on it would create tidal bulges. Since Earth would be trying to keep the bulges aligned with it (horizontal in Figure 7-4)—but a nonsynchronous lunar rotation would be trying to drag the tidal bulges around with it—the bulges would tend to exert a frictional drag, slowing the rotation until it became synchronized and one side faced the Earth at all times. This effect explains why most other satellites in the solar system, besides ours, keep one face toward *their* planet.

Tidal Forces Make the Moon Recede

The second effect is a low **tidal recession** of the Moon away from the Earth because of gravitational forces on tidal bulges. As shown in Figure 7-5, Earth's rapid rotation drags its tidal bulge slightly ahead of the Earth-Moon line by some angle θ. The gravitational effect of the bulge *M* exceeds that of *M′*, because *M* is closer to the Moon. Because *M* is in front of the Moon, the net force from *M* tends to pull the Moon ahead in its orbit. This accelerates the Moon forward and makes it spiral very slowly out from the Earth. Laser beam reflectors placed on the Moon by astronauts have allowed scientists to measure this slow outward movement directly. The Moon was much closer to Earth several billion years ago. Analyses do not indicate the exact date, but most researchers believe the Moon was closest to Earth about 4.6 billion years ago during the formation of the two bodies.

Tidal Forces Slow Earth's Rotation

The third effect of tides is that the Moon slows Earth's rotation, just as Earth has slowed the Moon to synchronous rotation. The effect can be seen in Figure 7-5, where the Moon pulls back on the tidal bulge *M*, acting like a brake on Earth. Studies indicate that billions of years ago, when the Moon was closer, the Earth's day was only about 5 or 6 h long. These theoretical results have been confirmed by an unexpected finding. Certain marine creatures daily and monthly create banded structures in their shells or other hard parts, allowing biologists to count the number of day bands in a monthly cycle. Fossil evidence suggests that whereas the present month is 29.5 d long, it was only about 29.1 d long 45 million years ago. Older fossils show that tides existed as long as 2.8 billion years ago and that the month was as short as 17 d (Kaula and Harris, 1975). We cannot extrapolate such data further back in time because the tidal effects depend on the configuration of terrestrial oceans and continents, and these configurations were different by unknown amounts in the past.

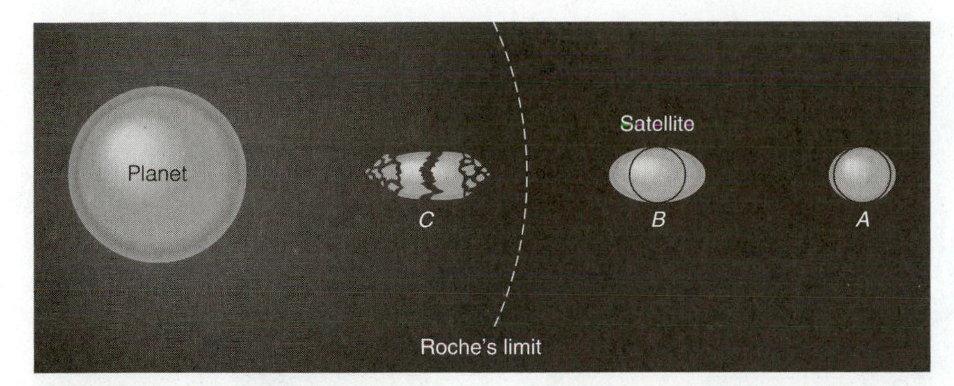

Figure 7-6 Roche's limit. If a satellite is located at *A*, a small tidal bulge develops. At closer distance *B*, a larger bulge develops. Within a critical distance, called Roche's limit, the stretching force differential between the near and far sides is so great that the satellite is torn apart (for example, at distance *C*).

a

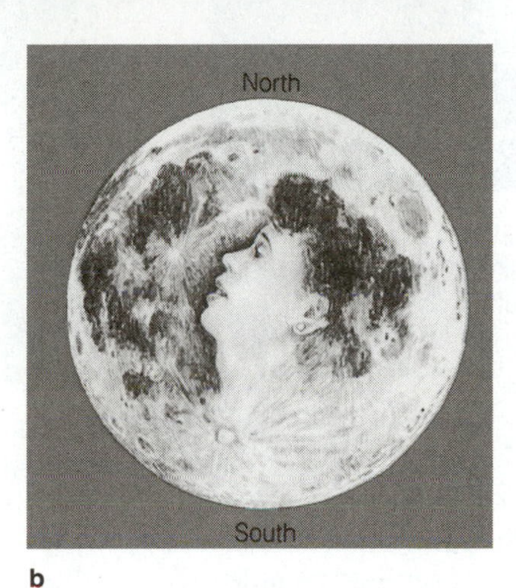

b

Figure 7-7 **a** At full moon, the rugged appearance of the terrain is minimized by the absence of shadows. Compare with Figure 7-9. Full lighting emphasizes different features, such as the bright rays emanating from the crater Tycho (bottom). **b** Several folklore figures, such as "the woman in the Moon," are formed by the pattern of dark lava plains. Squinting at a and b may help you see these features. (Photo from Lunar and Planetary Laboratory. University of Arizona)

As Earth slows and the Moon recedes in the future, it might eventually recede so far that Earth's gravitational pull would be very weak. Before the Moon can escape into an orbit around the Sun, however, the whole process will be stopped by small tidal bulges that are constantly being raised on Earth by the Sun. These tides will slow Earth's rotation so that the length of the day and the month will become equal, stopping the tidal recession. Earth and the Moon will then *both* be in synchronous rotation. Eventually, because of this small solar effect, Earth's rotation will slow so much that the day will exceed the month and the tidal process will reverse: The Moon will begin to approach Earth again.

Tidal Forces Create Roche's Limit

The fourth application of tidal theory shows that a small body, if close enough to a large body, can be torn apart by tides. Calculated around 1850 by French mathematician Edward Roche, **Roche's limit** is the distance between any two different-size bodies within which the tide-raising force exerted on the smaller body is sufficient to disrupt it. The effect occurs because the *difference* between the forces on the near and far sides of a satellite increases as the satellite moves closer to the primary body. This effect causes the satellite to stretch into a slight egg shape as it approaches the primary, as shown in Figure 7-6. The position of Roche's limit depends on the size, density, and strength of the satellite. For instance, a small metal spacecraft is not torn apart while orbiting near Earth. But a body as big as the Moon, if similarly placed for a long period, would develop internal stresses and fractures and eventually disintegrate into a cloud of orbiting particles like Saturn's rings. The Moon could not remain an intact body much closer to Earth than about 18,000 km (11,000 mi). (Its present distance is 384,000 km, or 240,000 mi.)

SURFACE FEATURES OF THE MOON

Until the telescope was invented around 1608, no one knew much about the lunar surface features except for the gray patches that make up the face of "the man in the Moon" and "the woman in the Moon" (Figure 7-7). Some thought the Moon was a polished sphere. Thomas Harriot and Galileo Galilei, the first

Figure 7-8 Comparison of one of Galileo's first lunar telescopic drawings (left), made in 1610, to a photograph of the Moon at the same phase. Letters show corresponding features. Galileo detected mountains, plains, and craters (bottom). (Courtesy Ewen A. Whitaker, University of Arizona)

Figure 7-9 Photograph of the Moon at first quarter phase shows dark, flat lava plains and brighter, cratered uplands. (Lunar and Planetary Laboratory, University of Arizona)

persons known to have seen the Moon's features through a telescope, made their early observations in 1609 and 1610 (see Chapter 3). They both recorded rugged regions with prominent shadows cast by mountains and crater rims. The shadows are most prominent along the **terminator**, the line dividing lunar day from lunar night (see Figure 7-8) and less prominent under high lighting (full moon) or at the edge of the disk, called the **limb**. Galileo reported:

> *The moon certainly does not possess a smooth surface, but one rough and uneven, and just like the face of the earth itself, is everywhere full of vast protuberances, deep chasms, and sinuosities.*

★ These features were later recorded in better telescopic photographs (Figure 7-9). Most of the roughness was caused by thousands of **impact craters,** or circular depressions caused by meteorite impacts (Figures 7-10), ranging up to 1200 km in diameter. Smaller ones (up to a few kilometers) are bowl-shaped, while larger ones have a more complex structure, such as central mountains or concentric rings of cliffs. Galileo found the dark gray patches that are visible to the naked eye and form "the man in the Moon" to be much smoother than the brighter, cratered areas, as seen in his sketch in Figure 7-8. He mistook these dark patches for seas. Using Latin, as scientists did in the 1600s, he called them *maria* (singular **mare** [MAH-ray]). These "seas," as Galileo himself probably eventually realized, are actually vast plains, which we now know are covered with dark lava. Galileo's work greatly strengthened the conception that the Moon is an Earth-like, planetary body having familiar features such as mountains. He used this as a proof against those who argued that the Earth was not to be included among the planets.

★ Mare lava plains cover much of the front side of the Moon, but only 15% of the whole Moon. The bright regions are much rougher and more cratered than the lava plains and cover the other 85%. Some of the bright uplands have traces of older lava flows, partly chewed up by the cratering. More views of the lunar surface from above can be found in this chapter on our web site.

The first reasonably accurate lunar map was produced by the German Johannes Hevelius in 1647. (A modern lunar map with names of certain prominent features is seen in Figure 7-11.) In 1651, an Italian priest, Riccioli, started the present practice of naming craters after well-known scientists and philosophers, such as Copernicus, Tycho, and Plato. The maria were given poetic and fanciful names, such as Mare Imbrium (Sea of Rains). Lunar mountains were named for prominent terrestrial ranges, such as the Alps. These mountains, however, are unlike terres-

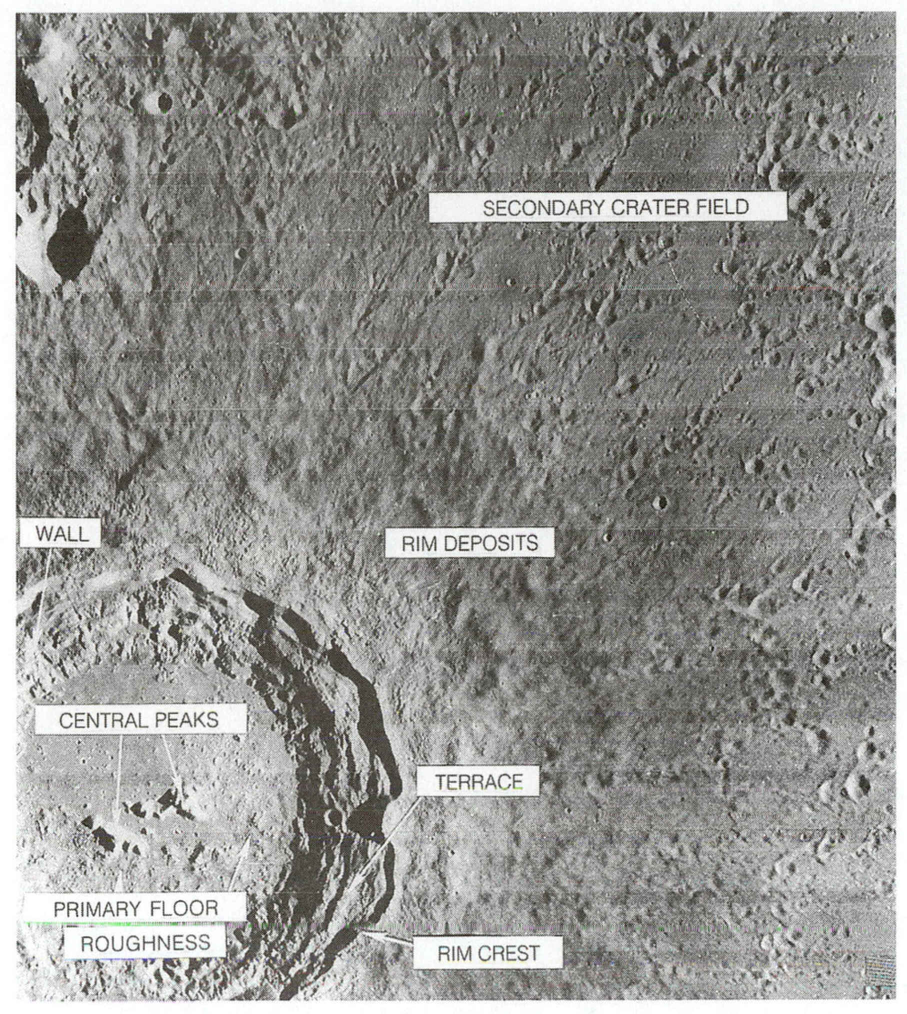

a

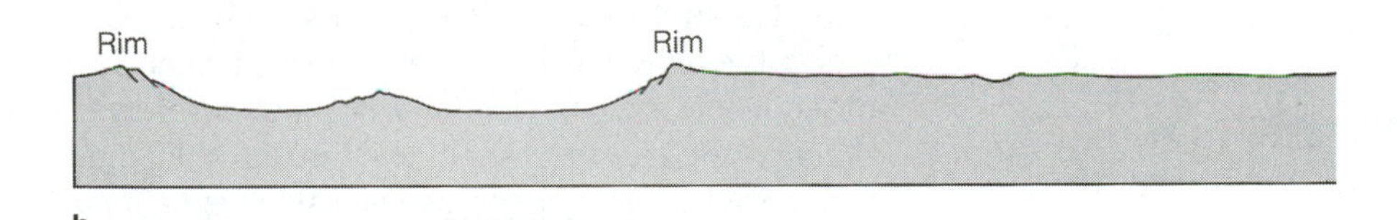

b

Figure 7-10 a The typical structures in a large lunar crater are exemplified by this Orbiter photograph of the crater Copernicus, which is about 90 km across. The main bowl was excavated by the impact of a meteorite a few kilometers in diameter. Central peaks and terraces are rebound and slump features. Rim deposits and satellite craters are caused by debris blasted out of the crater during its formation. (Courtesy James W. Head III, Brown University.) **b** An approximate cross section of such a crater.

trial folded ranges; they are the rims of vast multiringed craters, called **basins,** which in turn contain the mare plains. Figure 7-12 illustrates one of these huge impact features and its probable subsurface structure. Bright streaks called **rays,** which radiate from various craters but show no relief, are fine debris blasted out of the craters. Most maria are merely seas of lava that have flooded ancient basins.

Astronomers of the 1700s and 1800s undertook a tantalizing endeavor, searching with ever larger telescopes with ever greater magnification for diagnostic details that would show how the Moon formed. Were there any signs of changes—new craters forming or volcanic activity? Were there any artificial structures that might have been built by the inhabitants that popular writers continued to place on the Moon?

★ Despite years of searching, no changes or artificial structures were found. Lunar photographs, first made in 1849 (see our web site), showed the same structures year after year. Because there seemed to be nothing new to discover, interest in the Moon dwindled, and many astronomers saw it as a sterile nuisance that lit up the night sky and blotted out the faint objects they wanted to observe.

Nonetheless, evidence for minor geological activity on the present-day Moon came in 1963, when on two occasions observers at Lowell Observatory saw

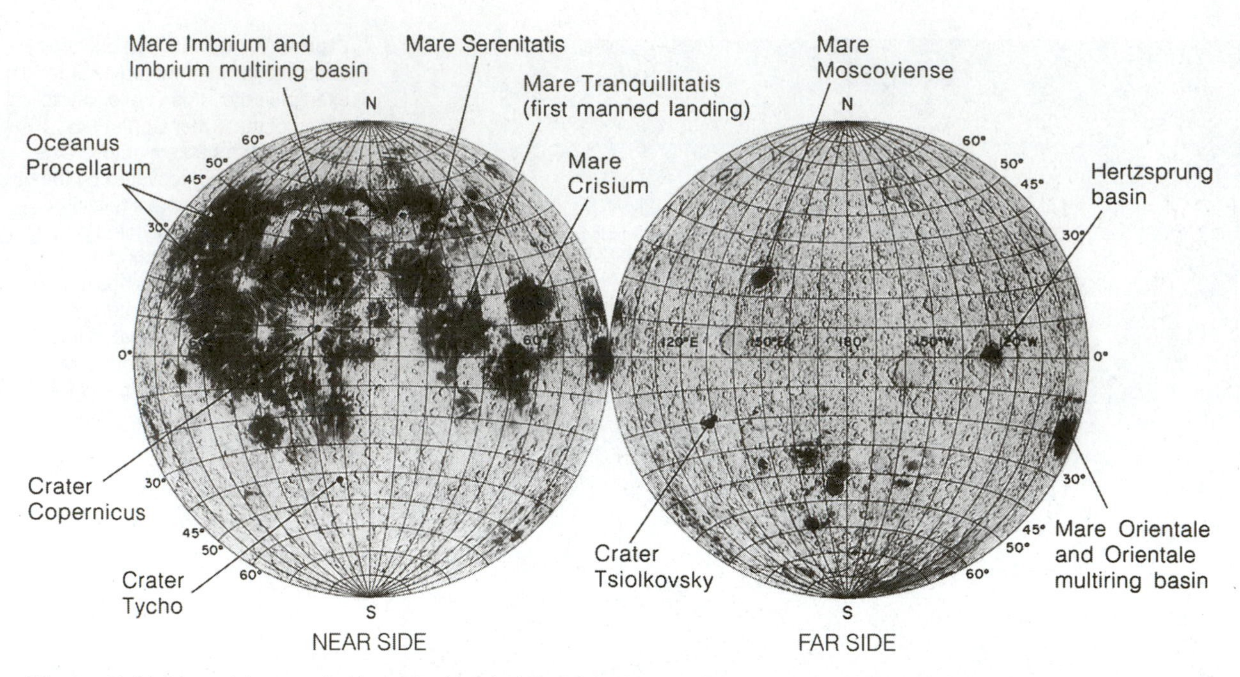

Figure 7-11 Map of the near (left) and far (right) sides of the Moon, showing some prominent features. (U.S. Geological Survey, courtesy R. M. Batson)

a red glow near the crater Aristarchus. This glow may have been a volcanic eruption or gas discharge; later data from Apollo flights indicated occasional gas emission in this region.

Why is the Moon's geological character different from Earth's? Why do craters dominate? Why are there no major mountain ranges? Where did the Moon come from? What would it be like to stand on the Moon? Such questions motivated actual journeys to the Moon.

FLIGHTS TO THE MOON

The first human device to reach the Moon was a Russian spacecraft that carried little scientific equipment and crashed into the Moon in 1959. The first close-up photos of the surface, from the American probe Ranger VII in 1964, showed that the surface was not craggy but covered by a gently rolling layer of powdery soil, scattered rocks, and shallow craters of various sizes. This type of soil cover, shown well in Figures 7-13 and 7-14, is present nearly everywhere on the Moon, and is called the **regolith** (rocky layer). The lunar regolith is typically 3 to 30 m (10 to 100 ft) deep and made primarily of debris blasted out of lunar craters as they were formed. Each well-preserved lunar crater is surrounded by a sheet of such debris, called an **ejecta blanket.** The regolith is therefore said to be composed of overlapping ejecta blankets. The powdery surface of the regolith is due to small meteorites (some microscopic), which are so abundant they have "sandblasted" most of the upper few meters into fine dust.

★ Table 7-1 lists the six Apollo lunar landings and two earlier test flights. Twelve men walked on the Moon during the Apollo program from 1969 to 1972. Figures 7-13 and 7-14 (and several other figures on our web site) show typical scenes during these missions. The first two Apollo missions were cautious tests that touched down on flat, smooth plains. The subsequent four Apollo landings sampled a variety of complex and rugged sites. Astronauts collected many samples and placed various instruments in position to make measurements.

★ As seen in Table 7-1, these flights proved that the lunar surface features are very old. The maria—the dark regions visible to the naked eye from Earth—turned out to be vast flows of basaltic lava, 3 to 4 billion years old. One such lava sample is shown on our web site.

The light-colored bright uplands are still older; they formed during the early years of the solar system about 4 to $4\frac{1}{2}$ billion years ago, although most of the oldest rocks are heavily fragmented or pulverized. Chemical evidence from the rocks shows that around $4\frac{1}{2}$ billion years ago, the surface layers of the Moon were molten, forming a vast sea of lava called a **magma ocean.** Low-density feldspar crystals then solidified in the magma ocean. Because of their low

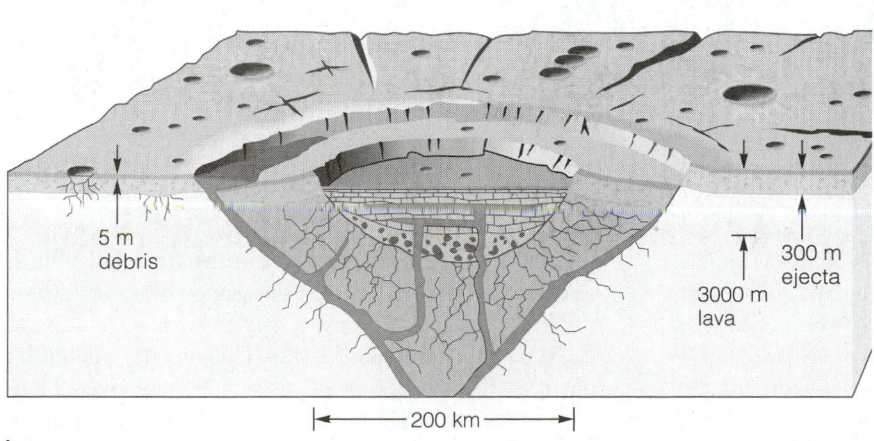

Figure 7-12 **a** The Orientale basin is the youngest and most dramatic multiring basin on the Moon. The outermost ring of cliffs is nearly 1000 km in diameter. Lava erupted to form dark "ponds" along fractures at the base of some cliffs. Map shows scale. (NASA photo from unmanned orbiter spacecraft) **b** Hypothetical cross section.

Figure 7-13 Apollo 16 astronauts drove their battery-powered "rover" (rear) to the rim of this 40-m-diameter crater in 1972. Note that the crater is old enough to have been smoothed by the sandblasting effect of innumerable small meteorite impacts that pock it with smaller pits and blanket it in regolith. (NASA)

density, they floated in the magma ocean, accumulating into an ever-thickening lithosphere of basaltlike rock called anorthosite. This is the rock type that now composes most of the uplands.

Most of the ancient anorthosite rocks did not survive intact. Numbers and ages of craters prove that the rate of meteorite impact in the first half-billion years of lunar history was thousands of times higher than today. The high cratering rate pulverized most of the primordial lunar rocks. Upland sites visited by astronauts revealed overlapping layers of ejecta blankets from many craters, composed of dust, glass

Figure 7-14 The landing module at the Apollo 15 site at the foot of the Apennine Mountains. The tracks of the astronauts' battery-powered "rover" vehicle are visible in the powdery regolith. (NASA)

droplets, rock chips, and **breccia,** or rocks composed of cemented rock fragments. These results explain why rocks older than about 4.2 billion years are rare or heavily pulverized and why rocks younger than 3.0 billion years are also rare. The oldest rocks were destroyed by impacts, and by 3 billion years ago, the Moon had cooled enough, and the lithosphere was thick enough, that subsequent volcanism or rock-forming activity was infrequent. Because 3 to $4\frac{1}{2}$ billion-year-old rocks are rare on Earth, lunar rocks have given geologists a welcome insight into conditions in the Earth-Moon system during that era. More detailed descriptions of the advances resulting from Apollo exploration are given by Taylor (1982).

LUNAR ROCKS: IMPLICATIONS FOR THE MOON AND EARTH

Rocks are cosmically significant. They are solid materials that contain many clues to their long histories of crystallization, melting, recrystallization, and so on. Rocks also tell the histories of their parent planets. Rocks can be analyzed in various ways:

1. By their elements and isotopes, which indicate the material from which the parent planet originally formed
2. By their minerals, which indicate the degree of differentiation that produced different chemical compounds inside the planet
3. By their structure, which indicates their environments through history
4. By their radioisotopic ages, which indicate when the planet-forming, differentiating, and rock-altering processes occurred

If the concepts in this list are not clear, the reader should review Chapter 6.

Many kilograms of lunar rock were brought back from the Moon. Once their appearance was known, an additional dozen were discovered on Earth between 1982 and 1991. Evidently, they were occasionally blasted out of lunar impact craters and then drifted through space until they eventually hit the Earth. All of the lunar materials have taught us a great deal about the Moon.

The elements and isotopes of lunar rocks are similar to those of Earth's rocks, though the Moon

TABLE 7-1

Manned Apollo Explorations of the Moon

Mission	Date	Landing Site	Results	Typical Rock Ages (billions of years)[b]
Orbital Missions				
Apollo 8	Dec. 24, 1968[a]	—	First lunar orbit. Orbital mapping. 115-km minimum altitude.	
Apollo 10	May 21, 1969[a]	—	Test of approach to approximately 17-km minimum altitude.	
Landing Missions				
Apollo 11	Jul. 20, 1969	Mare Tranquillitatis	First landing. Samples of mare material.	3.5–3.7
Apollo 12	Nov. 18, 1969	Oceanus Procellarum	Samples of mare material.	3.2–3.4
Apollo 14	Feb. 5, 1971	Fra Mauro (ejecta from Imbrium basin)	Samples of ejecta from Imbrium basin	3.9
Apollo 15	Jul. 30, 1971	Edge of Mare Imbrium at foot of Apennine Mts.	Samples of material from Apennine Mts., forming rim of Imbrium basin. Samples of mare material. First use of roving vehicle.	up to 4.3 (upland) 3.3–3.4 (mare)
Apollo 16	Apr. 20, 1972	Lunar uplands near crater Descartes	First landing in uplands. Samples of upland materials.	3.8–4.4
Apollo 17	Dec. 11, 1972	Taurus Mts.; edge of Mare Serenitatis	Samples from a region suspected of recent volcanism.	3.8 (mare) 4.2–4.4 (upland fragments)

Note: In addition to samples mentioned above, lunar soil samples were returned to Earth by three unmanned Soviet probes, Luna 16, 20, and 24. Ages were 3.3–3.4 billion years for basalt lava and about 4.4 billion years for an anorthosite upland rock chip. Additional studies of isotopes in all the rock samples indicate that the moon as a whole formed around 4.5 billion years ago.

[a]Date lunar orbit began.

[b]Date of solidification of crystalline rocks, or formation of breccias, based primarily on radioisotopic results of Wasserburg, Papanasstasiou, Tera, and colleagues at the California Institute of Technology; of Carlson and Lugmair (1988); and on a summary by Taylor (1982).

has much less iron. No new elements or bizarre compounds were found on the Moon, and the proportions of elements suggest that the Moon formed from material similar to that of Earth's mantle. Relative to Earth's mantle, however, the Moon is strongly depleted in **volatile elements** and compounds—substances that are driven off by heating, such as water. Similarly, the Moon is enriched in **refractory elements** and compounds—substances with high boiling points, such as aluminum and titanium. These findings indicate that the lunar material may have been strongly heated before the Moon formed, driving off the water and other volatiles into space.

Although the minerals of the lunar rocks are similar to those of Earth's mantle and crustal basalts, there is a very interesting difference. The Moon's rocks are not as differentiated as Earth's, and there are no granitic continental blocks or plates in the lunar crust. These facts are among many lines of evidence suggesting that the Moon has not been nearly as geologically active as Earth. This is because the Moon is so much smaller than Earth that it cooled more quickly, formed a thicker lithosphere, and

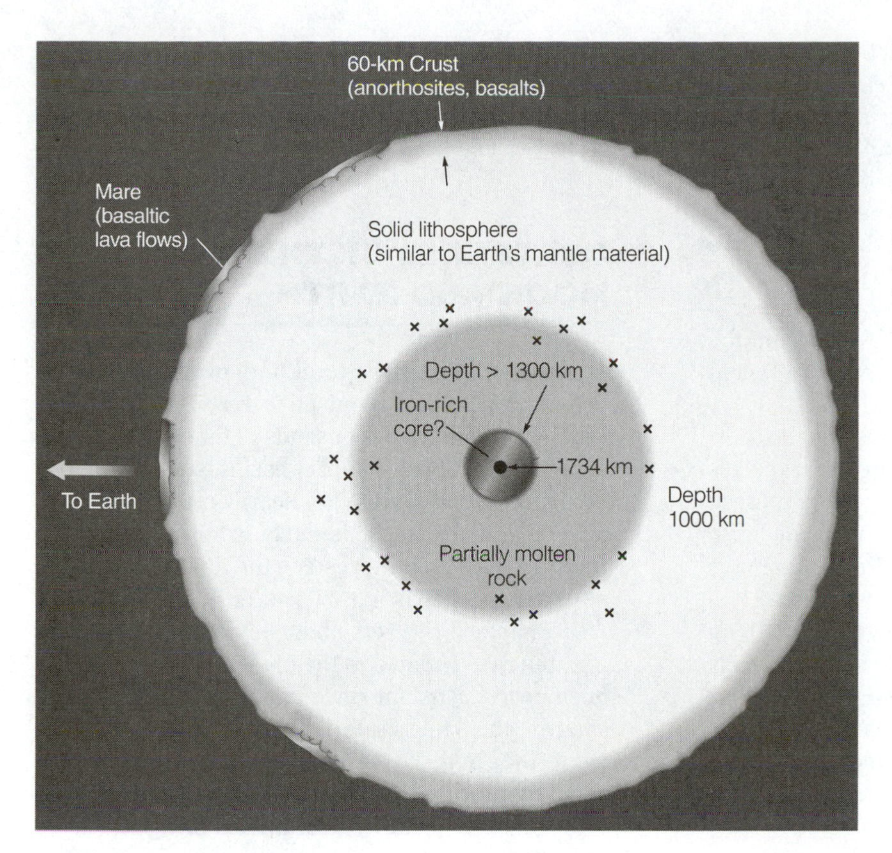

Figure 7-15 Cross section of the Moon as revealed by Apollo and other sources. The crust is thinner on the front side than on the far side. Fractures under large impact craters on the front side allowed lava to reach the surface and create more mare lava flows than on the far side. A moonquake zone (x) marks the bottom of the lithosphere. The existence of an iron-rich core is uncertain.

never developed the plate movements or perpetual volcanic and tectonic activity that make Earth's geology so complex.

The evidence for a lunar magma ocean has led to a new view that many or all planets may initially have had magma oceans that allowed low-density minerals to concentrate near their surfaces. This influenced the rock types, minerals, ores, and so on that we now find on the surfaces of these bodies.

All of the huge impact basins formed as part of the intense cratering that occurred before or around 4 billion years ago. The biggest impacts created fractures that allowed basaltic lavas to erupt from depths of around 300 km (a depth determined from certain mineral properties), and lunar lava flows filled many of the impact basins between 4 and 3 billion years ago. Because the Moon is smaller than the Earth, however, it cooled much faster than the Earth and formed a thick, solid, rigid lithosphere. It has been relatively quiet since then.

It is extraordinary to realize that astronauts have walked and driven across landscapes whose mountains and craters have lain still and silent since trilobites crawled on our own seafloors and the first lizardlike creatures crawled out on our land. Perhaps the most important fact to remember is that the lunar lithosphere formed very early and was very thick, as deep as 1000 km. Thus it was too rigid to allow mountain-building activity or plate tectonics, and it blocked magmas from reaching the surface. Thus the Moon's surface preserves craters and rocks formed 3 to 4 billion years ago. In contrast, the Earth has a thin lithosphere that is only 100 km thick and easily broken by earthquakes and volcanic eruptions. It is so active that it has destroyed most landforms and rocks formed during this period. Thus the Moon supplies some heretofore missing information about the early history of the planets and the early environment of the Earth-Moon system.

THE INTERIOR OF THE MOON

We can understand the Moon better by considering its interior structure, shown in Figure 7-15.

First, the **mean density** (total mass divided by total volume) of the Moon is much less than that of the Earth—3300 vs. 5500 kg/m^3 (3.3 vs. 5.5 g/cm^3). This proves that the Moon is mostly rocky, like Earth's mantle, and that it lacks a big iron core. Second, measurements indicate that the Moon has virtually no magnetic field. This again suggests the lack of a large molten iron core, because scientists believe that the planets' magnetic fields originate in currents in such cores. Nonetheless, magnetic measurements on lunar rocks indicate that when they solidified billions of years ago, there was a lunar magnetic field roughly 4% the strength of the Earth's field. Because no such field exists today, the implication is that the Moon had a molten iron core only in the past, when its interior temperatures were higher. A small core might still exist but reach no farther than about 350 km from the Moon's center, or about 20% of the radius (Taylor, 1982, 1992).

Third, seismic data show much less quake activity on the Moon than on Earth. Large lunar quakes rank only 0.5 to 1.3 on the Richter scale, compared with 5 to 8 for major earthquakes. In a year, the Earth expends about 100 million million times as much seismic energy as the Moon.

Nonetheless, moonquakes tell us something about the Moon's interior. The quakes occur mostly at depths of 700 to 1200 km. Just as Earth's quakes are concentrated at the bottom of the brittle lithosphere and the top of the sluggishly moving asthenosphere, the Moon's quakes are believed to mark the bottom of a 1000-km-thick lunar lithosphere. Many moonquakes occur in monthly cycles associated with tidal flexing as the Moon's slightly elliptical orbit brings it toward and then away from the Earth (Toksoz and others, 1974). Some quakes detected by Apollo instruments had a different source—the impacts of modest-sized meteorites (too small to make craters visible from Earth).

CRATERING OF THE MOON AND EARTH

As noted earlier, the Moon underwent an intense bombardment from about 4.6 to 4.0 billion years ago, with a cratering rate thousands of times higher than today's rate. All large impact basins, such as the one in Figure 7-12, as well as the heavily cratered uplands formed at that time. This **early intense bombardment** of the Moon represents the final stages of the sweep-up of debris left over after planet formation (see Chapter 14). From about 4.0 to 3.0 billion years ago, the rate declined to the present level, which has been nearly constant since that time. Earth has had roughly the same cratering rate, and scientists assume that both Earth and the Moon experienced similar cratering histories. The Apollo data from the Moon thus help clarify the impact history of Earth, paving the way for recent theories that giant impacts may have altered climates and biological evolution (see Chapter 6).

OPTIONAL BASIC EQUATION V

The Definition of Mean Density

When an astronomer begins to get data on a new object, such as a newly measured planet, asteroid, or star, one of the first things he or she wants to know is the mean density of the object because this helps clarify the composition of the object. The mean density is defined as

$$\text{Mean density} = \frac{\text{total mass of object}}{\text{total volume of object}}$$

This is called the *mean* density because it gives only the average density, averaging over the whole object. Parts of the object's interior might have higher or lower densities. The Earth's mean density, for example, is about 5500 kg/m^3, but the iron core has much higher values, in excess of 8000 kg/m^3, while surface rocks have values closer to 2500 kg/m^3. Thus, for example, if you sighted a round object in space, it would make quite a difference to your interpretation if the mean density turned out to be 8000 kg/m^3 instead of 2000 kg/m^3!

The value of 1000 kg/m^3 is of particular interest, since this is the density of water and the approximate density of ice. (When the metric system was set up, the kilogram was defined such that 1000 kg equaled the mass of a cubic meter of water.)

Since most objects in astronomy are roughly spheroidal in shape, we can replace volume by the expression for the volume of a sphere, $4\pi R^3/3$. Thus, using R as the radius, M as the mass, and the Greek letter ρ (rho) traditionally used as the density, we get the equation

$$\text{Mean density} = \rho = \frac{3M}{4\pi R^3}$$

Figure 7-16 Explosion tests on Earth simulate many features of meteorite impacts. This explosion of 100 tons of TNT produced a crater 39 m (128 ft) across and 7 m (23 ft) deep. Secondary impact craters formed as far as 110 m (360 ft) away, and the farthest ejecta went 200 m (660 ft). The jets of material may simulate material that formed lunar rays; the turbulent cloud at the base of the explosion expands across the landscape and deposits much of the ejecta blanket. (Photo by WKH)

Craters are the most important landforms on many other planets and moons as well. The photo in Figure 7-16 shows how meteorite impact explosions may have looked. The meteorites ranged from abundant microscopic particles, through numerous kilometer-size chunks, to objects over 100 km in diameter. The latter made craters about 1000 km across.

Repeated cratering by a rain of meteorites explains lunar topography. At the scale witnessed by humans on the surface, features of characteristic size of 10 to 100 m are shaped by the sandblasting effects of countless small impacts. These smoothed out the original crags, producing the powdery soils, rolling topography, and scattered boulders typical of the lunar landscape.

Thus, if an astronomer or astronaut can measure the mass M of an object (perhaps by its gravitational force on another object) and its radius R, he or she can quickly calculate the mean density and make some comments about the types of materials that might compose the object.

Note that in the SI system of units, M is measured in kilograms and R in meters, so the density is expressed as kilograms per cubic meter. While water and ice have densities of about 1000 kg/m^3, rock has densities around 2500 to 3000 kg/m^3, and gaseous objects have densities less than 1000 kg/m^3, unless they are compressed by strong gravity. In many scientific and other books, density is more commonly expressed in units of grams per cubic centimeter, which are 1/1000 of the value in kg/m^3. For example, water has density of 1 g/cm^3.

Sample Problem Voyager spacecraft and Earth-based measurements showed that the small moon of Saturn, Enceladus, has a surface of frozen water, a radius of 250 km, and a mass of 8.4×10^{19} kg. Find the mean density of Enceladus and comment on whether it might be largely icy or rocky throughout. *Solution:* Converting R to meters and inserting the given values for R and M, we compute a density of 1283 kg/m^3. Since R and M are given only to two significant figures, we should round off our answer to two significant figures, or 1300 kg/m^3. Note that this is only 30% greater than the density of ice! Thus Enceladus could not be made mostly of rock, but it could be mostly icy throughout.

ICE DEPOSITS AT THE LUNAR POLES?

Among the objects that hit the Moon are comets, which are made mostly of ice. As early as the 1960s and 1970s, a few scientists suggested that if water molecules were released on the Moon during comet impacts, some of them might collect in shadows at the lunar poles (Watson and others, 1961; Arnold, 1979). The reasoning is that the lunar polar axis is 90° to the sunward direction, so the floor of a deep crater at the pole is always shadowed. If you stood on the floor, the Sun would never quite rise above the crater wall. This means that such crater floors would be intensely cold, and water molecules that found their way there would condense into ice, trapped on the crater floor, never to be warmed or melted by the Sun.

For some years, this theory seemed far-fetched. But in 1998, the inexpensive Lunar Prospector robotic spacecraft flew to the Moon and observed evidence that such deposits do exist! The spacecraft monitored neutrons ejected from the lunar surface when cosmic rays hit the surface; their energies are modified by the presence of hydrogen or ice (Feldman and others, 1998). The strong modifications at each pole suggested the possibility of 0.3 to 1% ice in the polar soils. This ice has yet to be confirmed, but it could offer an important resource for explorers on an otherwise waterless Moon!

WHERE DID THE MOON COME FROM?

The Moon's origin has long frustrated theorists. Prior to the Apollo missions, astronomers debated various theories about the Moon's formation. It seemed not unreasonable that the Moon might have formed alongside the Earth at the same time Earth formed. However, a major problem was the lack of iron. How could the Moon have formed alongside Earth from the same general materials without getting as much iron as Earth did? To resolve this, some scientists proposed other theories, but none of the theories fitted the lunar data. The Moon had become a real puzzle, and one scientist quipped that if our theories couldn't explain it, maybe it was an illusion that didn't really exist!

Summing up pre-Apollo and post-Apollo evidence, there are several clues illuminating the Moon's origin:

1. The Moon has much less iron than Earth.

2. Except for the possible deposits of cometary ice at the poles, the Moon has virtually no water and few other volatiles—the substances easily driven off by heat.

3. The gross composition of the Moon resembles that of Earth's mantle rocks, except for the lack of water and volatiles.

4. The proportions of different isotopes composing oxygen in lunar rocks are exactly the same as in Earth's oxygen (whereas the proportions of oxygen isotopes in rocks formed at different distances from the Sun—that is, in meteorites—are different).

In the 1970s and 1980s, a new theory of lunar origin emerged to explain these facts. Called the **giant impact hypothesis,** it states that during the final stages of Earth's formation, but after Earth's iron core formed, our fledgling planet was hit by a large interplanetary body—perhaps as large as Mars. As will be clearer in Chapter 14, planets formed from aggregation of interplanetary debris, and as Earth formed, other sizable interplanetary bodies were growing at the same time. According to the giant impact hypothesis, one of them hit Earth. The resulting catastrophic impact, the biggest ever suffered by Earth, probably occurred 4.45 to 4.54 billion years ago, only 10 to 100 million years after Earth started forming. This was shortly after Earth's iron core had formed. Thus the impact blasted hot material out of the mantles of both the Earth and the impactor, into a cloud of debris around primordial Earth (Figures 7-17 and 7-18). The Moon aggregated from those mantle debris.

The giant impact theory fits into models of how the planets formed from interplanetary bodies and explains a number of facts about the Moon. For example, to return to the list above:

1. The scarcity of iron on the Moon is explained by the fact that Earth's iron was safely hidden in its center, and only iron-poor mantle material was ejected by the giant impact.

2. The lack of water and volatiles on the Moon (except for possible cometary ice deposits at the poles) is explained by the fact that the ejected material was so intensely heated that water and other volatiles were driven off into space before the Moon could form.

3. The overall similarity of composition of the Moon and Earth results because the ejected material came from the mantles of Earth and the impactor.

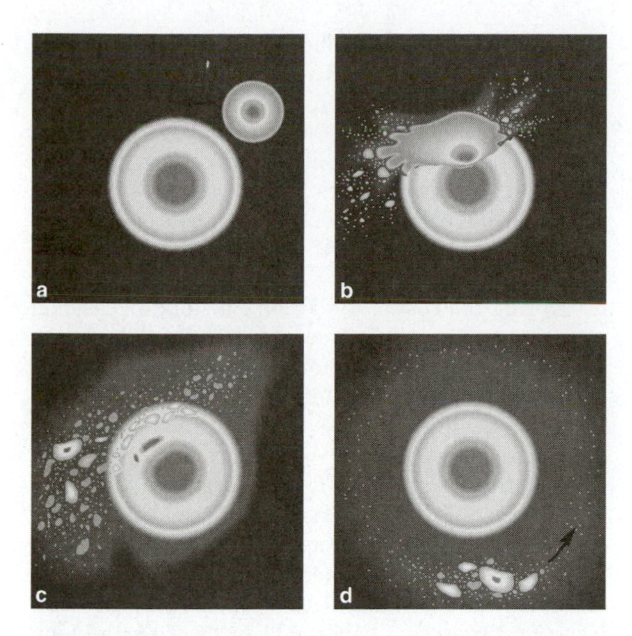

Figure 7-17 Schematic view of the giant impact hypothesis for lunar origin. A large interplanetary body approaches (**a**) and collides with the primordial Earth (**b**), following formation of iron cores (dark) in both bodies. Hot gas and condensing debris from the mantles of both bodies are thrown into near-Earth space (**c**). Part of the debris forms an orbiting cloud around Earth, in which the Moon begins to aggregate (**d**). (Adapted from computer models by A. Cameron, W. Benz, W. Slattery, M. Kipp, and J. Melosh)

4. The similarity of isotopic composition of oxygen is explained by the fact that the impactor was a body that also grew in Earth's part of the solar system and thus had the same isotopic composition.

The theory has the advantage that it can explain why Earth has a large moon, but the other terrestrial planets do not; giant impacts were chance occurrences that affected some planets and not others. In science, a useful theory is one that can be tested; the giant impact theory is currently being tested by computer models, which confirm that such a collision could blast enough material into orbit to form a Moon-size satellite. The theory is also being tested by chemical studies of Earth's mantle rocks, which support the probability that Earth's iron core formed very early as Earth grew—and before the impact occurred (Newsom and Sims, 1991).

RETURN TO THE MOON?

Ever since Apollo 17 blasted off the Moon in 1972, the Moon has been deserted by humans. However, there has been interest in returning to the Moon as part of a program to extend human capability in the solar system. For example, a program to measure dates of lunar craters could not only clarify lunar history, but could also reveal whether the impact that wiped out the dinosaurs on Earth was part of a periodic wave of impacts 65 million years ago, as some have suggested, or just a solitary statistical fluke. Thus, strangely enough, a lunar base could reveal important clues about the history of life on Earth. Such a lunar research station could also be an effective site for astronomical observations at a variety of wavelengths.

Figure 7-18 The giant impact that blew mantle material out of primordial Earth's mantle. This view, based on computer models of the event, show luminous matter, some as hot as the Sun's surface, spraying outward about $\frac{1}{2}$ h after the impact. (Painting by WKH)

The logistics of such a base are beginning to be understood. Permanent manned lunar stations could produce oxygen from lunar rocks for breathing and for fuel (Thomsen, 1986; Lewis and others, 1993). Polar ice deposits might supply water. Supplying liquid oxygen fuel from lunar rocks for spaceships and space stations might ultimately be cheaper than hauling it up from Earth, because so much less energy is needed to launch material off the Moon than off the Earth. Construction techniques for future lunar bases have already been tested on Earth. Although modules might be transported ready-made from Earth, 1986 tests on lunar soil showed that it is ideal for making concrete! Various lunar resources are discussed in the book "Resources of Near Earth Space" (Lewis and others, 1993).

SUMMARY

The Moon is an ancient planetary body, little disturbed since the formative days of the solar system. Much of the information in this chapter yields a chronological history of the Moon. Because the Earth shared much, if not all, of this history, we summarize the history of the Earth-Moon system in Table 7-2.

Lunar rocks and meteorites reveal that the Moon and planets were formed 4.6 billion years ago. The Moon's formation probably involved a giant impact that blew rocky material off Earth's surface. Formation of the Moon from this material explains why the Moon has no large iron core, as Earth does. Earth and the Moon were close together shortly thereafter. Differences in ages among lunar and meteorite specimens indicate that the Moon and planets reached approximately their present sizes a few million to 90 million years after the solar system's formation began.

Analyses of lunar rocks indicate that the lunar surface was initially covered with a molten magma ocean several hundred kilometers deep. An initial magma ocean may also have formed on Earth, but evidence here has been destroyed. Nearly all lunar rocks that formed before about 4 billion years ago have been pulverized by the intense early meteorite bombardment. According to Apollo data, the cratering rate 4 billion years ago was 1000 times the present rate, and before that it was still higher. The Earth was presumably also bombarded at that time. Many large craters formed on the Earth and the Moon during this period. The cratering rate declined to the current value by 3 billion years ago.

Tidal analysis proves that after its formation near Earth, the Moon moved out quickly, reaching about half its present distance in only 100 million years, or 2% of Earth's age. It probably approached its present orbit between 4.4 and 4 billion years ago, and the Earth's day approached its present length at that time. As the Moon moved away, and after its surface cooled, basaltic lavas erupted in many places on the front side of the Moon, especially where the crust was thinnest because of large-scale impacts. Successive lava flows formed the mare plains from 4 to 3 billion years ago. Because the Moon is smaller than the Earth, it cooled more quickly, and most volcanism died out around 3 billion years ago. This cooling produced a 1000-km-thick lithosphere that shielded its surface from internally caused changes. Therefore, the Moon developed no plate tectonics. On Earth, in contrast, convection currents continued, breaking its 100-km-thick lithosphere into moving plates.

During the rest of the Moon's history, occasional large impacts, involving meteorites a few kilometers across, excavated major craters, throwing bright rays of pulverized ejecta across the dark maria and old uplands. On Earth some large craters formed by meteorites during the last billion years have also been preserved, though most of Earth's craters have been destroyed by Earth's intense erosion and geological activity, as described in Chapter 6. The parts of terrestrial history studied by most geologists are only the tail end of Earth-Moon history. Most of Earth's surface rocks represent only the last 12% of Earth-Moon history, whereas most lunar surface structures date back through 84% of it. The Moon shows more clearly than the Earth the early and middle parts of geological history and the combined results of internal geological processes (for example, partial melting) and external astronomical processes (impacts). The Moon has given us new understanding of processes that shaped our own world, and we have left our mark on it.

CONCEPTS

satellite
phases
new moon
first quarter
gibbous
full moon
third quarter
sidereal period
synchronous rotation
tide
body tide
ocean tide
tsunami
tidal recession
Roche's limit
terminator
limb
impact crater
mare (maria)
basin
ray
regolith
ejecta blanket
magma ocean
breccia
volatile elements
refractory elements
mean density
early intense bombardment
giant impact hypothesis

TABLE 7-2

History of the Earth-Moon System

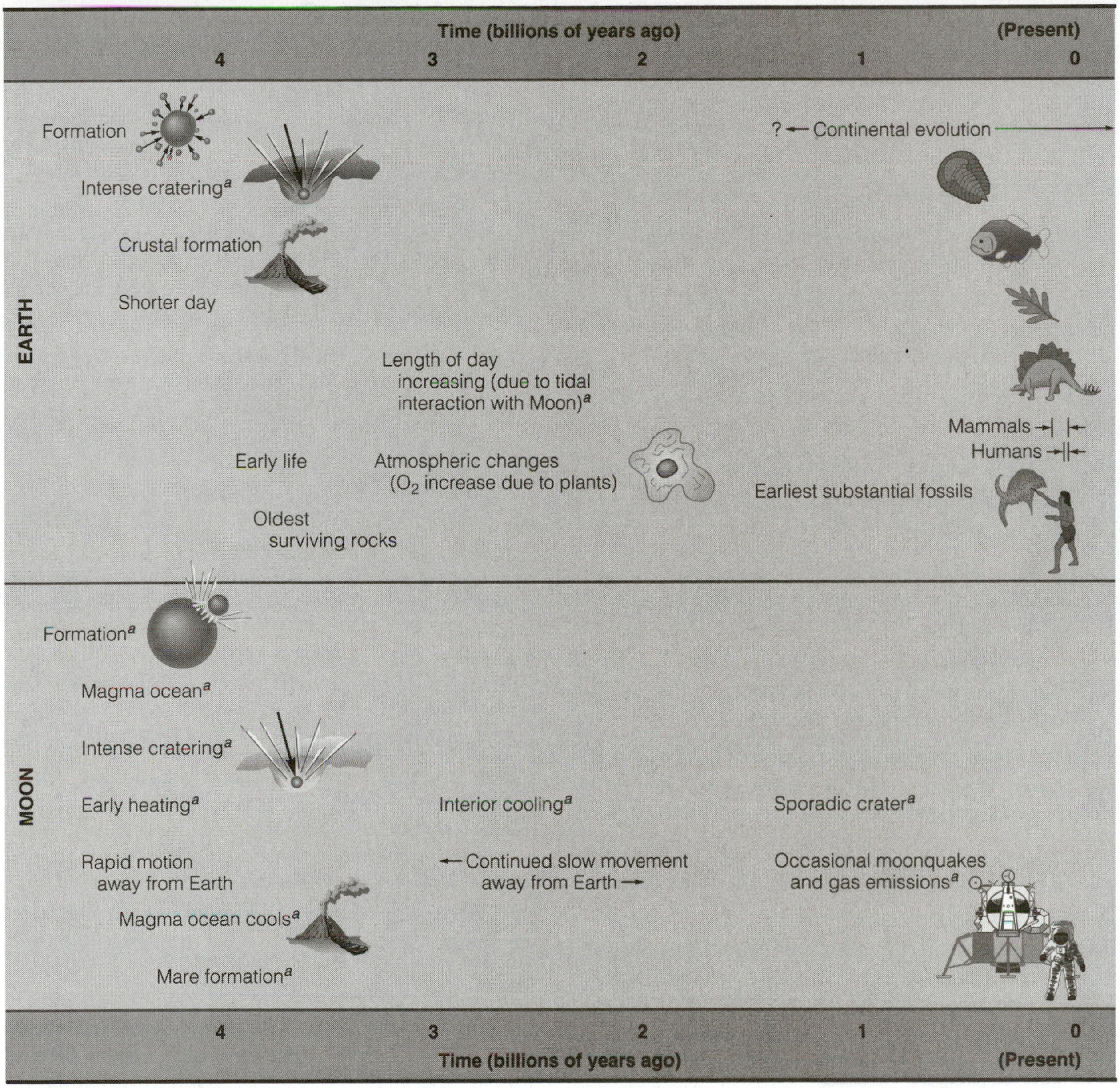

Time (billions of years ago): 4, 3, 2, 1, 0 (Present)

EARTH

- Formation
- Intense cratering[a]
- Crustal formation
- Shorter day
- Length of day increasing (due to tidal interaction with Moon)[a]
- Early life
- Atmospheric changes (O_2 increase due to plants)
- Oldest surviving rocks
- ? ← Continental evolution →
- Earliest substantial fossils
- Mammals →| |←
- Humans →||←

MOON

- Formation[a]
- Magma ocean[a]
- Intense cratering[a]
- Early heating[a]
- Rapid motion away from Earth
- Magma ocean cools[a]
- Mare formation[a]
- Interior cooling[a]
- ← Continued slow movement away from Earth →
- Sporadic crater[a]
- Occasional moonquakes and gas emissions[a]

Time (billions of years ago): 4, 3, 2, 1, 0 (Present)

[a]Information discovered or improved through Apollo-related lunar research.

PROBLEMS

1. At what time of day (or night) does:

a. The first quarter moon rise?

b. The full moon?

c. The last quarter moon?

d. The new moon?

2. When a terrestrial observer is recording a new moon, what phase would the Earth appear to have to an observer on the Moon?

3. Draw a diagram like Figure 7-2 and show approximate locations from which Figures 7-3a and 7-3b could have been taken.

4. Explain why the Moon keeps one side toward the Earth.

5. Why might one expect the highest tides to occur at noon or midnight on the date of the new moon or the full moon? Why don't the highest tides always occur at these times?

6. Suppose an astronaut orbiting just above the atmosphere releases into an orbit of their own two Ping-Pong balls just in contact with each other. Would you expect them to stay in contact with each other indefinitely? Why or why not?

7. Imagine you are selecting a lunar landing site.

a. What type of feature might offer fresh bedrock where regolith layers have been stripped away?

b. Would you expect the landscape inside a young 100-km-diameter crater to be rougher or smoother than inside an old crater of the same size?

c. Where might astronauts land to seek evidence of recent volcanic activity?

8. Imagine you are an astronaut exploring the Moon.

a. Would an ordinary compass work on the Moon? Why or why not?

b. What celestial object could serve as a directional aid (like the North Star) for astronauts hiking on the Moon's front side?

c. What property of this object's apparent motion would make it especially useful as a navigational aid during a lunar stay of several months?

9. If ages of the Earth and Moon are nearly identical, as believed, why are most rocks found on the Moon so much older than Earth rocks?

10. Suppose sedimentary rocks had been discovered on the Moon. How would this affect our beliefs about the Moon's history?

11. By comparing pictures of lunar maria (3 to 4 billion years old) and uplands (4 to 4.5 billion years old), prove that the meteoritic cratering rate was much higher during the first few hundred million years of lunar history than during the last 3 billion years.

ADVANCED PROBLEMS

12. The Moon has about 0.012 of the Earth's mass and 0.27 of the Earth's radius.

a. Using Newton's law of gravity, show that the Moon's surface gravity is about one-sixth that of the Earth.

b. How much would an 82-kg (180-lb) person weigh on the Moon? Give your answer in pounds.

c. Note that pounds are a measure of weight, whereas kilograms are a measure of mass. Would an 82-kg person have a different mass on the Moon?

13. Prove that the Moon subtends an angle of about $\frac{1}{2}°$. It is 3476 km in diameter and averages 384,000 km away.

14. How fast must a projectile move to escape from the Moon? (Mass of Moon = 7.35×10^{22} kg; radius = 1.74×10^{6} m) Compare this with the escape velocity from Earth.

15. From planetary data given in Table 8-1, confirm that the Earth's mean density is roughly 5500 kg/m^3 and that the Moon's is roughly 3300 kg/m^3. Explain how this result alone proves that the Moon cannot contain as much iron as the Earth.

16. If you have a small telescope that resolves details 2 seconds of arc across, what is the smallest crater you can see on the Moon.

PROJECTS

1. Observe the Moon at different phases with a telescope of at least 5-cm (2-in.) aperture. Locate and compare the texture of upland regions with maria (dark plains). Compare visibility of detail near the terminator and away from the terminator, and explain the difference. Sketch examples of craters, ray systems, and mountains.

2. With a telescope, find an example of a bright-ray crater, such as Tycho or Copernicus, and compare its appearance at full moon (high lighting) with its appearance near the terminator (low lighting). Why do the rays disappear under low lighting? Make a simulation of this effect by scattering a thin dusting of white flour or powder on a slightly darker, textured surface with a ray-like pattern. Illuminate with a bright light bulb from above (full moon) and from the side at a low angle (low lighting), and compare the appearance.

3. Prepare a box with white flour several centimeters deep and a light dusting of darker surface powder (flush with the top edge of the box). Throw different-sized dirt clods into the box to make craters. Illuminate with a bright light bulb from various angles and compare with the appearance of the lunar surface. Compare the number of craters needed to simulate a mare region and an upland region. Can the surface be saturated with craters if enough "meteorites" are thrown at the target? What physical differences exist between this experiment and lunar reality? (Example: These dirt clods hit at a few meters per second, whereas meteorites hit the Moon at several kilometers per second and cause violent explosions.)

PART

4

The Solar System

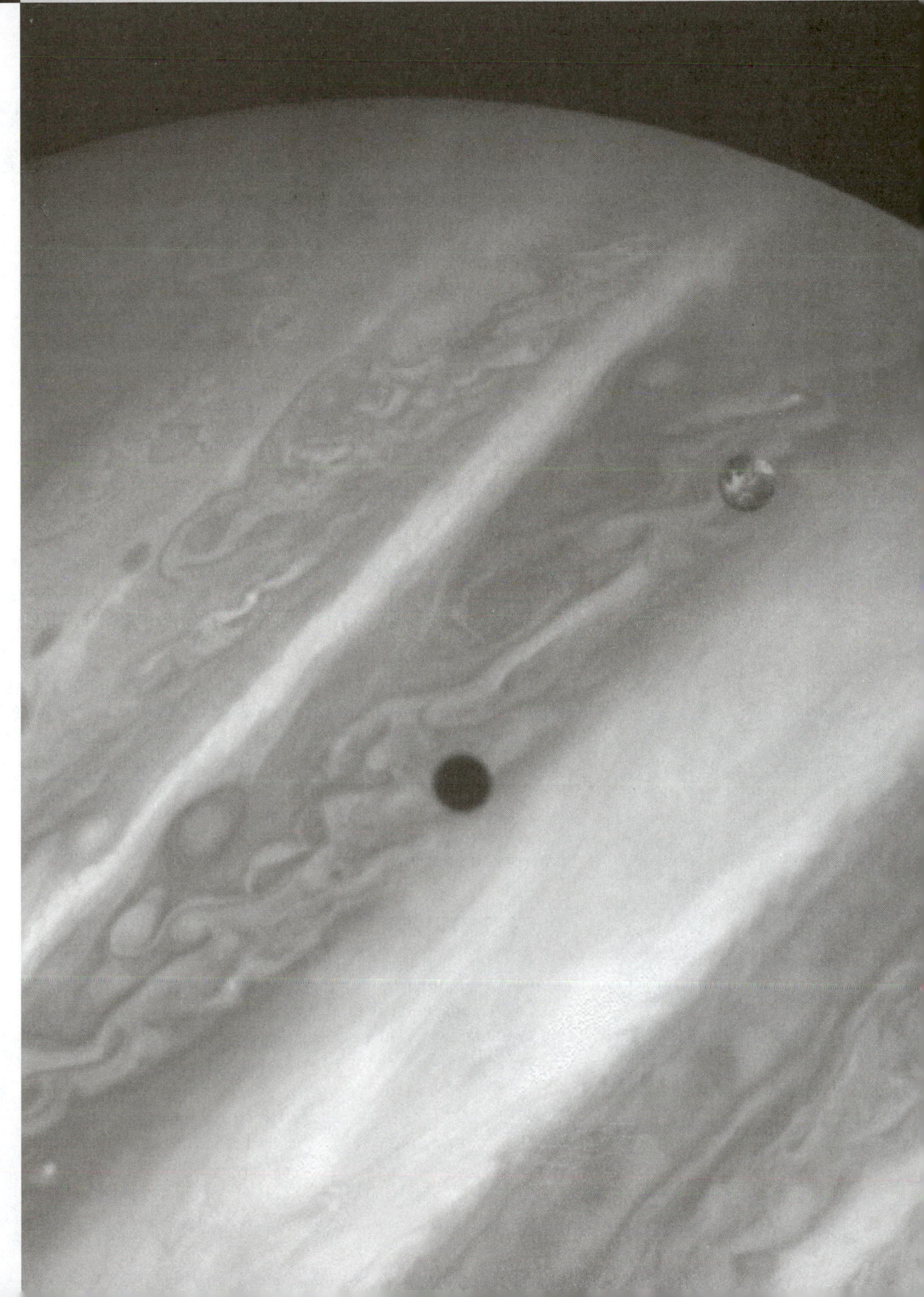

Two worlds in one photo. The Hubble Space Telescope made this image of Jupiter and its satellite, Io (upper right). The black spot is Io's shadow cast upon the giant, cloud-banded planet. (J. Spencer, Lowell Observatory, NASA, and Hubble Space Telescope Institute)

CHAPTER 8

Introducing the Planets—Mercury

WHAT THE READER SHOULD WATCH FOR IN THIS CHAPTER

This chapter presents an overview of the arrangement of the solar system. The student should learn the order of planets from the Sun and the difference between the terrestrial planets (inner solar system) and gas giant planets (outer solar system). Distances in the solar system are given in terms of astronomical units, or AU, where 1 AU is defined as the average distance from the Earth to the Sun. Bode's rule gives an easy way to learn the distances of planets from the Sun. The planet Mercury is the closest planet to the Sun. About one-third the size of Earth, it strongly resembles the Moon in general appearance and heavily cratered surface. ■

If we could journey far beyond the Moon and look back, we would see the Sun and its family of planets—the **solar system.** We would discover that the Earth is only the fifth largest of many worlds that orbit around the average-sized star we call the Sun. This is a far cry from the conception of 20 generations ago, when the Earth was viewed as a kind of imperial capital of the universe—a unique stationary scene of human activities around which the Sun, Moon, planets, and stars moved. The exciting transition from the older idea to the modern conception began what astronomer Carl Sagan has called "the cosmic connection," the realization that we are only one part of a larger system of worlds—an idea still growing in our consciousness even today.

Chapter 3 described how the arrangement of the planets' orbits was discovered, and Figure 3-10 showed our modern knowledge of that arrangement. In this chapter we begin describing the other planets, their moons, and the interplanetary bodies. We will focus here not so much on their orbits as on their properties as worlds. The idea that we live in a system of worlds suggests a new conception of a large theater in which we may travel; investigate many new examples of geological, meteorological, and biological processes; and perhaps exploit new sources of energy and materials.

A SURVEY OF THE PLANETS

The solar system is defined as the Sun, its nine orbiting planets, their own satellites, and a host of small interplanetary bodies, such as asteroids and comets. Starting in the center of the solar system, the major bodies and their symbols are:

⊙ **Sun**

☿ **Mercury**

♀ **Venus**

⊕ **Earth**

♂ **Mars**

♃ **Jupiter**

♄ **Saturn**

♅ **Uranus**

♆ **Neptune**

♇ **Pluto**

★ The symbols, mostly derived from ancient astrology, are sometimes used as convenient abbreviations today. A traditional memory aid for this outward sequence is "*M*en *V*ery *E*arly *M*ade *J*ars *S*tand *U*pright *N*icely, *P*eriod." Surely today's students can do better! Many students get mixed up about the order in the outer solar system; to avoid confusion about the order of Saturn, Uranus, and Neptune, remember that the *SUN* is a member of the system, too.

Some simple facts about the bodies of the solar system are useful to remember. For example, the Sun is about 10 times the diameter of Jupiter, and Jupiter is about 10 times the diameter of Earth. Whereas the Sun is a **star,** composed of gas and emitting radiation by its own internal energy sources, **planets** are bodies at least partly solid, orbiting the Sun, and known to us primarily by reflected sunlight. **Satellites,** in turn, are solid bodies orbiting the planets.

The planets divide into two groups. The **terrestrial planets** are the four inner planets, Mercury through Mars. They most nearly resemble Earth in size and in rocky composition. The **giant planets** are the four large planets of the outer solar system, Jupiter through Neptune. Much bigger than the terrestrial planets, they also have a different composition, being rich in icy or gaseous hydrogen compounds such as methane (CH_4), ammonia (NH_3), and water (H_2O). Because they are so rich in hydrogen and other gases, they are often called the gas giant planets. Pluto falls into neither category, being a special case.

Table 8-1 presents data on 77 bodies. Notice that some satellites are bigger than some planets! The largest satellites in the solar system are Jupiter's moon Ganymede and Saturn's moon Titan, with diameters over 5000 km. They are both bigger than the planets Mercury and Pluto. Many additional, known, kilometer-scale moons are too numerous to list in this table. Planets and moons are not the only large bodies in the solar system. The large asteroid Ceres (diameter 1020 km), which orbits the Sun in a planetlike orbit between Mars and Jupiter, is bigger than half the known satellites of the solar system. As Figure 8-1 shows, there are probably 26 worlds (in addition to the Sun) in the solar system larger than 1000 km across.

Until the decade of the 1970s, even the best telescopic views of Uranus, Neptune, Pluto, and all satellites beyond the Moon revealed only poorly perceived disks, like pinheads held at arm's length. Virtually nothing was known about the features of these worlds. The satellites were especially anonymous. Because these moons were no bigger than Mercury (and mostly smaller than our Moon), astronomers assumed they had cooled off quickly, like our own Moon, and were cratered globes, geologically dead. The most astonishing discovery of solar system exploration so far has been the unforeseen variety among these worlds. As we will see in the next chapters, these distant moons include a sulfur-orange world with dozens of active volcanoes; a world with an icy crust and an ocean of liquid water underneath, like our Arctic sea; a world with one blackboard-black hemisphere and an opposing snowy-white hemisphere; a world with smoking geysers; and a cloudy world where gasolinelike compounds may rain out of the sky! The solar system is not just nine planets and a Sun; it has dozens of worlds, each with its own personality.

COMPARATIVE PLANETOLOGY: AN APPROACH TO STUDYING PLANETS

Planetology is the study of individual planets and systems of planets. In the early years of planetary studies through telescopes, each planet tended to be characterized as a world unto itself: Certain markings could be glimpsed on Mars; Jupiter had a different type of markings; Saturn had rings; and so on. But in recent decades, as spacecraft and sophisticated astronomical instruments have enabled us to learn more about planets' surfaces, atmospheres, interiors, and evolution, a new style of planetology has come into being. Called **comparative planetology,** it is a systematic study of how planets compare with each other, why they are different, and why certain planets have certain similarities.

In comparative planetology, each planet and each moon is regarded as an experiment that teaches us what type of environment evolves if you start with a certain mass, with a certain composition, at a certain distance from the Sun. A comparison of Earth and Venus provides a good example of this approach.

TABLE 8-1

Objects in the Solar System
(Including the Sun, all planets, major satellites, the six largest asteroids, and Chiron)

Object	Equatorial Diameter (km)	Mass (kg)[a]	Rotation Period[b] (d)	Orbital Period (days unless marked)	Distance from Primary (10^3 km unless marked)	Orbit Inclination[b,c] (degrees)	Orbit Eccentricity	Escape Velocity (km/s unless marked)	Known or Probable Surface Material
Sun	1,391,020	1.99 (30)	25.4	—	0	—	—	617	Ionized gas
Mercury	4880	3.30 (23)	58.6	89	0.387 AU	7.0	0.206	4.2	Basaltic dust & rock
Venus	12,104	4.87 (24)	243R	225	0.723 AU	3.4	0.007	10.4	Basaltic & granite rock
Earth	12,756	5.98 (24)	1.00	365	1.00 AU	0.0	0.017	11.2	Water, granitic soil
Moon	3476	7.35 (22)	S	27	384	18–29	0.055	2.4	Basaltic dust & rock
Mars	6792	6.44 (23)	1.02	687	1.52 AU	1.8	0.093	5.0	Basaltic dust & rock
Phobos	27 × 19	9.6 (15)	S	0.32	9.4	1.0	0.015	11 m/s	Carbonaceous soil
Deimos	15 × 11	1.0 (15)?	S	1.26	23	2.8	0.001	6 m/s	Carbonaceous soil
Asteroids									
1 Ceres	1020	1.2 (21)?	0.38	4.6 y	2.77 AU	10.6	0.08	0.6	Carbonaceous soil
4 Vesta	549	2.4 (20)?	0.22	3.6 y	2.36 AU	7.1	0.09	0.3	Basaltic soil
2 Pallas	538	?	0.33	4.6 y	2.77 AU	34.8	0.24	0.3	Meteoritic soil
10 Hygiea	443	?	0.75	5.6 y	3.15 AU	3.8	0.10	0.2	Carbonaceous soil
511 Davida	341	?	0.21	5.7 y	3.19 AU	15.8	0.17	0.2	Carbonaceous soil
704 Interamnia	338	?	0.36	5.4 y	3.06 AU	17.3	0.15	0.2	Unidentified soil
Jupiter	142,984	1.90 (27)	0.41	11.9 y	5.20 AU	1.3	0.048	60	Liquid hydrogen?
J16 Metis	40	?	?	0.29	128	0.0	0.0	20 m/s	Rock?
J15 Adrastea	25 × 15	?	?	0.30	129	0.0	0.0	10 m/s	Rock?
J5 Amalthea	270 × 150	?	S	0.50	181	0.4	0.003	0.13	Sulfur-coated rock?
J14 Thebe	120? × 90	?	?	0.67	222	0.0	0.0	60 m/s	Rock?
J1 Io	3630	8.94 (22)	S	1.77	422	0.0	0.000	2.6	Sulfur compounds
J2 Europa	3138	4.80 (22)	S	3.55	671	0.5	0.000	2.0	H_2O ice
J3 Ganymede	5262	1.48 (23)	S	7.16	1070	0.2	0.001	3.6	H_2O ice, dust
J4 Callisto	4800	1.08 (23)	S	16.69	1883	0.2	0.008	2.4	Dust, H_2O ice
J13 Leda	8?	?	?	239	11,094	26.7	0.146	4 m/s?	Carbonaceous rock?
J6 Himalia	180	?	0.4	251	11,480	27.6	0.158	90 m/s?	Carbonaceous rock?
J10 Lysithea	40	?	?	259	11,720	29.0	0.130	20 m/s?	Carbonaceous rock?
J7 Elara	80	?	?	260	11,737	28.0	0.207	40 m/s?	Carbonaceous rock?
J12 Ananke	30	?	?	631	21,200	147R	0.17	16 m/s?	Carbonaceous rock?
J11 Carme	44	?	?	692	22,600	163R	0.21	20 m/s?	Carbonaceous rock?
J8 Pasiphae	70	?	?	735	23,500	148R	0.38	40 m/s?	Carbonaceous rock?
J9 Sinope	40	?	?	758	23,700	153R	0.28	20 m/s?	Carbonaceous rock?
Saturn	120,536	5.69 (26)	0.43	29.5 y	9.54 AU	2.49	0.056	36	Liquid hydrogen?
S15 Atlas	38 × 28	?	?	0.60	138	0.0	0.000	13 m/s?	Ice?
S16 Prometheus	140 × 74	?	?	0.61	139	0.0	0.002	50 m/s?	Ice?
S17 Pandora	110 × 66	?	?	0.63	142	0.0	0.004	35 m/s?	Ice?
S11 Epimetheus	140 × 100	6 (17)?	S	0.69	151	0.3	0.009	50 m/s?	Ice?

Objects in the Solar System, *continued*

(Sources and notes listed on page 160)

Object	Equatorial Diameter (km)	Mass (kg)[a]	Rotation Period[b] (d)	Orbital Period (days unless marked)	Distance from Primary (10^3 km unless marked)	Orbit Inclination[b,c] (degrees)	Orbit Eccentricity	Escape Velocity (km/s unless marked)	Known or Probable Surface Material
S10 Janus	220 × 160	2 (18)?	S	0.69	151	0.1	0.007	70 m/s?	Ice?
S1 Mimas	394	3.8 (19)	S	0.94	186	1.5	0.02	0.2	Mostly H_2O ice
S2 Enceladus	502	8.4 (19)	S	1.37	234	0.0	0.00	0.2	Mostly H_2O ice
S3 Tethys	1048	7.6 (20)	S	1.89	295	1.1	0.00	0.4	Mostly H_2O ice
S13 Telesto	≈25 × 11	?	?	1.89	295[d]	0	0	7 m/s?	?
S14 Calypso	30 × 16	?	?	1.89	295[d]	0	0	9 m/s?	?
S4 Dione	1118	1.0 (21)	S	2.74	377	0.0	0.00	0.5	Mostly H_2O ice
S12 Helene	36 × 20	?	?	2.74	377[d]	0.2	0.00	11 m/s?	Mostly H_2O ice
S5 Rhea	1528	2.5 (21)	S	4.52	527	0.4	0.00	0.7	Mostly H_2O ice
S6 Titan	5150	1.3 (23)	?	15.94	1222	0.3	0.03	2.7	Ices, liquid NH_3 & CH_4
S7 Hyperion	350 × 200	?	chaotic	21.28	1481	0.4	0.10	0.1	Ices?
S8 Iapetus	1436	1.9 (21)	S	79.33	3560	14.7	0.03	0.6	Ice and soil
S9 Phoebe	230 × 210	?	0.4	550.5	12,930	150R	0.16	0.1	Carbonaceous soil
Asteroid/Comet 2060 Chiron[e]	200?	?	0.25	50.7 y	13.70 AU	7.0	0.38	0.1?	Carbonaceous soil (?) and volatile ices
Uranus	50,800	8.76 (25)	0.72R	84.0 y	19.18 AU	0.8	0.05	21	?
U6 Cordelia	26	?	?	0.34	49.3	0	0	14 m/s?	Ice and soil
U7 Ophelia	32	?	?	0.38	53.3	0	0	17 m/s?	Ice and soil
U8 Bianca	44	?	?	0.44	59.1	0	0	23 m/s?	Ice and soil
U9 Cressida	66	?	?	0.46	61.75	0	0	35 m/s?	Ice and soil
U10 Desdemona	58	?	?	0.48	62.7	0	0	31 m/s?	Ice and soil
U11 Juliet	84	?	?	0.49	64.35	0	0	44 m/s?	Ice and soil
U12 Portia	110	?	?	0.52	66.09	0	0	58 m/s?	Ice and soil
U13 Rosalind	58	?	?	0.56	69.92	0	0	31 m/s?	Ice and soil
U14 Belinda	68	?	?	0.62	75.10	0	0	36 m/s?	Ice and soil
U15 Puck	160 × 150	?	?	0.76	85.89	0	0	126 m/s?	Ice and soil
U5 Miranda	484	7 (19)	S	1.41	130	3.4	0.02	0.4	H_2O ice, soil
U1 Ariel	1160	1.4 (21)	S	2.52	192	0	0.00	0.7	H_2O ice, soil
U2 Umbriel	1190	1.2 (21)	S	4.14	267	0	0.00	0.6	H_2O ice, soil
U3 Titania	1600	3.4 (21)?	S	8.71	438	0	0.00	1.1	H_2O ice, soil
U4 Oberon	1550	2.9 (21)	S	13.46	586	0	0.00	1.0	H_2O ice, soil
Neptune	48,600	1.03 (26)	0.67	164.8 y	30.07 AU	1.8	0.01	24	?
N3 Naiad	54	?	?	0.30	48.2	4.5	?	26 m/s?	Ice and soil
N4 Thalassa	80	?	?	0.31	50.0	<1	?	48 m/s?	Ice and soil
N5 Despina	150	?	?	0.33	52.5	<1	?	74 m/s?	Ice and soil
N6 Galatea	180	?	?	0.40	62.0	<1	?	84 m/s?	Ice and soil
N7 Larissa	190	?	?	0.55	73.6	<1	?	106 m/s?	Ice and soil
N8 Proteus	400	?	?	1.12	117.6	<1	?	222 m/s?	Ice and soil
N1 Triton	2705	2.1 (22)	S	5.88	354	159R	0.00	2.5	CH_4 ice
N2 Nereid	400	?	?	360.2	5515	27.6	0.75	0.2?	CH_4 ice
Pluto	2300	1.5 (22)?	6.4	247.7 y	39.44 AU	17.2	0.25	0.9?	CH_4 ice
P1 Charon	1190	2 (21)?	S	6.39	19	0R	0.00	0.6?	CH_4 ice?

TABLE 8-1

Sources and Notes

Sources: Data from IAU announcements 3463–3476; Voyager team reports (*Science,* 1979, *204,* 964ff.; 1979, *206,* 934ff.; 1981, *212,* 159ff., *246,* 1422ff.); Reitsema, Smith, and Larson (*Icarus,* 1980, *43,* 16); Dunbar and Tedesco (on Pluto; *Astron. J., 92,* 1201); data tables in *Satellites,* ed. J. Burns, 1986; Thomas and others, 1989, on small Uranus satellites. S18 discovered in 1990; designation "Pan" is suggested but not yet official.

Notes: Numbers assigned to asteroids and outer planets' satellites indicate order of discovery, except for largest satellites. The tables in this book use a system of symbols to indicate data that is approximate (~), uncertain (?), and not available or not applicable (—).

[a]Numbers in parentheses are powers of 10.

[b]An R in this column indicates retrograde motion; S indicates that synchronous rotation has been confirmed.

[c]To ecliptic for planets; to planet equator for satellites.

[d]S12 is in the leading Lagrangian point of Dione's orbit. S13 and S14 are in following and leading Lagrangian points of Tethys' orbit, respectively. Lagrangian points are stable points for small bodies in larger bodies' orbits, and are discussed further in Chapter 13.

[e]D. Tholen, W. Hartmann, and D. Cruikshank discovered anomalous brightening of this strange object in 1988, probably due to cometary activity. Karen Meech and M. Belton found a coma (a gas and dust cloud emitted by a comet) in 1989. Though it is cataloged as an asteroid, Chiron turns out to be the largest known comet nucleus!

★ Here are two planets with nearly the same mass and size, but they have radically different atmospheres and climates. This is well shown by the color views on our web site. The comparative planetologist tries to understand what processes control the similarities and differences of the various planets and moons. Are the properties controlled mostly by distance from the Sun, size, composition, or what? Researchers then try to use this knowledge to clarify our understanding of the Earth. For example, Venus and Earth are about the same size, but the atmosphere of Venus has much more carbon dioxide than Earth's atmosphere, so Venus may offer some clues about the effects of the increase in CO_2 in Earth's atmosphere due to pollution.

★ More possibilities for comparative planetology may be glimpsed from Figures 8-1 and color images of planets. Go to our web site and study the different worlds shown at the same scale. Orange deserts on Mars contrast with the gray lavas of the Moon. Jupiter's icy-gray moon Ganymede (lower left) contrasts amazingly with Saturn's same-size moon, orange smog-covered Titan. Explaining such differences clarifies not only the individual worlds but also the whole system of worlds, including our own. Each world tells its own story and a story of connections as well. In the next few chapters, we will explore these stories one at a time.

THE PLANET MERCURY

We will start our survey of the planets with the planet closest to the Sun—Mercury. Mercury is the second smallest planet in the solar system. It is only about 40% the size of Earth and about 40% bigger than the Moon. Figure 8-2, showing Mercury passing in front of the Sun, emphasizes its tiny size seen from Earth. For this reason, little was known about Mercury prior to the era of space exploration. A telescope with an aperture larger than about 15 cm (6 in.) reveals that Mercury has phases (Figure 8-3), very faint dusky markings, and a pinkish-gray cast.

In 1974 and 1975, the American spacecraft Mariner 10 sailed past Mercury on three different occasions. Due to Mariner 10's orbit, its cameras were able to record details on only about half of Mercury's surface. But this was sufficient to reveal that Mercury is a world much like the Moon, pocked with craters, marked with giant multiring basins and lava flows, and with virtually no atmosphere. From the standpoint of comparative planetology, therefore, Mercury provides a valuable example of a planet intermediate in size between the Moon and the Earth. Before we go on to show how Mercury's size affects its surface features, we pause to consider some rather peculiar properties of Mercury's motions.

Rotation and Revolution of Mercury

A curious relationship exists between Mercury's rotation and its orbital revolution around the Sun. Due to complex tidal effects (similar to those that hold the Moon facing the Earth), Mercury has gotten locked into a 59-d period of rotation, which is just two-thirds of its 88-d orbital revolution period. The combination of these two rates means that the Sun moves very slowly across Mercury's sky, taking 176 d to go from noon until the next noon!

This can be seen better from Figure 8-4, in which we visualize Mercury at position 0, where the noontime Sun is overhead at a certain mountain. At position 1, Mercury has completed a quarter of its rotation relative to the stars. By position 3, Mercury is halfway through its orbit, 44 d have elapsed, and the Sun is just setting on the mountain. By position 4, Mercury

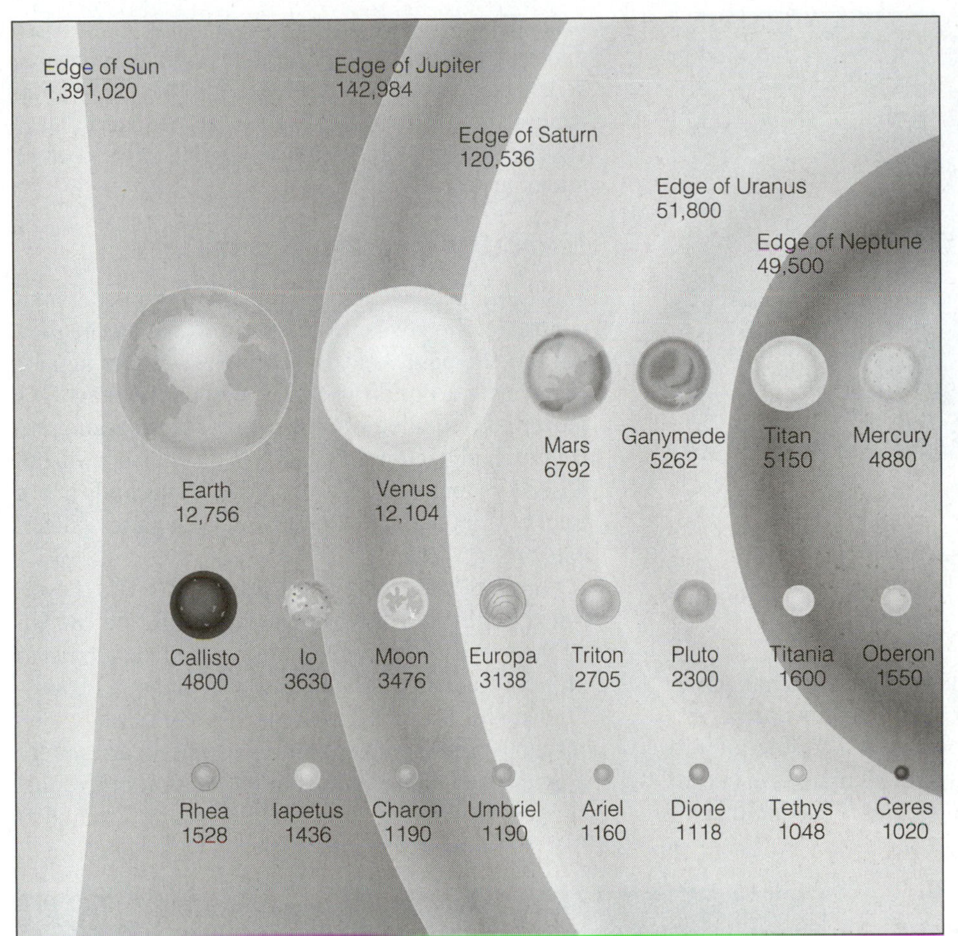

Figure 8-1 The 27 largest bodies in the solar system, shown to true relative scale. They include 1 star, 9 planets, 16 satellites, and 1 asteroid. Diameters are in kilometers.

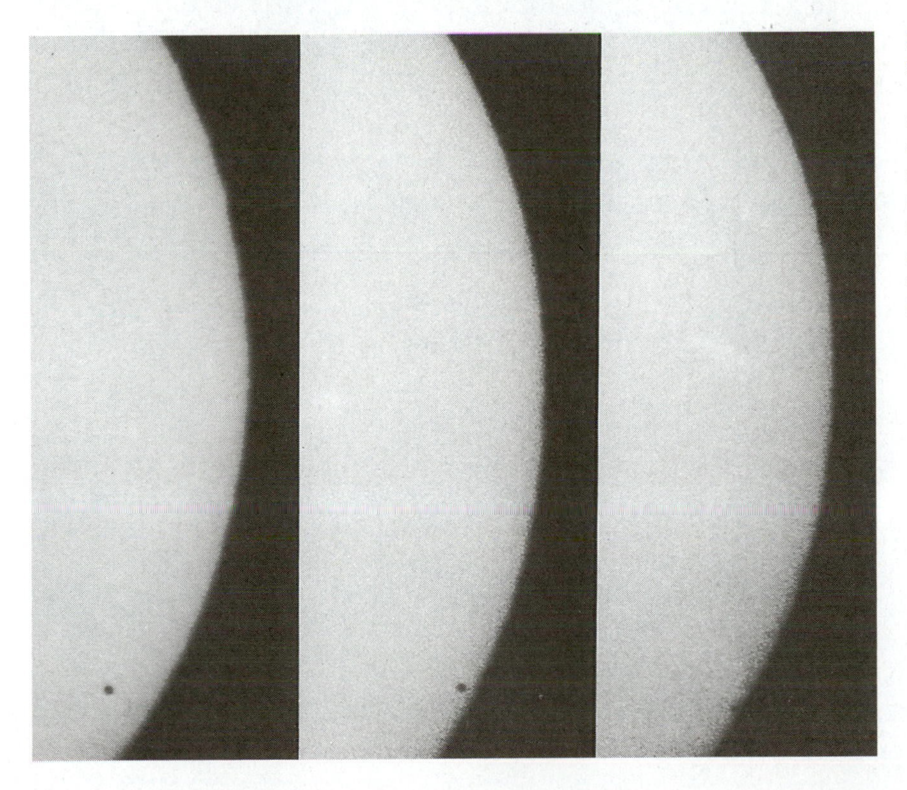

Figure 8-2 Three photos of Mercury as it transited the Sun on November 10, 1973. Mercury (the small black disk at bottom) is moving off the disk of the Sun. These photos show the small size of Mercury's disk during its closest approach to the Earth, illustrating how poor an earthbound observer's view of the planet is. (Photos with a Questar telescope by W. A. Feibelman)

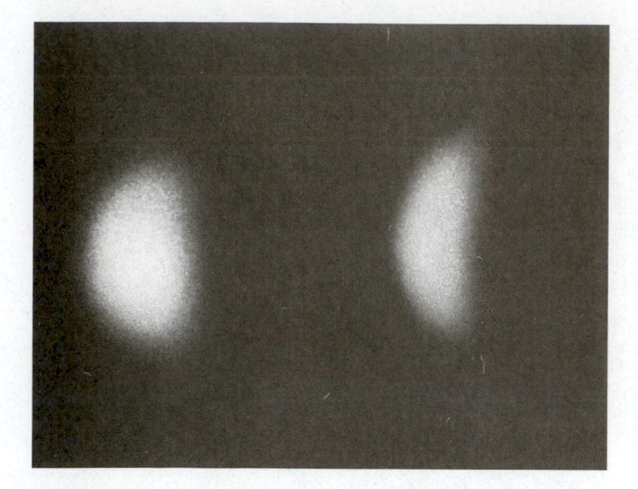

Figure 8-3 Two photographs of Mercury (June 7 and 11, 1934) showing the changing phases of the planet. Virtually no surface detail can be seen; very few photos from the Earth show reliable detail. (Lowell Observatory)

is two-thirds of the way around its orbit, but it has completed one rotation, since the mountain now points in the same direction as in position 0. By position 6, Mercury has gone once around the Sun in 88 d, but it is now midnight on the mountain, since the mountain faces directly away from the Sun. Another 88 d are necessary to bring the mountain back to noontime, as you can confirm by sketching in the next orbital trip, positions 7 through 12. Thus the Mercurian "day" (from sunrise to sunrise) is not the 59-d period of rotation but 176 d.

Mercury and Dr. Einstein

At first glance, Mercury's orbital motion seems to contradict the laws of Kepler and Newton: The *perihelion*—the point nearest to the Sun—shifts in position slowly around the Sun from year to year. This movement is called **orbital precession.** Some precession had been predicted from Newton's laws, but observers found an excess shift of 43 seconds of arc each century—a tiny amount, but enough to consternate orbital theorists!

Around 1860, the French astronomer U. J. Leverrier thought that the excess shift might be caused by the gravity of a small, undiscovered planet inside Mercury's orbit. He even gave the planet a name—Vulcan (Moore, 1954). Leverrier had already successfully predicted Neptune's existence from similar gravitational disturbances in the motion of Uranus, but he was wrong in the case of Mercury. Twentieth-

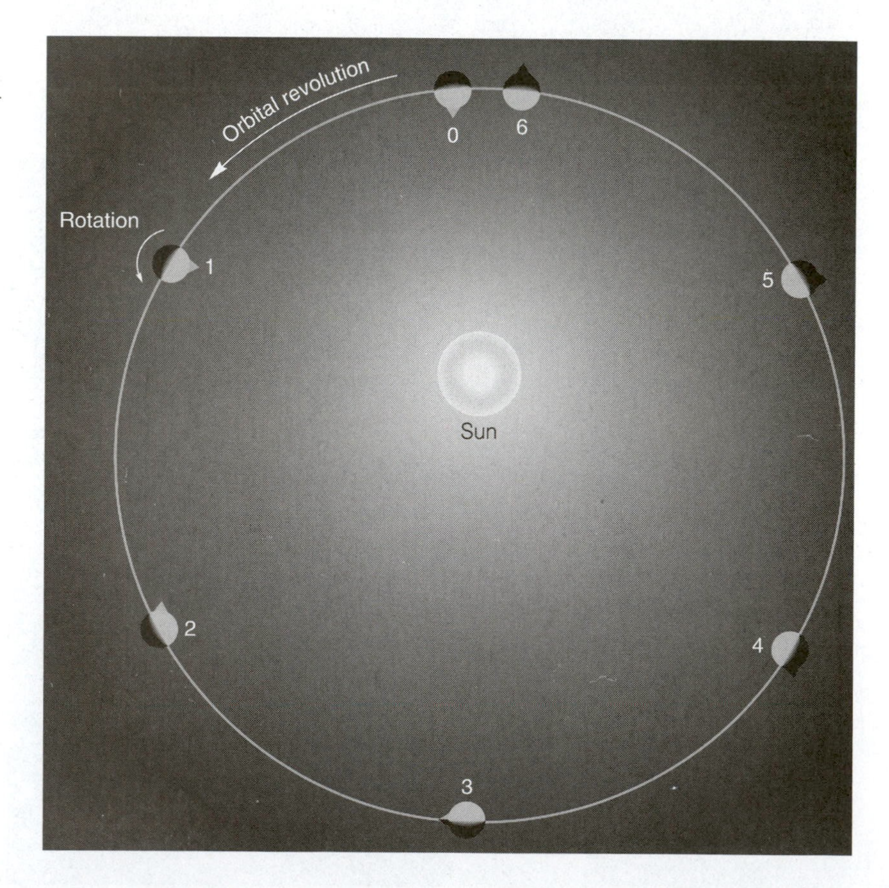

Figure 8-4 A view of Mercury's orbit showing the relationship between its rotation and revolution. The cartoons represent Mercury as a globe with one high mountain marking the rotation position. Shading shows the nighttime side. See the text for a discussion.

Figure 8-5 Mariner 10 spacecraft photo of Mercury shows a heavily cratered planet that resembles the Moon. Compare with Figure 7-9. (NASA)

century observations revealed no planet inside Mercury's orbit.

But how can Mercury's orbital precession be explained? The solution came in 1915, when Albert Einstein showed that the great mass of the Sun disturbs the orbits of nearby planets in a way unpredicted by Newton's laws. Einstein's theory of relativity predicted almost exactly the excess precession observed—43.03 seconds of arc per century. Einstein predicted smaller excesses for Venus and Earth, and these too were confirmed by observation. Thus Einstein's contribution to solving the puzzle of Mercury's precession played a major role in the acceptance of his theory of relativity.

Surface Properties of Mercury

Because Mercury is so close to the Sun, its daytime surface is much hotter than Earth's or the Moon's. Measurements by infrared detectors of the thermal radiation from both the day and night side of Mercury show that temperatures in the upper few millimeters of soil range well above 500 K (441°F) in the "early afternoon" near perihelion to lows of about 100 K (–279°) at night. In some areas, depending on soil type, the temperature might exceed 600 K (621°F). Early science fiction used to describe hypothetical pools of molten metal on the daylight side, but this now seems unlikely for several reasons. First, the melting point of lead is about 600 K; that of aluminum is 832 K. Second, because of the insulating effect of the overlying soil, the temperature slightly below the surface is only 314 to 446 K.

Mariner 10 produced the best available data about Mercury, including photographs of the surface (Figure 8-5) and magnetic measurements. The surface resembles the Moon's in its abundance of rugged craters and in its huge, multiringed craters known as basins. A few vague features glimpsed from Earth correlate with certain bright and dark features photographed

Figure 8-6 The Caloris basin on Mercury. The center of the basin lies in shadow out of the frame to the left. The left half of the frame is dominated by curved cliffs and fractures surrounding the impact site. The photo from top to bottom covers an area of about 1300 km; the cliffs are believed to be about 2 km (6000 ft) high. Compare with similar lunar features in Figure 7-12. (NASA)

by Mariner 10. The most prominent impact basin, shown in Figure 8-6, is named the Caloris basin (in keeping with the high Mercurian temperatures). Its concentric ring structure, more than 1200 km in diameter, strongly resembles the Orientale ringed basin on the Moon (see Figure 7-12). Evidently, the same impact and lava flow processes occurred on Mercury as on the Moon some 3.5 to 4.5 billion years ago. From this discovery, most scientists believe all planets suffered an intense bombardment by interplanetary debris at the close of the planet-forming process.

★ Mercury's landscape probably superficially resembles the lunar landscape: Rolling, dust-covered hills have been eroded by eons of bombardment by meteorites and covered by a regolith. Dusty lava plains and fault-cliffs remind us of violent ancient volcanism. Fresh impact craters might display rugged boulders and outcrops of craggy rocks.

Mercury's Internal Properties and Tectonic Activity

Another comparison with the Moon relates to Mercury's lithosphere structure. As we indicated in the last two chapters, the lithosphere's thickness is a key to a planet's surface evolution. If the planet is small, like the Moon, it cools rapidly, and a thick, rigid lithosphere forms, preventing any internal activity from breaking through to disturb the surface. If the world is as big as the Earth, however, it takes a long time to cool, and only a thin lithosphere has had time to form. In the case of Earth, the lithosphere is so thin that it is readily broken by faulting and volcanism associated with plate tectonics.

Scientists were thus interested to study the surface features of Mercury, because it is in between the size of the Moon and Earth (but closer to lunar size).

Using the philosophy of comparative planetology, we might predict that Mercury would have a thinner lithosphere and more signs of surface tectonic disturbance than the Moon. This turns out to be correct. Cliffs such as those in Figure 8-7 appear to mark huge faults, suggesting that the lithosphere was just thin enough to fracture when Mercury contracted as it cooled. As shown in Figure 8-7b, the faults appear to be compressional faults, formed because the surface was growing smaller as Mercury contracted. Aside from these faults, which are not found on the Moon, the rest of Mercury's surface is very lunarlike, with occasional patches of smooth lava filling the larger craters.

The dense crowding of impact craters over most of Mercury indicates that, like the Moon, its surface was partly molded by the period of **early intense bombardment** from about 4.5 to 4.0 billion years ago. Mercury's cratered uplands may contain somewhat more ancient lava flows than the Moon's, reaffirming that the larger the world, the more geologic activity it has. However, the main thing to remember about Mercury is that in terms of most surface and crustal properties, it more or less resembles our Moon.

a

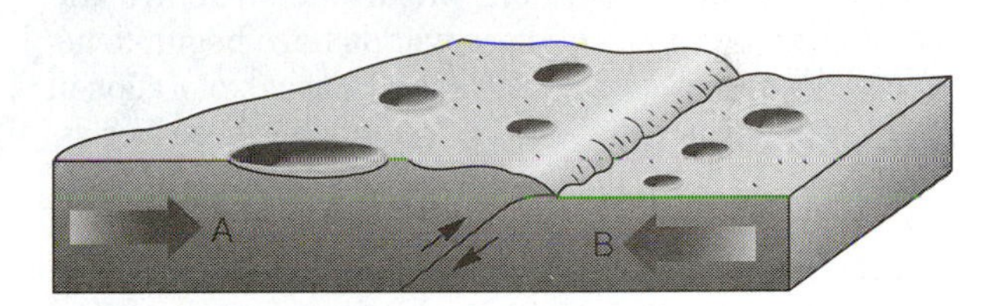

b

Figure 8-7 **a** A heavily cratered portion of Mercury resembles the lunar uplands. The cliff running from the upper left to the lower right, called Discovery Scarp, may be a fault caused by contraction of the planet following early heating and volcanism. The largest crater (left center) is about 125 km across. (NASA, Mariner 10) **b** Schematic diagram showing probable geometry of Discovery Scarp.

Mercury's Magnetism and Large Iron Core

Mercury has a weak magnetic field—about 1% as strong as Earth's. The field's magnetic axis is tilted about 7° to the planet's rotation axis, and the general form matches that for a field generated from an iron core. Thus scientists believe that Mercury has an iron core, which also accounts for its high mean density (5500 kg/m^3)—high compared to that of the Moon (3300 kg/m^3), which lacks much of an iron core. Scientists disagree, however, as to whether the core is currently molten or has cooled to a mostly solid metal state.

Atmospheric Vestiges and a Surprising Polar Cap

Mercury has essentially no atmosphere, but the planet is surrounded by a local concentration of gas atoms from several sources. Most abundant are sodium atoms, discovered in 1985 by astronomers in Texas. These atoms are probably knocked off the surface continually by gases streaming from the Sun. This does not mean that sodium is the most abundant material on Mercury's surface, but only that sodium atoms are easy to dislodge from the rock minerals because of their chemical properties. The solar gases (mostly hydrogen and helium) are also concentrated near Mercury. All these gases are so thin that an astronaut would not even notice them. They are interesting mainly in that they reveal the interaction of solar gases with the planet's surface.

★ An amazing discovery came in 1991. Because Mercury is the closest planet to the Sun's inferno, any possibility of ice on its surface had scarcely been considered. But Caltech scientists bounced radar waves off Mercury and discovered a spot exactly at its north pole that reflected radar much better than any other spot on the planet—a property that could be explained by ice on (or just under) the surface. How could this be? Because Mercury's polar axis is perpendicular to its orbit, the Sun never rises fully above the horizon, as seen from the pole—just as in the case of the Moon's pole. This means that many polar crater floors are shadowed and cold, and the polar soil temperature may be as low as 125 K (–148°C or –234°F). Careful mapping of the unusual radar spots shows that they match up with the positions of specific polar crater floors. As on the Moon, water vapor delivered to Mercury by a comet impact could condense into ice on such cold surfaces.

SUMMARY

According to present data, the solar system includes about 26 planetary worlds (in addition to the Sun) larger than 1000 km and a host of smaller bodies. These worlds show a great deal of variety, depending primarily on their mass but also on other conditions, such as distance from the Sun. Details of surface structure and composition for most of these worlds have begun to become clear only since the advent of space exploration in the 1960s. A current approach to studying these worlds, called comparative planetology, is to examine them not as isolated cases but in comparison with each other and especially in comparison with the Earth. In this way, we learn which forces determine surface environmental properties. Mercury is the closest planet to the Sun. It is the most Moon-like of the planets and is only 40% larger than the Moon. It has a heavily cratered surface that is broken in a few places by large, faulted cliffs produced by contraction. In spite of some gaseous atoms knocked off its surface and probable polar ice deposits, Mercury has virtually no atmosphere.

CONCEPTS

solar system
star
planet
satellite
terrestrial planets
giant planets
planetology
comparative planetology
orbital precession
intense early bombardment

PROBLEMS

1. Which planet can come closest to Earth? (*Hint:* See Table 8-1.)

2. How many times bigger in diameter than Earth is the largest planet? How many times more massive than Earth is the most massive planet?

3. Which planet has the largest satellite?

4. Which planet has the largest satellite measured in terms of the diameter of its planet—that is, which satellite's diameter is the largest fraction of its planet's diameter?

5. **a.** Which planet can come closer to Saturn: Earth or Pluto?

b. Advanced students might consider this same question for Uranus. *Hint:* For elliptical orbits, the closest distance to the Sun is given by $a\ (1 - e)$.

6. Using Table 8-1, list some systematic differences in physical properties between terrestrial planets and giant planets. Consider size, temperature, density, number of satellites, and orbital properties.

7. List processes of planetary evolution that have occurred on both Mercury and the Moon. List the processes, if any, that are unique to each. Is there any indication of plate tectonic activity on Mercury? What might this indicate about heat flow and mantle convection inside Mercury?

8. Some of the best telescopic observations of Mercury are made during midday instead of after sunset or before sunrise. Why? Why are none made at midnight?

9. Suppose astronauts plan to land on Mercury wearing spacesuits of the Apollo type used on the Moon. What modifications, if any, might have to be made to use the suits on either the daytime side or the nighttime side of the planet?

ADVANCED PROBLEMS

10. Calculate and compare the orbital velocities of the Earth (1.5×10^{13} cm from the Sun) and Pluto (about 5.9×10^{14} cm from the Sun). The Sun's mass is 2.0×10^{33} g.

11. Suppose the Sun were replaced by a star four times as massive, and the Earth were moved to a new orbit 2 AU from this star. Compare the gravitational force on this "new earth" to that in the present solar system.

12. When earthbound observers can see Mercury partly illuminated, it is at a distance of about 0.9 AU, or 135 million kilometers.

a. If you used a telescope that revealed details 1 second of arc across, what would be the smallest features you could see on Mercury? (*Hint:* Use the small-angle equation.)

b. Why can virtually no detail be seen on the surface of Mercury when it is closest to Earth, about 0.61 AU away?

c. With the telescope in part (a), what would be the thinnest cloud layers visible on Venus at its closest approach to Earth?

13. Of the terrestrial planets, Mercury is most like the Moon in size and surface character. Using the data in Table 8-1, calculate the mean density of Mercury and use this to confirm that Mercury has an iron percentage much more like the Earth's than the Moon's. Comment on how this might affect a lunar origin theory in which the Moon is viewed as an ordinary planetary body captured into orbit around the Earth.

14. Using Mercury's high mean density of 5500 kg/m^3, cited in the text, suppose that Mercury's mantle has a density like the Moon or Earth's mantle, around 3300 kg/m^3, and that the core is made of iron alloys with a density of 8000 kg/m^3.

a. Describe how you might calculate the radius of Mercury's core from this information.

b. Do the calculation if you can. (*Hint:* Recall that the volume of a sphere is $\frac{4}{3}\pi R^3$, where R is the radius of the sphere.)

PROJECTS

1. Try to see Mercury. This is likely to be harder than it sounds, because Mercury is prominent only during its period of elongation, which lasts only a few days. During that period, it is visible for an hour or less each day, just after sundown or before sunrise. Determine from an almanac or an astronomy magazine when Mercury comes to evening elongation and find a site with a very clear western horizon. It is best to start a few days before the elongation so you can become familiar with the background stars in the appropriate region of the sky.

2. If Mercury becomes visible during your studies, observe it with a telescope of at least 20 cm (8 in.) aperture. Try to detect the phase of the planet. Observe on several successive days, and see if you can detect a change in phase as Mercury circles the Sun.

3. If a planetarium is nearby, arrange a demonstration of the motions of Mercury as seen from space or from Earth.

CHAPTER

9

Venus

WHAT THE READER SHOULD WATCH FOR IN THIS CHAPTER

Venus is the nearest planet to Earth and is nearly the same size as Earth, but its environment is very different. Venus' thick carbon dioxide atmosphere causes a strong greenhouse effect that results in surface temperatures of around 750 K (891°F), both day and night. Several Soviet-era Russian probes have landed on the surface and revealed barren landscapes or rock and gravel under a cloudy sky. The crustal geology does not reveal plate tectonic motions like Earth's, but rather a mysterious global resurfacing by lavas that occurred about 500–800 million years ago. Comparisons between conditions on Venus and Earth help us understand our own planet better. Venus helps reveal several important rules of comparative planetology, which state that, on average, larger planets tend to have more geologic activity, younger surface features, and more atmosphere than smaller planets. ■

Venus is widely regarded as Earth's sister planet because its size is so similar to Earth's. Venus has about 95% of the Earth's diameter (see Chapter 8 on our web site) and about 82% of its mass. It has no moons.

From the viewpoint of comparative planetology, the similarities between Venus and Earth make us eager to compare the geological and atmospheric properties of the two planets. Since Venus comes closer to Earth than any other planet, you might think that it would be one of the best-observed worlds. Yet its surface remained a mystery until space probes arrived, because Venus is completely covered by clouds. The brilliant, yellowish-white, nearly blank cloud layer is the first feature to impress a telescopic observer. It prevented early astronomers from observing a single detail on the surface, so space probes were needed to reveal the planet's secrets.

From the 1960s to the 1990s, unmanned probes orbited Venus, dangled from balloons in its atmosphere, and landed on its surface. They revealed an environment wildly different from Earth's. For example, the atmosphere has a completely different composition. Why should our neighbor planet, so similar to ours in size, be so radically different? This chapter will explain some of the differences, which are caused by the different evolutionary histories of the two planets.

THE SLOW RETROGRADE ROTATION OF VENUS

Radar signals, bounced off Venus from Earth, reveal that Venus has an unusual rotation unlike that of Mercury, Earth, or Mars. Those planets all have **prograde rotation**—they spin from west to east. Venus has **retrograde rotation**—it spins from east to west (Figure 9-1). Venus' spin is also unusual in being very slow, taking 243 d to make a complete turn on its axis. These properties were discovered in 1962, when

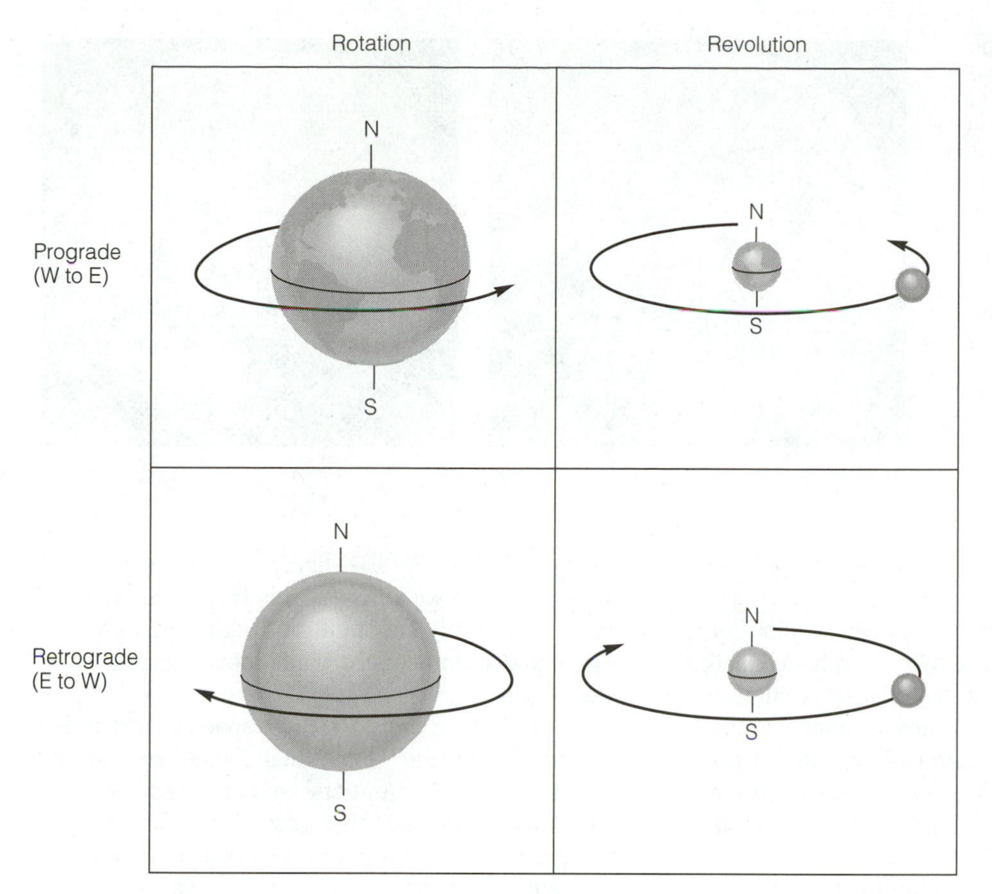

Figure 9-1 Rotation (motion around an internal axis) and revolution (motion around an external body). Prograde motion (top) is west to east. Most planets, including Earth, have prograde rotation and revolution. The rotation of Venus is retrograde, as diagrammed in the lower left figure. Remember the difference between rotation and revolution by noting that the common handgun should be called a *rotator,* not a *revolver!* Venus has retrograde rotation but prograde revolution.

radar signals were first bounced off the planet. The cause of the unusual reverse spin may involve tidal forces between the Earth and Venus or an ancient collision with a body larger than our Moon, of which there may have been many in the early solar system.

Just as was the case with Mercury, the length of the day (from sunrise until the next sunrise) on Venus is complicated by the combination of the rotation (243 d, retrograde for Venus) and the period required to go around the Sun (225 d for Venus). The net result is that the time from one sunrise until the next on Venus is 117 Earth days. As experienced on the surface of Venus, daylight would last $58\frac{1}{2}$ Earth days, followed by a hot night of $58\frac{1}{2}$ Earth days. Daybreak would be eerie and gradual, with the sky brightening slowly for a week as the Sun rose hidden beyond the cloud-blanketed sky.

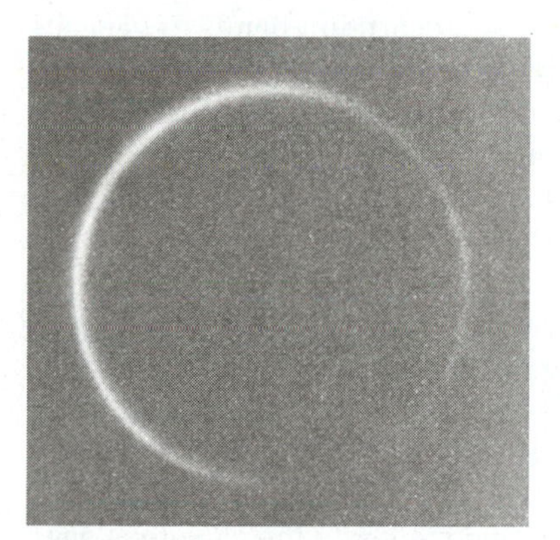

Figure 9-2 A telescope view of Venus as it passed nearly between the Earth and the Sun. Backlighting by the Sun illuminates the atmosphere of Venus all the way around the disk. This effect, first recorded in 1761, not only proves the existence of an atmosphere but gives information on its structure. (The Observatory, New Mexico State University)

VENUS' INFERNAL ATMOSPHERE

The first proof that Venus has an atmosphere came as long ago as 1761, when the Russian scientist M. Lomonosov observed the backlit atmosphere extending around the disk, as shown in Figure 9-2. This phenomenon occurs when Venus is approximately between the Earth and the Sun so that sunlight backlights the

a

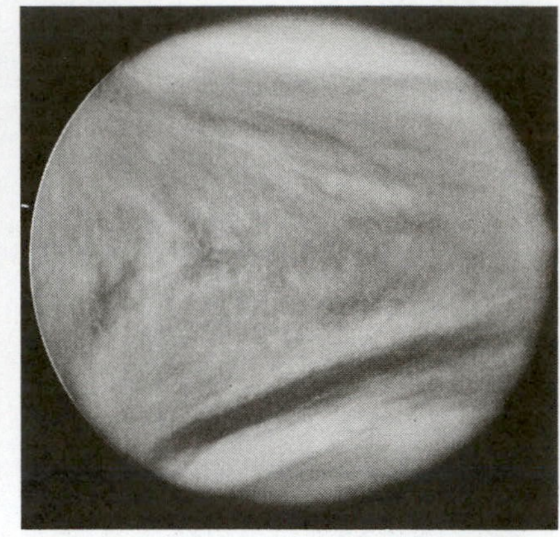
b

Figure 9-3 Circulation of Venus' atmosphere is shown in these two views taken 5 h apart. In **b,** the features have shifted to the west (left; note V-shaped dusky marking in left center). While the planet itself takes 243 d to turn, 200-mhp winds blow the clouds around the planet in about 4 d. This is a contrast-enhanced image made in ultraviolet light. In visible light, the contrast is so low that the yellowish-white clouds look nearly featureless, although telescope observers have sometimes reported the bright polar cloud cap and adjacent dark band. (NASA, Pioneer Venus Orbiter)

hazy upper atmosphere of Venus, revealing the haze layer ringing the planet.

The Venusian atmosphere intrigued scientists for two centuries after it was discovered.[1] What caused this whitish shroud? What kind of planetary surface did it hide? In 1928, the American astronomer Frank Ross photographed dusky patterns by using plates and film sensitive to ultraviolet. These patterns, shown in Figure 9-3, are formed by cloud layers differing in composition, particle size, or altitude.

In 1932, Mt. Wilson astronomers W. S. Adams and T. Dunham studied the spectrum of Venus and detected very strong absorption bands of gaseous carbon dioxide (CO_2, the same gas that is dissolved in carbonated soft drinks). This revealed that Venus' atmosphere has much more CO_2 than Earth's atmosphere. Later data showed that Venus' atmosphere is about 96% CO_2, as seen in Table 9-1.

Still undiscovered was the composition of the opaque cloud layer and the air below it. In the 1940s and 1950s, some writers imagined stormy clouds of water droplets, a surface swept by torrential rains, and vegetation like that of a Brazilian rain forest. Some supposed that the highly reflective clouds would shade the surface and moderate the climate in spite of Venus' closeness to the Sun. Around 1958–63, however, strong thermal radiation was measured at far-infrared and radio wavelengths, and Wien's law revealed that the lower atmosphere has a temperature of about 750 K (891°F)—hardly conducive to liquid water or life as we know it. This torrid temperature remains roughly constant, both day and night, because upper atmosphere winds move the air around the planet.

During the era of the Soviet space program, Soviet scientists made Venus "their planet," so to speak, while American scientists concentrated more on the Moon, Mars, and other targets. The Soviet probe Venera 4, which crashed on Venus in 1967, was the first human artifact to touch another planet. In 1970, **Venera 7,** a Russian probe, became the first spacecraft to land successfully on another planet. It transmitted data from the Venusian surface for 23 min. The data confirmed the high temperature and revealed an atmospheric pressure about 90 times as great as Earth's! Instead of our 101,000 N/m^2 (14.7 lb of force pressing on every square inch of surface), the pressure on Venus is about 9,000,000 N/m^2 (1320 lb/in.^{2}), equivalent to that endured by a diver nearly a kilometer (3000 ft) below the terrestrial ocean surface! Later Soviet spacecraft landings confirmed these results. Venus is a stranger world than most humans had imagined!

Venus was soon revealed to be stranger yet. In 1972–73, astronomers discovered that the clouds of Venus consist not of water droplets, like Earth's clouds, but rather of tiny droplets of sulfuric acid (H_2SO_4)! In 1978, probes dropped by the American spacecraft Pioneer Venus showed that the clouds lie primarily in a high layer 48 to 58 km above the surface.

In 1985, two balloons (dropped by Soviet probes on their way to Halley's comet) floated in the clouds for 46 h, measuring hurricanelike winds (150 mph) but relatively pleasant conditions at this altitude (T = 95°F and pressure like that on Earth's surface).

[1]Adjectives for planets are controversial. Some writers, claiming that *Venusian* is an ugly word, use *Cytherean* (from a name of Aphrodite), which is merely confusing. *Venereal* is already preempted by other areas of human endeavor. While *Venusian* (ve-NOO-sian) has the sanctity of science fiction tradition, many astronomers use *Venerian* or the noun *Venus* as an adjective.

TABLE 9-1

Atmospheres of Venus and Earth

Venus		Earth	
Gas	**Percent Volume**	**Gas**	**Percent Volume**
Co_2	96.5	N_2	78.1
N_2	3.5	O_2	20.9
SO_2	0.015	H_2O	0.05 to 2 (variable)
H_2O	0.01	Ar	0.9
Ar	0.007	CO_2	0.03
CO	0.002	Ne	0.0018
He	0.001	He	0.0005
O_2	≤0.002	CH_4	0.0002
Ne	0.0007	Kr	0.0001
H_2S	0.0003	H_2	0.00005
C_2H_6	0.0002	N_2O	0.00005
HCI	0.00004	Xe	0.000009

Note: Compositions are for near-surface conditions, with terrestrial data other than H_2O tabulated for dry conditions. CO_2 on Earth is probably increasing by 2% to 3% of the listed amount in each decade, because we are burning so much fossil fuel. This activity may be modifying Earth's climate.

Sources: Oyama and others (1979); von Zahn and others (1983).

Venus' clouds are higher than Earth's, which are mostly less than 10 km high.

Recall that in the case of Earth, Coriolis drift associated with the 24-h rotation causes a shearing of north- and south-moving cloud masses, which helps produce spiral-shaped cyclonic cloud whorls (see page 125). On Venus, with its slow rotation and fast east-west jet-stream winds, there is negligible Coriolis drift and virtually *no spiral cloud patterns.* Of course, the white cloud masses look nearly featureless to the naked eye in any case, but ultraviolet photos show that the winds shear the clouds into fairly straight east-west bands that are nearly parallel or slightly inclined to the equator, as seen in Figure 9-3.

★ In spite of the differences, the clouds of Venus and Earth form in a similar way, in atmospheric layers where the temperature and pressure cause condensation of some relatively minor atmospheric constituent. On Earth, this constituent is H_2O, condensed either into droplets (lower clouds) or ice crystals (high cirrus clouds). On Venus, it is H_2SO_4 droplets, which begin to fall as they grow. If a droplet gets big enough, it falls out of the cloud deck, where it encounters much higher temperatures and evaporates. Thus Venus' weird H_2SO_4 rain never reaches the ground. This explains why the clouds have a well-defined bottom surface, as detected by the Pioneer probes, and why the lower atmosphere and surface are clear, as found by the Soviet Venera landers.

The Greenhouse Effect on Venus

Chapter 6, about the Earth, described the **greenhouse effect.** Review Figure 6-16 and recall that sunlight comes into Earth's atmosphere and heats the ground (producing outgoing thermal infrared radiation), but that CO_2 absorbs the outgoing infrared rays, causing a warming effect.

Because Venus has an incredibly dense CO_2 atmosphere, *Venus has an extremely strong greenhouse effect, explaining the extraordinarily high temperature of the air near Venus' surface* (Figure 9-4). Studies of Venus are thus important, because they confirm that the greenhouse effect is a real planetary effect, not just a wild theory of environmental pessimists (as is sometimes charged). Without Venus' thick CO_2 atmosphere, its surface would be much cooler.

We are fortunate that nature has provided us with a CO_2-rich planet next door, because it allows us to check our theoretical understanding of the greenhouse effect on planets. Carl Sagan (later to become the twentieth century's most effective popularizer of astronomy) and others first called attention to the important role of CO_2 as a greenhouse gas on Venus around 1963. Of course, Earth has much less CO_2 than Venus, but human activity is adding significant amounts of the gas to the atmosphere. The important issues today are to understand how much heating occurs for a given amount of added CO_2 and whether other effects help offset the heating. In effect, Venus acts as a valuable natural laboratory for testing our understanding of potential environmental changes on Earth.

Why Venus Has a CO_2 Atmosphere

Compare the columns in Table 9-1. Why should the atmosphere of Venus—a nearby planet nearly the size of ours—be mostly CO_2 instead of N_2/O_2 like Earth's? The answer is clearer if we rephrase the question: Why does Earth *not* have a massive CO_2 atmosphere?

The groundwork for answering this question was laid in Chapter 6, where we saw that Earth's early

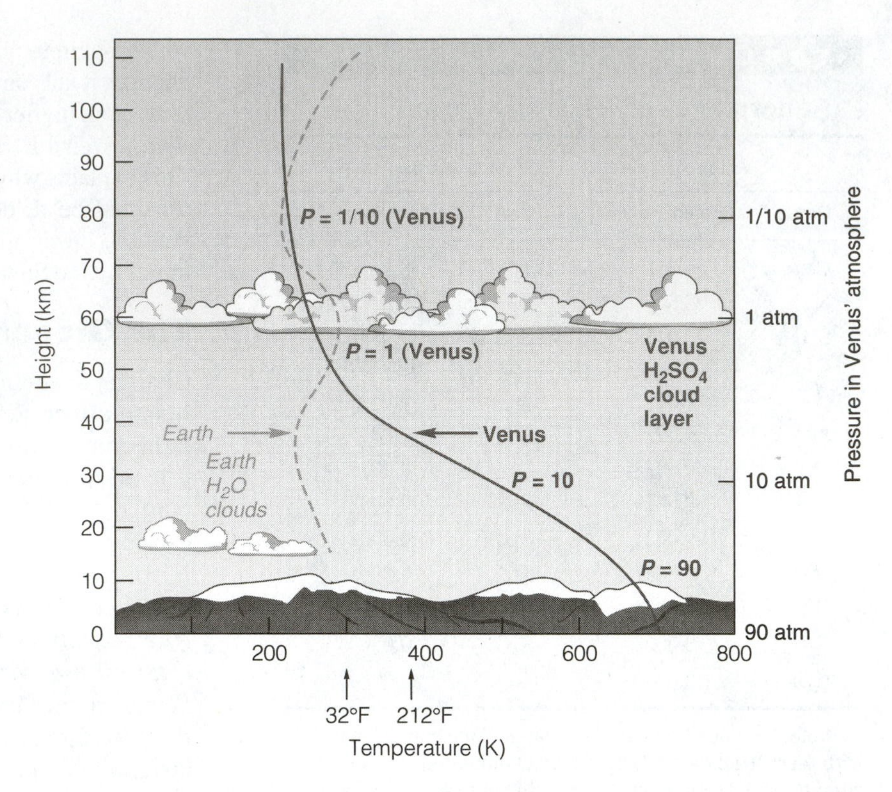

Figure 9-4 Temperature and pressure of Venus' atmosphere compared to Earth's atmosphere. Italic features refer to Earth; bold, to Venus. The intense greenhouse effect on Venus strongly heats its atmosphere below 50 km. Pressures (*P*) are measured in units of Earth's surface pressure (known as bars); they are much higher on Venus. Note the interesting coincidence that a layer just above Venus' clouds has close to room temperature and room pressure (but CO_2 instead of N_2/O_2 in the atmosphere).

dense atmosphere was mainly H_2O, CO_2, and N_2, all from volcanoes. On primordial Earth, the H_2O condensed into oceans. The CO_2 dissolved in the oceans, ending up primarily in the form of carbonate rocks. After plant life evolved, CO_2 was also consumed by plants, buried in deposits of plant and animal remains, and ended up in carbonate-rich sediment deposits, such as limestone. The important point is that *Earth has just as much CO_2 as Venus, but instead of being in the atmosphere, it is tied up chemically in the form of carbonate (CO_3-bearing) rocks.* Earth was left with a much thinner, N_2-rich atmosphere, to which O_2 was added by plants.

Imagine an Earth-like planet that lacks an ocean or plants; the CO_2 emitted by its volcanoes would not dissolve or be consumed; it would be left as the dominant atmospheric gas, along with H_2O molecules. If the H_2O molecules then disappeared in some fashion, an atmosphere of nearly pure CO_2 would be left.

That is probably what happened on Venus long ago. According to most researchers, Venus' primordial volcanoes emitted CO_2 and H_2O, but because Venus was closer to the Sun, it was too hot for the H_2O to condense into rain and form oceans. Therefore, the CO_2 could not dissolve in oceans and simply accumulated in the atmosphere, building up to a crushing 90 times our surface pressure. The H_2O molecules, however, were easily broken into H and O atoms by intense ultraviolet photons from the Sun. The H atoms, being light, floated to the top of the atmosphere and escaped into space, a process well documented by atmospheric chemists. With the loss of the hydrogen long ago Venus lost its chance to have water.

There is some interesting direct evidence for this. Hydrogen comes in two forms, or isotopes: ordinary hydrogen and *heavy hydrogen,* whose atoms have one extra neutron. The lighter hydrogen, of course, floats to the top of the atmosphere sooner and escapes faster than the heavier hydrogen. Thus more of the heavy hydrogen is left behind. If the water breakup and hydrogen escape process really happened on Venus, we should see more of the leftover heavy hydrogen in Venus' atmosphere than in Earth's atmosphere, where the water and its H atoms were never lost. Sure enough, probes on Venus have measured 150 times as much heavy hydrogen (relative to ordinary hydrogen) as on Earth. Using this information, many geochemists believe Venus originally had water when it formed; if the water had condensed into rain, there would have been enough to make oceans from 0.5 to 25 m deep, according to various theoretical reconstructions (Cowen, 1993; Bullock and Grinspoon, 1999). There probably never were such seas, however, because of the heat; the water molecules remained in the form of vapor until they were broken by solar ultraviolet photons and lost long ago.

★ What happened to Venus' leftover oxygen? Russian and American scientists, studying the data on rocks and soils returned by the various Venus landers, believe that most of the oxygen, being chemically very reactive, combined with rock minerals and disappeared from the air. Some scientists have a different version of the same story, because chemical details of how Venus lost its water have been hard to explain. These scientists believe the material from which Venus formed, being closer to the Sun, had few water molecules to begin with. According to this hypothesis, Venus formed dry, and its volcanoes emitted CO_2 alone, so Venus never had to get rid of its water or oxygen molecules. In this view, the extra heavy hydrogen came from material in comets, which have crashed into Venus during the eons of geologic time.

Future spacecraft measurements of the amount of oxygen in Venus' rocks and of the exact chemistry of the atmosphere may tell us which interpretation of the planet's history is correct. In any case, Venus is very dry and hot today!

Notice a general rule that emerges from this discussion. Terrestrial planets are likely to have volcanically emitted CO_2 atmospheres, and the CO_2 will be retained as the main atmospheric gas unless liquid water oceans can gain enough depth and permanence to dissolve the CO_2—in which case, it will disappear into carbonate rocks.

THE ROCKY LANDSCAPES OF VENUS

Following their 1970 success with Venera 7, the first landing probe on Venus, the Soviets landed a number of additional probes. As expected, the intense heat caused instrument failures after a matter of minutes, but the probes sent back much interesting information about the hidden surface. For example, photos were sent back from Veneras 9 and 10 in 1975 and from Veneras 13 and 14 in 1982. As shown in Figure 9-5, these panoramic photos stretched at an angle from the horizon to the soil near the spacecraft. The clarity of the photos surprised scientists and proved that although Venus has opaque high clouds, the surface is free from haze. The photos revealed stark, dramatic landscapes under a high, bright sky. Of course, there were no plants or life-forms, because of the heat. Angular boulders, gravel, flat outcrops of slabby rock, and fine soil appeared at various sites. The varied states of erosion, with weathered soils at some sites, suggest ongoing geologic erosion processes. Very similar rock formations can be found among lightly volcanic lava flows on Earth, as shown on our web site.

Chemical measurements made by seven Soviet landers (Veneras 8, 9, 10, 13, and 14 and VEGAS 1 and 2) confirm that the surface is primarily volcanic lava. Compositions at five sites resemble basaltic lava found on Earth's seafloor crust and somewhat like that of the lunar lava plains. A sixth site showed a slightly different chemistry, but still close to basaltic composition. The seventh site (Venera 8) was the most provocative. It showed a more granitic composition, similar to that of Earth's continental rocks (Saunders and others, 1991). This raises the question of whether Venus, like its sister, Earth, has a certain fraction of its surface covered by granitic continents. The answer is still uncertain (see below).

Although jet-stream winds (up to 185 mph) were measured near the cloud layer at altitudes around 40 km, four lander probes found only gentle breezes of $\frac{1}{2}$ to 3 mph at the surface (Kerzhanovich and Marov, 1983). Nonetheless, more recent orbiting probes, which mapped the surface by radar, detected dunes in scattered locations, proving that occasional winds are strong enough to blow dust around.

★ The landscape of Venus has orange-brown tones (as seen on our web site) because it is bathed in orangish light that filters through the clouds. Direct sunlight never falls on Venus' rocks because of the high overcast. Detailed studies of the color images, however, reveal that the *intrinsic* color of the rocks is a more neutral gray. The orangish colors come from the orangish light, just as your friend's gray shirt looks red in the light of a red neon sign. These same rocks would look gray if the astronaut turned a normal, white-light spotlight on them at night.

Mapping Venus by Radar

Even after the Soviet probes returned photos from a few surface sites, scientists had no idea of the planet-wide topography. Detecting and mapping Venus' surface features was a real challenge, because it is totally obscured by clouds. Were there mountain ranges, plains, raised continental areas, or unfamiliar features? This question was answered by sending radar through the clouds and bouncing it off the surface. Imagine bouncing a radar wave off the surface from

Figure 9-5 Six landscapes on Venus: views toward the horizon (top) from Russian Venera landers. Parts of landers appear in the bottom of views *a, b,* and *f.* **a** First photo from the surface of Venus, showing loose boulders near Venera 9. **b** and **c** Boulders and gravel near Venera 13. **d** Platey rock outcrops and gravel near Venera 13. **e** and **f** Platey rock surfaces near Venera 14. (Photos *b-f* courtesy C. Florensky and A. Basilevsky. Vernadsky Institute, Moscow)

an orbiter flying over the surface at a constant altitude. If the radar wave comes back quickly, the surface is closer to you and thus has higher altitude. If the radar echo takes a longer time to return, the surface must be lower. Thus it is easy to map the surface relief by means of radar. Moreover, the character of the radar wave varies with different materials, such as solid smooth rock or loosely packed soils. Thus it is also possible to map the surface geology by radar.

Radar mapping provided remarkable images of diverse topography. Some areas show similarities to Earth, but there are also intriguing differences. The American Pioneer Venus mission made the first global radar map in 1978–80; it was followed by a 1983 Soviet mission and detailed U.S. mapping by the Magellan mission in 1990–92. The radar map is shown in Figure 9-6. It shows that about 60% of the planet is covered by low, rolling plains, with only about a kilometer (3000 ft) of relief. The other 40% is covered by

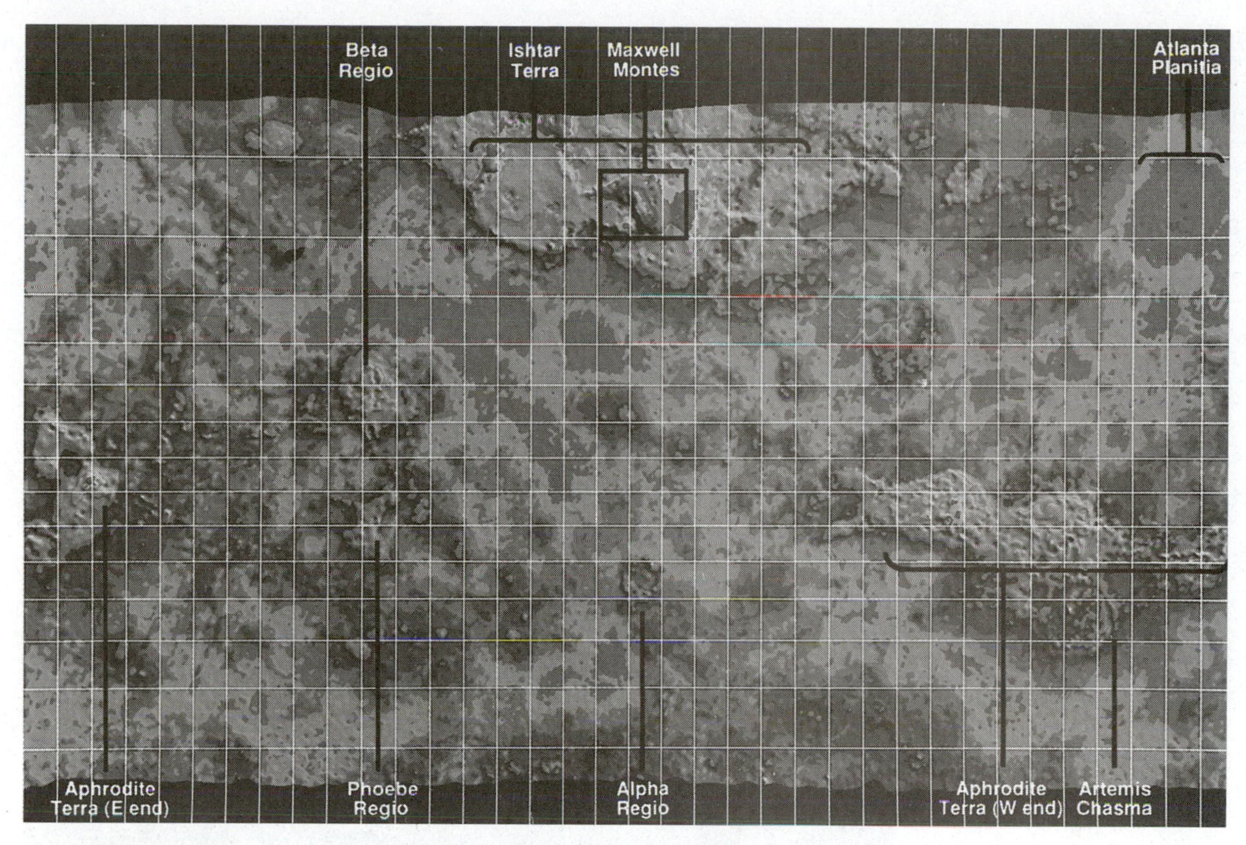

Figure 9-6 Radar map of the surface of Venus. Radar signals penetrated the clouds and measured altitudes of landforms underneath. Dark shows lowest terrain; light, highest. Topography is reminiscent of Earth's, with "seafloor" interrupted by continentlike blocks containing mountain ranges. Latitude and longitude intervals of 10° are shown. (NASA; courtesy M. Kobrick, Jet Propulsion Laboratory)

uplands, including a few Australia-size continentlike plateaus (24% of the planet) standing a few kilometers above the surface. A third type of surface structure comprises a few huge volcanic peaks (16% of the planet) that tower as high as 10.6 km (34,500 ft). These peaks rise higher than our Mt. Everest rises above sea level.

★ The Magellan probe produced amazing images of familiar and unfamiliar geologic features down to sizes of a few hundred meters (Figures 9-7 and 9-8). The images show that Venus is virtually covered by lava flows less than a billion years old, and in many areas it is intensely fractured.

Venus is named for the goddess of beauty and love, and its symbol is also the biological symbol for *female,* so scientists named most of its surface features after historical and mythological women. The largest plateau is called Aphrodite Terra. Other features include an impact crater named Eve, which marks the zero meridian on maps.

At first, the map of Venus reminded researchers of Earth without an ocean. The low volcanic plains resemble our ocean floors; Earth's ocean floor crust is also composed mostly of basaltic lavas. In support of this idea, some trenches have been found on Venus that resemble seafloor trenches off the coast of Japan and elsewhere on Earth, where crustal plates are being dragged back down into Earth's mantle (Sandwell and Schubert, 1992). The plateaus on Venus suggest small, poorly developed continents, which may be associated with the granitic rock indicated by the Soviet probe Venera 8.

An important difference between Earth and Venus, however, is that Venus lacks long, folded mountain ranges like the Himalayas, Andes, and Rockies. On Earth, such mountain chains are produced by collisions between the moving plates. Their absence on Venus convinces most scientists that Venus does not have as much active plate tectonic activity as Earth. In support of this, we see few horizontal displacements

Figure 9-7 A large impact crater on Venus. The crater is 50 km (30 mi) across and was named Barton after Clara Barton, founder of the Red Cross. The beautiful double ring structure, typical of impact craters in certain size ranges, is caused by mechanical processes during the impact explosion. (NASA, Magellan radar image, Jet Propulsion Laboratory)

along faults on Venus, even though we see many fractures with vertical movement. In short, there seems to be less continental drift on Venus than on Earth.

Why? At 95% of Earth's diameter, Venus' interior has probably cooled a little faster than Earth's. Thus it may lack enough convection currents in its mantle to drive plate tectonic activity. Alternatively, the high surface temperatures may make the surface layers more plastic, making for less well-defined, less rigid tectonic plates (Solomon and others, 1992).

★ One interesting feature, found on Venus but on no other planet, is called a **corona.** Coronas, as shown in Figure 9-9, are depressed or raised circular rings of ridges and fractures. About 360 coronas are known on Venus; most range from 200 to 400 km in diameter. Some are flooded with lavas. They are believed to be sites where upwelling currents in the mantle have ascended under the crust and pushed up the surface, allowing lavas to break out. Some of them have subsided to create depressions. Coronas support the idea that Venus' mantle does have thermal currents and rising mantle plumes, though they may not be strong enough to produce full-fledged continental drift (Kaula, 1990; Phillips and others, 1991; Saunders and others, 1991; Smrekar and Stofan, 1997).

Impact Craters and the Age of Venus' Surface Features

Venus has far fewer impact craters than the Moon or Mercury, but more than Earth. An example is shown in Figure 9-7. The crater statistics suggest that, on average, Venus' surface has been exposed to meteorite impacts for less time than the Moon's or Mercury's surface, but longer than Earth's. By counting the total number of craters and dividing by the estimated rate of crater formation, scientists estimate that the surface of most of Venus averages 500 to 800 million years old (Phillips and others, 1991; Schaber and others, 1992). This young age is believed to result from continual resurfacing of the planet by lava flows. Remember, of course, that the planet *as a whole* is far older; it formed 4.5 to 4.6 billion years ago, at the same time as the Earth. Here, we are dis-

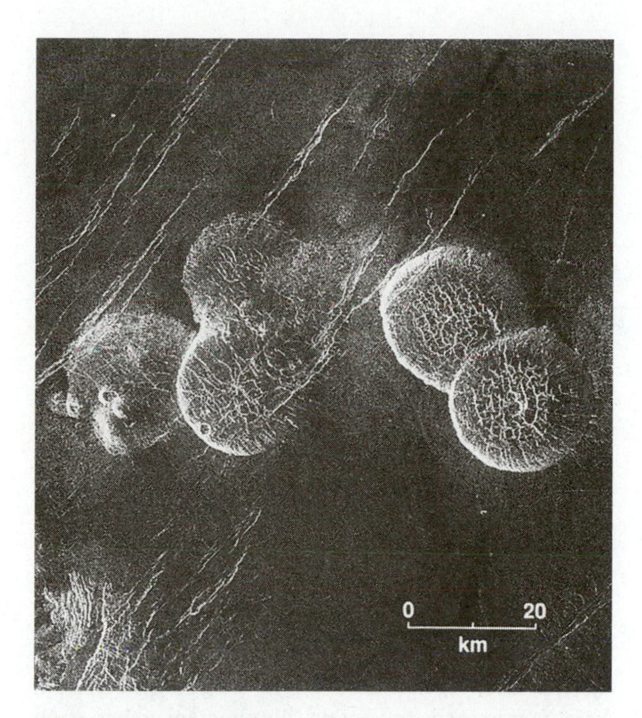

Figure 9-8 Magellan radar image revealed strange circular domes on the east side of Alpha Regio, averaging about 25 km (15 mi) across and about 750 m (2500 ft) high. They are believed to be pancakelike eruptions of viscous lava. Cinder cones can be seen at the left edge of the left domes. Many tectonic fractures cross the area (bright lines). Vague shadings of light and dark at the top mark different lava flows. Smallest details are about 150 m across (NASA)

cussing only the average age of the landscape features, such as mountains and lava plains. By comparison, Earth's surface measured by this technique might average around 500 million years old, but it would show much more variety of ages, ranging from fresh lava flows and 10-million-year-old sediments to 2000 million-year-old continental core areas, like central Canada.

★ The relatively uniform surface age of Venus creates a scientific mystery. It suggests that most of the planet was resurfaced by lavas around 500–800 million years ago. Could there have been one episode of planetwide volcanic activity so recently in terms of planetary history (the last 10 to 15% of the planet's age)? Some calculations suggest that in the presence of strong lava outflows, and in the absence of strong continental drift to churn crust back into the mantle, Venus' crust grew thicker and thicker until about 800 million years ago. At that point, the heavy lava crust began to break up and sink into the mantle, causing widespread eruptions of new lava and creating a "new" surface during the period from 800 to 500 million years ago. This hypothesis is controversial. Other researchers favor a more gradual evolution (see the extended discussion on our web site). In any case, more work will be needed to understand why Venus, with its Earth-like size, seems to have differed from Earth in its style of crustal evolution (Kaula, 1995; Phillips and Hansen, 1998).

Are Venus' Volcanoes Active Today?

★ The images of fresh-looking volcanic calderas and lava flows on Venus raise the question of whether some of the planet's volcanoes are still active. The answer is probably yes but there has been no firm proof. Among the suggestive evidence: fluctuations in sulfur

a

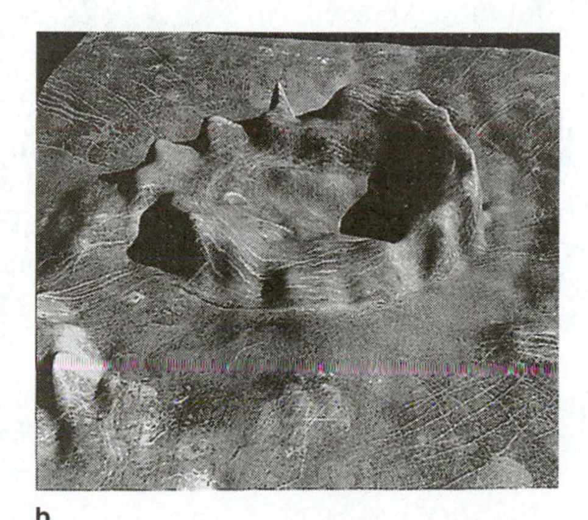

b

Figure 9-9 A typical corona on Venus is a roughly circular pattern of fractures, with some additional radial fractures, 100 km or more in diameter. These features have not been found on other planets and seem to be caused by upward forces causing doming, followed by collapse. **a** Normal Magellan radar image, looking "straight down" on surface. (NASA) **b** Use of Magellan-measured altitudes to construct an artificial, computer-generated oblique view with highly exaggerated relief. This view shows the raised edges and collapsed central region. (Courtesy D. M. Janes, Cornell University, and Jet Propulsion Laboratory)

dioxide (SO_2) gas in the atmosphere of Venus in recent decades. Since SO_2 is emitted by volcanoes on Earth, its recent increases on Venus may be due to volcanic outbursts there (Esposito, 1984). More exotic evidence came from reanalysis of data from the Pioneer Venus probe that crash-landed on Venus in 1978. Some kilometers above the surface, it apparently passed through a methane-rich cloud of unusual composition, which some scientists suspect was a plume of gas from an active Venusian volcano (Donahue, 1992)!

LESSON 1 IN COMPARATIVE PLANETOLOGY: SURFACE FEATURES VS. PLANET SIZE

On all planets there has been a competition between the construction of *internally* derived features, such as volcanoes, and their destruction by the *external* bombardment of meteorites. Primordial surfaces were probably hot with much volcanism and many lava flows, because the formative processes and early intense meteorite bombardment caused heating.

As we noted above, the smaller a planet, the quicker it cooled. If a planetary body is very small, the surface layers cooled and formed a solid lithosphere relatively early, ending volcanism long ago. In that case, early lava flows were destroyed by the intense early bombardment of meteoritic debris, which lasted until about 4.0 billion years ago. Such bodies' surfaces are covered mostly with impact craters. The Moon is an example; only a few "last-gasp" lava flows have occurred there in the last 3.5 billion years, creating the few lava plains that we can still see. On a still larger world, like Mercury, we see more lava and evidence of other internally driven geologic activity, such as faulted cliffs. On worlds as large as Venus and Earth, internally generated surface features won out over cratering, and the ancient craters were destroyed by volcanism and other processes. In the words of science writer Richard Kerr (1999), "For a planet, volcanism is the secret to a youthful appearance," because volcanism obliterates the oldest features, often creating fresh fracture systems and covering the ancient, rugged impact craters that accumulated from ancient times.

Thus we see a rule of thumb: The larger a world, the younger and more active its surface. In the next chapters, we will see this rule more or less confirmed on worlds of many sizes.

LESSON 2 IN COMPARATIVE PLANETOLOGY: WHY DO SOME PLANETS LACK ATMOSPHERES?

Our discussions of Earth and the Moon showed that the planets were probably heated by impacts as they formed, and that resulting volcanism emitted volcanic gases such as CO_2. Why, then, is there so much variation among planetary atmospheres? We've already seen the reasons for the different evolution of Venus' and Earth's atmospheres, but let's look at the larger picture. Why do some worlds lack atmospheres while other planets have dense ones? The explanation comes from three principles that govern the motions of gas molecules in atmospheres:

1. The higher the temperature, the higher the average speed of the molecules.
2. The lighter the molecules, the higher their average speeds. Light gases like hydrogen and helium have faster average speeds than heavier gases such as oxygen, nitrogen, carbon dioxide, or water vapor.
3. The bigger the planet, the higher the speed needed for a molecule to escape into space.

If you could heat a planet's atmosphere, more and more molecules would move faster than escape velocity, and fast-moving molecules moving upward near the top of the atmosphere would shoot out into space, never to return. First hydrogen, then helium, and then heavier gases would leak away into space. Cold, massive planets are most likely to retain all the gases of their primitive and secondary atmospheres; hot, small planets (with weak gravity and low escape velocity) are most likely to lose all their gases. Calculations based on these principles show that planets as small as Mercury and the Moon have lost virtually all of their gases. Venus and Earth have lost most of their hydrogen and helium but have kept heavier gases.

These principles help explain why Venus could rapidly have lost its water if water molecules broke into hydrogen and oxygen atoms: The light hydrogen atoms would quickly escape into space. (Oxygen atoms would oxidize surface rocks and thus be removed from the atmosphere as well.)

SUMMARY

Comparing features on different planets helps explain planetary phenomena, particularly on Earth. The study of planets, their origins, and their development has come to be called *planetology. Comparative planetology* is the comparison of different planets to understand what makes them alike or unlike. Only a few decades ago, knowledge of phenomena on Earth, such as rock chemistry, weather circulation, and mountain building, had to be derived from knowledge found on our planet alone. Today we can gain insights from other planets, treating them as special "laboratory examples" of other conditions. Venus, for instance, is an Earth-size planet nearer the Sun with a different rotation rate. From it we learn that an Earth without oceans could have retained much more CO_2 gas in its atmosphere, instead of in rocks. Venus also proves that atmospheric CO_2 causes atmospheric heating through a strong greenhouse effect.

From Venus we have also learned that a more slowly rotating Earth would have more linear weather circulation and less developed cyclonic spiral systems. The Moon, Mercury, and Venus show that smaller planets apparently do not develop enough internal energy to drive the plate tectonic activity that has broken and reformed the Earth's original, cratered crust. Venus is largely a volcanic planet whose surface lavas are about 500–800 million years old on average, somewhat older than typical land surfaces of Earth. From our studies of the Earth, Moon, Mercury, and Venus, we can derive three general principles that will clarify phenomena of other planets as well:

1. *Larger planets are more likely to have internal geological activity.* Internal heat is the energy source that drives geological activity such as tectonic faulting, earthquakes, and volcanism; the larger a planet, the more radioactive minerals it contains and the more radioactivity there is to release heat. Also, the larger a planet, the better insulated the interior and the harder it is for the heat to escape. Small planets, on the other hand, cool rapidly and lose whatever heat they may have generated. The Earth, unlike Mercury and the Moon, has enough internal energy to drive plate tectonics.

2. *The larger a planet is, the younger its surface features are likely to be.* This principle follows from the one above. The more internal heat, the thinner the lithosphere and the more likely that the lithosphere has been broken by recent geological activity. Small planets that cooled long ago retain very ancient surface features. The Earth and probably Venus retain fewer ancient craters than Mercury and the Moon.

3. *The larger and cooler a planet is, the more likely it is to have an atmosphere, and the more likely this atmosphere is to have retained its original gases.* This rule comes from the fact that a larger planet has more gravity, and the additional gravity helps prevent gas from leaking off into space.

These rules not only explain general features of the Earth, Moon, Mercury, and Venus, but also apply to Mars and other planets and satellites, as we will see in the next few chapters.

CONCEPTS

prograde rotation
retrograde rotation
Venera 7
greenhouse effect
corona

PROBLEMS

1. Which is hotter, Mercury or Venus? Why?

2. Venus and Earth are about the same size and mass, and degassing volcanoes on each probably produced both CO_2 and H_2O gas. Why is CO_2 a major constituent of the atmosphere only on Venus, while H_2O is not a *major* constituent of the atmosphere on either planet?

3. If astronauts are to walk on Venus, what sort of space suit design might be needed? Would it need to *contain* pressure or *resist* pressure?

4. Compare the cloud patterns on Earth and Venus (see Figures 5-14, 9-3, and Chaper 8 on the web site).

- **a.** What types of features are similar?
- **b.** These two planets have nearly the same radius, mass, and surface properties, but different rotations. What might be learned about circulation (wind patterns) of planetary atmospheres by comparing patterns on Venus and Earth?
- **c.** Would Coriolis drift be greater or smaller on Venus? How would this affect airflow and cloud patterns.

5. State which of the following characteristics of Venus suggest a primitive surface (little disturbed since planet formation) and which suggest an evolved surface (affected by geological processes such as erosion, differentiation, and plate tectonics):

- **a.** Craters
- **b.** A large, rifted canyon
- **c.** The lack of long, folded mountain ranges
- **d.** Basaltic surface rocks
- **e.** Granitic surface rocks (if any)

6. Compare the state of Venus' geological evolution with the evolution of the Moon, Mercury, and Earth.

7. What are the chances that life as we know it exists on Venus? Why?

a. If Venus were to have a surface temperature of about 300 K and an abundance of H_2O in its clouds, how would you rate the chances for life? Why?

b. If Venus were exactly like the Earth, would life necessarily exist there?

8. How close is Venus during its nearest approach to Earth (see Table 8-1)? How many times farther is this than the distance to the Moon?

9. If a telescope shows Venus to be a thin crescent, where is Venus relative to the Earth and Sun?

ADVANCED PROBLEMS

10. Suppose you have a telescope that resolves $\frac{1}{2}$ second of arc. What would be the thinnest cloud layers visible on Venus in a view such as that in Figure 9-2, during its closest approach to Earth?

11. If Venus has a surface temperature of 750 K, at what wavelength is its strongest radiation emitted? Is this ultraviolet, visible, or infrared radiation?

12. Calculate the mean density of Venus and compare it with that of Earth. Would you expect Venus to have a larger or smaller amount of iron than Earth?

PROJECTS

1. Determine whether Venus will be prominent in the evening or morning sky during this semester, and observe its motions and brightness from day to day. Observe on which date it is farthest from the Sun and estimate this angle.

2. Observe Venus in a telescope of at least 5-cm (2-in.) aperture on several dates a few weeks apart. Observe and sketch the changes in phase, and explain them in terms of Venus' motion relative to the Earth and the Sun.

3. If a planetarium is convenient, arrange a demonstration of the motions of Venus.

CHAPTER

10

Mars

WHAT THE READER SHOULD LOOK FOR IN THIS CHAPTER

Mars is the next planet beyond Earth and the planet most like ours. It is about half the size of Earth. Because it is further from the Sun, it is colder than Earth. Mars apparently has abundant water, but most of it is frozen (in the polar ice caps and in the soil), so Mars is a cold, dry desert. Close-up photos from spacecraft show various features including volcanoes, canyons, sand dunes, ice deposits, scattered clouds, and even dry riverbeds similar to those in the American Southwest. A century ago, many astronomers believed Mars had vigorous plant life or even intelligent civilizations! Today this has been ruled out, but some scientists believe that microbial life might have evolved in the ancient past and that microbes may still be living in the soils and rocks in areas where liquid water may occasionally be accessible. Although Mars generally supports the rules of thumb about planetary evolution described in the last chapter, two major mysteries have yet to be solved: (1) If the water on Mars is frozen under today's conditions, why did water flow in the past to form riverbeds. Did the climate change in the past? (2) Did life ever form on Mars? If so, what was it like? Was it like terrestrial microbial life? If not, why not? ■

On a Martian summer morning in 1976, the Sun came up as usual on a rock-strewn plain. The dawn temperature was around –84°C (–120°F or 189 K), but by afternoon the air warmed to about –29°C (–20°F or 244 K). The wind was light, and the rust-colored rocks lay about as they had for the last few million years. The first sign of something unusual came at about 4:11 in the afternoon when a tiny, white, starlike object appeared high in the dusky red Martian sky. It was the 16-m-(53-ft-) diameter white parachute of the Viking 1 landing craft.

Within a few moments, the contraption would have been clearly visible from the surface. At an altitude of 1.2 km (4000 ft), the lander's three engines fired with a pale, transparent flame. Within seconds the parachute cut loose and drifted away. Slowed by its rocket engines, the spacecraft dropped for another 40 s or so. As it dropped the last few meters, reddish dust swirled into the air. The first of three lander legs hit the ground about as hard as you would if you jumped off a chair, and the jolt automatically switched off the engines. As the ungainly spacecraft bumped to rest, Viking 1 became the first human-built machine to gather data on the surface of Mars.

Until the instant that Viking's cameras clicked on, no one knew what existed on the surface of Mars; many scientists expected plants or other life-forms. As might be hoped for any successful exploration, Viking's voyage confirmed some existing ideas but forced surprising revisions of other ideas.

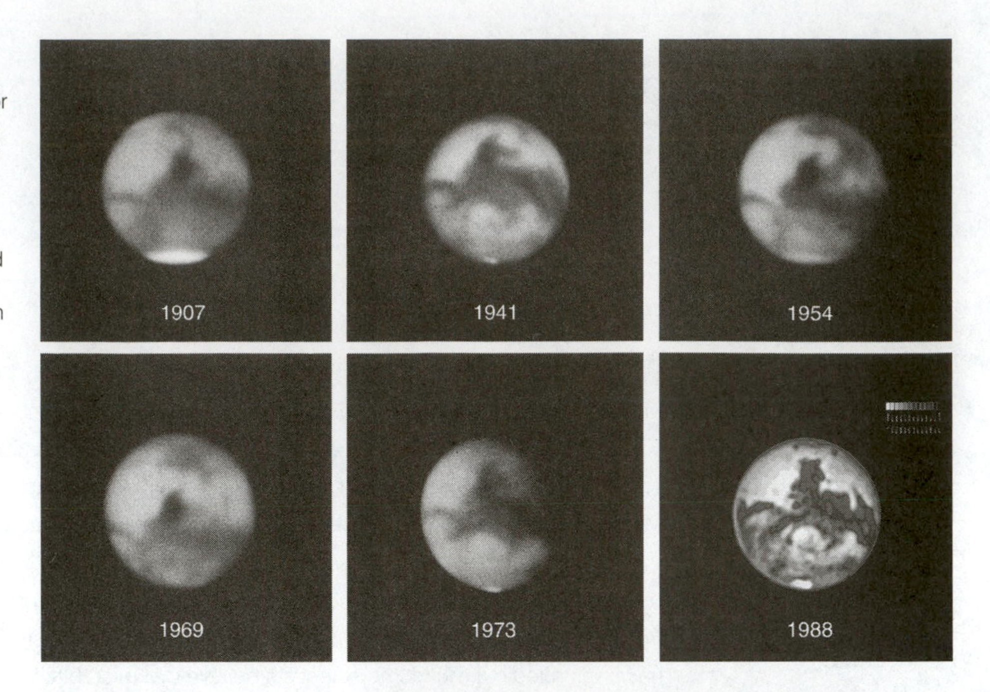

Figure 10-1 Syrtis Major region of Mars as photographed from 1907 to 1988. (Syrtis Major is the north-pointing dark triangular area in the central part of each image.) Variations in the patterns of the dark markings can be seen. A dark "wing" to the right of the Syrtis Major and a brightening of the nearly circular Hellas bright region south of Syrtis Major are evident in the 1941 photo. Note the large winter south polar ice cap in the 1907 view and the small summer cap in the 1941, 1973, and 1988 views. (1907–1973, Lowell Observatory. 1988 view uses new CCD technology, courtesy P. Pinet, S. Chevral, C. Buil, and E. Thouvenot, Toulouse OMP Observatoire, France)

MARS AS SEEN WITH EARTH-BASED TELESCOPES

At its closest, Mars comes within about 56 million kilometers (35 million miles) of the Earth—closer than any other planet but Venus. When it is that close, a telescope with an aperture of only 7 to 10 cm (3–4 in) will show features on Mars' reddish surface, including polar ice fields, clouds, and dusky markings, and a larger telescope will show still more details (Figures 10-1 and 10-2). However, no telescope on Earth's surface could see enough detail to distinguish clearly any mountains, canyons, or other topographic features, so the nature of the markings remained uncertain until spacecraft arrived in the 1960s.

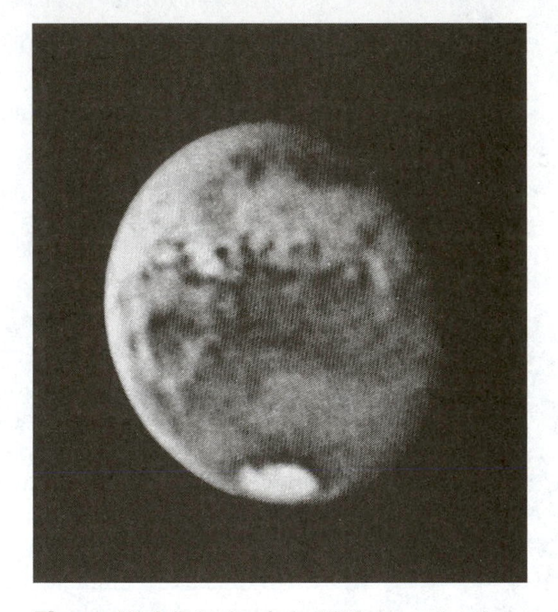

Figure 10-2 This view from 1988 shows the sharper modern pictures that are possible with ground-based telescopes using CCD technology (see Chapter 5). The short, stubby projections of dark areas northward into the bright deserts are the features that were once interpreted as artificial canals. (Catalina Observatory photo, courtesy S. M. Larson, University of Arizona)

The Markings of Mars

The first person to see Mars through a telescope was probably Galileo, who wrote in 1610 that he could see its disk and phases. These observations proved that Mars was a spherical world illuminated by the Sun. When telescopes improved, observers began mapping dusky and bright patches that we can still recognize today. The Dutch physicist Christian Huygens first clearly sketched these markings in 1659, and a French-Italian observer Giovanni Domenico Cassini tracked them a few years later to determine that **Mars' rotation period** is $24^{h}37^{m}$, only a bit longer than the Earth's. Our web site shows some of the drawings of Mars made by these early telescopic observers.

Seasonal Changes on Mars

Mars has seasons just like Earth, though each season lasts about twice as long as on Earth, because the Martian year is nearly twice as long as ours. Telescopic observations in the 1700s and 1800s revealed **seasonal changes in the features of Mars.** In the Martian hemisphere experiencing summer, the bright, white polar cap shrinks away and may disappear from view, while the dusky markings darken and grow more prominent.

Year-to-Year Changes on Mars

★ While retaining roughly constant shapes, the markings also change slightly from year to year, as shown in Figures 10-1 and 10-2. Once these changes were established, many observers thought that the dark areas were regions of vegetation, perhaps losing their leaves in winter and turning dark and lush in summer, as on Earth. To understand the studies of Mars, it is important to realize that many observers in the late 1800s erroneously believed that Mars had climate and vegetation like the Earth's. This opinion evolved even further as a result of the celebrated affair of the Martian **"canals."** By 1900, some astronomers, especially Percival Lowell, who had built a new Mars-dedicated observatory in Arizona, claimed that Mars had a system of faint, straight, narrow lines, which they called canals. Some observers thought these lines actually *were* canals—artificial canals built by Martians to carry water from the polar ice caps to the warm equator, where they lived. For more details on the "mystery of the canals," see our web site. Figure 10-2, a modern photo, shows that while some of the markings of Mars are definitely streaky, no network of straight, artificial lines exists.

★ The possible existence of Martian civilizations made philosophers and theologians face the possibility that humans might not be the only intelligent beings in the universe, and it gave science fiction writers, film makers, and artists fertile grounds for speculations. Our web site gives more information about the mythic Mars of science fiction.

★ Gradually, during the twentieth century, scientific data painted a more and more forbidding view of conditions on Mars. In 1947, the Dutch-American astronomer Gerard Kuiper used newly developed infrared spectroscopic equipment to detect the carbon dioxide absorption bands on Mars. He found very low atmospheric pressure. The normal air pressure on Earth at sea level is defined as "1 bar," which could also be written as 1000 millibars, or 1000 mb. Estimates of the Martian atmosphere by mid-century pictured a surface pressure of around 1 to 85 mb, only 0.001 to 0.085 of the surface pressure on Earth, with subfreezing temperatures—a forbidding environment indeed! The idea of intelligent civilizations or even abundant vegetation was finally disproved by the first Martian space probes in the 1960s and 1970s, which confirmed a surface air pressure typically around 7 mb.

Names of Martian Features

★ Giovanni Schiaparelli, an Italian astronomer and classical scholar, was one of the important observers of Mars in the 1800s. He was the first to emphasize the streaky character of many Martian markings, and he also named the larger Martian dark and light regions (as well as the illusory canals) after historical, mythological, and geographic features of his native Mediterranean area. These names, such as Hades, Arabia, and Libya, are still used for the major dark and light areas. Bright areas were often named after deserts, which was appropriate! In 1971, when the Mariner 9 spacecraft revealed actual geological structures such as craters, mountains, and canyons (Figure 10-3)—all too small to be seen from Earth—these were assigned additional names. As on the Moon, craters were named after deceased scientists. The largest canyon complex, big enough to stretch across the United States from coast to coast, was named Valles Marineris (Valleys of Mariner). A modern map with some of the markings and geological structures is shown in Figure 10-4. The tallest volcano was named Olympus Mons (or Mt. Olympus, after the peak in Greece).

VOYAGES TO THE SURFACE OF MARS

The first three human-made devices to reach the surface of Mars were unsuccessful Soviet probes. Mars 2 crashed in November 1971; Mars 3 landed in December 1971 but failed after 20 s on the surface; another probe sent back data while parachuting through the atmosphere in 1974 but failed moments before touchdown. The failures may have been caused by design problems or hostile Martian conditions; a dust storm was raging during the 1971 landings.

The first successful landing on Mars was by the **Viking 1** spacecraft, which touched down on July 20,

Figure 10-3 Mars without clouds. This image was constructed by taking high resolution mapping photography and constructing a global photo-map, reprojected digitally onto a globe. Many or most of the dark and light features are associated with windblown dust deposits. The map clearly shows the streaky character of many markings. The giant canyon, Valles Marineris, is at lower left. The residual north polar water ice cap is at the top (the transient winter CO_2 frost cap is still larger). X marks the spot of the Pathfinder landing in 1997, at the mouth of channels draining from the southern uplands. The real Mars is generally more hazy than this, with occasional veils and cloud patches, especially at dawn and dusk. (NASA and U.S. Geological Survey, print courtesy Dan Britt, University of Arizona)

1976 (7 y to the day after the first human landing on the Moon). It was followed on September 3, 1976, by a duplicate spacecraft, Viking 2. Both landings were on plains that looked relatively smooth from orbit but turned out to be rock-strewn, as seen in Figures 10-5 and 10-6. No more probes landed on Mars for 21 y, until the Pathfinder probe landed in 1997. It took photos at the mouth of a Martian riverbed and sent a small rover out to measure rock compositions within a few dozen meters of the lander (Figure 10-7). Desolate but beautiful landscapes with sand dunes and scattered boulders as wide as 3 m were photographed at all three sites. Disappointingly, no Martians, deserted cities, canals, or strange vegetation were found!

★ Mars turned out to be an awesome desert. The reddish color of Mars is vividly shown in color photos from the various landers, as seen on our web site. Though some rocks appeared dark gray, like terrestrial lavas, most rocks and soil particles were covered with a coating of rustlike, reddish iron oxide minerals. Similar iron minerals give terrestrial deserts their familiar red-to-yellow coloration, especially when moisture is present only occasionally. Viking scientists were surprised to find the daytime sky of Mars reddish-tan instead of blue. The sky color is caused by much fine red dust stirred from the surface into the air by winds and deposited even on rock tops as it settles out of the air. (See color photos of the various landing sites on our web site.)

Not surprisingly, the most Mars-like landscapes on Earth are found in extremely dry, volcanic regions. As shown in Figure 10-8, some of these provide amazingly Mars-like vistas.

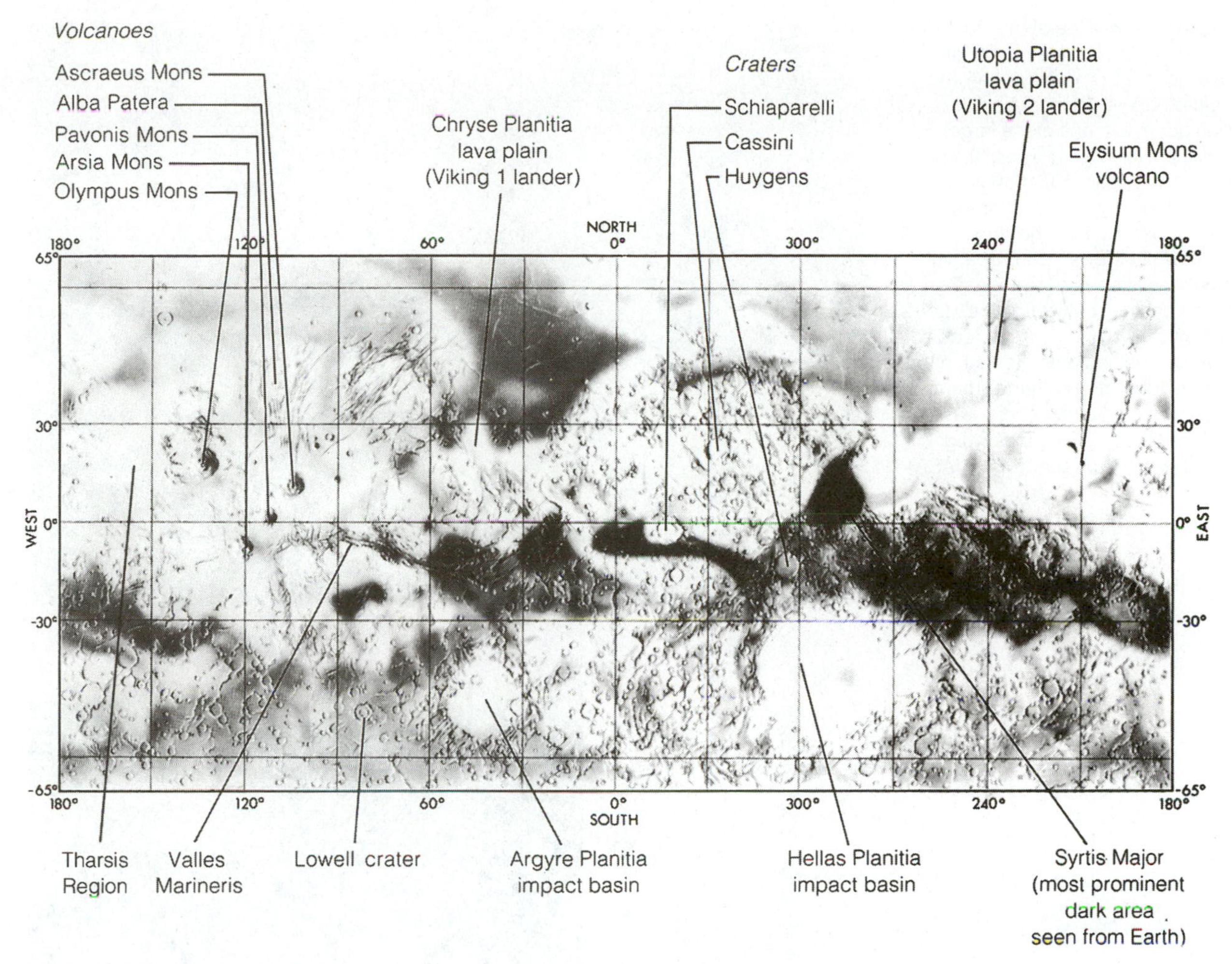

Figure 10-4 A map of Mars. Dark shadings are the somewhat changeable markings visible from Earth, probably associated with windblown dust. Topographic features such as craters and volcanoes are too small to see from Earth. The names of some features are given. (Base map courtesy R. M. Batson, U.S. Geological Survey)

Figure 10-5 The first close-up photo of Martian soil was sent from the Viking 1 lander on July 20, 1976. Fine dust and rock chips are shown. Landing leg is at right. The landing dislodged soil, which settled in the center of the concave pad at the end of the leg. The larger chips are about 5 cm (2 in.) across. (NASA)

Figure 10-6 One of the first vistas photographed on Mars. The late afternoon Sun reveals dunes and scattered foreground rocks. The boulder at left, one of the larger rocks photographed by either Viking lander, is about 3 m (10 ft) wide and is located only 8 m (26 ft) from the Viking 1 lander. Had the lander come down on it, the vehicle probably would have been destroyed. Detailed photos show a frothy texture of basalt lava and a cap of reddish dust that settled onto the rock after the last dust storm. At the time of this photo, the atmosphere was particularly dusty, giving a hazy light with low-contrast shadows. (NASA)

Figure 10-7 The Sojourner Rover backs up against a boulder to measure its composition. The rover's aft end carried a composition measuring device, which bombarded the rock with alpha particles from a radioactive source and measured particles and X rays scattered back off the rock. This boulder had basaltic composition. The rover landed on Mars in 1997 with the Mars Pathfinder. (NASA, courtesy Sara Smith, Pathfinder project)

Figure 10-8 Close terrestrial analogs to Martian landscapes occur in very arid regions with a history of weathering. Glacier-dropped boulders in volcanic plains, Iceland. (Photo by WKH)

CONDITIONS ON THE SURFACE OF MARS

The various landers, plus measurements made from telescopes on Earth and orbiters circling Mars, have given us a great deal of information about the environment on this cold, desert planet.

Atmospheric Composition

The composition of the **Martian atmosphere,** is basically carbon dioxide. This is shown in more detail in Table 10-1. The atmosphere is extremely thin, and the **Martian air pressure** only about 5 to 12 mb, depending on the elevation and the season. It is something like the thin air encountered by a high-flying jet at 100,000 ft above Earth, so a spacesuit would be essential to survive on Mars.

There is a clue here to planet history. As on Earth and Venus, the atmosphere on Mars has been generated chiefly by planetary degassing of CO_2 and H_2O through volcanic activity. The H_2O froze. The CO_2 was left in the atmosphere instead of being trapped in carbonate rocks, as it was on Earth. The low gravity allowed much of the early atmosphere to escape, leaving only a thin remnant atmosphere of CO_2.

TABLE 10-1

Composition of Martian Atmosphere

Gas	Percent Volume
CO_2 (carbon dioxide)	95
N_2 (nitrogen)	2.7
Ar (argon)	1.6
CO (carbon monoxide)	0.6
O_2 (oxygen)	0.15
H_2O (water vapor)	0.03
Kr (krypton)	Trace
Xe (xenon)	Trace
O_3 (ozone)	0.000003

Note: Amounts of gases vary slightly with season and time of day. H_2O is especially variable. Some CO_2 condenses out of the atmosphere into the winter polar cap; changing cap sizes cause small changes in the total Martian atmospheric pressure.

The Martian Climate

The Martian climate is very cold and dry by Earth standards. During the weeks after landing in summer, the two Vikings measured **Martian air temperatures** ranging from 187 K (–123° F) at night to 244 K (–20° F) in the afternoon. The soil is warmer, because it absorbs sunlight, which mostly passes through the air. The soil temperature gets above freezing on some summer afternoons.

Winds measured at the landing sites were usually gentle, less than 17 kph (11 mph), but gusts exceeded 50 kph (31 mph). Much higher winds occur at certain seasons, raising clouds of dust that can be observed from Earth.

The Polar Ice Caps

The polar ice caps (Figure 10-9) play an important role in Mars' climate and are more complex than Earth's polar ice caps. Frozen H_2O exists at each pole all year-round (as on Earth). These caps are small, extending only 400 to 600 km from the pole. In 1997–98, the Mars Global Surveyor made careful measurements of surface elevations at the north pole. The north polar cap lies in a broad lowland; its ice has an average thickness of about 1 km, but increases to 3 km thick in some places, forming a deposit that rises above the surroundings. If this volume of ice melted, it could fill the northern polar lowland with a sea averaging about 270 m deep, or if spread uniformly around the whole planet, it would make a layer 9 m deep (Zuber and others, 1998). More water is locked in the south polar cap, and much more water is believed to be frozen underground in soils around the planet.

In the hemisphere that is having winter, the pole grows so cold that CO_2 freezes out of the atmosphere and forms an additional frost layer of white CO_2 frost

Figure 10-9 The south polar ice cap of Mars seen from orbit. This picture shows the small cap left near the south pole during the southern summer. It is roughly 360 km across and consists of CO_2 ice (possibly mixed with H_2O ice). Spiral breaks mark warmer, Sun-facing ridges in the layered sedimentary deposits around the pole. In winter, condensation of CO_2 frost and snow makes the cap expand to 10 times this size. Compare with Earth-based views of the cap in Figure 10-2. The night side of the planet is at the left. (NASA Viking orbiter photo; courtesy Tammy Ruck and Larry Soderblom, U.S. Geological Survey)

and ice. (This is the material popularly known as dry ice.) This produces a frost layer ranging from centimeters to no more than 10 m deep, stretching 2500 km from the pole to mid-latitudes, just as snow cover stretches from the North Pole to the northern United States during winter on Earth. This frost made a scenic addition to winter morning landscapes at the Viking 2 site, when white frost deposits hid in the shadows of rocks, as seen in the color photos on our web site. During spring, the polar frost cap recedes as the CO_2 frost changes back to gas, but at this time, the other hemisphere is growing colder, and CO_2 begins to freeze on the other pole. A significant fraction of the atmosphere's CO_2 changes from gas to solid during this cycle, causing fluctuations in the Martian air and climate.

A peculiarity of both polar caps is that ice and dust are deposited in thin layers, which have been photographed from orbit. Researchers believe these layers represent some sort of climatic cycles. For example, dust grains get blown into the air by global dust storms, arrive over the pole, and serve as condensation nuclei for condensing snowflakes or frost grains that condense in the winter. These are deposited on the surface. Then when spring comes, the ice *sublimes* (goes from solid to gas, without passing through a liquid phase), leaving the thin dust layer behind. The polar layers could preserve a record not only of annual climate cycles, but much grander, million-year cycles in the planet's climate history, which might be caused by orbital changes, volcanic eruptions, or other effects. The first attempt to study these layers on the ground ended in failure in 1999 when the Mars Polar Lander mission failed to deploy properly.

Rock and Soil Types

One of the goals of Martian exploration is to understand the types of rocks and soils in order to interpret the geological history of the planet. We have two sources of information on this: observations from landers on Mars' surface, and—amazingly enough—actual rock samples that were blown off Mars during meteorite impacts, drifted through space, fell on Earth, and were collected. The observations from landers revealed mostly igneous rocks, fragments of basaltic lava flows, as judged by textures and chemical analysis carried out by the landers. The Pathfinder lander in 1997 included a small rover, which traveled out from the lander, nuzzled up against several rocks, and measured their compositions. They all turned out to be generally basaltic, with compositions more specifically listed as basalts and andesites. Missing were quartz-rich rocks, granites, sandstones, limestones, and other such rocks that commonly form on Earth due to the endless cycles of crustal melting and differentiation that are associated with plate tectonic activity and deposition of seafloor sediments. These results are consistent with other evidence that Mars has had a less complex tectonic history than Earth.

The Martian soils, as measured by the various landers, show a certain uniformity of composition at different sites, because they are mixed when the wind blows dust around the planet. They are basaltic, derived from the erosion of these rocks plus the addition of carbonate minerals, salts, sulfates and other minerals associated with the presence and evaporation of water in the past. (In the same way, arid regions of the American West are plagued by the accumulation of surface salts, as water carries dissolved salts through the soil but then leaves them on the surface as it evaporates.) This evidence of water flow is important, because it confirms the evidence from the dry riverbeds that liquid water has been active on the Martian surface.

Martian surface soils still contain water today, not in the form of moisture or ice, but as H_2O molecules in the crystal structures of the mineral grains. When the soils were heated to around 700 K (800°F) in the Viking landers, water was driven off. This shows that future Martian explorers could get water by heating the soil. Most investigators believe that underground ice is present among the soil particles below the surface, especially at high latitudes. This would be similar to **permafrost** in arctic tundra regions of Earth.

The second source of information is the rocks that came from Mars. As will be clearer in Chapter 13, most meteorites, or rocks that fall from space, are fragments of asteroids. During the 1980s, researchers realized that certain rocks, cataloged as ordinary meteorites, were actually from Mars. The proof was that they have bits of gas trapped in them, and this gas exactly matches the composition of the atmosphere measured on Mars. They are basaltic lavas or igneous rocks, and most have much younger ages than other meteorites. One group—apparently blasted out of a single crater on Mars—is basalt that solidified only about 1.3 billion years ago. This is young, compared to the planets' age of 4.6 billion years. Another group may be even younger.

The value of these **Martian meteorites** is that their chemical properties provide direct evidence about

Mars. For example, they show that Mars has basalt lavas, a mantle composition somewhat similar to Earth's, and large amounts of subsurface water or ice (McSween, 1985; Nyquist and others, 2001; Bridges and others, 2001). They also show that Martian lavas erupted within the last 1.3 billion years, which proves that Mars is geologically a much more active world than the Moon. These rocks may also tell us something about the history of water activity on Mars. In the oldest Martian meteorite, a single specimen that has an age of 4.53 billion years, carbonates that were probably deposited by water have been dated to an age of 3.90 billion years ago—implying that water flowed very early on Mars, perhaps during a period when the primordial atmosphere was thicker and the climate was warmer (Borg and others, 1999). However, in one of the 1.3 billion-year-old Martian basalt specimens, researchers found minerals produced by interaction with liquid water, which dated from only about 670 ± 91 million years ago. This result seems to prove that liquid water may have been active on the surface in episodes that occurred in relatively recent geologic time.

The limitation of these Martian meteorites is that we don't know exactly where on Mars they came from. To unravel the mysteries of Mars' past, scientists still would like to obtain samples from specific features, such as Olympus Mons or the polar ice fields, so that chemical properties and ages could be matched with specific Martian formations.

MAJOR GEOLOGICAL FEATURES

Detailed mapping of Mars from orbit has led to additional discoveries. Some of the discoveries clarified old mysteries, but others raised new mysteries.

Martian Dust: The Markings Demystified

★ The evidence that early observers cited for vegetation on Mars—the dark patches that changed shape with seasonal and yearly cycles—turned out to have a more prosaic explanation. They are due to windblown dust deposits. The seasonal and yearly changes in shape are caused by seasonal winds that move the dust from place to place. Some of Lowell's streaky "canals" turned out to be windblown dust deposits aligned in swaths on the downwind side of craters and mountains. Stubby examples can be seen in Figure 10-3. Such deposition has been found in desert regions of Earth and has been simulated in lab experiments.

Martian Impact Craters: Clues to Surface Processes and Ages

Impact craters on Mars, such as the one shown in Figure 10-10, are interesting for three reasons. First, they confirm that impacts of interplanetary bodies have been common on Mars as well as on Earth, the Moon, Venus, and Mercury. Second, the craters have varied states of degradation, indicating that erosive processes have been much more active on Mars than on the Moon or Mercury. Of course, this is partly due to processes of Mars' atmosphere; for example, some craters are partly obliterated by massive accumulations of sediments or windblown dust. Third, as on Venus, the craters give a way to estimate the age of the surface. Mars has a wide variety of surfaces with different ages. Much of the southern hemisphere is covered by ancient, heavily cratered plains, probably 2 to 4 billion years old. Much of the northern hemisphere, however, is covered by young lava flows, averaging perhaps 1 billion years old. A few of these flows are much younger. Counts of impact craters suggest that some of them are much less than 100 million years old. This means that volcanoes have been active in the last 1 or 2% of the history of Mars, which, in effect, means that volcanoes are still active on Mars (Hartmann, 1999). All these results agree with the finding that of the three or four Martian impact sites that we sampled via Mars meteorites, one produced lavas 1.3 billion years old, and another produced igneous rocks that may be a few hundred million years old. This is a radical change from the suspicion of many Mars scientists in the 1970s and 1980s, who thought that all volcanism on Mars had ended around 2.5 billion years ago.

Mars confirms our rule of thumb about planet size vs. geologic activity (page 178). Midway in size between Mercury and Venus, its surface is also midway in geologic age. It has many regions younger than Mercury's surface, but about half of it is covered by cratered uplands older than Venus' surface.

Martian Volcanoes: Clues to Lithospheric Stability

In addition to its many lava flows, Mars also has huge volcanic mountains. These **Martian volcanoes** are

Figure 10-10 Example of a young impact crater on Mars. In this and many other Martian craters, the ejected material outside the rim is not a thin blanket of dust, but a thick mass of congealed material. The interpretation is that the asteroidal projectile hit an area of ice-rich soil and blew out a muddy slurry that piled up outside the rim. (Mars Global Surveyor image; NASA, Jet Propulsion Lab, and Malin Space Science Systems)

the largest known in the solar system. They were discovered by Mariner 9 in 1971. The highest is Olympus Mons (Figure 10-11a), which looms 24 km (78,000 ft) above the lower deserts. In contrast, Mt. Everest rises only 9 km above sea level and 13 km above the greatest ocean depths. The huge base of Olympus Mons volcano is about 500 km across and would nearly cover the state of Missouri. Its summit caldera, or volcanic crater, is about 65 km across and could contain the Los Angeles urban sprawl.

Olympus Mons is in the middle of a huge region of young lava plains, including regions called Tharsis, Amazonis, and Elysium. Tharsis has a slightly dome-shaped profile and dominates one hemisphere of Mars. Three other slightly smaller volcanoes are in the central Tharsis plain, and one has a summit caldera 140 km across. Another, older volcano, cut by channels, is well shown in Figure 10-11b. The Martian lavas were very fluid, spreading into broad flows and dome-shaped volcanoes, like Mauna Loa volcano in Hawaii, rather than steep-sided cones like Mt. Shasta or Mt. Fuji. Mauna Loa, for example, rises 9 km above the seafloor, stretches 120 km across at its seafloor base, and has a summit caldera a few kilometers wide. The Elysium region has some of the youngest lava flows known on Mars. As shown in Figure 10-12, they have few impact craters and may thus have an age 30 million years or less—less than 1% the age of Mars.

Figure 10-11a Oblique view from orbit looking east across the south flank of the volcano, Olympus Mons. The summit caldera can be seen at left, and the distant horizon is mostly covered with clouds and haze. (Mars Global Surveyor image; NASA, Jet Propulsion Lab, and Malin Space Science Systems)

Figure 10-11b A highly eroded ancient Martian volcano about one-tenth the size of Olympus Mons. This orbital view is a vertical image of a gently sloped volcano whose summit caldera is at left center. Radiating from the summit are eroded channels. Some geologists believe this mountain was formed by ash flows, which would be more easily eroded than basaltic lava flows. Erosion channels may have been formed by water released during the eruption when hot ash melted icy permafrost layers in the soil. Photo covers a region about the size of Rhode Island. (NASA Viking photo; courtesy A. S. McEwen, U.S. Geological Survey)

★ Comparison with volcanoes on other planets, such as Mauna Kea on Earth and Maxwell Montes on Venus, shows that tall volcanoes are common on the larger terrestrial planets (all but Mercury). The mantles of these planets apparently have ascending currents of hot material, known as **mantle plumes.** A plume rises and hits the underside of the lithosphere, causing an up-doming—probably the source of the coronas on Venus and perhaps the whole Tharsis dome on Mars. Where the lavas break through the lithosphere and erupt, they pile up into huge volcanic mountains, such as Mauna Loa on Earth, Maxwell Montes on Venus, and Olympus Mons on Mars.

TWO GREAT MYSTERIES OF MARS: ANCIENT CLIMATE AND POSSIBLE ANCIENT LIFE

Mariner 9 shattered previous conceptions of a Moonlike Mars by photographing so-called **channels,** or dry riverbeds, on Mars. (These should not be confused with the so-called canals, which turned out to be nonexistent). Channels are shown in Figures 10-13 and 10-14. Thus the greatest surprise of Martian exploration was to discover that although Mars today is frozen and very arid, it once had flowing rivers of liquid water!

The number of recent impacts superimposed on the channels indicates that the channels are perhaps 1 to 3 billion years old. The duration of the flow episodes is unknown. Generally, the channels cut across older terrain. They are older than the most recent lava flows; they do not cut across the younger lava plains. Some channels emanate from chaotic collapsed areas believed to have formed when ice deposits melted and released water onto the surface. Other channels—fine channels near the equator and large channels with tributaries—form a drainage network across the old deserts and suggest that water came from some other source, possibly rainfall from a once denser atmosphere. Some scientists support the idea of ancient rainfall and even ancient lakes and oceans filling the low areas of Mars. The best evidence was announced in 1999: A northern basin is surrounded by a shelf that looks like an ancient shoreline, and mapping by the Mars Global Surveyor orbiter showed that the possible shoreline is at a constant altitude (as a water surface would be) and has smooth deposits below it (Head and others, 1999). Other scientists argue against

Figure 10-12 Example of a young lava flow in the Elysium Planitia area of Mars. The young flow (dark, coming from bottom of picture) has split and flowed around the rim of an impact crater, creating a long tongue of lava at upper left. Age differences can be seen because the light-toned background surface is moderately cratered, while the lava flow has almost no impact craters and must therefore be much younger. (Mars Global Surveyor image; NASA, Jet Propulsion Lab, and Malin Space Science Systems)

ancient rainfall or oceans; the whole subject is controversial. In the next section, we will look at some of the arguments in more detail.

Mystery 1: The Ancient Martian Climate

The mystery of the ancient climate is simply stated: How could Mars ever have had a climate that allowed flowing rivers, not to mention possible rain or lakes? It is clear that the ancient Martian climate was different from today's climate—but how different?

The Viking landers and other spacecraft have provided important evidence proving that the atmosphere was actually denser in the ancient past. One technique involves measuring chemically inert gases such as argon in the present-day atmosphere. These gases are called inert because they don't combine readily with other elements. Once they are injected into the atmosphere by volcanoes, they remain there. Thus argon and other inert gases accumulate in the air and are good indicators of the total volcanic outgassing during a planet's history. By measuring the present argon content—and by knowing the typical amounts of nitrogen and carbon dioxide that are emitted by volcanoes along with the argon—Viking scientists estimated the total N_2 and CO_2 dumped into Mars' atmosphere during its history. They concluded that Mars once had 10 to 100 times as much of these gases as are now present in the atmosphere—meaning that the early Martian atmosphere was surely much denser than it is now and could have been nearly as dense as Earth's present atmosphere (Owen, 1992, Fanale and others, 1992).

Similarly, geologists studying the total H_2O ejected by volcanoes during Mars' history conclude that the total would have been enough to make a water or ice layer from tens of meters to a few hundred meters deep over the whole planet. Other workers studying orbital photos point to features that resemble scars of glaciers and shores of ancient lakes or oceans, up to hundreds of meters deep in some areas.

All this leads to the conclusion that Mars in the ancient past had a thicker atmosphere, more liquid water, and a more clement environment than it does today. Where did the ancient atmosphere go? What could have caused the transition from earlier, more clement conditions to today's arid, freezing conditions? This mystery is especially interesting in view of our concerns about possible changes in Earth's climate due to increases in CO_2 and other factors.

The answers are partially known. Some of the lighter molecules leaked off into space, a process that is aided on Mars by the planet's low gravity. This means that the total pressure of any early thick atmosphere would have slowly declined. If the early CO_2 atmosphere was thicker and if there was liquid water, then (as we saw on Venus and Earth) the CO_2 would dissolve in the water and form carbonate-bearing rocks. Indeed, Viking landers found carbonate minerals in Martian soils. Some researchers think that this process removed vast amounts of CO_2 from the Martian air and that the early surface pressure of

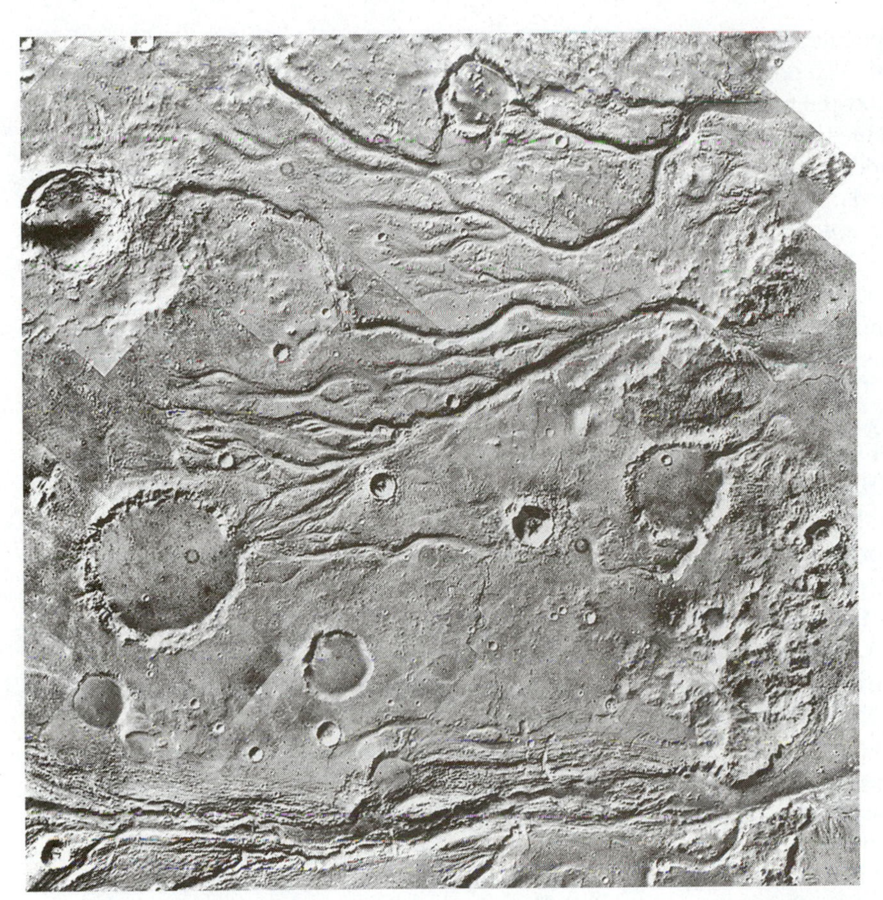

Figure 10-13 Dramatic evidence of ancient riverbeds on Mars is found in these channels and tributary systems. The area is about 180 km (110 mi) wide and drops about 3 km in the direction of flow, from the west (left) to the east. Water apparently cut into some old craters, but predated others. (NASA Viking photo from orbit)

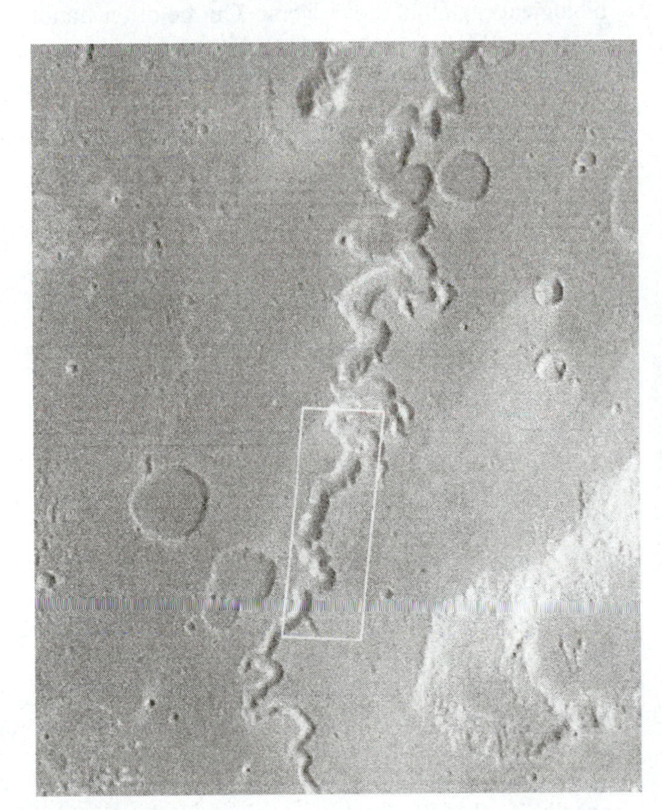

a

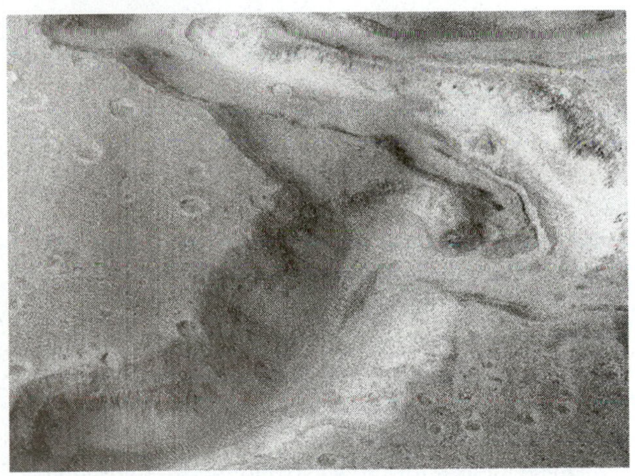

b

Figure 10-14 Possible evidence of a long-time duration of water flow on Mars? **a** Nanedi Vallis is a meandering channel with cutoff loops, or "oxbow" formations, which are often evidence of long, slow flow. Upper end of box is the region of b. (NASA, Viking orbiter). **b** Oblique close-up shows a portion of the winding channel with a much smaller channel down the center, marking a final flow. It is partly obscured by dust drifts. Outcrops halfway down the walls may mark resistant rock layers cut through by the flow. (NASA, Jet Propulsion Lab, and Malin Space Science Systems, Mars Global Surveyor)

Mars' CO_2 atmosphere could have equaled the pressure of Earth's present atmosphere (Warren, 1987; Fanale and others, 1992). Also, any oxygen produced from breakup of CO_2 or H_2O molecules would have combined chemically with rock minerals, oxidizing the iron minerals to rustlike compounds. *The intensely rusted state of Martian iron minerals explains the reddish color of the Martian soils.*

The main constituent of Mars' atmosphere was probably always CO_2, but the total amount may have varied due to loss of the early atmosphere and also due to sporadic volcanic eruptions. If the CO_2 pressure was once, say, 100 times today's value (60% of Earth's current surface air pressure), then the CO_2 greenhouse effect and/or certain light-scattering effects of the CO_2 clouds would have helped ancient Mars have warmer temperatures (Forget and Pierrehumbert, 1997). The warmer temperatures, plus the higher pressure, would have ensured that water could be liquid and not evaporate too fast. This explains conditions that could have produced rivers. However, as the amount of CO_2 in the atmosphere dwindled, Mars would have gone through a stage when all the water began to freeze, forming the permanent polar ice caps and the underground permafrost layer. In this sense, there was a lot of truth behind the idea popularized around 1900 by Percival Lowell: Mars is a drying, dying planet, evolving from a more habitable ancient state to a less habitable state.

★ Thus, understanding the history of Mars' atmosphere and water helps us understand many properties, such as the Mars' color, its ancient riverbeds, and its polar caps.

A best guess about the history of Mars is that it had cold but more Earth-like conditions during its first billion years or so. In the middle one or two billion years, perhaps until one billion years ago, there may have been intermittent periods of water flow, separated by periods of freezing drought. Following the major volcanic activity in the Tharsis region, probably for the last one billion years, the planet has been mostly arid and frozen.

These ideas show how astronomical exploration of other planets can illuminate conditions on our own Earth. Everybody is familiar with the ice ages that occurred in the last hundred thousand years; geological records indicate even greater climate changes in the last three billion years. Martian studies help us realize that planetary climate changes are common and that the Earth's climate, too, may be subject to astronomically caused changes. Finally, Martian studies may help us understand the magnitude of changes we may cause by adding CO_2 to our own atmosphere.

Mystery 2: Where Are the Martians?

Chemical experiments and other data prove that complex **organic molecules,** or massive molecules containing carbon-hydrogen bonds, form easily wherever conditions are suitable. These experiments suggest (but do not prove) that life may evolve easily when conditions are Earth-like, especially when liquid water is available. (We will discuss this issue in more detail in the last chapter.) Therefore, the evidence for an ancient Mars that was more Earth-like, with running water, suggests that life might have evolved on Mars in the ancient past. So, the second mystery of Mars is: Did life evolve there? If not, why not? If so, are there any traces today?

Much of the excitement of Martian exploration has revolved around this question. At the time the Viking landers were launched, scientists thought the first surface pictures might reveal alien animals, strange desert bushes, or at least primitive mosses and lichens.

The question of **life on Mars** is important scientifically and philosophically. Just as the Copernican revolution showed that Earth is not the center of the solar system, the discovery of life on Mars (or elsewhere) would show that Earth is not necessarily the biological capital of the universe. On the other hand, proof that life never evolved there would put some important limits on our ideas of life's origins.

★ The Viking mission was specifically designed to look for life on Mars. Viking soil analysis showed that the soil in the upper centimeters of the surface of Mars was absolutely sterile, with no organic material to an accuracy of a few parts per billion. The cameras, of course, showed no traces of life on the surface. Three other experiments were designed to put nutrients in soil samples and look for signs of life. They gave curious results that indicated chemical reactions in the soil, but probably not life. The chemical reactions are discussed in more detail on our web site. In retrospect, we understand why the Martian surface soils are sterile. Mars lacks an ozone layer. Remember that Earth's ozone layer blocks the energetic ultraviolet photons of sunlight, which can break up organic molecules and kill microorganisms. With no ozone layer, Mars cannot protect its hypothetical surface organisms from the Sun.

So, contrary to the dreams of generations, the surface of Mars "is not teeming with life from pole to pole" (to quote a memorable understatement by astronomer Carl Sagan). But does that rule out the possibility that life evolved on the red planet? The

a

b

Figure 10-15 **a** Overhead view of a cliff face on Mars where water has apparently seeped out of the lower rocky layers at the top, run down the hill, and left fan-shaped deposits of debris at the bottom. Because the deposits are on *top* of dunes, the gullies must be geologically very young. **b** Ground view of nearly identical features in a cliff face in Iceland. Straw bales in foreground give scale. (**a** Mars Global Surveyor; NASA, Jet Propulsion Lab, and Malin Space Science Systems. **b** Photo WKH)

surface soils are sterilized by solar ultraviolet light, but what about the deep subsurface regions?

The emerging new view of Mars is quite exciting in this regard. As we will discuss in more detail in the final chapter, many experiments suggest, but do not prove, that life will form on a planet if conditions are right—in particular, if there is abundant liquid water. No lab experiment has ever produced a living cell, but amino acid molecules, the building blocks of protein, form easily in a variety of lab conditions simulating early Earth, and extraterrestrial amino acids formed inside certain types of water-bearing asteroids. Mars' surface apparently did have abundant liquid water in the past. Furthermore, Mars seems to have widespread permafrost layers of mixed ice and soil, and there is considerable evidence that underground aquifers of water have circulated at the underside of the permafrost layer. In 2000, Mars Global Surveyor scientists announced discovery of gullies in many Martian cliff sides, where water seems to have seeped out of the cliff face and run downhill. Many of these look geologically very young, as seen in Figure 10-15. Therefore, scientists remain excited about searching for microbial life that might have evolved in the deep layers of melted underground ice, or for fossil evidence of ancient Martian life that might have evolved and then become extinct. In fact, this quest to ascertain whether life evolved on Mars is a major thematic driver of the modern Martian exploration program.

It is a perfect scientific question, because any answer would be profound! If Martian life ever did exist, what was it like? Did it base itself on DNA and RNA molecular chemistry like Earth's life? Or is it more alien? If Martian life never appeared, why not? Would that tell us that life on Earth is unique, or that life is much rarer in the universe than we thought?

All these ideas and questions were given a shot in the arm by new research. First, starting around 1977, scientists in Antarctica, discovered that in Mars-like areas where the soils are relatively sterile, algae and other simple organisms live in the fractures and microscopic pore spaces *inside rocks,* shielded from

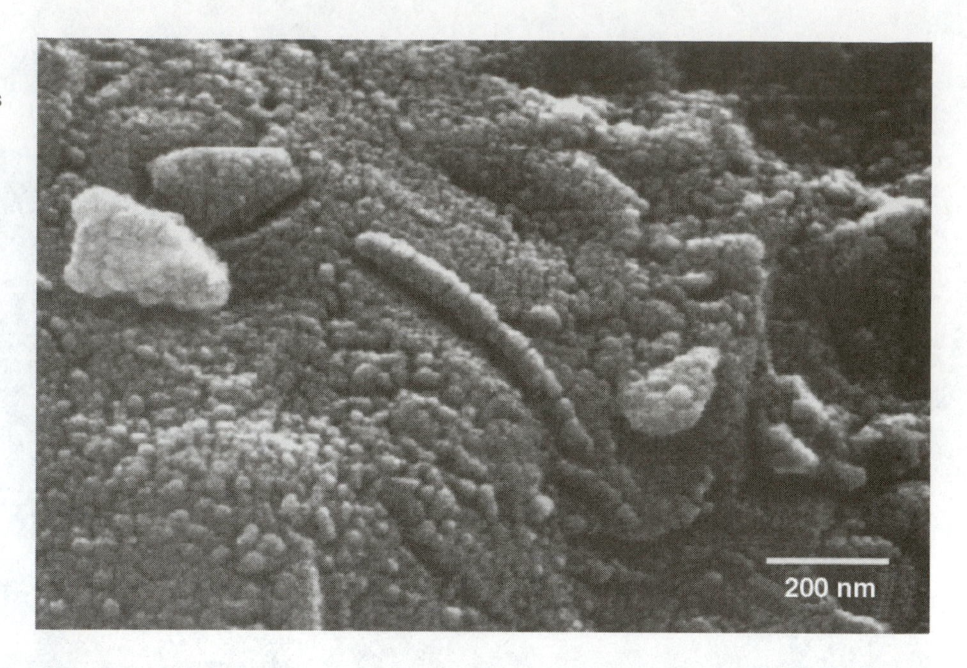

Figure 10-16 Example of the microbelike structures found in carbonate globules inside cracks in a 4.5 billion-year-old Martian meteorite, Allan Hills 84001. Controversy has swirled around whether these are fossil Martian organisms. (NASA, courtesy Everett Gibson)

the harsh external conditions. Second, researchers have found colonies of strange organisms clustered around geothermal vents that emit hot water and minerals on the ocean floor; these organisms derive at least some of their sustenance from geothermal energy and minerals emitted by the vents, rather than depending on the Sun as do all life-forms involved in the food chain on Earth's surface. Finally, scientists in the 1990s pointed out that microbe communities exist in sediments nearly 3000 m below Earth's surface. Perhaps we need to seek primitive life inside Martian rocks or in buried sediments, protected from the solar ultraviolet light of the hostile surface!

These findings set the stage for a stunning development in 1996, when NASA researchers David McKay and Everett Gibson announced that they had found some evidence of fossil microbes inside one of the Martian meteorite rocks—a fragment that formed 4.5 billion years ago in the primordial Martian crust! The microscopic structures, shown in Figure 10-16, were in the right place; they were inside concentrations of organic compounds, which were inside 3.9 billion-year-old carbonate globules deposited by liquid water that had seeped into the rock when it was on Mars (McKay and others, 1996; Gibson and others, 1997). Their announcement initiated violent arguments and controversies. Philosophically, some scientists criticized McKay and Gibson for even announcing their results, arguing that such an extraordinary result required much more extraordinary proof. Others praised them for alerting other researchers about a number of intriguing indications that should be followed up. In terms of the evidence, some said that the structures found by McKay and Gibson were too small to be microbes, because they could contain barely enough DNA molecules to operate a living cell. McKay and Gibson pointed out that some of the structures were larger, and their supporters pointed to unusually small microbes found on Earth. The McKay-Gibson team later found larger possible structures, and proposed that the small structures might be remnant fragments of larger microbes. Some critics said that the materials found must be contamination introduced into the meteorite after it fell on Earth. McKay and Gibson countered that the materials were not concentrated near the surface of the rock, as would be the case for terrestrial contaminants, but were distributed deep inside the rock. The controversy is not settled. Isotopic research showed that some of the organic compounds were terrestrial, but seemed to indicate that another fraction of the organics was not terrestrial (Jull and others, 1998). The quest to find out if life evolved on Mars continues!

Some scientists have proposed that the place to look for fossil (or modern?) life on Mars is at the underground base of the permafrost layer, where the ice might be melted by geothermal heat, and materials would be shielded from the solar ultraviolet rays. This idea became all the more intriguing in 1999 when two teams of scientists reported viable microbes living 3590 m below Antarctica in ice that has been isolated from Earth's atmosphere and surface for more than a million years (Priscu and others, 1999; Karl and others, 1999)! This thick ice covers Lake Vostok, a buried lake that probably also contains organisms.

Figure 10-17 First on Mars. In this concept of a Martian landing a backup supply ship has landed first (background), followed by astronauts in a second ship. (Painting by WKH from nature in a Mars-like volcanic region of cinder cones and dunes in Mexico)

More recently, biologists have claimed discovery of terrestrial microbes that can go dormant for 100 million years or more, and then be revived! Such behavior might allow for viable underground microbial life even on today's Mars!

Do similar buried ice deposits and aquifers carry microbial life on Mars? Only further exploration will tell the tale. Perhaps someday a Mars expedition will ferret out the answer (Figure 10-17). Lunar astronaut (and later U.S. senator) Harrison Schmidt has remarked that today's students will be the parents of the first Martians!

MARTIAN SATELLITES: PHOBOS AND DEIMOS

Not all of the mysteries of Mars are on its surface. In 1877, the American astronomer Asaph Hall became the first human to see a satellite of Mars. Shortly afterward, he charted the positions of two Martian moons, naming the inner satellite Phobos ("fear") and the other one Deimos ("terror"), after the chariot horses of Mars in Greek mythology. Close-up photographs by spacecraft have revealed these moons to be strange, potato-shaped, cratered chunks of rock (Figure 10-18). Their dimensions are given in Table 10-2. The craters of Phobos and Deimos were caused by collisions with small bits of meteoritic debris. The largest crater on Phobos is 8 km (5 mi) across and can be seen at the top of Figure 10-18a. It is named Stickney, the maiden name of Mrs. Hall, who encouraged her husband's successful search for the satellite. A collision violent enough to create such a crater would release as much energy as 100,000 atom bombs of the Hiroshima size, or about 1000 hydrogen bombs of megaton size.

A peculiar feature of Phobos was revealed by close-up photos from Viking orbiters. Networks of grooves reach widths of around 100 m and lengths of as much as 10 km. Some are rows of adjoining craters. The exact nature of the grooves is unclear, but they radiate roughly from the large crater Stickney and may be fractures (somewhat masked by surface dust) caused by the mighty Stickney impact, which nearly shattered Phobos.

The origins of Phobos and Deimos are unclear. Their surfaces are dark in color and apparently of a composition resembling a certain type of black asteroid and also a carbon-rich type of meteorite (called carbonaceous chondrite), which represent fragments of that type of asteroid. In 1989, the Soviet Phobos-2 spacecraft went into an orbit that circulated around Phobos and improved earlier Viking measurements of its density—giving a value of 1.95 g/cm^3. This is lower than the density of ordinary rock and matches the density of the carbon-rich asteroidal material mentioned above. For these reasons, many researchers believe Phobos and Deimos originated as asteroids but were later captured into orbit around Mars. This could have happened 4.5 billion years ago, when primordial Mars had a more extensive atmosphere, which could slow asteroids happening to pass through

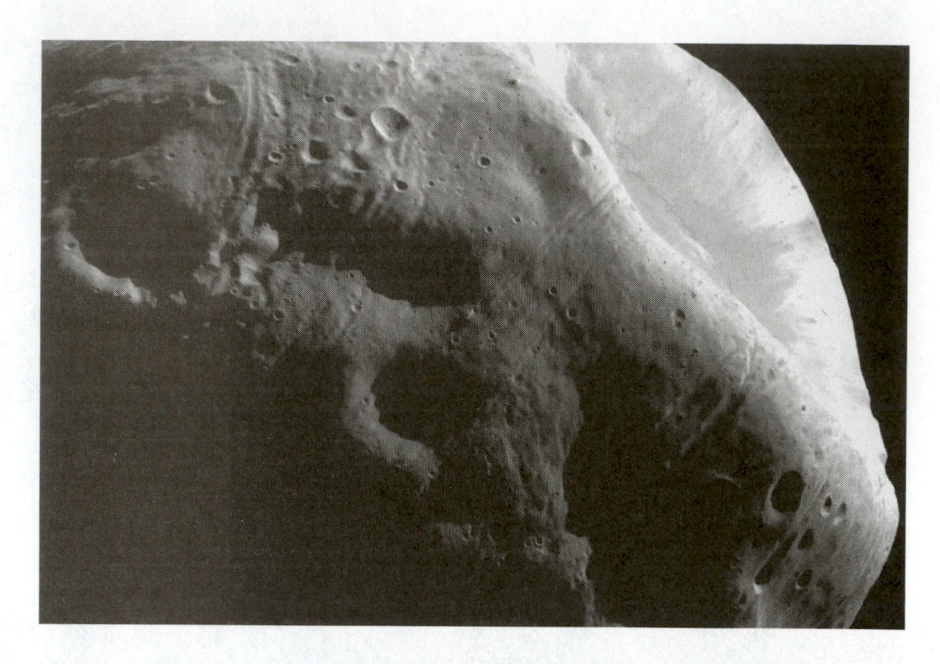

Figure 10-18a Close-up view of a portion of Mars' larger satellite, Phobos. This moon is 28 km long. The largest crater, Stickney dominates the top third of the picture and is 8 km (5 mi) in width. Grooves may be fractures produced by the impact-explosion that created Stickney. Note several scattered house-sized boulders on Stickney's rim (for example upper center). (NASA, Jet Propulsion Lab and Mars Global Surveyor)

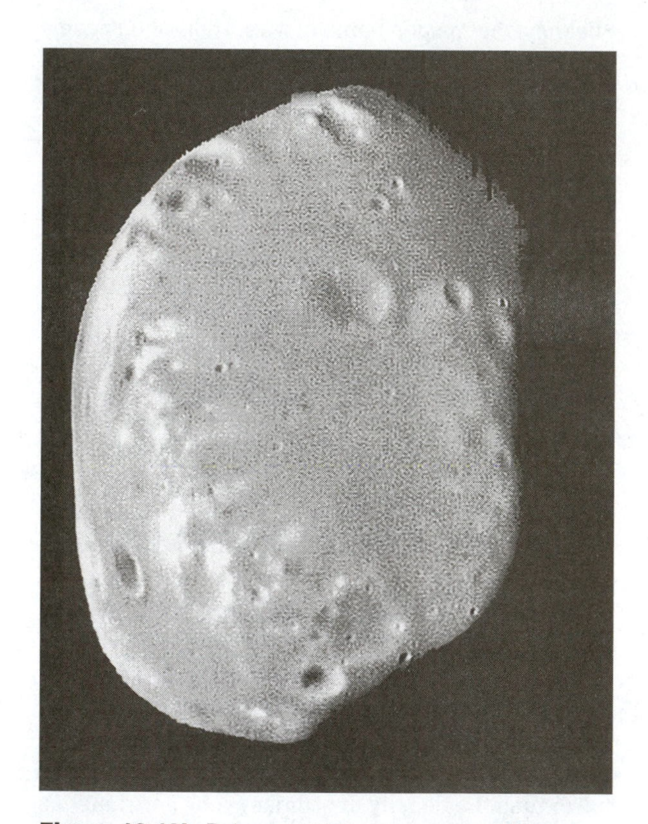

Figure 10-18b Deimos, a 16-km-long moon of Mars. Its surface texture is smoother than that of Phobos, perhaps because of a different duration of microcratering since the most recent large impacts. Streaky markings are believed to be due to slipping of loose material in locally downhill directions in Deimos' weak, asymmetric gravity field. (NASA)

TABLE 10-2

Dimensions of Phobos and Deimos

	Phobos		Deimos	
Diameter	km	mi	km	mi
Longest	28	17	16	10
Intermediate	23	14	12	$7\frac{1}{2}$
Shortest	20	12	10	6

its outer fringes. This slowing could have caused a passing asteroid to be captured into Martian orbit. Phobos and Deimos, according to some astronomers, might be pieces of a single asteroid broken during the capture process. Further drag from the primitive, extended atmosphere may have altered the moons' orbits, but the early atmosphere soon dissipated, leaving the moons stranded in their present orbits.

A LESSON IN COMPARATIVE PLANETOLOGY: THE TOPOGRAPHY OF EARTH, VENUS, AND MARS

Orbital mapping of the planets has allowed researchers to prepare topographic maps showing the altitudes of points across the surfaces of Earth, Venus,

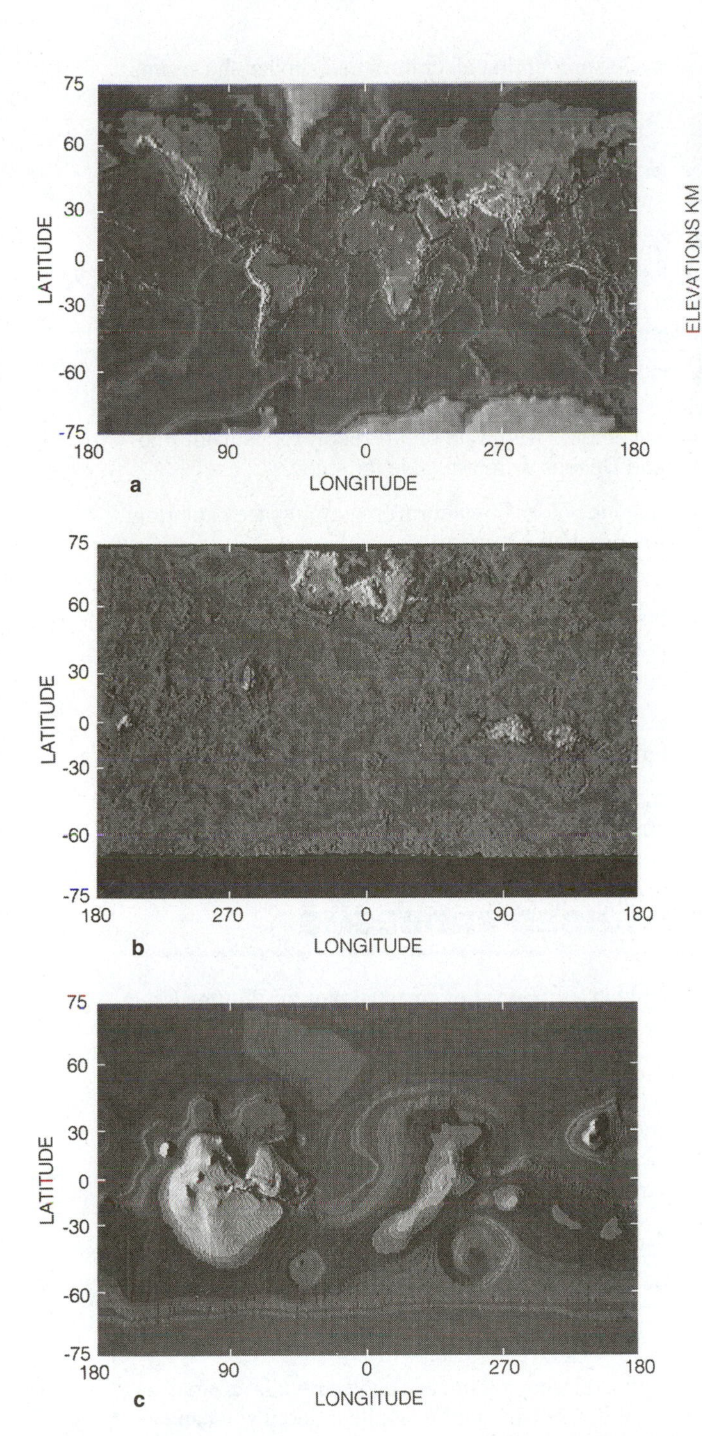

Figure 10-19 Topographic maps of the three largest terrestrial planets; altitude scale (shown by bar) is the same in all three cases. Longitude scale at bottom of each decreases in the direction of rotation of the planet. **a** Earth, showing continental blocks with arc-shaped mountain chains formed by plate collisions. **b** Venus, showing lesser development of "continental" masses. **c** Mars, showing a more primitive surface with circular impact basins and highest spots created by giant volcanic peaks. (Computer-generated topographic maps from orbital radar and other data; courtesy M. Kobrick, Jet Propulsion Lab)

and Mars. Figure 10-19 compares three such maps. A comparison of the maps reveals interesting differences in the geological "styles" of the planets related to their sizes. Earth, the largest, is dominated by rolling seafloor plains interrupted by continental blocks (Figure 10-19a). The same could be said for Venus, although there is not as much elevated "continental" land on Venus, perhaps because smaller Venus did not have as well-developed plate tectonic motions (Figure 10-19b). Indeed, we can see that the Earth's major mountains are arc-shaped ranges that develop when plates collide. Because a smaller planet loses its internal heat faster, Venus did not have as much internal energy to drive such plate motions.

These ideas are affirmed as we turn to the map of still-smaller Mars (Figure 10-19c). Here the map shows a quite different style. This planet was too small to generate even enough tectonic energy to destroy all its original cratered topography. Thus we see that some of the deepest depressions are well-defined circular impact basins, not rolling seafloor plains. The Hellas basin (lower right center) is a prominent example. The uplands (southern $\frac{1}{3}$ of map) are old and heavily cratered. There is a link with Venus, however. The highest Martian mountain areas, such as the broad Tharsis dome (left center), are simply piles of volcanic lavas surmounted by the mighty Olympus Mons volcano and other volcanic peaks, not unlike Venus' Maxwell Montes.

In other words, the maps confirm our rule of thumb at the end of Chapter 9: Worlds smaller than Mars preserve ancient surfaces dominated by the *external* forces of cratering during the early intense bombardment. For worlds around the size of the Moon and Mars, internal energy is significant enough that volcanic forces break through the lithosphere and resurface parts of the planet with lava plains and volcanic mountain peaks. Worlds larger than Mars have surfaces dominated by these *internal* forces, including volcanism and tectonic restructuring, which reshape the surface continually, obliterating all but the youngest craters.

SUMMARY

Mars, the planet once thought to have fields of vegetation or even a dying civilization, has been revealed by spacecraft to be a barren but beautiful desert lacking any visible life-forms. Orbital photos reveal a wide variety of landscapes including lava flows, sand dunes, grand canyons, landslides, polar snowfields, eroded strata, impact craters, dust devils, dry river channels, and the solar system's largest volcanoes.

Evidence about Mars' ancient climatic history seems to contradict its present-day barrenness. Nearly all water on Mars today is locked in polar ice, frozen in the soil, or chemically bound in the soil; virtually no liquid water exists. But liquid water apparently once flowed and eroded the surface, indicating different past climates. Some Mars researchers believe that lakes or even oceans of water occupied some of the low areas of Mars during certain intervals in the past.

One of the most exciting questions is whether life ever evolved on Mars. The conditions required for formation of life, according to our best (but imperfect) understanding, seem to have been present on early Mars. Some researchers believe they have found microscopic fossil Martian microbes and Martian organic compounds inside carbonates left by evaporating water in fractures in Martian rocks. Other researchers dispute this. The jury is still out, and the quest to unravel the secrets of Mars continues.

CONCEPTS

Mars' rotation period
seasonal changes in Mars' features
canal
Viking 1
Martian atmosphere
Martian air pressure
Martian air temperature
permafrost
Martian meteorites
Martian volcanoes
mantle plume
channels
organic molecules
life on Mars

PROBLEMS

1. Describe a day on Mars as a future astronaut might experience it. Include the length of the day, the appearance of the landscape, possible clouds and winds, possible hazards, and objects visible in the sky.

2. What scientific opportunities for long-term exploration does Mars offer compared with the Moon? What qualities of the Martian environment might make operating a long-term base or colony easier on Mars than on the Moon once initial materials were delivered to the site?

3. Compare photos of craters on Mars and on the Moon.

a. Assuming that all craters had similar sharp rims when fresh, which craters have suffered most from erosion?

b. What does this say about lunar versus Martian environments?

4. Give examples of how the Martian environment and geology are midway between those of the smaller planet Mercury and the larger planet Earth. Comment on atmosphere, craters, volcanism, and plate tectonics.

5. What scientific knowledge might be gained from close-up investigation of Phobos and Deimos? What measurements would be of interest if rocks from Phobos and Deimos were available for study?

6. Imagine you are a visitor from outer space exploring the solar system.

a. If two Viking-type spacecraft landed at random places on the Earth, took photos, measured the climate, and took soil samples, what might they reveal about the Earth?

b. How many landings might be needed to characterize the Earth adequately?

c. To characterize Mars to the same degree, how many might be needed?

ADVANCED PROBLEMS

7. What is the weight of an astronaut on Phobos if he or she weighs 130 lb on Earth?

8. In a gravitational field, objects fall through a height h in a time $t = \sqrt{2h/a}$, where a is the acceleration due to gravity. Suppose you are an astronaut exploring Deimos, where a is about 0.0042 m/s^2. Suppose you drop a tool from eye level.

a. How long does it take to reach the ground?

b. How long would it take on Earth, where a is about 10 m/s^2? (*Hint:* Because we are using the SI system, express h in meters.)

9. With a telescope that can resolve details as small as 1 second of arc, what is the smallest detail you can see on Mars during its closest approach to Earth at a distance of about 56 million kilometers?

10. Calculate the velocity needed to launch an object into circular orbit around Deimos from a point on the highest "mountain" on Deimos, assumed to be 8 km from the center. Assume the mass of Deimos is 4×10^{15} kg. Many people can throw an object at about 30 m/s. Would you need a rocket to launch a satellite from Deimos?

PROJECTS

1. Observe Mars with a telescope, preferably within a few weeks of an opposition (the date when Mars is closest to us and opposite the Sun) and with a telescope having an aperture of at least 15 cm (6 in.). Magnification around 250 to 300 is useful. Sketch the planet. Can you see any surface details? Usually, the most prominent detail is one of the polar caps, a small, brilliant white area at the north or south limb contrasting with the orangish disk. Can you see any dark regions? Compare the view on different nights and at different times of the night. (Because Mars turns about once in 24 h, the same side of Mars will be turned toward Earth on successive evenings at about the same hour.) If no markings can be seen, three explanations are possible: Observing conditions are too poor; the hemisphere of Mars with very few markings may be turned toward Earth; or a major dust storm may be raging on Mars, obscuring the markings.

2. For the previous observations, determine which side of Mars you were looking at. (Your instructor may need to assist you.) First determine the date and Universal Time of your observation (UT = EST + 5 h = PST = 8 h. Thus 10 P.M. EST on April 2 = 03^h00^m on April 3, Universal Time). In *The Astronomical Almanac,* the table "Mars: Ephemeris for Physical Observations" gives the central meridian (or longitude on Mars of the center of the side facing Earth) at 0^h00^m Universal Time on each date. From these tables you can find the Martian central meridian for the time of your observation. (Mars turns about 14.7°/h.) Compare your observations with a map of Mars, locating the part of Mars that you observed.

CHAPTER

11

Jupiter, Saturn, and Their Moons

WHAT THE READER SHOULD WATCH FOR IN THIS CHAPTER

The outer solar system has a different kind of planet than the inner solar system. In the outer solar system, we find giant planets, much larger than Earth, with massive atmospheres rich in hydrogen gas. Jupiter and Saturn are the largest and best examples. Their deep atmospheres have dense clouds arranged in bands with various color tones of white, tan, and reddish-brown. All four giant planets are surrounded by rings composed of millions of small particles, but only the ring system of Saturn is visually prominent; the other three systems are thinner and fainter. Both Jupiter and Saturn have extensive systems of over a dozen moons, ranging from bigger than the planet Mercury (!) to kilometer-scale moonlets. Most noteworthy among Jupiter's moons are Ganymede (the biggest moon in the solar system), Io (with erupting volcanoes), and Europa (with a thin ice crust over an ocean of liquid water). Most noteworthy among Saturn's moons are Titan (the second biggest in the solar system, with a thick atmosphere of nitrogen and orangish smog), Iapetus (one black side and one side of white ice), and several midsize icy moons. ■

The terrestrial planets, which we have been studying, are huddled relatively close to the Sun. Now we leave them behind and move to *the outer solar system*—the part of the solar system beyond the asteroid belt.

INTRODUCING THE OUTER SOLAR SYSTEM

The outer solar system contains four **giant planets**—Jupiter, Saturn, Uranus, and Neptune—and a small planet, Pluto. The name of the monarch of the Roman gods is fitting for Jupiter. The biggest planet, it contains 71% of the total planetary mass, nearly $2\frac{1}{2}$ times as much as all other planets combined. All four of the giant planets also have large families of satellites, at least 41 in all. The four giant planets together contain $99\frac{1}{2}$% of the total planetary mass and harbor about 91% of the known satellites.

Much of the best data on the four giants and their satellites was gathered by two space probes, Voyagers 1 and 2, which flew past Jupiter and its satellites in 1979 and then flew on through the Saturn system in 1981. Voyager 2 did the same for the other two giants later in the 1980s. Additional data on the Jupiter system were gathered by the Galileo probe, which went into orbit around Jupiter and gathered stunning pictures of the various moons in the late 1990s.

The four giant planets have much lower mean densities than the terrestrial planets—700 to 1600 kg/m^3 as compared with 3900 to 5500 kg/m^3. Saturn, at 700 kg/m^3, would float like an ice cube if we could find a big enough ocean. (Water's density is 1000 kg/m^3.) This simile is particularly apt. The giant planets evidently *are* made largely of ices, as well as low-density liquids such as liquid hydrogen. Jupiter's diameter measures a little over 10 times Earth's diameter, and Saturn just under 10 times. Uranus and Neptune have diameters about four times the Earth's. Placed on the face of Jupiter, Earth would look like a dime on a dinner plate (see Figure 8-1).

Perhaps the most important principle to remember about the outer solar system is that because it is farther from the Sun and much colder than the inner solar system, it contains much more ice than the inner solar system. Because the gases that formed the Sun and its planet-spawning surroundings were mostly hydrogen, the ices that formed in the outer solar system are frozen compounds of hydrogen, such as water (H_2O), methane (CH_3), and ammonia (NH_3). Thus, instead of forming worlds of rock like the terrestrial planets, the outer solar system formed worlds of rock *plus* ice, often in roughly 50–50 mixtures. This explains many properties of worlds in the outer solar system. The giant planets are giant for two reasons: (1) They had ices in addition to rock, and (2) once they reached a size several times more massive than Earth, during their formation, their gravity was so great that they began to pull in the surrounding hydrogen-rich gases. (The gravity of Earth and smaller planets was too weak to hold these light gases.) Hence the giants are huge, hydrogen-rich worlds. Their surfaces are hidden by colored clouds. Because of the immense, deep atmospheres, the four giant planets are often called *gas giants.*

Again because of the cold temperatures and abundance of compounds such as H_2O, most of the moons in the outer solar system have icy or dirty-ice surfaces, rather than rock surfaces. On some of them, geological processes tended to evaporate the ice off the surface, leaving darker, soil-rich surfaces. On others, internal heating melted the ice, producing watery "lava" that erupted and formed bright ice patches. These moons are fascinating worlds with many unexpected properties.

a

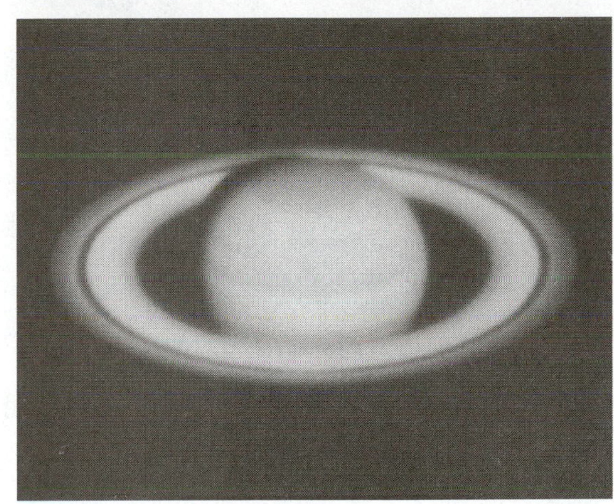

b

Figure 11-1 Comparison of examples of the best Earth-based telescopic imagery of **a** Jupiter and **b** Saturn. Note the dark cloud bands contrasting with lighter cloudy zones. The oval below center on Jupiter is its Great Red Spot, a storm system. Color versions of these pictures are on the web site, where you can see a contrast between the yellow-orange clouds of Saturn and the whitish tones of the rings, which are composed of ice particles. (Catalina Observatory 61-in. telescope; courtesy S. M. Larson, University of Arizona)

JUPITER AND SATURN: THE PLANETS

Jupiter and Saturn look radically different from each other at first (Figure 11-1) because of Saturn's rings. The more we study these planets, though, the more similarities we see and the more natural it is to group them together. They are the largest and most colorful of the giant planets, and they have similar cloud patterns and atmospheric compositions. They both have many satellites. Jupiter has the largest known satellite, and Saturn has the second-largest satellite.

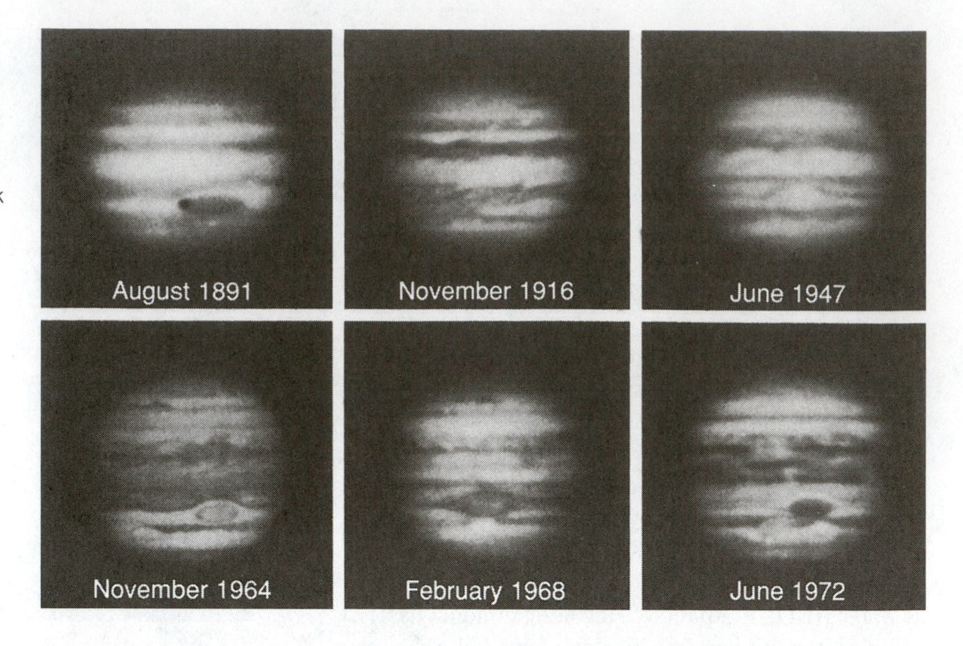

Figure 11-2 Photographs of Jupiter spanning 81 y, showing the changing array of Jupiter's belts and zones. The dark North Temperate Belt is relatively permanent, but the Equatorial Zone changes from bright (1891) to dark (1964). Many of these images show the Great Red Spot. (Lowell Observatory)

Both satellites are larger than the planets Mercury and Pluto, as seen in Figure 8-1.

When we look through a telescope at Jupiter or Saturn, the "surfaces" we see on the globes are not surfaces at all, but merely the tops of the thick cloud decks that cover both planets. These cloud decks have interesting patterns, though the contrast and coloration are stronger on Jupiter. Even a week's observations with a backyard telescope will reveal not only the swirling cloud patterns of Jupiter and the yellowish, blander disk of Saturn, but also the rings of Saturn and a few moons of both planets. The clouds on both planets maintain their general features for many years, but details of these features change from month to month and year to year because of turbulent movements of the clouds, just as Earth's cloudscapes change from day to day (Figure 11-2).

★ The clouds of both Jupiter and Saturn are organized in systems of dark **belts** and bright **zones,** running parallel to their equators (see Figures 11-1 and Figure 11-2). On both planets, the general features are named, as shown in Figures 11-3 and 11-4. The bright, whitish Equatorial Zone and the brownish North and South Temperate Belts are the most prominent features on each planet. Generally, the zones are whitish or yellowish, and the belts are darker, with gray-brown and reddish tinges. The colors in the clouds probably come from colored compounds made from minor constituents. These include polysulfides, phosphorus, and organic compounds. (*Organic* does not imply the presence of life, but merely complex chemistry involving carbon-hydrogen bonds created by the complicated mix of chemicals in the clouds of these planets.) Close-up photos from the Voyager 1 and 2 space probes vividly recorded the colors, even in the smaller cloud formations, as seen in figures on our web site.

The reason for the changes and contortions of smaller clouds on Jupiter and Saturn has to do with high-speed, high-altitude wind motions, similar to Earth's jet streams. The Voyager probes discovered eastward jet-stream winds on Jupiter blowing along the equatorial and temperate zones at speeds 150 m/s (338 mph) faster than the movements of the clouds in the dark equatorial belts. Saturn has even higher differential wind speeds, with the eastward jet stream along the equator moving up to 450 m/s (1010 mph) faster than the neighboring belts.

★ Jupiter has an additional famous feature, the amazing **Great Red Spot,** which is an oval storm system that has existed for at least 300 y, since it was first reported by G. D. Cassini in 1665. It can be seen in Figures 11-1a and 11-2 and is shown in color on our web site. It became very prominent in 1887 when it was rediscovered and given its current name. The Great Red Spot and other smaller, transient spots are probably hurricanelike systems. Small clouds approaching the Red Spot get caught in a counterclockwise circulation like leaves in a giant whirlpool. Close-up Voyager photos give a sense of how neighboring clouds get contorted into fantastic patterns as they get caught in the Red Spot vortex.

★ The cloud features of these planets may look small through a telescope, but they are huge. The Great Red Spot averages three times the width of the entire

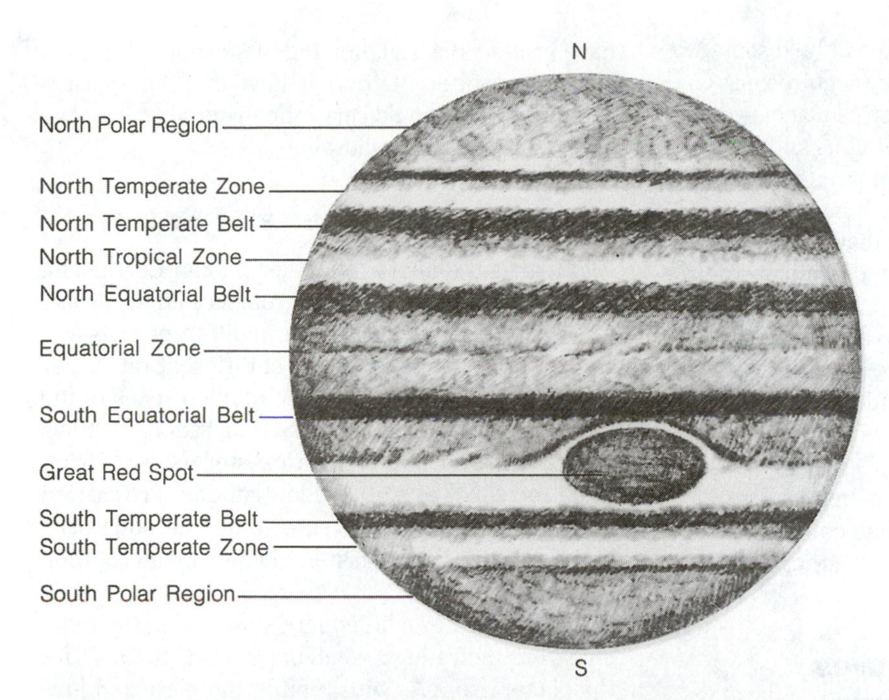

Figure 11-3 Semipermanent cloud formations of Jupiter, visible from Earth when viewed through small telescopes.

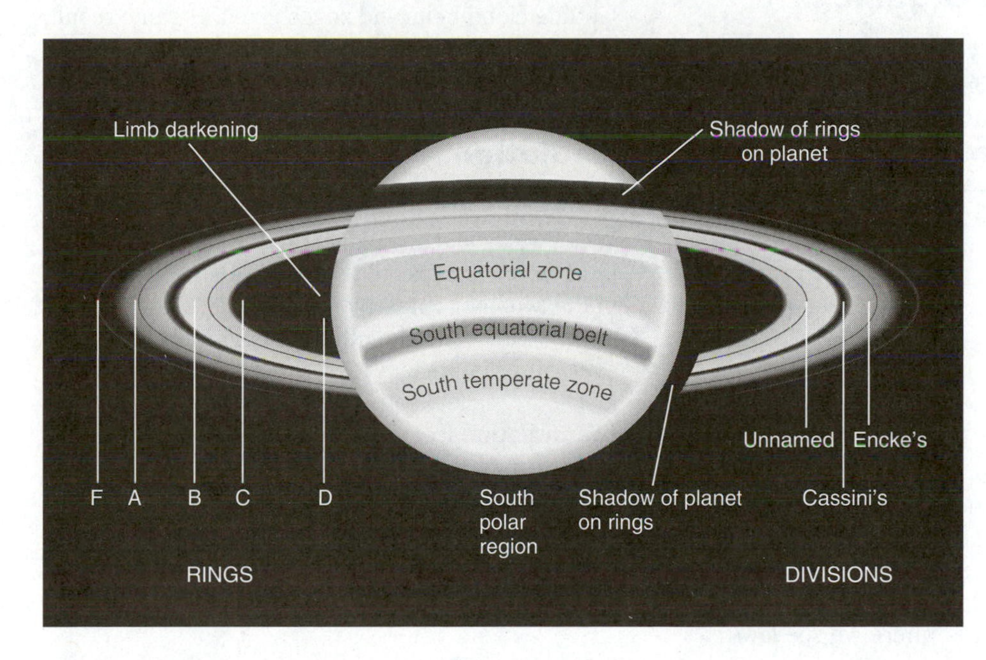

Figure 11-4 Features of Saturn. Telescopes with apertures as small as 5 cm (2 in.) will show the rings; telescopes larger than about 25 cm (10 in.) will sometimes show all these features. *Limb darkening* is a darkening of the edge (or *limb*) of the planet, due to absorption of light in the planet's atmosphere.

planet Earth! It is interesting to speculate on the awesome cloudscapes that would confront a visitor who made a flight into Jupiter's or Saturn's atmosphere, and such visualizations are shown on our web site.

Atmospheric Composition of Jupiter and Saturn

It is very significant that the **atmospheric composition of Jupiter and Saturn** is a mixture of mostly hydrogen and less helium, with only tiny traces of other compounds, and that this hydrogen-helium mixture is about the same composition as the Sun. This gives us an important clue about the formation of the giant planets, as mentioned earlier. The hydrogen-helium mix is roughly the same basic composition of all stars and gas in our region of the galaxy, and it was the composition of gaseous debris that filled the solar system as the planets formed. Jupiter and Saturn got so massive that they had very strong gravity, and their gravitational attraction captured and held some of this gaseous debris. This is how they acquired massive atmospheres. None of the terrestrial planets got big enough to do this. As we study Uranus and Neptune in the next chapter, we will compare the atmospheric compositions of all the gas-rich giant planets and the Sun in more detail, showing that all four gas giants trapped hydrogen and helium in similar proportions as the Sun did.

★ On both Jupiter and Saturn, the clouds are believed to be composed primarily of ice crystals of ammonia, ammonium hydrosulfide, and frozen water. These cloud crystals represent minor constituents that have condensed out of the atmospheres of Jupiter and Saturn. They float in the hydrogen-helium gas of the atmosphere, just as water droplets condense in Earth's atmosphere and form clouds floating in our nitrogen-oxygen air.

Atmospheric Temperatures, Cloud Colors, and Organic Chemistry

Saturn's clouds are less active and less colorful than Jupiter's. The reason for this difference is that Saturn is colder, because it is farther from the Sun (and also because Saturn radiates less internal heat; see section on radiation, below).

To be specific, the daytime and nighttime cloudtop **temperatures of Jupiter and Saturn** average about 100 K to 140 K (–207°F to –279°F); this temperature defines the cloudtops because this is the temperature at which ammonia (NH_3) and certain ammonia-based compounds condense and form clouds. However, because of Jupiter's greater heat, it has more energy to drive storm systems; it has more atmospheric disturbances, such as the Red Spot, and more upwelling of colorful clouds from the warmer regions below. On these planets, the zones usually represent higher, brighter clouds, and the belts represent gaps in the high clouds where we see lower, darker, and more colorful clouds.

The farther down we go in these atmospheres, the warmer they become. The Voyager probes glimpsed gaps in Jupiter's midlevel clouds where deeper clouds could be seen at temperatures as high as 250 K (–9°F). Scientists believe that lower regions, about 60 km below the cloudtops, may reach room temperature at pressures nearly 10 times greater than on Earth—relatively benign conditions except that the gas would be poisonous to Earth's organisms!

Some scientists have speculated that reactions among organic chemicals at these conditions might have produced aerial, floating life-forms in Jupiter's lower atmosphere. Others believe that the updrafts and downdrafts would carry the materials to regions too hot or too cold to allow life.

Rotation of Jupiter and Saturn

All four giant planets rotate faster than Earth. The cases of Jupiter and Saturn are complicated by the fact that the different wind speeds in different belts and zones carry the clouds along at different rates. For Jupiter, the best estimate of the rotation speed of the underlying planetary body, based on radio radiations from deep-atmosphere electrical storms, is $9^h55.5^m$. Belts and zones at high and low latitudes are carried around the planet in this same time, but equatorial belts and zones move faster, giving a mean rotation period of 9^h50^m.

Saturn rotates a little more slowly, with the radio emissions indicating a rotation period of $10^h39.4^m$ for the planet itself. As with Jupiter, the high- and low-latitude cloud belts and zones give this same result, but the equatorial belts and zones rotate a bit faster, giving an average period of about 10^h14^m.

Radiation and Magnetism of Jupiter and Saturn

★ When we look at any planet through a telescope, the light we see is reflected sunlight, just as in the case of the Moon or a nearby mountain on Earth. On Jupiter, Saturn, and the other two giant planets, however, other important kinds of radiation come from inside the planet itself. These include **thermal infrared radiation** from heat of the planet itself and **synchrotron radiation,** emitted when electrons interact with the strong magnetic fields of each planet. These forms of radiation and the strong magnetic fields of Jupiter and Saturn are described in more detail on our web site.

Internal Structure and "Surfaces" of Jupiter and Saturn

What are these planets like under their clouds? Contrary to our experience on terrestrial planets and moons, Jupiter and Saturn probably have no well-defined surface on which we could land. If we descended far below the visible clouds, we would find the atmosphere growing thicker and thicker, at higher and higher pressure. If the temperature were cool enough, a well-defined ocean surface of liquid hydrogen might exist, perhaps 100 km below the clouds; but the temperature is too high, so the gas simply gets denser at lower depths, turning into a mush resembling a thick, hot liquid, but with no well-defined surface. The term "pea-soup fog" might be appropriate on these planets!

Perhaps it is just as well that there is no surface on Jupiter: The gravity is so strong that a 150-lb person from Earth would weigh 400 lb on Jupiter. It would be hard to walk around!

The deeper **internal structures of Jupiter and Saturn** are even stranger. An easy test proves that Jupiter and Saturn are very different from Earth in bulk composition. As mentioned in the opening paragraphs of this chapter, Earth's mean density is about 5500 kg/m^3, whereas Jupiter's is 1300 kg/m^3, and Saturn's is only 700 kg/m^3, which is even less than ice. Clearly, Jupiter and Saturn are not made mostly from rock, like Mercury, Venus, Earth, the Moon, or Mars!

Then what are Jupiter and Saturn made from? By considering the bulk density and other properties, scientists have concluded that Jupiter and Saturn have a bulk composition of roughly 66% hydrogen, with the rest being helium mixed with small amounts of silicates and other impurities. To determine the state of matter inside these planets, scientists use laboratory data on the properties of hydrogen, helium, and other elements at high pressure. By the late 1990s, hydrogen had been subjected to laboratory pressures as high as 1 to 2 megabars—1 million to 2 million times Earth's atmospheric pressure (Guillot, 1999). As was predicted by theory, the experiments indicate that at about 1 to 1.4 megabars, pressure and temperature are so great that hydrogen atoms collide frequently and their electrons' structures break down. The electrons are stripped away from the atoms, so that we have protons surrounded by loose electrons. This state of hydrogen is called liquid **metallic hydrogen,** because the material acts somewhat like a metal. For Jupiter, this result means that at a depth perhaps 9000 km below the clouds, the pressures and temperatures are great enough to transform the gaseous hydrogen into metallic hydrogen. Metallic hydrogen conducts electricity, perhaps explaining the electrical currents that generate Jupiter's strong magnetic field. The structure of Saturn's interior is generally similar to that of Jupiter, though the transition to metallic hydrogen is closer to the center, because the interior pressure is less.

The heavier elements have sunk to the centers of the giant planets. Thus a core of rocky material, or mixed rock and high-pressure ice, resembling a buried terrestrial planet, exists near the centers of Jupiter and Saturn. According to some models, these cores in Jupiter and Saturn may be about 1.5 times the diameter of Earth and contain 10 to 15 Earth masses of rocky or rock/ice material.

Thus, in a real way, Jupiter and Saturn can be visualized as cold, buried super-Earths, surrounded by vast oceans of high-pressure metallic hydrogen, mushy hydrogen fog, and deep hydrogen atmospheres full of clouds, stretching to about 10 times Earth's diameter. A sketch of the structures of the interior of Jupiter and Saturn, compared with Earth and also Uranus and Neptune, is seen in Figure 11-5.

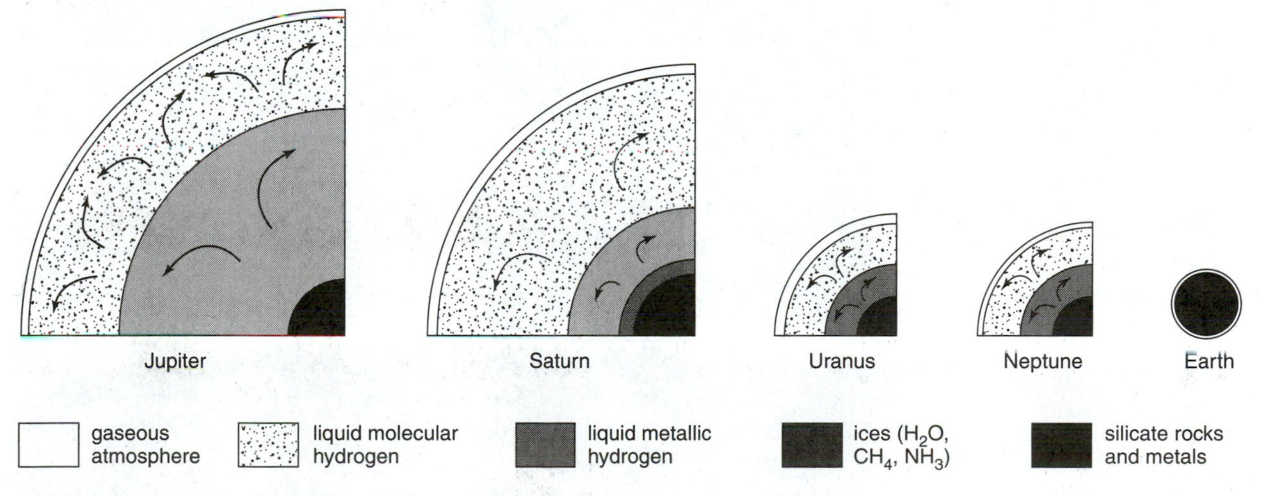

Figure 11-5 Schematic models of the giant planets. Some variations can be found, depending on model assumptions. (Adapted from Hartmann, *Moons and Planets*, 4th ed.; Hubbard, 1990)

RINGS OF JUPITER AND SATURN

Until the late 1970s, Saturn was the only planet known to have rings around it. Then the flight of the Voyager revealed that all four giant planets have **ring systems.** There are many questions about the rings, and we understand only some of the answers. For example, what are they made of, and how did they form? Why is only Saturn's ring system prominent? What explains differences in properties of individual rings from one planet to another?

The Nature of Ring Systems

★ The beautiful ring system of Saturn, as shown in Figures 11-1 and 11-6, has intrigued scientists since 1610, when Galileo first turned his telescope on Saturn. With his poor optics, he could make out only a fuzzy blob on either side of Saturn's disk, and he drew Saturn as a triple planet. Not until 1655 did Christian Huygens realize that a ring system encircled the planet over its equator. Seen with a modern backyard telescope, Saturn's rings present a changing appearance from year to year in a 29-yr cycle. This is because the orientation of the ring plane relative to Earth changes as Saturn orbits around the Sun every 29 y (Figure 11-6).

Contrary to appearances, Saturn's rings are not a solid flat plate but are composed of billions of separate particles, too small to be seen individually from Earth. This was proved by a stepwise combination of theory and observation. In 1859, Scottish physicist James Clerk Maxwell showed that a solid disk would

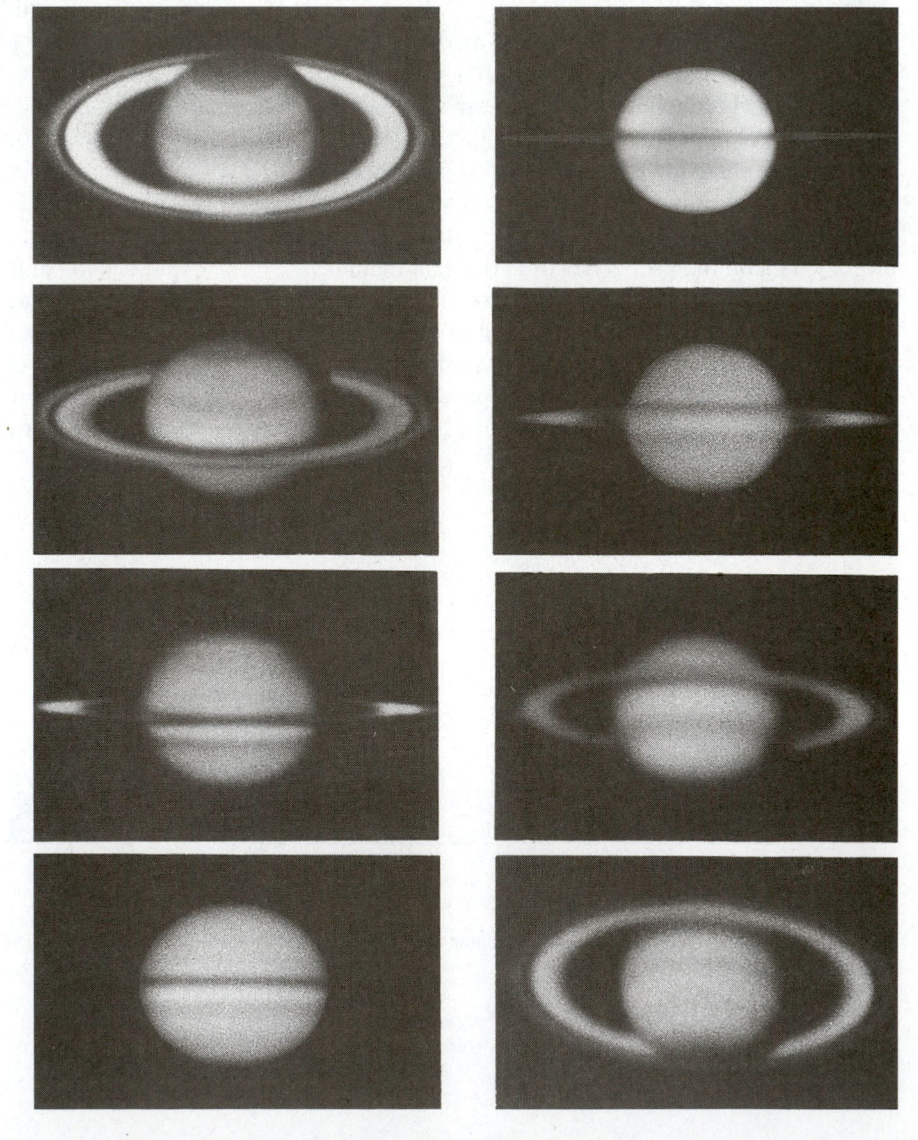

Figure 11-6 Varying aspects of Saturn during its 29-y orbit around the Sun, as photographed with Earth-based telescopes. Because the rings maintain a fixed relation to the ecliptic, the Earth-based observer sometimes sees from "above" and sometimes "below" the ring plane. This series shows half the 29-y cycle, including two views that show the disappearance of the rings as Earth passes through the ring plane. (Lowell Observatory)

Figure 11-7 An unusual view of Saturn in the crescent phase, obtained by Voyager 2 as it left Saturn's system. Such backlighting of the globe can never be seen from Earth because Earth is always closer to the Sun than Saturn is; from Earth we see nearly full lighting on the globe. The shadow of the globe makes a black swath across the rings, and the rings cast a narrow shadow on the globe. Note that part of the globe can be seen through the rings, proving that the rings are not solid. (NASA photo)

not be stable so close to Saturn because it would be torn apart by tidal forces. Then, in 1895, American astronomer James Keeler confirmed spectroscopically that parts of the rings orbit around Saturn at different speeds, proving that it is not one solid ring. Modern spacecraft photos confirm that the rings are not solid because we can see through parts of them, as shown in Figure 11-7.

These discoveries about Saturn's rings apply to all ring systems: They are made of countless tiny particles orbiting each giant planet over its equator. Each giant has its own form of rings, and only Saturn's ring system is dense enough and bright enough to be prominent.

Jupiter's Ring

Our conception of the solar system was changed forever in the late 1970s when scientists discovered rings around planets other than Saturn. There was suspicion of a ring around Uranus in 1977, but Jupiter's dramatic ring was the first one photographed at close range. Voyager 1 took the first pictures of it, unexpectedly; and Voyager 2 showed it more clearly (Figure 11-8). It is a single ring with a sharp outer edge and a diffuse inner edge. Much fainter and narrower than Saturn's rings, it is composed of dark, microscopic dust particles, probably of rocky composition. The ring is so tenuous that it shows up well only from the far side of Jupiter, from a position looking back toward the Sun so that the ring is backlit. Just as cigarette smoke is more prominent when backlit, the tiny particles in the ring make it stand out when the light is behind it. These properties of thinness and lighting explain why we cannot see Jupiter's ring from Earth.

Saturn's Rings

Saturn's rings' structure is very different from Jupiter's and, in fact, shows several rings of different brightness, separated by narrow gaps. As seen in Figure 11-9, the rings have a complex structure, being divided into brighter rings and darker rings, separated by divisions including one gap, wider than the others, called **Cassini's division.** This is a narrow "cleared" zone in the rings, containing only a very low concentration of tiny ring particles, similar to that in Jupiter's ring. Even the broad, seemingly uniform rings named from Earth, such as Ring B, are subdivided into thousands of well-defined ringlets. Wavy ring edges, gentle twists in ringlets, and transient radial shadings were also found.

★ Saturn's rings are not made of dust particles, like Jupiter's, but of chunks of ice. This was discovered in 1970 by American astronomers using spectroscopic techniques (Lebofsky, Johnson, and McCord, 1970). Common particles in the rings range in size from that of a Ping-Pong ball to that of a house. Some larger and smaller particles probably also exist. A visitor to the ring system would be surrounded by an amazing swarm of floating hailstonelike bodies, as shown in an artist's conception on our web site. Although this system is 274,000 km (171,000 mi) from tip to tip, dynamical forces probably keep the rings less than 100 m thick. In terms of relative proportions

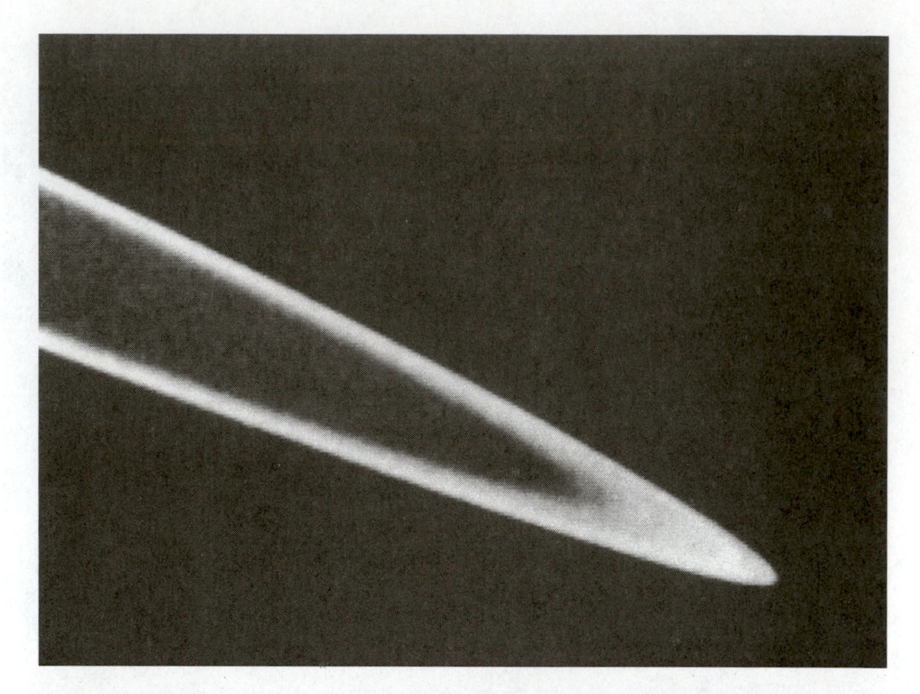

Figure 11-8 Jupiter's ring is believed to be fine material spiraling in toward Jupiter from small moons near the outer edge of the ring. Jupiter is out of the picture at left. (NASA Voyager photo)

Figure 11-9 Close-up photo of Saturn's rings showing the unexpectedly intricate structure—including thousands of individual ringlets. This is a general view of the rings, with overexposed disk of Saturn in background. Disk can be seen through the thinner parts of the rings, but is blocked by the more crowded parts of the rings. Shadow of rings falls on disk at top. (NASA Voyager 2 photo)

of thickness vs. width, Saturn's rings are thinner than a piece of paper!

Tidal Forces Prevent Ring Particles from Aggregating into Moonlets

As explained in Chapter 7, about the Moon, tidal forces tend to pull apart bodies that get too close to a planet. Study Figure 7-6. Look at the fragments making up body *C.* The planet's gravity pulls harder on the pieces closer to the planet than on the ones farthest away. Recall that inside **Roche's limit,** this difference in force is enough to pull the bodies apart. Thus, if two ring particles inside Roche's limit began to clump together due to their mutual gravity, the tidal force would soon pull them apart again.

This is an important principle. It says that if once a lot of small particles are injected into the region inside Roche's limit for any particular planet, they will

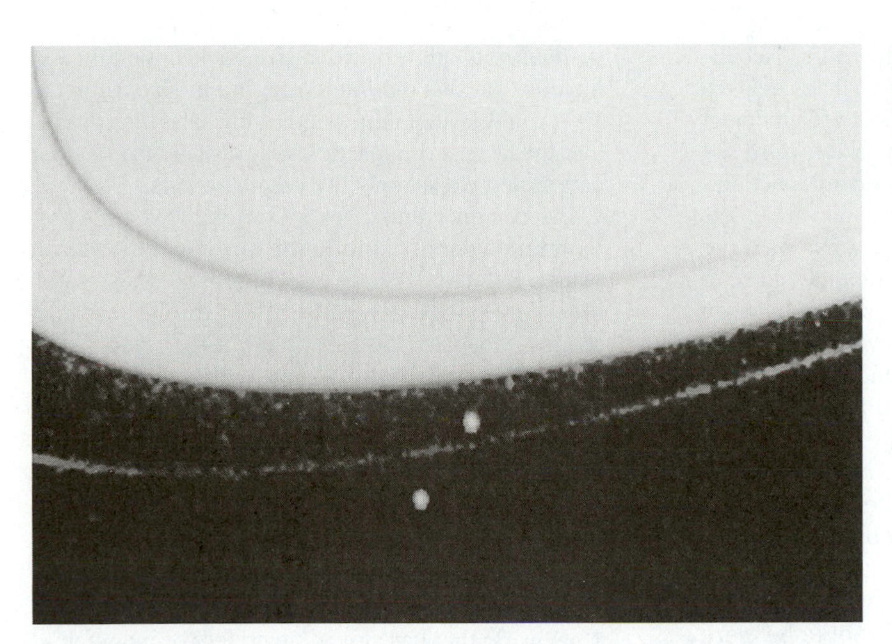

Figure 11-10 Two "shepherd satellites" straddle the very thin F ring of Saturn, which lies just outside the edge of the A ring. In this view, the shepherd moons are less than 1800 km apart and will pass each other in about 2 h. (NASA Voyager 2 photo)

stay separated. Roche's limit for a given planet depends on the planet's density and the ring particles' density, but it is typically between two and three planetary radii from the center of the planet. These seemingly abstract principles are supported by direct observation: *All ring systems are inside Roche's limit for their planets.*[1] Some ring systems, such as Saturn's, have their outermost edges very near Roche's limit.

This permits us to understand why rings persist once particles are established in orbit close to a planet: The particles are too close to the planet to aggregate into moons. Well-understood forces bring such particles into a thin disk over the equator of the planet.

Structure and Origin of Ring Systems

If the rings are made of swarms of millions of particles, where did the particles come from? What clusters them into ringlets with well-defined edges or wavy edges? What forces keep particles out of the almost-empty gap known as Cassini's division?

Scientists are still debating these questions, but it now seems clear that *most of the ring structure is caused by forces acting on the particles from nearby moons and also from smaller, unseen moonlets that may exist within the rings.* Supporting this theory was Voyager 1's discovery of two 200-km moonlets straddling the narrow F ring, just outside Ring A, confining it to a narrow zone as shown in Figure 11-10. Moons of this type, which confine rings into narrow zones, are called **shepherd satellites.** Like a sheepdog, they confine the flock of ring particles into an orderly group.

★ Another support for this idea comes from the biggest gap in Saturn's rings—Cassini's division. Particles in that location orbit around Saturn in about half the orbital period of the nearest large moon, Mimas. This means that those particles would feel a repeated gravitational tug from Mimas, at the same position, every two trips around Saturn. This phenomenon, when one body has an orbital period equal to a simple fraction of the orbital period of another—such as $\frac{1}{2}$, $\frac{1}{3}$, or $\frac{2}{3}$—is called **resonance.** The repeated resonant tug of the larger body disturbs the orbit of the smaller body, often kicking the smaller body clear out of its original orbit.[2] This phenomenon helps explain how nearby moons could clear gaps in the rings.

As for the source of the particles in ring systems, the consensus is that they are debris knocked off

[1]Since the exact position of Roche's limit depends on the density of the ring particles, it would be more correct to say that this is true for plausible densities of icy or rocky particles.

[2]A more familiar example is that of a child on a swing. If you push on a swing at random times, you don't "pump up" the swing's motion because the pushes tend to cancel each other. But if you push at exactly the resonant interval that matches the swing's natural period, the motion of the swing gets higher and higher. The swing is like the small particle in the ring, getting pushed resonantly at just the right intervals to "pump up" its velocity and change its orbit.

satellites just at the outer edge of the rings. Calculations of gravitational forces on such particles help explain the rings. First, particles knocked off a moonlet close to the planet tend to spiral inward toward the planet, thus spreading into a ring. Second, particles close to the planet can't aggregate into new moonlets because of tidal forces. For both reasons, if a large cloud of debris were knocked off a moonlet close to the planet, it would spread into a ring system between the moonlet and the planet, as observed.

In the case of Jupiter, the dark microscopic particles are believed to be debris "sandblasted" off the innermost small moons by meteorites. The ring has a sharp edge near the orbits of the inner moonlets but diffuses inward. In a sense, Jupiter's ring system is like a river: It is not a fixed object but is composed of material flowing through it toward the planet.

Because Saturn's ring system is so much denser and is full of larger bodies, it seems not to be mere "sandblasted" debris. Rather, its particles are thought to be a swarm of fragments of a moderate-size icy moon (like some of Saturn's other inner moons; see below) that was blown apart eons ago by a meteorite impact. According to this view, Saturn's ring system may be a fairly recent addition to the solar system, as some calculations suggest that much of the ring material would spiral inward and disappear into Saturn's upper atmosphere on time scales less than the age of the solar system. Perhaps we live in an era a few hundred million years after a moon was fragmented—an era when Saturn's rings are unusually dramatic! In other eras of solar system history, Saturn may have had thinner rings, and one of the other giant planets may have had a thicker ring system due to a breakup of one of its own moons.

SATELLITE SYSTEMS OF GIANT PLANETS: GENERAL PROPERTIES

Voyagers 1 and 2, which made close-up studies of all four giant planets in the decade between 1979 and 1989, revolutionized our view of the satellites of the giant planets. For one thing, the Voyagers discovered many new small moonlets in each system, bringing the total for Jupiter to 16 and the total for Saturn to 18. Several more small moonlets have been discovered since then. Whereas, in the past, the number of moons of a planet seemed very fixed and well defined, it now appears that the moons in these large systems grade downward in size, in a continuum that includes various moonlets ranging in size from 1 to 10 km on the outer edges of the rings and then grades into the largest ring particles. For this reason, in our data tables, we list only the largest moons.

More importantly, Voyagers 1 and 2 revealed astounding diversity among the large moons of each planet. Before 1979, even the biggest of these satellites were only fuzzy pinheads in the largest telescopes. Most astronomers thought they would turn out to be merely cold, boring iceballs with thousands of craters and nothing more. How wrong this idea was! Voyager data transformed the satellites into a variety of worlds with distinctive personalities. Among them:

- A sulfur-dominated world with the most active volcanoes in the solar system
- A world that has almost no craters or mountains, but only an ice sheet, probably covering an ocean of water
- A world with a nitrogen atmosphere denser than Earth's and smoggy methane-rich clouds
- A world with fractured canyons of ice, kilometers deep
- A world with a thin atmosphere and columns of smoke rising from geyserlike vents
- Several worldlets that go around their planets in a "backward" direction—opposite to the direction of all planets and most other moons

As stressed earlier, two of the moons of the giant planets are bigger than the planets Mercury and Pluto. Clearly, the moons of the outer planets should not be demoted to minor status. It is fair to say that the solar system is composed not just of nine planets, but rather of two dozen world-class objects (not to mention countless smaller bodies); some are planets and some are moons—with individual geologic and astronomical personalities.

Satellite systems of the giant planets are very different from satellites of the terrestrial planets in another regard. Earth has one moon spawned by a giant collision, and Mars has two tiny moons consisting of captured asteroids. But each of the giant planets is like a miniature solar system, with a massive central body surrounded by a family of orbiting smaller bodies of different sizes. As we will see in a later chapter, the formation of these satellite systems apparently involved the same processes as the formation of planets around the Sun (Peale, 1999).

To explain our approach to satellite systems, let us point out that each giant planet is surrounded by an imaginary spheroidal volume of space called its **sphere of gravitational influence,** in which the planet's gravity dominates the Sun's gravity in terms of controlling satellite motions. For example, if you tried to put a satellite in orbit around Jupiter, but *outside* Jupiter's sphere of gravitational influence, the satellite would eventually drift away from Jupiter into an orbit around the Sun.

Thus satellites occur only inside the sphere of gravitational influence. Inside this imaginary spheroidal volume, we find that as a rough rule of thumb, each giant planet's family of moons can be divided into four groups:

- Countless minimoonlets and dust particles involved in the ring systems. We have already discussed these.
- Small moons close to the planet, on the outskirts of the ring systems (or, in some cases, inside the ring system). These are sources of ring material. Most of them were too small to be seen from Earth and were discovered by the Voyager probes.
- Large moons at intermediate distance from the planet. These have the most distinctive geology.
- Small moons on the outskirts of the planet's sphere of gravitational influence. These moons seem not to be native to the planet, but rather passing interplanetary bodies that were captured by the planet's gravity.

In summary, each giant planet has a satellite family composed of several classes of satellites plus a ring system of tiny particles.

The Voyager data revealed an interesting trend among satellites: The larger satellites closest to the planets tend to be the most modified by geological processes, while the distant satellites tend to be more pristine. There are several causes of active geological processes close to the giant planets. In decreasing order of importance, these are:

1. The inner satellites are not only closest to the planet, but also closer to each other. For this reason, the tidal stretching force of the planet on the moon is greatest for the innermost moons, and the tidal and other gravitational forces of each moon acting on its neighbor are stronger for inner moons. In certain cases, these interactions produce strong and variable tidal stretching forces on some inner moons, leading to internal heating, which in turn powers geological activity on those moons. This internal **tidal heating** of satellites turned out to be much more important in the solar system than anyone had dreamed before the close-up study of these moons.

2. The giant planets attract interplanetary meteorites, so the rate of impacts is much greater on inner moons. This means that the innermost moons have the greatest chance of having suffered catastrophic impacts.

3. Particles blasted off moons by meteorite impacts tend to spiral *inward.* Therefore, certain inner moons are affected by material from their outer neighbors.

4. When a planet forms, it gives off heat; thus some inner moons may have been affected by early heat radiated from the planet itself.

These processes produce an important rule of thumb for understanding satellite systems: While most outer moons are rather inert, cratered worlds, some moons closer to their planets have undergone heating and have distinct "geological personalities."

For these reasons we will discuss different moons in order outward from the planets, but first we will consider their basic compositional materials.

Two Basic Satellite Materials: Bright Ice and Black Soot

When we dealt with the worlds of the inner solar system, we were dealing with ordinary silicate rock materials, like familiar gravel or the basaltic rocks from lunar and terrestrial lava flows. However, the two most common materials of the worlds of the outer solar system are not silicate rocks, but ices (of several compositions) and black, carbon-rich dirt called **carbonaceous material,** which probably resembles black, sooty dust and gravel. Thinking about these two materials, bright ice and dark soot, will help in understanding the surfaces of the moons of the outer solar system.

The reason ices and soot dominate is that the outer solar system is so cold that, in addition to silicate materials, icy and carbonaceous materials also condensed there to form solid compounds. The ice and sooty material overwhelmed the smaller amount of rocky material. The situation is like going to Antarctica and finding that the landscape is composed mostly of ice, not familiar soil.

Therefore, a simple but useful picture can be obtained by thinking in terms of a "salt and pepper" model with two main types of materials: bright ices and black carbonaceous soils. The ices include familiar frozen H_2O at very low temperatures, along with frozen carbon dioxide (CO_2, also called dry ice), frozen methane (CH_4), and other frozen material. All the ices are white, and the pure carbonaceous material is black, as black as black velvet. This color comes from graphitelike carbon compounds. At greater distances from the Sun, the carbonaceous material has a dark chocolate-brown or even reddish-brown color, apparently due to organic compounds that condense at lower temperatures farther from the Sun.

A small amount of the dark material can be very effective in discoloring the ice, so a mixture of mostly ice and a little carbonaceous material can be very black in color. The primordial, undisturbed mixtures of ice and sooty carbonaceous dirt are believed to have this character. Scientists often refer to such material as **dirty ice,** but it should be visualized as blackish in color in its original form, not like a white snowbank sprinkled with dirt.

These ideas help explain what we see on most of the moons. The more pristine moons (usually, the outer, unheated moons) have dark surfaces. But certain geological processes can separate the black and white materials. For example, heating melts the ices, causing the dark soils to sink and watery material to freeze on the surface, leaving bright white areas. On the other hand, countless impacts can tend to vaporize the ices and concentrate the dark soils.

Watch for these processes as we discuss the moons. They help explain why the major moons have distinct personalities. During our discussion, you can refer to Table 8-1 for specific physical properties of these bodies.

SATELLITES OF JUPITER

A partial map of Jupiter's system of satellites is shown in Figure 11-11. It is partial because, in order to show the outermost moons' orbits, the scale must be too small to show the innermost moons' orbits, which are crowded close to Jupiter. Jupiter is the gray dot at the center of the system (lower left), and the innermost moons are very close to it.

Moving outward from Jupiter, the next four circles show the near-circular orbits of the four large intermediate moons. Like the inner moons, their orbits lie in the same plane, over Jupiter's equator.

★ These large moons are named Io, Europa, Ganymede, and Callisto (after mythological companions of the god Jupiter). They are easily visible in small telescopes and were discovered on two consecutive nights in 1610 by Galileo and the German astronomer Marius as they viewed Jupiter through the newly invented telescope. Recall that this discovery helped Galileo advance the Copernican revolution (see Chap-

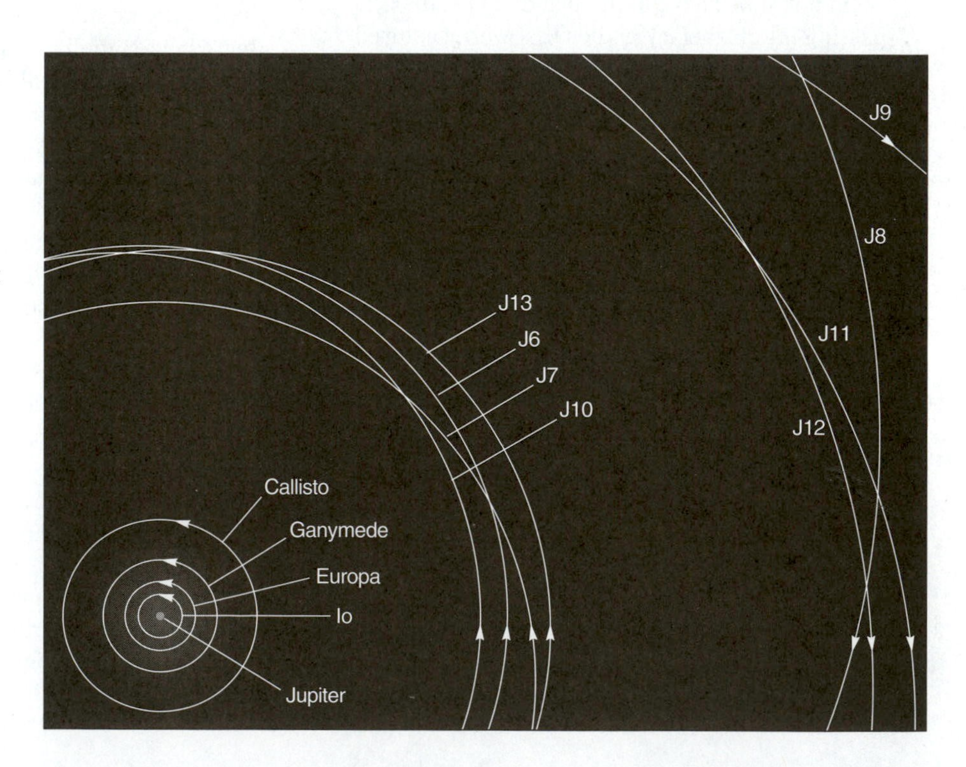

Figure 11-11 The "miniature solar system" of Jupiter and its satellites. Orbits of the four large Galilean moons are shown close to Jupiter (lower left). Still closer, smaller moons and the rings are too close to show on this scale. Two groups of outer moons lie in closely clustered orbits. Note that the outer group moves in retrograde direction—opposite to the usual sense of orbital motion in the solar system. See text for further discussion.

ter 3). These moons are referred to as the **Galilean satellites,** in honor of Galileo's observations.

Just as our Moon keeps one side toward Earth, the Galilean satellites keep one side toward Jupiter. The reason is the same: tidal forces. If you don't remember how this works, review the section on the Moon's synchronous rotation in Chapter 7.

Finally, Figure 11-11 shows two groups of small outer moons, numbered J6 through J13. Note the arrows that show the directions of orbital motion, and note that the four outermost moons move in a "backward" direction—a curious fact that we will discuss below. The orbits of these eight moons, unlike the others, do not lie over Jupiter's equator but are inclined at about 30° off the equator.

Among the moons in Figure 11-11 are good examples of the separation of the bright ices and dark sooty material. For example, Europa, the second Galilean satellite, has a surface of relatively pure, uncratered white H_2O ice, while Callisto, the outer Galilean satellite, is dark and cratered. Our goal in exploring the satellites is not only to reveal and experience these alien worlds, but to explain their differences. We will find some surprising effects.

The Small Innermost Moons near Jupiter's Ring

At least four small moonlets exist just outside Jupiter's faint ring. The largest, designated J5 Amalthea, was discovered in 1892 and is the fifth largest of Jupiter's moons. It is a cratered lump 270 km long and 155 km wide (roughly the size of New Jersey). Amalthea violates our black-and-white salt-and-pepper theory. It is orangish. The color is believed to come from sulfur atoms ejected from the next outer moon, Io. These atoms spiral inward and coat Amalthea. Thus Almalthea probably does not reveal its true color.

The three other small inner moons, all discovered by the Voyagers, are around 40 km across. One orbits between Amalthea and Io, and the other two are nearly on the outer edge of Jupiter's ring. Little is known about them.

Io: Jupiter's Volcanic Moon

★ We will now discuss the four large Galilean satellites one at a time, because their properties are very different and each is an important world in its own right. We include some striking black-and-white pictures in the book, but many more vivid color images can be found on our web site.

★ The inner Galilean moon, **Io,** is probably the most bizarre world in the solar system. It is about the size of our Moon, but what a difference! In the first place, Voyagers 1 and 2 discovered a unique, mottled surface of pale yellow, orangish, tan, and white sulfur deposits (Figure 11-12). To appreciate the colors of Io, look at the color images on our web site. When the pictures of Io first arrived from Voyager 1, one researcher joked that it looked like the kind of world we used to laugh at when it appeared outside spaceship windows in grade-B science fiction movies. But there it was, real!

Io's second claim to fame is that the Voyagers also discovered active volcanoes shooting ashy debris into umbrella-shaped plumes 100 km above the moon's surface, as seen in Figure 11-13. Once they were discovered by the Voyagers, Io's volcanoes were also detected by infrared telescopes from Earth, which picked up their thermal infrared radiation. This technique, plus later studies in the 1990s by the Galileo spacecraft orbiting Jupiter, showed that Io's main volcanoes are long lived. The Galileo probe saw many of the same vents in 1999 that had been seen by the Voyagers in 1979. The eruptions are ongoing: one volcanic vent may shut down, but another one may start up. During the Galileo flight, twelve major volcanoes were erupting at once on Io! Thus Io is the most volcanically active world in the solar system! Although Jupiter's satellites generally have very cold surfaces (measured dawn and afternoon temperatures range between 80 and 155 K, or −315 and −180°F), Io's volcanoes are local hot spots. Typical measured lava temperatures are as high as 420 to 620 K (297 to 657°F) (Carlson and others, 1996), with some vents erupting material as hot as 1000 to 2000 K (McEwen and others, 1998).

In a realm of ice and cold carbonaceous compounds, what source of heat could explain the erupting volcanoes and their sulfur flows? This would have been a total mystery when Voyager arrived, except for some brilliant detective work by California geophysicists S. J. Peale, P. Cassen, and R. Reynolds (1979). While the two Voyager probes were on their way to Jupiter, Peale and his colleagues discovered the tidal heating effect mentioned earlier. They calculated that although Io's orbit is nearly circular, gravitational pulls of neighboring satellites cause Io's distance from Jupiter to vary slightly during its orbital motion. As the distance varies, the tidal stretching of Io changes. This flexing heats Io's interior, just as rapid flexing of a tennis ball makes it heat up from friction. The calculations showed that this heating

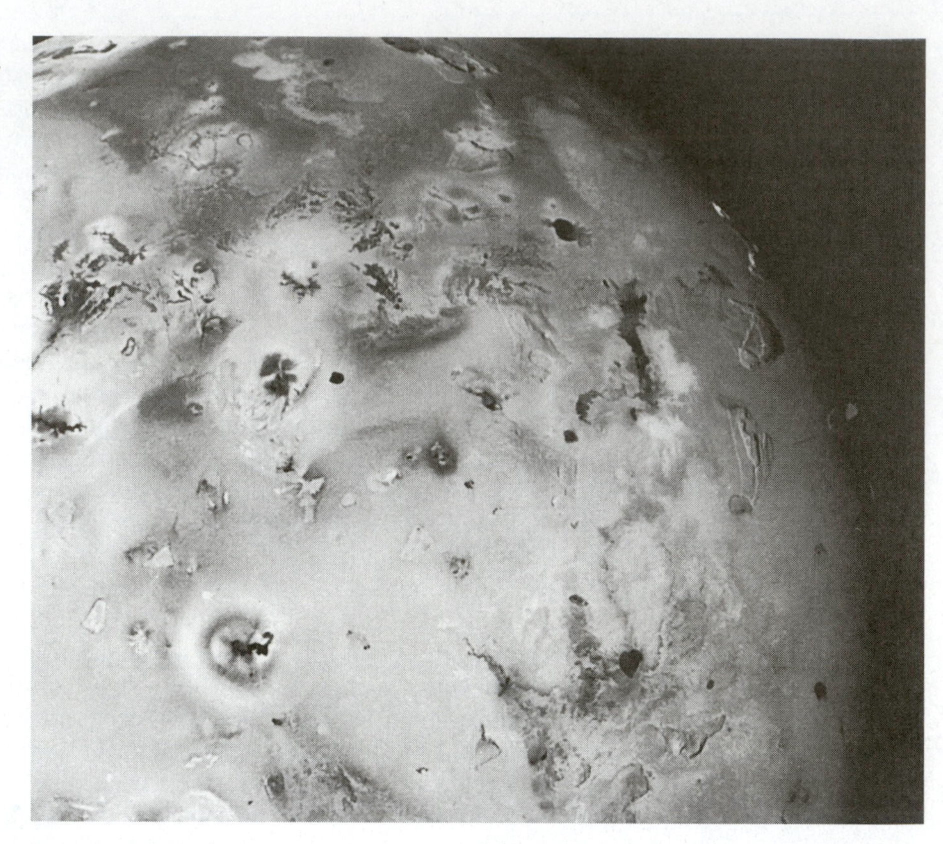

Figure 11-12 The mottled surface of Io. The surface is relatively flat, although mountain masses can be seen along the terminator (right). Small black spots are volcanic calderas, possibly containing molten sulfur. Dark flows emanate from some of them. Vertical view through an erupting volcanic plume produces the doughnut-shaped feature surrounding a dark caldera, in lower left center. (NASA, Galileo spacecraft view)

effect is stronger in Io than in any other satellite and causes Io's interior to be molten. In a beautiful example of the scientific method at work, Peale and his colleagues published their prediction that volcanoes might be found on Io in a journal that came out just a few days *before* Voyager 1 reached Io. A few days later, Voyager discovered the solar system's most active volcanoes, spewing 100-km-high plumes of debris!

★ What are the compositions of the colored compounds and the hot lavas? Since we just explained that most giant planets' satellites are mixtures of ices and black carbonaceous materials, isn't it strange that Io is surfaced by colorful sulfur compounds? The explanation comes from the intense heating of the moon. The heating and volcanism have driven off all the ices and easily gasified carbonaceous compounds. (Io may have been considerably bigger before it lost those materials.) Only sulfur-rich materials and silicates remain. The lower-temperature volcanoes are too cool to be silicate lavas like Earth's; they are probably erupting molten sulfur. Molten sulfur is black but cools to reddish-orange and yellowish colors. As can be seen in Figure 11-12 and the web site figures, many of the volcanic calderas have black spots on their floors and orangish-brown flows around them; these are probably black pools of molten sulfur lava, surrounded by cooling sulfur flows. The highest-temperature volcanoes on Io may be erupting silicate lavas like those of Earth. The white areas in the photos are believed to be sulfur dioxide condensates.

Europa: Jupiter's Icy Billiard-Ball Moon

Europa, the next moon outside Io's orbit, is a bit smaller than Io or our Moon. It is almost as astonishing as Io, but for the opposite reason. Instead of orange lava flows and erupting volcanoes, it has a bland white color and few contrasting features! Seen from a distance, as in Figure 11-14, Europa has a relatively smooth surface. There are no big craters, mountainous peaks, or dark markings—hence the comparison to a billiard ball.

Spectroscopic studies show that the white material is frozen water, about as bright as a sheet of paper. The relatively high bulk density of 2970 kg/m^3, however, shows that Europa cannot consist entirely of ice, which has a density of 1000 kg/m^3. There must be a large, rocky core.

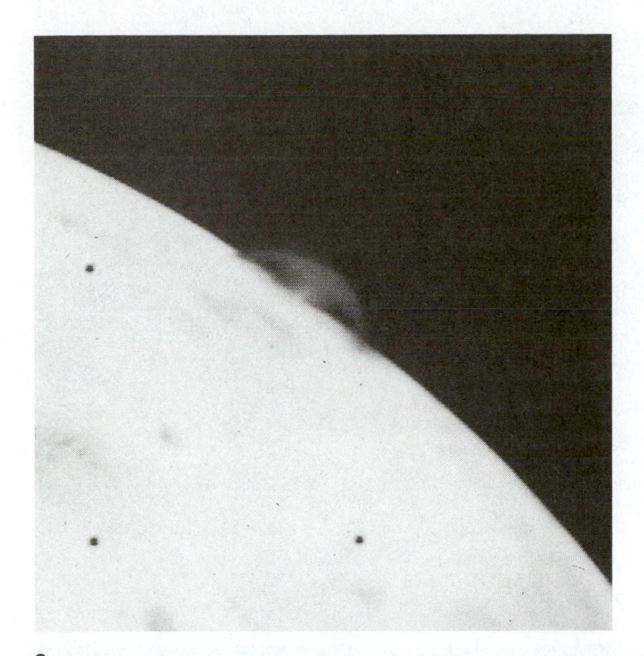

a

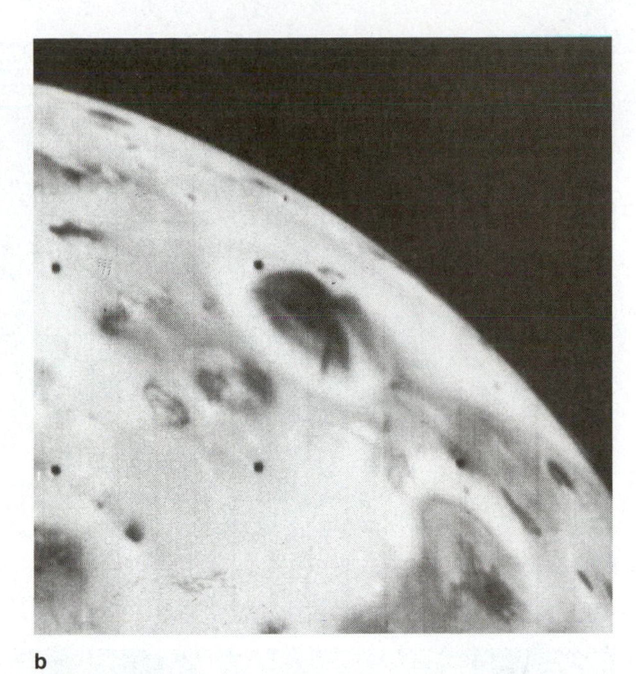

b

Figure 11-13 An erupting volcanic plume on Io. The plume is about 70 km high and 250 km wide. It was seen for 18 h by Voyager I and was still active 4 mo later, as revealed by Voyager 2. **a** Plume silhouetted on horizon. **b** Plume seen obliquely against Io's disk. Note distinct arcs of ejecta, indicating complex jetting at vent. (NASA Voyager 1 photo)

Closer examination and contrast enhancement (Figure 11-15) show that the surface is broken by delicately shaded, shallow grooves and cracks. The cracks separate large plates that have drifted apart. How did such a strange surface form? Probably a modest tidal flexing, like that which acted on Io, heated the interior and melted the ice components. Water erupted onto the surface and froze into a clean, smooth ice layer, obliterating the original crust. Fractures formed due to changing tidal forces as Europa moved around Jupiter (Hoppa and others, 1999). This fracturing and resurfacing must be continuing even today, because the lack of impact craters shows that the surface is very young, geologically speaking. The mean surface age is estimated at a few hundred million years, perhaps even younger than the average age of Earth's land surface. Some Voyager scientists believe that the final Voyager 2 photo of Europa shows a transient cloud, marking an eruption. Perhaps Europa has occasional giant geysers, although no such eruptions were confirmed by the Galileo orbiter in the 1990s.

Models of Europa suggest that the ice crust may be only 100–150 km (60–90 mi) thick, overlying a liquid water ocean (Carr and others, 1998). In his novel

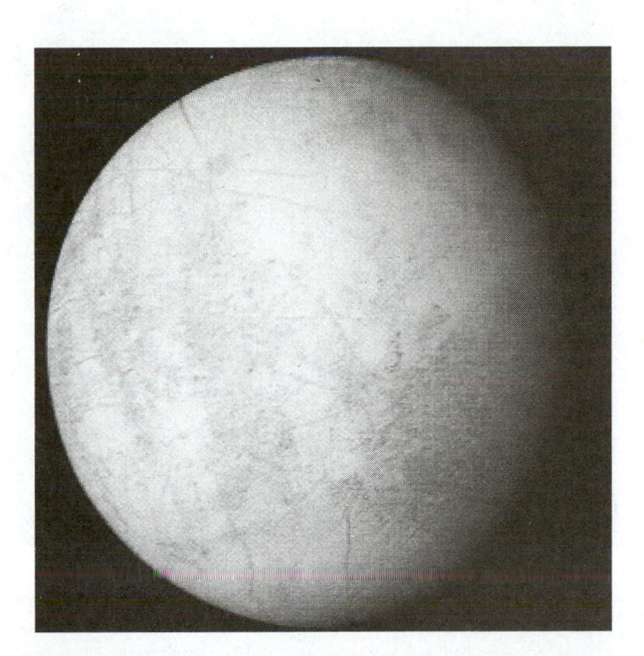

Figure 11-14 Jupiter's icy billiard-ball moon Europa. This world is nearly the size of our Moon but is covered by nearly smooth, whitish ice. Pale dark streaks appear to mark fractures. (NASA Voyager 2 photo)

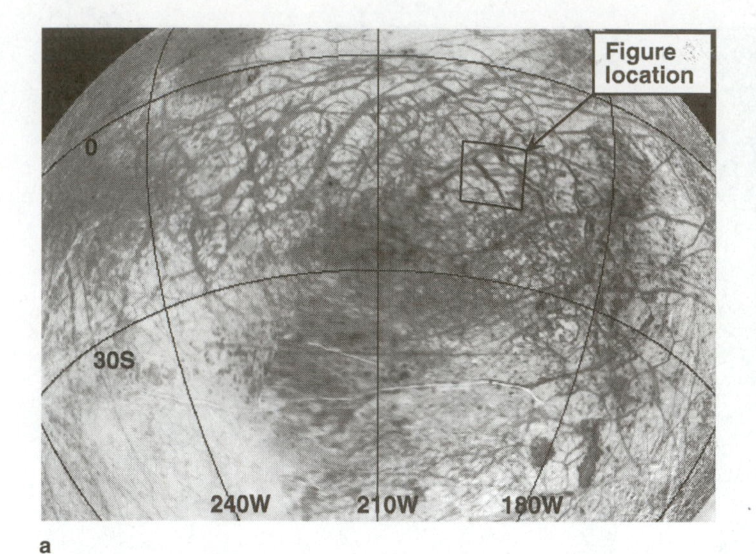

a

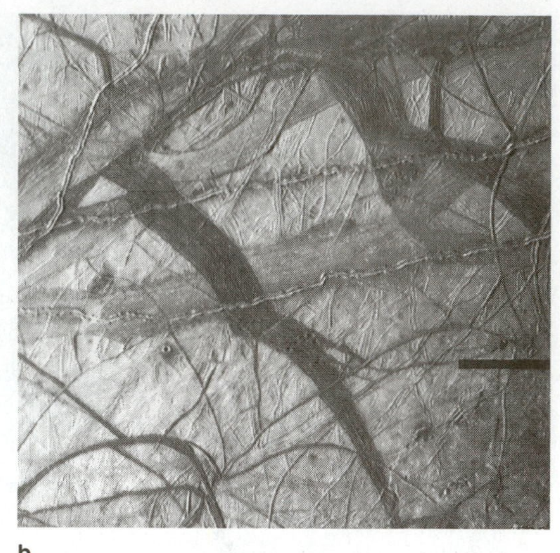

b

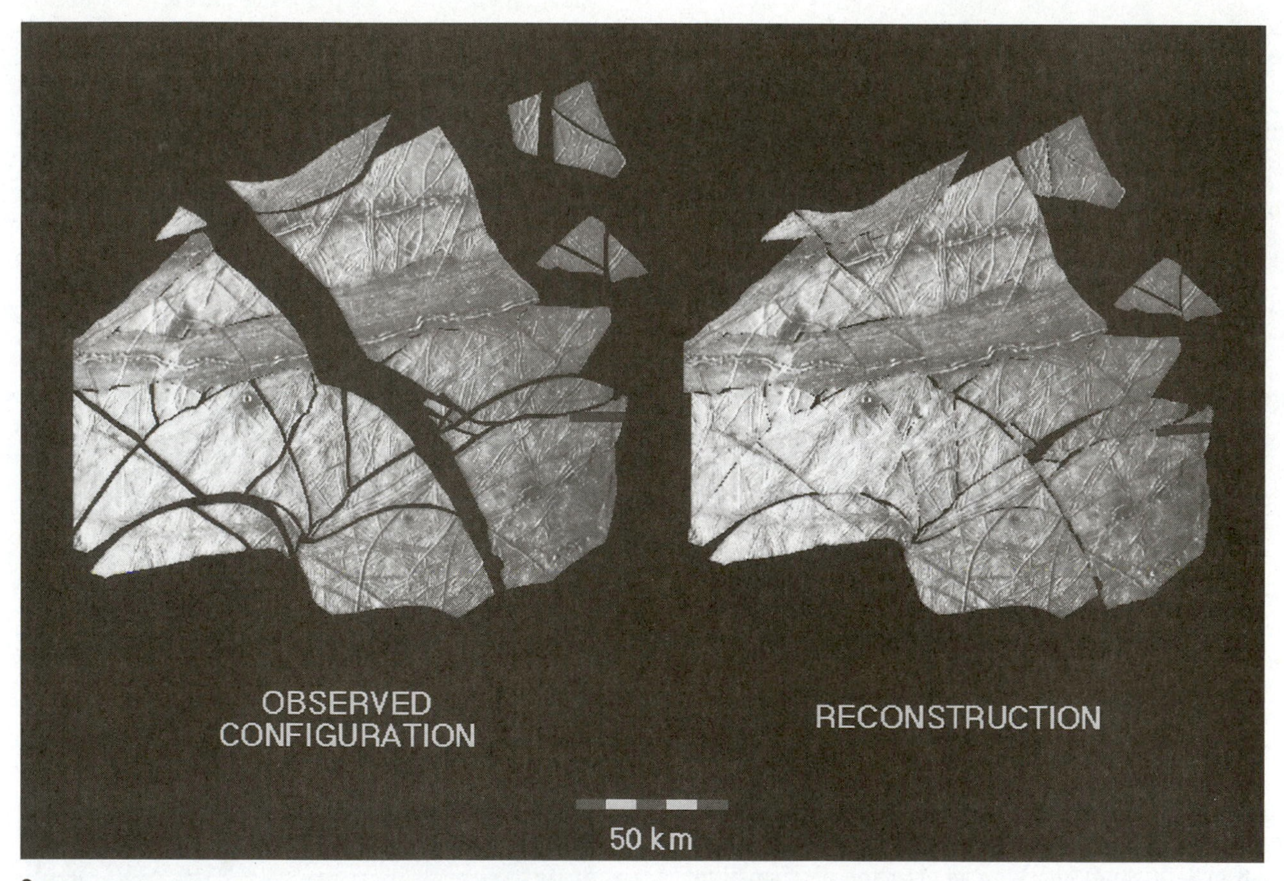

c

Figure 11-15 Solving the Europa jigsaw puzzle. **a** Global view of Europa showing location of detailed figure. **b** Close-up view of present-day fractured system. The darker bands mark fractures that have separated the brighter platelike masses of ice. Contrast is enhanced in a and b. **c** Reconstruction of ice places in b. Fracture spaces have been filled in with black (left), and then the plates have been reconstructed into the original surface (right). This shows dramatically how Europa's ice crust is constantly fracturing and ice plates separating into new positions, with fresh ice filling in the fractures. (NASA, Galileo images courtesy Robert Sullivan, Cornell University; see Sullivan and others, in press)

2010 (sequel to *2001*), science fiction writer Arthur C. Clarke speculated that life might have evolved in Europa's buried oceans. The plausibility of underground life is buttressed by the discovery on Earth of microbe communities living in sediments and ices as much as 3 km below the surface, as mentioned in our discussion of Mars. Europa may harbor interesting secrets for future exploration!

Ganymede: Jupiter's Giant, Fractured Moon

Ganymede, with a 5262-km diameter, is the largest moon in the solar system. It is shown in Figure 11-16. Ganymede is 8% bigger than the planet Mercury and nearly twice as big as Pluto. Its average density of 1900 kg/m^3 implies an ice-rock composition. A thin, bright polar cap is believed to be water frost. This would indicate that water vapor has been emitted from the interior, circulated poleward in an almost negligible atmosphere, and frozen onto the ground at the poles.

★ The ancient parts of Ganymede's surface are heavily cratered and covered with dark, dusty soil (see images on our web site). The abundance of impact craters shows that those parts of Ganymede's surface are much older than Io's surface or Europa's surface. The cratered areas are divided by swaths of light-toned, grooved terrain, shown in Figure 11-17. The bright terrain is younger than the dark areas and has not accumulated as many impact craters. Photos from the Galileo orbiter reveal that the intense fracturing exists at all scales down to a few meters, to the extent that it has obliterated old craters in some areas.

Such splitting may imply internal heating that melted at least the icy component and caused expansion and cracking, reminiscent of that on Europa. The bright swaths may mark fracture systems along which the water erupted, though the eruption of water and formation of ice flows are clearly not as ongoing as on Europa. Scattered impact craters on the old, grooved terrain suggest that it formed 1 to 4 billion years ago, after which Ganymede's geological activity declined. The source of heat that drove the activity is uncertain. Tidal heating and Jupiter's primordial heat might have been involved, but they would not have been as strong as on Io or Europa because of the greater distance from Jupiter and greater separation from neighbor moons. This may explain why the surface of Ganymede is less altered than the surfaces of Io or Europa.

When the Galileo space probe flew close to Ganymede in 1996, it surprised scientists by revealing that the moon has a magnetic field. This in turn indicates a hot, metal-rich core, which in turn supports the idea that the interior of Ganymede has been heated, perhaps tidally, and that this is the engine that drives some of the geologic complexity seen on the surface (Stevenson, 1996; Showman and Malhotra, 1999).

Figure 11-16 Global view of Ganymede showing old, dark, cratered crust and bright, highly fractured regions. The freshest impact craters, in general, are the brightest. The surface is mostly ice. (NASA, Galileo spacecraft)

Ganymede tells an important story that may help us understand Earth better. Ganymede offers a case of incipient plate tectonics, since its brittle lithosphere of ice split into platelike blocks as it floated on a warmer, watery subsurface. Although Ganymede's interior was not active enough (hot enough?) to cause full-scale continental drift among the plates, offsets do show that the plates have moved slightly. Ganymede thus provides a transitional example from cold, inactive planets to larger, geologically active worlds.

Callisto: Jupiter's Cratered Moon

The outermost of the four Galilean moons, **Callisto,** is almost as big as Ganymede and is the third-largest moon in the solar system. At 4800-km diameter, it is only 2% smaller than the planet Mercury. Callisto's density, 1790 kg/m^3, is the lowest of the Galilean

a

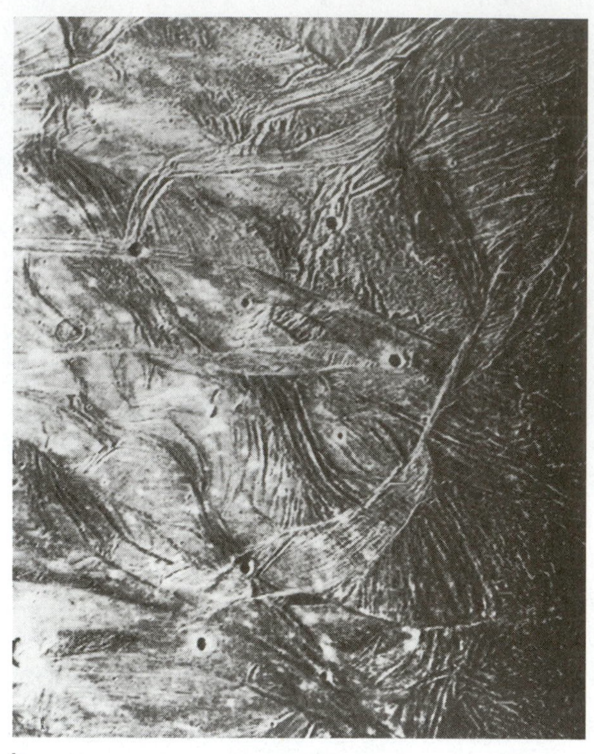

b

Figure 11-17 Fracture systems on Ganymede. **a** Bright, 150-km-wide band of fractures crossing an older Callisto-like cratered terrain. Note offset in broad fracture band by a narrower fracture band in lower left. **b** Close-up of fracture patterns. The frame width is about 580 km, and the smallest features are about 3 km across. (NASA, Voyager)

satellites, implying the highest ratio of ice to soil. This implies the least heating, since heating drives off the water and leaves the soil behind. The density corresponds to a mixture of half ice and half carbonaceous material. Properties of the gravity field suggest a somewhat higher central density, with a large rocky core, a rock/ice mantle, and an outer ice crust a few hundred kilometers thick.

The surface geology also reveals little melting or fracturing and no fresh ice flows. Callisto is covered by meteorite impact craters. Apparently, countless microimpacts have vaporized the ice from the upper meters of the surface, leaving a surficial layer of dark, tannish-gray soil; wherever a larger impact has occurred, brighter, cleaner subsurface ice is exposed, as shown in Figure 11-18.

Of the four Galilean moons, Callisto seems to be the least altered by internal heating or geological evolution. Judging from its high number of craters, the surface is very old and has changed little since the planets' formative era, 4 to 4.5 billion years ago. Nonetheless, the Galileo orbiter discovered a weak magnetic field. The field's properties and the lack of evidence for a large iron core suggest that the field may not be produced in an iron core (as the fields of Earth, Ganymede, and other worlds are), but by electric currents in a buried "ocean," or layer of melted ice, perhaps 10 km thick (Showman and Malhotra, 1999).

The Eight Outermost Moons of Jupiter: Captured, Black Interplanetary Bodies

The eight outermost moons are on the outskirts of Jupiter's sphere of gravitational influence. If they were much farther from the planet, the Sun's gravity would be more important than Jupiter in controlling their motions. (Indeed, a "moon" briefly observed in this region some years ago subsequently disappeared and is believed to have been a comet nucleus that was temporarily captured by Jupiter, but then wandered off again into an orbit around the Sun.)

These moons are too far from Jupiter to be influenced by tidal forces. Hence they have probably not been heated, and they do not keep the same side facing Jupiter.

Physically, the sizes and black surfaces of these moons resemble the sizes and surfaces of interplanetary bodies in the outer solar system—that is, comets and remote asteroids. For these reasons, astronomers believe these moons are not native to Jupiter but are interplanetary bodies that originally orbited around

Figure 11-18 Heavily cratered surface of Callisto. Multiringed structure (top) is believed to be a remnant of a huge impact feature, the largest impact feature known in the solar system. Colors are about the same as in the dark regions of Ganymede. (NASA Voyager photo)

the Sun, came close to Jupiter, and became **captured moons**—interplanetary bodies that were captured into orbit around the planet. This could have happened in the early history of the solar system when Jupiter may have had a more extensive atmosphere; asteroids passing through the fringes of that atmosphere could have been slowed in a way that resulted in capture.

Satellites J6, J7, J10, and J13 have orbits around 12 million kilometers from Jupiter and inclined 27° to Jupiter's equator. All these moons, as well as the other moons mentioned so far, orbit in a **prograde** direction like our own Moon. (This is the "normal" direction of orbital motion in the solar system, counterclockwise as seen from the north celestial pole. See Figure 11-11.) But satellites J8, J9, J11, and J12 all orbit in a **retrograde** direction (clockwise as seen from the north celestial pole), 23 million kilometers from Jupiter with inclinations of around 52°. How did the outer satellites come to be clumped into two such groups? Analysts have suggested that during the capture, when the asteroids passed through the fringes of an early atmosphere, they broke into pieces as they were decelerated. Thus there may have been only two capture events, each contributing four large pieces. One object approached Jupiter in a prograde sense and fragmented to produce prograde moons in similar orbits; the other approached in a retrograde sense and produced retrograde moons. Perhaps there are smaller fragments still undiscovered in each of the two orbital groups.

Because there are no close-up photos of any of these moons, we don't know their surface features. They are probably cratered like Mars' moons, Phobos and Deimos. Indeed, they may be quite similar to Phobos and Deimos in appearance, size, and origin: Recall that Phobos and Deimos are also believed to be dark asteroids captured into satellite orbits.

SATELLITES OF SATURN

Saturn's satellite system has somewhat less variety than Jupiter's. Most of the moons are fairly heavily cratered bodies. They are somewhat like Callisto and Ganymede, but with brighter, more ice-rich surfaces. Almost all Saturn's satellites have lower density than any of Jupiter's Galilean moons. This implies that they have a higher percentage of ice and also explains their lighter color. The fact that Saturn's system is farther from the Sun and colder than Jupiter's system may explain why more ice condensed in Saturn's system.

The Small, Innermost Moons near Saturn's Rings

The Voyagers discovered six small satellites close to the outer edge of Saturn's rings.[3] The largest is a respectable 220 × 160 km (140 × 100 mi, or about the size of New Hampshire and Vermont), and all are more or less potato-shaped. They are fairly bright and believed to consist of fairly pure ice, like the ring particles. An example of one of these moons is shown in Figure 11-19.

Many meteorites pass through this region because of the gravitational attraction of massive Saturn, and these moons—together with the ring particles—may be fragments of more massive moons that were shattered by large impacts long ago.

Saturn's Midsize Icy Moons

Beyond the small inner moons are five midsize moons that were discovered telescopically from Earth. They are bright-colored, icy, cratered moons, ranging from about 300 to 1500 km (190 to 940 mi) in diameter. Their bright surfaces and low densities—only 20–40% greater than pure ice—imply that they are made mostly of ice, without much rocky material. Examples of these moons are shown in Figure 11-20 and on our web site.

All of these moons keep one side facing Saturn all the time, in the same way that the Moon keeps one side toward Earth—and for the same reason: tidal forces. The moons of this group—Mimas, Enceladus, Tethys, Dione, and Rhea—give evidence of mild degrees of internal heating. Running across their cratered surfaces are some large canyons that look like fractures, perhaps caused by expansion.

Of these five moons, Enceladus seems to have had the most active geologic history (Figure 11-21). It is like a missing link between the other moons and Jupiter's smooth ice moon, Europa. Enceladus has had more internal geologic activity than Saturn's four other nearby moons. Part of its surface is cratered, like the surfaces of the other moons, but part is covered by lightly cratered plains of ice, as water erupted and covered some regions before freezing. The paucity of impact craters shows that these surfaces are geologically young. Enceladus' surface is the brightest in the solar system and resembles clean white snow.

★ Curiously, there is a thin ring of tiny particles scattered along Enceladus' orbit; some scientists think eruptions may have occurred on Enceladus and blown debris into space, where it spread along Enceladus' orbit. What caused the heating of this icy, cold moon? Calculations indicate that Enceladus may have experienced mild tidal heating in the past, though perhaps not enough to melt it. Thus the exact history of Enceladus remains one of the mysteries of Saturn's system.

[3] Two had been glimpsed from Earth in 1966, but their orbits were uncertain.

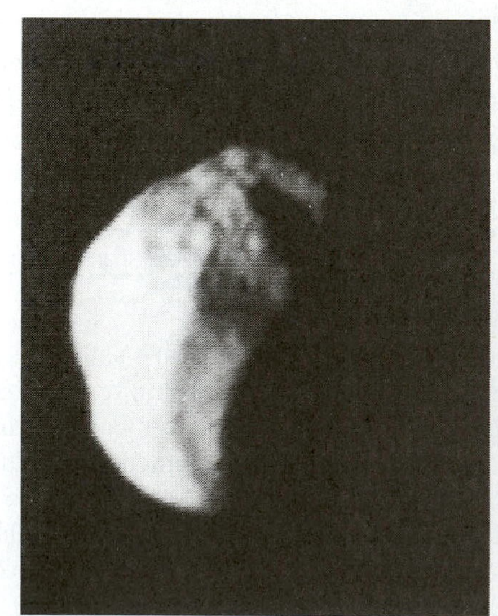
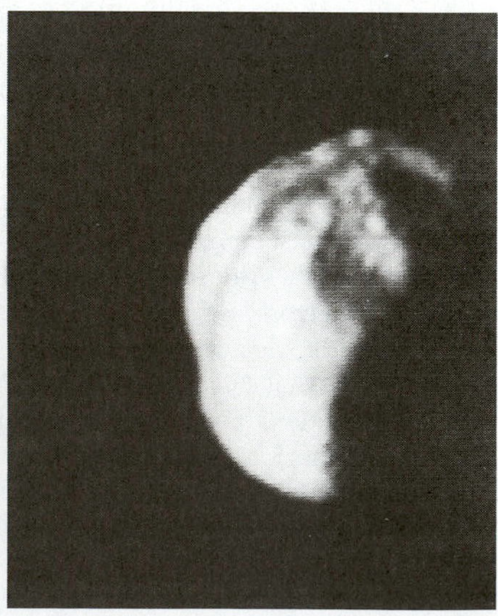

Figure 11-19 Saturn's satellite Epimetheus is a 100 × 140-km object, probably eroded into its irregular shape by intense cratering. These two views, taken 13 min apart, show the shadow of Saturn's nearby rings moving across the satellite. (NASA Voyager 1 photo)

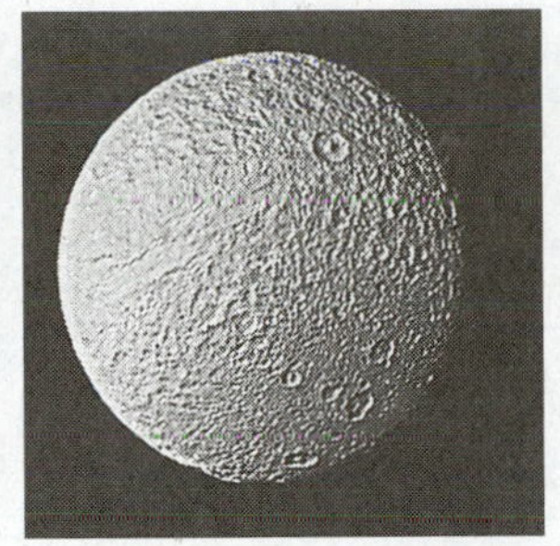

Figure 11-20 A rogue's gallery of Saturn's major icy satellites. **a** Mimas, 390-km diameter, showing a very large crater one side, caused by an impact almost large enough to have shattered the satellite. **b** Tethys, 1048-km diameter, showing craters and a canyonlike fracture or groove (left center). **c** Rhea, 1530-km diameter, showing craters and bright streaks. The bright streaks on some moons may mark frost condensed along fractures. (NASA Voyager photos)

Titan: The Saturnian Moon with a Thick Atmosphere

Beyond the icy midsize moons lies the unique moon, **Titan.** We called Io the most bizarre moon, but Saturn's largest moon, Titan, is a close runner-up. With a diameter of 5150 km (3200 mi), it is probably the second-largest moon in the solar system and the only moon with a thick atmosphere, seen in Figure 11-22. Like all large moons, Titan keeps one side toward its parent planet because of tidal forces.

Methane (CH_4) was discovered in spectra of Titan in 1944. Observations in 1973 showed that Titan's sky is not clear but is filled with reddish-orange haze. Later observations showed that this haze is photochemical smog produced by reactions of the methane and other compounds when they are exposed to sunlight—like the smog produced by the action of sunlight on hydrocarbons over Los Angeles. Titan is the smoggiest world in the solar system. The Voyagers showed that the methane and smog are not more than 10% of the atmosphere. The main constituent is nitrogen, meaning that the atmosphere is denser than was realized before the Voyager flights. Voyager 1 discovered that the surface air pressure is 1.6 times that on Earth! Since the Earth's air is also mostly nitrogen, Titan's atmosphere offers fascinating comparisons to present or primeval conditions on Earth. The main difference is that Titan is very cold, around 93 K (–292° F). Minor constituents detected in Titan's air include ethane, acetylene, ethylene, and hydrogen cyanide. The abundance of these organic molecules suggests that Titan offers a good natural laboratory for research on the origins of life.

Based on the measurements of temperature and pressure at Titan's surface, researchers visualize dramatic weather conditions, illustrated in artists' conceptions on our web site. Methane not only may exist as a gas but also may be able to rain out of the clouds and exist as snow or ice, playing the same triple role of gas, liquid, and icy solid as water does on Earth. In addition to methane (CH_4), meteorologists conclude that ethane (C_2H_6) would form in the atmosphere, slowly drizzle out, and accumulate on the ground. Thus the surface of Titan may have oceans or ponds of liquid methane and liquid ethane. Other chemical calculations about Titan's environment suggest raindrops or snowflakes of additional complicated, gasolinelike compounds. Radar waves bounced off Titan, together with *Hubble Space Telescope* infrared images, suggest a fairly solid surface in at least some areas. During the Voyager mission, one researcher characterized Titan as a "bizarre murky swamp."

The most interesting thing about Titan is that it may teach us about the consequences of complex organic chemistry on other planetary surfaces. Titan is so cold that life is unlikely to have evolved, though we cannot rule out local Yellowstone-like geothermal areas, which might host very complex organic compounds. In any case, Titan may be a natural laboratory for studying primordial biochemical evolution on Earth!

Saturn's Satellites beyond Titan

★ Beyond Titan are three moons, starting with Hyperion, which is notable for having a biscuit-shaped body about 350 × 200 km (220 × 125 mi) in size. (See the Voyager photos on our web site.) Hyperion is also noted for its unique, irregular rotation rate. Instead of a steady spin, it has a wobbling motion, like a

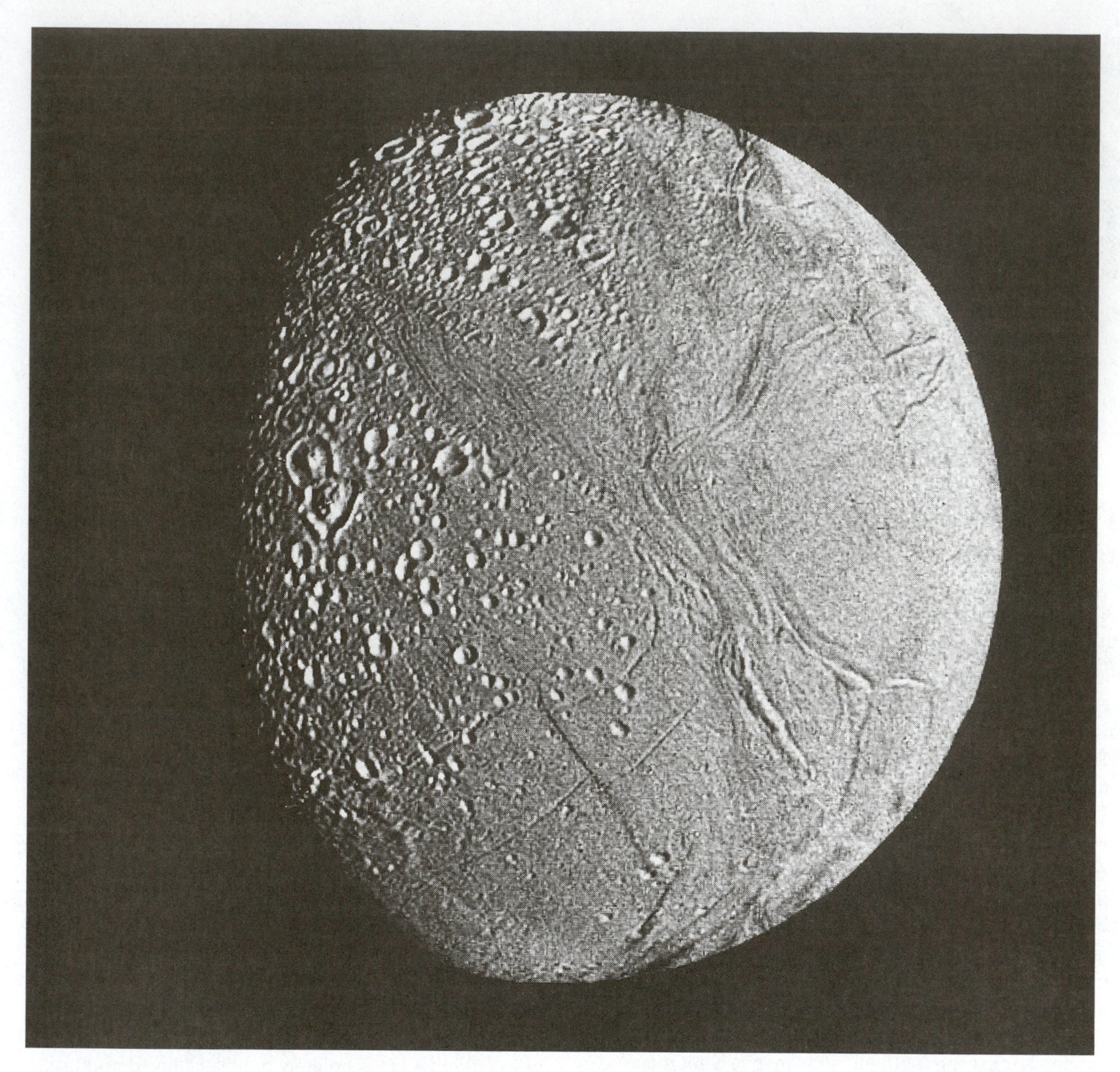

Figure 11-21 Saturn's moon Enceladus, 502 km across, offers a missing link between the "cracked billiard-ball" surface of Jupiter's moon Europa and the cratered, fissured surface of Jupiter's moon Ganymede—though Enceladus is only one-sixth the size of Europa. Portions of the ancient cratered surface seem to have been cracked as water erupted and froze into smooth ice plains, obliterating older craters—for example, the bottom central region. This suggests that Enceladus may have been heated enough to melt portions of its icy interior at some time in its history. (NASA Voyager 2 photo)

blackboard eraser tossed at random; this may be the result of an off-center hit by a large meteorite in recent geologic history.

The next satellite is one of the major moons, 1440-km **Iapetus.** It is the strangest of Saturn's mid-size icy moons, because it is literally two-faced. One side is black, and the other side is white, as shown on our web site. Like the other large moons, Iapetus keeps one side facing Saturn because of tidal forces. That means that one hemisphere always faces forward as Iapetus moves around Saturn, and the other hemisphere trails. Interestingly, the leading hemi-

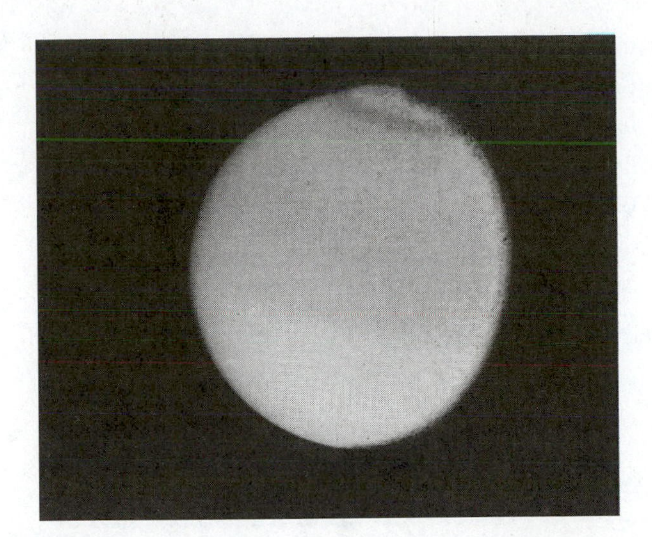

a b

Figure 11-22 Saturn's giant moon, Titan, about as big as Mercury. Titan has a smoggy atmosphere of nitrogen, colored reddish-orange by organic compounds. **a** In this gibbous lighting view, a dark cloud band can be seen around the north polar region, and the southern hemisphere haze is brighter than the northern hemisphere. **b** Under crescent lighting, the illuminated atmosphere can be traced all the way around the moon. As seen in the color version on the web site, the color is bluer where sunlight has been scattered farther through the atmosphere—for the same reason that scattered sunlight colors our own sky blue. (NASA Voyager photos)

sphere is brownish-black, and the trailing hemisphere is white. The difference between the two sides is so pronounced that it was discovered in the 1600s, when observers with early telescopes could see Iapetus only on one side of Saturn (when the bright, trailing side faced Earth). The cause of the dark hemisphere is almost certainly black dust hitting Iapetus as it moves around Saturn, like bugs hitting the front windshield of a car. Where does the black dust come from? It is apparently blown off a black captured satellite named Phoebe (a captured moon that we will discuss next) on the outskirts of Saturn's system. Dynamical studies confirm that dust knocked off that moon by meteorite impacts would spiral in toward Saturn and crash on Iapetus.

Saturn's outermost moon, 220-km **Phoebe,** is unique in the Saturn system in several ways. First, its orbital motion around Saturn is in the backward, retrograde direction, like that of the outermost four moons of Jupiter. Second, while the large intermediate moons follow a sort of Saturnian Bode's rule, with each one being about $1\frac{1}{2}$ to $2\frac{1}{2}$ times farther from Saturn than the preceding one, Phoebe is isolated on the outskirts of Saturn's sphere of gravitational influence, nearly $3\frac{1}{2}$ times as far from Saturn as its inner neighbor Iapetus. Third, instead of being bright and icy, like most other large moons, it is coal-black. This black color roughly matches the color of most interplanetary bodies in the outer solar system. These three factors all suggest that Phoebe is not a native moon of Saturn, but rather an interplanetary body that approached Saturn and was captured into orbit on the fringes of the sphere of influence, where capture can be affected by slight changes in velocity. We know little about Phoebe's surface because it was too far from Saturn and the path of Voyagers 1 and 2 to yield any good photographs.

★ As meteorites hit the carbonaceous surface of Phoebe and blow off black dust, this dust spirals inward toward Saturn. As was remarked earlier, much of it hits the leading side of Iapetus, explaining Iapetus' black hemisphere.

SUMMARY

Jupiter and Saturn are the largest planets in the solar system. Their massive atmospheres, which consist of about four-fifths hydrogen and one-fifth helium by mass, are probably remnants of gas from which the Sun and planets formed. The atmospheres were accumulated by Jupiter and Saturn after those planets grew to a size of about 10–15 Earth masses, when their gravity was so strong it trapped surrounding gas.

Both planets are covered by dense cloud systems, arranged in dark belts and bright zones parallel to the equator. Probably neither planet has a well-defined surface, because the gas and clouds grade into a thick fog or slush as one goes downward. The 10–15 Earth-mass rocky core is hidden deep in the center of each planet.

Each planet has a ring system. Jupiter's is an extremely thin ring of microscopic dark particles, discovered by Voyager 1. Saturn's is a much more massive ring system (but only about 100 m thick), easily visible in small telescopes. Saturn's rings consist of ice bodies; many are centimeters to meters in size, and a few larger ones probably exist. The ring particles in both systems may be debris knocked off the small inner satellites.

Jupiter's and Saturn's systems of moons provide us with "miniature solar systems" to compare with the system of planets. Each has a group of small moons on the edge of the rings, a group of major moons at intermediate distance, and at least one black, captured moon on the outskirts of its sphere of gravitational influence.

Jupiter has four major moons with different properties. Io is volcanic; Europa is a smooth iceball; Ganymede, with craters and fractures, is the largest moon in the solar system; and cratered Callisto is the third-largest moon in the solar system.

Saturn has many major icy moons of intermediate size. Most are heavily cratered, but some have distinctive surface qualities. Saturn also has the second-largest moon in the solar system, Titan, which is the only moon in the solar system with a thick atmosphere and opaque clouds.

CONCEPTS

giant planet
belts
zones
Great Red Spot
atmospheric composition of Jupiter and Saturn
temperature of Jupiter and Saturn
thermal infrared radiation
synchrotron radiation
internal structures of Jupiter and Saturn
metallic hydrogen
ring systems of Jupiter and Saturn
Cassini's division
Roche's limit
shepherd satellites
resonance
sphere of gravitational influence
tidal heating
carbonaceous material
dirty ice
Galilean satellites
Io
Europa
Ganymede
Callisto
captured moons
prograde
retrograde motion
Titan
Iapetus
Phoebe
Galileo space probe (web site)
Cassini space probe (web site)

PROBLEMS

1. Answer the following problems:

a. How might studying cloud patterns on Jupiter, Mars, and Venus help us understand terrestrial meteorological theory?

b. What planetary or environmental characteristics might figure in such a theory?

c. How do these factors vary among these planets?

d. Why are other planets not included on the list?

2. How would gravity on the Galilean satellites compare with gravity on the Moon (see data in Table 8-1)? Which three bodies in the solar system would you expect to have general environments most like the Moon's?

3. Describe a landscape on Io.

4. Two large planets have the same size and mass but different orbits.

a. Which would you expect to have more hydrogen? Why?

b. If they have had different amounts of volcanism, which would you expect to have more carbon dioxide? Why?

5. Discuss why a probe to the surface of Titan might be interesting to biologists trying to understand how life started on Earth.

6. What difficulties might be met in sending a spacecraft through the rings of Saturn to examine the ring environment? Consider a pass *perpendicular* to the ring plane vs. a pass *in* the ring plane. (Note that without elaborate retrorockets, such a spacecraft would probably travel about 20 to 30 km/s relative to the rings.)

ADVANCED PROBLEMS

7. Juptier's four Galilean satellites revolve in orbits about 400,00 to 2 million kilometers from the planet.

a. What would be their maximum angular separation from the planet when Jupiter is at its closest distance of about 4 AU from Earth?

b. These satellites have a brightness about equal to the faintest stars visible to the unaided eye. The eye can normally distinguish details as little as 2 minutes of arc apart. Comment on the feasibility of seeing the Galilean satellites with the naked eye.

8. If you had a telescope that could reveal details as small as $\frac{1}{2}$ second of arc, would you be able to see dark markings on Ganymede (diameter of 5270 km)?

9. Calculate the weight of a 140-lb person on the "surface" of Jupiter. Use Table 8-1.

10. Calculate the mean density of Saturn (lowest of any known planetary body) and discuss its implications for a dense core of rock or metal inside Saturn. Use Table 8-1.

11. Assume the mass of Saturn is 5.7×10^{26} kg.

a. What is the orbital velocity of a particle orbiting around Saturn in Saturn's rings, about 250,000 km from the center of Saturn?

b. Suppose this particle hits another one that orbits 1 km farther away from Saturn. Estimate how fast they come together, using whatever mathematical techniques you know.

12. Calculate the weight of a 140-lb person on the "surface" of Titan. Use Table 8-1.

PROJECTS

1. Observe Jupiter with a telescope with at least an 8-cm (3-in.) aperture. Sketch the pattern of belts and zones. Which are the most prominent belts? Which zones are brightest? Compare these results with photos in this book and on the web site. Is the Red Spot or other dark or bright spots visible on the side of Jupiter being observed?

2. Observe Jupiter with large binoculars or a telescope with at least a $2\frac{1}{2}$-cm (1-in.) aperture. How many satellites are visible? Observe the satellite system at different hours over a period of several days and try to identify the satellites. (This could be done as a class project, with different students making sketches at different hours.)

3. With a telescope having at least a 150-cm (6-in.) aperture, determine the rotation period of Jupiter by recording the time when the Red Spot is centered on the disk. Note that intervals between appearances must be an integral number (1, 2, 3, and so on) of rotation periods.

4. With a telescope having at least an 8-cm (3-in.) aperture, observe Saturn. Sketch the rings. Estimate the angle by which the rings are tilted toward Earth during your observation. Does this angle change much from day to day? From year to year? (Your teacher might save drawings by students of past years for comparison.) Can you see any belts or zones? (Usually, they are less prominent than on Jupiter.) Can you see Cassini's divisions? Sketch nearby starlike objects that may be satellites and track them from night to night. Identify Titan, the brightest satellite.

CHAPTER

12 The Outermost Planets and Their Moons

WHAT THE READER SHOULD WATCH FOR IN THIS CHAPTER

Uranus and Neptune are similar in size. Though they are giant planets, they are much smaller than Jupiter or Saturn, both being about four times the size of Earth. Uranus and Neptune have deep hazy atmospheres and are blue in color due both to absorption of red light by methane and to the same scattering process that produces the blue skies and bluish color of Earth. Both planets have systems of narrow rings and families of satellites. Two satellites stand out as unusual. Uranus' moon, Miranda, has unusual fracture patterns. Neptune's largest moon, Triton, has smoking vents and a "backward" orbit. Both of these satellites must have had interesting and unusual geologic histories. Pluto, traditionally named as the ninth planet, is smaller than our Moon. If it had been discovered recently, it would probably have been listed as an asteroid instead of a planet. The basic distinction between giant planets and terrestrial planets in our solar system (and perhaps in other solar systems) has to do with the fact that once a terrestrial planet grows to two to three times the size of Earth, its gravity is strong enough to trap and retain a massive hydrogen-rich atmosphere, which would leak off a smaller planet. ■

Beyond Jupiter and Saturn are two more giant planets, Uranus (U′-ra-nus) and Neptune. At first glance, if seen through a large telescope or from a nearby spaceship, they seem quite different from the other two giant planets. Instead of displaying the tan colors of Jupiter's and Saturn's clouds, they are strikingly blue. They are both about one-third the size of Jupiter and Saturn or, to put it in other terms, about four times the diameter of Earth.

Closer examination of Uranus and Neptune reveals some of the same features as Jupiter and Saturn—**features common to all four giant planets:** (1) They are all cloud-covered. (2) They all have massive atmospheres composed mostly of hydrogen and helium. For this reason these four planets are sometimes called the *gas giants.* (3) They all have ring systems, though each ring system has distinct features. (4) They all have many satellites. In this chapter we will discuss some of the distinctive features of Uranus and Neptune as compared to Jupiter and Saturn, but we will also stress the significance of common traits among the four giant planets.

We will also discuss Pluto, which is very different from the giant planets. It was hailed as the ninth planet when it was found in 1930, but it is only the sixteenth largest world in the solar system and today seems more related to a family of interplanetary bodies that occupy the region of Neptune and beyond.

URANUS AND NEPTUNE—THE PLANETS

Uranus and Neptune are the first planets that were discovered in historic times by telescopes, instead of by prehistoric observers.

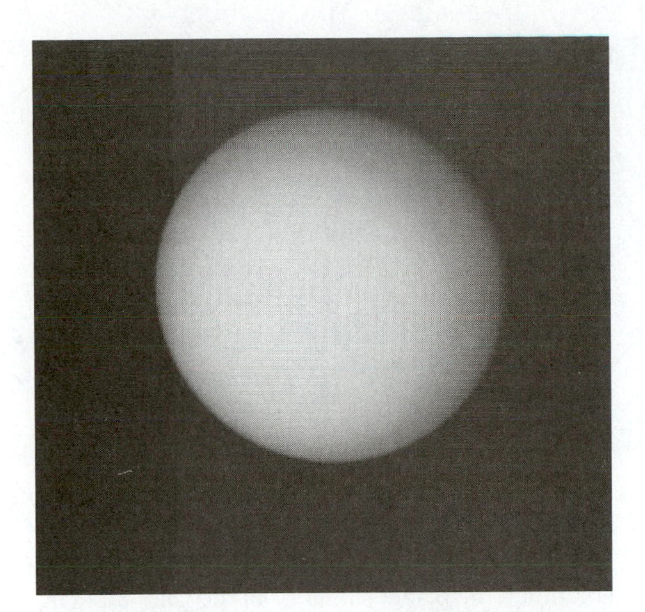

Figure 12-1 The hazy sky-blue planet, Uranus. As Voyager 2 approached Uranus from the inner solar system and the Sun's direction, it took this photo with Uranus at a gibbous phase under nearly full sun. The view looks down on the pole, which is pointed nearly toward the Sun due to Uranus' unusual axial tilt. Cloud features are nearly invisible, though a faint dark band can be seen along the upper right edge. (NASA Voyager 2 photo)

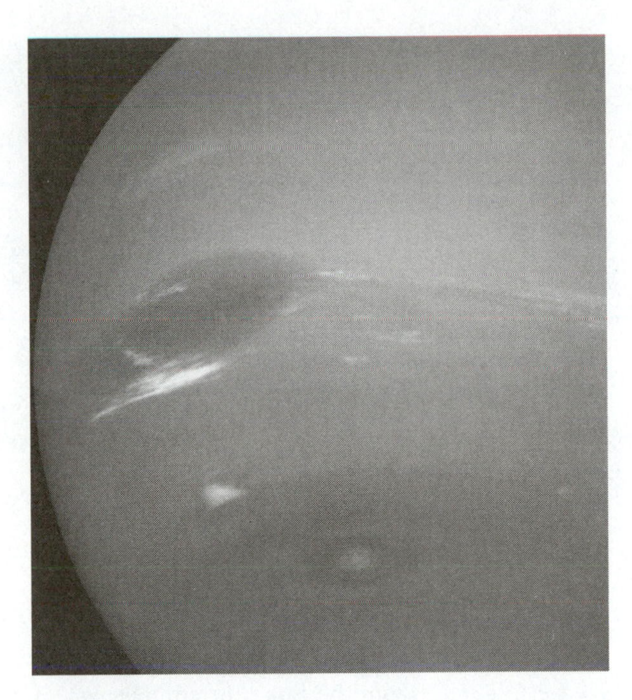

Figure 12-2 This view of a portion of Neptune shows some of the cloud structures in its atmosphere. Dark oval clouds, such as the prominent "Great Dark Spot," mark circulating storm systems, capped by white condensed clouds higher in the atmosphere. (NASA Voyager 2 photo)

Discovery of Uranus and Neptune

Uranus was discovered in 1781, when the German-English musician-turned-astronomer William Herschel accidentally detected the planet during an ambitious star-mapping project. Herschel thus became the first known human to recognize a new planet. The planet is so far from Earth that it has a very tiny apparent angular size. To a visual observer looking through even a large modern telescope, it appears as a bland, bluish disk without any markings. The planet's name was suggested by Johann Bode (of Bode's rule) because in mythology Uranus was the father of Saturn, who in turn was the father of Jupiter. Because of Uranus' long 84-y period of revolution around the Sun, it has made only two orbital circuits since its discovery.

★ The discovery of Uranus led to the discovery of **Neptune.** After 1800, theorists tried unsuccessfully to fit observations of Uranus' position into Kepler's laws of planetary motion. Uranus seemed to have its own somewhat irregular motions. A few scientists though this might signal a breakdown of Newton's law of gravity at great distances from the Sun. Others correctly suggested that Uranus was being attracted by a still more distant planet. In the 1840s, an English astronomer and a French mathematician set out independently to predict where the new planet should be. After a race among European observers to find the new world (described on our web site), it was discovered by German observers in 1846 and given the name Neptune.

The Physical Nature of Uranus and Neptune

After the two Voyager probes flew past Saturn, Voyager 1 zoomed "upward" out of the plane of the solar system, but Voyager 2 went on to Uranus and then to Neptune. It provided the first good data on the **atmospheric properties of Uranus and Neptune.** Both planets are blue, with a slight greenish tinge. Uranus has virtually no belts and zones of dark and bright clouds, like those of Jupiter and Saturn. Rather, its atmosphere has a deep layer of almost featureless hazy gas almost obscuring a deeper cloud deck, as can be seen in Figure 12.1. Neptune turns out to have more pronounced cloud structures than Uranus, including dark belts and a dark oval cloud system, called the **Great Dark Spot.** The Great Dark Spot is an Earth-size storm system, reminiscent of Jupiter's Great Red Spot, circulating in an anticyclonic direction. These features are shown in Figure 12-2.

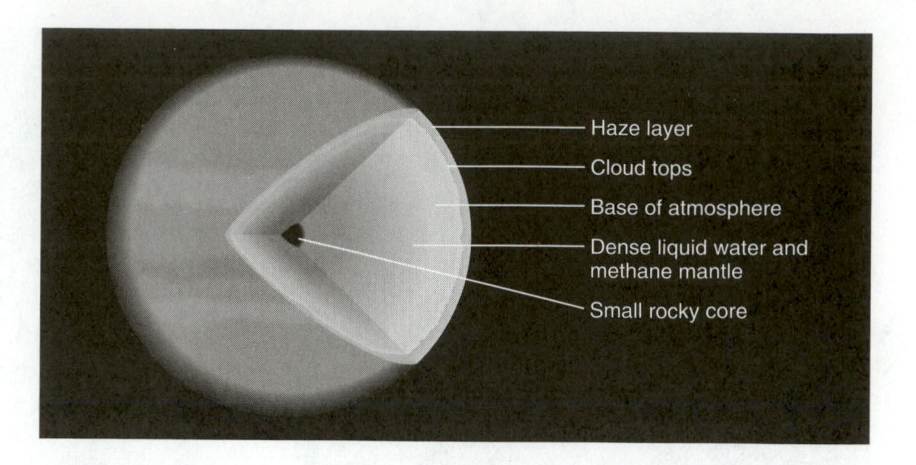

Figure 12-3 Schematic diagram of the interiors of Uranus and Neptune. Each world has a central rocky core smaller than those of Jupiter and Saturn, surrounded by a deep layer of compressed liquid mixtures of water, methane, and minerals, and topped by a deep atmosphere.

Voyager measurements showed that, like the other two giant planets, Uranus and Neptune have thick atmospheres composed mostly of hydrogen (H_2) and helium (He).[1] The cloudtop temperatures are colder than on Jupiter and Saturn, because of the greater distance from the Sun. For comparison, at the level where the pressure is equal to sea-level conditions on Earth, the temperatures of Earth, Jupiter, Saturn, Uranus, and Neptune are, respectively, roughly 290 K, 170 K, 150 K, 78 K, and 69 K (63°F, –153°F, –189°F, –319°F, and –335°F).

Why is Uranus nearly featureless? The ultimate reason is a lack of heat from the interior. The largest giant planets, Jupiter and Saturn, have a large amount of internal heat: they radiate nearly $2\frac{1}{2}$ times more heat than they get from the Sun. This means that their atmospheres are heated and churned up from the bottom, just as a pot of soup on the stove is churned by heat from below. This helps stir up storm systems and raises cloud patterns to high altitudes, where they can be seen from space. Uranus radiates only about the same internal heat as it gets from the Sun (the figure is somewhat uncertain). Because of the reduced outward flow of internal heat and the huge distance from the Sun, the clouds of Uranus are cold; they do not billow upward but remain veiled by the upper haze layer.

If Uranus is nearly featureless, why does Neptune have strong cloud features? Surprisingly, Neptune does have a large internal heat source and radiates 2.7 times more energy than it receives from the Sun. The heat source is mysterious, but it may relate to radioactivity inside Neptune or to large impacts that stirred up the planet's interior in the last half of its history. The net result is that the atmosphere of Neptune has fairly strong heating from below, which in turn explains why its clouds billow up into the haze layer to produce visible cloud patterns.

Why are the two planets blue? There are several reasons. Remember that on the giant planets, sunlight passes through a thick haze layer before reaching the clouds and reflecting back to us. The haze layers of Uranus and Neptune are richer in methane gas than the haze layers of Jupiter and Saturn, and the methane absorbs red and orange light, producing an overall blue color with a faint greenish tinge. A second effect is that the thick haze layer scatters sky-blue light, just as our own atmosphere does. Third, because of the intense cold at their great distances from the Sun, Uranus and Neptune probably have less chemical activity—and hence less colorful compounds in their upper clouds—than Jupiter or Saturn, so there are fewer other colors to compete with the sky-blue color of the haze.

The Interiors of Uranus and Neptune

As shown in Figure 12-3, the **internal structures of Uranus and Neptune** are somewhat like those of Jupiter and Saturn, but without the deep mantle of metallic hydrogen. This is because, according to calculations, the pressures in the two smaller giant planets do not reach high enough values to force hydrogen into the high-pressure metallic state. Taking into account the mean density of the planets and the behavior of material under high pressure, theorists believe Uranus and Neptune have rocky cores smaller than those of Jupiter and Saturn, surrounded by indis-

[1]The similarity of the actual percentages of hydrogen among the four giant planets and in the Sun (averaging around 75% by mass), and the percentages of helium (averaging around 20% by mass), are clues about the origin of the solar system. They are therefore tabulated in the chapter on that topic, Chapter 14, on page 282.

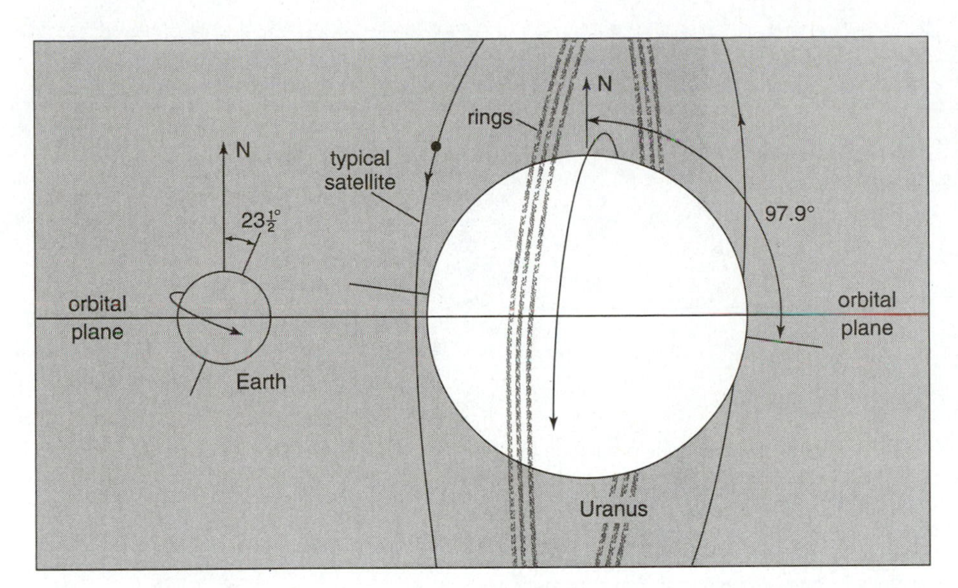

Figure 12-4 Comparison of the sizes and rotations of Earth and Uranus. Earth has an obliquity (or axial tilt) of $23\frac{1}{2}$ and a prograde (west-to-east) rotation. Uranus has a much steeper obliquity and retrograde rotation.

tinctly layered mantles—dense, sluggish oceans—of very compressed liquid water and methane, containing rocky minerals. About 5000 km below the clouds is the boundary between this liquid ocean and the dense lower atmosphere. This boundary—the surface of the ocean—is probably not as distinct as the surface of our ocean, because the high pressures and densities at this depth on Uranus and Neptune create a mushy, murky boundary between the cold, dense liquid phase and the compressed, high-density gas.

The Extreme Axial Tilt and Peculiar Seasons of Uranus

A unique property of Uranus is that its axis of rotation is highly tilted to the plane of the solar system. This is illustrated in Figure 12-4. Most planets, like Earth, have their north pole pointing more or less "upward," at a steep angle to the plane of the solar system. Uranus' pole lies nearly *in* the plane. The angle between the polar axis and the perpendicular to the plane of the orbit is called the **obliquity.** Thus another way to express the last two sentences is to say that the obliquity of most planets is low, but that of Uranus is high.

The rotation of Uranus, as shown in the sketch, is from east to west, which is "backward" compared to Earth's rotation. This is called **retrograde rotation.** Due to the high obliquity of Uranus, the planet's north pole or south pole can point nearly at the Sun at different times during its 84-y trip around the Sun, as shown in Figure 12-5. This means that Uranus has a unique seasonal sequence. When the north pole points almost directly toward the Sun, the southern hemisphere is plunged into a long, dark winter, lasting for about a quarter of the planet's 84-y revolution. After this 21-y south polar winter and north polar summer, the Sun shines on the equatorial regions. Each point on the planet now goes from day to night during the planet's 17-h rotation. During this season, the situation is much like that on the Earth or Mars, but much colder. After 21 more years, the south pole points approximately toward the Sun, and the southern hemisphere experiences a 21-y summer.

What causes the extreme axial tilt of Uranus? Scientists speculate that just as Earth was hit by an unusually large planetesimal to create its moon, Uranus was hit by an unusually large planetesimal that tilted its rotation axis. The difference in outcome between the two collisions was a result of different geometries of the collisions. Earth may have been hit more straight-on in a way that blasted out mantle material; Uranus was hit in a way that affected its axial tilt. In spite of the regularities of the planets (for example, most planets have low obliquity, prograde rotation, and near-circular orbits), a few random large collisions may have given distinct "personalities" to some planets.

A LESSON IN COMPARATIVE PLANETOLOGY: WHY GIANT PLANETS HAVE MASSIVE ATMOSPHERES

At the end of Chapter 9, we gave three principles that explain why some planets lack atmospheres. Generally, of course, bigger planets have stronger gravity

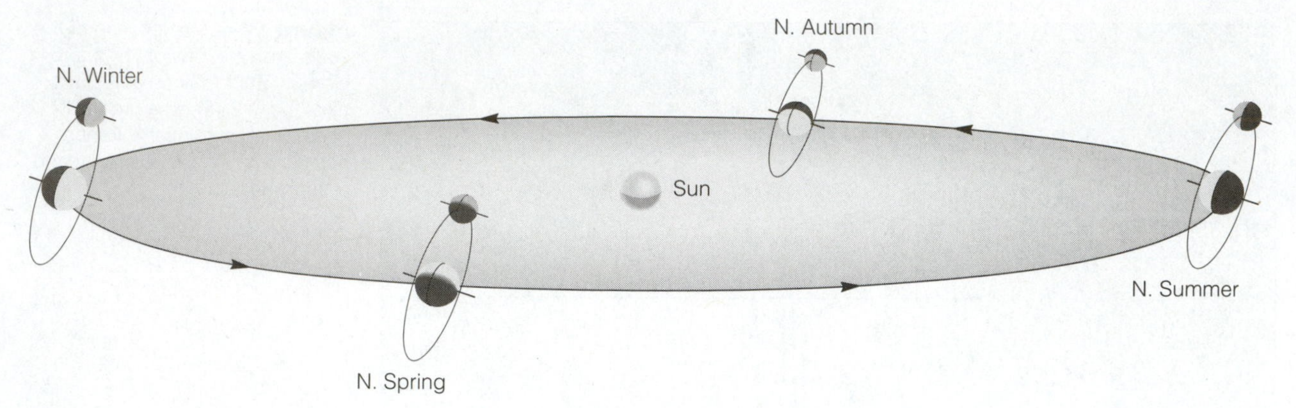

Figure 12-5 Schematic drawing of Uranus' unusual seasons (not to true scale). The extreme axial tilt of Uranus produces extreme seasons during its 84-year trip around the Sun. For example, during northern summer (right) Uranus' north pole points almost exactly toward the Sun, and the equator remains in twilight all day. In spite of many years of direct Sun at the pole, the temperature remains frigid because of Uranus' great distance from the Sun. Note the season effect on the satellites, one of which is shown schematically. Because each satellite revolves above the equator and keeps one face toward Uranus, the satellites undergo the same extreme seasons, as can be seen here.

and thus are better able to keep gas from drifting away into space. But we can use the three principles to explain the situation in more detail.

The first principle says that the higher the temperature, the higher the molecules' speed. (For more detail, see accompanying Optional Basic Equation VI.) This means that molecules will generally be moving at lower speeds in the outermost, cold atmospheres of giant planets than in the atmospheres of warmer terrestrial planets. The second principle says that the lighter the molecule, the higher its speed. This means that on any given planet, hydrogen will be the fastest-moving molecule. The third principle says that the bigger the planet, the more gravity and hence the higher the speed needed for a molecule to escape the planet's gravity and shoot off into space. Thus hydrogen is the gas molecule most likely to escape from any planet. The combination of the three principles means that giant planets will hold hydrogen and whatever other gases they have much better than smaller planets like Earth.

Let's compare Earth and a giant planet. On small, warm Earth with its modest gravity, the fast-moving hydrogen molecules will move more than fast enough to escape into space. Thus Earth has lost virtually all its original hydrogen, which was by far the most abundant primordial gas. The slower nitrogen and oxygen did not escape. Conversely, on any of the four cold giant planets the fast-moving hydrogen and all other gases have been retained. Since the original hydrogen was more abundant than all other gases put together, hydrogen dominates on all four of these planets, and the hydrogen-rich atmospheres today are very massive and very deep.

RINGS OF URANUS AND NEPTUNE

The **ring systems of Uranus and Neptune** were not even suspected until a few years before the Voyager 2 space probe reached the planets. The rings are almost too faint to detect from Earth. Once Voyager 2 reached each planet, it made magnificent photos of the rings and revealed that each system has distinct features. The reason for such different features in the ring systems of the four giant planets is still a puzzle for dynamical astronomers, who try to understand the ring structures in terms of the motions of the ring particles as they orbit around each planet.

Uranus' Rings

In 1977, Uranus passed in front of a relatively bright star. A number of astronomers watched, expecting to see the star dim as it passed behind Uranus' upper atmosphere and thus to learn about the haze layer's structure. They were astonished to see, in addition, the star dim several times at some distance from the Uranian disk, but symmetrically on each side of the disk. These observations marked the discovery of narrow rings of dark material around the planet. Uranus' rings blocked the starlight as Uranus moved.

Voyager 2 photos revealed that the rings are much narrower than those of Jupiter or Saturn, as seen in Figure 12-6. Most of the ring material is confined in nine narrow bands. Scientists do not fully understand the forces that confine the ring particles into such tight bands, although **shepherd satellites,** such as those on each side of Saturn's narrow F ring, are probably involved. Their gravitational force apparently keeps the ring particles from straying outside the narrow ring, just as a shepherd and his dog keep their sheep confined in a flock. Figure 12-6 also shows Voyager 2's discovery of two such shepherd moons straddling the thickest Uranian ring. Gravitational

OPTIONAL BASIC EQUATION VI

Typical Velocities of Atoms and Molecules in a Gas

A useful result about the motions of atoms or molecules in a gas was worked out around 1860 by physicists, particularly the Scottish researcher James Clerk Maxwell and the Austrian researcher Ludwig Boltzmann. Like our other Basic Equations, it is a simple law that gives us many insights throughout astronomy. It comes from the recognition that any gas is made of particles (atoms or molecules or even microscopic dust grains), all moving at various speeds determined only by the temperature T. In fact, T is defined as a measurement of the average velocity of energy of a particle in the gas. The higher the T, the higher the mean velocity (or energy) of any representative particle, as noted at the end of Chapter 9. Now recall from basic physics that the kinetic energy of any moving particle is $\frac{1}{2}mv^2$, where m is the particle mass and v is its velocity. The discovery of Maxwell and Boltzmann is that the mean kinetic energy is proportional to T. This statement is usually written

$$\frac{1}{2}mv^2 = \frac{3}{2}kT$$

where

$$k = \text{Boltzmann constant}$$
$$= 1.38(10^{-23}) \text{ joules/degree}$$

The joule (abbreviated J) is the SI unit of energy. Solving for v, we have a value that may be thought of as the typical velocity of an average particle:

$$\text{Typical velocity} = v = \sqrt{3kT/m}$$

(To be more precise, this value is the square root of the mean squared velocity and is about 10% higher than the true mean velocity. Physicists widely use this *root mean square* velocity as an all-purpose indicator of gas atom speeds, however.)

Note that the mass of an atom or molecule is conveniently figured from

$$m = (\text{atomic weight of atom or molecule}) \times (\text{mass of a hydrogen atom})$$
$$= (\text{atomic weight}) \times 1.67(10^{-27}) \text{ kg}$$

Sample Problem 1. What is the typical velocity of a molecule of air in your room? *Solution:* Assume $T = 68°\text{F} = 20°\text{C} = 293$ K. (We have to convert to SI units, which for T are kelvin degrees.) Assume the air molecule is the nitrogen molecule, N_2. A check in a chemistry book shows that these two nitrogen atoms have total atomic weight 28 (28 × the weight of a hydrogen atom), so that $m = 4.68(10^{-26})$. Therefore

$$v = \sqrt{\frac{3(1.38)(10^{-23})(293)}{4.68(10^{-26})}}$$
$$= 509 \text{ m/s}$$

Thus the atoms and molecules around you are typically hitting your skin at about $\frac{1}{2}$ km/s! You don't feel the individual hits because they are too small, but you do feel the sustained pressure exerted by the ensemble of them.

Sample Problem 2. Prove that a hydrogen atom would move much faster than a nitrogen atom in the same atmosphere. *Solution:* Since v is proportional to $\sqrt{1/m}$, we confirm that the smaller the mass, the greater the velocity. Since an N atom is 14 times more massive than an H atom, it would move $1/\sqrt{14}$ as fast. Or, to say the same thing, the H atom moves 3.7 times faster. You should also show that an H *atom* would move 5.3 times faster than an N_2 *molecule* in the same gas.

As illustrated in the advanced problems at the end of this chapter, this basic equation can easily be used to show why hydrogen has escaped from the Earth and not from the giant planets.

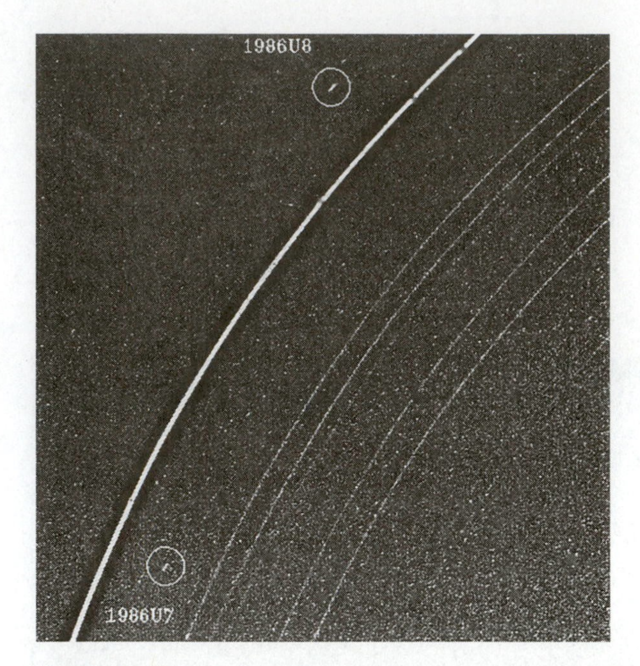

Figure 12-6 Portion of the rings of Uranus and two shepherd moons (circled), whose gravitational pulls help keep ring particles confined to narrow widths. The width of the widest, outer ring is only 100 km (60 mi). All nine rings appear; the inner three may be glimpsed by holding up the book and sighting along the line of the rings. (NASA Voyager 2 photo)

forces of other undiscovered moonlets among the rings may help confine the other rings. Further research is under way to clarify the curious "shepherding" effects.

Neptune's Unique Arc-Rings

During the 1980s, prior to Voyager 2's flight past Neptune, Earth-based observers attempted to repeat the successful discovery of Uranus' rings by watching Neptune pass in front of stars. They got puzzling results. Sometimes the star dimmed before or after the planet passed it (indicating a ring), but at other times it did not. They concluded that Neptune might have arc-shaped ring segments that go only part way around the planet, with the rest of the ring being thin or nonexistent—a puzzling conclusion because no one knew how such arc-rings might form.

As shown in Figure 12-7, Voyager 2 confirmed that the rings contain thicker arclike segments. These have been called **arc-rings.** Generally, the rings of Neptune are similar to those of Uranus in being extremely narrow, separate rings; but several of these contain concentrated arc-rings. Much fainter, and almost invisible in the Neptune system, are broad, dispersed rings like the ring of Jupiter. As is true of the other ring systems, researchers believe that gravitational effects of small moonlets moving within or near the rings may force Neptune's ring particles into narrow rings and arcs. The ring particles themselves may originate in debris knocked off the shepherd moons by meteorite impacts.

The origin, age, evolution, and dynamics of all four ring systems—of Jupiter, Saturn, Uranus, and Neptune—remain a subject of lively research.

THE SATELLITE SYSTEM OF URANUS

The satellite systems of both Uranus and Neptune are so far away from us that only the largest moons in each system were discovered by telescopes from Earth. It remained for Voyager 2's flight through each system to discover many smaller moons and some surprising facts about the bigger ones.

Uranus' satellites number 15. Before Voyager, five moons of Uranus were known and named after

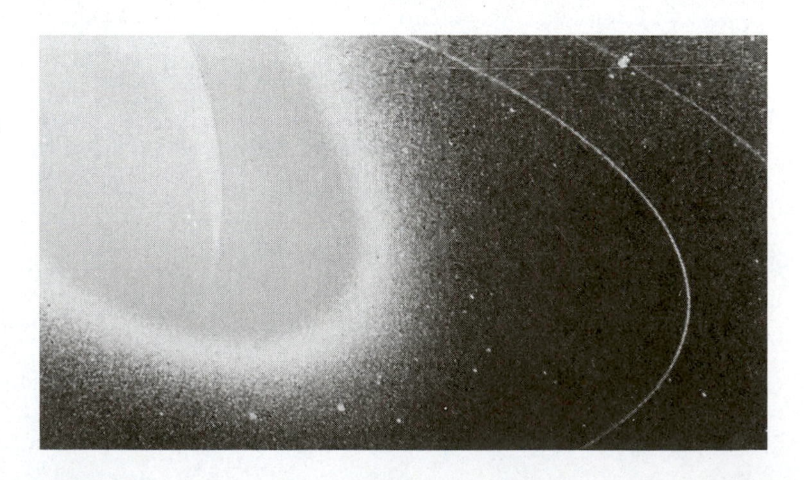

Figure 12-7 The rings of Neptune. Two rings exhibit denser arclike segments (bottom), but are continuous around Neptune. Fainter broad rings were also found inside each of these rings. (NASA Voyager 2 photo)

characters in Shakespeare's *Midsummer Night's Dream* (see Table 8-1). But 10 smaller moons were discovered in 1985 and 1986 as astronomers studied Voyager 2's photos! The newly found moons were named after additional characters in Shakespeare's plays and in a poem by Alexander Pope. These 10 moons are quite small, ranging from about 30 to 110 km (19 to 70 mi) in diameter. Voyager did not get good photos of them but did show that they are much darker-colored than Saturn's innermost moons, probably containing black, carbonaceous soil. In general, they fit the giant planets' pattern of small moonlets orbiting close to the outer edge of the ring system.

Voyager did get good photos of the five larger moons. Scientists had assumed that at such large solar distance, they would be little more interesting than inert iceballs. But once again, as in the cases of Jupiter and Saturn, Voyager revealed more complex geology than expected.

Of the five larger moons, the outer four are the largest, all about a third the size of our Moon. True to our rule of thumb that small worlds show primarily external cratering without internal disturbance, these moons are fairly heavily cratered and somewhat resemble Saturn's moons of similar size, as exemplified by Figure 12-8. But there are complications. Several of the outer four are strongly fractured by canyons, as seen in the figure. These fissures may indicate internal heating from unknown sources or dynamic forces acting on each satellite from gravitational interaction with other moons.

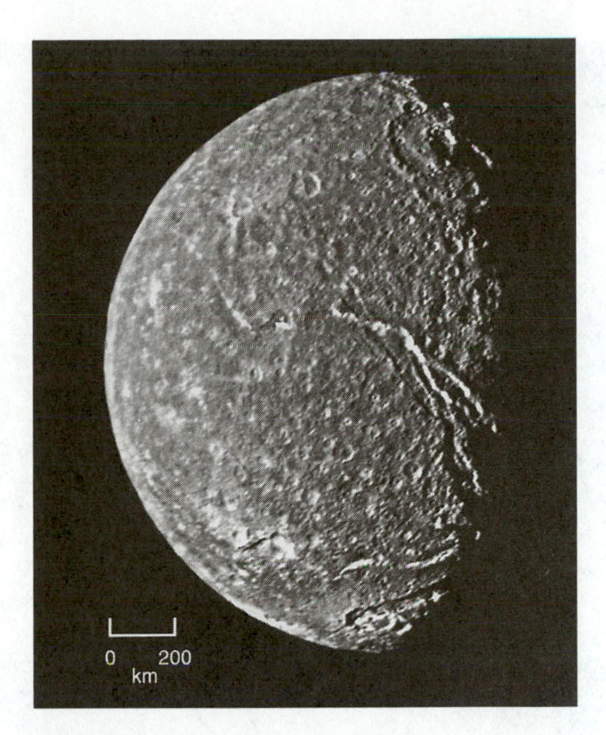

Figure 12-8 Uranus' largest moon, 1580-km Titania. With a cratered, icy surface split by fracturelike canyons. Titania resembles the icy moons of Saturn. (NASA Voyager 2 photo)

Miranda: Uranus' Fractured Moon

★ The greatest surprise came from the close-up photos of the innermost of the five large moons—modest-size **Miranda.** At a mere 470-km diameter, it is the smallest of the five main satellites. In accord with our rule of thumb that smaller size signifies less internal activity, scientists expected it to have an old surface, saturated with eons of impact craters. But unexpectedly it is the most fractured and resurfaced of the five moons, as seen in Figure 12-9. How could such a small, asteroid-size moon have generated enough internal energy to create these impressive features? Some external energy source must have been involved. Voyager scientists suggested that, as with Saturn's inner large moon Mimas, Miranda might have been smashed by impact and reassembled, with the fractures forming as the new moon adjusted its shape while the fragments settled. Another suggestion is that external forces from nearby satellites may have forced Miranda into a period of chaotic rotation during its past history, and the resulting stresses might have caused heating and fractures (Marcialis and Greenberg, 1987). Whatever its history, Miranda rotates smoothly and synchronously now, and its unusually fractured face is a cause for awe.

THE SATELLITE SYSTEM OF NEPTUNE

★ Of **Neptune's satellites,** only two were known before Voyager 2's flight through the system in 1989, but the spacecraft discovered five more. As was the case at Uranus, the five newly discovered moons have orbits close to the planet, on the outskirts of the ring system. The inner four are small, ranging from about 50 to 190 km (30 to 120 mi) in diameter. The fifth, named Proteus, is a little farther from Neptune and is a respectable 400 km (250 mi) in diameter. It is the only one of these five that was well photographed; Proteus turned out to be a not-quite-spherical black moon, as seen on our web site. A large impact crater on one side of the moon is nearly big enough to have shattered it. This testifies again to the strong impact

Figure 12-9 Uranus' 470-km moon, Miranda, surprised scientists with its strange swaths of fractures and linear ridges, cutting through older cratered terrain. Miranda had been thought too small and cold to have geologic resurfacing activity. This image has been reprojected by computer to place one of the fracture arrays at the center; faint latitude and longitude lines reveal that the pole (left) was pointing nearly toward the sun during Voyager 2's encounter. (NASA Voyager 2 photo)

rate suffered by many of the small moons close to the giant planets—a result of the giants' tendency to attract meteoroids with their strong gravity.

The two outer satellites, the ones discovered from Earth, constitute a peculiar system because of their orbits. The largest satellite, **Triton,** was discovered in 1846. With a diameter of about 2700 km (1690 mi), it is slightly smaller than our Moon and is the seventh-largest satellite in the solar system. One might expect Triton to be in a normal orbit like all other large moons, but, surprisingly, it has an orbit inclined at a high angle to Neptune's equator. Moreover, its motion is in the backward, or retrograde, direction compared to all the other moons. The tidal forces operating on Triton are causing it to spiral in toward Neptune, and it will eventually crash onto Neptune's surface, probably many millions of years in the future.

The most distant moon, 340-km (210-mi) Nereid, discovered in 1949, is not in the usual circular orbit of a satellite, but in a very elliptical orbit on the outskirts of Neptune's gravitational sphere of influence.

The strange orbits of Nereid and Triton suggest that both may not be native Neptunian satellites, but rather captured interplanetary bodies—or that they may have been disturbed by the passage of some other large interplanetary body through Neptune's system. Combining the high impact rates of the inner satellites and the evidence for capture or disturbance of the outer moons, we can infer that the satellite systems of the giant planets have not evolved in isolation but have been influenced by passing bodies, reminiscent of the evidence that Earth and Uranus were each influenced by a giant collision with a passing body. In spite of its quiet appearance, the solar system has experienced some significant disruptions!

Figure 12-10 Neptune's largest moon, Triton, shows a strangely mottled surface of bright ice and frost deposits. The brighter top area surrounds the south pole and is believed to mark a polar frost cap. Dark streaks are surface deposits of sootlike material blown downwind from geyser vents. (NASA Voyager 2 photo)

Triton: Neptune's Erupting Moon

Voyager photos revealed Triton to be a cream-colored icy moon with many unusual features. Photos such as Figure 12-10 reveal that Triton has few impact craters, which means that its surface must be, geologically speaking, very young. A closer view is shown on our web site. This in turn means that Triton must undergo geologic activity that renews the surface—a fact totally unanticipated by scientists.

Triton has other unusual properties. Earthbound spectroscopic observers in the late 1970s discovered that Triton has a thin atmosphere (Cruikshank and Silvaggio, 1979). Voyager 2 instruments showed that the atmosphere is mostly nitrogen, with some methane. This is reminiscent of the only other satellite atmosphere, Titan's thick, nitrogen-rich atmosphere. (It's a handy aid to memory that the two moons with atmospheres have such similar names.) Triton's atmosphere, however, is minimal; the surface pressure of about 0.01 millibar is barely a thousandth of that of Mars. The atmosphere is far too thin to cause enough erosion to explain Triton's young surface. Nonetheless, Voyager 2 cameras showed thin clouds and a haze layer 5 to 10 km above the surface (Figure 12-11).

Triton may have seasonal patterns of atmospheric change (Brown and others, 1991). The high inclination of the orbit leaves first one pole and later the other in darkness for 41-y winters during Neptune's 165-y trip around the Sun. As a result, nitrogen frost (frozen N_2) and perhaps methane frost (frozen CH_4) condense in thin layers on the winter pole. Then they are exposed to the Sun in spring, sublime into gas, and transfer to the other winter pole. This transfer of gas may cause variable atmospheric hazes and density at different times during Triton's year. The bright south polar frost cap can be seen at the top of Figure 12-10, where it was probably shrinking due to its exposure to sunlight. The same picture shows a bluish-white frost deposit along the terminator, the cold zone dividing night and day.

With a measured surface temperature of 38 K (–391°F), Triton is the coldest body yet observed at close range. Analysis of the surface materials indicates that the surface is composed of somewhat dirty ices

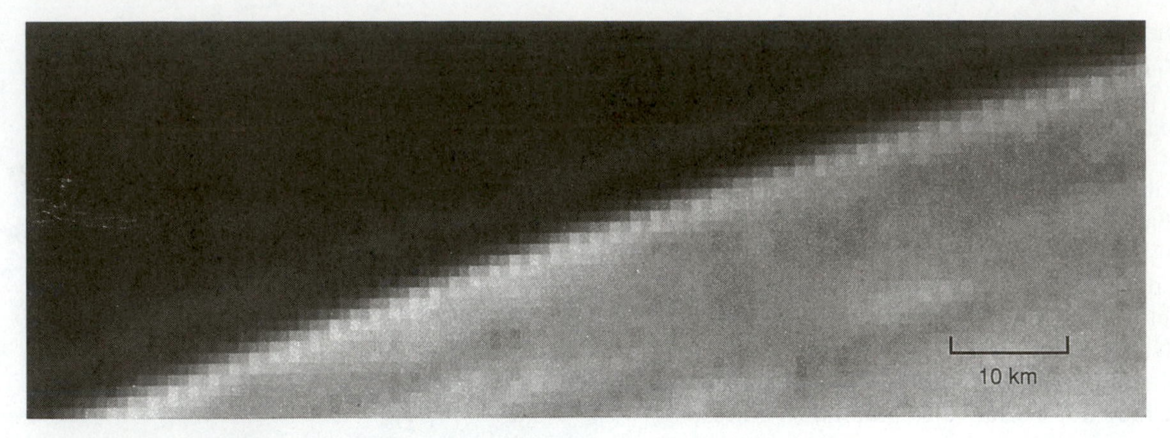

Figure 12-11 Triton's haze layer, a few kilometers thick, floats about 6 km above the curved horizon, proving the existence of an atmosphere. (Enlarged detail from NASA Voyager 2 photo. Note that individual pixels are visible at this enlarged scale.)

of frozen nitrogen (N_2) and methane (CH_4), which form at those frigid temperatures. In some regions, these are overlain by the seasonal nitrogen frosts.

The greatest surprise about Triton came when highly processed Voyager camera stereo images revealed active, erupting geysers or volcanoes with columns of dark smoke rising vertically 8 km (26,000 ft) into the sky, where they were then sheared off by jet-stream winds. These eruptions confirm the conclusion drawn from the young, uncratered surface: Triton is geologically active! Dark streaks on the ground emanate from some of the vents; apparently, these are deposits of sooty material blown downstream from the eruption by prevailing winds.

The big question is, what type of geologic energy produces the underground heat that drives the geyserlike eruptions and creates a young surface? Triton's "backward" orbit indicates an unusual history that may explain the geologic energy source. If Triton was captured by Neptune into an elliptical orbit, tidal forces at that relatively close distance to Neptune would have circularized the orbit to its present circular shape. At the same time, however, these forces could have heated Triton to be as hot as Jupiter's volcanic moon Io, according to calculations by Washington University researcher William McKinnon and others. This process, and/or a large impact involved in the capture process, might be a source of Triton's present internal heat (Goldreich and others, 1989; Smith and others, 1989). Final answers will probably await future generations' explorations of this unusual moon. Perhaps the main lesson to be learned, or repeated, from Neptune's moons is that diverse historical circumstances, from one system to another, have produced many unique moons.

PLUTO: NINTH PLANET OR INTERPLANETARY BODY?

After Neptune was discovered, scientists expected that its gravitational pull would explain irregularities in Uranus' motions. But Neptune's gravity did not account for all of them, and Neptune itself has some unexplained irregular motions. These irregularities suggested at least one more planet beyond Neptune. For this reason, Percival Lowell began a search for the ninth planet in 1905. As shown in Figure 12-12, this search led to the discovery of **Pluto** in 1930, when Clyde Tombaugh found it on Lowell Observatory photos. After it was named for a god of the underworld, a new planetary symbol ♇ was created from the first two letters, which were also Lowell's initials.

Though Pluto was hailed as the ninth and final planet in 1930, its status as a planet is now dubious, for at least five reasons:

1. It is much smaller than any other full-fledged planet. At an estimated diameter of 2300 km, it is smaller than our Moon and only 47% as big as Mercury. It ranks only sixteenth in size among the complete list of planets and moons. It is 2.3 times as big as the largest asteroid in the asteroid belt.

2. It fails to continue the trend of the giant planets in size; among them, it is an anomaly.

3. Its orbit is not close to twice the size of Neptune's, in keeping with other planet spacings; instead, it overlaps Neptune's orbit, ranging from solar distance 29.7 AU to 49.3 AU. It is the only "planet" to overlap another's orbit, whereas many asteroids and comets do this.

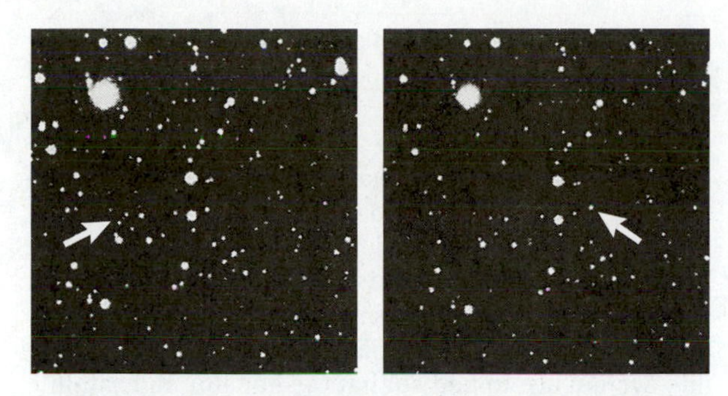

Figure 12-12 Portions of the photographs on which Pluto (arrow) was discovered. The planet was distinguished from stars by its motion. (Lowell Observatory)

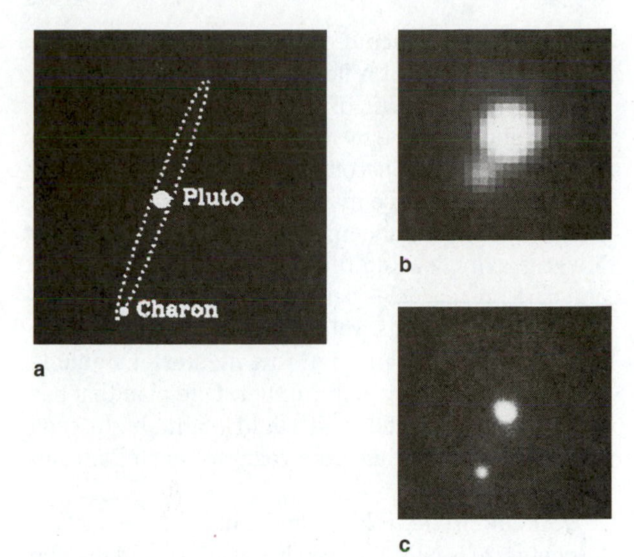

Figure 12-13 Pluto, the double planet, as seen by the *Hubble Space Telescope* (HST). **a** Map of Pluto, Charon, and Charon's orbit as they appeared at the time of the photography. **b** Earth-based telescopes show only this blurry level of detail. **c** Image of Pluto and Charon as photographed with the HST, using computer corrections to compensate for the telescope's flawed optics. This was the first long-exposure photo made with the HST on a moving object; the telescope had to track Pluto during the exposure. (NASA)

4. In the early 1990s, several interplanetary bodies roughly $\frac{1}{10}$ the size of Pluto were discovered orbiting in the general region from Neptune to somewhat beyond Pluto. This proves that a whole population of objects exists in the region of Pluto. Moreover, we know that many more comet nuclei (not yet discovered individually) must exist beyond Neptune's orbit (see next chapter). Pluto could be categorized as merely the largest of such interplanetary bodies.

5. Pluto's moon is much larger in diameter with respect to its planet (57%) than any other moon.

For these reasons astronomers are beginning to think of Pluto (and its moon) as more like a giant double asteroid than a normal planet (Figure 12-13). The distinction may be more than semantic: Pluto may indeed be more closely related to asteroids and comets than to full-fledged planets, because a group of small bodies was apparently left in that region after the giant planets finished forming. Nonetheless, most writers still list Pluto as the ninth planet, and the International Astronomical Union has resisted suggestions that the status of Pluto be changed to an asteroid or comet designation.

Pluto's light variations and eclipses of its moon establish that it rotates every 6.39 d on an axis tipped about 120° to its orbit. Its density, or, more accurately, the mean density of Pluto and the satellite, is close to 2000km/m^3, less than normal rock density. This suggests a mixed composition of about 70% rock and 30% ice (Stern, 1992a)—a somewhat lower percentage of icy material than has been found among the icy satellites of the giant planets.

Pluto's satellite, named **Charon,** was discovered in 1978. The name comes from a mythological figure associated with Pluto. Charon (pronounced KEHR-on) circles Pluto in 6.39d, keeping one face toward Pluto. Observations with the *Hubble Space Telescope* suggest that Charon has a lower density than Pluto—as low as 1300±230 kg/m^3—and thus is more ice-rich. Other reports give a density closer to Pluto's.

According to Washington University planetary scientist William McKinnon (1984), a giant impact on primordial Pluto might explain some of the properties of the Pluto/Charon system. In the same way that Earth's Moon is now believed to have formed from material blown off primordial Earth in a giant impact, Pluto may have suffered an impact that blew icy material out of its mantle. Charon may have aggregated from that material. This idea is consistent with the growing view that a few catastrophic impacts and close encounters have affected the solar system's character. It would explain not only Charon's existence but also its relatively ice-rich composition.

You might expect that Pluto and Charon would have the same ices on their surfaces, but spectroscopic studies show unexpected differences. Observations from American and Japanese telescopes at Mauna Kea Observatory in Hawaii from 1976 to 1999 showed that Pluto is covered mostly with frozen nitrogen ice, mixed with a few percent of frozen methane, carbon monoxide, and ethane (C_2H_6). The ethane may be produced by the action of sunlight on the methane (CH_4). Pluto's icy surface is bright, reflecting about 60% of the incoming sunlight. Charon is surprisingly different. It reflects about 40% of the light—a figure characteristic of slightly dirty ice—and the spectra show that the methane component on Charon is virtually absent. Charon's surface is primarily frozen water. The difference may involve a loss of the lightweight methane molecules from Charon, due to its lower gravity, leaving the heavier water molecules and dark soil particles behind (Stern, 1992a).

Clearly, scientists would like to know more about the origin and evolution of this mysterious double world of the solar system frontier. One plan that has been discussed would send a lightweight, high-speed space probe that would take years to reach Pluto, arriving around 2015.

On cold, remote Pluto and Charon, the surface temperature is only about 50 K (–369° F), and the Sun would be too far away to be perceived as a disk. From this outpost of the solar system, the Sun would look like a bright streetlight across the street, reminding us that it is only one star out of many.

"PLANET X"?

★ Several astronomers have sought dynamical or photographic evidence of a planet beyond Pluto, sometimes called **"Planet X."** Clyde Tombaugh, the discoverer of Pluto, conducted a long search of the region beyond Pluto and ruled out any planet as large as Neptune near the plane of the solar system out to a distance of around 100 AU. However, as mentioned above, starting in 1992, observers Jane Luu and David Jewitt began finding many modest-size objects on orbits similar to Pluto's. So far, none is as large as Pluto, but many astronomers thus suspect that additional Pluto-size bodies may lurk in the outermost solar system waiting to be discovered.

SUMMARY

The four giant planets—Jupiter, Saturn, Uranus, Neptune—and their satellite systems show some systematic similarities as well as differences. All four of the giant planets have dense, cloudy atmospheres, averaging around four-fifths hydrogen and one-fifth helium by mass. All have rings composed of billions of small particles ranging in scale from microscopic to house-size. All have systems of satellites. And most of the moons contain abundant ice as well as black soil, because this part of the solar system is so cold that ices were a main constituent of the material available for building moons and planets.

Among differences, we find that the colder, more remote planets have more hazy, less colorful cloud patterns. Uranus and Neptune are smaller than Jupiter and Saturn and have a blue color caused by a methane-rich haze, instead of the reddish browns and tans of the larger bodies. Uranus has narrow rings; Neptune has rings concentrated in unusual arcs. Uranus and its satellite system are tipped so that the equator and satellite orbits lie nearly perpendicular to the plane of the solar system. While Uranus has cratered and fractured airless satellites, Neptune has a nearly uncratered moon, Triton, with a young surface and erupting geyserlike vents.

Pluto, usually classed as the ninth planet, may also be viewed as only the largest of many interplanetary bodies in that region. It is smaller than our Moon and is unique in having a moon more than half its own size.

CONCEPTS

features common to all four giant planets
Uranus
Neptune
atmospheric properties of Uranus and Neptune
Great Dark Spot
internal structures of Uranus and Neptune
obliquity
retrograde rotation
ring systems of Uranus and Neptune
shepherd satellites
arc-rings
Uranus' satellites
Miranda
Neptune's satellites
Triton
Pluto
Charon
"Planet X"

PROBLEMS

1. Why do Uranus and Neptune have more Earth-like colors (bluish) than Jupiter and Saturn?

2. Discuss how rings may have formed around giant planets.

3. Why doesn't Earth have a hydrogen/helium atmosphere as the giant planets do?

4. Describe the seasons and other effects that would occur if Earth had the same obliquity as Uranus.

5. Do orbits of planets ever cross, as seen from far north or south of the plane of the solar system?

6. Where would Bode's rule, if extended, predict a tenth planet? Would it be easier to detect if it was covered with snow or with rock? Why?

ADVANCED PROBLEMS

7. What is the typical velocity of a hydrogen atom:

a. In the 1500-K gas at the top of Earth's atmosphere?

b. In a thin 1500-K gas at the surface of the Moon?

c. In a 78-K gas in Uranus' atmosphere?

d. The fastest atoms in a gas may move three or more times faster than the typical velocity calculated above, due to random collisions in the gas. Calculate the escape velocities of the Earth, Moon, and Jupiter, and use these facts to explain the absence of hydrogen on the Earth or Moon but its retention on Jupiter.

8. Calculate the weight of a 140-lb person on the surface of Pluto. Use Table 8-1.

PROJECT

1. With a telescope having at least a 15-cm (6-in.) aperture, observe Uranus or Neptune. Can you see the greenish-blue color? Can you see any detail? Any rings? Any satellites?

CHAPTER

13 Comets, Meteors, Asteroids, and Meteorites

WHAT THE READER SHOULD WATCH FOR IN THIS CHAPTER

Many small planetary bodies orbit the Sun in the vast stretches of space between the planets. They range in size from innumerable dust grains and boulders to dozens of objects between 100 and 1000 km in diameter. The bodies with significant ice in them are called comets, and the bodies of rocky material are called asteroids, although the distinctions are somewhat blurred in some cases. When a comet comes close enough to the Sun (typically, inside the orbit of Jupiter or Mars), the Sun's heat causes the ice to turn into gas, and the cometary body emits gas and dust; this is what causes the tail of a comet. Fragments of asteroids and comets can hit the Earth and other planets. Bigger ones create impact craters, and the biggest ones have affected the evolution of life on Earth. Fragments that hit the ground are called meteorites. Meteorites can be seen in museums. Some are rock, others are pure nickel-iron alloy, and still others are made of black, carbon-rich material. The carbon-rich ones may offer some clues about the origin of life. ■

★ On June 30, 1908, a mysterious explosion occurred in Siberia. English observatories 3600 km (2200 mi) away noted unusual air pressure waves. Seismic vibrations were recorded 1000 km (600 mi) away. At 500 km, observers reported "deafening bangs" and a fiery cloud. The explosion was caused by an unknown object that struck the Earth's atmosphere from space. An artist's reconstructions of this event, based on Russian eyewitness reports, are shown on our web site.

Some 200 km (120 mi) from the explosion, the object was seen as "an irregularly shaped, brilliantly white, somewhat elongated mass . . . with [angular] diameter far greater than the Moon's." Carpenters were thrown from a building and crockery knocked off shelves. An eyewitness 60 km from the blast reported that

> *the whole northern part of the sky appeared to be covered with fire. . . . I felt great heat as if my shirt had caught fire . . . there was a . . . mighty crash. . . . I was thrown onto the ground about [7 m] from the porch. . . . A hot wind, as from a cannon, blew past the huts from the north. . . . Many panes in the windows were blown out, and the iron hasp in the door of the barn was broken.*

Probably, the closest observers were some reindeer herders asleep in their tents about 80 km (49 mi) from the site. They and their tents were blown into the air, and several of the herders lost consciousness momentarily. "Everything around was shrouded in smoke and fog from the burning fallen trees."

The cause of this remarkable explosion was a collision between the Earth and a relatively modest bit of interplanetary debris. Many such objects circle the Sun. The Siberian event was merely the fall-in of the largest object to hit the Earth in the last century or so. Even larger objects have hit the Earth in earlier eras, as proved by the 20,000-y-old Arizona crater and by the evidence for a giant impact 65 million years ago that ended the reign of the dinosaurs, as discussed in Chapter 6.

Smaller impacts have been recorded more often; interplanetary stones fall from the sky in various locations every year (sometimes remote and sometimes inhabited—houses and cars have been hit in recent decades!). Tiny dust grains that burn up in the atmosphere before hitting the ground can be seen every night; these are sometimes called shooting stars.

★ From these observations we know that interplanetary space contains many small bodies of different sizes and that some of them occasionally hit Earth, as well as the Moon and other planets.

COMETS AND ASTEROIDS: RELATIONSHIPS

Just as the outer and inner parts of the solar system have produced planets of two different types (hydrogen-rich giants and rocky terrestrial planets), they have produced two different kinds of interplanetary bodies. **Comets** are ice-rich bodies that formed primarily in the cold, outer solar system beyond Jupiter's orbit; **asteroids** are rocky bodies that formed primarily in the inner solar system inside Jupiter's orbit. As we will see, though, the terminology is a bit fuzzy; some objects cataloged as asteroids have turned into comets and vice versa. The original distinctions in the 1800s had to do more with appearance than with composition or origin (as discussed further on our web site). Comets were usually fuzzy (because they can give off gas), and asteroids looked like faint stars.

The terms *comet* and *asteroid* refer mostly to the larger bodies moving through interplanetary space—ranging in size from hundreds of meters to 1000 km (600 mi) across. The reason comets can look fuzzy and give off gas has to do with their composition. They contain abundant ices as well as dark carbon-rich soil; when comets come into the inner solar system and are warmed by the Sun, their ices evaporate away into space, or **sublime,**[1] causing the comets to give off gas and dust. For example, water (H_2O) ice, when warmed, changes to gaseous H_2O molecules, hydroxide (OH) molecules, and hydrogen (H) molecules. Because asteroids are composed mainly of rocky and metallic material, they do not give off significant amounts of gas. Comets and asteroids have been studied through telescopes and by spacecraft visits. Smaller bits of debris (either icy or rocky) cannot be studied in this way and are known only when they hit Earth. They typically range in size from microscopic grains to 100 m or so in diameter. Since their physical character is less well known, they are given the generic name, **meteoroids.**

When a meteoroid hits the atmosphere or surface of Earth or another planet, it is typically traveling at 10 to 40 km/s. At this speed, hitting the atmosphere creates friction and a strong shock pressure in the material; the meteoroid heats up and becomes a glowing object, and it may break into many pieces. **Meteors** are the pea-size and smaller meteoroids that burn up in the atmosphere and do not hit the ground. **Meteorites** are larger meteoroids, or pieces of them, that do hit the ground.

A note on terminology: The term *comet* comes from the Greek root *coma,* for hair, because a comet's tail fanning out reminded the ancients of long hair blowing in the wind. *Asteroid* comes from the Greek root *aster,* referring to a star, because an asteroid, as seen through a telescope, looks like a faint star. The prefix "meteor," as in meteoroids, meteors, and meteorites, comes from the Greek root for "air" or "atmosphere." The same root is used in the term *meterology,* the study of the atmosphere. This root was used because in ancient times meteors were known as streaks of light high in the air, and meteorites were rocks that seemed to fall out of the atmosphere. The concept of interplanetary space beyond the atmosphere was not well understood, and meteors and meteorites were initially considered phenomena associated with the air itself. These old names don't have much to do with the physical nature of the objects, as revealed in the twentieth century!

Studies of meteorites prove that they are fragments of asteroids and, perhaps in some cases, of comets. Similarly, many of the smaller asteroids are also believed to be fragments of once-larger asteroids. When a fragment of this sort is discussed, the hypothetical body of which it is a piece is called its **parent body.** Meteorites come in many different compositions, from metal to rock. One unsolved mystery is which types of asteroids or comets are the parent bodies of different types of meteorites.

One way of answering such questions is to study the orbits of the objects, illustrated in Figure 13-1. Table 13-1 compares some orbits for the

[1] *Sublime,* which means to change from solid form directly into gaseous form, is a more correct technical word than *evaporate. Evaporate* technically means to change from liquid form into gaseous form.

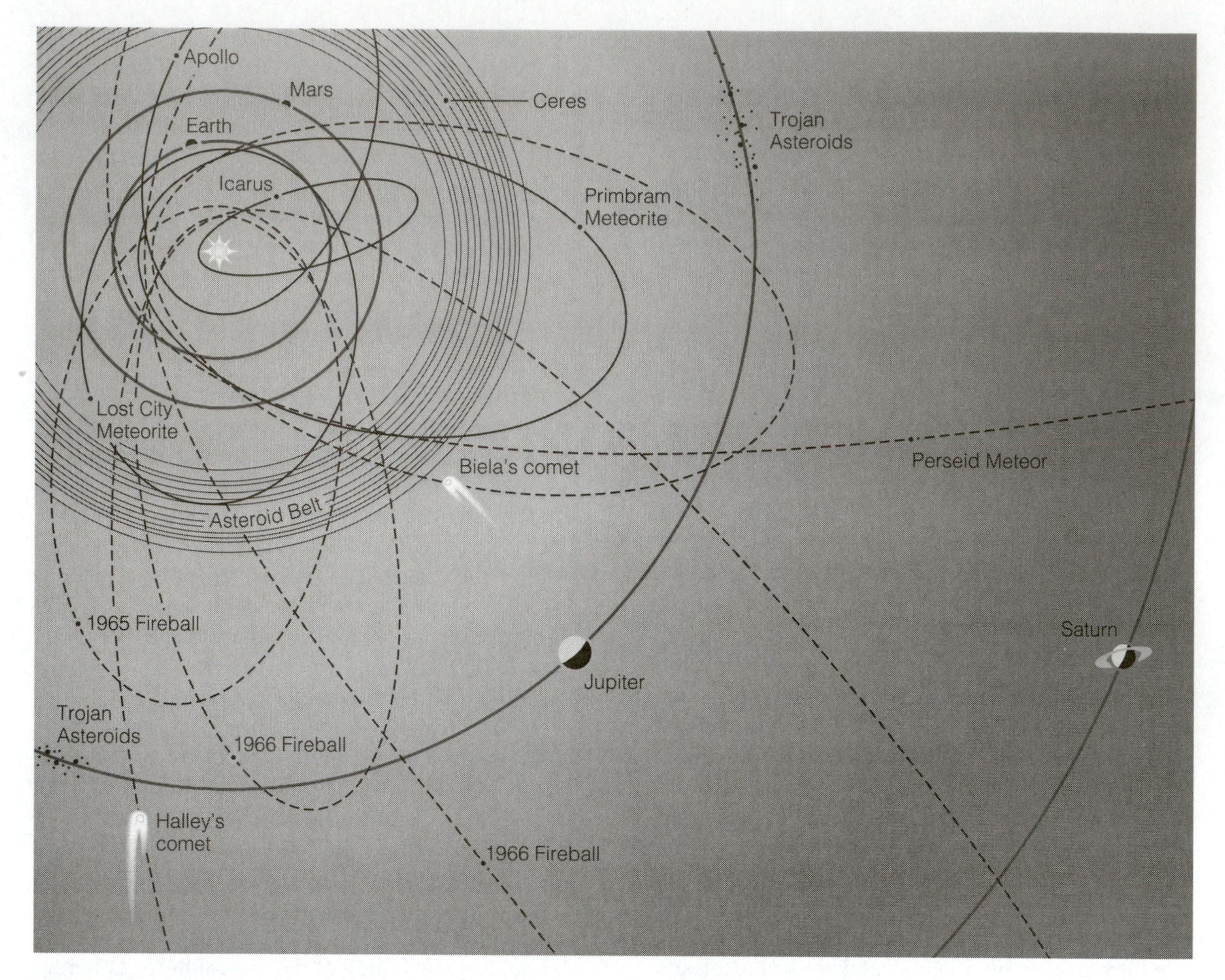

Figure 13-1 A portion of the solar system illustrating a small fraction of the crowded population of interplanetary bodies. Planet orbits are shown; orbits of selected comets, main-belt asteroids, Trojan asteroids, Apollo asteroids, meteorites, and large meteors (fireballs) are shown in dashed and solid lines.

four classes discussed in this chapter: comets/meteors and asteroids/meteorites. Notice, for example, that asteroids tend to be concentrated in the region between Mars' orbit and Jupiter's orbit, from about 2 to 5.2 AU. Specifically, most of the larger ones are crowded into the **asteroid belt,** a region from about 2 to 3 AU. Notice the table column labeled "Max. Distance from Sun"; this is the farthest each body can get from the Sun as it travels in its elliptical orbit. As you can see, true asteroids rarely get outside the 2-to-5 AU region (except for a few, such as the so-called Apollo group, whose orbits have been disturbed by the gravity forces of the planets).

Now look at the maximum distances from the Sun for comets and meteors. As you can see, most comets have traveled from much vaster distances, far beyond Pluto in many cases. Most meteors have similar orbits. On the other hand, all the meteorites listed come from no farther than the asteroids. This was one of the first clues that established that *most meteors are bits of debris from comets,* while *most meteorites are fragments of broken asteroids.*

Comets, asteroids, and meteoroids of every size are all debris left over from the origin of the solar system 4.6 billion years ago. As we'll see in the next chapter, astronomers believe that the present planets formed from such bodies. At the time the planets formed, the solar system was filled with innumerable small, cometlike and asteroidlike bodies, ranging up to 1000 km across. To distinguish them from modern comets and asteroids, they are known collectively as **planetesimals.**

TABLE 13-1

Properties of Selected Small Bodies in the Solar System

Class	Example	Diameter (km)	Orbit: Semimajor Axis (AU)	Orbit: Max. Distance from Sun (AU)	Orbit: Eccentricity	Orbit: Inclination	Remarks
Comets							
Short-period	Encke	1–8	2.2	4.1	0.85	12°	Probably dirty ice
	Halley	Few?	18	35	0.97	162°	Probably dirty ice
Long-period	Kohoutek	8 × 15	Very large	Very large	1.0	14°	Probably dirty ice
Meteors							
Shower	Perseid	10^{-6}	40	79	0.97	114°	Cometary debris
	Taurid	10^{-6}	2.2	4.0	0.80	2°	
Fireballs	July 31, 1966	?	32	63	0.98	42°	May be related to comets or asteroids
	May 31, 1966	?	3.0	5.4	0.80	9°	
Asteroids							
Main Belt	1 Ceres	1020	2.8	3.0	0.08	11°	Carbonaceous rock surface
	2 Pallas	538	2.8	3.4	0.23	35°	Rocky surface
	3 Juno	248	2.7	3.4	0.26	13°	Rocky surface
	4 Vesta	549	2.4	2.6	0.09	7°	Lavalike surface
	14 Irene	170	2.6	3.0	0.16	9°	Rocky surface
Trojan	624 Hektor	100 × 300	5.1	5.2	0.02	18°	Unusual shape
Apollo	433 Eros	7 × 19 × 30	1.5	1.8	0.22	11°	Elongated
	1862 Apollo	1.4	1.5	2.3	0.56	6°	
Comet/asteroid	2060 Chiron	100–320?	13.7	19	0.38	7°	Erupted in 1988[a]
Meteorites (chondrites)							
	Pribram	10^{-4}	2.5	4.2	0.68	10°	
	Lost City	10^{-4}	1.7	2.4	0.42	12°	
	Leutkirch[b]	10^{-4}?	1.6	2.2	0.40	2.5°	

[a]Chiron was cataloged as an asteroid after its discovery in 1977. It brightened and developed a coma in 1988, proving it is really an icy body. It illustrates the "fuzziness" of comet/asteroid nomenclature.

[b]An object photographed over Europe in 1974. No fragments were recovered, but believed to be a stone meteorite.

Sources: Data from Hartmann (1975, 1983); Binzel and others (1989); Beatty and Chaikin (1990). The diameter of Comet Encke was measured with radar by Kamoun and others (1982).

To put it in other words, comets, asteroids, and meteoroids are all leftover planetesimals and planetesimal fragments. Thus they provide a direct link to the conditions under which the planets and the Sun itself formed. Some of them even contain grains that are believed to have formed in *interstellar* space, before the Sun formed. By implication, they provide clues to the formative conditions of other stars and hence provide an additional link with the most distant stellar regions, which will be a topic of much of the rest of this book.

COMETS

Comets are the most spectacular of the small bodies in the solar system. When they pass through the inner solar system, for example, between the orbits of Earth and Mars, they can be seen drifting slowly from night to night among the stars. (Writers sometimes incorrectly describe comets as "flashing across the sky" like shooting stars. They do not. They seem to hang motionless and ghostly among the stars. Their motion relative to the stars can be detected by the naked eye only after a few hours.) Thus a typical comet moves slowly among the stars and may be visible for many weeks as it passes through the inner solar system, around the Sun, and back out to the obscurity of the outer solar system, where it is too faint for us to see.

Figure 13-2 The motion of Comet Ikeya-Seki in the dawn sky during an interval of a few days in 1965. The comet's head is the sharply defined tip nearest the horizon; the tail extends diffusely upward. The comet in the predawn sky, photographed with an ordinary, stationary 35-mm camera (20-s exposure at f1.9 on Tri-X film). (Photo by S. M. Larson, University of Arizona)

★ A bright comet has several parts, as seen in Figures 13-2 and 13-3. The brightest diffuse part is the **comet head,** sometimes called the coma. The **comet tail** is a fainter glow extending out of the head, usually pointing away from the Sun. Although a typical comet tail can be traced for only a few degrees by the naked eye, binoculars or long-exposure photos may reveal fainter extensions of the tail extending tens of degrees. Some historic comets in past centuries have been close enough to Earth to be spectacular and have tails extending clear across the night sky. A telescope reveals a brilliant, starlike point at the center of the comet head. At the center of this bright point is the **comet nucleus.** This is the only substantial, solid part of the comet, but it is too small to be resolved by telescopes on Earth. Studies reveal that a typical comet nucleus is a worldlet of dirty ice only about 1 to 20 km (a few miles) across—tiny compared to most planets and moons! The gas and dust that make up the rest of the comet's head and tail are material emitted from the nucleus. The origin of the comet's head and tail is shown in Figure 13-4. As the comet nucleus moves through the inner solar system, roughly inside the orbit of Jupiter or Mars, the sunlight warms it and causes the ice to sublime into the form of gas. This gas, together with dislodged dust grains from the dirt in the nucleus, is then carried away from the nucleus by the pressure of radiation and thin gas rushing outward from the Sun. This outrushing solar gas is called the **solar wind.** Only microscopic grains

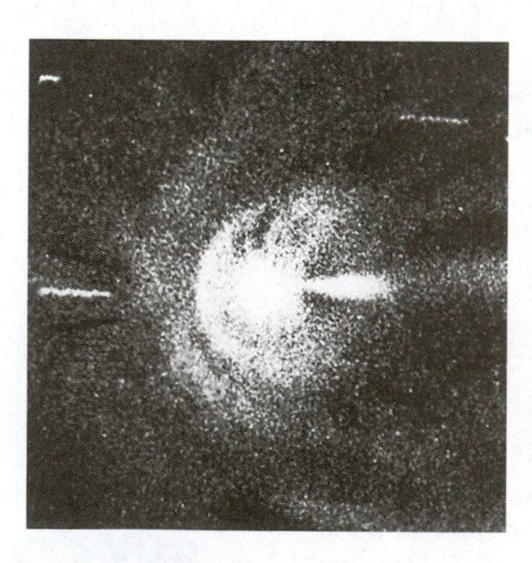

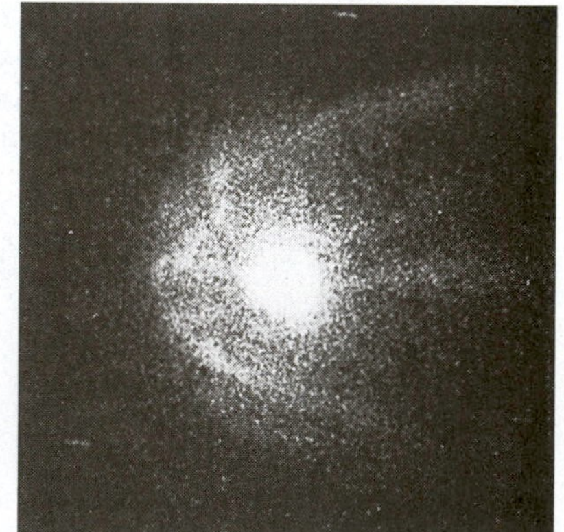

Figure 13-3 A new look at old photos of Halley's comet. The head of Comet Halley was photographed on different days as it passed unusually close to the Earth in June 1910, but photographic prints made by standard techniques of that era showed only vague structure. In 1984, astronomers Z. Sekanina and S. M. Larson reprocessed some of the old plates with modern equipment that digitizes the image and enhances low-contrast detail. The nucleus is in the brightest region at the center of each image. These images clearly show individual curved jets of dusty debris blown off the nucleus by individual outbursts. Sun is on left; tail is to right. Streaks are star images trailed by long-exposure tracking on comet. (Photos courtesy S. M. Larson, University of Arizona)

get blown outward by it; larger grains are too heavy to be caught in the solar wind. (Similarly, if you throw a handful of dust and gravel into the air, the dust gets blown by the wind, but the pebbles do not.)

The gas and dust from the nucleus form the comet's tail, often stretching more than an astronomical unit. Caught in the solar wind, the tail streams out behind the comet as the comet approaches the Sun, but leads

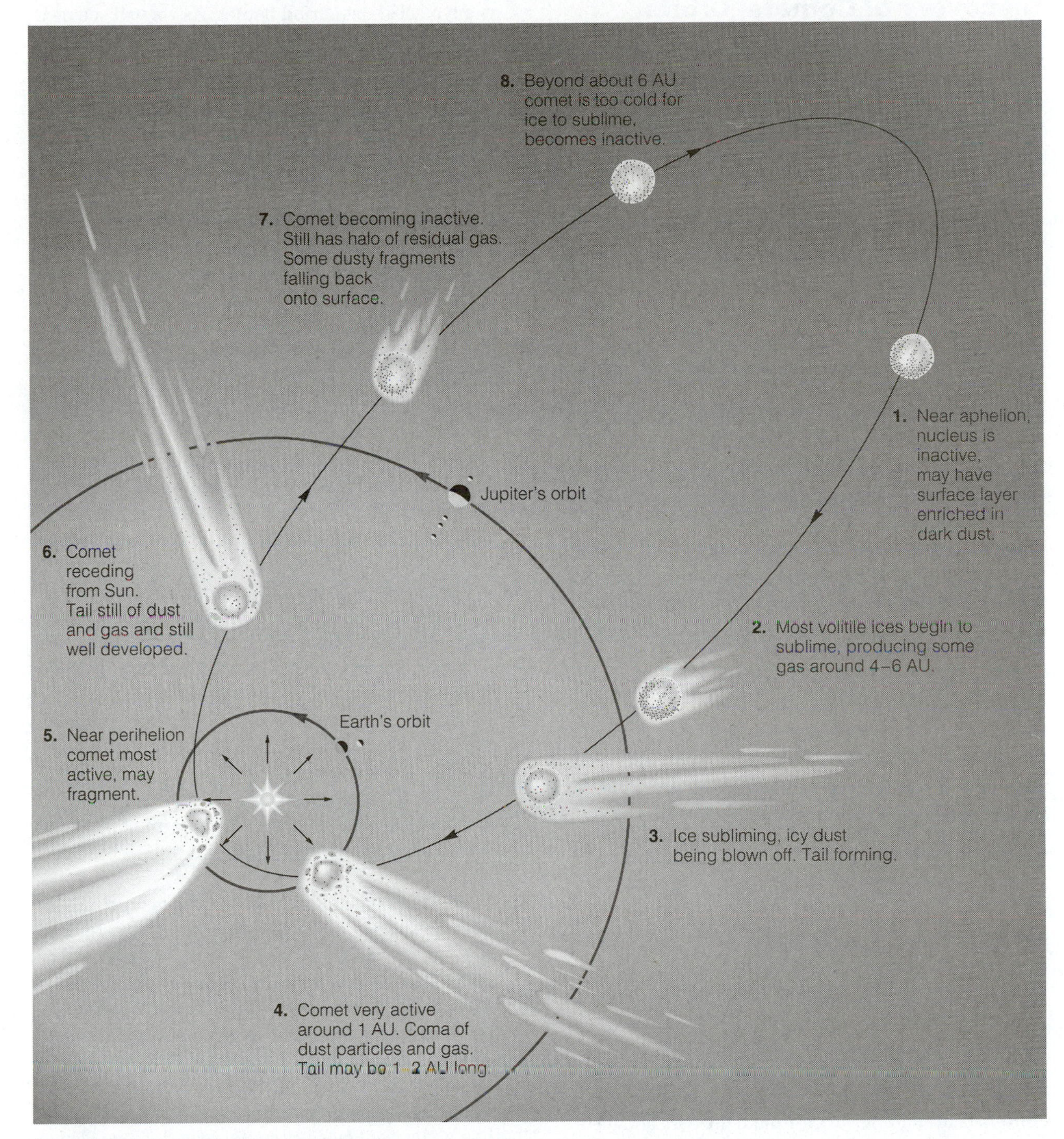

Figure 13-4 Stages in the development of a typical comet as it travels from the outer solar system into the inner solar system, where it loops rapidly around the Sun and moves outward again. The tail develops only within a few astronomical units of the Sun, where sunlight is warm enough to sublime the ice in the nucleus. In order to show comet detail, the figure is not drawn to scale.

as the comet recedes from the Sun. It is like the long hair of a woman streaming out behind her as she walks into the wind, but in front of her if she reverses direction.

Discovery of Comets' Orbits

★ In ancient days of superstition, comets were believed to be evil omens, as discussed on our web site. No one knew how far away comets were, and many people thought they were phenomena in our own atmosphere. Seneca, a Roman contemporary of Jesus, wrote: "Some day there will arise a man who will demonstrate in what regions of the heavens the comets take their way." That man was Tycho Brahe, who in 1577 arranged observations of a bright comet from two different locations. Finding no parallactic shift in the coment's angular position relative to the stars (see our earlier discussion of parallax in Chapter 2), Tycho correctly concluded that it was more distant than the Moon and thus not terrestrial.

Comets' true nature as interplanetary bodies became even more clear after Newton explained the laws of orbital motion. In 1704, the English astronomer Edmond Halley applied newly developed methods of computing orbits and discovered that comets travel on *long, elliptical orbits* around the Sun and that certain comets reappear each time they loop around the Sun and pass near Earth's orbit. Calculating the orbits of 24 well-recorded comets by Newton's methods, Halley found that four comets (seen in 1456, 1531, 1607, and 1682) had the same orbit and a periodicity near 75 y. Halley correctly inferred that these appearances were by a single comet and that the slight irregularities in periodicity were caused by gravitational disturbances from the planets, especially Jupiter. Halley predicted that the comet would return about 1758. It did so on Christmas night. It was named **Halley's comet.**[2]

★ Halley's comet became very famous due to its prominence in 1910, when the Earth passed through its tail. It was not so prominent during its most recent approach in 1986, because it did not pass as close to the Earth (see the photograph on our web site). Even so, the 1986 approach was historic because it marked the first close-up study of a comet by space probes.

As shown in Table 13-1, comets are divided into two groups, according to their orbits. **Long-period comets** take centuries to go around the Sun, and their orbits can extend to thousands of AUs from the Sun. **Short-period comets** take about a century or less to go around the Sun, and their orbits are mostly inside the region of Pluto. Comet Halley, with its period of 76 y and maximum distance of about 18 AU, is a typical short-period comet. As we will see next, these two groups of comets probably had different histories.

[2]Comets discovered today are first named for their discoverers but are later given scientific names in order of their passage around the Sun—for example, Comet 1995 II.

Where Comets Come From: The Oort Cloud and the Kuiper Belt

When most comets are discovered, they are looping into the inner solar system from somewhere in the outer solar system. In the outer solar system they are too faint to be seen—remember they brighten only as they come inside the orbit of Jupiter or Mars. The Dutch astronomer Jan Oort began the work of mapping comet source regions in 1950, when he showed that most long-period comets must come from a vast region surrounding the solar system, ranging from 50,000 to 150,000 AU from the Sun. This region is far beyond the giant planets or Pluto. Oort found that a typical long-period comet has an elliptical orbit with one end in the very distant region and the other end among the terrestrial planets. Comets on such extremely elongated orbits take 10 to 60 million years to go around the Sun, which explains the name *long-period comets.* From statistics of comet orbits and their rate of appearance, astronomers estimate that roughly 100 billion inactive comet nuclei, invisible from Earth, lie in this frigid, distant region. These nuclei are not distributed in a flattened disk, like the planets of the solar system; rather, they comprise a huge, spheroidal swarm of objects, like bees buzzing around a hive. This swarm of comet nuclei is called the **Oort cloud** (see review by Weissman, 1998).

How do comets get from the Oort cloud into the inner solar system? Disturbances, such as the passage of a star near the solar system, can alter the orbits of Oort cloud comets, deflecting them into new orbits where they fall toward the Sun and pass among the terrestrial planets. As we will discuss in more detail in the next chapter, astronomers believe that the comets in the Oort cloud formed among the giant planets as the planets themselves were forming, but then underwent close encounters with those planets and were deflected out of the planetary region by their strong gravitational fields. In the same way that the two Voyager space probes were flung beyond Pluto by the gravitational fields of Saturn and Neptune, comet nuclei were also flung beyond Pluto, that is, into the Oort cloud.

Statistics of short-period comets' motions indicate that they have a different distribution and a different history. A concentration of these comets—barely detectable from Earth—exists just beyond the region of Neptune in the general region of Pluto, from about 30 to 100 AU from the Sun. This concentration began to be recognized in the 1980s and is known as the **Kuiper belt,** or sometimes the Edgeworth-Kuiper belt, after the astronomers who first surmised its existence between 1943 and 1951. The Kuiper belt is not a spheroidal swarm like the Oort cloud, but a flattened disk of comets, lying in the plane of the planets, closer to the Sun and planets. These comets were never deflected into the Oort cloud, but rather were stranded where they formed in the earliest years of the solar system (Luu and Jewitt, 1996). The Kuiper belt may contain around a billion small comet nuclei; it has been estimated that 35,000 of them might be larger than 100 km (Jewitt and Luu, 1995). Only a handful of the largest ones are bright enough to be seen from Earth with large telescopes. These large examples have diameters about 10–15% the diameter of Pluto. They are like little Plutos, and since they orbit in the same general region, they have often been called **Plutinos.** Here are some examples, compared to Pluto:

	Distance of Orbit from Sun
Pluto	30–49 AU
Plutino 1996 TL	35–134 AU
Plutino 1996 TP	26–53 AU
Plutino 1997 CQ	41–47 AU

In the same way that gravitational disturbances in the Oort cloud can deflect Oort cloud comets into the inner solar system, disturbances from Neptune and other outer planets can deflect Kuiper belt comets into the inner solar system. The typical Kuiper belt comet first is deflected into an orbit among the giant planets, then may be further deflected into the inner solar system if it passes near one of them. The objects temporarily hung up among the giant planets began to be discovered in the late 1970s and 1980s—before the Plutinos. They were named after Centaurs in Greek mythology and are often called Centaurs, or Centaur objects. It has been estimated from discovery statistics that 2600 of them may be larger than 75 km across (Jewitt, Luu, and Chen, 1996). Here are some examples, compared to Saturn, Uranus, and Neptune:

	Distance of Orbit from Sun
Saturn	close to 9 AU
Uranus	close to 19 AU
Neptune	close to 30 AU
Centaur 2060 Chiron	between 8.5 and 19 AU
Centaur 5145 Pholus	between 8.8 and 32 AU
Centaur 1995 GO	between 6.6 and 22 AU

From these examples, you can see that Centaurs move across the orbits of giant planets and are likely to be disturbed. Their orbits are typically unstable, so the Centaurs' lifetimes in these orbits are only a few million years or sometimes less. The supply is constantly replenished from the Kuiper belt.

Chiron is an interesting example. It was discovered in 1977 and cataloged as an asteroid, before the Kuiper belt was even known. It is roughly 250–300 km across. In 1988, it was found to be twice as bright as expected! Further observations the next year proved that it had developed a coma of dust. Chiron was not an asteroid; it was an active comet. It has cometary activity as far as 11 AU from the Sun—one of the greatest distances for which cometary activity has ever been found, implying that Chiron has gases that sublime at the slightest temperature increase. This in turn suggests that Chiron has never been closer to the Sun, because if it had, these ices would have burned off. Thus Chiron is probably a very primitive, pristine object that has been preserved in the deep freeze of the outer solar system. It might make an interesting target for future exploration.

★ Unexpected implications about the Kuiper belt come from observations of Sun-like stars. In the 1980s, telescopes revealed that a number of these stars have disks of dust—probably comet debris—surrounding them, with the dimensions extending out several hundred AU from the star. (We will discuss these further in Chapter 18.) These disks may mark *those stars'* Kuiper belts of cometary debris.

Comet's Nuclei: Dirty Icebergs in Space

By the 1700s, when people discovered that comets were not wispy atmospheric phenomena or evil visitors but objects orbiting the Sun, the next question became: What is the physical nature of the comet? Progress on this question came in 1868 when English astronomer William Huggins first studied comets through a spectroscope. As discussed in Chapter 5, the spectroscope allows astronomers to measure properties of the material in a celestial body. Huggins found three bright emission bands, which he could identify with gaseous carbon. This proved comets were giving off gases, rich in carbon. Modern astronomers have added many other kinds of atoms, molecules, and ions to the list, including H (hydrogen), O (oxygen), CN, CH, OH, H_2O^+, CN^+, CH^+, OH^+, N_2^+, CO^+, and CO_2^+, and HCO. Recent studies have made exciting additions to the list: complex organic molecules such as CH_3CN and H_2CO. This discovery is exciting because it proves that organic molecules—the building blocks of life—form elsewhere in the universe besides Earth.

These results have been supported by spectroscopic observations of various Centaur and Kuiper belt objects, which have provided direct confirmation of water ice and CH-rich ices such as methane (CH_4), ethane (C_2H_6), and other CH-based molecules (Brown and others, 1997, 1998).

Note that all these atoms (H, O, C, N) and molecules are just the ones that would be expected if the gases were formed by the sublimation of common solar system ices such as H_2O (water), CH_4 (methane), CO_2 (carbon dioxide), and NH_3 (ammonia).

★ Apart from revealing gases, the spectroscope shows that some particles in comet tails are much bigger than molecules. They are dust grains. Thus a comet tail contains both gas and dust coming off the nucleus. Indeed, these two materials can easily be distinguished in many color comet photos, as shown on our web site. The gas tends to blow straight back from the Sun in the solar wind, but the heavier dust grains move partly under the influence of the solar wind and partly under the influence of Keplerian motions, like microplanets. Thus two slightly separated but distinct tails often develop. The gas tail tends to scatter blue light just as the gas in our atmosphere does. The dust tail is composed of dark, reddish-brown dust and thus tends to have a pinkish cast. Thus comets can often be seen to have a straight, bluish gas tail and a slightly curved, pinkish dust tail. These colors can't be sensed with the eye because comet tails are so faint, but they do show up in color time-exposure photos.

Evidence of this kind proved that the nucleus, whatever it is, can slough off both gas and dust when warmed by the Sun. Around 1950, Harvard astronomer Fred Whipple first put together all this evidence to give an essentially correct prediction of the nature of a comet nucleus. He argued that the gases do indeed come from subliming ice and that the ice must be full of dust grains. As the ice sublimes, the dust grains are dislodged and blown away by the solar wind along with the gas. He predicted the ice to be mostly frozen water with certain amounts of frozen methane (CH_4), carbon dioxide (CO_2), frozen ammonia (NH_3), and other materials to explain the various C, H, O, and N atoms in the gas. Thus Whipple's theoretical picture of a comet nucleus came to be called the **dirty iceberg model,** and it has been confirmed by modern data.

The dirty iceberg that makes up the comet nucleus must be fairly weak in many cases. Some comets have been seen to break up spontaneously, especially as they get close to the Sun and have a large degassing rate. Examples are shown in Figure 13-5. The degassing may open up fractures that allow the nucleus to fall apart. Alternatively, the nucleus may be made up of discrete chunky objects, which come apart (Weidenschilling, 1997).

A Comet Nucleus at Close Range

The greatest advance in studying comets came in the 1980s, when sensitive telescopes began to measure properties of comet nuclei and comet behavior at large distances. This work culminated when an international fleet of five spacecraft (two Japanese, two Soviet, and one European) probed Halley in 1986. The first close-up data and pictures supported the earlier work.[3] After the first four probes tested the environment at moderate distance, the European probe flew by closest, only about 600 km from the nucleus. The probe was named Giotto, after the Italian artist whose 1304 painting of the comet may have been the first to show it.

The heart of Halley's comet was a hazy environment where the probes found more fine dust than expected. About 960 km from the nucleus, Giotto was hit by a dust grain, knocking it partially out of commission for 32 min during the closest approach! As seen in Figure 13-6, the photos revealed bright jets of illuminated gas and dust coming off "active" spots on the black, peanut- or potato-shaped nucleus, about 15 km long and 7 to 10 km wide. The haze made the

[3] The encounter dramatized that while our news media may be adept at reporting tragedies, they fail in dealing with the adventure of human exploration. During the night of the first Soviet encounter, a TV hookup with the Soviet Union allowed live coverage of humanity's first close look at a comet; but my local affiliate carried an Evel Knievel movie instead. The next day the papers ran a front-page picture of the comet, reported from the Soviet spacecraft. A few days later, however, on an inner page they reported sheepishly that this was not the new Soviet picture at all but a computer processed 1910 photo that had been misidentified by a news agency.

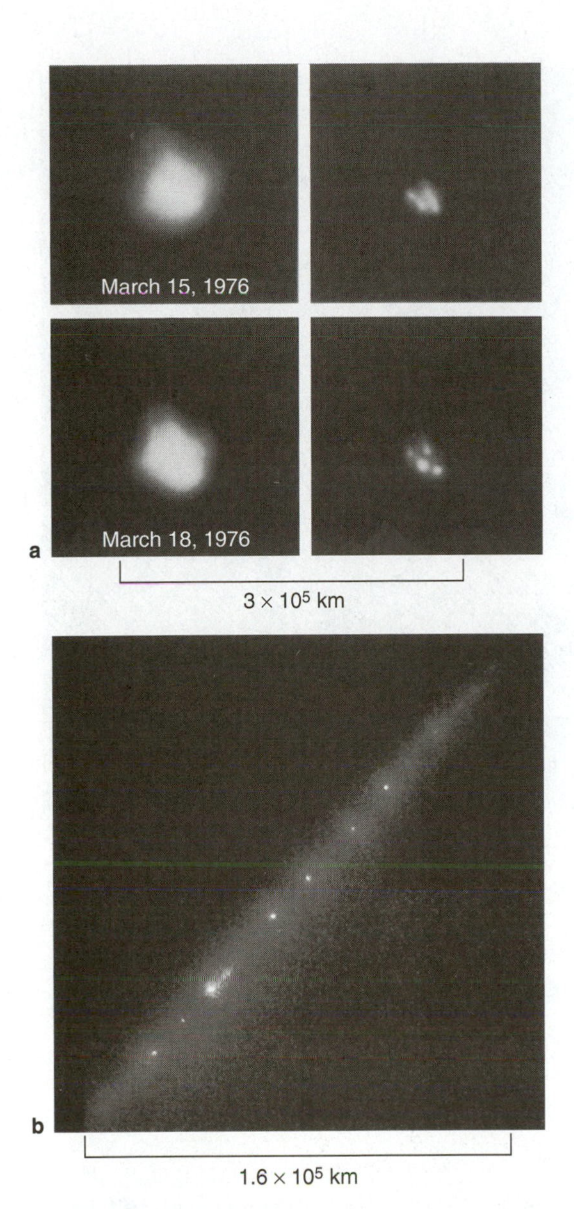

Figure 13-5 Several comets have been seen to break into pieces. **a** These four photos show the breakup of Comet West (1975n) into four pieces in 1976. The photos on the left are normal exposures, which incorrectly suggest a single bright head and nucleus. However, the shorter exposures on the right show four nuclei separating at relative speeds of about 1 to 5 m/s (2 to 11 mph) over a period of three days. One piece lasted only two weeks; others were followed for over six months. (Lunar and Planetary Laboratory, University of Arizona, courtesy S. M. Larson) **b** The extraordinary comet Shoemaker-Levy was already broken into a string of fragments spread along its orbit plane when it was discovered and imaged in 1993. The breakup probably happened due to tidal forces exerted by Jupiter during a close pass by the planet a year before this image was made. Tail material spreads from the nuclei toward the upper right. The string of fragments collided with Jupiter in 1994, the largest impact event in the solar system ever witnessed by humanity. (**a:** CCD image by J. Luu and D. Jewitt with University of Hawaii Mauna Kea Observatory telescope; courtesy Jane Luu. **b:** Hubble Space Telescope image by H. Weaver and T. Smith, Space Telescope Science Institute and NASA)

pictures fuzzy. The gas in the head, or *coma,* was found to be about 80% water vapor by volume. The dust particles were found to be rich in C, H, O, and N and to somewhat resemble carbon-rich meteoritic particles collected on Earth. There was so much carbon, hydrogen, oxygen, and nitrogen in these particles that they came to be called CHON particles. They are believed to be rich in organic molecules (large, carbon-based molecules) and are different from familiar terrestrial dust, which is richer in silicon, iron, other metals, and their oxides.

Although a few astronomers had predicted dark, asteroidlike dust on comets, Halley's nucleus turned out to be even blacker than many astronomers expected, reflecting only 4% of the light that strikes it. (Compare with 60 to 80% for clean ice.) It is as dark as black velvet! Comet nuclei are very dirty icebergs! Earth-based observers have found similar results for the nuclei of other comets (Campins and others, 1987). Since abundant carbon was observed in the dust shed from Halley, the black material is likely to be very carbonaceous. Thus scientists are becoming increasingly interested in this cometary dust, which is apparently rich in the building blocks needed for life. No life is expected to have evolved in the frozen interiors or surface layers of comets, but perhaps comets contributed organic materials as they struck primordial planets.

The Origin of Comets

From the evidence described so far, astronomers have theorized a five-stage history that explains the origin of comets. First, comets must have formed by aggregations of ice crystals and dust grains in the cold, outer regions of the solar system where the ice-rich planets and moons were forming. We know from studying the gases in the atmospheres of Jupiter and other giant planets that hydrogen, carbon, oxygen, and other elements were available as the outermost planets formed. In the environment of the outer solar system, these would have condensed into ice grains (such as H_2O) that could accumulate into icy planetesimals—the comet nuclei. The comet nuclei that formed in the region of Pluto had no giant planets to disturb their orbits and remained in place as part of the Kuiper belt.

As a second stage, many of the nuclei that were building among the giant planets approached close to the giant planets and were flung by their strong gravity into highly elliptical orbits that took the nuclei hundreds or thousands of astronomical units from the Sun. There, beyond the Kuiper belt, they formed the distant Oort cloud.

a

b

Figure 13-6 Giotto space probe's close-up views of the nucleus of Halley's comet. **a** Distant view gives a sense of the dusty haze surrounding the 15-km- (9-mi-) long nucleus, and of the bright jets of gas and dust venting from the surface. **b** The best available image. In the weeks after the encounter, computer processing yielded this view based on the closest images. Bright jets shoot out of localized active vents on the sunlit side. Smallest details are about the size of a football field. The nucleus is dark gray, and its dark side can be seen silhouetted against the softly glowing comet head (right side). Because of the journalistic tradition of reporting only "today's news," fuzzy initial images were widely published during the week of the encounter, but this much improved image has rarely been published in the popular media. (Giotto photo courtesy Harold Reisema and Alan Delamere, Ball Aerospace; copyright 1986 MPAE)

Third, comets were "stored" in the Kuiper belt and Oort cloud during much of solar system history. As a result of the process of ejection by giant planets, orbits in the Oort cloud included high inclinations and both prograde and retrograde directions, but they were further randomized by the passage of nearby stars. This explains why long-period comets' orbits are randomly distributed in inclination, direction, and other orbit properties.

Fourth, comets in the Kuiper belt and Oort cloud occasionally found themselves on trajectories back into the inner solar system. There, their ice sublimed, and they formed the familiar heads and tails that give comets their traditional appearance.

★ Fifth, a tiny fraction of those comets had new encounters with planets as they passed through the solar system, resulting in new changes in their orbits; in a few cases, the aphelion distance was greatly reduced, creating comets that spend most of their time among the planets—such as Halley's comet, with its 76-y period, or Encke's comet, with its 3-y period. Note in Table 13-1 that Halley's comet makes it back to its original "home" in the Kuiper belt, with an aphelion distance of 35 AU, but that the orbit of Encke's comet has been modified so much that now it never gets farther than 4.1 AU from the Sun.

METEORS, METEOR SHOWERS, AND THE COMET CONNECTION

"Shooting stars" that flash momentarily across the sky might seem totally unrelated to the ghostly appearance of a comet. However, a study of their frequency on different nights reveals a direct connection. On an average night, you may see about 3 meteors per hour before midnight and about 15 meteors per

TABLE 13-2

Dates of Prominent Meteor Showers

Shower Name (After Source Constellation)	Date of Maximum Activity[a]	Associated Comet
Lyrid	April 21, morning	1861 I
Perseid	August 12, morning	1862 III
Draconid[b]	October 10, evening	Giacobini-Zinner
Orionid	October 21, morning	Halley
Taurid	November 7, midnight	Encke
Leonid	November 16, morning	1866 I
Geminid	December 12, morning	"Asteroid" 1983 TB[c]

[a]Showers can last several days before and after the peak activity on the listed date. Observations are best when the constellation in question is high above the horizon, usually just before dawn.

[b]The Draconids are now weak because their orbits have been disturbed by the gravity of planets, but further disturbances may again strengthen the shower in the future.

[c]This object was discovered in 1983 by the IRAS satellite and cataloged as an asteroid, but it is probably a "burnt-out" comet nucleus.

hour after midnight. (You see more meteors after midnight because you are then located on the leading edge of the Earth as it moves forward in its orbit, sweeping up interplanetary debris.) But on certain dates each year, you may see **meteor showers** of 60 meteors or more per hour, all radiating from one direction in the sky (Table 13-2).

The best-known example is the Perseid shower, which occurs every year around August 12, when bright meteors streak across the sky every few minutes from the direction of the constellation Perseus. (A shower is named for the constellation most prominent in the area of the sky from which the shower radiates.) Occasionally, the showers are so intense that meteors fall too fast to count. The Leonid shower of mid-November is normally only a modest shower. But during the Leonid shower of November 17, 1966, meteors fell like snowflakes in a blizzard for some minutes, at a rate estimated to be more than 2000 meteors per minute (see Figure 13-7)!

What is the connection between the showers and the comets? In 1866, G. V. Schiaparelli (of Martian canal fame) discovered that a meteor shower occurred whenever the Earth crossed the orbit of Comet 1862 III, now known as Comet Swift-Tuttle. The Perseids, then, must be spread out along the orbit of that comet. Other relationships were soon found between specific meteor showers and specific comets, as Table 13-2 shows. In 1983, an infrared astronomical telescope in orbit (called IRAS) discovered the thermal infrared

Figure 13-7 A rare meteor shower; the Leonids of November 17, 1966. The rate of meteors visible to the naked eye was estimated to exceed 2000 per minute. The brightest star (upper left) is Rigel, in the constellation Orion. This exposure of a few minutes' duration was made with a 35-mm camera. (D. R. McLean)

emission from a swarm of meteor dust spread along the orbit of Comet Tempel 2. Here, then, was a direct detection of the dust scattered by a comet. Even the orbits of individual sporadic meteors, tracked photographically, often resemble long- or short-period cometary orbits. *Therefore, most meteors must be small bits of debris scattered from comets.*

The Leonid shower gives a good example of this connection. It occurs because Earth passed through a dense clump of dust grains spread along the orbit of Comet Tempel-Tuttle (1866 I). By studying statistics of showers, astronomers mapped this clump and predicted another intense shower in 1999. The prediction was correct, and observers under dark skies in rural areas saw the meteor counts rise to several thousand per hour as Earth passed through the swarm of debris!

Since the 1960s, rockets and balloons have collected microscopic fragments believed to be meteoroids, shown in Figure 13-8. These fragments are irregular glassy silicate and metallic particles, supporting the theory that comet nuclei are dirty icebergs with bits of entrapped grit. Their compositions are similar to those of certain types of meteorites. Most meteors are far too small to reach the ground, "burning" at altitudes around 75 to 100 km. Occasional large ones, called **fireballs,** are very bright and spectacular. They generally explode in the air instead of hitting the ground, again indicating that they are too fragile to survive atmospheric entry. Those have sometimes been reported as UFOs.

In 1986, scientists discovered a "missing link" between comets and meteor showers. An infrared orbiting telescope in space had mapped the sky in infrared light coming from all sorts of celestial materials. Among the asteroids, nebulae, and other sources, researchers found trails of dust spread along the orbits of active comets (Sykes and others, 1986). The solar system is ringed with these tenuous **dust trails** of microscopic particles blown off the comets' surfaces when they are active. Visible meteors may be just the larger particles in these trails. As we will see in later chapters, these rings of dust may relate to rings of dust found around other stars.

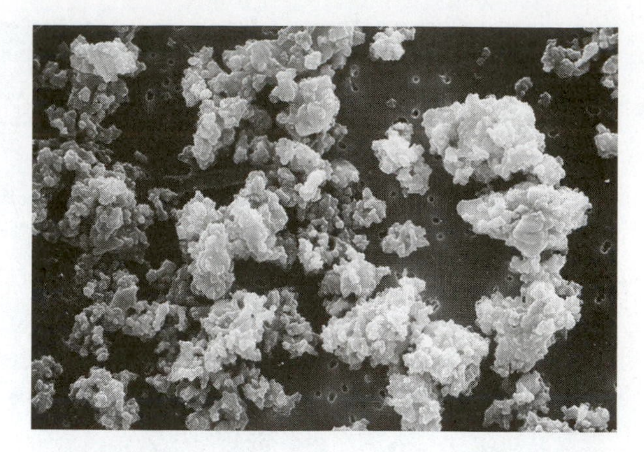

Figure 13-8 Clumps of microscopic meteroidal particles collected by high-flying aircraft. They may have originated in comets. The largest clump is about 60 μm across. (Courtesy D. E. Brownlee, University of Washington, Seattle)

ASTEROIDS

Asteroids are rocky and metallic interplanetary bodies. As noted earlier, they are distinguished from comet nuclei by having no ice (or, in some cases, insufficent ice to sublime and blow off enough gas and dust to make a coma). Asteroids include the largest interplanetary bodies, ranging from the biggest, Ceres (about 1000 km or 600 mi across), to objects only meters across. Spectral studies show that they are mostly rocky, sometimes with appreciable metal. Detailed images from passing spacecraft, radar techniques, and other studies show that those smaller than 100 km or so are irregularly shaped, cratered objects (Figure 13-9). The larger asteroids are more spheroidal, and Ceres, at 1000 km, is relatively spherical like the Moon.

Asteroids appear throughout the solar system but are most abundant in the region between the orbits of Mars and Jupiter.

Discovery of Asteroids

Asteroids played an interesting role in the mapping of the solar system. Normally too small to be seen by the naked eye, they were unknown before 1800. Bode's rule, confirmed by the discovery of Uranus in 1781, called for a planet at 2.8 AU from the Sun in the large space between Mars and Jupiter. Therefore, in 1800, astronomers set out to find the "missing planet"; success came on the first night of 1801, when Ceres was discovered—at 2.8 AU, just as predicted.

Arthur C. Clarke pointed out an interesting aspect of Ceres' discovery. It came just when the philosopher Hegel had claimed to "prove" philosophically that there could be no more than the seven then-known planetary bodies, Mercury through Uranus. This is yet another demonstration that going out and looking at the real universe is worth more than sitting at home speculating on the basis of supposed rules. Observation and theory must always go hand in hand.

Between 1802 and 1807, three more small, planet-like bodies turned up in the interval from 2.3 to 2.8 AU

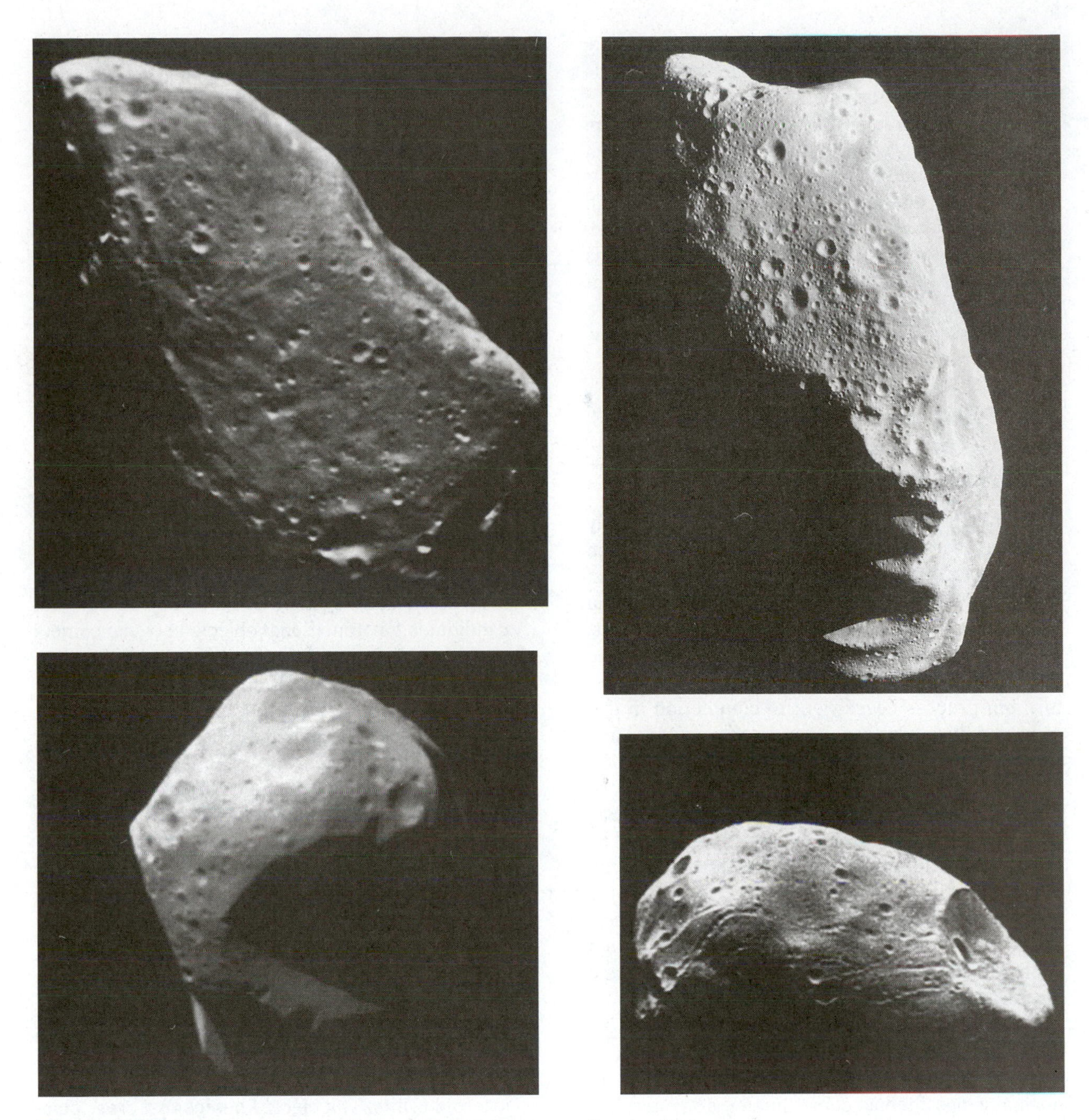

Figure 13-9 The first three close-up photos of asteroids by spacecraft and a close-up photo of Mars' satellite, Phobos. **a** Asteroid 951 Gaspra, about 12 × 16 km in size. **b** Asteroid 243 Ida, about 52 km long. **c** Asteroid 253 Mathilda, about 59 × 47 km. Scientists were somewhat surprised by the large craters on the asteroids, especially Mathilda; the impact explosions needed to make these craters are almost big enough to have fragmented the asteroid. **d** Mars' satellite, Phobos, 19 × 27 km in size, is believed to possibly be an asteroid captured into orbit around Mars. The similarity in appearance gives some support to this idea. (Source: NASA missions, **a** and **b** are from the Galileo spacecraft; **c** from the NEAR spacecraft; and **d** from the Viking orbiter)

from the Sun. Because of their small size, they came to be called minor planets, or *asteroids*. This name came to be applied to any interplanetary body that is not a comet (that is, is not known to emit gas/dust and look fuzzy in a telescope). By 1890, some 300 asteroids were known. In 1891, German astronomer

Max Wolf began searching for them photographically by time exposures and detected many new asteroids by their telltale motion among the stars.

★ Asteroids are known by numbers (assigned in order of discovery, but only after the orbit has been accurately identified) and a name (chosen by the discoverer)—1 Ceres and 2 Pallas, for example. The names cover a wide range of human interests, including cities, mythology (1915 Quetzalcoatl), politicians (1932 Hooveria!), spouses, and other lovers. About 6000 are now cataloged, and from statistics of discovery rates, astronomers estimate that 100,000 observable ones remained uncharted.

The Main Asteroid Belt and Other Groupings of Asteroids

All of the first asteroids to be discovered in the early 1800s were located in a belt between Mars and Jupiter, known as the **main asteroid belt,** or sometimes just the asteroid belt. An early theory was that a planet in that zone had somehow exploded, leaving fragments. But today we know that the variety of materials is too great to represent a single planet. Rather, the planet-forming process never went to completion in this zone, and these are debris that never aggregated into a planet.

Asteroids are not distributed uniformly across the belt. In Chapter 11, we describe how satellites of Saturn clear gaps in the rings through gravitational interactions called resonances. In the same way, asteroids with $\frac{1}{2}$, $\frac{1}{3}$, and some other simple fractions of the period of Jupiter are cleared out of the asteroid belt by resonant gravitational effects of Jupiter. Through such resonances, Jupiter has shaped the width and structure of the belt.

Additional observations have revealed other groupings of interplanetary bodies in the solar system. As we have seen, the Oort cloud and the Kuiper belt are normally classified as comet nuclei, not asteroids. Moving inward toward the Sun, we find a grouping called the **Trojan asteroids,** lying in two swarms in Jupiter's orbit, 60° ahead of and 60° behind the planet, called **Lagrangian points.** The astronomer Joseph Louis Lagrange discovered that particles can be held in the two swarms by the combination of gravitational forces from the Sun and Jupiter. This effect is yet one more example of a gravitational resonance in the solar system. The name "Trojans" comes from the tradition of naming these particular asteroids after heroes in the Homeric epics about the Trojan War. Many dozens of Trojans are known, mostly bigger than 50 km across. Taken together, the two Trojan swarms compose the solar system's "second asteroid belt," which is nearly as populous as the main belt between Mars and Jupiter, although most Trojans are too far away to see easily. The largest Trojan is Hektor, estimated to be 100 km wide and 300 km long, tumbling end over end in Jupiter's orbit. Trojan asteroids and their neighbors, the outer satellites of Jupiter (J6 through J13), are similar in having a very dark color. Jupiter's outer satellites may even be Trojan-like asteroids that were captured by Jupiter's strong gravity into orbits around the giant planet.

Trojans are nearly always called "asteroids," and their composition may be largely the dark, carbonaceous material common in the outermost astronomical belt and among comets. However, Trojans may also have a certain amount of ice content and thus mark a transition between asteroids and comets. Recall that they are at the same solar distance as Jupiter, and Jupiter's moons (except for hot, volcanic Io) are rich in ices. Thus ices may be native to this zone of the solar system. Trojans may be our best candidates for transitional objects between comets and asteroids. If one of the Trojans were brought into the inner solar system, it might turn into an active comet. However, since the Trojans are trapped around the Lagrangian points, this "experiment" may never be performed.

Moving in toward the Sun, the next major group of asteroids is the main asteroid belt itself, home of the classic asteroids discovered early in the 1800s. Just as the Kuiper belt acts as a reservoir of comets beyond the giant planets and occasionally feeds comets inward to create the Centaurs among the giant planets, the main asteroid belt acts as a reservoir of asteroids beyond the terrestrial planets, and it occasionally feeds asteroids inward to cross the orbits of Mars, Earth, Venus, and even Mercury. Asteroids are ejected from the belt primarily by the disturbing force of gravity from nearby Jupiter, which can alter asteroid orbits. The ejected bodies that cross planets' orbits are called Mars-crossing asteroids, **Earth-crossing asteroids,** and so on. Asteroids that cross the orbit of a planet generally have some chance of hitting that planet. Typically, once an asteroid is ejected from the main belt, it lasts only about 1 to 10 million years before crashing into a planet and making a crater. The supply is replenished as other asteroids are thrown out of the belt by the gravitational force of massive Jupiter.

Another name for the Earth-crossing asteroids is **Apollo asteroids,** named after 1862 Apollo, the first of the group to be discovered. These are the aster-

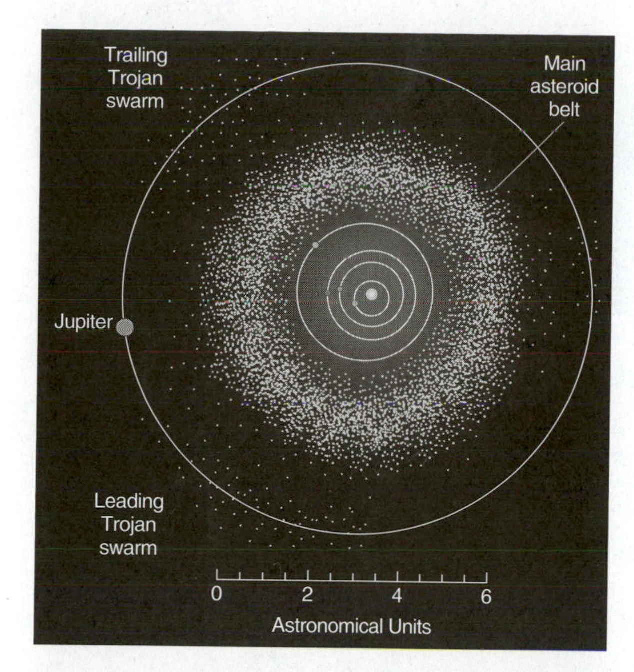

Figure 13-10 A map of the positions of thousands of cataloged asteroids on a specific day (March 19, 1993). Each small dot represents one asteroid. Large gray dots show the positions of five planets that happened to be on the same side of the Sun on this date. (Adapted from diagram courtesy of Mark Sykes, Steward Observatory)

oids that have a chance of hitting our planet. Many dozens are known, including some only tens of meters across. While most of the Apollos are true, rocky asteroids from the belt, others may be remnants of burnt-out comets that have lost their ices and are no longer active. Because they come close to Earth, Apollo asteroids are the easiest bodies beyond the Moon to get to from our planet, either by robotic spacecraft or future human voyages of exploration.

Perhaps a sense of the welter of asteroids in the solar system can best be obtained from a map showing the actual positions of thousands of cataloged asteroids on a specific, randomly chosen date. This is shown in Figure 13-10. The main asteroid belt stands out clearly. Additionally, the figure shows the two somewhat vaguely defined swarms of Trojans in Jupiter's orbit and a scattering of **Earth-crossing asteroids** among the terrestrial planets.

★ In spite of the apparently dense crowding in the main asteroid belt in Figure 13-10, remember that distances are so vast and asteroids so small by comparison that if you were riding on an asteroid (or a spaceship in the belt), you would rarely see another asteroid passing close by, contrary to Hollywood movies where belt asteroids are shown as densely crowded navigational hazards. From any given asteroid, other asteroids would usually look like faint stars at best. Several unmanned spacecraft have already flow through the belt with no serious consequences. Nonetheless, during the immense duration of geologic time, collisions occasionally occur between asteroids, as we will see later in this chapter.

Rocky, Metallic, and Carbonaceous Asteroids: A Zonal Distribution of Compositional Classes

As was indicated in Figure 5-13, the spectra of asteroids can be used to estimate their mineral properties. This technique, which is possible for much more distant asteroids than the radar technique, has been used to study the compositions of asteroids throughout the solar system. In the 1980s, astronomers discovered that asteroids are divided into different compositional classes and that different classes are concentrated at different distances from the Sun (see review by Binzel and others, 1991). Spectra of some classes match spectra of meteorites, providing one more piece of evidence that most meteorites are fragments of certain types of asteroids. Figure 13-11 shows this distribution of asteroid classes among the two main groups of asteroids, the asteroidal belt and the Trojan swarms. We will discuss the five major classes S, M, C, P, and D. Remember that the main asteroid belt runs from about 2.1 AU to about 3.4 AU. The inner half is dominated by S-class stony objects (like stony meteorites—see discussion of meteorites later in this chapter) and M-class objects, which are believed to contain a high percentage of metal (like iron meteorites and stony-iron meteorites. For example, 1986 DA, the 2-km "piece of nickel-iron" measured by radar, had earlier been cataloged spectroscopically as an M-class asteroid. The S-class and M-class objects are fairly light-colored, reflecting 10 to 20% of the light that strikes them, like many familiar rocks on Earth. A few other minor classes of asteroids have been found in this region.

In the middle of the belt, at about 2.7 AU, lies a *soot line* beyond which black carbon-rich minerals are formed. They dominate the asteroids beyond 2.7 AU, which are mostly very black in color, reflecting only 4% of the light that strikes them. They are blacker than a blackboard and probably rich in carbonaceous materials (like carbonaceous meteorites). The two groups of Trojan asteroids, clustered in Jupiter's orbit, also have this dark color, as do comet nuclei associated with the Kuiper belt and Oort clouds, even farther from the Sun. These black objects are divided

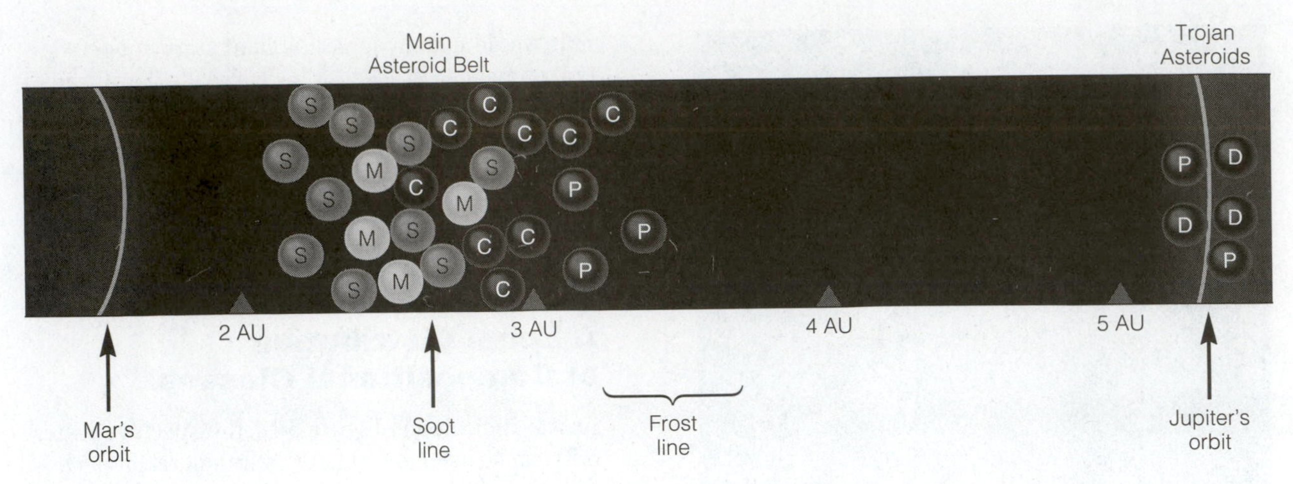

Figure 13-11 Schematic diagram of the zonal distribution of asteroid classes in the solar system. The inner asteroid belt consists of light-colored S-class stony objects and M-class metal-rich objects, while the outer belt and Trojan asteroids consist of black, carbonaceous C-, P-, and D-class objects. See text for further discussion.

into three groups: the C class is neutral black, the P class is slightly reddish or brownish in color, and the D class is still more strongly reddish-brown. The reddish-brown colors are believed to be due to organic molecules that form from the carbonaceous material in the cold outer part of the solar system under certain conditions.

Somewhere between 3 and 4 AU is another important dividing line that does not show up so clearly in asteroid appearance. This is the so-called *frost line,* where frozen water is stable. In other words, asteroids beyond this distance may contain substantial amounts of ice and might turn into active comets if they were deflected into new orbits that took them closer to the Sun.

Spectra have revealed many interesting properties of individual asteroids. For example, the largest asteroid, 1 Ceres—which is a somewhat modified C-class object—was found to have minerals that contain chemically bound water molecules (Lebofsky, 1981) and ammonium (NH_4) molecules (King and others, 1992). Many carbonaceous meteorites contain water and may be similar to Ceres' material. The water is chemically bound and can be driven off by mild heating—implying that the surface of Ceres was probably never strongly heated, since it still has its bound water.

On the other hand, the second-largest asteroid, 4 Vesta, has a surface consisting of lavalike rock similar to basalt; it must have melted. Certain other asteroids, notably a rare class called class A, are rich in the mineral olivine—a major constituent of Earth's mantle. To consider another example, when the Galileo spacecraft flew past the rocky, S-class asteroid 951 Gaspra, it measured a magnetic deflection that may indicate a significant proportion of iron in Gaspra (perhaps as large as that in some stony meteorites). Taken together with the evidence for nickel-iron asteroids and meteorites (discussed in the next section), these data show that at least some asteroids melted, forming nickel-iron cores like Earth's core, olivine-rich mantles of dense rock like Earth's mantle, and surfaces of basaltic lava as well as unmelted primitive rock. We know from meteorite samples that all this happened about 4.5 billion years ago, soon after the asteroids formed; after that time, many asteroids were shattered in collisions, giving us samples of all these types of material.

Earth-crossing asteroids are a mixture of the various classes of objects. Indeed, Earth-crossers are probably a mixture of true asteroids from the main asteroid belt and burnt-out comets (which have sublimed all their ice) from the outer solar system. A review by Binzel and others (1992) concludes that the fraction of comets among the Earth-crossers is between 0 and 40%.

Satellites of Asteroids

Asteroids sometimes pass between stars and Earth, causing the starlight to dim out. Asteroid orbits are so well known that these events are predictable and can be observed from Earth. When this work started in the 1970s, occasional cases were noted where the star dimmed twice, suggesting that not only the asteroid but a small satellite of the asteroid passed in front of the star. Such events were controversial, but in 1993, when the Galileo spacecraft made the second spacecraft encounter with an asteroid, the 52-km asteroid 243 Ida, a 1-km moonlet was found orbiting around the asteroid about 100 km away. Other asteroid satellites have been found since by Earth-based astronomers. Studies of certain double craters on planets suggest that they are caused by impacts of asteroid-satellite pairs, and that about 10–15% of all asteroids have such moonlets. Hartmann (1979) suggested these might be formed when pairs of fragments, blasted out of asteroids during collisions, end up orbiting around each other. Computer models by Durda (1996) support this view.

Origin and Collision of Asteroids

Asteroids are probably planetesimals that never finished accumulating into a planet. According to dates derived from meteorites, they all formed about 4.6 billion years ago. Possibly disturbed by nearby Jupiter, the planetesimals in the belt region collided too fast to coalesce into a planet. Instead, they collided and broke into thousands of fragments, as depicted in Figure 13-12. Most smaller asteroids in the main belt are such fragments. In 1983, IRAS discovered distinct rings of dust circling the Sun in the main asteroid belt, each interpreted as debris from a relatively recent, individual asteroid collision. Thus the collisions continue, sporadically, even in modern geologic time.

The zonal structure of the main asteroid belt (Figure 13-11) shows that the asteroids have not been strongly mixed; they lie in their original zones. The zonal structure also reveals that the asteroids of the inner belt, closer to the Sun, never reached cold enough temperatures to form carbonaceous and ice-rich compositions. Many of them got so hot that they melted and produced iron cores. On the other hand, asteroids farther from the Sun contained sooty carbonaceous material, water molecules, and even ice. Indeed, comet nuclei are probably nothing more than asteroids that formed so far from the Sun, in such cold regions, that they contained much ice mixed with the carbonaceous material.

Although asteroids on the average are far apart in the main asteroid belt, enough of them are there that they occasionally collide, down through geologic time. When a small one hits a big one, it just makes an ordinary impact crater in the side of the larger one. When two comparable-size asteroids hit, however, they can both be shattered into thousands of fragments. This adds to the population of small asteroid fragments and dust in the belt. Small asteroidal fragments, from brick size to mountain size, are very numerous in the main asteroid belt, and examples are constantly being ejected from the belt, primarily because of disturbances of their orbits by nearby, massive Jupiter. Some of these ejected fragments may approach Earth and become meteorites.

Figure 13-12 Imaginary view of the collision of two asteroids. Fragments created by the collision may ultimately fall on the Earth and other planets as meteorites. Since two asteroids are of different mineralogical types, some of the meteorites produced will be mixtures of different rock types. The larger body is about 30 km across. (Painting by WKH)

METEORITES

★ Meteorites are stony and metallic objects that fall from the sky. Modern evidence suggests that they are fragments of Apollo-type asteroids that cross the Earth's orbit and eventually collide with Earth.

Scientific Discovery of Meteorites

★ As described on our web site, meteorites were often venerated by ancient peoples, because stones from the sky seemed like an act of the gods. Until the late 1700s, most naturalists did not believe folk tales about stones falling from the sky. Then, in 1794, a German physicist, E. F. F. Chladni, reported that the Ensisheim and other supposed celestial stones seemed similar to each other and different from normal terrestrial stones. He concluded that these "meteorites did indeed fall from the sky." This conclusion started a controversy among naturalists. Upon hearing that a meteorite had fallen in Connecticut, Thomas Jefferson, himself an accomplished naturalist, is supposed to have said, "It is easier to believe that Yankee professors would lie than that stones would fall from heaven."

The French Academy, the scientific establishment of the day, dismissed the stones as a superstition. As luck would have it, a meteorite exploded over a French town in 1803, pelting the area with stones. Cautiously, the academy sent the noted physicist J. B. Biot to investigate. His report, one of the historic documents of science, methodically constructed an irrefutable chain of evidence from eyewitness accounts, measurements of the 2 × 6-km area of impacts, and specimens of the meteorite themselves. This report established that stones can indeed fall from the sky.

Meteorite Impacts on Earth

Interplanetary bodies collide with the Earth at very high speeds, usually 11 to 60 km/s (24,000 to 134,000 mph). At such speeds, material is heated by friction with the air. Dust grains and pea-size pieces burn up before striking the ground. Larger pieces are usually slowed by drag, although at least one has hit a car and two grapefruit-scale specimens have punched through house roofs, one bruising a woman occupant! Since they pass through the atmosphere too fast for their interiors to be strongly heated, stories of meteorites remaining red-hot for hours after falling are untrue.

Large meteorites are rare. Only a few brick-size meteorites are recovered each year. In 1972, an object weighing perhaps 1000 tons just missed the Earth, skipping off the outer atmosphere; it was filmed from the ground and detected by Air Force reconnaissance satellites. Objects weighing 10,000 tons, like the Siberian object of 1908, which are large enough to cause nuclear-scale blasts, fall every few centuries. Larger blasts, thousands of years apart, may form multikilometer-scale craters, such as that in Figure 6-14.

Since meteorites are "free samples" of planetary matter, recovering and reporting a meteorite is a rare honor. Meteorite discoveries should be presented, or at least lent, to research institutions—for example, the Smithsonian Institution in Cambridge, Massachusetts, and Washington, D.C., and the Center for Meteorite Studies, Arizona State University, Tempe.

Types and Origins of Meteorites

Meteorites as a group are among the most complex rocks studied by geologists. There are many different types, and sometimes they appear as **brecciated meteorites**—meteorites made of mixed fragments of different meteorite types all jumbled into one rock! After decades of study, scientists began to understand how meteorites formed and what they have to tell us about the ancient history of the solar system. The key to the story is that meteorites are fragments of asteroids (and sometimes mixtures of fragments of dissimilar asteroids), created when asteroids collided and blew apart at various times during the history of the solar system. Their fragments were ejected in various directions, often getting thrown into orbits that eventually intersected the Earth's.

The different types of meteorites thus tell us about conditions inside the asteroidal parent bodies. Some are samples of surface rock layers, and others are samples of deep cores inside these bodies. Table 13-3 summarizes the major types and their abundances. The first type is called **chondrites** (KON-drites), or sometimes "ordinary chondrites." They are so named because they contain peculiar BB-like glassy spherules called **chondrules** (from the Greek term for seedgrains), as shown in Figure 13-13. Chondrules are puzzling objects. They date from the earliest days of the solar system. Many researchers believe they started as clumps of dust that melted when shock waves spread through the early solar system, possibly close to the early, unstable Sun (Glanz, 1997; McKeegan and others, 1998; Connolly and Love, 1998). The preservation of the ancient chondrules in these meteorites proves that the meteorite rocks themselves have never been melted or severely altered since their formation long ago. Chondrites are by far the most common type of meteorite, which indicates that, whatever their history, the majority of meteoritic rocks have not been altered very much since the early days of the solar system.

The second type of meteorites is called **carbonaceous chondrites.** These are a type of chondrites,

TABLE 13-3

Types of Meteorites

Stony Meteorite Type	Percentage of All Falls[a]	Remarks	Parent Asteroid Class
Chondrite	80.0	Commonest type. Defined by millimeter-scale spherical silicate inclusions, sometimes glassy, called *chondrules.*	Some S's?
Carbonaceous chondrite	5.7	Most primitive, least altered material available from early solar system	C? P? D?
Achondrite	7.1	Most nearly like terrestrial rocks. Defined by lack of chondrules, which have been destroyed by a heating process. Some resemblance to terrestrial lunar igneous rocks of basaltic type.	Basaltic types and some S's?
Stony-Iron	1.5	Contain stony and metallic sections in contact with each other.	M?
Iron	5.7	Nickel-iron material. Museum specimens are often cut, polished, and etched to show interlocking crystal structure.	M?
Total	100	—	—

[a] This table is based on meteorites called *falls*—those actually seen to fall. Meteorites found by chance in the soil, called *finds,* are more numerous but less valuable statistically because they are biased toward iron meteorites, which attract attention whenever found in the ground, while stony meteorites eventually weather to resemble ordinary stones.

a

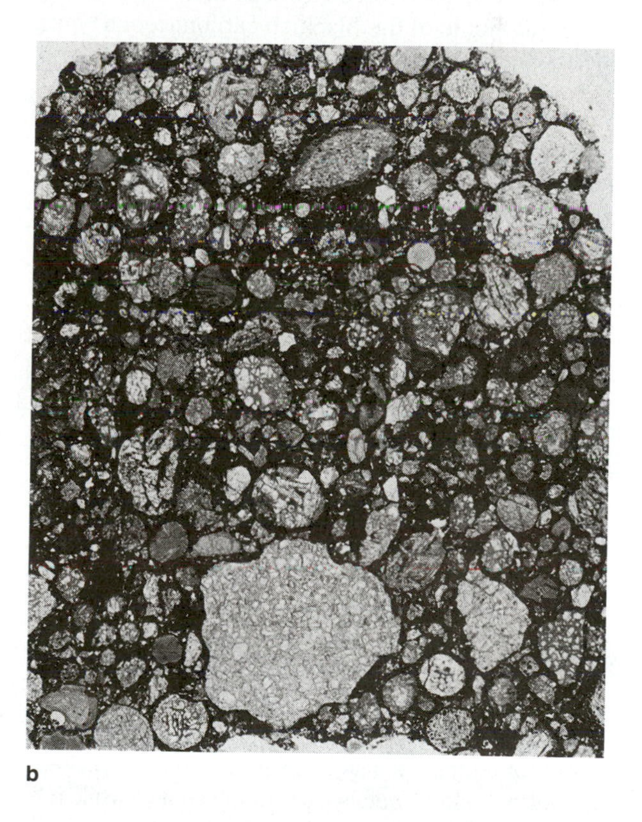

b

Figure 13-13 Meteorites with chondrules. **a** A 3-cm portion of the chondritic meteorite Björbole. Protruding from the surface are chondrules of various sizes, including a large one at lower left. **b** A cross section of the highly chondritic meteorite Chainpur, which consists mostly of chondrules embedded in a black matrix. (**a:** Center for Meteorite Studies, Arizona State University; **b:** Courtesy Laurel Wilkening, University of Arizona)

too, because they contain chondrules. However, instead of being made from ordinary minerals, as in ordinary chondrites, the rest of the rock is a black-colored mixture of poorly formed carbon-rich material. This blackish matrix contains water chemically bound in minerals; and this water can be driven off by mild heating to only a few hundred degrees. The preservation of such water in the matrix shows that these meteorites have not been heated above a few hundred degrees since the water-rich matrix formed. Apparently, we are seeing these rocks preserved exactly as they formed billions of years ago. Interestingly, cracks in some carbonaceous chondrites have carbonate deposits left by evaporating water (similar to "bathtub ring" deposits left by evaporation of mineral-rich "hard" water). Carbonaceous chondrites were thus probably formed in the cold, outermost asteroid belt or outer solar system where ices were abundant. Slight heating of these materials, perhaps to room temperatures, probably allowed some of the ice to melt and water to percolate through the material. Based on all evidence, these meteorites seem to be fragments of the blackish carbonaceous types of asteroids or even comet nuclei found in these regions—the classes C, P, and D.

Achondrite meteorites, on the other hand, are rocky meteorites with no chondrules. They resemble basaltic lavas found on the Earth, Moon, and Mars. Thus, in contrast to the two types of chondrite meteorites, achondrites are fragments of materials that were once melted and then resolidified inside parent asteroids or in lava flows on their surfaces. Spectroscopic matches show that they resemble the surface materials on the large asteroid 4 Vesta, and on certain other rare spectral classes. A certain group of achondrites, which resemble basalt lava, were probably formed as lava flows on Vesta and then blasted off by an impact that made a crater on Vesta (Asphaug, 1997).

Stony-iron meteorites (mixtures of stone and iron alloy) and **iron meteorites** (pure iron alloy metal) are probably samples from the central regions of asteroids that once melted, probably due in part to heat-producing radioactive minerals incorporated in them when they formed. Figure 13-14 shows the mixture of rock and metal in a stony-iron. As asteroids melted, the heavy molten metals sank to the center, while the lighter silicate magmas floated. As the asteroids later cooled, the central regions formed nickel-iron cores (like that believed to exist in the Earth), while the surface regions formed rocky lithospheres. When the asteroids were smashed in collisions, fragments of the metal regions, or mixed fragments of metal and achondrite rock, were released. Thus we have samples of the metallic cores of small worlds. Iron meteorites are especially interesting, since they are chunks of pure metal, falling out of the sky! They indicate that there must be many more such chunks—perhaps kilometers across—floating in space! Which asteroids are the metal ones? Spectral evidence suggests that they are likely to be the M class of asteroids.

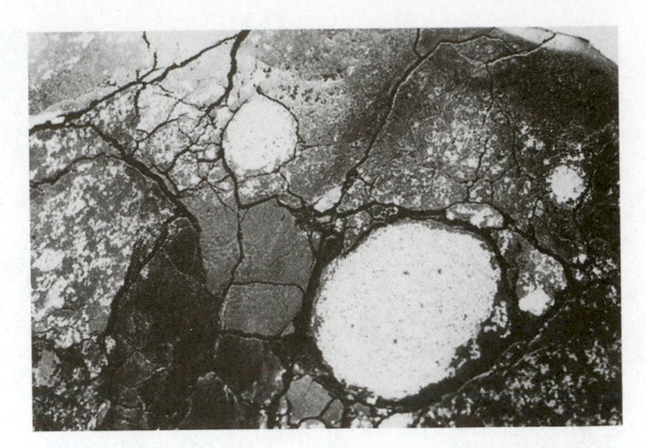

Figure 13-14 Cut and polished section of a stony-iron meteorite from the Bondoc Peninsula, Philippines. The dark fractured matrix is silicate rock, possibly fractured in collisions. The bright nodules are nickel-iron nuggets a few centimeters across. (Photo by WKH)

Studies of minerals in several types of meteorites show that they formed inside parent bodies with diameters of tens to hundreds of kilometers. For example, the large crystals in iron meteorites indicate that they needed a long, slow cooling in order to form, and this in turn means they could not have been in the surface layers of an asteroid or they would have cooled too fast. They had to be insulated by being buried about 100 km down. This supports the theory that they formed in the core regions of the larger asteroids, at least 200 km across. Those asteroids must have been broken up by later collisions in order to release the iron fragments.

Ages of Meteorites

One of the most important aspects of meteorites is that they can be dated by the techniques discussed in Chapter 6, and this gives us a history of events in the early solar system. Dating by various techniques shows that virtually all meteorites formed during a "brief" interval of only 20 million years, 4.6 billion years ago. This is an extremely important fact, because it gives us the date of the formation of the solar system. *"Formed"* means that the various objects made the transition from dispersed dust grain and gas to solid, rocklike objects. This interval, then, marked the birth of planetary material.

Other types of dating confirm that many major collisions happened at this time, 4.6 billion years ago, and sporadic additional collisions smashed some meteorites' parent bodies at other times scattered through solar system history. We will discuss some of these important events in more detail in the next chapter.

ZODIACAL LIGHT

With your own eyeball, you can observe the swarm of dusty debris in interplanetary space! The smallest interplanetary particles are microscopic dust grains and individual molecules and atoms spread out along the plane of the solar system and concentrated toward the Sun. This swarm of material is replenished primarily by dust and gas shed by comets, as they loop around the Sun. If you look west in a very clear rural sky as the last glow of evening twilight disappears (or look east before sunrise), you can detect the diffuse glow of sunlight reflecting off the cloud of these particles. It appears as the **zodiacal light**—a faint glowing band of light extending up from the horizon and along the ecliptic plane. It is brightest at the horizon and fainter as you look further from the sunset direction. A sequence of color photos starting at sunset and showing the emergence of the zodiacal light is shown on our web site. Measurements made by astronauts indicate that it merges with the bright glow of the Sun's atmosphere.

The zodiacal light is the visible effect of countless dust grains—once distributed in trails along comets' orbits but eventually spread out into a uniform disk-shaped cloud of dust around the Sun.

ASTEROID THREAT OR ASTEROID OPPORTUNITY?

As discussed in Chapter 6, in the 1950s researchers recognized that asteroids and comets have hit the Earth throughout geologic time; since 1980, we have learned that they have influenced the extinction of some species. Because there are more small asteroids and comets than big ones, the small ones hit frequently, whereas large impacts are rare. This knowledge penetrated popular awareness only in the late 1980s and 1990s, as astronomers' telescopes began discovering many potential "killer asteroids" passing Earth at distances not much farther than the Moon and as Hollywood produced films like *Deep Impact* and *Armageddon.* (For example, see Trefil, 1989; a 1993 *Newsweek* issue featured a supposed killer comet on its cover—drawn at a grossly wrong scale relative to Earth!) So-called killer asteroids and killer comets are the ones larger than a few hundred meters across—big enough to make catastrophic explosions on Earth that would disrupt civilization. Any Apollo asteroid or comet on an orbit that passes close to Earth's orbit has a fair chance of eventually hitting Earth (or some other planet), usually within 10 million years. Because new asteroids and comets are constantly being injected into the inner solar system from the main asteroid belt and from the Oort cloud, there is a constant supply of these objects throughout geologic time.

On the other hand, "killer asteroids" capable of disrupting civilization are extremely rare. Obviously, none has hit Earth in several thousand years. Figure 13-15 summarizes modern data about this threat, based on statistics about asteroids, comets, and lunar craters. The scale at the bottom gives the size of the explosion crater caused by an impact; the scales at the top give the size of the original meteroid and the equivalent size of the explosion in terms of nuclear bombs. The scale at the left gives the average frequency of impact of various sizes. The graph shows the division between meteors and meteorites: To the left of the dotted vertical line, most bodies burn up or break up in the atmosphere and do not reach the surface. This corresponds to about the size of the object that exploded over Siberia in 1908.

This same dotted line gives one example of how to read the chart: It shows that about every century or two, somewhere on Earth, we can expect an A-bomb- or H-bomb-scale explosion like the 1908 Siberian event. (Of course, most of these would happen over oceans and have minimal effect. Such an explosion over populated land may happen only once very 500 to 1000 y; it might be enshrined in local legends, but would have little other long-term effect.) The diagram predicts smaller explosions more frequently. In support of this, a 70-ton meteorite exploded over a different region of Siberia in 1947, scattering iron meteorites. Brilliant but smaller explosions were observed in the high atmosphere over Greenland, El Paso, and the South Pacific by Air Force satellites and/or ground observers in the mid-1990s. No meteorites were found (Gibbs, 1998). As the number of surveillance devices continues to increase, we are likely to detect more of these explosions in coming years.

The dashed line gives a more sinister example: It shows that every 100 million years or so we may get

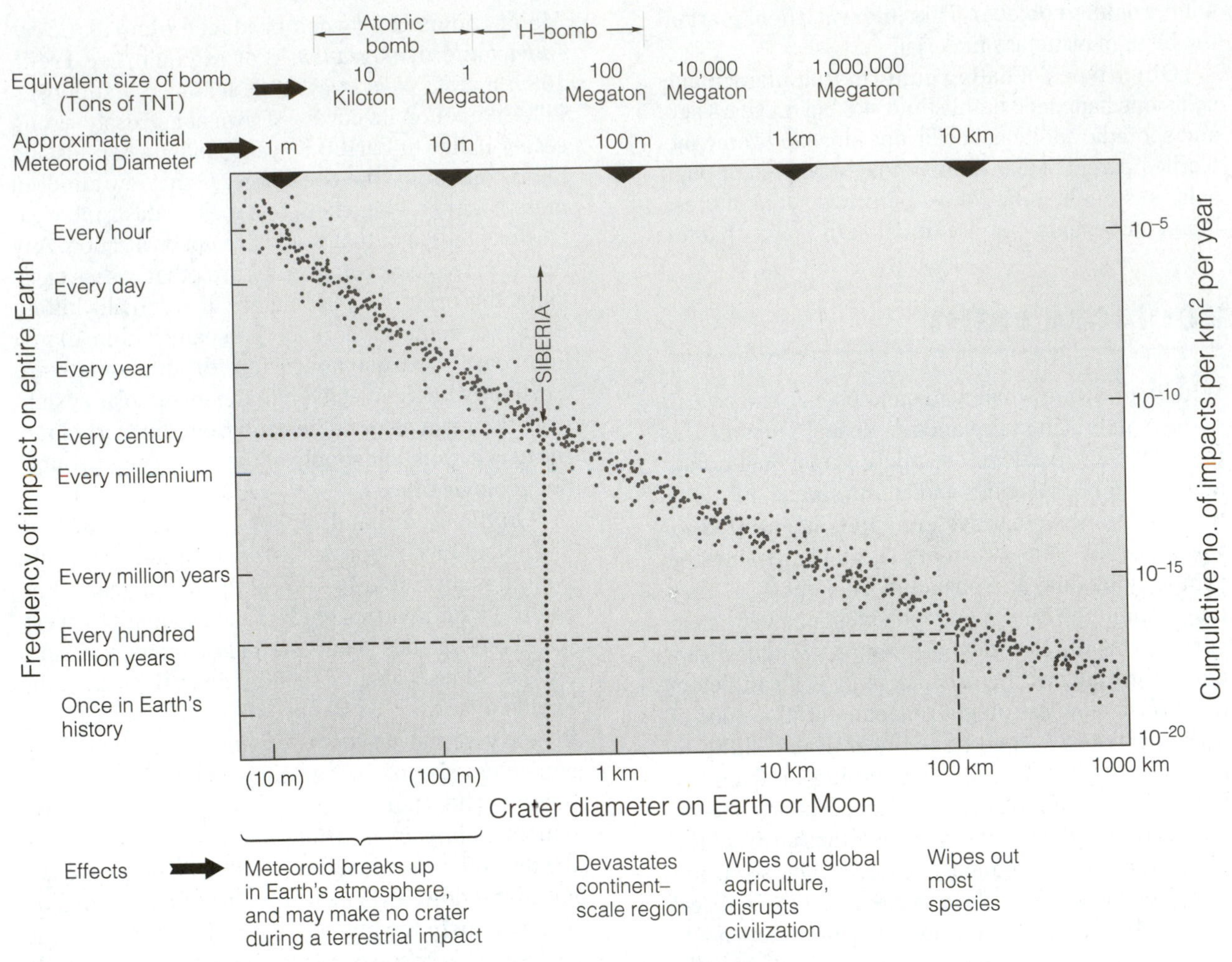

Figure 13-15 The *asteroid threat.* The frequency of impacts of asteroids and comets of various sizes. The bottom scale gives the size of the resulting explosion crater; the top scales give the size of the original meteoroid and the equivalent size of a nuclear bomb. The stippled band shows the approximate rate of impact of different sizes, according to the best available data. (See text for discussion.) (From diagram by Hartmann, 1984, 1993; and from 1993 data by G. Shoemaker, C. Chapman, D. Morrison, G. Neukum, and others)

an impact big enough to wipe out most species and reset the evolutionary clock. In fact, as discussed in Chapter 6, we now know that such an impact happened 65 million years ago and that it caused temporary climate changes big enough to wipe out the dinosaurs and about 75% of the species that existed at that time.

Many asteroid scientists have popularized the *asteroid threat.* Do we need a system to protect civilization against the chance of major disasters? There is a good chance of a local disaster (on the scale of a state or country) within a few centuries, of a regional disaster by impact within a few thousand years, and of an impact big enough to end civilization within every 100,000 or million years.

One response is to keep up asteroid surveys. Within a few decades, especially if new search telescopes can be brought on line, we should have a complete inventory of the larger Apollo asteroids and know if one is on a collision course with Earth.

Although the chances of this are small, a certain community of astronomers and Defense Department experts have suggested experiments with missile systems designed to detect, deflect, or blow up any approaching asteroid before it can hit Earth. Edward Teller, known as the father of the H-bomb, and members of the Strategic Defense Initiative ("Star Wars") project have supported these discussions. Critics have suggested that such ideas are merely an attempt to keep up SDI funding in a post-cold-war world.

Figure 13-16 Beyond the Moon, the closest planetary bodies are Apollo asteroids. Small Apollos, a few hundred meters across, may approach close enough to the Earth-Moon system (lower right) to be visited relatively easily by spaceships. Astronauts transfer from a "parked" ship to the asteroid surface. Some researchers believe asteroid materials could be economically exploited for use on or near Earth. (Painting by WKH)

The Asteroid Opportunity

A more positive approach, given the visionary nature of this issue, is to emphasize the *asteroid opportunity.* All reasonable projections of consumption rates indicate that in a closed, earthbound society, depletions of many natural resources will begin to alter society[4] in the middle of the 2100s. This conclusion was already apparent decades ago (for example, see review by Hartmann, Miller, and Lee, 1984). In the long term, asteroids might provide a way out. Asteroid research indicates that interplanetary bodies near Earth contain a variety of resources. Based on meteoric samples, chondrite-type asteroids (possibly the very common S class) are believed to contain ores of economically important platinum-group metals. M-class asteroids are probably pure nickel-iron. Such asteroids could be mined in space using the "free" solar energy streaming from the Sun 24 h a day. Scientists and technicians have realized for some decades that masses of metal worth hundreds of billions of dollars—enough to supply Earth's needs for certain metals for decades—should be obtainable from kilometer-scale and smaller asteroids (Gaffey and McCord, 1977; O'Leary and others, 1979; Hartmann, 1982). Sober, international technical conferences have already been held to begin analyzing the natural resources available in near-Earth space (Lewis and others, 1993).

Studies are already under way to examine the feasibility of flying to Apollo asteroids for economic and scientific exploration, as suggested by Figure 13-16. Many are easier to visit than Mars, and in terms of energy expenditure, some of them are actually easier to reach and return from than the Moon (Davis, Friedlander, and Jones, 1993)! The formation of a venture capital company to build technology to pursue such resources was announced in the late 1990s.

If such resources could be "harvested" and processed in space, then mining and industrial processing and pollution on Earth could decline, and our planet could begin to return to a more natural state. Thus, instead of a negatively oriented defense against the asteroid threat, we could invest in a positive (if admittedly visionary) program for the 21st century. The ability to reach asteroids and deflect them from one orbit to another would emerge as a by-product of such a program and would thus solve the problem of the asteroid threat. At the same time, such a program would help to resolve Earth's environmental problems.

Siberia Revisited

The mysterious Siberian explosion of 1908, which we described at the beginning of this chapter, is now

---

[4]Or catastrophically disrupt society, depending on how well the present generation responds to these issues and prepares for future changes in resource consumption and supply.

Figure 13-17 The moment of the explosion over Tunguska, Siberia in 1908. This view was reconstructed from descriptions by eyewitnesses at a trading station 60 km south of the blast. One of these witnesses was blown 20 ft off a porch. One witness stated, "The sky was split in two and high above the forest the whole northern part of the sky appeared to covered by fire. . . . I felt great heat, as if my shirt had caught fire." (Painting by WKH)

easier to explain. Interplanetary space contains debris of many sizes. Rocky and icy chunks up to a few kilometers in size cross the Earth's orbit and are likely to collide with the Earth from time to time. One of these objects apparently struck the atmosphere over Siberia on June 30, 1908.

Other dramatic events in Russia at this time kept Russian scientists from visiting the site until 1927. During the 1927 expedition and later expeditions, scientists found that, surprisingly, the object did not reach the ground and form a crater. It apparently exploded in the atmosphere (Figure 13-17). Trees at ground zero were still standing, but their branches had been stripped by the blast forces in a downward direction. Trees had been knocked over by the blast out to 30 km from ground zero. A forest fire was started, and trees were scorched by the blast out to about 14 km. Researchers found carbonaceous chondrite dust but no meteorites. The explosion injected so much dust into the high atmosphere that on June 30 and July 1, sunlight shining over the North Pole illuminated the dust all night, and newspapers could be read at midnight in western Siberia and Europe.

Because the object blew up in the atmosphere and left little stony residue, Russian investigators concluded that it may have been a piece of an icy comet. Others suggested a weak stony asteroid some 90 to 200 m across (Sekanina, 1983). A 1993 review of the data by Christopher Chyba suggests that the object was a weak carbonaceous asteroid, only 30 m across, hitting Earth at about 15 km/s.

Such was the effect of a relatively small bit of interplanetary debris striking the Earth. To put this event in perspective, if the same object had exploded over New York City, the scorched area would have reached nearly to Newark, New Jersey. Trees would have been felled beyond Newark and over a third of Long Island. The man knocked off his porch could have been in suburban Philadelphia. "Deafening bangs" might have been heard in Pittsburgh, Washington, D.C., and Montreal.

SUMMARY

The spatial and compositional relationships among the various types of asteroids and comets are summarized in Figure 13-18, which is arranged roughly in order of distance from the Sun.

All of the solar system's small bodies—comets, asteroids (and their fragments), meteors, and meteorites—began as planetesimals, which were the preplanetary bodies that formed in the solar system 4.6 billion years ago, during a relatively short formative period that lasted about 100 million years. The fact that meteorites of many different types, together with the Moon and Earth, share a common age of 4.6 billion years is one of the most important observational facts of solar system astronomy. It reveals that the solar system itself formed 4.6 billion years ago.

The evolution of these bodies from icy and rocky planetesimals to today's comets, asteroids, meteors, and meteorites is summarized in Figure 13-19. This figure shows how icy bodies in the outer solar system were mostly perturbed into the Kuiper belt or Oort cloud and became today's comets. Dusty debris from them created meteors and meteor showers. Rocky bodies from the main asteroid belt are occasionally perturbed toward the terrestrial planets, making meteorites and meteorite craters.

The meteorites are extremely valuable planetary samples because they represent the interiors of many different kinds of ancient parent bodies. They show that the parent asteroids—at least many of those larger than a few hundred kilometers across—melted soon after they formed, producing iron cores and rocky mantles. Some other asteroids survived strong surface heating and preserve primitive materials. The existence of water-bearing minerals on some of these asteroids proves that they were never strongly heated.

Around 65 million years ago, an impact of an asteroid or comet roughly 10 km across was involved in killing off the dinosaurs and the majority of Cretaceous-period species that existed at that time. Thus debates about the importance of asteroids have tended to emphasize the threat posed by their occasional collisions with Earth. At the same time, however, asteroids may offer an opportunity for future resource gathering that

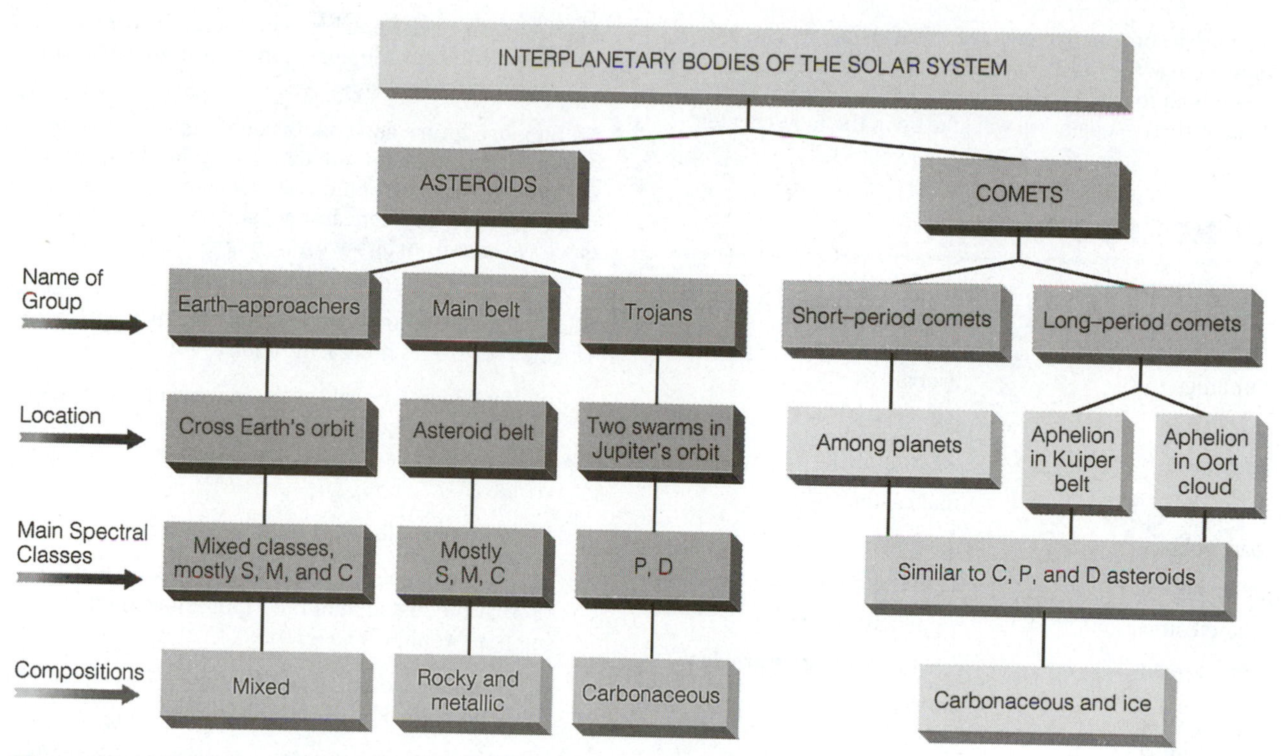

Figure 13-18 Schematic diagram of nomenclature and relationships among asteroids and comets and their different subgroups.

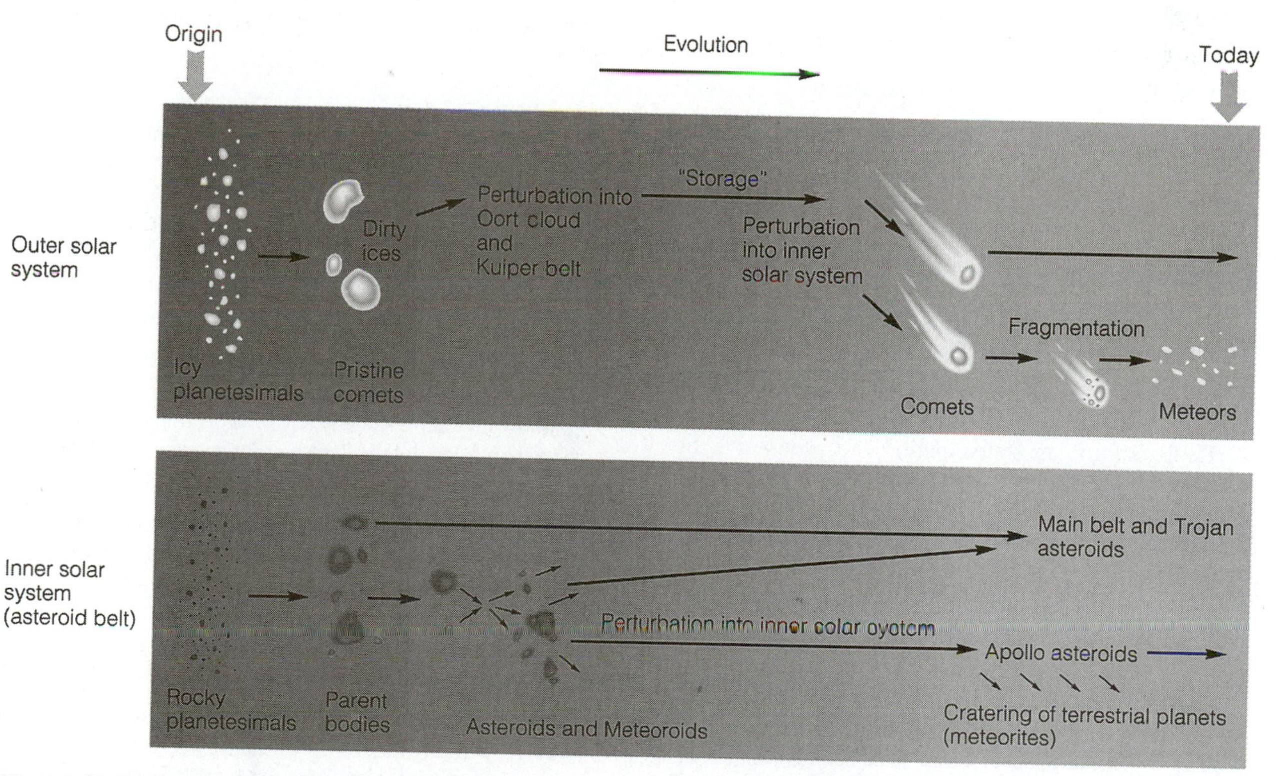

Figure 13-19 Schematic histories of cometary and asteroidal material, showing condensation into multikilometer bodies (left) and subsequent perturbation and fragmentation to explain phenomena now observable. The diagram illustrates names used for the various types of objects.

may solve some of the environmental problems of Earth. In any case, the asteroids and comets offer some of the best clues to the ancient history and formation of the solar system—a point we will take up in the next chapter.

CONCEPTS

comet
asteroid
sublime
meteoroid
meteor
meteorite
parent body
asteroid belt
planetesimal
comet head
comet tail
comet nucleus
solar wind
Halley's comet
long-period comet
short-period comet
Oort cloud
Kuiper belt
Plutinos
dirty iceberg model
meteor shower
fireball
dust trails
asteroid belt
main asteroid belt
Trojan asteroids
Lagrangian points
Apollo asteroids
Earth-crossing asteroids
brecciated meteorite
chondrite
chondrule
carbonaceous chondrite
achondrite
stony-iron meteorite
iron meteorite
zodiacal light

PROBLEMS

1. If a comet should happen to pass through Saturn's satellite system, why would it probably not be detected from Earth?

2. Kepler's third law states that $a^3 = P^2$, where a is the semimajor axis of a body orbiting the Sun (expressed in astronomical units) and P is the period (expressed in years). Should this result apply to comets? If a typical comet in Oort's cloud has a semimajor axis of about 100,000 AU (10^5 AU), how often would it return to the inner solar system?

3. In terms of measuring and reporting useful scientific information, what actions would be appropriate if you observed an extraordinarily bright meteor or fireball? In two columns, list examples of useful and nonuseful descriptions of the fireball's speed, brightness, and apparent size. What actions would be appropriate if you saw a meteorite strike the ground?

4. Summarize relations between the bodies and particles responsible for meteors, comets, the zodiacal light, asteroids, and meteorites.

5. Suppose future astronauts could match orbits with a comet and reach its nucleus. Describe the possible surface appearance of a comet. Consider gravity, surface materials, sky appearance, and so on. Would it be easier to match orbits with a long-period or a short-period comet?

6. Typical interplanetary material may move at about 15 km/s relative to the Earth-Moon system.

a. If a kilometer-scale asteroid were discovered on a collision course with the Earth when it was 15 million kilometers away, how much warning time would we have?

b. What would be the potential dangers?

c. If a much smaller asteroid were similarly discovered at the distance of the Moon, how much warning time would we have?

d. How long would the objects take to pass through the 100-km thickness of the atmosphere?

7. Based on everyday experience, what is the danger of an event such as that in Problem 6 compared with the danger of other natural disasters, such as earthquakes? Is a large-scale meteorite disaster a plausible source of myths during the 10,000-year history of humanity? Defend your answer.

ADVANCED PROBLEMS

8. Suppose a comet on a parabolic orbit is closest to the Sun at a point very near the Earth and moves in the same direction as the Earth. How fast does it move (in km/s) relative to the Earth?

9. Use the small-angle equation to calculate how close a 2-km-diameter asteroid would have to come to Earth to allow its shape to be resolved by a telescope that can reveal angular details 1 second of arc across. How does this distance compare with the Moon's distance?

10. Suppose a small asteroid of pure nickel-iron, with radius r = 100 m, could be located and exploited. If the value of the alloy were 90¢ per kilogram, what would be the potential economic value of the asteroid? Compare this with the $20 billion cost of the Apollo program. Assume that the density of the material is 800 kg/m^3.

11. How many asteroids would be required to provide enough material to make one planet the size of Earth? Assume a typical asteroid is 120 km across and the Earth is about 12,000 km across.

PROJECTS

1. Use a large piece of cardboard to make a model of the inner solar system out to the orbit of Jupiter. Assume that the planets travel approximately in the plane of the cardboard. Use orbital properties listed in Table 13-1 to cut out scale models of the orbits of various interplanetary bodies. (A slit through the first cardboard could be used to show how comet or asteroid orbits penetrate through the ecliptic plane. Note that the Sun must always occupy one focus of each orbit.)

a. Show how the geometry of passage of a comet (or other body) through the ecliptic plane, especially for highly eccentric orbits, depends on the angle between the perihelion point and the ecliptic plane, measured in the orbit. (This angle is fixed for each body but is omitted from Table 13-1 for simplicity.)

b. Show how the prominence of a given comet may depend strongly on where the Earth is in its orbit as the comet passes through the solar system.

2. Visit Meteor Crater, Arizona. Why is this feature misnamed? Observe the blocks of ejecta and deformation of rock strata, as explained in museum signs and tapes. What would prehistoric observers, if any (estimated impact date was 20,000 y ago), have witnessed at various distances from the blast that formed this crater nearly a mile across? (Other impact sites are known in various states, but they are eroded or undeveloped.)

3. Examine meteorite specimens in a local museum. Compare the appearance of stones and irons. Are chondrules visible in any stones? Heat damage? In iron samples that have been cut, etched, and polished, look for the crystal patterns that give information about the cooling rate and the environment when the meteorite formed inside its parent body.

CHAPTER

14

The Origin of the Solar System

WHAT THE READER SHOULD WATCH FOR IN THIS CHAPTER

The formation of planets around the Sun is believed to be a fairly common by-product of the formation of stars like the Sun—not a rare accident of the universe. Theory and direct observations combine to show that such a star forms with a surrounding, rotating disk of gas and dust grains, about the size of our solar system. This disk is called the solar nebula. The dust grains in the inner solar nebula, near the Sun, were rocky and metallic in composition, and those in the colder, outer solar nebula were more ice-rich. Asteroid- and comet-size bodies arose as the dust grains aggregated. The aggregation process continued until most of the material had been swept up into a few planets. The Sun's formation and the aggregation of planets happened within an interval of roughly 20 million years, about 4.6 billion years ago, as determined from dating of meteorites and lunar samples. This scenario explains many features of the solar system and also accords with the discovery that planetary bodies are circling in orbits around certain other stars. Scientists are still working to explain many details of the system, as well as certain differences between our system and planetary systems circling other stars. ■

The planets, satellites, meteoritic material, and Sun did not exist before about 4.6 billion years ago. Samples from the Earth, Moon, and meteorites suggest that the solar system formed from preexisting material within an interval of about 20 to 100 million years. Prior to that time, the atoms in the Earth, in this book, and in your own body were floating in clouds of thin gas in interstellar space. It is remarkable to realize that this interstellar material somehow aggregated not only into a dazzling star but also into a family of surrounding planets, as symbolized in Figure 14-1. How did the Sun and its planetary system form? (Note: Do not confuse this with the formation of the universe as a whole, which occurred around 14 billion years ago, or the origin of our galaxy, which probably occurred around 12–14 billion years ago. We will take up those topics in later chapters.)

FACTS TO BE EXPLAINED BY A THEORY OF ORIGIN

★ Nobel laureate Hannes Alfvén, who has spent years researching the solar system's origin, once said: "To trace the origin of the solar system is archaeology, not physics." He meant that our ignorance of the initial conditions forces us to work backward through time, reasoning from whatever clues we can find. The most

Figure 14-1 Earth, almost lost in the glare of its star, symbolizes the question of how planets formed around the Sun. (NASA photo from Apollo 14 on the way to the Moon)

important clues are *facts about the solar system that have no obvious explanation from present-day conditions* but must have arisen from initial conditions as the solar system formed. Table 14-1 lists some of these clues. In this chapter, we will account for them one by one. This process leads to an interesting realization. As we sift through clues found in meteorites, lunar rocks, and orbits in our own solar system, we will see that we are learning about the formation of not only our own system but also the stars themselves. Thus this chapter provides a link to the ensuing chapters.

THE PROTOSUN

Before the Sun existed, its material must have been distributed in interstellar space in a large cloud like the interstellar clouds we see today, where other stars are actually forming before our eyes. Based on the dates of materials in meteorites, which in turn sample many types of asteroids, the Sun began to form 4.6 billion years ago when this cloud (or part of it) became so dense that its own inward-pulling gravitational forces became stronger than the outward force of the pressure generated by the motions of its atoms and molecules. The cloud thus began to shrink.

Early Contraction and Flattening

Even if gas in the cloud had been randomly circulating initially, the contracting cloud would have developed a net rotation in one direction or another. To show this, you can do an experiment with a cup of coffee or a pan of water. If you stir the liquid vigorously but as randomly as you can and then wait a moment and put a drop of cream in it, the cream will usually reveal a smooth rotation in one direction or the other, because the sum of the random motions will usually be a small net angular motion in one direction or the other. Because the interstellar cloud was relatively isolated from the rest of the universe (as such clouds are today), this rotating motion, or angular momentum, could not be easily transferred anywhere and thus was *conserved* as the cloud shrank. (The term *angular momentum* refers to a quantity used by physicists to measure the total rotary motion

TABLE 14-1

Solar System Characteristics to Be Explained by a Theory of Origin

1. All the planets' orbits lie roughly in a single plane.
2. The sun's rotational equator lies nearly in this plane.
3. Planetary orbits are nearly circular.
4. The planets and the Sun all revolve in the same west-to-east direction, called prograde (or direct) revolution.
5. Planets differ in composition.
6. The composition of planets varies roughly with distance from the Sun: Dense, metal-rich planets lie in the inner system, whereas giant, hydrogen-rich planets lie in the outer system.
7. Meteorites differ in chemical and geological properties from all known planetary and lunar rocks.
8. The Sun and all the planets except Venus and Uranus rotate on their axis in the same direction (prograde rotation). Obliquity (tilt between equatorial and orbital planes) is generally small.
9. Planets and most asteroids rotate with rather similar periods, about 5 to 10 h, unless obvious tidal forces slow them (as in the Earth's case).
10. Distances between planets usually obey the simple Bode's rule.
11. Planet-satellite systems resemble the solar system.
12. As a group, most comets' orbits define a large, almost spherical swarm around the solar system (the Oort cloud). Other comets reside in the Kuiper belt, near Pluto and just beyond it.
13. The planets have much more angular momentum (a measure relating orbital speed, size, and mass) than the Sun. (Failure to explain this was the great flaw of the early evolutionary theories.)

of a system—it increases with the amount of rotating material and with the speed and diameter of the rotating system.) This is called the principle of **conservation of angular momentum.** It predicts that the cloud would have rotated faster as its mass contracted toward its center, just as a figure skater spins faster when she pulls in her arms. Mathematical studies show that centrifugal effects associated with the faster spin caused the outer parts of the cloud to flatten into a disk, while material in the center contracted fastest, forming the Sun, as shown in Figure 14-2. The planets formed in this disk, accounting for fact 1 in Table 14-1: *the single plane of all planets' orbits.* Because the Sun itself was an integral part of this disk, fact 2, *the Sun's rotation in the same plane,* is also explained.

At first, when the cloud was large, its atoms were far apart and could have fallen more or less freely toward the center of gravity in the middle of the cloud. This state is called **free-fall contraction.** If free-fall had persisted, the cloud could have completely contracted into the Sun in only a few thousand years. But eventually, because of their growing concentration and random motions, the atoms in the cloud began to collide with each other. Atoms interacting in a gas cause outward *pressure.* A buildup of pressure in the cloud would have slowed the collapse. Thus the inward force of gravity and the outward force of gas pressure competed. Because the properties of gases are well understood, astrophysicists can analyze the later contraction and evolution of the cloud under these competing forces. Even with the outward gas pressure, the Sun went from a cloud hundreds of solar radii across to a well-defined condensation surrounded by a dusty disk in only a million years or so.

Helmholtz Contraction

Gravitational contraction in which the shrinkage is slowed by outward pressure is called **Helmholtz contraction** (after Hermann von Helmholtz, the German astrophysicist who first studied it). In 1871, Helmholtz showed how contraction would have caused heat to accumulate in the contracting protosolar cloud. Helmholtz noted the following (quoted in Shapley and Howarth, 1929):

If a weight falls from a height and strikes the ground, its mass loses . . . the visible motion which it had as a whole—in fact, however, this is not lost; it is transferred to the smallest elementary particles of the mass, and this invisible vibration of the molecules is [what we call] heat.

In the contracting protosun, atoms or swarms of atoms would have fallen toward the center until they collided with other parts of the gas cloud. Temperature would have increased inside the cloud. Wien's law

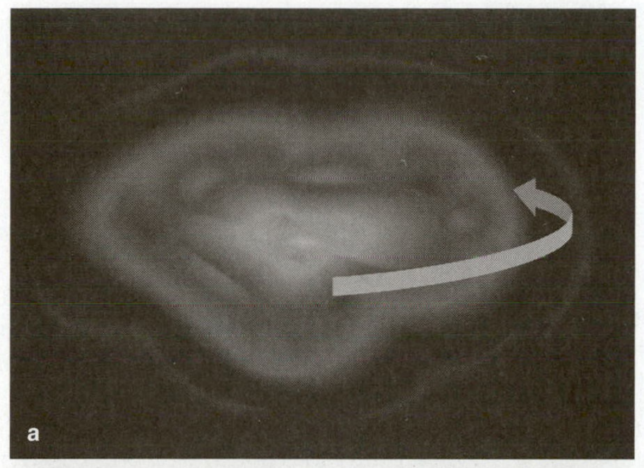

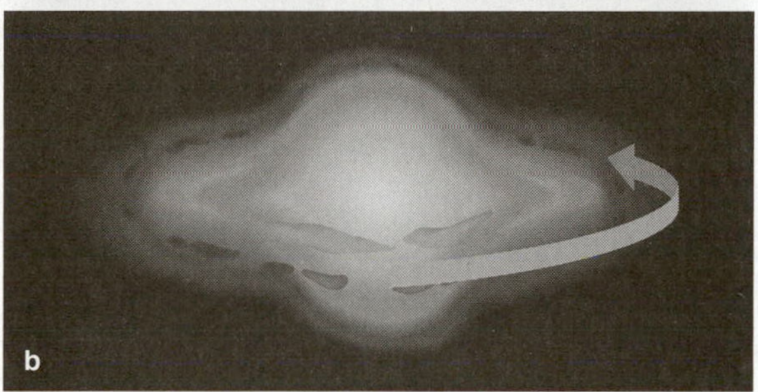

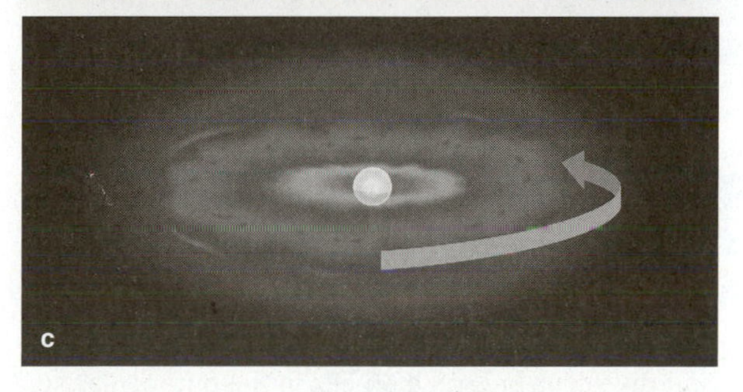

Figure 14-2 Three stages in the evolution of the protosun. **a** A slowly rotating interstellar gas cloud begins to contract because of its own gravity. **b** A central condensation forms, and the cloud rotates faster and flattens. **c** The Sun forms in the cloud center, surrounded by a rotating disk of gas.

and other physical laws guarantee that the protosun would have radiated energy and warmed the surrounding cloud.

Calculations based on the Helmholtz theory indicate the conditions: Eventually, the cloud's central temperature rose to 10 million kelvins or more, starting the nuclear reactions that made it a star and not just a ball of inert gas. Meanwhile, and more important for the formation of planets, the outer parts of the cloud, shown in Figure 14-2, formed a disk of gas as big as the solar system, at temperatures of a few thousand kelvins. By 10 million years after the collapse started, the Sun was a glowing star in the middle of a dusty disk. The precise dating of meteorites suggests this was about 4.56 billion years ago, or, to put it another way, 4560 million years ago. *As we will see in Chapter 18, these theorized steps have been confirmed by actual observations of newly formed stars surrounded by clouds of gas and dust.* This whole process of contraction is thus very important because it turns out to apply to the formation of all stars, not just the Sun.

THE SOLAR NEBULA

A cloud of gas and dust in space is called a **nebula** (plural: *nebulae*) from the Latin term for mist. The disk-shaped nebula that surrounded the contracting Sun is called the **solar nebula.** Molecules of gas or grains of dust must have moved in circular orbits, because noncircular orbits would have crossed the paths of other particles, leading to collisions that would have damped out the noncircular motions—in the same way that noncircular eddies get damped, or canceled, in the cup of coffee mentioned earlier. Thus, neglecting small-scale eddies in the gas, broad-scale motions in the cloud were in parallel circular orbits, accounting for facts 3 and 4 in Table 14-1. As the nebula stabilized, its gas began to cool.

Condensation of Dust in the Solar Nebula

As remarked above, Helmholtz contraction initially heated the solar nebula to at least 2000 K. At such a temperature, virtually all elements were in gaseous form. As is the case with other cosmic gas, most solar nebula atoms were hydrogen, but a few percent were heavier atoms such as silicon, iron, and other planet-forming material.

How did solid particles form in this gas? The answer can be seen on the Earth. When air masses cool, their condensable constituents form particles: snowflakes, raindrops, hailstones, or the ice crystals in cirrus clouds. Similarly, as the solar nebula cooled, condensable constituents formed tiny solid particles of dust. Various mineral compounds appeared in a sequence known as the **condensation sequence**.

Chemical studies show that as the temperature in any part of the nebula dropped 1600 K, certain metallic elements such as aluminum and titanium condensed to form metallic oxides in the form of **grains,** or microscopic solid particles, as shown in Table 14-2. At about 1400 K, a more important constituent, iron, condensed. Microscopic bits of nickel-iron alloy

TABLE 14-2

Condensation Sequence in the Solar Nebula

Approximate Temperature (K)	Element Condensing	Form of Condensate (with Examples)	Comments
2000	None		Gaseous nebula
1600	Al, Ti, Ca	Oxides (Al_2O_3, CaO)	
1400	Fe, Ni	Nickel-iron grains	Parent material of planetary cores, iron meteorites?
1300	Si	Silicate and ferrosilicate minerals [enstatite, $MgSiO_3$; pyroxene, $CaMgSi_2O_6$; olivine, $(Mg, Fe)_2\ SiO_4$] in form of microscopic grains	First stony material, combined to form meteorites; some still preserved in primitive meteorites
300	C	Carbonaceous grains	Forms black, carbonaceous soils and rocks
300 to 100	H, N	Ice particles (water, H_2O; ammonia, NH_3; methane, CH_4)	Large amounts of ice; still preserved in outer planets and comets

Sources: Data from Lewis (1974), Grossman (1975), and others.

formed as grains or perhaps coated existing grains. Still more important, at about 1300 K abundant silicates began to appear in solid form. For instance, a magnesium silicate mineral, enstatite ($MgSiO_3$), formed at about 1200 K (Lewis, 1974). These silicate minerals are the common rock-forming materials, so at this point the solar nebula was acquiring a large quantity of fine dust with rocky composition.

In the outer, colder regions of the nebula, some grains that formed in interstellar space before the solar nebula existed may have survived. These could have been mixed with the newly condensing grains. This may explain some unusual grains that chemists have found in ancient meteorites.

Complex mixtures of magnesium-, calcium-, and iron-rich silicates condensed, depending on the temperature, pressure, and composition of the gas at various points in the solar nebula. Since local conditions were determined by the distance from the newly formed Sun, different compositions of mineral particles may have dominated at different locations, explaining facts 5 and 6 in Table 14-1. Exploration of the planets has helped clarify conditions in the nebula. For example, comparison of Pioneer probe results on the atmosphere of Venus in 1979 with earlier Viking results on Mars shows that the amount of primordial argon gas incorporated from the solar nebula into the planets decreased by a few hundred times from Venus, past the Earth, to Mars. This suggests that the nebular gas further out from the Sun had lower pressure (Pollack and Black, 1979). Similarly, the outer nebula must have been cooler than the inner nebula, being further from the Sun and more shaded by the inner dusty nebula (Cameron, 1975). In these colder regions, at about 300 K, water was trapped in many minerals.

This agrees with what we saw when we studied comets and asteroids in the last chapter. Beyond the main asteroid belt, snowflakes of H_2O ice condensed, and the surviving small bodies in the outer solar system are comets, with high ice content. At 100 to 200 K, in the outermost nebula, ammonia and methane ices also condensed. In the outer solar system, these ices survive even direct sunlight; they survive today on comets and icy satellites of the giant planets. Closer to the Sun, in the main asteroid belt and among the terrestrial planets, we see that bodies were made primarily out of rock.

Meteorites as Evidence

★ The compositions of meteorites strongly support this theory of condensation. Carbonaceous chondrites contain inclusions of materials believed to be among the earliest solid particles in the solar system. These pebblelike inclusions, shown in Figure 14-3, are rich in elements that would have condensed first (at the highest temperatures), such as osmium and tungsten. Their minerals formed at temperatures of 1450–1840 K, consistent with fact 7 in Table 14-1. Yet

Figure 14-3 A piece of the Allende carbonaceous chondrite, showing white inclusions. These demonstrate how material that formed in one environment was later trapped in other material. The black matrix is composed of microscopic dust grains condensed at a few hundred kelvin. The white inclusions contain aluminum-rich minerals condensed at high temperatures. As recognized in the 1970s, the inclusions shed light on the earliest formative conditions in the solar system. (Courtesy R. S. Clarke, Smithsonian Institution)

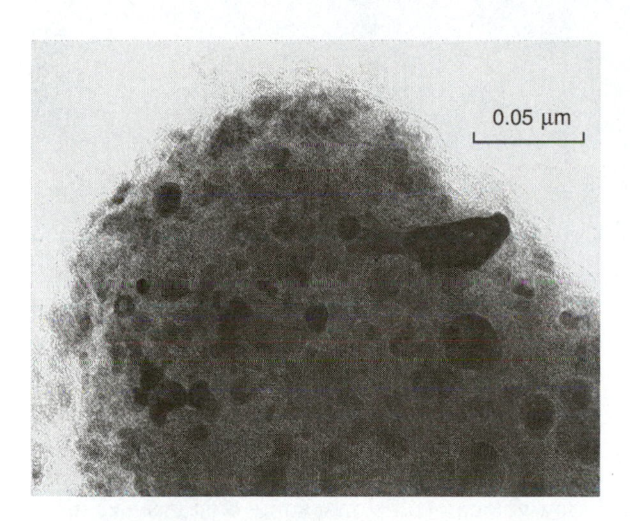

Figure 14-4 A bit of interplanetary dust. In this transmission electron microscope image, larger dark blobs are mineral grains, probably magnetite (Fe_3O_4) or iron-nickel sulfide. They are surrounded by an extremely fine-grained carbon-rich matrix, probably consisting of amorphous carbon particles formed at lower temperatures than the iron grains. This microscopic structure supports the theory that tiny mineral grains condensed and then accreted into larger bodies as planetary material formed in the dusty nebula around the primordial Sun. (Courtesy Roy Christoffersen and Peter Buseck, Arizona State University)

the bulk of carbonaceous chondrites contain microscopic grains formed at low temperatures. The high-temperature minerals must have condensed first and aggregated into pebblelike objects, only to be overwhelmed and surrounded by the sooty carbonaceous material, which condensed later after the nebula had cooled (see Figure 14-4). Water is a common constituent in minerals in many carbonaceous chondrites; because the water would have been driven off at high temperature, this shows that they formed and remained at low temperature. Also, as predicted by the theory (Table 14-2), enstatite is a common mineral in meteorites.

FROM PLANETESIMALS TO PLANETS

Although the preplanetary particles may have formed as microscopic grains, they clearly grew bigger. (Otherwise there would be no planets!) The hypothetical intermediate bodies, from millimeters to many kilometers in size, are usually called **planetesimals,** as noted in Chapter 13. Evidence that they existed includes the following:

1. Craters on planets and satellites indicate impacts of planetesimals with diameters of at least 100 km (Figure 7-12).

2. Meteorites are their surviving fragments, and their microstructures reveal how the dust grains clumped together. Iron meteorites' crystal structure indicates that they formed inside bodies around 30 to 400 km across.

3. Asteroids and comet nuclei up to 1000 km in diameter survive today throughout the solar system.

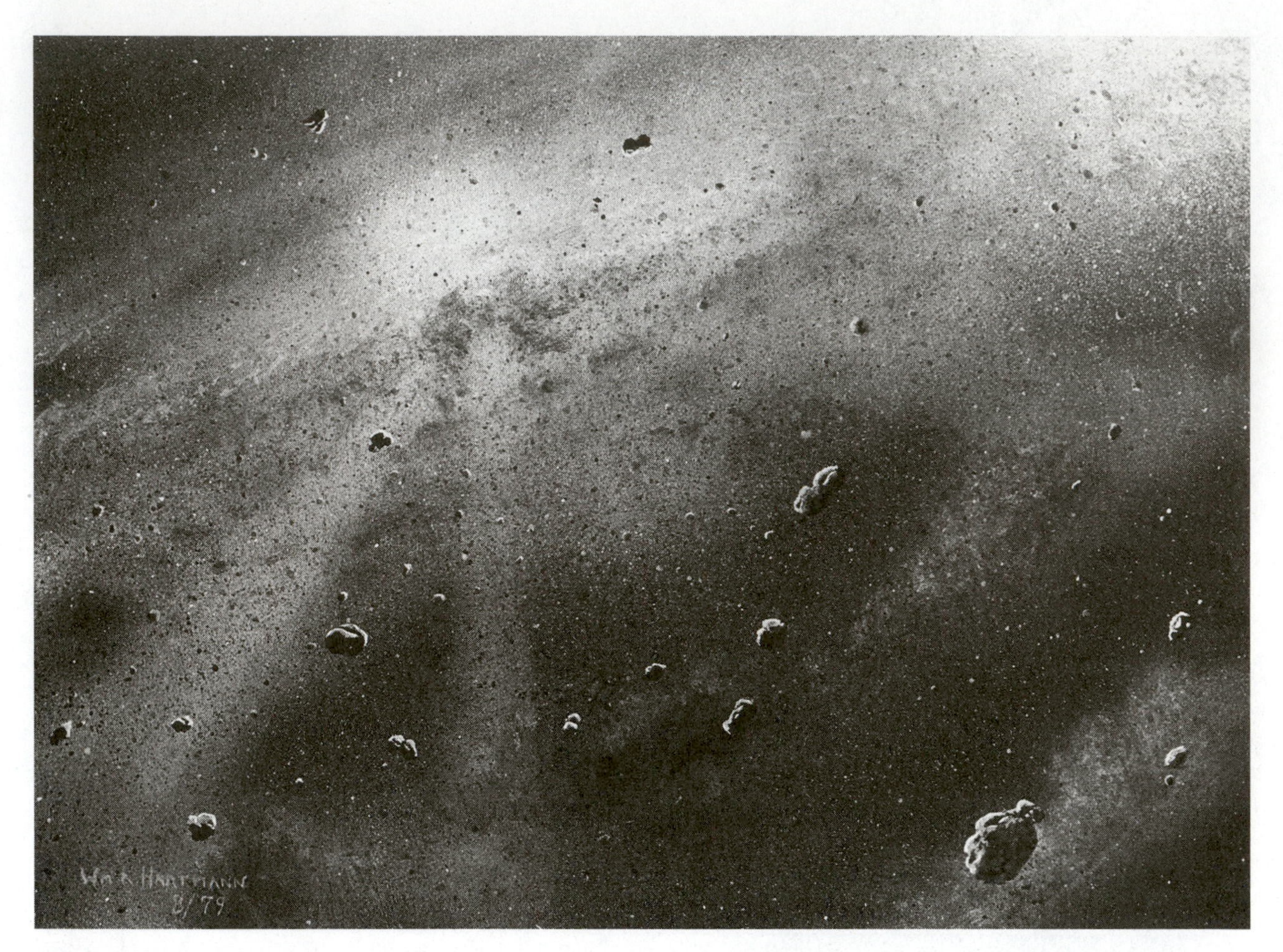

Figure 14-5 A scene in the early solar nebula. Planetesimals of rocky and icy materials orbit in the foreground. The sun is partially obscured by dust and gas in the inner nebula. (Painting by WKH)

But exactly how did microscopic grains aggregate to produce 100-km-scale planetesimals? If they had circled the Sun in paths comparable to present asteroidal and cometary orbits, they would have collided with one another at speeds much faster than rifle bullets. They would have shattered (as seen in Figure 13-12), and the solar system would still be a nebula of dust and grit.

Did dust particles have such high speeds in the early solar nebula? According to dynamical analyses, they collected in a swarm of particles with nearly parallel, circular orbits in the central plane of the disk. Because the orbits were nearly parallel, planetesimals approached each other gently, and collision velocities were low. To visualize this, picture race cars moving around a circular track at 150 mph. A head-on collision would be very violent, but if two neighboring cars on near-parallel courses bump against each other, the actual collision will be much slower. In low-velocity collisions, some dust grains simply stuck together, held by weak adhesive forces such as gravity and electrostatic attraction. Mutual gravity probably caused groups of planetesimals to clump together. Once a single clump grew, it would have a good chance of sweeping up and absorbing neighboring particles, in the same way that the front end of a car gets plastered with snow as it moves through a storm of snowflakes.

Computer simulations suggest that many of the dust aggregates grew to multikilometer size within only a few hundred thousand years—a mere moment in cosmic time. Radioisotope studies of meteorites provide direct evidence that the initial assembly of asteroid-size bodies was very fast—within about a million years (Lee and Halliday, 1996; Cowen, 1999). We say "fast," because the time interval is so short compared to the age of the solar system. For example, if the Sun formed 4560 million years ago, it already had asteroid-size planetesimals by about 4559 million years ago.

At this point, about 4.56 billion years ago, the solar nebula would have resembled the scene in Figure 14-5, with rocky planetesimals orbiting in dusty clouds that dimmed the central Sun. The growing planetesimals probably resembled the asteroids whose photos were shown in the last chapter.

Dynamical studies indicate that coalescing particles tend to form bodies rotating in a prograde motion (fact 8 in Table 14-1) with similar rotation periods (fact 9). Additional computer modeling of the collision and aggregation process suggests that as the planetesimals grew toward planet size, one large body came to dominate each doughnut-shaped zone around the Sun, and the planetesimal swarm naturally partitioned itself into zones with one large planet in each zone. For instance, imagine the planetesimals growing in Earth's zone, 1 AU from the Sun. Whichever planetesimal is the largest has the most cross-sectional area and the most gravity, so it has the greatest chance of sweeping up others. Thus it grows the fastest and dominates that zone and will probably eventually become the Earth. The second-largest body, perhaps at 0.9 AU from the Sun, is likely eventually to be swept up by the largest one. However, the proto-Earth's gravity does not extend as far as 0.7 AU from the Sun. Therefore, in that zone, the largest planetesimal grows into a separate planet, Venus. Each doughnut-shaped zone was about twice the width of the next inner zone, and this principle roughly explains Bode's rule (fact 10).

Some planetesimals, the *parent bodies* of meteorites, were heated, melted, and differentiated into metal and rock portions. Some were shattered by collisions with other large or fast neighbors, freeing iron, stony-iron, and achondrite meteorites from their interiors. As some shattered, others grew to replace them.

This scenario of gradual growth through collisions allows us to explain the major classes of bodies in the solar system—the terrestrial planets, giant planets, asteroids, and comets—as follows.

Terrestrial Planets

In the inner part of the solar system, this process simply continued until Mercury- to Earth-size planets were formed. Most of the planetesimals in this region were composed of silicate rocky material, familiar to us from the rocks of the Earth and Moon. Thus all the terrestrial planets grew from this material until most of it was swept up; some remaining gas and dust were blown away by outrushing gas and radiation from the newly heated Sun.

In support of this picture, calculations show that if you put a planetesimal (or asteroid) halfway between Mercury and Venus, or between Venus and Earth, or between Earth and Mars, the gravity of the planets would disturb its path and it would be swept up in a few million years by one of the planets. In other words, the terrestrial planets partitioned all the available space in the inner solar system. No more planets could grow there.

Giant Planets

There are two competing theories of how the giant planets formed. The first theory is just an extension of the terrestrial planet model. In this view, more material was available in the giant planet zone because ices as well as rocky material had condensed in this cold region, augmenting local planetesimal masses. Thus the "embryo planets" that were to become Jupiter, Saturn, Uranus, and Neptune grew to Earth's size and beyond. By the time they reached about 10–15 times the mass of present-day Earth (or two to three times the diameter of Earth), they had such strong gravity that they began to pull in gas from the surrounding solar nebula. Thus they accreted not only planetesimals to make a solid/liquid planet, but also massive atmospheres of gas whose composition approximately equaled that of the nebular gas. This brought Jupiter from 10–14 Earth-masses up to about 300 Earth-masses. In this theory, the giant planets can be thought of as two-phase planets. Their cores are "giant terrestrial planets," averaging around 15 times more massive than Earth; and these cores are surrounded by giant atmospheres. The terrestrial planets never accumulated these giant atmospheres of hydrogen-rich nebular gas because they never attained enough mass to pull in the nebular gas.

The second theory starts further back, with the solar nebula, before the planetesimals had reached a large size. In this view, the outer part of the nebula had some regions of high density. Just as the Sun collapsed gravitationally from an interstellar cloud, due to its self-gravity, regions of the nebula may have collapsed into Jupiter- or Uranus-size objects, forming directly from huge masses of hydrogen, helium, dust, and ice particles (Sincell, 2000). Thus, in this view, the planets formed all at once, whereas in the former view the hydrogen-helium atmospheres were added

TABLE 14-3

Comparison of Solar Nebula and Giant Planet Atmospheres

	Composition Percentages by Mass				
Gas	Solar Nebula (Sun's Present Composition)	Jupiter	Saturn	Uranus	Neptune
H_2 (hydrogen)	73	81	93	65	63
He (helium)	25	19	16	23	25

Sources: Gautier and others (1992); Ingersoll (1990); Anders and Grevesse (1988); Voyager reports (Stone and others, 1979, 1982, 1986, 1989); Hubbard and Stevenson (1986).

Note: Uncertainties are a few percent. Scientists debate the significance of the lower helium content reported for Saturn.

onto rocky-icy cores. Researchers are trying to choose between these theories by getting more information on the size of Jupiter's core (does it really have a 10-Earth-mass core of rock and ice?).

Both scenarios explain the gross composition of the giant planets' atmospheres. Although we don't have samples of the nebular gas, we *can* measure the Sun's present-day composition, which is believed to be about the same as that of the nebular gas from which the Sun formed. Table 14-3 makes this comparison. Here we see that gases from which the Sun and nebula formed were mostly hydrogen and about one-fourth helium. In the atmospheres of the giant planets, we see similar proportions: They are mostly hydrogen, and the second-ranked constituent is helium. Given the uncertainties of the measurements, the atmospheres of all four giants roughly resemble the original nebular gas—although Uranus and Neptune, with slightly weaker gravity than Jupiter and Saturn, may have lost some of their hydrogen atoms.

Note how different the giants are from the terrestrial planets. Terrestrial planets' original atmospheres were mostly carbon dioxide, emitted by volcanoes from inside the planet. But gas giants' atmospheres retain the hydrogen and helium of the ancient nebula.

As the giant planets formed, each giant formed a miniature "solar nebula" around itself. And in this disk-shaped cloud of gas and dust, the accretionary growth process was repeated all over again. That is, each giant planet became an analog of the Sun, and moons grew around it analogous to planets. In these miniature planetary systems, the most abundant building materials were ice and the black carbonaceous dirt common to the outer solar system. Thus giant planets spawned systems of dirty-ice satellites in prograde, circular, coplanar orbits around them, accounting for fact 11 in Table 14-1.

Asteroids

The asteroids are easily understood in this scenario. They are planetesimals that never made it all the way to planethood. There is probably a specific reason why an asteroid belt was left stranded between Mars and Jupiter: This zone would normally have been a location where a planet should have grown. Ceres, the largest asteroid, had already grown to 1000-km diameter by the time the growth process stopped in that zone. It probably stopped because Jupiter had grown so huge that its gravity disturbed the motions of the asteroids in the zone we now know as the main asteroid belt. As mentioned in the last chapter, this disturbance increased the collision velocities of the asteroids—just as a disturbance in a car race causes the cars to go in zig-zag paths instead of on parallel circular paths around the track. The resulting collisions caused the asteroids to smash each other into innumerable fragments instead of coalescing. Thus the main asteroid belt is a belt of fragments instead of one large planet. Direct evidence of this comes from the structure of certain types of meteorites that are composed of fragments of one type of meteorite jammed up against completely different types. An example is shown in Figure 14-6, where we can see a fragment of a black carbonaceous chondrite jammed into a matrix of ordinary chondrite, full of round white chondrules. Here are pieces of two parent asteroids that formed in different environments millions of miles apart in space, fused together by a collision eons ago, before falling into a third environment—Earth!

As mentioned in the last chapter, the gravity of Jupiter also shaped the structure of the asteroid belt, kicking out asteroids in certain resonances (such as $\frac{1}{2}$ the period of Jupiter), thereby removing asteroids with a mean distance of 3.28 AU from the Sun. Ju-

Figure 14-6 A meteorite breccia. An angular carbonaceous chondrite fragment (dark mass) lies inside the chondrite Sharps, illustrating the mixture of rock chips from different parent bodies. The white "fish" in the upper corner of the dark mass is a white refractory inclusion of minerals formed at high temperatures. The dark rim on the carbonaceous chrondrite fragment is indicative of some heating at the time the entire mass was forming into a chondritic rock. (Courtesy Laurel Wilkening, Lunar and Planetary Laboratory, University of Arizona)

piter's gravity also defined the edges of the belt; one resonance defines the inner edge at about 2.0 AU. Calculations show how Jupiter could have cleared out most asteroids in the region from about the outer belt edge at 3.2 AU out to Jupiter's orbit at 5.2 AU within the first 10 million years of solar system history (Liou and Malhotra, 1997).

The Trojan asteroids are a different story. They are trapped at the Lagrangian points of Jupiter's orbit, but scientists have not agreed on how they originally got there. They may be planetesimals that formed in Jupiter's region and were trapped in the two Lagrangian points of Jupiter's orbit, 60° ahead of, and behind, the planet. Alternatively, they may be comet nuclei, originally formed at greater distances, but trapped by Jupiter when they ventured from the Oort cloud or Kuiper belt into the inner parts of the solar system. The Trojans may even include examples of both types of objects.

Comets

Our picture of planet formation gives a good explanation of comets. Beyond the main asteroid belt, planetesimals formed from the ice and carbonaceous materials that dominated the condensates in the present region of the giant planets. Most were consumed in growing the rocky-icy cores of the giant planets. But millions were left in interplanetary space as the giant planets grew. Calculations of the aggregation of comets suggest that they may have been assembled from individual chunks of 10- to 100-m size condensed ice, loosely bound into 1- to 10-km-size comet nuclei, like clusters of marbles weakly glued together (Weidenschilling, 1997). This model explains several properties of comets, such as why they are weak enough to break into pieces.

The gravitational forces of the giant planets were so strong that whenever an icy planetesimal experienced a near-miss with one of them, the planetesimal was thrown into the Oort cloud. Other icy planetesimals formed at the very outskirts of the solar system, in the region from Neptune to beyond Pluto. Their orbits were not disturbed very much, so these planetesimals were left in the region from 30 to 100 AU, where they formed the Kuiper belt. Thus we can account for the facts in item 12 in Table 14-1. The Kuiper belt and Oort cloud contain icy bodies that were never strongly heated; thus the remote bodies in these regions probably contain some of the most ancient and primitive materials in the solar system, including low-temperature compounds of C, H, O, and N. For example, a 1993 discovery showed that Pholus, a Kuiper belt asteroid, is the reddest yet observed, probably

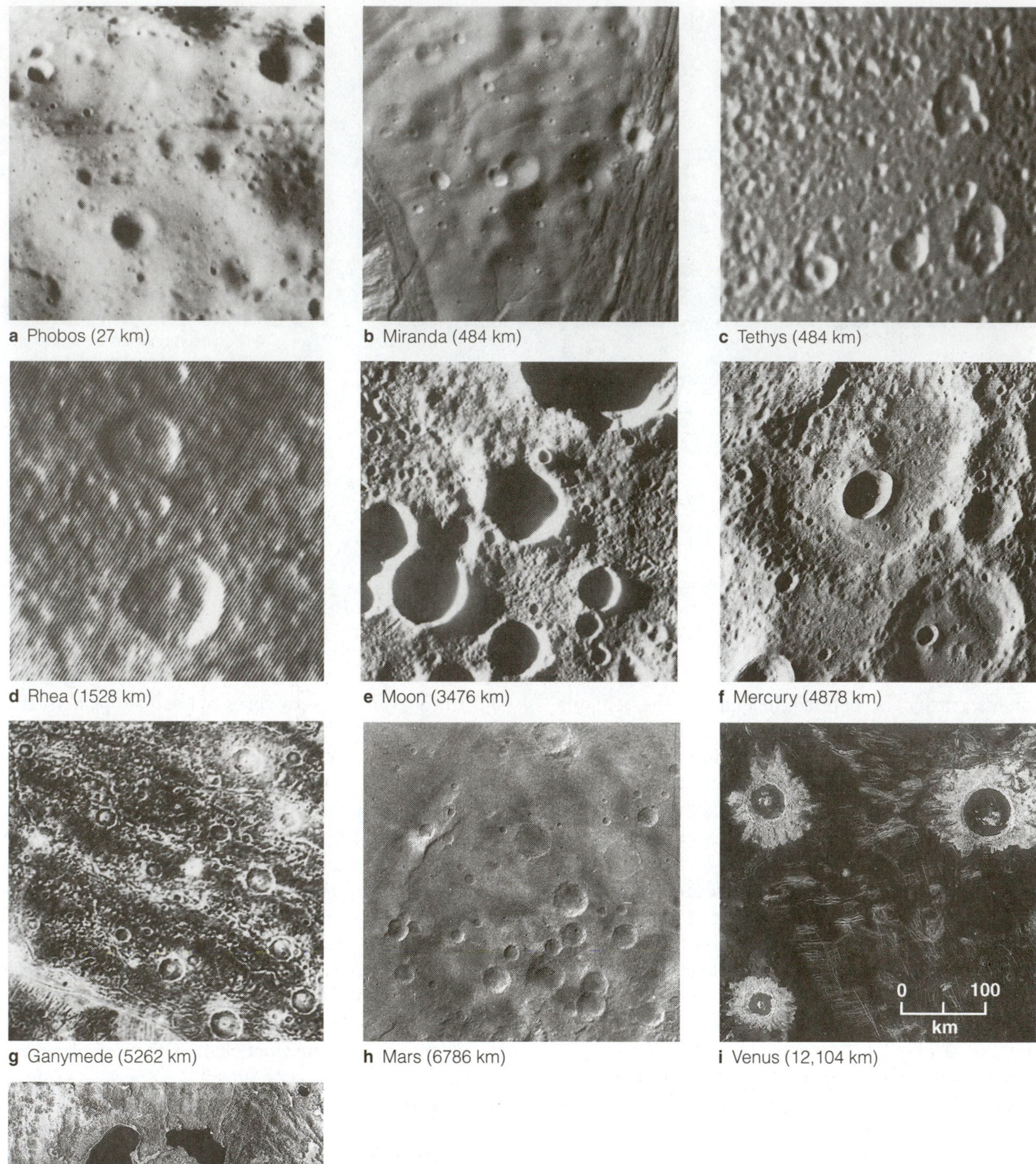

a Phobos (27 km)

b Miranda (484 km)

c Tethys (484 km)

d Rhea (1528 km)

e Moon (3476 km)

f Mercury (4878 km)

g Ganymede (5262 km)

h Mars (6786 km)

i Venus (12,104 km)

j Earth (12,756 km)

Figure 14-7 Impact craters throughout the solar system are evidence that countless planetesimals have crashed into worlds of all sizes. The diameter of each world is given. The arrangement by size shows that larger worlds (Mars, Venus, Earth) have fewer or more eroded craters. The crater shown on Earth is the 4-km, 450-million-year-old Brent crater in Ontario, Canada; it has been filled by glacial erosion. Scale varies from photo to photo (although the Phobos and Earth photos cover similar areas). (Photos **a-g** and **i,** NASA; **h,** Soviet Mars 5 orbiter photo, courtesy Vernadsky Institute, Moscow; **j,** Earth Physics Branch, Department of Energy, Mines, and Resources, Ottawa)

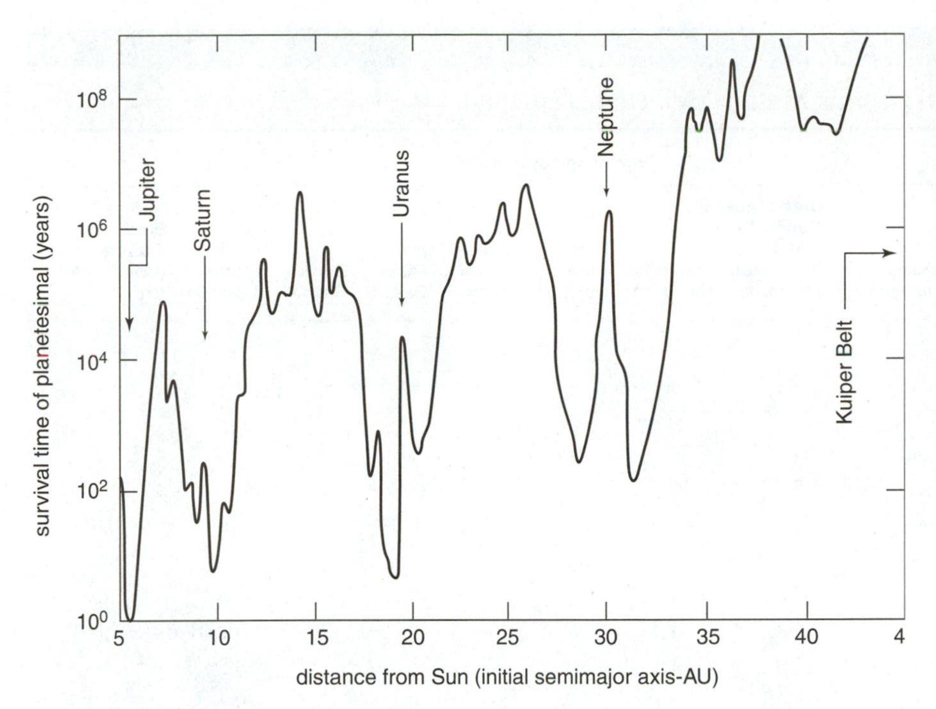

Figure 14-8 Calculated lifetimes of interplanetary bodies placed among the four giant planets. The graph shows the interplanetary bodies in this part of the solar system cannot be stable for the life of the system, but are swept up or ejected by the giant in a few million years or less. If a fifth planet had tried to grow in this space, it would have been swept up or ejected. The only stable region is beyond about 40 AU—where the Kuiper belt of leftover planetesimals actually exists. Spikes at the positions of the four planets represent Lagrangian point resonances. (After calculations by M. J. Holman and Jack Wisdom)

due to colored organic compounds that would have been destroyed had Pholus ever been heated. Many clues about the formation of our system may reside among such primitive, never-heated objects!

What Became of the Remaining Planetesimals and the Solar Nebula?

Most of the planetesimals were eventually accumulated by the planets, scattered into the Oort cloud, or left stranded in the Kuiper belt and the asteroid belt. This probably happened within 100 million years of the solar system's beginning. Each time a planetesimal crashed into a planet to add to the planet's mass, a crater was formed. For another 500 million years, as lunar rocks have taught us, the last remaining interplanetary planetesimal rained down onto planetary surfaces, creating still more craters on all planets and satellites, as shown in Figure 14-7. A few planetesimals were captured into orbits around the newly formed planets, explaining the dozen or so satellites with irregular orbits that seem to have been captured instead of growing in a nebular disk around the planet. Good examples include the outer eight satellites of Jupiter and the outmost satellite of Saturn; several of these have retrograde "backward" orbits.

Astronomers are beginning to understand the dynamical processes that cleared out these last planetesimals. Figure 14-8 shows computer results on the survival time of planetesimals placed in orbits between the giant planets. The point is that any small bodies in these zones would last only a few million years or less before the gravitational forces of the giant planets would disturb their motion and bring them close to one of the giants, either to be ejected from the solar system altogether or to crash into a planet. Note on the graph that only beyond Neptune are planetesimals "safe" enough to have lifetimes approaching the age of the solar system— and indeed this is exactly where the Kuiper belt objects have survived and are found today.

THE CHEMICAL COMPOSITIONS OF PLANETS

Why do the planets vary in their chemical composition? Why do meteorites differ from lunar and terrestrial materials? If Earth was made of meteoritic material, why aren't the rocks in your backyard the same composition as meteorites? The answers to these questions should now be clear from the earlier discussions. Silicate and metal-rich dust grains and planetesimals condensed in the inner solar system, but those that grew in the outer solar system were ice-rich because of the lower temperatures. Carbonaceous materials may have been produced partly from condensation or partly from chemical processing of CH-rich materials, such as methane ice. Since

TABLE 14-4

Elemental Abundances in Cosmic Matter (Selected Elements)

		Percentage by Weight								
Element	Element Behavior[a]	Sun ("Cosmic" Composition)[b]	Carbonaceous Chondrite (Most Primitive) Meteorite	Chondrite Meteorite	Earth: Ultrabasic Rock	Mars: Basaltic Rock	Moon: Mare Basalt	Earth: Basalt	Moon: Upland	Earth: Granite (Continental Crust)
Hydrogen (H)	V	78	—	—	—	—	—	—	—	—
Helium (He)	V	20	—	—	—	—	—	—	—	—
Oxygen (O)	L	0.8	34	35	43	—	43	44	45	48
Iron (Fe)	S	0.04	24	26	9.4	13	11	8.6	4.6	2.1
Silicon (Di)	L	—	15	18.5	20	21	21	23	23	33
Magnesium (Mg)	L	—	13	15	20	5	5.5	4.6	4.4	0.6
Sulfur (S)	—	—	8.6	2.3	0.03	3	0.2	0.3	—	0.3
Calcium (Ca)	L	—	1.5	1.3	2.5	4	8	7.6	7	1.5
Nickel (Ni)	S	—	1.4	1.4	0.2	—	0.0001	0.01	0.0001–0.01	0.001
Aluminum (Al)	L	—	1.2	1.1	2	3	8	7.8	13	8
Sodium (Na)	L	—	0.46	0.70	0.4	—	0.4	1.8	0.4	2.7
Titanium (Ti)	L	—	0.08	0.08	0.003	0.5	2.1	1.4	0.3	0.2
Potassium (K)	L	—	0.5	0.9	0.004	0.25	0.1	0.83	0.01–0.1	3

Sources: Fairbridge (1972); Taylor (1973); Mason and Melson (1970); Page (1973); Clark and others (1977).

[a]V = volatile (driven off by heating); L = lithophile (concentrated in silicaceous rocks; sulfur usually classified separately); S = siderophile (iron-affinity elements; concentrated in basic rocks and metallic minerals, as in the Earth's core and mantle).

[b]This composition is called cosmic because it is believed to be the same gas that formed not only the Sun but all nearby stars.

the interstellar cloud that formed the solar system, like most of the rest of the universe, was composed mostly of hydrogen and helium, the Sun itself had this composition, and when the high gravity of the giant planets began to capture gas from the solar nebula, the hydrogen-helium mix was mostly what they captured.

★ Chemical evidence for all this can be seen in Table 14-4. Massive bodies, such as the Sun, retained the original gases, as shown in the third column. Meanwhile, the solid materials condensed into dust grains, losing the hydrogen and helium gases; these dust particles aggregated into planetesimals, which we have sampled in the form of chondrite meteorites (columns 4 and 5). These columns show that chondrites were composed mostly of oxygen, iron, silicon, and magnesium. As soon as these materials melted, however, probably due to heat from radioactivity in certain minerals, the iron drained inward to form central cores in the planetesimals, asteroids, and planets. The cores were metal material, which we have sampled in iron meteorites. This differentiation process meant that the surface materials of planets became more iron-poor and more silicon-rich than the primitive materials. This *differentiation* process was discussed in Chapter 6. This can be seen by comparing the chondrite columns (4, 5) to the columns for crustal rocks of Earth, Mars, and the Moon (columns 6–11). Note that their surface and crustal rocks have only 2 to 13% iron, compared with 24–26% in the original solid planetary material. On the other hand, they have excess percentages of elements like silicon and oxygen, because those combined into lightweight silicate minerals that floated to the top and composed planetary surface layers.

A LESSON IN COMPARATIVE PLANETOLOGY: COMPARISONS AMONG MOON SYSTEMS

Knowledge about the origin of the solar system gives us a basis for making some interesting comparisons among satellite systems of the various planets. At first glance, satellites appear to be a chaotic mixture of different kinds of bodies, with no rhyme or reason: Giant planets have whole families of diverse moons that are small compared to the planet; some of the moons move in prograde directions, some retrograde. Earth and Pluto, on the other hand, are almost "double planets," with single moons that are a quarter to half the size of the planet.

In earlier chapters we discussed some general patterns among the satellite systems, but this chapter gives a better basis for explaining these patterns. It also provides a good opportunity to review some features that we have discussed in the last several chapters.

As we explained, the giant planets attracted so much gas from the surrounding nebula that they made miniature, disk-shaped "solar nebulae" around themselves. Their satellite systems formed within those disks, just as planets formed in the nebular disk around the Sun. The rings and small moonlets near the rings are remnants of the inner debris. Just as Jupiter formed roughly halfway out in the solar system, the largest moons of each giant planet, such as Ganymede, Titan, and Titania, formed roughly halfway out among the satellites. They are the Jupiters of their systems. Thus the world-building process was probably most efficient in the middle of the nebulae, at an intermediate distance from the central body. Tidal stretching forces, which cause heating on moderate- to large-size moons, are strongest close to the planet, and this is why some of the inner moons, like Io, Europa, and Enceladus, show evidence of volcanism or resurfacing.

As the gas giants' satellite systems finished forming, orbits of nearby planetesimals—were being disturbed by the strong gravity of the newly formed giant planets. A few errant passing asteroids or comet nuclei were captured into orbits around giant planets on the outskirts of their systems. This made some of the black outer moons, including all the retrograde ones, such as Jupiter's Pasiphae or Saturn's Phoebe. Mars probably also received two captured black moons. Neptune's Triton may be the largest of these captured moons, and it apparently came far enough into Neptune's system to disrupt some of the other moons' orbits.

By chance, in the case of Earth and possibly Pluto, very large intruding planetesimals apparently collided with the planet itself, blowing out enough debris to form satellites.

According to these ideas, we see that what appears at first to be a hodgepodge of moons actually shows some systematic patterns, and these are all a product of a few dynamical processes, especially the accretion of planetesimals. Accretion produced the planets themselves; gravitational attraction of gas and dust from the nebula added disks of nebular material around giant planets; accretion within those disks produced moons around giant planets; a few passing planetesimals were captured instead of accreting into the planet; in two cases out of nine, the growing planet suffered a large enough impact during accretion to create a debris cloud that could make a new moon.

GRADUAL EVOLUTION PLUS A FEW CATASTROPHES

We have stressed that this growth of planets by innumerable small planetesimals produced certain regularities of the solar system: planetary orbits that lie in the plane of the Sun's equator, the regular Bode's rule spacings of the planets' orbits, prograde orbital revolutions, prograde rotations, the mostly small obliquities of planets, and the regular systems of prograde satellites lying in their planets' equatorial planes. These regularities of the solar system require smooth evolutionary growth from a system of many small planetesimals, not a catastrophic creation in some chaotic system.

Nonetheless, one of the beauties of the modern picture of planet formation is that it requires a few catastrophic events that can explain some of the nonregularities of the solar system (Wetherill, 1985). These catastrophic events would involve the collisions or near-misses of the planets with the largest of their nearby planetesimal neighbors as the planetesimals were being swept up. As each planet grew, it experienced collisions of different sizes. If a collision was large enough, it could have had important effects.

We have already seen one example of this type of thinking: the theory that a Mars-size planetesimal hit Earth late in its growth and blew off material from which the Moon formed. Another example is that Uranus was probably hit by a relatively large planetesimal, tipping its rotation axis to lie nearly *in* the plane of the solar system instead of nearly perpendicular to it, as is the case for most other planets. Other properties of the solar system—such as the different styles of ring systems and the geological differences between hemispheres of some planets, such as Mars—may trace back to large impacts. The largest impacts were not big enough to randomize the characteristics of planets and their orbits, but they were large enough to give planets individuality of character!

Chaos and Orbit Changes

The effects caused by catastrophic collisions after the planets formed are related to another concept that has gained attention in recent years. Once the planets formed, the orbits of many interplanetary bodies became chaotic. "Chaotic" in this sense means that the orbits are essentially unpredictable, because a small change in the position of a body could have a huge effect during a planetary encounter. More specifically, a **chaotic orbit** is one that diverges exponentially from that of a neighboring body, as time goes on (Lissauer, 1999). Moving a body by a few hundred meters might not make much difference for many years, but later on, during a close encounter with a planet, it might make the difference between zipping past the planet, being pulled down into its atmosphere and crashing, or even being deflected out of the solar system altogether by the planet's gravity. Chaotic orbits can also be produced in certain kinds of resonances, even if the small body does not come close to a larger body. The concept of chaotic orbits means that it is difficult to predict the future history of any given planetesimal, even though we understand Kepler's laws of planetary orbital motion. This is part of the point of Figure 14-8, discussed above. Planetesimals wandering among the outer planets simply couldn't survive in stable orbits.

It is important to recognize that the concept of chaotic orbits does not override Kepler's laws and the general understanding of Newtonian gravity. Major planetary orbits are generally predictable, and even chaotic orbits are theoretically predictable in principle, but in practice the future path is so sensitively dependent on initial conditions that no computers are powerful enough to handle all the nuances (Lissauer, 1999).

For these same reasons, although astronomers feel that they have a good understanding of the general processes forming planetary systems, it is not possible to predict configurations of specific planetary systems. Small changes in a system might have big effects. If a large planetesimal had missed Earth, instead of hitting it, we might not have a Moon. What if a second Jupiter-size object had tried to grow near Jupiter and they had ended up colliding or disturbing each other's motions? A major scientific question thus remains: Do the processes sketched above produce a system of planets around most Sun-like stars? Or do things have to work out "just right" to produce a relatively stable system like ours instead of a chaotic mess?

STELLAR EVIDENCE FOR OTHER PLANETARY SYSTEMS

All stars formed from gravitationally unstable, collapsing nebulae. All stars should have some angular momentum and thus produce rotating, flattened clouds around themselves as they form. Such clouds should cool as their heat radiates away into space, and dust grains of varied compositions should appear. Many observations indicate that these processes occurred among distant stars, as well as in the Sun's case. Astronomers have actually detected disk-shaped "solar nebulas" of dust and gas around some newly forming stars, and have detected dust grains of ice and dust in these systems, as we will see in later chapters.

But does such material always accumulate into planet-size bodies? Probably not, according to simple observations of stars. Many, if not most, stars have one or two other *stars* going around them, rather than a system of planets. Such massive companions may interfere with planets forming in the system, just as Jupiter interfered with the formation of a planet in the main asteroid belt.

Nonetheless, one of the most exciting areas of astronomy in recent years has been the discovery of what seem to be planets orbiting around other stars, as we will discuss further in Chapter 21. Starting in 1995, astronomers made confirmed detections of planet-size objects moving around several stars. By 2000, the total had risen to 33 of these planets (Marcy and Butler, 2000). Based on the principles in this chapter, astronomers had expected that such planets, if detected, would have orderly circular orbits, with giant planets existing only in the cold outer regions of each system. However, many of the systems have planets in strongly elliptical orbits or Jupiter-scale objects closer to their star than Mercury is to our Sun! Why do these differences exist? Are many systems so chaotic that after Jupiter-size objects are formed at, say, 5 AU from the star, they then move inward toward the star? If so, it would argue against the existence of stable Earth-like planets in those systems!

You can see that a stimulating new merger is occurring between studies of our own planetary system and the study of other stars. A new interdisciplinary field is emerging: the study of planetary material around other stars. If only 0.1% of all stars had planets on which liquid water could exist, then 100 million habitable planets could plausibly exist in our galaxy! We will come back to this subject—and its provocative consequences for possible alien life—in later chapters.

SUMMARY

Information about the origin of the solar system has been culled from meteorites, lunar samples, the oldest terrestrial rocks, and chemical and dynamical analysis of planets and satellites. This information indicates that the Sun formed from a contracting cloud of gas about 4.6 billion years ago. As outer parts of this cloud cooled, solid grains of various minerals and ices condensed and accumulated into planetesimals during a relatively brief interval, lasting a few million years.

During this interval, neighboring planetesimals collided, often gently enough to allow them to hold together by gravity. In this way small planetesimals aggregated into a few larger bodies. Sometimes these bodies collided at high enough speeds to shatter each other, producing meteoritelike fragments. The largest bodies survived and grew into planets. Most moons (the moon systems of the giant planets) formed by a similar process. A few additional moons were added when interplanetary bodies were captured, and the moons of Earth and Pluto may have formed as the result of giant collisions.

Many of the last surviving planetesimals crashed into the planets and moons, forming craters. Other planetesimals were stranded in the main asteroid belt, the two Trojan swarms, and the Kuiper belt. When icy planetesimals in the outer solar system made near-miss encounters with giant planets, the planets' gravity flung many of them almost out of the solar system, forming the Oort cloud reservoir of comets (as described in Chapter 13).

The Sun and the outer planets, with their high masses and strong gravity, retained the light, hydrogen-rich gases of the original cloud. Lower-gravity terrestrial planets lost these gases.

Studies of solar system history are producing an interesting rerun of the old debates between catastrophist and evolutionary theories of planetary development. In general, we can say that the general evolutionary picture is correct, in the sense that planet-forming processes—formation of disk nebulas and silicate dust grains—seem to be ordinary developments, not unique anomalies. On the other hand, we can't be sure whether Sun-like systems, with planets in circular, stable orbits, are the "normal" outgrowth of the situation, or whether Earth-like planets can avoid orbit changes, climate changes, and catastrophic asteroid impacts long enough to produce life in the "normal" case. Do we have cosmic neighbors, or are we a unique accident?

To answer this question, we must keep the 500-y-old scientific adventure going through continued observations of other stars and exploration of our cosmic surroundings.

CONCEPTS

catastrophist theories (web site)
evolutionary theories (web site)
conservation of angular momentum
free-fall contraction
Helmholtz contraction
nebula
solar nebula
condensation sequence
grain
planetesimal
magnetic braking (web site)
chaotic orbit

PROBLEMS

1. The gas in the early solar system was about 73% hydrogen (Table 14-3). Considering the theory of thermal escape of gases from planetary atmospheres (see Chapter 9), explain the absence of abundant hydrogen in the terrestrial planets' atmospheres. Why does this not need to be listed as a fact to be explained by theories of solar system *origin* in Table 14-1?

2. Since all planetary material condensed from the same nebula, why do meteorites have different chemical and geological properties than rocks you might find in your own yard?

3. Because of heating by the Sun and by the contraction process, gases in the inner solar system were probably warmer than gases in the outer solar system when planetary solid matter formed. In terms of the condensation sequence, relate this to the estimated or observed composition of the planets.

4. Judging from planetary composition, where was the inner boundary of the part of the solar nebula where ices condensed?

5. If planets orbited the Sun in randomly inclined orbits, both prograde and retrograde, how might theories of such a solar system's origin differ from the theory described in this chapter?

6. List some observations that support the theory of solar system origin described in this chapter.

7. How are theories of solar system origin different in principle from theories of the origin of the universe? (*Hint:* In principle, do we have prospects for testing origin models of either the solar system or the universe by studying other examples of them?)

8. What reasons do we have for believing that the entire solar system formed during a single, relatively short interval?

9. If a five-year-old member of your family asked where the world came from, how would you answer?

ADVANCED PROBLEMS

10. Suppose that a planetary system was forming around a nearby star and the star was obscured by a "solar nebula" of dust grains at a temperature of 1000 K. How might such a dust nebula be detected? (*Hint:* Apply Wien's law and assume that a large telescope is available with infrared detectors.)

11. If dust grains in the early solar nebula collided at a speed equaling 0.1% of their orbital velocity, how fast would they have collided near the present orbit of Earth? How fast near the present orbit of Pluto?

12. A rule of thumb is that a planet can retain a gaseous atmosphere for a geologically long time if the typical velocities of the molecules are no more than about one-fourth the escape velocity of the planet. Assume that icy bodies with density 1000 kg/m^3 were growing in the outer solar system in a nebula of hydrogen at temperature 500 K.

a. Use this rule of thumb to show that if any primitive planets grew to around 38,000 km across they would form cores that would begin to retain hydrogen atoms and form massive gaseous envelopes, thus beginning to form giant planets.

b. Analysts have shown that the giant planets all have cores of ice and rock that have about 8 to 30 times the mass of Earth. Prove that the mass of the cores described in Problem 12a would fall between Earth's mass and the mass of these giant planet cores.

c. Comment on how these simple calculations help explain how the giant planets may have formed.

PART

5

Stars and Their Evolution

A huge, billowing pair of gas and dust clouds are captured in this image of the supermassive star Eta Carinae. Even though Eta Carinae is more than 8,000 light years away, structures only 10 billion miles across (about the diameter of our solar system) can be distinguished. (Jon Morse and NASA)

CHAPTER

15 The Sun: The Nature of the Nearest Star

WHAT THE READER SHOULD WATCH FOR IN THIS CHAPTER

The Sun is the nearest star and the engine for all life on Earth. This chapter discusses the properties of the Sun, starting with a summary of the laws of radiation that lead to the different types of spectra of a hot gas—continuous, emission line, and absorption line. These laws of radiation help us interpret the Sun's spectrum and learn the properties of the Sun. The origin of the Sun's enormous energy output lies deep within the Sun, where hydrogen nuclei (or protons) undergo fusion and release nuclear energy. Knowing the cause of the Sun's energy output, it is possible to calculate the physical conditions throughout the sphere of hot gas. For example, the core where fusion occurs must be at a prodigious temperature of 10 to 15 million kelvins, while the much cooler surface is at 5700 K. Energy flows out from the Sun's core in the form of radiation, and it travels through the outer regions by churning gas motions, called convection. The Sun is not smooth and featureless, but is marked by dark spots and arcs of gas. The gas below the surface is in constant, seething motion. This activity follows a cycle of 22 y, and variations in the Sun have profound effects on the Earth's climate. ■

Much of modern astronomy deals with stars—how they generate their energy, what kinds of light they radiate, how they form, and how they evolve. How can we study such remote objects as the stars? We can study the closest one, which is nearer to the Earth than five of the nine planets. Light from its surface reaches us in only 8 min. Our eyes are dazzled by it. The Earth is bathed in its flow of radiation, washed by the winds of its outer atmosphere, blasted by seething swarms of atoms blown out of it, bombarded by bursts of X rays and radio waves emitted by it. It is our Sun, a million-kilometer ball of hydrogen and helium in the center of the solar system. The Sun's million-kilometer diameter is roughly 10 times Jupiter's size and roughly 100 times the Earth's size (Edberg, 1995).

Humans pondered the stars for many centuries before they realized that the Sun is just another star, and the stars are suns. The Sun is one of very few stars on which we can see surface details. It is so close that we can watch storms develop on its surface and track them as they are carried around it by *solar rotation.* This rotation averages 25.4 d relative to the stars (and 27.3 d relative to the Earth, since the Earth's orbital motion is in the same direction as the solar rotation and must be added in). See Figure 15-1 for a daily sequence of images that show the rotation as tracked by sunspots. As in Jupiter's atmosphere, the equatorial region rotates faster (25 d) than the polar regions (33 d)—proof that the Sun has a gaseous, not solid, surface.

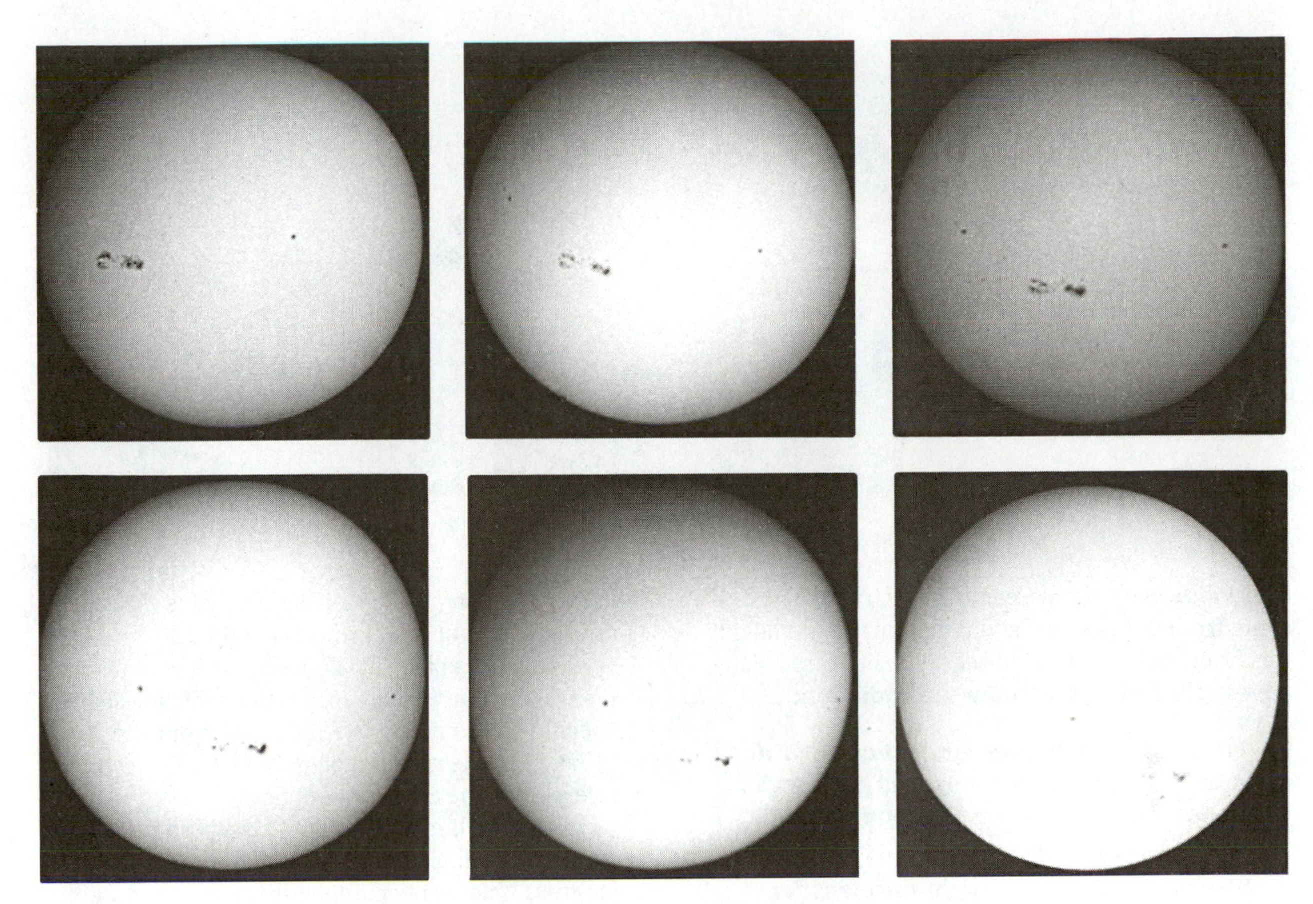

Figure 15-1 Six photos of the Sun in a daily sequence, taken August 21–26, 1971, showing the rotation of the Sun by the movement of sunspot groups. (W. A. Feibelman)

REMOTELY PROBING THE SUN

We have not yet sent space vehicles to land into the Sun's atmosphere, as we have done for planets. To study the Sun or other stars, we must for the present rely on interpreting their light, sampled by means of our telescopes.

The Solar Spectrum

As Newton showed, sunlight is a mixture of all colors. Light of all colors combined appears white to our eyes. However, a prism can array the colors in order of wavelength; this method of producing a spectrum was discussed in Chapter 5. In 1817, German physicist Joseph Fraunhofer found that certain wavelengths were missing from the Sun's spectrum, so the spectrum appeared to be crossed by narrow, dark absorption lines[1] (see Figure 5-10). What were these lines?

[1]For a good review of the early discoveries about solar radiation and its effects on Earth, see Meadows (1984).

Let us review and expand on what we learned in Chapter 5. By the mid-1800s, scientists discovered that when a given element is burned, it emits radiation of certain wavelengths and no others. These very narrow wavelength intervals, unique to each element and as unmistakable as a set of fingerprints, are called the emission lines. Researchers soon found that some of the emission lines exactly matched the position of Fraunhofer's solar absorption lines. Were these two sets of lines caused by the same element?

Kirchhoff's Laws of Radiation

The answer proved to be yes. The proof? In the 1850s, German physicist Gustav Kirchhoff discovered in the laboratory the conditions that produce the three different kinds of spectra described in Chapter 5: the continuum, absorption lines, and emission lines. When Kirchhoff looked through his spectroscope toward a sodium flame against a dark background, he saw a strong red line in emission. But when he changed the background to a brilliant beam of sunlight passing through the same flame, he saw a strong absorption

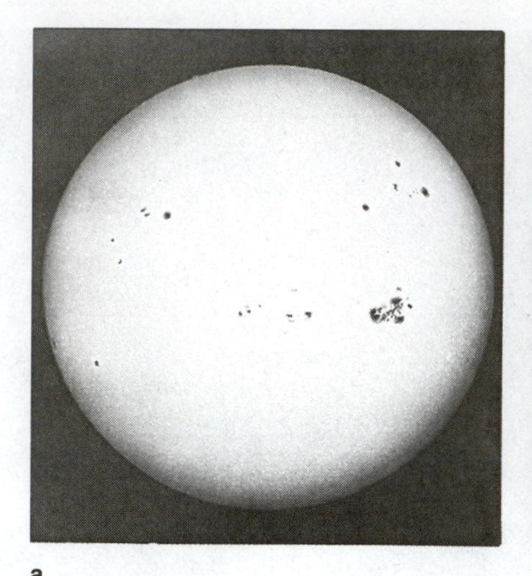

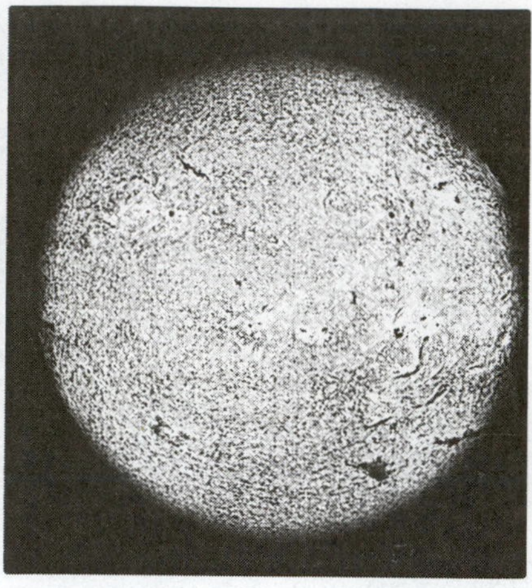

Figure 15-2 a The Sun in normal visible light, with sunspots. Shading around edge, called limb darkening, is caused by the solar atmosphere's absorption of light. **b** The Sun on the same date, photographed by a spectroheliograph in red light emitted by hydrogen. Bright areas involve intense hydrogen emission. (Hale Observatory/California Institute of Technology)

line at the same red wavelength. In each case, the lines came from the gaseous sodium atoms in the flame. Kirchhoff reduced such observations to three statements called **Kirchhoff's laws of radiation:**

1. **A gas at high pressure, a liquid, or a solid, if heated sufficiently, will glow with a continuous spectrum, or continuum.**
2. **A hot gas under low pressure will produce only certain bright wavelengths, called emission lines. Each element emits a characteristic set of emission lines.**
3. **A cool gas at low pressure, if placed between the observer and a hot continuous-spectrum source, absorbs certain wavelengths, causing absorption lines in the observed spectrum.**

We can now discuss physical examples of the electron transitions in atoms that were discussed in Chapter 5. The continuum arises when free electrons are available; this occurs in high-pressure gases, liquids, or solids. Since free electrons are not confined to the discrete or *quantized* energy levels of an atom, they can have any energy and so give a smooth featureless spectrum. The filament of a light bulb is an example. Emission lines arise from electrons inside atoms in an excited state, as in a hot gas. A neon sign is an example. The purity of the color reflects the fact that the radiation is concentrated in a few red emission lines. Absorption lines arise when atoms are in, or near, the ground state, as in a cooler gas. Continuum photons are absorbed at specific wavelengths, corresponding to electron transitions in the cooler intervening gas. Photons of these wavelengths are released again as the excited electrons drop back down into lower energy orbits, but in this case the emission takes place in all directions including back toward the source. The result is that radiation at these specific wavelengths is subtracted from the background source, resulting in *dark* lines. The cool surface layer of the Sun is an example.

Kirchhoff himself found that the absorption lines and emission lines of a given gas have identical wavelengths. What is seen depends on the temperature and density of the gas relative to the radiation coming from behind it, as indicated in the third law. Later an important modification was made to Kirchhoff's laws: An absorption spectrum need not originate *in front of,* or in a cooler gas than, the continuous spectrum. It can arise within the same gas as the continuous spectrum. This is because within a single gas, electrons may be jumping upward in some atoms (making absorption lines) and downward from the free state into other atoms (forming the continuum). The light-emitting layers that form the well-defined visible surface of the Sun are called the **photosphere.**

In summary, spectroscopy allows us to determine many things about the properties of the Sun. Most important, *identification of the spectral lines allows identification of the elements in the Sun.* Application of Kirchhoff's laws proves that the photosphere is a layer of hot gas. Application of Wien's law allows us to use the wavelength of the strongest solar radiation to determine the temperature of the photosphere. Other spectroscopic principles enable astronomers to measure temperatures and pressures at different depths in the gas near the Sun's surface.

Figure 15-2 shows the Sun as viewed in the normal yellow-white light of the continuum, as well as an image made over a narrow range of wavelengths centered on the hydrogen alpha emission line (Hα emission). The Sun contains hydrogen over its whole surface, but the bright regions in Figure 15-2b show where the temperature and density are just right to excite the particular transition in hydrogen that gives the Hα line. Note that while the examples we have given are all in the visible spectrum, emission and absorption lines can be found in the ultraviolet and infrared regions of the electromagnetic spectrum too.

COMPOSITION OF THE SUN

After more than a century of spectroscopic study, the **Sun's composition** is accurately known. It is about 76% hydrogen and 22% helium by mass—roughly the same H/He proportions we found in the giant planets' atmospheres. The heavy elements common in the Earth comprise only 2% of the Sun by mass. The most abundant elements are listed in Table 15-1, which is believed to reflect the bulk composition of material from which the solar system formed.

An interesting episode occurred in 1868, when the French astronomer Pierre Janssen and the English astronomer Norman Lockyer independently found solar spectral lines corresponding to an unknown element. This element, named *helium* (from the Greek *helios,* "sun"), was the first to be discovered in space instead of on the Earth (where it was not observed until 1891). After hydrogen, it is the second most abundant element in the Sun.

TABLE 15-1

Composition of the Sun

Element	Percent Mass of the Sun	Atomic Number
Hydrogen (H)	76.4	1
Helium (He)	21.8	8
Oxygen (O)	0.8	8
Carbon (C)	0.4	6
Neon (Ne)	0.2	10
Iron (Fe)	0.1	26
Nitrogen (N)	0.1	7
Silicon (Si)	0.08	14
Magnesium (Mg)	0.07	12
Sulfur (S)	0.05	16
Nickel (Ni)	0.01	28

Source: Adapted from Anders and Ebihara (1982).
Note: Based on spectroscopic measurements of the Sun and measurements of meteorites and other samples.

SOLAR ENERGY FROM NUCLEAR REACTIONS

Hermann von Helmholtz showed in 1871 that the energy output of the Sun corresponds to the burning of 1500 lb of coal every hour on every *square foot* of the Sun's surface. No ordinary chemical reactions can produce energy at this rate! Thus, Helmholtz realized, the Sun is not burning in the normal sense.

Then what *is* the source of the Sun's heat and light? In the 1920s, astrophysicists realized that the energy of the Sun and other stars comes from **nuclear reactions**—interactions of atomic nuclei near the star's center. Normally, nuclei are protected from interacting by their surrounding clouds of electrons and by the electrical repulsion between the positively charged nuclei. Familiar **chemical reactions,** such as coal burning, involve interactions only between *electrons* of different atoms, far outside the central nucleus. If the temperature and pressure are high enough, the nuclei can be forced to overcome their mutual electrical repulsion and interact. This happens in the Sun's central core.

States of Matter in Stars and in the Universe

When we deal with nuclear reactions and the material inside stars, we are no longer dealing with the familiar forms of matter that make up the solids, liquids, and gases of our planet and our bodies. We are about to make the leap from cold planetary matter to hot stellar matter. This is a good moment to pause and describe the way that matter behaves under a variety of conditions.

There is a simple way to arrange the states of matter—by temperature, as in Table 15-2. Although this arrangement oversimplifies certain effects of pressure and other variables, it is very useful in understanding the forms of material we are dealing with in this book. The table is arranged from the bottom up in order of increasing temperature.

At the lowest temperatures, matter is "frozen" and exists as a solid. This means that the atoms are bonded together, often in a lattice pattern. The bonds are formed by the sharing of electrons (tiny dots) between nuclei. The nuclei consist of protons (larger black circles) and neutrons (white circles). Most of the mass of each planet is in this solid state—the

TABLE 15-2

States of Matter in the Universe

Approximate Temperature Scale	Velocity of Typical Atoms and Molecules	Schematic Chart	State of Matter	Typical Location	Typical Radiation Emitted
60 billion K	Relativistic (appreciable fraction of speed of light)		NUCLEAR FRAGMENTS (nuclear particles collide hard enough to shatter)	Accretion disk near a black hole	Gamma rays X rays
10 million K	500 km/s		BARE NUCLEI (nuclei collide and fuse, causing nuclear fusion reactions)	Core of a star	X rays Ultraviolet light
5000 K	10 km/s		IONIZED GAS (electrons knocked free)	Atmosphere of a star	Visible light
500 K	1 km/s		GAS (separate atoms)	Atmosphere of a planet	Infrared light
300 K	$\frac{1}{2}$ km/s		LIQUID (some atoms linked in chains)	Water	Far infrared light
0 K	0		SOLID (atoms linked in lattice)	Rock	Radio waves

HUMAN REGIME (500 K to 0 K)

crystals that form rocks generally consist of atoms bound together in lattice patterns. Temperature is merely a way of measuring the rate of motions of the atoms. The faster the motions, the higher the temperature and vice versa. At absolute zero temperature, 0 K, atomic particles would have virtually no motion (except for a tiny residual predicted by quantum theory). But as we raise the temperature of a solid, its atoms vibrate faster and faster.

By the time room temperature is reached, around 300 K, the typical atoms in typical substances are moving at around $\frac{1}{2}$ km/s. This activity is sufficient to break many of the atoms loose from their lattice. We perceive this as the substance melting, or turning into a liquid. The liquid oceans of planet Earth are a good example. In the liquid state, chains or groups of atoms may move among each other. By the time we reach another 100 or 200 K higher, the atoms are moving at around 1 km/s and the chains are broken. Now we have created a gas. In a gas, the individual atoms (or molecules, such as H_2O in water vapor) are moving freely. This is the state of matter in the air we breathe. But all these forms of matter are cooler than matter in the Sun or in most stars. If we keep heating the gas, the speeds of the atoms increase. They hit each other harder and harder.

At a temperature of a few thousand degrees, the atoms are hitting each other so hard that they break the electrons free from their orbits. The Sun's surface, for instance, has a temperature a little over 5000 K. There the hydrogen nuclei, many of them stripped of their electrons, move at speeds around 10 km/s. Gas that has had its electrons knocked off is called **ionized gas,** or *plasma.* This is the form of matter not only in the Sun's surface layers but also in a familiar flame. Because the electrons are constantly being bumped off atoms and rejoining them, they are constantly changing energy states and giving off light. Following Kirchhoff's second law, the light of a candle flame is in the form of emission lines. But as we noted above, in accordance with Kirchhoff's first law, the light given off from the high-pressure gas in the surface of the Sun or a star is in the form of continuum.

The insides of stars are even hotter than their surfaces. The center of the Sun, for instance, is at about 15 million kelvins. What happens if we confine the gas (in a container or inside a star) and keep raising the temperature toward such a value? The electrons have all been stripped off, and the bare nuclei of atoms are colliding. As always, when the temperature increases, the particles collide harder and harder. Atomic nuclei might be compared to spitballs: If they just brush together at low speed, they merely bound apart; but if they hit hard enough, they fuse together. Among nuclei, it takes temperatures on the order of 10 million kelvins before the fastest hydrogen nuclei begin to fuse. Typical hydrogen nuclei at this temperature move at around 500 km/s. The merger of nuclei at high temperatures is called *fusion.* The fusion may be among individual protons (black circles in Table 15-2) and neutrons (white circles), or it may be between nuclei to build even larger nuclei with many protons and neutrons.

Fusion of atoms inside stars is one of the most important processes in the universe. We will be dealing with its consequences throughout the rest of this book. The fusion process has the important effect of giving off energy, both in the form of the kinetic energy of particles and in the form of radiation (represented as the wavy lines in Table 15-2). This energy heats the gas, maintains the temperature inside the Sun at 10 to 15 million kelvins, and thus keeps the reactions going. But what happens if we force the temperature still higher in a sample of ionized gas? Eventually, at temperatures of billions of degrees, the particles are moving at nearly the speed of light. Eventually, the collisions are violent enough that the repulsive force between more and more massive nuclei can be overcome. As temperatures increase, progressively heavier elements fuse.

The Nuclear Reactions inside the Sun

Nuclear reactions inside stars do two important things: They generate energy, and they gradually change the star's composition because they build up more and more heavy nuclei. The principal reactions inside the Sun are believed to be a three-part sequence that fuses four hydrogen atoms into a helium atom. Because this chain of reactions starts with two hydrogen nuclei—that is, two single protons—it is called the **proton-proton chain.**

The reaction occurs in three steps. In step 1, two protons collide and fuse. The fusion produces a form of hydrogen nucleus called deuterium, which is designated ^{2}H. Two additional particles are released. The positron is the antiparticle of the electron, identical except for having a positive charge. It is designated e^+ (or sometimes p^+). The **neutrino** is a ghostly particle, with no charge and with very small mass. It is designated by the Greek letter ν, pronounced *nu.* In step 2, the hydrogen nucleus hits another proton and fuses into a form of helium known as helium-3, designated ^{3}He. A photon of radiation is emitted. In step 3, two of the ^{3}He nuclei collide and fuse into the natural form of helium, helium-4, designated ^{4}He. Two protons are left over, and another photon is emitted.

Note that the reaction has done two important things: It has given off energy in the form of radiation of photons, and it has created helium out of the

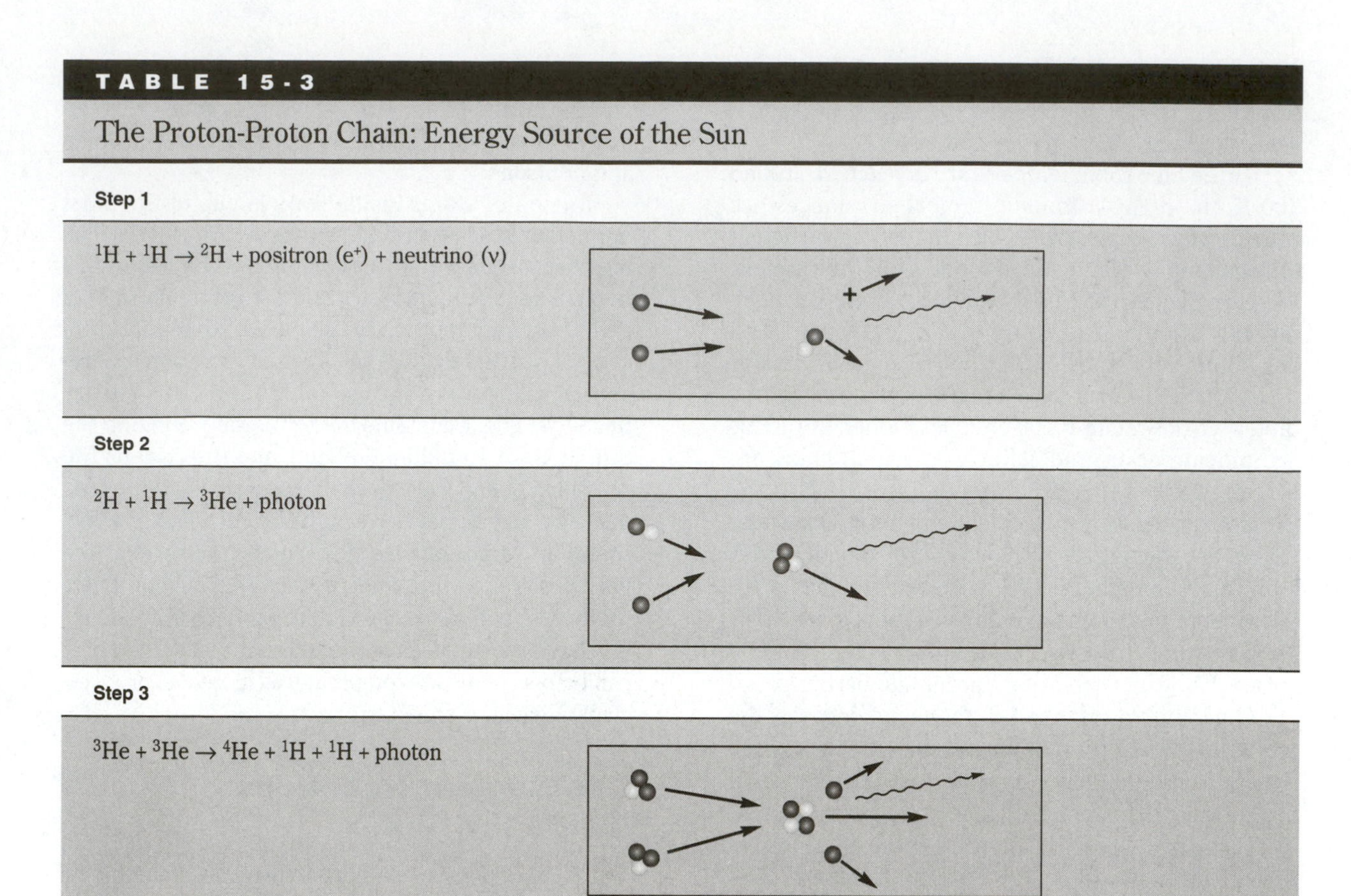

TABLE 15-3

The Proton-Proton Chain: Energy Source of the Sun

Step 1	$^{1}H + ^{1}H \rightarrow ^{2}H$ + positron (e^+) + neutrino (ν)
Step 2	$^{2}H + ^{1}H \rightarrow ^{3}He$ + photon
Step 3	$^{3}He + ^{3}He \rightarrow ^{4}He + ^{1}H + ^{1}H$ + photon

Note: There are other, less significant branches for fusion that are not shown here.

lighter element hydrogen. See Table 15-3 for a summary (see also Hathaway, 1995).

The total amount of mass left at the end of the three-step chain is slightly less than the mass of the initial hydrogen atoms. During the fusion, a small amount of mass m is converted to an amount of energy E, according to Einstein's famous equation $E = mc^2$. (The constant c is the velocity of light, 3×10^8 m/s. When the units are in the SI metric system, m is in kilograms and E is in joules.) In the Sun's fusion sequence, about 0.007 kg of matter is converted into energy for each kilogram of hydrogen processed. This liberates 4×10^{26} J/s inside the Sun, and the Sun radiates this much every second to maintain its equilibrium. This corresponds to 400 trillion trillion watts—which equals a lot of light bulbs!

Every second, the Sun converts 4 million tons of hydrogen into energy and radiates it into space. Long before the Sun can use up all its mass, the solar core will have converted so much hydrogen to helium that there will not be enough hydrogen left in the core to fuel further reactions, and the reactions will stop. According to a recent calculation, the Sun won't run out of hydrogen for about 4 billion years. The consumption of hydrogen inside stars proves the important point that stars are not permanent, but must evolve and run down.

THE SUN'S INTERIOR STRUCTURE

As a result of Helmholtz contraction (see Chapter 14), the temperature at the Sun's center reached 15 million kelvins as the proton-proton reactions became established. This remains the current temperature in the **solar core,** where nuclear energy is generated. Approximate conditions in the layers between the core and the surface can be calculated by using equations that describe the pressure at any depth, the properties of gas under different pressures, and energy generation rates at points inside the Sun. The results can be converted into a schematic view of the cross section of the Sun, as shown in Figure 15-3. However, for reminders that our knowledge of the nearest star is not perfect, see Lang (1996).

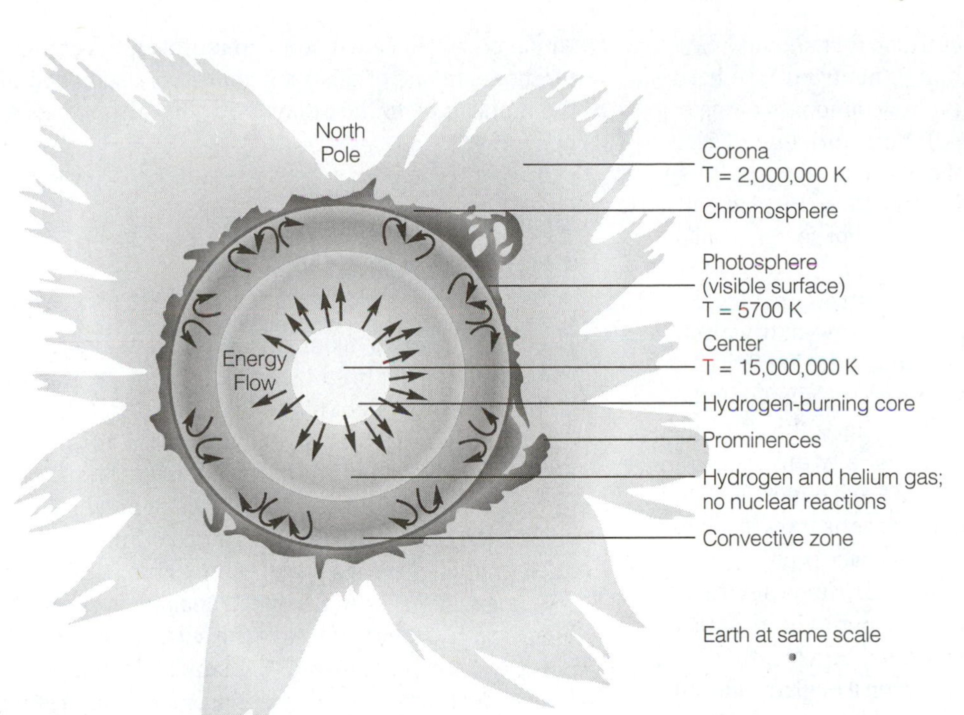

Figure 15-3 A cross section of the Sun and its atmosphere, showing (to approximate scale) the energy-producing core, the outer convective zone, and the tenuous corona. Note the relative size of Earth at the same scale.

These calculations indicate that the gas pressure at the Sun's core is about 250 billion times the air pressure at the Earth's surface. This high pressure compresses the gas in the core to a density of about 158,000 kg/m^3—158 times denser than water and about 20 times denser than iron. If brought to Earth, one cubic inch of this gas would weigh nearly 6 lb! The core of the Sun occupies about the inner quarter of the Sun's radius. This $\frac{1}{64}$ of the Sun's volume contains about half the solar mass and generates 99% of the solar energy.

How Energy Gets from the Core to the Surface

As heat energy always flows from hot to cool regions, solar energy travels outward from the hot core, through a cooler zone of mixed hydrogen and helium, toward the surface. Throughout most of the Sun's volume, this energy moves primarily by **radiation.** That is, the energy radiates through the gas in the form of light, just as electromagnetic radiation travels through our atmosphere. Very little moves by conduction, the mechanism by which a pan on a stove becomes hot.

In the outer part of the Sun, we find a third mechanism of energy transport—**convection.** Convection occurs when the temperature difference per unit length between the hot and cold regions is so great that neither radiation nor conduction can carry off the outward-bound energy fast enough. So-called cells of gas, having become heated enough to expand, become less dense than their surroundings and rise toward the surface, move across the Sun at about 20 m/s, cool by radiating their energy into space (sunlight!), and sink.

Note that energy always moves from the core of the Sun, where energy is being generated, to the surface. At a temperature of 15 million kelvins, the photons produced in the core have a thermal spectrum that peaks in X rays. As the photons travel out, they constantly collide and lose energy. Energy takes about a million years to diffuse to the surface, by which time the temperature has been reduced to a few thousand kelvins. The visible surface of the Sun, or photosphere, corresponds to a low enough density that photons no longer collide on their way out. At this point they travel freely through space. Thus energy takes a million years to diffuse out from the core of the Sun to the surface, but then takes only 8 min to travel the large distance to the Earth!

The Solar Neutrino Puzzle

Although solar astronomers believe they understand the basic structure and energy sources of the Sun, a frustrating problem has arisen in recent years. Step 1 of the proton-proton chain produces a neutrino. From the Sun's total energy production and the rate of reaction of the proton-proton chain, astronomers predicted the amount of neutrinos that should be released by the Sun. Even though neutrinos are hard to detect (they pass through huge amounts of material without being absorbed), detectors to measure neutrino numbers were first constructed in the 1970s. To the surprise of most scientists, these instruments have consistently detected fewer neutrinos than solar theories predict.

The techniques of neutrino astronomy are extremely difficult. Neutrinos interact so weakly with ordinary matter that they pass through the Earth as if it were not there. After neutrinos are produced in the core of the Sun, they leave the Sun within 2 s and streak toward the Earth at essentially the speed of light. Neutrinos therefore offer a great opportunity to "see" directly into the thermonuclear heart of the Sun. The neutrino flux at the Earth's surface is prodigious, about 10^{14} neutrinos/m^2/s. About 10 trillion neutrinos pass through your body every second! They interact so weakly that vast detectors must be assembled to catch the rare interactions between a neutrino and an atomic nucleus. The detectors must be placed deep underground to shield them from contaminating signals due to cosmic ray particles from interstellar space.

The first neutrino experiments were sensitive only to a type of neutrino produced in a minor branch nuclear reaction in the Sun, not the main proton-proton chain. Ray Davis of Brookhaven National Laboratory was the pioneer of this type of experiment, and his original experiment has been running for nearly 30 y deep in a gold mine in South Dakota. Recent experiments are directly sensitive to neutrinos from the Sun's principal nuclear reactions, and they have finally detected the number of neutrinos predicted by solar theories (Bahcall, 2001). However, the surprising result is that neutrinos oscillate or change from one variety to another. Agreement with solar theory has been reached but with the realization that our standard model of particle physics is incomplete.

Solar Oscillations

★ Another way to study the Sun's interior is to measure the way it vibrates. The Sun oscillates and vibrates at many frequencies, like an ocean surface or a bell. The wavelike solar oscillations can be used to infer the interior properties, just as geologists use seismic waves to study the structure of the Earth. The oscillations are seen as volumes of gas near the Sun's surface that rise and fall with a particular frequency. The most well studied oscillation has a 5-min period, during which portions of the Sun's surface move up and down by 10 km. This discovery has created a whole new field called **solar seismology.** A computer model of one of the many modes of oscillation of the Sun can be found at our web site (also see Kennedy, 1996).

THE PHOTOSPHERE: THE SOLAR SURFACE

Energy ascending from inside the Sun heats the photosphere—the bright surface layer of gas that radiates the visible light of the Sun—to a temperature of about 5700 K. The convective motions inside the Sun disturb the surface layers. A detailed view of the solar surface shows granules, each of which is a convection cell 1000 to 2000 km across (Figure 15-4). Each granule rises at a speed of 2 to 3 km/s and lasts only a few minutes. A terrestrial analogy is shown in Figure 15-5. Other features of the surface layer of the Sun are shown in Figure 15-6. Many of the masses of gas being churned on the seething surface of the Sun are as large as the Earth!

If the Sun is a giant ball of gas, why does it appear to have a sharply defined surface? The answer involves the **opacity** of the gas—its ability to obscure light passing through it. The density and temperature of the Sun decrease smoothly and continuously moving outward from the core. Photons are continually colliding with particles, but as they migrate outward, there is a point where the density is low enough

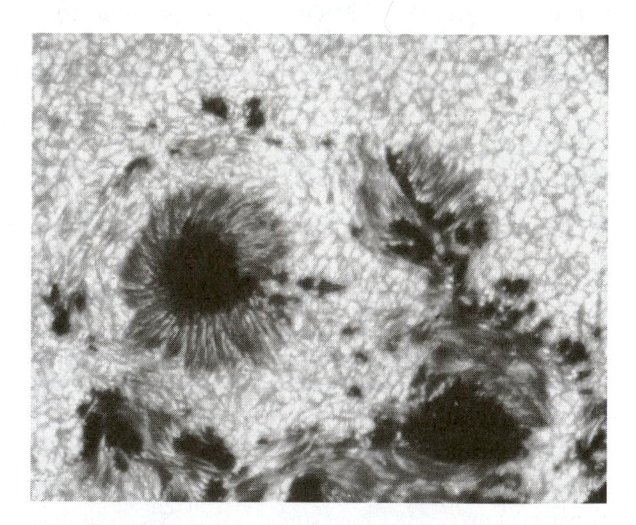

Figure 15-4 The solar surface in the region of a sunspot. Outside the sunspot, the surface is mottled by granules, convection cells in the solar gas. The main sunspots are similar to Earth in size, and the large granules are comparable to continents. (Project Stratoscope photo, supported by NSF, ONR, and NASA)

that no more collisions occur. The photons travel unimpeded to the Earth, and we see an "edge" at that point. A solid probe, if it could survive the 5700-K temperature, could drop directly through the photospheric "surface" and plunge into the Sun, like an airplane passing through the surface of a cloud. Inside a cloud, for example, light bounces off water droplets. The edge of the cloud corresponds to a region where the density of water droplets is low enough that light travels freely outward. As with a cloud, there is no sharp discontinuity in density or temperature at the photosphere of the Sun.

Figure 15-5 View from a plane of convection cells in cumulus clouds over Indiana, showing a convective pattern similar to that seen on a larger scale in the solar gas. (Photo by WKH)

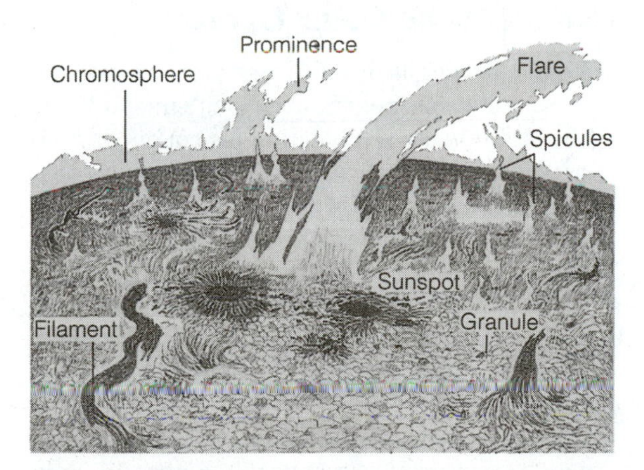

Figure 15-6 A schematic, oblique view of the solar surface showing various features, the most important of which are mentioned in the text. Granules are the result of convective "boiling" of the surface. Flamelike spicules, prominences, and flares shoot up above the surface.

CHROMOSPHERE AND CORONA: THE SOLAR ATMOSPHERE

★ The **chromosphere** (which means "color layer") is a pink-glowing region of gas just above the photosphere, at a temperature of 10,000 K. Its light is mainly the red Hα emission line. The chromosphere can be seen by the naked eye during a solar eclipse. When the Moon covers the rest of the solar disk, this thin outer layer is visible as a ring of small, intense red emission (see image at our web site).

Above the chromosphere is the rarefied, hot gas of the **corona.** Gas in the corona reaches the amazing temperature of 2 million kelvins, due to heating by violent convective motion in the photosphere and chromosphere. As a result of the extreme heat, the gas expands rapidly into space. Whereas the gas density is about 0.001 kg/m^3 within a few hundred kilometers of the photosphere, it drops to 10^{-7} kg/m^3 in the middle chromosphere and to less than 10^{-11} kg/m^3 in the lower corona. Clearly, the corona is only the outermost, tenuous atmosphere of the Sun.

Why are both the chromosphere and the corona hotter than the photosphere? After all, they are farther from the Sun's internal energy source. The answer is that magnetic effects and shock waves from the violent subsurface convection transfer a lot of energy to this gas. The Sun's magnetic field controls motions of gas in the corona, creating delicate streamers (see details in schematic Figure 15-6).

SUNSPOTS AND SUNSPOT ACTIVITY

Although sunspots sometimes can be seen with the naked eye when the Sun is dimmed by fog or a dark glass (Warning: **NEVER** point any telescope or binoculars at the Sun), their nature was not realized until 1613, when Galileo studied them and concluded that they are located on the solar surface and are carried around the Sun by solar rotation. Recall that Galileo spent the last years of his life blind due to his excessive observations of the Sun (for tips on how to view the Sun safely, see Mims, 1990).

The Nature of Sunspots

A **sunspot** is a magnetically disturbed region that is cooler than its surroundings. A sunspot looks dark only because its gases, at 4000–4500 K, radiate less than the surrounding gas at about 5700 K (see

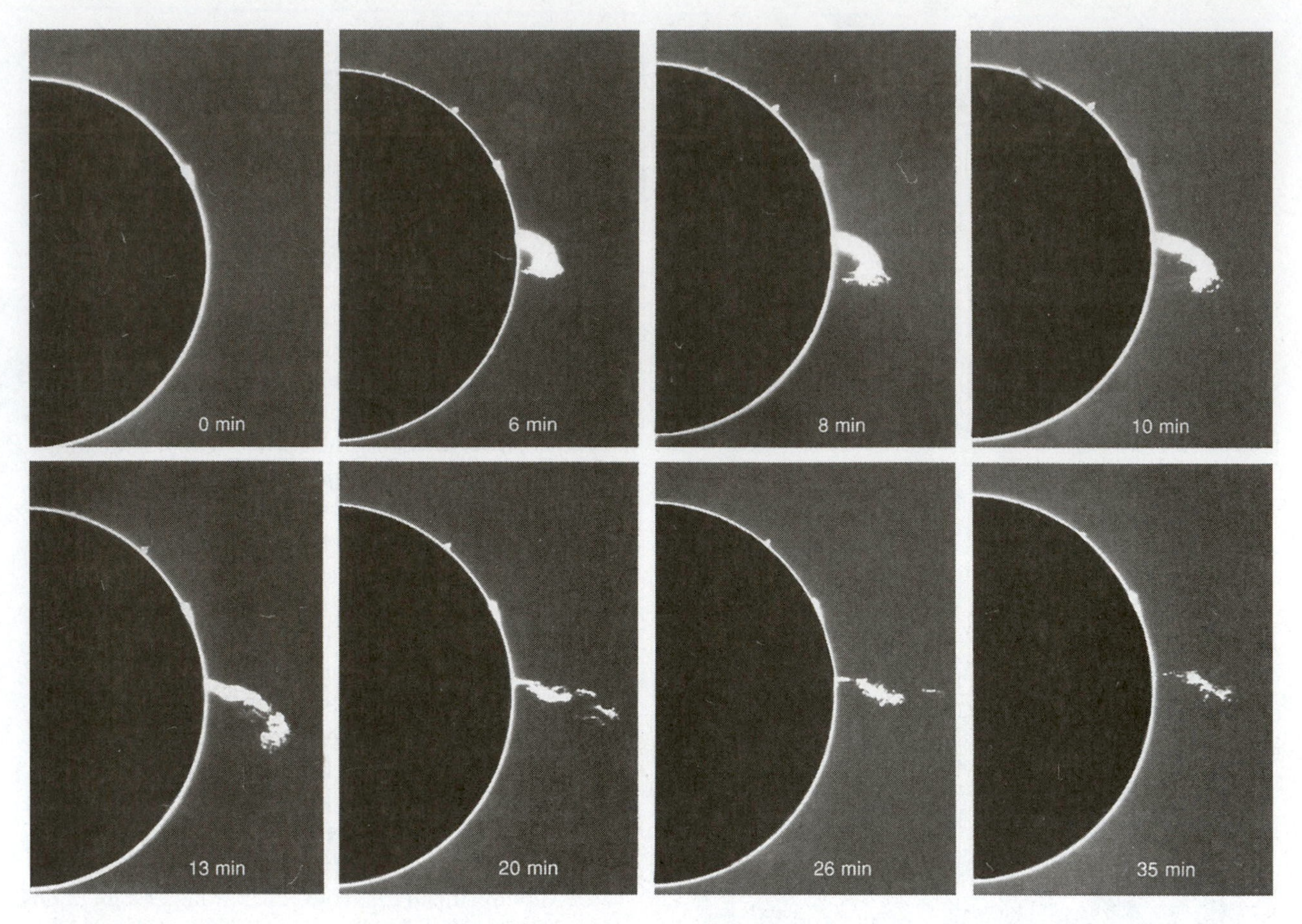

Figure 15-7 Sequence of photos showing an eruptive prominence, or jet of gas, blasting off the Sun over a period of 35 min. The photos were made with a coronograph, which obscures the bright solar disk and allows solar atmospheric activity to be monitored. (National Center for Atmospheric Research)

Figure 15-4). Motions of solar gas near sunspots are controlled not by atmospheric forces, as with terrestrial storms, but by magnetic fields of the Sun. *Ions* (charged atoms or molecules), which are common in the Sun, cannot move freely in a magnetic field, but must spiral around the magnetic field lines and stream from one magnetic pole to the other.

A gas with many ions is called a **plasma,** and unlike a neutral gas, its motions are strongly influenced by magnetic fields. For this reason, plasmas in the sunspots and elsewhere in the solar atmosphere move in peculiar patterns that indicate the twisted patterns of the solar magnetic field. Huge clouds of gas, larger than the whole Earth, erupt from the disturbed regions of sunspots. These *prominences* can be seen when silhouetted above the solar limb, or edge, as seen in Figure 15-7. The largest blasts of material and their very active sunspot sites are called *flares.*

The 22-Year Solar Cycle

Around 1830 an obscure German amateur astronomer, H. Schwabe, began observing sunspots as a hobby. After years of tabulating his counts, in 1851 he announced a **solar cycle:** The number and positions of sunspots vary in a cycle, as shown in Figure 15-8. After Schwabe's observations, previous cycles were recovered by careful study of historical records. This discovery, followed a year later by the discovery that terrestrial magnetic compass deviations exactly follow the same cycle, was a key step in understanding the Sun and its effects on the Earth.

The cycle's duration averages 22 y and consists of two 11-y subcycles, as shown in Figure 15-9. We are currently emerging from a peak in the sunspot cycle. Notice that the number of sunspots visible at any time exhibits large variations, from a typical

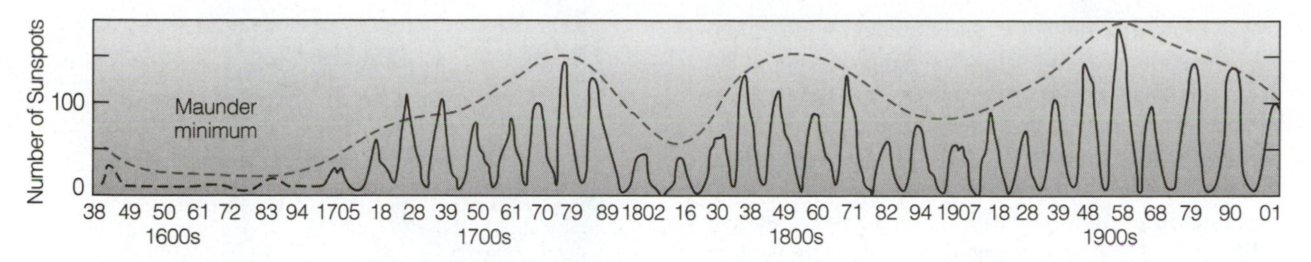

Figure 15-8 Sunspot counts since the 1600s show a cycle averaging 11 y for sunspot numbers (half the 22-y magnetic cycle), with evidence for a longer 80-y cycle (dashed line). The extensive period of low sunspot activity during the 1600s is believed to correlate with climate changes at that time. (Data from M. Waldmeier, Gibson; Pasachoff; National Solar Observatory)

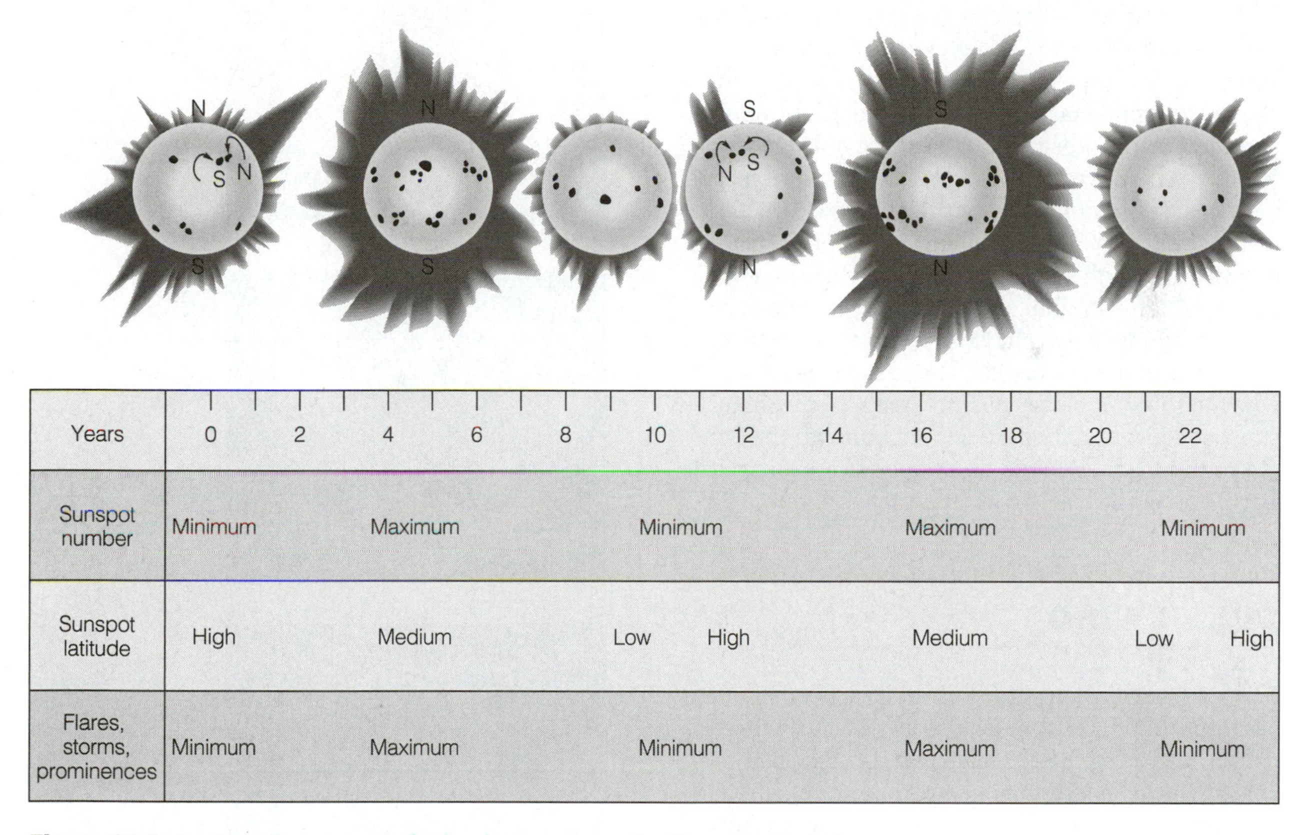

Figure 15-9 A schematic sequence of solar changes during the 22-y cycle. Typical coronal appearances are shown for the minimum and maximum of the cycle in the two drawings at left. (See text for explanation.)

mean of 5–10 up to a typical mean of 100–110. The extremes are even larger—the peak recorded number is 254 and the lowest recorded number is zero!

At a time of *sunspot minimum* (a minimum number of sunspots), the few visible spots are grouped within about 10° of the solar equator. When a new cycle begins in a year or so, groups of new spots appear at high latitudes, about 30° from the solar equator. The spots often appear in pairs, and the eastern spot in each northern hemisphere pair is of specific polarity—for example, a north magnetic pole. (This means that a compass placed in that location would point toward the spot with "north" polarity.) In the southern hemisphere, the polarity is reversed. After a few years, the sunspot number reaches a maximum, and the spots are at intermediate latitudes, about 20° from the solar equator. After about 11 y, the spots appear mostly about 10° from the equator, and a sunspot minimum occurs again.

Now the cycle begins to repeat, except for a noticeable difference. The new spots forming between +30° and −30° have reversed polarity. The eastern spots in the northern hemisphere are now south magnetic poles! Thus it takes another 11 y to complete the full cycle, when all features resume their initial pattern. The most recent sunspot maximum occurred in 2001, and we are currently entering a decline in solar activity.

Still more remarkable is the fact that the magnetic field of *the entire Sun* reverses during each 11-y subcycle; thus the entire Sun participates in the full 22-y cycle. Imaginary observers on the Sun would find their compasses pointing north in one direction for 11 y (subject to disturbances by frequent magnetic storms) and in exactly the opposite direction for the next 11 y. This behavior is not entirely unknown: The Earth's field reverses every few hundred thousand to few million years. Both patterns of reversal may involve cyclic flow patterns in the deep fluid cores of the two bodies. The sunspot cycle is important to us because during years of maximum sunspot activity, solar particles shooting off the Sun affect the magnetic field and upper atmosphere of Earth, disturbing radio communications and causing aurora.

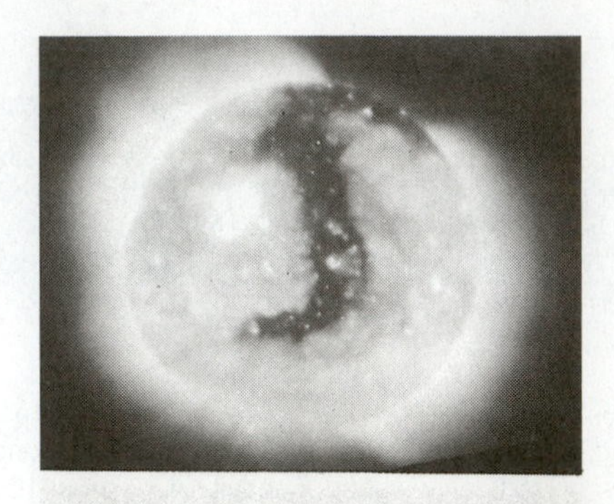

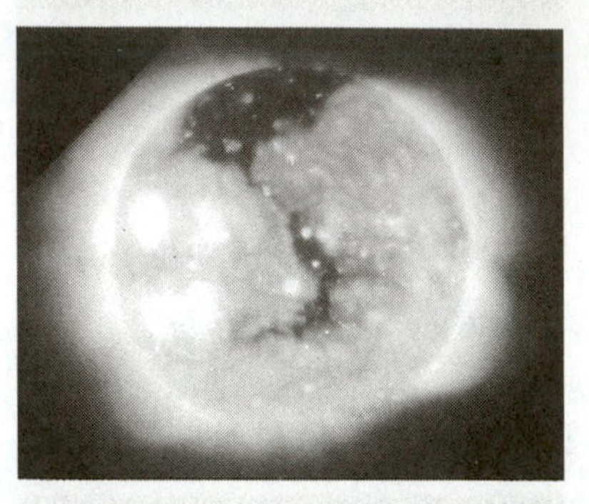

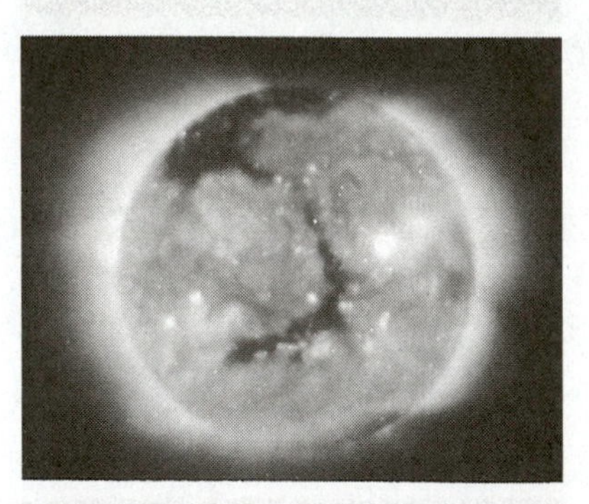

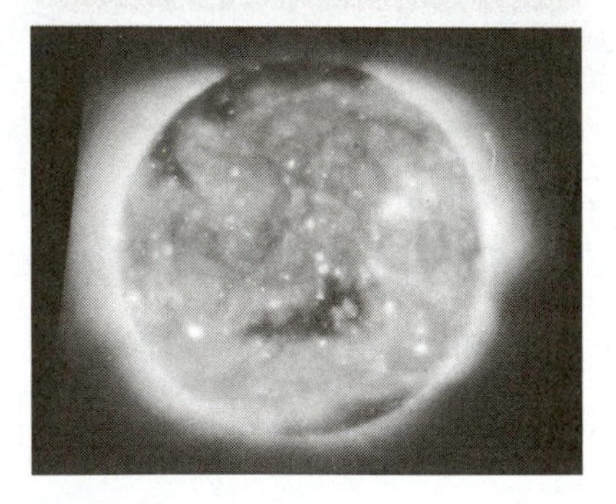

Figure 15-10 The coronal structure of the Sun's atmosphere, photographed by Skylab astronauts using an X-ray telescope. The technique revealed rifts in the coronal structure, through which the cooler (darker) surface gases can be seen. The evolution of such a rift is shown in the sequence. (NASA; American Science and Engineering, Inc.)

SOLAR WIND

Particles blasted out of flares and spots rush outward through interplanetary space. The solar coronal plasma, having been heated to nearly 2 million kelvins by the violence of photospheric convection, also expands rapidly into space (limited only by magnetic forces acting on charged particles). Together these effects cause the **solar wind:** an outrush of gas past the Earth and beyond the outer planets. Near the Earth, the solar wind travels at velocities near 400 km/s and sometimes reaches 1000 km/s. The gas has cooled only to 200,000 K, but it is so thin that it transmits no appreciable heat to the Earth. According to spacecraft data, the solar wind extends significantly farther than Saturn's orbit.

In addition to the solar wind, solar radiation itself exerts an outward force on small dust particles. This effect, which is greater on small particles, is called

Figure 15-11 Surface and atmospheric activity of the Sun. During the June 30, 1973 solar eclipse in Kenya, the outer corona of the Sun was photographed in white light (outer image). About an hour earlier, astronauts in the orbiting Skylab photographed the solar surface (circular inset) using X radiation, showing centers of flare activity. Major streamers in the outer corona are aligned with X-ray flares on the surface. (National Center for Atmospheric Research; American Science and Engineering, Inc.; NASA)

radiation pressure. Together these are the forces that blow comet tails away from the Sun.

AURORAE AND SOLAR-TERRESTRIAL RELATIONS

★ During a solar flare, the total visible radiation from the Sun changes by much less than 1%, but the X-ray radiation may increase by a factor of a hundred. The X-ray photons are energetic because they have very short wavelengths. When they strike Earth's upper atmosphere, they change its distribution of ions and affect radio transmission on the ground. Imaging of the Sun at X-ray wavelengths from above the atmosphere has permitted us to visualize the remarkable appearance of the X-ray Sun, as shown in Figure 15-10, which vividly shows flares and active areas emitting X rays (for a color image, see our web site). The flares shoot material upward into the corona, affecting the coronal structure, as seen in Figure 15-11.

Solar flares emit not only radiation, such as X rays, but also streams of atomic particles, such as protons and electrons. These join and enhance the solar wind. If a flare directs material toward Earth, the enhanced solar wind hits Earth after a few days' travel. Normally, the charged particles of the solar wind are strongly deflected by Earth's magnetic field, as shown on the left side of Figure 15-12. But during solar flares, the surge in the solar wind is often so strong that the Earth's magnetic field is seriously distorted, affecting distributions and motions of charged particles throughout the Earth's vicinity.

The ions near Earth are concentrated into doughnut-shaped regions around Earth, about over the equator. These **Van Allen belts** of radiation were discovered in 1958 by the first artificial satellites. They are shown by the x's in Figure 15-12. They are particularly affected by disturbances in the solar wind. Under normal conditions, as the solar wind sweeps around the outer limits of Earth's outermost atmosphere, or ionosphere, it builds up voltages of 100,000 volts or more between the outer regions of the magnetic field and the atmosphere, driving some charged particles along the magnetic field lines toward the poles (dashed lines in the middle of Figure 15-12).

★ These particles crash into the upper atmosphere, excite the gas atoms there, and cause them to glow. This glow can be seen from the ground; it is called the **aurora** (plural: *aurorae*). The ions crash into the atmosphere only near the magnetic poles, causing intermittent aurorae in the Arctic and Antarctica. These aurorae are also called the northern and southern lights (or aurora borealis and aurora australis, respectively). See our web site for examples.

IS THE SUN CONSTANT?

Solar radiation reaches the Earth at a rate of 1.37 kW/m^2, called the **solar constant.** That is, 1370 J of energy reach every Sun-facing square meter of Earth every second. This rate is called the solar constant because it was once assumed not to vary. Actually, it is not truly constant. Careful measurements of visible sunlight by space probes show variations of about

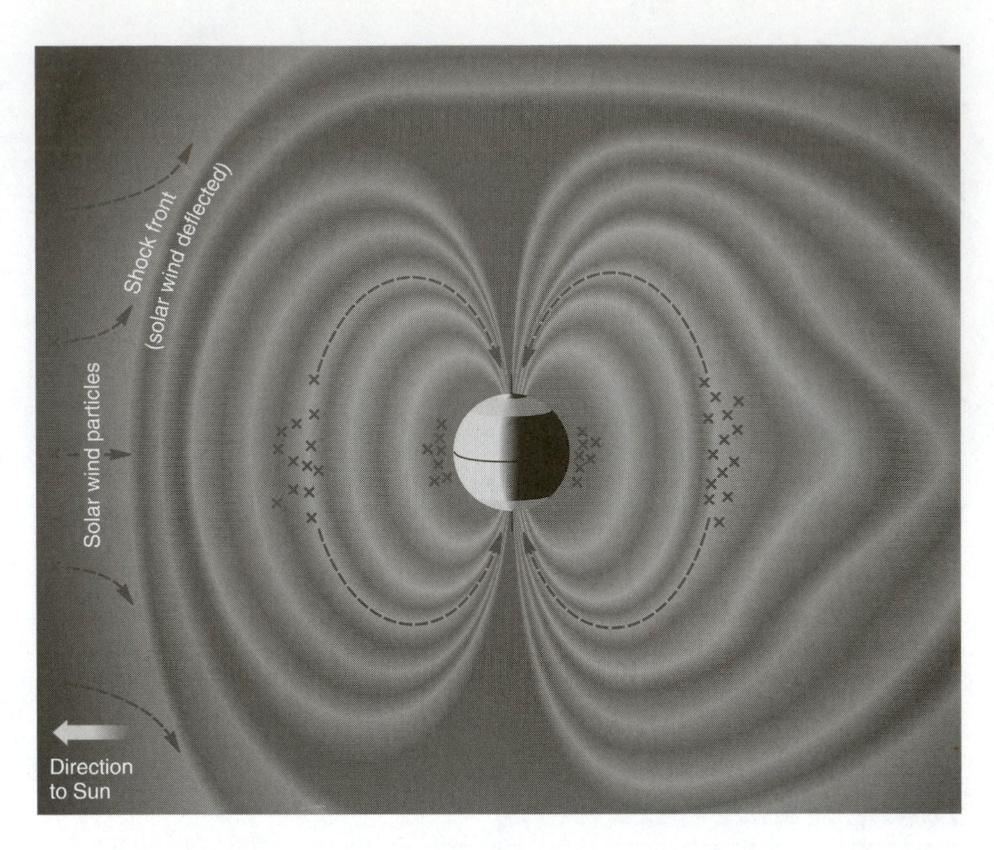

Figure 15-12 Schematic view of the interaction of solar wind ions (from left) with Earth's magnetic field. The shock front is analogous to the bow wave cut by a moving boat. Dashed lines show the typical paths of solar ions (x's). Those ions that penetrate Earth's field accumulate in doughnut-shaped Van Allen radiation belts around Earth. Concentrations of x's mark their positions. Ions in the belts eventually empty into Earth's polar atmosphere, colliding with air molecules and forming auroral zones near the north and south magnetic poles.

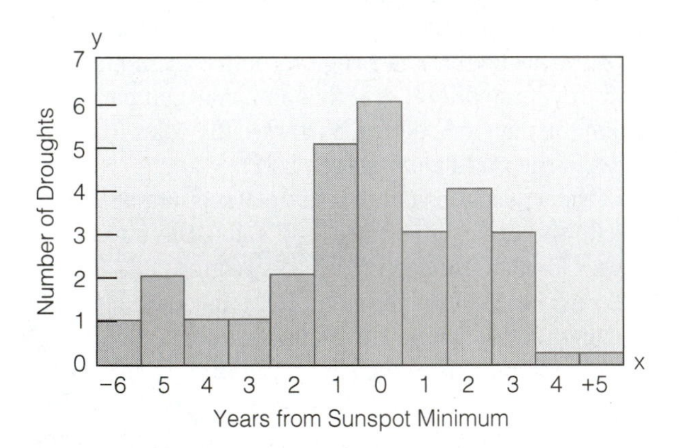

Figure 15-13 A graph of suspected correlation between terrestrial weather and the solar sunspot cycle. Major droughts in Ethiopia, recorded from 1540 to 1985, appear to correlate with years with minimum numbers of sunspots. Such studies may benefit agriculture by helping to predict weather cycles. (Data from Wood and Lovett)

a tenth of a percent per year (Foukal, 1990). During the few days of a single flare, X rays may change dramatically, even though the total solar output barely changes.

Researchers suspect that longer-term changes affect climatic conditions. For example, during the period 1645–1715, sunspot numbers were unusually low (see Figure 15-8). This phenomenon is called the Maunder minimum. Tree-ring patterns and other evidence suggest that the European climate was unusually cold during this period. Similarly, Figure 15-13 shows a correlation between the sunspot cycle and murderous Ethiopian droughts over more than four centuries. Monitoring of solar changes is growing more sophisticated. These changes may seem small, but small changes in the Earth's heat budget can have drastic effects. Changes of only a degree in mean annual temperature can cause dramatic changes in climate and food production. This is why signs of global warming are causing such concern. The whole question of solar influence on climate and agriculture is attracting new research. Solar studies have growing importance for astronomy, meteorology, agriculture, and world economics.

SOLAR ENERGY AND OTHER COSMIC FUELS

Solar energy has accumulated in the ground as coal and petroleum, originally stored in organisms. Industrial civilization has been built with this fossil solar energy, and a region's standard of living is correlated with its rate of consumption of such energy. But while it took about 300 million years to accumulate the fossil fuels, we have burned through a significant fraction of them in only 100 y.

With the Sun providing every daylit square meter in the Earth's vicinity with 1.37 kW, solar energy seems a likely alternative to fossil fuel. A giant test of such technology is the power station Solar One, near Barstow, California. It uses 1818 mirrors to focus sunlight on a central collector, generating 10,000 kW of electricity. More down-to-Earth is the fact that the roof area of a typical home intercepts roughly 1000 to 2000 kWh of energy per day, equaling the average daily energy consumption of an American household. Rooftop solar cell tiles, still under research, might make homes more self-sufficient and relieve energy expenses, which average more than $1500 per year for most households.

In contrast to fossil fuels, energy from basic planetary or astronomical sources might be called **cosmic fuels.** In addition to solar energy, cosmic fuels include nuclear fission energy (used in present-day reactors), nuclear fusion energy (use of solar-type fusion reactions, possibly using water as a source of hydrogen; still in the experimental stage), geothermal energy (heat from the interior of the Earth), and energy from tides and winds. Although nuclear fission energy is a current favorite to get us over the short-term energy hurdle, it has the overwhelming disadvantage of involving and creating deadly radioactive wastes that could wreak health havoc if they escaped into the environment, either by accident or through terrorism. The other cosmic fuels are cleaner than nuclear or fossil fuels. They have an additional subtle advantage: Since they can be collected anywhere in the world, they could make countries energy-independent, allowing a more stable planetary culture.

SUMMARY

The Sun can be studied both observationally and theoretically. Observational studies have yielded information about solar gas revealed by spectral absorption and emission lines created in the solar spectrum as radiation from the Sun's interior passes outward through the layers of solar gas. These studies reveal, among other things, that the Sun is about three-quarters hydrogen; most of the rest is helium, and a few percent is composed of heavier elements.

Theoretical studies reveal that as the Sun formed by contraction of an interstellar gas cloud, it got so hot that atoms at the center collided at high speeds. These collisions cause nuclear reactions in which hydrogen atoms are fused into helium atoms, releasing energy. This nuclear fusion is the source of the Sun's light and heat. The most important reactions in the present-day Sun are a series called the proton-proton cycle.

Transport of this energy from the Sun's center to the outer layers violently disturbs the surface, producing phenomena such as granules, prominences, flares, and sunspots. Particles are shot off the Sun in outward-moving gas called the solar wind, which interacts with the Earth, causing aurorae and other phenomena.

Comparing the Sun with other energy sources used on the Earth shows that we are rapidly consuming our planetary budget of fossil fuels and will soon have to convert to cosmic energy sources, such as solar energy, to maintain our present rates of energy consumption.

CONCEPTS

Kirchhoff's laws of radiation
photosphere
Sun's composition
nuclear reactions
chemical reactions
ionized gas
proton-proton chain
neutrino
solar core
radiation
convection
solar seismology
opacity
chromosphere
corona
sunspot
plasma
solar cycle
solar wind
Van Allen belts
aurora
solar constant
cosmic fuels

PROBLEMS

1. Why do astronomers infer that the Sun's energy comes from nuclear fusion reactions of the proton-proton cycle? How do we know it does not come from chemical burning?

2. If you see a cluster of sunspots in the center of the Sun's disk, how long would the spots take to reach the limb (apparent edge of the disk), carried by the Sun's rotation? How long would it take for the cluster to appear again at the center of the disk?

3. Why is the solar cycle said to be 22 y long, even though the number of sunspots rises and falls every 11 y?

4. Suppose you could make detailed comparisons of the appearance of the Sun at different times. What variations in appearance would you see if the intervals were:

a. 10 min?

b. 1 wk?

c. 5 y?

d. 10 y?

e. 100 million years?

5. Why is a radio disturbance on the Earth likely to occur within minutes of a solar flare near the center of the Sun's disk, whereas an aurora occurs a day or two later, if at all?

6. How much more massive is the Sun than the total of all planetary mass (see Table 8-1 for data)?

7. What is the most abundant element in the solar system? The second most abundant element?

8. If the Earth formed in the same gas cloud as the Sun, why is the Earth made from different material than the Sun?

9. Why will the Sun change drastically in several billion years?

10. Why is the Sun's energy generated mostly at its center and not near its surface?

ADVANCED PROBLEMS

11. Using the small-angle equation, calculate the following:

a. The angular size of a sunspot that has the same diameter as Earth.

b. The angular size of Earth as seen by an imaginary observer on the Sun.

12. At what velocity must particles move to escape from a region of the corona 2 solar radii out from the center of the Sun?

13. Use Wein's law to confirm the temperature estimate for the Sun's surface, based on a maximum energy emission at wavelength 510 nm.

14. a. Calculate the speed of a typical hydrogen ion in a portion of the solar corona with a temperature 2 million kelvins.

b. Compare this with the escape velocity in the corona (from Problem 12). Using the rule of thumb that the fastest atoms move more than three times faster than the typical thermal velocity, comment on the possibility of solar atoms escaping from the corona and thus supplying gas for the solar wind.

PROJECTS

1. According to the principle of the pinhole camera, light passing through a small hole will cast an image if projected onto a screen many hole diameters away. Confirm this by cutting a 1-cm hole of any shape in a large cardboard sheet and allowing sunlight to pass through the hole onto a white sheet in a dark room or enclosure several feet away (3 m—more if possible). Confirm that the projected image is round, an actual image of the Sun's disk. Are any sunspots visible?

2. Cut a 1-cm hole in a sheet of cardboard and use it as a mask over the end of a small telescope. *After* masking the telescope, point it toward the Sun and project an image of the Sun through an eyepiece onto a white card. **Under no circumstances should anyone ever look through the eyepiece at the Sun, since all the light entering the telescope is concentrated at that point and can burn the retina!** Professional supervision of the telescope is suggested. Note also that an unmasked large telescope may concentrate enough light and heat to crack eyepiece lenses.

Are sunspots visible? If so, trace the image on a piece of paper and reobserve on the next day. Confirm the rotation of the Sun by following sunspot positions for several days. Class records kept from year to year can be used to record the cyclic variation of the numbers of sunspots.

CHAPTER

16

Measuring the Basic Properties of Stars

WHAT THE READER SHOULD WATCH FOR IN THIS CHAPTER

Having described the characteristics of the Sun, a typical and well-studied example of a star, we move on to discuss the properties of stars in general. A star's distance from Earth is a key property. Stars have a very wide range of intrinsic brightness or luminosity, so a star that is far away and luminous may appear to have the same brightness as one that is nearby and has low luminosity. Distances are needed to distinguish between these two possibilities. Astronomers measure the distances of stars near the Sun by the parallax technique—a simple application of trigonometry. Spectroscopy yields further information, such as the star's temperature, chemical composition, and motion toward or away from the observer. Other important properties, such as mass and size, cannot be measured directly; they must be inferred from a model of the structure of the star. ■

Now we are ready to take the great leap out of the solar system into the realm of the stars. It is quite a leap—the next star beyond the Sun is 260,000 times as far as the Sun and 6800 times as far as Pluto. If we make a model solar system the size of a half-dollar, the neighbor stars would be dots smaller than a period in this book, scattered about 100 meters apart.

Because of optical limitations and turbulence in the Earth's atmosphere, the world's largest telescope cannot distinguish details smaller than a few hundredths of a second of arc (about 10 millionths of a degree). But the stars with the largest apparent angular size are no larger than about 0.06 arc second. So the disks and surface details of nearly all distant stars are hidden from us, in contrast to the great detail we can study on the Sun.

In spite of these limitations, we know there are giant stars bigger than the whole orbit of Mars, stars the size of the Earth, and stars the size of an asteroid. There are red stars and blue stars. There are stars of gas so thin you can see through parts of them, and stars with rocklike crusts with properties similar to diamonds. There are stars that are isolated spheres, stars with rings around them, and stars that are exploding. All of these bodies fit the definition of stars: **Stars** are objects held together and powered by gravity, with so much central heat and pressure that energy can be generated in their interiors by nuclear reactions (Kaler, 1992). The most familiar stars visible in the night sky are balls of gas with solar composition and sizes usually a few times smaller or larger than the Sun.

How can we know all these details about stars if we cannot even see their disks in telescopes? In the next chapters, we will describe the details of familiar and unfamiliar types of stars, but first we will describe *how* we learn about them.

NAMES OF STARS

Stars are named and cataloged by several systems. Because Ptolemy's *Almagest* was passed on by Arab astronomers, many of the brightest stars ended up with Arabic names. Since *al-* is the common Arabic article, many star names start that way: Algol, Aldebaran, Altair, Alcor. Other scientific *al-* words also have Arabic origins: *algebra, alchemy, alkali,* and *almanac.* Stars in constellations are cataloged in approximate order of brightness using Greek letters. Thus the brightest star in the constellation of the Centaur is called Alpha Centauri (α Centauri). Fainter stars or stars with unusual properties are often known by English letters followed by constellation names or by catalog numbers, such as T Tauri or BD +4° 4048.

DEFINING STELLAR DISTANCES: LIGHT-YEARS AND PARSECS

The vast distances that separate stars and make them so hard to observe are awkward to express in ordinary units. Astronomers use units appropriate to these distances. The easiest to understand is the **light-year** (abbreviated *ly*), the distance light travels in 1 y, which is about 6 million million miles, or 10^{16} m. Remember that the light-year is a unit of *distance,* not time. The common mistake of using light-year as if it were a unit of time is like saying that the ball game lasted for 2 mi. (Even noted space pilot Han Solo confused distance and time units when he claimed in *Star Wars* that the Millennium Falcon could "make the Kessel run" in less than 12 parsecs.)

The nearest star beyond the Sun, Proxima Centauri (which is in orbit around Alpha Centauri), is about 4.3 ly away. The Sun could be said to be 8 light-minutes away. The North Star, Polaris, is about 650 ly away. Polaris' light takes about 650 y to reach us, so we are seeing it now as it was in the 1300s! If Polaris had suddenly exploded in 1950, we would not know it until about A.D. 2600. Astronomers more commonly use a still larger unit of distance called the **parsec** (abbreviated *pc*):

$$1 \text{ pc} = 3.26 \text{ ly} = 3 \times 10^{16} \text{ m}$$

The term *parsec* refers to a distance that produces a *par*allax shift of 1 arc *sec*ond. Parallax was introduced in Chapter 2 and its use to measure stellar distances will be discussed later. In this book, we will use the parsec and its multiples (kiloparsecs and megaparsecs) to express cosmic distances as we move to more remote parts of the universe. You can convert parsecs to light-years approximately by multiplying by 3. Another convenient fact to remember is that near the Sun, stars are roughly a parsec apart. For instance, Alpha and Proxima Centauri, the closest stars to the solar system, are about 1.3 pc away.

DEFINING A BRIGHTNESS SCALE

A nearby flashlight may *appear* to be brighter than a distant streetlight, but in *absolute* terms the flashlight is much dimmer. This statement contains the essence of the problem of stellar brightness. A casual glance at a star does not reveal whether it is a nearby glowing ember or a distant great beacon. Hence astronomers distinguish between **apparent brightness** (the brightness perceived by an observer on Earth) and **absolute brightness** (the brightness that would be perceived if all stars were magically placed at a standard distance).

Apparent Brightness

Let us look first at apparent brightness. In this book, we will use a decimal system to calculate units of brightness difference. Astronomers, however, are victims of history. When Hipparchus cataloged 1000 stars in about 130 B.C., he ranked their apparent brightness on a *magnitude* scale of 1 to 6, with 1st-magnitude stars the brightest and 6th-magnitude stars the faintest visible to the naked eye. A difference of one magnitude corresponds to a factor of roughly 2.5 in apparent brightness; a difference of five magnitudes represents a factor of 100 in brightness. This nonlinear scale roughly matches the response of the eye, which is a logarithmic detector. The 2100-y-old magnitude system is so ingrained that astronomers continue to use it; however, we will use a more sensible system based on a linear and decimal brightness scale.

Table 16-1 lists the relative brightness of various objects, referenced to the bright star Vega. As the table shows, the brightest star in the night sky is 11 billion times fainter than the Sun, and the best telescopes in space can detect objects 800 million times fainter than the eye can see!

TABLE 16-1

Apparent Magnitudes of Selected Objects

Object	Apparent Magnitude	Apparent Brightness (relative to Vega)
Sun	−26.5	4×10^{10}
Full Moon	−12.5	100,000
100-W light bulb at 100 m	−11.1	27,700
Venus (at brightest)	−4.4	58
Mars (at brightest)	−2.7	12
Jupiter (at brightest)	−2.6	11
Sirius (brightest star)	−1.4	3.6
Canopus (second brightest star)[a]	−0.7	1.9
Vega	0.0	1.0
Spica	1.0	0.4
Naked-eye limit in urban areas	4	0.025
Uranus	5.5	0.0063
Naked-eye limit in rural areas	6.5	0.0025
Bright asteroid	6	0.0040
Neptune	7.8	0.0008
Limit for typical binoculars	10	10^{-4}
3C 273 (brightest quasar)	12.8	8×10^{-6}
Limit for 15-cm (6-in.) telescope	13	6×10^{-6}
Pluto	15	1×10^{-6}
Limit for visual observation with largest telescopes	19.5	2×10^{-8}
Limit for CCDs with largest telescopes	28	6×10^{-12}
Limit for Hubble Space Telescope	29	3×10^{-12}

Source: Data from Allen (1973).

[a]This lesser-known southern Hemisphere star is used as a prime orientation point for spacecraft. A small light detector on spacecraft is called the *Canopus sensor.*

To be meaningful, apparent brightness must be specified at a particular wavelength. Stars have different colors, which means that the apparent brightness depends on the wavelength. Also, light detectors (the eye, photographic films, and electronic CCDs) have different sensitivities to different colors or wavelengths. For this reason, astronomers specify exactly what wavelength any set of measurements refers to. Standards have been derived to express apparent brightness measured in blue light, red light, infrared light, radio waves, X rays, and so on. Here we will usually be referring to a system having the same color sensitivity as the human eye, which is sometimes called *visual apparent brightness* (centered on the green part of the visible spectrum).

Absolute Brightness

We have just discussed the *apparent* brightness of stars and other objects as seen from Earth. This depends on the star's distance and thus doesn't express the star's true energy output. The true, or intrinsic, energy output is called the star's *absolute brightness* or, equivalently, the **luminosity.** Astronomers compare the brightness of stars using a standard reference distance of 10 pc (Figure 16-1 with d set to 10 pc). The exact reference distance is not important, just as we could have chosen to refer all the values of brightness in Table 16-1 to an object other than Vega. The relationship among apparent brightness, absolute brightness, and distance is fundamental to astronomy.

Note that the distinction between apparent and absolute brightness would not be important if all stars were the same. Imagine a large darkened room scattered with 100-W light bulbs. Assuming you had measured the distance and absolute brightness (the "wattage") of just one reference light bulb, you could deduce the distances of all other light bulbs by observing their apparent brightness and applying the inverse square law. For instance, a light bulb that *appeared* to be four times brighter than the reference bulb must be two times closer, and a light bulb that *appeared* nine times fainter than the reference bulb must be three times further away. In this situation, apparent brightness is a good indicator of distance.

In reality, stars differ enormously in luminosity or the amount of energy they emit each second. Imagine now that the large room is scattered with bulbs of widely different wattage, from 1-W nightlights to 10,000-W arc lamps. In this case, a 1-W bulb would have the same apparent brightness as a 100-W bulb 10 times further away, or a 10,000-W bulb 100 times further away. These calculations are simply applications of the inverse square law. Apparent brightness is therefore a very poor indicator of distance. Conversely, the intrinsic energy output of a star cannot be calculated without knowing the distance. Astronomers must deal with this situation when studying stars as different as red giants and white dwarfs that are widely distributed through space.

Figure 16-1 A schematic figure that illustrates the difference between the apparent and intrinsic brightness of a star. The arrows represent light rays emerging from the star in all directions. The total number of rays gives the absolute brightness or luminosity of the star. The apparent brightness of the star is given by the number of rays that penetrate a particular area (usually a unit area or m^2) at a distance *d* from the star. This number of rays, and therefore the apparent brightness, diminishes with the square of the distance *d* (see Figure 4-1 on the inverse square law). At a fixed distance, the apparent brightness of a star varies with the absolute brightness. If the distance is known, and the apparent brightness is measured, the intrinsic brightness of the star can be determined using the inverse square law.

MEASURING DISTANCES TO NEARBY STARS

From what you have read so far, can you think of a way to measure the distance of a star? The question is not far-fetched, since star distances were first measured more than a century ago. A crude method may be applied to the information in Table 16-1. The brightest stars have values of *apparent* brightness about 10 billion (10^{10}) times fainter than the Sun (Kaler, 1991). In 1829, the English scientist William Wollaston used this simple fact to estimate that most typical stars must be at least 100,000 (10^5) times more distant than the Sun, since the inverse square law indicates that dimming 10 billion times corresponds to increasing distance 100,000 times.

Because this gives only a typical distance, we need a better technique to measure distances of individual stars. The most important such technique is *parallax* measurement. You will recall that parallax is the apparent shift in the position of an object caused by a shift in the observer's position.

The principle is easy to understand. Hold your finger in front of your face and look past it toward some distant objects (see Figure 16-2). Your finger represents a nearby star; the objects represent distant stars. Your right eye represents the view on one side of the Sun. Your left eye represents the view 3 mo later, after the Earth has traveled to a point lined up with the Sun (a shift in the Earth's position by 1 AU as seen from the star). First wink one eye and then the other. Your finger (the nearby star) seems to shift back and forth. Hold your finger only a few centimeters from your eyes; the shift is large. Hold your finger at arm's length; the shift is smaller. Likewise, the farther away the nearby star, the smaller the parallax. The parallax in this experiment may be measured in degrees, but the parallaxes of actual stars are all less than a second of arc. Parallaxes as small as $\frac{1}{100}$ second of arc have been reliably measured by spacecraft, and the distance of such stars is 100 pc.

Knowing accurate distances is the prime requirement of measuring most other properties of stars. Thus the estimated 20,000 to 25,000 stars that lie within 100 pc are our main statistical sample for mea-

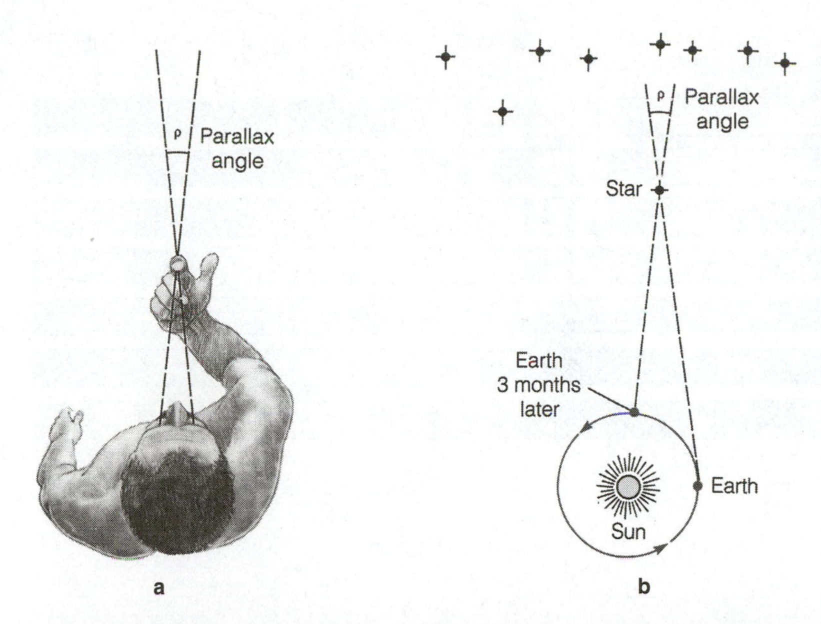

Figure 16-2 The principle of parallax determination applied to one's finger **a** and to stars **b.** Nearby objects observed from two positions appear to shift by a parallactic angle, ρ, compared with background objects; measurement of ρ gives the distance of the object, once the separation of the two positions is known. (By astronomical convention, the angle cataloged as a star's parallax corresponds to the angle diagrammed here, but astronomers actually observe from two positions 6 mo apart. This yields twice as large a parallactic shift, which is correspondingly easier to measure.)

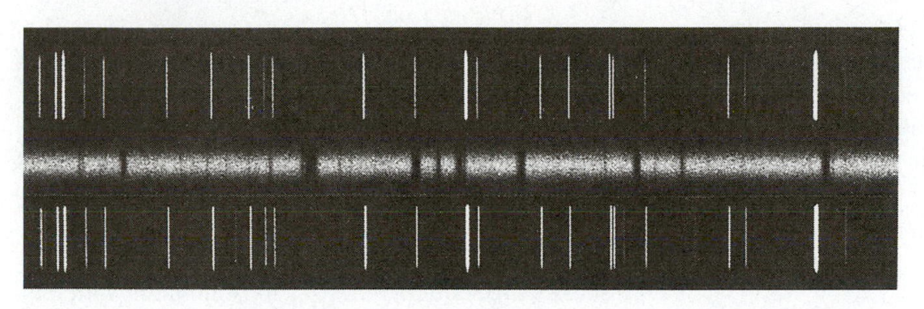

Figure 16-3 The stellar spectrum is the fuzzy horizontal band in the center, crossed by dark vertical absorption lines. It is flanked at the top and bottom by the specific set of emission lines of a known element (usually neon or iron) produced by a discharge lamp inside the instrument. These lines have known wavelengths (wavelength is the horizontal scale, with blue usually at the left, as in this case), so they form a reference scale that permits measurements of the wavelengths in the stellar spectrum.

suring stellar properties. Note that even this large sample is not sufficient for us to be able to measure the distance to certain rare stellar types. Other techniques have been devised for estimating distances to more remote stars, but they all ultimately depend on the accuracy of the parallax measures of nearby stars.

BASIC PRINCIPLES OF STELLAR SPECTRA

The last chapter showed that the Sun's **spectrum,** or distribution of light into different colors (*wavelengths*), gives information about the Sun's atmosphere, surface layers, and interior. The same is true of other stars. **Spectroscopy**—the study of spectra—is a vital tool for understanding the physical properties of astronomical objects, and many astronomers devote their entire careers to it. Chapter 5 gives more details on the principles of spectra, and you may wish to review that material while studying this chapter. Let us see how spectra can be used to reveal the nature of stars.

★ As seen in Figure 16-3, a *spectrograph* produces an image of the spectrum. The intensities of the image represent intensities of light, or radiant energy. Usually, the blue end is to the left, as in the figure. In research applications, photographic spectra have been superseded by spectra taken with electronic detectors, or CCDs. **Absorption lines** appear as dark vertical lines on an image of the spectrum

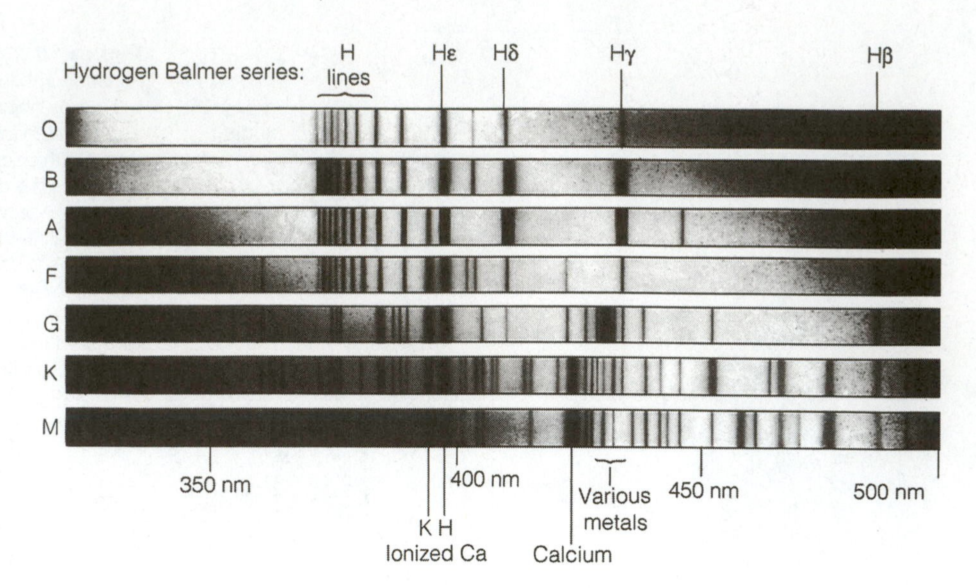

Figure 16-4 Representations of spectra of stars in the seven main spectral classes. Classes A and B show the Balmer series of hydrogen lines (see the caption to Figure 16-5) especially well (compare with Table 16-2). The recently discovered classes L and T are too cool to produce substantial flux in the visible spectral range.

and as notches or valleys on a graph. **Emission lines** would appear as bright vertical lines or as sharp peaks on the graphical version of the spectrum (most stars do not show emission lines). The general level of brightness between absorption or emission lines is the **continuum.** (See our web site for an example of the spectrum of a sun-like star in color and in the form of a chart.)

SPECTRA OF STARS

In 1872, Henry Draper, a pioneer in astronomical photography, first photographed stellar spectra. This represented a tremendous advance. Instead of sketching or verbally describing spectra, astronomers could directly record, compare, and measure them. Spectra of thousands of stars became available for precise analysis.

Such massive amounts of data required a classification scheme. This was begun in the 1880s by Harvard astronomer Edward C. Pickering and completed by Annie J. Cannon and a group of young women assistants who invented a system of **spectral classes** based on the number and appearance of spectral lines (Welther, 1984). The classes—A, B, and so on—started with spectra with strong hydrogen lines. When Annie Cannon published the Henry Draper Catalog (1918–24), it contained spectral data on 225,320 stars and became the basis for all modern astronomical spectroscopy.

Further work showed that the classes had to be rearranged to bring them into a true physical sequence based on temperature, as shown in Figure 16-4. The sequence finally adopted begins with the hottest stars—class O—which show ionized helium lines in their spectra. The sequence of classes is O, B, A, F, G, K, and M. The M stars are the coolest in the traditional classification (MacRobert, 1992), although L and T stars have recently been identified using infrared techniques. (The rearrangement of the letters makes the sequence harder to remember, which is another example of the historical baggage that astronomers carry around.) The coolest spectral types emit very little energy in visible light; most have been found by the 2MASS (2-Micron All-Sky Survey), which selects stars at wavelengths four times longer than visible light. L and T stars have molecules and dust grains in their atmosphere. They can be thought of as transitional objects to planets.

About 99% of all stars can be classified into these groups. For finer discrimination, the classes are subdivided from 0 to 9; in this case, a B9 star much more closely resembles an A0 star than it does a star of spectral class B0. The Sun, for example, is classified as a G2 star.

Spectral Lines: Indicators of Atomic Structure

In Chapter 5, we gave a simplified description of how emission and absorption lines depend on the orbital structure of electrons in atoms. Now we look at the process in more detail by considering the simplest and most abundant type of atom in the universe: hydrogen. As shown in Figure 16-5, hydrogen atoms' orbits can be numbered. Since a neutral hydrogen

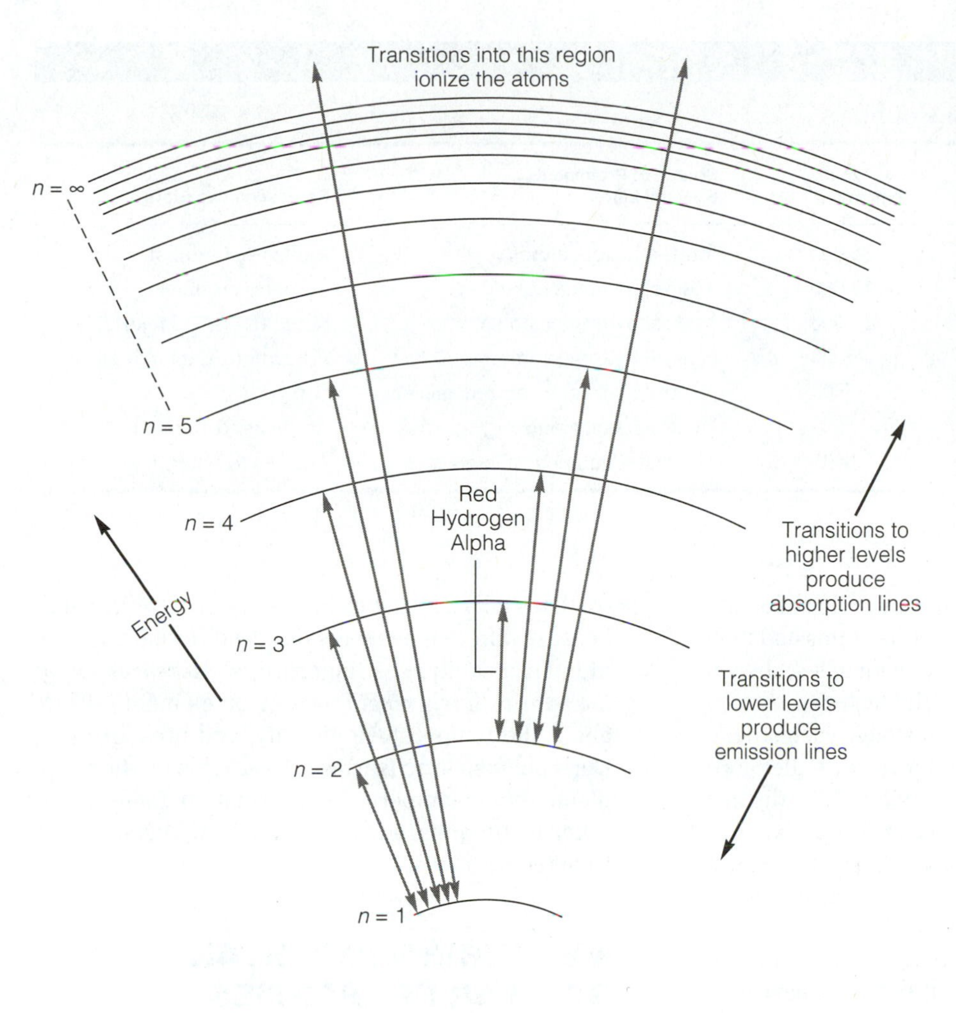

Figure 16-5 Schematic diagram of a hydrogen atom showing different possible electron orbits or energy levels ($n = 1, 2, 3 \ldots$). Each change of orbit by an electron produces a spectral line, because energy is removed or added to the light beam. The series of lines starting from or ending on $n = 1$ occur in the ultraviolet part of the electromagnetic spectrum and are called the Lyman series. Lines starting from or ending on $n = 2$ occur in the visible part of the electromagnetic spectrum and are called the Balmer series. The well-known red Hα line of the Balmer series involves transitions between $n = 2$ and $n = 3$.

atom has just one electron, an atom in the ground state would have one electron in the orbit $n = 1$. If the atom had been bumped by other atoms or if it had absorbed radiation, the electron might be in the orbit $n = 2, 3$, and so on. Further absorption of energy might cause it to jump from $n = 3$ to $n = 4$, creating an absorption line, or it might spontaneously revert from $n = 3$ to $n = 2$, creating an emission line. Each possible transition (1 to 2, 2 to 3, 4 to 2, and so on) creates a different line. As Figure 16-5 shows, transitions between the $n = 2$ level and higher levels create the lines prominent in the visible part of the electromagnetic spectrum. The famous line *hydrogen alpha* lends its brilliant red color to many astronomical gases, including the solar chromosphere and many nebulae.

Very small amounts of energy are required to raise the energy level of an electron in a hydrogen atom or to ionize it completely. The hydrogen within a star is completely ionized. In diffuse hydrogen in interstellar space, electrons can change energy level either due to radiation that permeates the gas or due to collisions between hydrogen atoms. The calculation of the expected strength of each emission line in atoms of each element can be very complex. Emission line strengths depend on the temperature and density of the gas, on the presence of dust, and on details of the atomic physics.

Spectral Classes as a Temperature Sequence

Each spectral class corresponds to a different temperature in the gases in the light-emitting layers of the stars. **Temperature** is merely a measure of the

TABLE 16-2

Principal Spectral Classes of Stars

Type	Spectral Class	Typical Temperature (K)	Source of Prominent Spectral Lines	Representative Stars
Hottest, bluest	O	30,000	Ionized helium atoms	Alnitak (ζ Orionis)
Bluish	B	18,000	Neutral helium atoms	Spica (α Virginis)
Bluish-white	A	10,000	Neutral hydrogen atoms	Sirius (α Canis Majoris)
White	F	7000	Neutral hydrogen atoms	Procyon (α Canis Minoris)
Yellowish-white	G	5500	Neutral hydrogen, ionized calcium	Sun
Orangish	K	4000	Neutral metal atoms	Arcturus (α Boötes)
Coolest, reddest	M	3000	Molecules and neutral metals	Antares (α Scorpii)

average velocity of the atoms or molecules of a substance. The hotter a gas, the faster its atoms and molecules move. If the temperature quadruples, the velocities of all particles double. The faster the atoms collide, the more they disturb or dislodge each other's electrons. Furthermore, the hotter the gas, the more radiation it emits; the resulting photons also disturb electron structures of atoms. Because spectral lines depend on electron structures, and because spectral classes depend on spectral lines, spectral classes therefore form a temperature sequence.

Consider what happens as we reduce the temperature of matter. The hottest stars, the O stars, have temperatures of 40,000 K or more, although a more typical temperature is 30,000 K, as shown in Table 16-2. Here the atoms and radiation are so energetic that even the most tightly bound atoms, such as helium, have had their electrons knocked off, as the table indicates. By class B, the temperature has dropped to around 18,000 K, and helium atoms keep their electrons. By classes A and F, even relatively weakly bound hydrogen keeps its electrons. Many metals have at least one electron that can be very easily knocked off the outer part of the atom, and in G-type stars like the Sun, at around 5500 K, lines of ionized metals are prominent along with Balmer lines of hydrogen. (Recall that astronomers refer to all elements heavier than the two most simple and common, hydrogen and helium, as "metals.") By class K, at 4000 K, most metals are neutral. By class M, at 3000 K, energies are so low that different atoms can stick together into molecules. Even water molecules have been identified in spectra of such cool stars, where, of course, they take the form of steam!

Note the power of the spectroscopic techniques we have discussed so far. *Identification* of the spectral lines tells us what elements are present in a star, because different elements display different patterns of lines, as unique as fingerprints. *Measurement of the relative line strengths* of a given element tells us about the temperature, density, and pressure in a star's photospheric (surface) layer. This method supplements the temperature information gained from other techniques, such as Wien's law, discussed in Chapter 5.

MEASURING IMPORTANT STELLAR PROPERTIES

We now can apply simple physical principles, such as the Stefan-Boltzmann law, to explain how properties of remote stars can actually be measured. These applications, to be discussed in the following sections, are summarized in Table 16-3 (see also Davis, 1991).

Measuring Stellar Luminosity

The luminosity of a star is its absolute brightness—the total amount of energy it radiates each second. Unlike apparent brightness (also discussed earlier in this chapter), luminosity is intrinsic to a star. The most useful concept of luminosity is **bolometric luminosity**—the total amount of energy radiated each second in all forms at all wavelengths. Since many stars radiate primarily visible light, visual luminosity and bolometric luminosity are often roughly the same. Note however, that objects much hotter than the Sun radiate primarily at ultraviolet wavelengths, and objects much cooler than the Sun radiate primarily at infrared wavelengths. In these cases, visual luminosity represents only a small part of the bolometric luminosity.

TABLE 16-3

12 Basic Properties of Stars and How They Are Measured

Property	Method of Measurement
Distance	Trigonometric parallax
Luminosity (absolute magnitude)	Distance combined with apparent brightness
Temperature	Color or spectra
Diameter	Luminosity and Temperature
Mass	Measures of binary stars and use of Kepler's laws
Composition	Spectra
Magnetic field	Spectra, using Zeeman effect
Rotation	Spectra, using Doppler effect
Atmospheric motions	Spectra, using Doppler effect
Atmospheric structure	Spectra, using opacity effects
Circumstellar material	Spectra, using absorption lines and Doppler effect
Motion	Astrometry or spectra, using Doppler effect

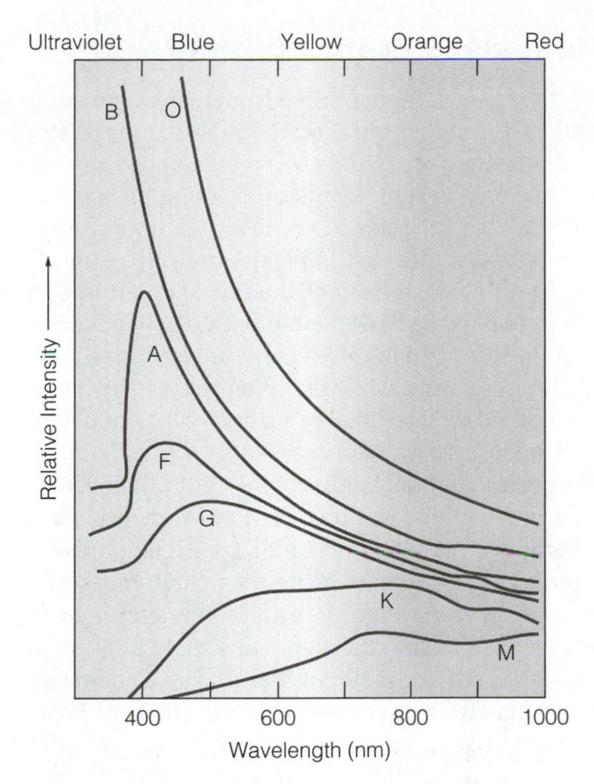

Figure 16-6 A schematic comparison of the energies emitted by various stars. Curves show intensity of radiation in the continuous spectrum of each star, neglecting absorption lines. The hottest and brightest are O stars, which emit most of their energy as blue and ultraviolet light. G stars, like the Sun, emit most of their radiation in the visible part of the spectrum, especially as yellow light. Class M stars emit most light in the red part of the spectrum, and classes L and T emit primarily in the infrared part of the spectrum and so are not represented in this figure. Absorptions cause irregular structure in the curves. Compare with a similar diagram for an idealized radiator at different temperatures in Chapter 5.

In this book, *luminosity* will generally mean *bolometric luminosity,* abbreviated *L.*

The most basic method of estimating luminosity derives from the measurement of distance. A faint light in the night may be a candle 100 m away, a streetlight a few kilometers away, or a brilliant lighthouse beacon 100 km away. Once we know the distance to an object, we can determine its absolute brightness. This method becomes less accurate for very distant stars. For one thing, as mentioned earlier, parallax distance measures become unreliable beyond about 100 pc. More important, the space between stars contains a thin haze of dust that dims starlight over larger distances. Just as smog might make it difficult to estimate the distance to a distant mountain range, interstellar dust in uncertain amounts complicates estimates of stellar luminosity. Since interstellar dust reduces the apparent brightness of a star, it results in an *underestimate* of the absolute brightness or luminosity.

★ As discussed previously, the luminosity, the distance, and the apparent brightness of an object are all interrelated. If we know any two of these quantities, we can estimate the third. For a figure that shows you how to look up the relationship between apparent magnitude, absolute magnitude, and luminosity, see our web site.

Measuring Stellar Temperature

We have seen that spectral classes offer a rough guide to temperature. For more accuracy, we can study stellar spectra, measure which color is the most strongly radiated, and then use Wien's law to calculate the temperature (Steffey, 1992). Stellar temperatures range all the way from about 2500 K for cool M-type stars to 40,000 K or more for hot O-type stars. The temperature only applies to the outer layer or photosphere of the star from which visible light is emitted. See the range of energy distributions in Figure 16-6.

Measuring Stellar Diameter

Most diameters are measured from temperature and luminosity by applying the **Stefan-Boltzmann law.** By this law, a star radiating a certain number of watts and having a certain temperature must have a certain area A. The values can be inserted in the Stefan-Boltzmann equation and the equation solved for A. The mathematical form of the law states that L is proportional to AT^4. If the temperature of a source doubles, the amount of energy radiated increases by 2^4, or 16. Thus, while doubling the area of a star would increase its output by a factor of two, doubling its temperature would increase its output 16 times. Hotter stars not only radiate bluer light than cooler stars (a result that was predicted by Wien's law) but also radiate *more* light per unit area. The star's diameter can be found from its area. Such measures indicate a vast range of stellar diameters, from less than Earth size, through Sun size, to huge stars bigger than the diameter of Mars' orbit! The Stefan-Bolzmann law is discussed further in Optional Basic Equation VII.

Measuring Stellar Mass

Mass turns out to be one of the most difficult stellar properties to measure. Astronomers estimate the mass of most stars in terms of a *model* of how stars evolve and radiate their energy, as we will see in the next chapter. In some cases where stars are in orbit around each other, we can calculate stellar masses by an application of Kepler's Laws. Binary stars are discussed in further detail in Chapter 21.

THE DOPPLER EFFECT

Probably the most important physical phenomenon in astronomical spectroscopy is the Doppler effect (named after the Austrian physicist Christian Doppler, who discovered it in 1842). The **Doppler effect** is a shift in wavelength of a spectral absorption or emission line away from its normal wavelength, caused by motion of the light source toward or away from the observer. If the source approaches, there is a **blueshift** toward shorter wavelengths, or bluer light. If the source recedes, a **redshift** occurs toward longer wavelengths, or redder light (Figure 16-7). The amount of the shift is just proportional to the approach or recession speed of the source (as long as that speed is well below the speed of light). As a mnemonic aid, remember that *re*cession produces *red*shifts.

In the shorthand of mathematics, the shift in wavelength divided by the rest wavelength equals the ap-

OPTIONAL BASIC EQUATION VII

The Stefan-Boltzmann Law: Rate of Energy Radiation

Before describing how to measure specific properties of stars, we need the seventh of the 11 basic equations to be used in this book. This is the Stefan-Boltzmann law, which describes the total amount of energy radiated in each second from any hot surface. The total energy radiated per second by a star is called its luminosity, abbreviated L.

In studying Wien's law, we saw that every warm body radiates. The higher the temperature, the bluer the radiation. About a century ago the Austrian physicists Josef Stefan and Ludwig Boltzmann discovered another characteristic: *The higher the temperature, the more energy is radiated each second,* as shown in Figure 16-6. Although the subjects of their study were radiating objects in the laboratory, the law applies to stars and all other bodies in the universe. The law gives the total energy L radiated per second from a body with temperature T and surface area A. For bodies that radiate efficiently, like stars, the law is

$$L = \sigma T^4 A$$

Sigma (σ), called the Stefan-Boltzmann constant, equals 5.67×10^{-8} W/m^2·K^4, T is given in kelvins, A in square meters, and L in the SI metric units of watts.

This equation shows that if the temperature or area of a star increases, the total energy radiated every second will increase. Suppose the star has some radius R; then its area is $A = 4\pi R^2$. Thus

$$L = 4\pi\sigma T^4 R^2$$

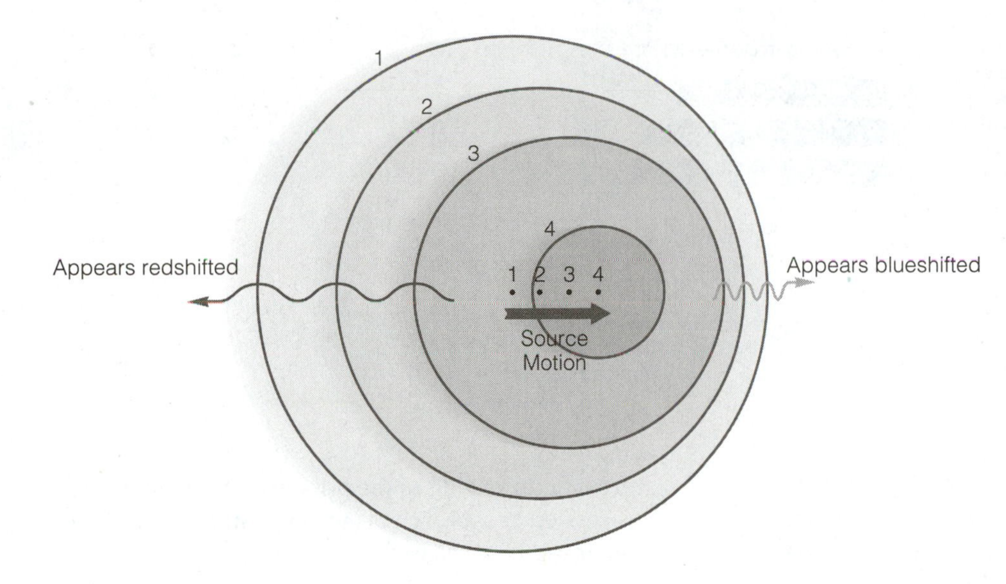

Figure 16-7 An illustration of the Doppler effect, the way that a moving source of waves can affect the observed wavelength. If the source of waves is moving and emits waves successively at positions 1, 2, 3, and 4, the waves are illustrated by the circles centered on each of the four positions. Viewed along the direction of motion, the wavefronts are bunched up, and so the waves are shifted to bluer wavelengths. Viewed away from the direction of motion, the wavefronts are stretched out, and so the waves are shifted to redder wavelengths. There is no Doppler effect when the source of waves is moving transverse to the direction of observation.

proach or recession speed divided by the velocity of light. For example, if the light source is receding at 10% the speed of light, the light will be redshifted by 10% of its normal wavelength; a line normally found at wavelength 500 nm would appear at 550 nm. By measuring such wavelengths, velocities of stars moving toward us or away from us can be studied (for further details, see Optional Basic Equation VIII).

Doppler discussed the effect in the light from distant stars orbiting around each other, but his effect also applies to any signal transmitted by waves, including familiar sound waves on Earth (Figure 16-7).

This variation of the Stefan-Boltzmann law allows us to calculate a star's radius if the luminosity and temperature are known.

Sample Problem Suppose a certain B-type star is found to have a temperature of 18,000 K and a luminosity of 10,000 times that of the Sun. What is its radius relative to the sun's radius? *Solution:* The straightforward way to do the problem is to solve the equation for R:

$$R = \sqrt{\frac{L}{4\pi\sigma T^4}}$$

We then look up the luminosity of the sun, $L_\odot$, in Appendix 2 (Table A2-2), multiply it by 10,000 to get L, and plug in the appropriate numbers. The answer, in meters, can be compared to the Sun's radius, using the diameter listed in solar system data (Table 8-1). Solve the problem this way for experience. However, also try the following simpler way. Since we asked for R relative to the Sun, write the equation again for the Sun, with solar subscripts on R, L, and T. Then divide it into the preceding equation to get the ratio of stellar R to solar R. Many factors cancel, and we get:

$$\frac{R}{R_\odot} = \sqrt{\frac{L}{L_\odot}\left(\frac{T_\odot}{T}\right)^4} = \sqrt{10{,}000\left(\frac{5700}{18{,}000}\right)^4} = 10.0$$

(Ask your instructor for help if this is not clear.) In this method the bothersome constants simply cancel out, since we are solving for the ratio, and we quickly see that the star is 10 times bigger than the Sun.

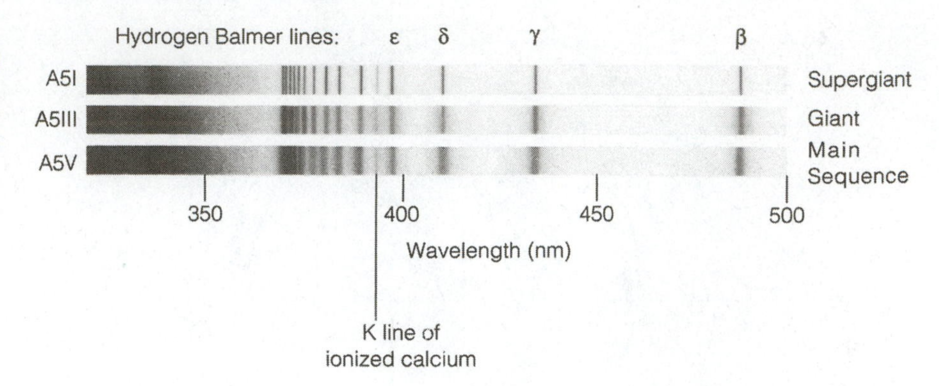

Figure 16-8 Comparison of the spectra of a supergiant, giant, and main-sequence star of the same spectral class (see Chapter 17). In the larger stars, pressures are less and atoms are less disturbed by collisions, resulting in slightly narrower, better-defined absorption lines.

The most familiar example of the Doppler effect is the shift in pitch of sound from an approaching or receding source. As a car or train passes by and recedes in the distance, the pitch of the sound dramatically decreases (longer wavelength means lower pitch). As the source approaches, the sound waves rush past the observer apparently more closely spaced than normal, decreasing the observed wavelength. As the source recedes, the waves are perceived to be more spread out, increasing the wavelength. The effect modifies whatever property of the signal is determined by wavelength—the color of light or the pitch of sound.

Note that the observer of a Doppler shift cannot say whether it is the source or the observer that is moving relative to any absolute external frame of reference. According to the principle of relativity, only recession or approach of one body with respect to another can be measured.

OTHER INFORMATION AVAILABLE FROM SPECTRA

Spectroscopy reveals more than just the luminosity and size of a star. The pattern, shape and position of spectral lines can be used to deduce remarkably detailed information. In particular, very luminous stars have slightly broader spectral lines than low-luminosity stars of the same spectral class. This useful fact allows astronomers to use spectra to classify a star's luminosity, as shown in Figure 16-8.

OPTIONAL BASIC EQUATION VIII

The Doppler Effect: Approach and Recession Velocities

As explained in the text, the Doppler effect is a change in wavelength proportional to any line-of-sight velocity between observer and source. (That is, the wavelength of a spectral line from a fast-approaching body will differ from that of a fast-receding body *and* that of one slowly approaching.)

The equation that describes the Doppler effect is the eighth of the 11 major equations we will use in this book. The equation is commonly written

$$\Delta\lambda/\lambda = v/c$$

In the equation λ designates wavelength, Δ designates change, v designates velocity, and c designates the speed of light read as "change in wavelength divided by the original wavelength equals v divided by c." Thus, if the source of light is not approaching or receding, the Doppler shift is zero. The faster the source approaches or recedes, the greater the Doppler shift in wavelength of the light.

Suppose a certain infrared spectral line normally has a wavelength of 1000 nm. If the light source is *receding* from the observer at $\frac{1}{1000}$ the speed of light, the line would appear at the slightly longer wavelength of 1001 nm, a little toward the red end of the spectrum. If the source *approached* at $\frac{1}{1000}$ the speed of light, the line would appear at 999 nm.

Measuring Stellar Composition

As described earlier, compositions are revealed by spectra. There are two problems: detecting the *presence* of an element and measuring its *amount.* The presence of an element is detected by identifying at least one—or preferably several—of its absorption lines or emission lines in the spectrum of the star.

The amount of the element is indicated by the appearance of the spectral line. Generally, the wider and darker the absorption lines, the more atoms of the element are present. Likewise, the wider and brighter the emission lines, the more atoms. Studies of this type were first made around 1925 at Harvard College Observatory by Cecilia Payne-Gaposhkin, who was a pioneer in many ways. She was at the vanguard of a group of women astronomers who were making important discoveries yet were not considered worthy of employment as research astronomers at the time. She was also the first to demonstrate that nearby stars are composed primarily of hydrogen and helium—and so are approximately Sun-like in composition.

Measuring Stellar Rotation

If a rotating star is seen from any direction except along the rotation axis, one edge will be approaching the observer and one edge will be receding. Light emitted or absorbed at the approaching edge will be blueshifted. Light from the other edge will be redshifted. Consider an absorption line being formed in all parts of the star's atmosphere. If the star were not rotating, neither shift would occur and the line would be very narrow. Because the star rotates, the line is broadened. The faster the rotation and the closer our line of sight to the equatorial plane, the more line broadening occurs.

The rotations of stars can thus be inferred to some extent from measurements of the broadening of spectral lines. In general, the fastest rotators are the hot, bluish, massive stars of class O, which have equatorial rotation speeds of around 300 km/s. Slightly cooler stars rotate more slowly, with a sharp break in the trend occurring at stars a little hotter and more massive than the Sun. The Sun's equatorial speed is only 2 km/s.

Detecting Atmospheric Motions of a Star

If masses of stellar gas rise and fall in convection cells (as they do in the Sun), the various cells will display a range of approach and recession velocities. Therefore, blue and red Doppler shifts (respectively) would occur, broadening the star's spectral lines. In practice, astronomers have trouble distinguishing this effect from rotational line broadening. Especially strong turbulence has been found in the atmospheres of certain cool, large-diameter stars known as red

When source and observer are getting closer together, the wavelengths of all light coming from the source are *shifted toward the blue* (shorter wavelengths). But when source and observer are getting farther apart, the spectrum of light from the source is *shifted toward the red* (longer wavelengths). Hence the two kinds of shifts are loosely referred to as *redshifts* and *blueshifts,* and the source is said to be *redshifted* or *blueshifted.*

Sample Problem The hydrogen alpha absorption line in a certain star is carefully measured and found to lie at 656.4 nm instead of the normal 656.3. Describe the motion of the star in the line-of-sight direction toward or away from the observer. Does this observation tell us anything about motion perpendicular to this direction? *Solution:* Solving the equation for v, and looking up c in Appendix 2 (Table A2-2), we have

$$v = \frac{\Delta\lambda}{\lambda}c = \frac{0.1}{656.3} \times 3(10^8) = 4.6\ (10^4)\ \text{m/s}$$

The Doppler shift is toward longer wavelength, or redward, so the star must be moving away. The Doppler shift is sensitive only to motion toward or away from us, so we cannot say anything about a possible additional component of lateral motion. We can only say that the star is moving away from us at 46 km/s in the line of sight.

giants. In these stars, masses of gas rise and subside with speeds as high as 40 km/s.

Detecting Circumstellar Material

Unstable stars, such as newly forming or dying stars, may throw out clouds of gas and dust (see our web site). These surround the stars and are called *circumstellar nebulae.* Because many of these clouds are nearly transparent, we can see light arising from both the near side and the far side of the cloud. As such a cloud expands away from the star, absorption lines arising on the near side (where the starlight passes through the gas) are blueshifted compared with lines in the star because the gas is approaching the observer. Emission lines from glowing gas on the near side are also blueshifted, but emission lines from the far side of the cloud are redshifted (see the discussion of Kirchhoff's third law in Chapter 15). These shifted lines are telltale clues to expanding circumstellar nebulae.

Gas around dying stars is believed to have been blown off the star in explosions related to the exhaustion of the star's energy supplies. In certain newly formed stars, circumstellar material has been found in the form of dust grains, rather than glowing gas. These dust grains are warmed by the star and detected by the thermal infrared radiation they give off.

Measuring Stellar Motion

The Doppler shift is a very simple way to detect part of a star's motion—its **radial velocity,** or *motion along the line of sight.* A consistent blueshift or redshift of all of a star's spectral lines proves that the star is moving toward or away from us. Among the 50 nearest stars, about half the radial velocities measured are more than 20 km/s toward or away from us.

The other component of a star's motion is its *tangential velocity—the motion perpendicular to the line of sight.* It cannot be measured as simply as radial velocity. To measure tangential velocity, we must measure the distance of the star and its rate of angular motion across the sky, called **proper motion.** (Values of a few seconds of arc per year are common for nearby stars.) These measurements of stellar motion are often lumped together with parallax in a branch of astronomy called **astrometry.**

If both radial and tangential velocities are known, they can be combined to give the star's **space velocity**—its true speed and direction of motion in three-dimensional space *relative to the Sun* (Figure 16-9).

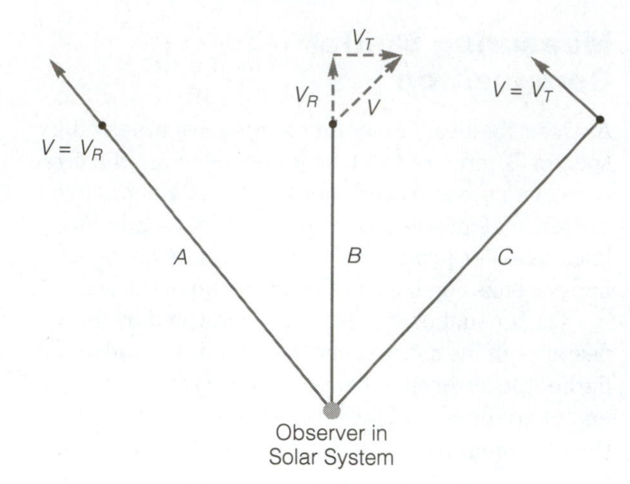

Figure 16-9 Three examples of a star's space velocity (V) compared with its radial velocity (V_R) and tangential velocity (V_T). In *A*, the star's space velocity is aligned with the line of sight; no proper motion would be seen. *B* is the most common case, in which both V_R and V_T are appreciable. In *C*, the space velocity is perpendicular to the line of sight; no radial velocity or Doppler shift would be seen.

Space velocities of most stars near the Sun are a few tens of kilometers per second and are nearly random in direction.

SUMMARY

This chapter emphasizes methods of studying stars, rather than the nature of the stars themselves. The chapter has two basic sections. The first section describes concepts: the system of naming stars; the visual appearance of stellar images; the definitions of light-year and parsec; the scale for measuring brightness; distances to nearby stars; and features of stellar spectra. The second section describes methods of determining basic stellar properties, as summarized in Table 16-3, using physical principles such as the Stefan-Boltzmann law and the Doppler effect. The key point is that once we have a good model for the structure of a star, we can estimate quantities that would be impossible to measure directly, such as stellar diameter and the existence of atmospheric motions in a star.

Clearly, our journey outward from the Earth has taken us far beyond the places that we can investigate in the near future by manned or robotic space missions. Even at the speed of light (which is 10,000 times greater than speeds our spacecraft have achieved), it would take years to reach the nearest stars. Thus, instead of sending instruments to the stars, we must rely on the messages in the starlight coming to us. Our abilities to interpret starlight are rapidly increasing, not only with construction of new large telescopes but also with the launching of such telescopes into orbit. In the coming chapters, we will interpret starlight in order to understand the nature and histories of the stars that surround us.

CONCEPTS

star	spectral class
light-year	temperature
parsec	bolometric luminosity
apparent brightness	Stefan-Boltzmann law
absolute brightness	Doppler effect
luminosity	blueshift
spectrum	redshift
spectroscopy	radial velocity
absorption line	proper motion
emission line	astrometry
continuum	space velocity

PROBLEMS

1. If a telescope of 25-m (1000-in.) aperture could be put in orbit or on the Moon, it could resolve an angle of only about 0.005 second of arc. Compare this resolution with the maximum angular size known for stars (0.03 second). Why would this telescope perform better on the Moon than on Earth? Why do astronomers never use the highest magnifications theoretically possible with earthbound telescopes?

2. A star is 20 pc away. How many years has its light taken to reach us?

3. What is the chemical composition of most stars?

4. A reddish star and a bluish star have the same radius. Which is hotter? Which has higher luminosity? Describe your reasoning.

5. A reddish star and a bluish star have the same luminosity. Which is bigger? Describe your reasoning.

6. How does the magnitude scale differ from the scales by which we measure lengths (centimeters or inches), temperature (degrees Celsius or kelvins), weight (kilograms or pounds), or other familiar properties?

7. A certain star has exactly the same spectrum as the Sun but is 30 magnitudes fainter in apparent magnitude. List some conclusions you would draw about it and describe your reasoning.

8. A star has a parallax of 0.05 second of arc. How far away is it?

9. The spectral lines of metals, such as calcium, are prominent in the solar spectrum. Hydrogen lines are less prominent. Why does this not indicate that the sun consists mostly of these elements instead of hydrogen?

10. Each of a certain star's spectral lines is found to be spread out over a wide range of wavelengths. What might cause this?

ADVANCED PROBLEMS

11. How many times brighter is a daylit landscape than the same landscape lit at night by the full moon? (*Hint:* See Table 16-1; note that 5 magnitudes equals a factor of 100 in brightness; 1 magnitude, a factor of $2\frac{1}{2}$.) If your cameral needs a $\frac{1}{100}$-s exposure in daylight, what exposure might record the moonlit scene? (Actually, about twice the calculated exposure will be better, because film sensitivity falls off in weak light.)

12. A certain star has a spectrum similar to the Sun's, but all the spectral lines are shifted 0.1% of their wavelength toward the red.

a. What do you conclude about this star's motion toward or away from the observer?

b. Is its speed in this direction unusually fast, average, or slow?

13. A star 10 pc away is observed to have a proper motion of 1 second of arc per year. Use the small-angle equation to derive its tangential velocity.

14. If the stars in Problems 13 and 14 were the same star, what would its space motion be like?

15. A star has the Sun's luminosity (4×10^{26} J/s) but a temperature of only 2850 K. How big is it?

16. a. Using Optional Basic Equation VI, prove that the mean energy of atoms and ions in a star will increase as the temperature increases. (*Hint:* See discussion of Optional Basic Equation VI.)

b. Discuss how this explains the pattern of increasing ionization as temperature increases in the spectral sequence of stars.

c. In previous problems we have used a rule of thumb that the fastest atoms of any element move more than three times faster than the typical velocity of the same element in the same gas. Note that in this case they would have nine times as much kinetic energy. Assume that these fastest atoms dominate in ionizing a mass of hydrogen gas and calculate the temperature at which these fastest atoms would have enough energy to ionize adjacent hydrogen atoms. The ionization energy for ground-state hydrogen is about 2×10^{-18} J.

d. Comment on where a star of such temperature would fall in the sequence of spectral classes, and show that hydrogen is indeed ionized in hotter stars. (*Note:* The model for ionization in problems *c* and *d* is overly simplistic. First, ionization in a star is also caused by photons of ultraviolet radiation, as well as thermal jostling by neighboring atoms. Second, the percentage of atoms ionized depends on gas density and pressure, as well as temperature, as worked out by Megh Nad Saha in 1920. Nonetheless, these problems give a rough idea of how ionization sets in as temperature increases in a gas, due to increasing energy of the gas particles.)

PROJECT

1. Using a telescope of fairly high magnification (such as 300× or 400×), examine the image of a bright star on several different nights and sketch it. Are there differences from night to night? Are the images smooth and featureless, or can you identify any rings or spikes around the core of the image? If so, label them on your sketches. Note any shimmering due to atmospheric turbulence or air currents in and around the telescope. Run the eyepiece inside or outside the focus point, making the image a round blob. Shimmering and other turbulent effects are often more evident in this way.

CHAPTER

17

Behavior of Nearby Stars: The H-R Diagram

WHAT THE READER SHOULD WATCH FOR IN THIS CHAPTER

Astronomers can classify stars according to their surface temperature and luminosity. The primary tool for stellar classification is a plot of temperature and luminosity called the Hertzsprung-Russell diagram, or H-R diagram. Stars like the Sun have a clear relationship between temperature and luminosity—these are called main-sequence stars. Like the Sun, main-sequence stars are all converting hydrogen into helium by the process of nuclear fusion. Some stars have sizes and luminosities that differ greatly from the Sun's. There are cool stars hundreds of times larger than the Sun and hot stars hundreds of times smaller than the Sun. The most prominent stars in the sky are mostly hot main-sequence stars and giants, much more luminous than the Sun. However, a careful census of nearby regions of space shows that most stars are actually cool main-sequence stars and dwarfs, much less luminous than the Sun. Stars are not unchanging, as was once believed. All stars must evolve as they use up their cosmic fuel supply. Models show that the diverse life stories of stars are dictated primarily by their mass. ■

As soon as astronomers could begin to measure different properties of stars, they began to categorize the stars. This chapter will outline three interesting discoveries that resulted: (1) Stars come in a wide variety of forms: massive and not so massive, large and small, bright and faint. (2) All stars pass through a variety of evolutionary stages. (3) Stars of different initial mass evolve at different rates and into different final forms.

In this sense, the population of stars can be likened to that of people. If you travel from country to country and observe people, you tend to see mostly ordinary people—youngsters, middle-aged people, and elderly people. The creation of new people—the birth of babies—is secluded in special places such as hospitals and is not readily visible to the tourist. In addition, the tourist may see tombstones, but the actual deaths of people are usually brief events not commonly encountered. (Unlike people, however, stars generally do not grow in mass.) This metaphor explains the approach of this chapter and the next two chapters. In this chapter we will tour the local population of stars. We will find that the great majority of these stars are ordinary stars—that is, youthful, middle-aged, or becoming elderly. The next chapter will show how stars come into existence and cite some observed examples of star birth. Chapter 19 will examine some of the extraordinary phenomena of stellar old age.

CLASSIFYING STAR TYPES: THE H-R DIAGRAM

When astronomers realized that there is such a variety of star forms, their first step was to devise a way to arrange and study the data about stars, in hopes of finding relationships among the various forms. This could have been done in various ways, but one method has become traditional. This method was introduced around 1905 to 1915 by Danish astronomer Ejnar Hertzsprung and American astronomer Henry Norris Russell. They constructed diagrams plotting the spectral class of stars vs. their luminosity. This type of plot is usually called the **H-R diagram** in honor of Hertzsprung and Russell.[1] The first published H-R diagram is shown in Figure 17-1.

As we saw in the last chapter, temperature is the principal factor that governs the differences among spectra of O, B, A, F, G, K, and M stars. For this reason, many astrophysicists make H-R diagrams by using a temperature scale instead of a scale showing spectral class. In this book, we identify temperature, spectral class, and luminosity along the edges of the diagram as an aid to interpretation. (Following Russell's original version, the spectral classes are arranged with *cooler* stars to the right.)

A schematic version of the H-R diagram, showing stars of all kinds, is shown in Figure 17-2. Different locations on the H-R diagram correspond to different types of stars. It is important to understand why this is true. First, remember from Wien's law that stars of different temperature have different color. The right part of Figure 17-2 corresponds to redder, cooler stars, and the left part to bluer, hotter stars. Similarly, the upper part corresponds to more luminous stars, and the lower part to less luminous stars.

Note that the Sun is near the middle of the diagram (luminosity 1 $L_{\odot}$, spectral type G). Since the Sun has typical properties, it is convenient to express stellar properties in terms of the Sun. Thus we will often use "solar units." The sign of solar units is a subscript circle with a dot in the center. The symbols for solar luminosity, solar mass, and solar radius are $L_{\odot}$, $M_{\odot}$, and $R_{\odot}$, respectively. A star of 4.6 $M_{\odot}$ is 4.6 times the mass of the Sun. A star of 0.3 $R_{\odot}$ is 0.3 times (30% of) the size of the Sun. By expressing stellar properties in this way, we avoid having to use very large numbers to express luminosity in watts or radius in kilometers.

If we move upward from the Sun's position on the diagram, we encounter stars more luminous than the Sun, even though they have the same temperature. These stars must be bigger than the Sun, since

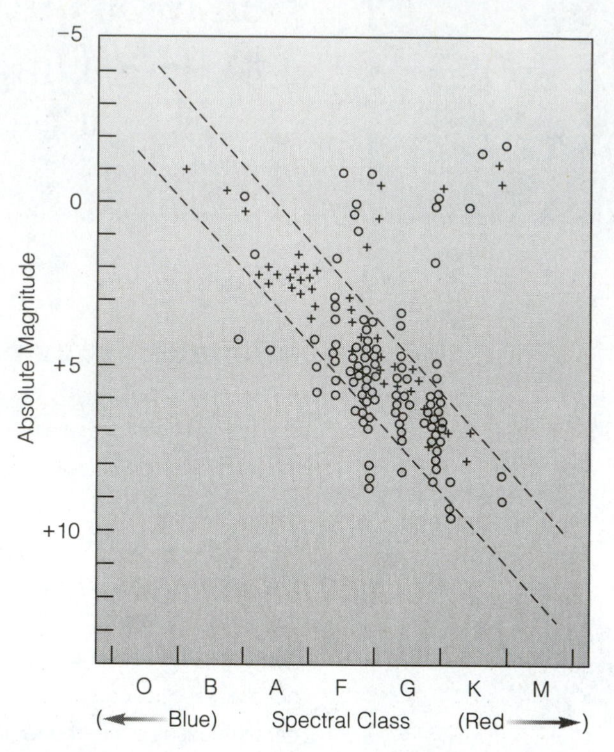

Figure 17-1 The first H-R diagram, redrawn from Russell's original 1914 publication. Circles represent stars with directly measured parallaxes; crosses represent estimated data from stars in four clusters. Dashed lines enclose Russell's identification of the main sequence—a discovery made from this diagram.

the Stefan-Boltzmann law tells us that each square meter on a star at the same temperature as the Sun will radiate as much energy as a square meter on the Sun. Thus, to get more total luminosity, we must have more square meters of surface area and hence larger size. Similarly, a star below the Sun on the diagram must be smaller than the Sun.

Do stars appear *throughout* the diagram? Or are there only certain types such as main-sequence stars, giants, and dwarfs? The answer is that only certain parts of the H-R diagram are crowded with stars. Stars are rare or nonexistent in other parts of the diagram. Even the earliest H-R diagram revealed an important discovery about stars: *Among most stars there is a smooth relation between spectral class and luminosity.* Hertzsprung called stars obeying this relation (falling on the diagonal band from upper left to lower right in Figure 17-1) **main-sequence stars.** The main sequence runs from hot luminous blue stars (upper left) to cool faint red stars (lower right). It turns out that main-sequence stars all have something in com-

[1]Sometimes this graph is called the spectrum-luminosity diagram or the color-magnitude diagram, because plots of spectral type vs. luminosity or color vs. absolute magnitude produce equivalent information.

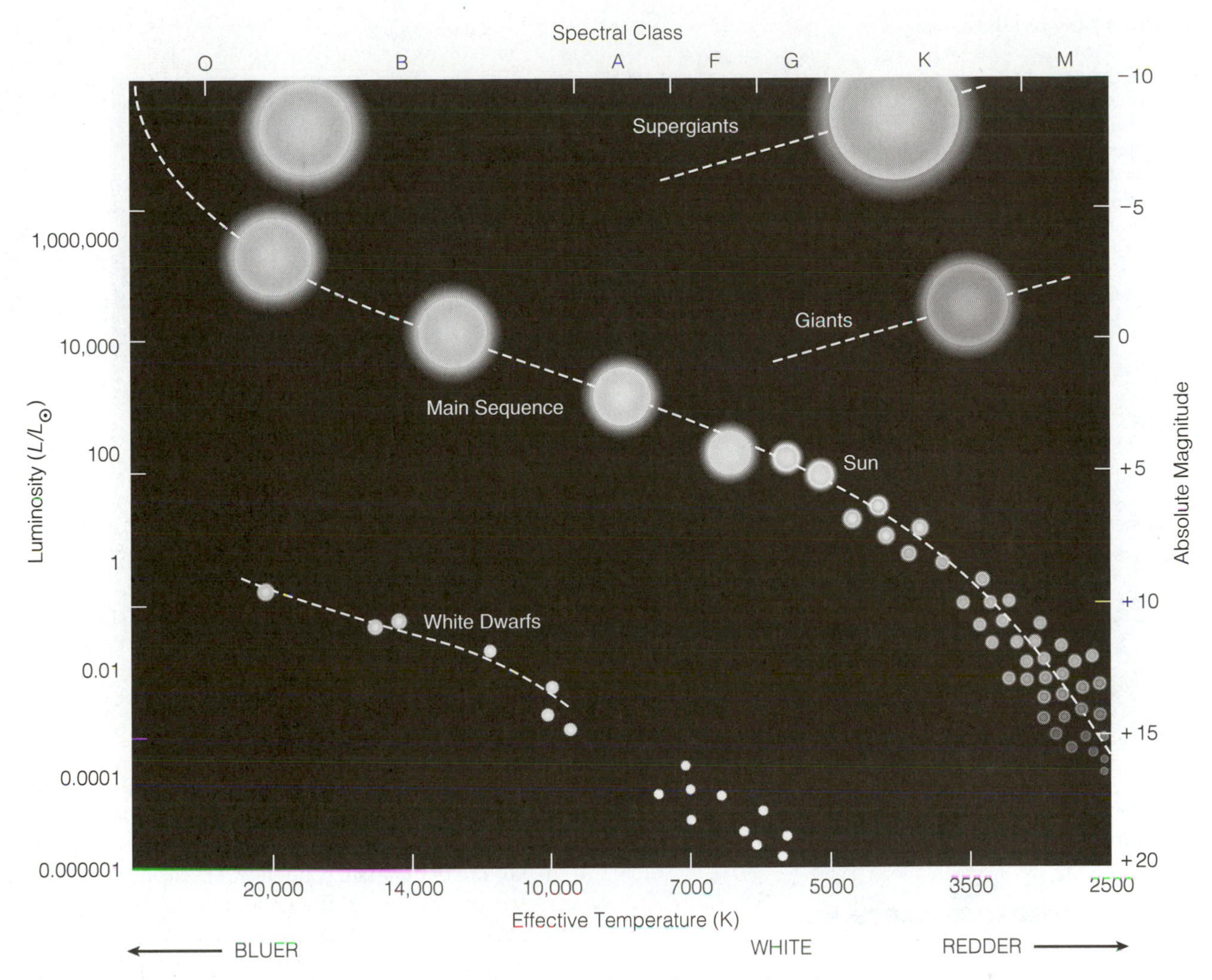

Figure 17-2 An H-R diagram for a selection of different types of stars. Stars are plotted schematically to show relative sizes, but not to true scale (which would require much larger red giants). Main-sequence stars, which release energy by fusing hydrogen into helium, run diagonally across the diagram.

mon: They use the same energy-generation mechanism as the Sun.

A Representative Sample of Nearby Stars

Figure 17-3 shows a modern H-R diagram for the 100 stars closest to the Sun. Stars at this range are near enough for us to measure accurate distances and to detect examples of very low luminosity. Stars in such a sample are believed to be representative of all stars in the neighborhood of the Sun (Crosswell, 1987). The figure shows that over 90% of a representative sample of stars in the solar neighborhood fall on the main sequence. The Sun is plotted near the center, at spectral class G, luminosity 1 $L_{\odot}$. Figure 17-3 also reveals the important fact that *most representative stars are fainter, cooler, and smaller than the Sun.*

★ Table 17-1 lists more detailed properties of the 24 stellar systems within 4 pc of the Sun. The census of the solar neighborhood was produced by the Hipparcos satellite, which measured parallax distances for a large number of stars. Seventeen of the total of 34 stars are inhabitants of systems in which two or three stars orbit around one another. The table also makes clear that the nearest stars are quite unlike the Sun. Most of them are 10–100 times less luminous than the Sun; 26 out of the 34 are K- or M-type dwarfs with very red colors. Even though these are the nearest stars to us, they are so intrinsically faint that only

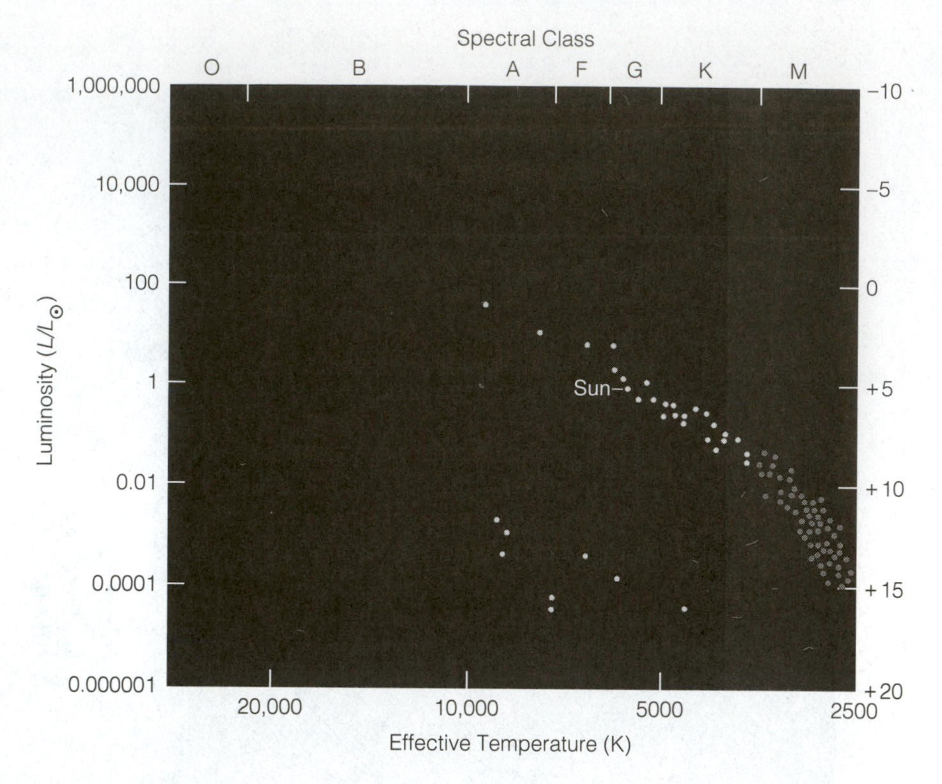

Figure 17-3 An H-R diagram for the 100 nearest stars—a representative sample from a volume of space around the Sun. This diagram shows that most stars are on the main sequence. Note also that most are fainter and cooler than the Sun. Therefore, they fall in the lower right corner.

8 out of the 34 are visible to the naked eye. (Our web site shows the nearest star system to the Sun, Alpha Centauri.)

Representative Stars vs. Prominent Stars

If we try to expand our statistical sample of stars by tabulating the lower portion of more distant ones, we run into a problem. At great distances the less luminous stars, like those in Figure 17-2, are too faint to see (Kaler, 1991). The only distant stars we see are unusually luminous ones. In fact, the *prominent stars* in our night sky are mostly distant, unusually luminous stars. These stars are quite rare in three-dimensional space. Hertzsprung aptly called them the "whales among the fishes."

Thus the statistics of prominent stars do not give us a sample of representative stars (Crosswell, 1992). Rather, they are biased toward the "whales." Nonetheless, a tabulation of such stars is important because it reveals that some of the "whales" don't belong to the main sequence.

Non-Main-Sequence Stars

If we make an H-R diagram by adding a sampling of the most prominent stars in our sky (as ranked by apparent brightness) to an equal-size sampling of the nearby stars, we get an H-R diagram similar to Figure 17-2. The nearby stars are the small red stars at the bottom of the main sequence; the prominent stars are mostly "whales" in the upper part of the diagram.

Let's examine the types of non-main-sequence stars more closely. Some red K and M stars have much higher luminosity than the faint, main-sequence K and M stars. They radiate more light than the fainter stars of the same temperature. Thus, according to the Stefan-Boltzmann law, they must have larger areas. Hertzsprung named them **giant stars.** Most are called *red giants,* since they are reddish K and M stars.

Other stars were found to be brighter than the giants or the main-sequence O stars; they came to be called **supergiant stars.** As more data came in, still other stars appeared on the H-R diagram below the main sequence. By the same logic used for giants, these stars were recognized as unusually small. They came to be called **white dwarf stars,**[2] because their colors were intermediate, like the Sun's, being neither strongly blue nor red.

The work of Hertzsprung, Russell, and their followers was important because it showed systematic

[2]Unfortunately, main-sequence stars are sometimes called dwarf stars to distinguish them from giants, but for clarity we reserve the term *dwarf* for white dwarfs.

TABLE 17-1

The Nearest Stellar Systems (Stars within 4 Parsecs)

Star Name	h	′	Parallax π π (″)	Distance (1/π) (pc)	Spectral Type	Apparent Magnitude (m_V)	Absolute Magnitude (M_V)	Luminosity ($L_\odot = 1$)
1 Sun					G2 V	–26.75	4.82	1.
2 Proxima Cen	14	41	0.772	1.29	M5.5 e	11.05	15.49	0.00005
α Cen A	14	5	.742	1.35	G2 V	.02	4.37	1.51
α Cen B					K0 V	1.36	5.71	0.44
3 Barnard's star	17	42	.549	1.82	M4 V	9.54	13.24	0.0004
4 Wolf 359 (CN Leo)	10	1	.419	2.39	M6 V	3.45	16.56	0.00002
5 BD +36°2147 (HD95735)	11	58	.392	2.55	M2 V	7.49	10.46	0.006
6 Sirius A	6	43	.379	2.64	A1 V	–1.45	1.44	22.49
Sirius B					DA2	8.44	11.33	0.0025
7 L 726-8, BL Cet = A	1	57	.374	2.68	M5.5 V	12.41	15.27	0.00007
UV Cet = B					M6 V	13.25	16.11	0.00003
8 Ross 154 (V1216 Sgr)	18	50	.337	2.97	M3.5 V	10.45	13.08	0.0005
9 Ross 248 (HH And)	23	10	.316	3.16	M5.5 V	12.29	14.79	0.0001
10 ∊ Eri	3	27	.311	3.22	K2 V	3.72	6.18	0.286
11 CD –36°15693 (HD217987)	23	51	.304	3.29	M2 V	7.35	9.76	0.01
12 Ross 128 (FI Vir)	11	48	.300	3.34	M4 V	11.12	13.50	0.00034
13 L 789-6 (EZ Aqr) = A	22	18	.290	3.45	M5 V	12.69	15.00	0.00008
= B						13.6	15.9	0.00004
14 61 Cyg A	21	45	.286	3.49	K5 V	5.22	7.51	0.084
61 Cyg B					K7 V	6.04	8.32	0.0398
15 Procyon A	7	14	.286	3.50	F5 IV-V	0.36	2.64	7.45
Procyon B					DA	10.75	13.03	0.0005
16 BD +43° 44 A (GX And)	0	1	.280	3.57	M1.5 V	8.08	10.32	0.006
BD +43° 44 B (GQ And)	0	2	.280	3.57	M3.5 V	11.05	13.29	0.0004
17 BD +59 1915 A	18	38	.280	3.57	M3 V	8.90	11.14	0.00
BD +59 1915 B	18	38	.280	3.57	M3.5 V	9.69	11.93	0.0014
Struve 2398AB								
18 ∊ Ind	22	47	.276	3.63	K5 Ve	4.69	6.89	0.148
19 G 51-15 (DX Cnc)	8	47	.276	3.63	M6.5 V	14.79	16.99	0.00001
20 τ Ceti	1	56	.274	3.65	G8 Vp	3.49	5.68	0.45
21 L 372-58	3	31	.270	3.70	M5.5 V	13.01	15.17	0.00007
22 L 725-32 (YZ Cet)	1	0	.269	3.72	M4.5 V	12.05	14.20	0.0002
23 BD +5°1668	7	14	.263	3.80	M3.5 V	9.85	11.95	0.0014
24 Kapteyn's star	5	1	.255	3.92	M1 p V	8.85	10.89	0.004
25 CD –39°14192 (AX Mic)	21	52	.253	3.95	M0.5 V	6.68	8.70	0.028

Source: Stellar data from the Hipparcos satellite, the Catalog of Nearby Stars (Jahreiss, 2000), and the General Catalog of Trigonometric Stellar Parallaxes (van Altena, Lee, and Hoffleit, 1995).

groupings of stars: one group fitting on the main sequence, a distinctly different group of giants, a group of dwarfs, and a group of supergiants. Further work showed that these groups occur in all known regions of space. This meant that universal physical processes could be sought to explain the groups.

Table 17-2 lists the 17 most prominent stars in our sky and reveals that nearly half are giants or

TABLE 17-2

The Stars of Greatest Apparent Brightness

Star Name	Apparent Magnitude (m_v)	Bolometric Luminosity ($L_\odot$)	Type[a]	Radius[b] ($R_\odot$)	Distance (pc)
Sun	−26.7	1.0	Main sequence	1.0	4.85×10^{-6}
Sirius (α Canis Majoris)	− 1.4	23	Main sequence (primary)	1.8	2.7
Canopus (α Carinae)	− 0.7	(1400)	Supergiant	30	34
Arcturus (α Boötis)	− 0.1	115	Red giant	(25)	11
Rigel Kent (α Centauri)	0.0	1.5	Main sequence (primary)	1.1	1.33
Vega (α Lyrae)	0.0	(58)	Main sequence	(3)	8.3
Capella (α Aurigae)	0.1	(90)	Red giant (primary)	13	14
Rigel (β Orionis)	0.1	(60,000)	Supergiant (primary)	(40)	(280)
Procyon (α Canis Minoris)	0.4	6	Main sequence (primary)	2.2	3.5
Achernar (α Eridani)	0.5	(650)	Main sequence	(7)	37
Hadar (β Centauri)	0.7	(10,000)	Giant (primary)	(10)	150
Betelgeuse (α Orionis)	0.7	(10,000)	Supergiant	800	160
Altair (α Aquilae)	0.8	(9)	Main sequence	1.5	5
Aldebaran (α Tauri)	0.9	125	Red giant (primary)	(40)	21
Acrux (α Crucis)	0.9	(2500)	Main sequence (primary)	(3)	(110)
Antares (α Scorpii)	0.9	(9000)	Supergiant (primary)	(600)	(160)
Spica (α Virginis)	1.0	(2300)	Main sequence (primary)	8	84

Source: Data from Burnham (1978).
Note: These are stars brighter than the first apparent visual magnitude.
[a]The designation *primary* indicates data for the brighter companion in binary pairs.
[b]Parentheses indicate estimates.

supergiants. Although the whales among the fishes are rare, they account for many of the familiar stars.

★ Occasionally, we find groups of stars at about the same distance from us and formed at about the same time. These *star clusters* offer us, so to speak, living H-R diagrams in the sky. Look on the web site for an image of the Jewel Box cluster, where the brightest stars offer a startling color contrast. Just by studying the colors and brightness of stars in clusters, we can see the pattern of the H-R diagram. Most of the stars are on the main sequence; the brightest ones are blue-white, and the faintest ones are red.

Luminosities and Radii on the H-R Diagram

★ Figure 17-4 shows the positions of the major stellar types on an H-R diagram. Lines of constant luminosity are horizontal in this plot. Figure 17-4 also has lines added to show positions of constant radius. The important concept is this: *Any point on the H-R diagram corresponds to a star of a certain radius.* The reasoning should be familiar. Any point on the diagram corresponds to a certain temperature T. According to the Stefan-Boltzmann law, any surface at temperature T must radiate a certain amount of energy per square meter each second. But any point on the diagram also corresponds to a particular luminosity L, which is the total amount of energy radiated each second. This luminosity fixes the number of square meters involved: hence the total area; hence the radius. (To get a sense of the enormous size of the supergiants on the diagram, see the image on our web site.)

STARS OF DIFFERENT MASSES AND DIFFERENT AGES

What causes the distinctive types of stars found on the H-R diagram, such as main-sequence stars, giants,

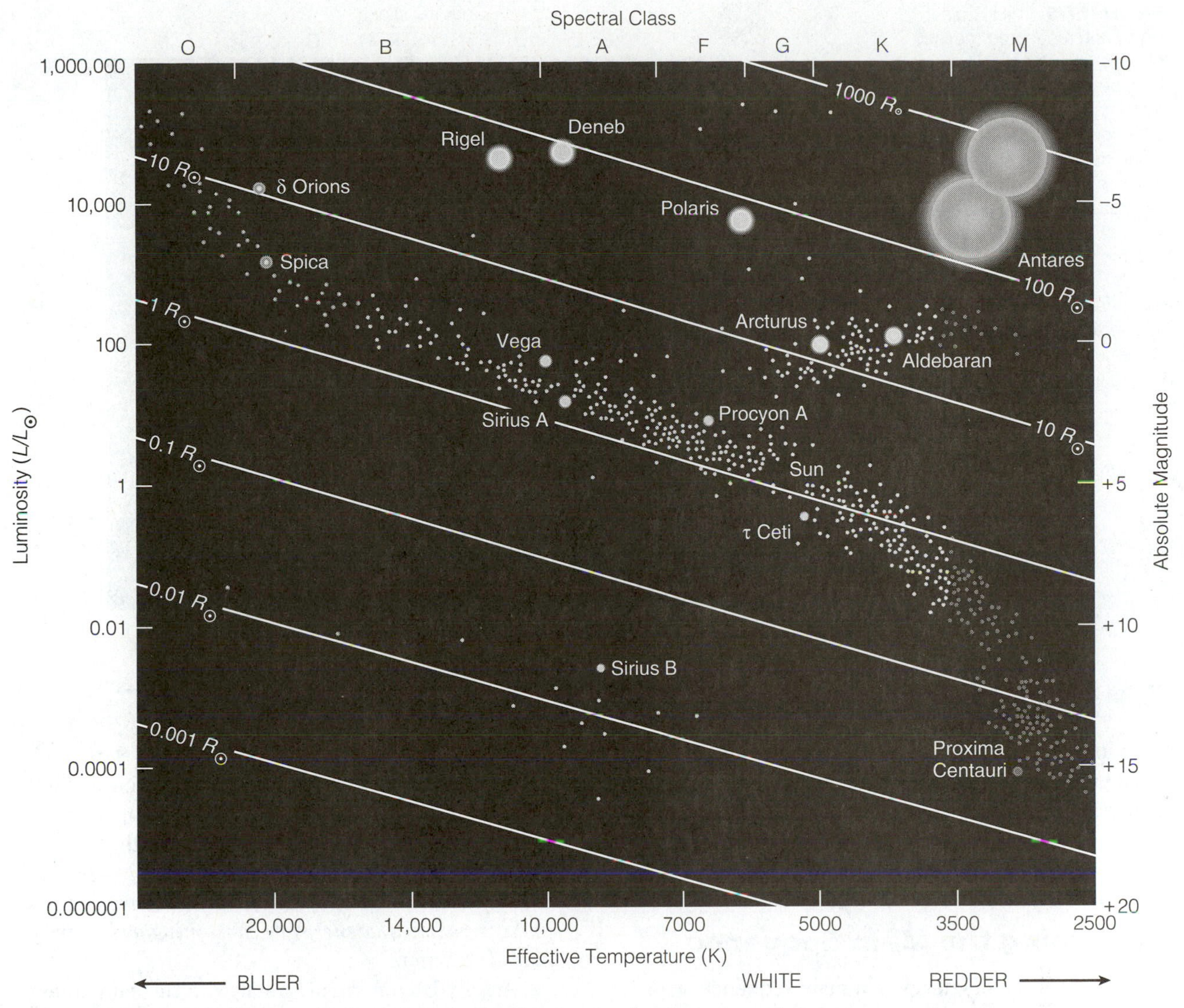

Figure 17-4 An H-R diagram with lines of constant radius showing the dimensions of stars in different parts of the diagram, in units of the radius of the Sun. Selected well-known stars are marked. Their relative sizes are schematically indicated, but there is no room to show them to true scale.

and white dwarfs? First, stars have different masses, as shown in Figure 17-5. *High-mass* main-sequence stars are hotter and brighter and bigger than *low-mass* main-sequence stars. Second, stars evolve from one type to another, and since we see stars with different ages, we see stars with different forms.

How did we discover the principles that control the structure and evolution of stars? Much of this work was begun by the English astrophysicist Arthur Eddington in the 1920s, and was carried on by other scientists through the 1950s. Eddington showed that the same two opposing influences at work in the Sun are competing in any star: Gravity pulls *inward* on stellar gas while gas pressure and radiation pressure push *outward*.[3] This is the principle of hydrostatic equilibrium (see also Chapter 14). In any stable star, these forces are just balanced.

As we have already seen in the discussion of Helmholtz contraction (Chapter 14), heat is produced during gravitational contraction of a star. A stable main-sequence star has contracted until the inside is hot enough to start nuclear reactions among hydrogen atoms. At this point the interior becomes a stable

[3]As mentioned in Chapter 15, radiation exerts a distinct pressure on material through which it passes. In stars, especially massive ones, this pressure is important.

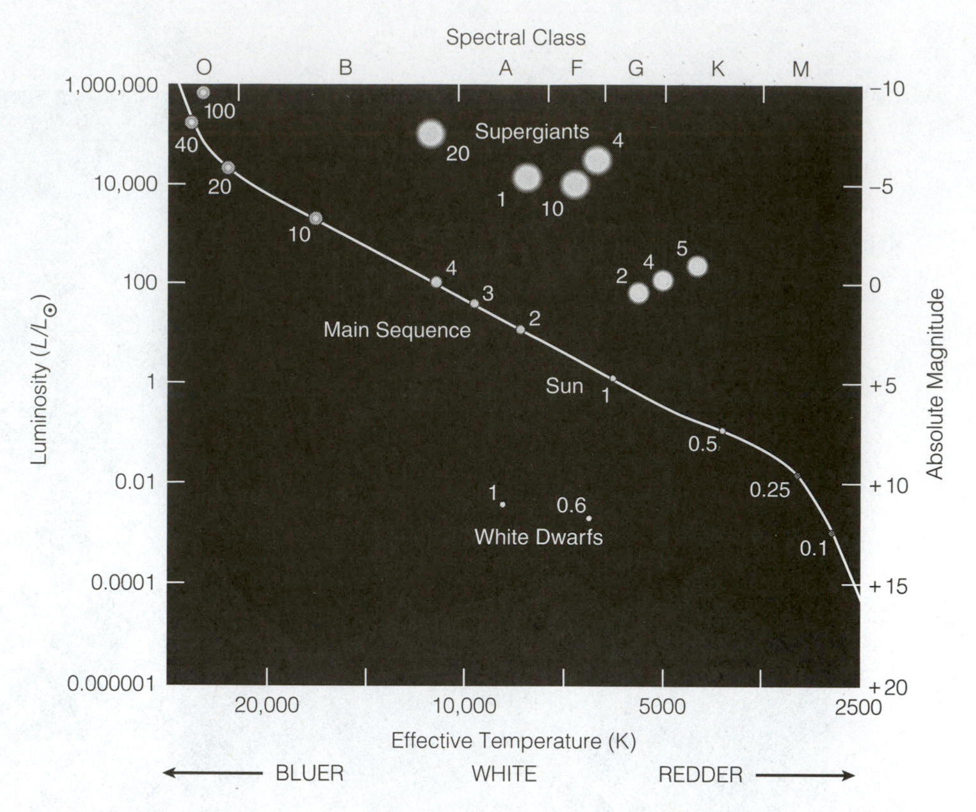

Figure 17-5 An H-R diagram with measured masses of stars (in solar masses) marked in different parts of the diagram. A smooth progression of masses is found along the main sequence, but masses in other parts of the diagram are mixed because stars of different mass evolve into the same regions—in other words, evolved stars of different masses can have the same temperature and luminosity. Schematic sizes are shown (not to true scale).

heat source, radiating light and creating enough outward pressure to counterbalance the inward force of gravity. What had been a contracting ball of gas now becomes a star with a stable size, governed by the release of energy in the interior (Furth, 1995).

Explaining the Main Sequence

The main source of energy in a main-sequence star's interior is nuclear reactions in which hydrogen is consumed. Because stars form with a huge supply of hydrogen, stars remain stable on the main sequence for a relatively long time at a fixed size. If a stable star were magically expanded, its gas would cool and the reactions would decline, reducing the outward pressure, and the outer layers would fall back to their original state. If it were magically compressed, the inside would get denser and the reactions would increase, raising the outward pressure and expanding the star. A main-sequence star tends to stay in a stable state due to the hydrostatic equilibrium that exists at each point within the star. The star remains stable *as long as its internal composition and energy production rate stay the same.*

Eddington did calculations that revealed a **mass-luminosity relation.** A hydrogen-burning star more massive than the Sun has a higher luminosity and surface temperature than the Sun. Hence, on the H-R diagram, it would lie to the upper left of the Sun. Similarly, a star of lower mass would lie to the lower right of the Sun. These stars therefore fall along a line in the H-R diagram—the main sequence. In other words: *The main sequence is explained as the group of stars of different masses that have reached stable configurations and are generating energy by consuming hydrogen in nuclear reactions.*

Any hydrogen-burning star with the same mass as the Sun and solar composition must look like the Sun. Any hydrogen-burning star with twice the mass of the Sun and solar composition must be brighter and bigger than the Sun. (Note: The word *burning* technically refers to chemical reactions involving only electrons. But astrophysicists use the term informally for nuclear reactions that convert small fractions of the mass of atomic nuclei into energy.) The properties of main-sequence stars are summarized in Table 17-3.

Figure 17-5 shows this trend more clearly by charting masses of some stars on the H-R diagram. Note the smooth trend of increasing mass and luminosity as we ascend the main sequence. We can now follow the physical chain of reasoning that explains the main sequence of stars. More massive stars have greater gravity, which creates higher pressure in the stellar interior. The higher pressure results in higher temperature, which causes higher energy output by

TABLE 17-3

Properties of Main-Sequence Stars

Spectral Type	Mass ($M_\odot$)	Radius ($R_\odot$)	Effective Temperature (K)	Luminosity ($L_\odot$)	m_v	Bolometric Correction
O5	40	18	40,000	5×10^5	−5.8	−4.0
B0	18	7.4	28,000	2×10^4	−4.1	−2.8
B5	6.5	3.8	15,500	800	−1.1	−1.5
A0	3.2	2.5	9900	80	+0.7	−0.4
A5	2.1	1.7	8500	20	+2.0	−0.12
F0	1.7	1.4	7400	6	+2.6	−0.06
F5	1.3	1.2	6580	2.5	+3.4	0.00
G0	1.1	1.1	6030	1.3	+4.4	−0.03
G5	0.9	0.9	5520	0.8	+5.1	−0.07
K0	0.8	0.8	4900	0.4	+5.9	−0.19
K5	0.7	0.7	4130	0.2	+7.3	−0.60
M0	0.5	0.6	3480	0.03	+9.0	−1.19
M5	0.2	0.3	2800	0.008	+11.8	−2.30

Note: Mass is in units of solar mass (2×10^{30} kg), radius is in units of solar radius (7×10^8 m), effective temperature is a measure of surface temperature, luminosity is in units of solar luminosity (3.8×10^{26} watts), bolometric correction is in magnitudes. The coolest L and T types continue this sequence to even smaller masses and luminosities.

the fusion process, giving both higher luminosity and higher surface temperature. The same reasoning does not apply to stars that lie off the main sequence; a star's position on other parts of the H-R diagram is not a simple indicator of mass. In other words, giants, supergiants, and dwarfs are stars that do *not* have the same energy-generation process, composition, or energy-transport processes as the Sun.

The Russell-Vogt Theorem

As Eddington reached these conclusions, the astrophysicists H. N. Russell and H. Vogt independently derived a related result in 1926 known as the **Russell-Vogt theorem:** The equilibrium structure of an ordinary (main-sequence) star is determined uniquely by its mass and chemical composition.

This says that a certain mass of material with fixed composition—for example, one solar mass of solar composition—can reach only one stable configuration. This is represented by the temperature and luminosity of the point on the H-R diagram actually occupied by the Sun. But if the composition were altered to $\frac{1}{2}$ hydrogen and $\frac{1}{2}$ helium, the rate of nuclear fusion would be different, and so the configuration of the star and its location on the H-R diagram would be different. This principle, plus a lot of detailed physics of nuclear reactions, is used to make models of stars. Astronomers have actually calculated what the stars look like—they have made tabulations of the pressures, temperatures, and other characteristics of the interiors, surfaces, and atmospheres of stars to accurately classify them.

The equilibrium states of stars are shown in Figure 17-6. Main-sequence stars of a given mass tend to have the same radius. In contrast, giants, supergiants, and white dwarfs of the same mass, because they have evolved and have different compositions, have different equilibrium structures and radii.

The Rate of Evolution of Stars

Although a main-sequence star generates energy in a stable configuration, it is gradually using its fuel supply. The Russell-Vogt theorem explains the cause of **stellar evolution** from one form to another form: A star converts hydrogen into helium and changes its composition; therefore, the star must change to a new equilibrium structure. All nuclear reactions cause changes in composition, and all changes in composition cause evolution to a new structure.

Stars evolve at different rates. The more massive a star, the higher its interior temperatures and the faster it uses its nuclear fuel. A Sun-like star of 1 $M_\odot$ stays

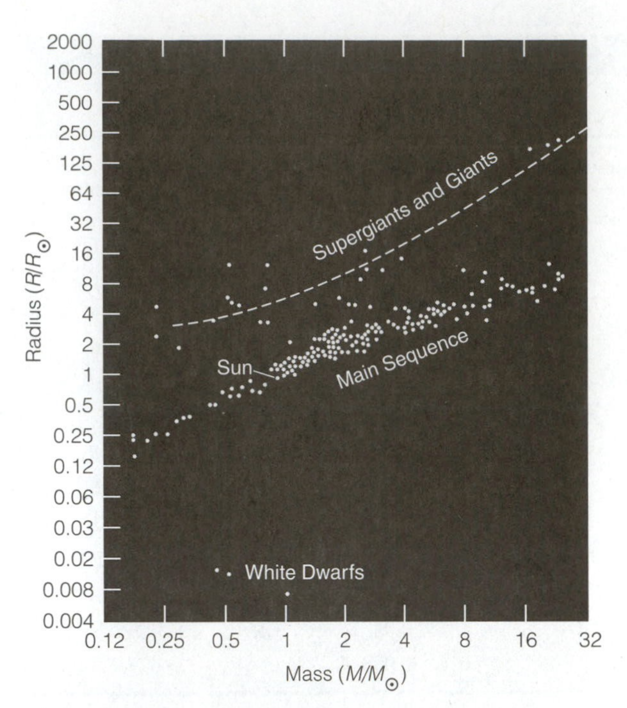

Figure 17-6 This figure reveals that distinctly different radii exist among stars of a given mass. This is a result of evolutionary changes. Most stars in the mass range shown here remain at nearly constant mass during most of their lives (main sequence). However, later evolution takes them to giant size, then to dwarf size. (Data are derived from double stars' orbits from Popper, 1980; Aller; 1971; and Liebert, 1980)

on the main sequence about 9 billion years (9×10^9 y), but a star of 10 $M_\odot$ stays there only about 20 million years (2×10^7 y). The entire history of a star of 1 $M_\odot$ (protostar to white dwarf) takes about 11 billion years, whereas a 10 $M_\odot$ star lasts only about 24 million years. In spite of the differences in total time, the largest fraction of each star's life is still spent on or near the main sequence.

Since a star's hydrogen-burning lifetime depends mostly on its mass and luminosity, a simple formula gives the time the star will spend on the main sequence—which is most of the star's lifetime: *Hydrogen-burning lifetime = M/L times 9 billion years,* where M and L are the star's mass and luminosity in solar units.

★ The most massive stars stay on the main sequence for only the twinkling of a cosmic eye. Some of them evolve into the supergiant region, and some less massive ones become ordinary giants. All of them quickly evolve to unstable configurations; many may explode; and all disappear from visual prominence. For a detailed description of the evolution of a star like the Sun, see our web site (also Kaler, 1994).

Determining the Ages of Stars

Consider a group of stars that formed at the same time. After about 10 million years, stars larger than 20 $M_\odot$ will have disappeared from the main sequence. That is, the H-R diagram of the group will contain no main-sequence O stars. After about 100 million years, stars more massive than 4 $M_\odot$ will have evolved off the main sequence, and the H-R diagram will contain scarcely any main-sequence B stars. The older the cluster, the more main-sequence stars will be gone. The missing stars will have been transformed into giants, white dwarfs, or even fainter terminal objects.

Thus we reach an important conclusion: The H-R diagram can serve as a tool for dating groups of stars that formed together. This principle will be applied often in later chapters as we probe the **ages of stars** in our galaxy.

Isolated individual stars are harder to date. The Sun's age was measured at 4.6 billion years by dating planetary matter—unavailable to us in the case of other stars. Certain indicators, such as the amount of "unburned" light elements (lithium, for example) in a star's atmosphere, can also be used to estimate a star's age.

PHILOSOPHICAL IMPLICATIONS OF THEORETICAL ASTROPHYSICS

Philosophically, the achievements of Eddington and other astrophysicists were profound. Eddington pointed out that humans (or other intelligent creatures), reasoning purely from elementary principles even without telescopic observation, could show that the universe must be populated by objects like stars because gravity causes gas to "clump" into star-size masses. Smaller or larger objects have too little gravity to assemble themselves from interstellar gas, and larger objects would produce so much radiation that they would blow themselves apart. These suppositions by Eddington have been proved correct by further observation and theory: The most common stars in the universe have a mass around 0.1 to 1.0 $M_\odot$, as shown in Figure 17-7. Stars smaller than 0.1 $M_\odot$ are less common, and stars bigger than 85 $M_\odot$ are very rare.

Eddington wrote in 1926:

We can imagine physicists working in a cloud-bound planet such as Jupiter who have never seen the stars. They should be able to deduce by [these methods] that

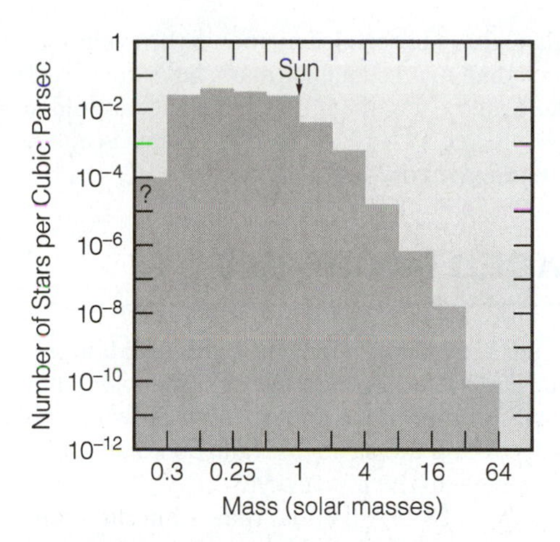

Figure 17-7 A histogram showing the frequency of occurrence of different stellar masses in the solar neighborhood. The most common stars have about $\frac{1}{4}$ $M_\odot$. The abundance of the lowest-mass objects is uncertain.

if there is a universe existing beyond the clouds, it is likely to aggregate primarily into masses of the order of 10^{27} tons. They could then predict that these aggregations will be globes pouring out light and heat and that their brightness will depend on the mass in the way given by the [mass-luminosity relation].

Perhaps Eddington was too optimistic; he was not giving enough credit to the interplay between observers and theorists. Nonetheless, he became fascinated by the idea that intelligent creatures could mentally derive the appearance of the universe without actually observing more than a few fundamental relations.

SUMMARY

The nearby stars, within about 100 pc of the Sun, display a range of masses, luminosities, temperatures, and compositions. These properties are not randomly distributed but are grouped into distinct types. Some stars are on the main sequence, some are giants, some supergiants, and some dwarfs. Differences among these types are conveniently displayed on the H-R diagram, which plots stellar luminosity against temperature.

To explain these different types of stars, this chapter developed three important principles, mentioned in the introduction. First, stars form with a variety of masses ranging from as low as 0.1 to 85 $M_\odot$ and more. As the stars form and begin to fuse hydrogen in nuclear reactions, they settle onto the main sequence, but the more massive the star, the more luminous its main-sequence position in the H-R diagram.

Second, stars evolve from one form to another. They begin with pre-main-sequence configurations, settle onto the main sequence for a relatively long time to convert hydrogen into helium, and then evolve off the main sequence. Post-main-sequence forms include giants, variables, and dwarfs, which will all be discussed in Chapter 19. Third, stars evolve at different rates. The more massive a star, the faster it goes through its sequence of life stages. This is because more massive stars have stronger gravity, which leads to higher pressures and higher central temperatures, and they therefore consume their hydrogen faster.

The great majority of stars, including most of those in the Sun's neighborhood, have masses in the range of 0.1 to 1 $M_\odot$. They are relatively faint and cool; they lie in the lower right part of the H-R diagram. The most massive stars are much brighter and can be seen from much farther away. They lie in the upper part of the H-R diagram. Thus many of the stars prominent in the night sky are prominent not because they are close but because they are the very luminous "whales among the fishes," far away among myriads of distant fainter stars.

CONCEPTS

H-R diagram	mass-luminosity relation
main-sequence star	Russell-Vogt theorem
giant star	stellar evolution
supergiant star	ages of stars
white dwarf star	

PROBLEMS

1. Suppose you could fly around interstellar space, encountering stars at random.

a. Describe the stars you would encounter most often. Mention masses and H-R diagram positions.

b. About what percentage of stars would be as massive as or more massive than the Sun? (*Hint:* Use statistics from either Table 17-1 or 17-2, as appropriate.)

c. Would the stars encountered by similar to bright stars picked at random in our night sky?

2. Do giant stars necessarily have more mass than main-sequence stars? Why are they called *giants*?

3. Four stars occupy the four corners of an H-R diagram. Which one has:

a. The highest temperature?

b. The greatest luminosity?

c. The largest radius?

4. Two stars lie on the main sequence in different parts of the H-R diagram. Which is:

a. Larger?

b. More luminous?

c. More massive?

d. Hotter?

5. Two stars form at the same time from the same cloud in interstellar space, but one is more massive than the other. Describe differences or similarities at later moments in time.

6. What would an imaginary terrestrial observer see as the Sun runs out of hydrogen in the future? If life is confined to Earth when this happens, would life perish from heat or from cold? (See web site material on evolution of the Sun.)

7. Which part of Problem 4 could not be answered just from location in the H-R diagram if both stars were not on the main sequence?

8. Consider newly forming cosmic objects of stellar composition. What would you expect to happen in terms of stars' life cycles as the masses decrease from 1 $M_{\odot}$ toward values comparable to Jupiter's mass?

9. Show that astrophysical estimates of the Sun's lifetime are consistent with meteoritic and lunar data on the age of the solar system. Why would the data be inconsistent if astrophysicists calculated that solar-type stars stay on the main sequence only 1 billion years?

10. Why don't normal stars all collapse at once to the size of asteroids or tennis balls under their own weight, since gravity always pulls their material toward their center?

11. How was the chemical composition of the Sun 3 billion years ago different from what it is now? (See web site material on evolution of the Sun.)

12. Red giants have proceeded further in the evolutionary sequence than main-sequence stars, but is it correct to make the blanket statement that red giants are older than main-sequence stars? Explain. Which star is older, the Sun or a red giant of 10 $M_{\odot}$?

ADVANCED PROBLEMS

13. Use Wien's law to show that an M star with temperature about 2900 K is strongly red in color and that an O star with temperature about 29,000 K would be strongly blue. Would the strongest radiation emitted by each star be visible to the human eye?

14. For a group of stars of equal radius but different temperatures T, the Stefan-Boltzmann law shows that the total luminosity will be proportional to T^4.

a. Among such stars, how many times brighter would a 20,000-K star be than a 2000-K star?

b. Using Figure 17-4, confirm your results by reading luminosities on one of the lines of constant radius.

15. a. As shown in Figure 17-6, a typical white dwarf could have a radius of 0.015 $R_{\odot}$ and a mass of 0.8 $M_{\odot}$. Calculate the mean density of matter in such a star and compare it with the density of familiar material such as water (1000 kg/m^3), rock (about 3000 kg/m^3), and lead (11,350 kg/m^3).

b. How many pounds would a matchbox full of such material weigh?

PROJECT

1. Locate or construct spherical objects that have the same proportional sizes as selected different types of stars, such as an Antares-type supergiant, the Sun, and a white dwarf.

CHAPTER

18

Stellar Evolution I: Birth and Middle Age

WHAT THE READER SHOULD WATCH FOR IN THIS CHAPTER

Star birth and death occur quickly on a cosmic time scale, so most of the stars we see in the night sky are in the middle of their lives. We know that stars must form continuously because there are stars that are older than the Sun and the solar system, as well as massive stars whose ages must be far less than that of the Sun. Star birth occurs in regions of space shrouded by gas and dust. Stars form by a complex process, and astronomers do not yet have a comprehensive theory of star formation. Young stars are energetic, emitting copious amounts of infrared radiation and often spewing jets of material from the poles. Stars form with a wide range of masses, up to about a hundred times the mass of the Sun and down to objects less than one-tenth of the Sun's mass. Cool substellar objects have been discovered with properties similar to those of the giant planets. After its birth, every aspect of a star's life is controlled by its mass. Massive stars can use nuclear fusion to build heavier elements than less massive stars. Starting with hydrogen, stars create helium and then heavier elements. ■

Just as new living individuals are being formed from chemical materials against a background of slowly changing genetic pools, new stars are being formed from interstellar gas and dust against a background of slowly changing elemental abundance of the cosmic gas. Locally—and by this we now mean a substantial part of our galaxy—the environment is constantly changing, with new stars forming and old stars blasting material into interstellar space. From a philosophical point of view, identifying and understanding newly forming stars are challenges in our long quest to comprehend the role of stars in our universe.

THREE PROOFS OF "PRESENT-DAY" STAR FORMATION

Here are three lines of evidence for "present-day" star formation—that is, star formation in the current part of astronomical history.

Youth of the Solar System The solar system is only about 4.6 billion years old, whereas the entire Milky Way galaxy—our system of 100 billion stars—is at least 10 billion years old. Thus our own Sun formed much more recently than the galaxy, and stars did not all form in one burst at the beginning.

Short-Lived Star Clusters Many young, massive stars are grouped in *open star clusters*. The stars in the cluster formed at about the same time. Because of tidal disruptive forces and the tendency of each star in the cluster to follow its own orbit around the galactic center, most clusters dissociate into isolated stars in only a few hundred million years. This is confirmed by analyses of H-R diagrams of clusters, which show that few of these clusters are much older than this. The famous Pleiades cluster, for example, is only about 50 million years old. The existence of such clusters shows that star clusters did not all form at the beginning, but continue to form today.

Short-Lived Massive Stars Chapter 17 showed that massive stars evolve fastest. Calculations show that stars of 20 to 100 $M_{\odot}$ can last only a few million years in a visible state, yet we see them shining. They must have formed less than a few million years ago.

Thus star formation has been a continuing process during the whole history of the galaxy, including the last million years. There are stars in the sky younger than the species *Homo sapiens.*

THE PROTOSTAR STAGE

Stars form from tenuous material between other stars. Typically, a near vacuum with very thin, cold gas exists between stars. Atoms and molecules number only about 10^6 per cubic meter. About three-quarters of them are hydrogen atoms, and most of the rest are helium—as is the case with the atoms in most nearby stars. There are scattered dust particles. The temperature is typically around 10 to 20 K.

Molecular Clouds

In certain regions of space, however, the density of gas and dust may be 10,000 times greater. With some 10^{10} atoms per cubic meter, the atoms are close enough to collide frequently. The most important molecular species, molecular hydrogen (H_2), forms on dust grains. These regions are called *molecular clouds.* They are rich in atoms, dust grains, and diverse molecules such as molecular hydrogen, carbon monoxide (CO), water (H_2O), and more complex forms such as formaldehyde (H_2CO) and ethyl alcohol (C_2H_5OH). Such complex molecules are virtually absent in most parts of interstellar space. Although a molecular cloud is denser than most interstellar gas, it is still a nearly perfect vacuum compared with ordinary room air, where there are about 2 trillion trillion (2×10^{25}) molecules per cubic meter.

Molecular clouds are important because they are the only active regions of star formation. In these regions, the atoms and molecules and dust grains are crowded close enough to begin to attract each other gravitationally—a key step in forming stars from gas. The rich variety of molecular transitions in star-forming regions is illustrated by Figure 18-1, which shows a spectrum of the Orion molecular cloud made at millimeter wavelengths. The spectral features are caused by rotational and vibrational transitions in molecules. (A photon is released with energy corresponding to the difference between two energy states.) The low energy of the transitions means that the spectral lines are seen at long wavelengths, in the millimeter or submillimeter part of the electromagnetic spectrum. The spectrum indicates the complex chemistry of molecular clouds.

Toward an Astrophysical Theory of Star Formation

Chapter 14 explained how the Sun formed when one of these diffuse interstellar clouds contracted and produced a central star surrounded by a dusty nebula. Now we need a more general theory of this process that can explain other stars with different masses.

When scientists use the word *theory,* they usually mean a well-tested body of related ideas, often with a mathematical formulation that can be applied to a variety of cases and often backed up by observations. A less complete or less tested idea is called a *hypothesis,* or a working hypothesis. An untested idea is often called *speculation.* Unfortunately, newspapers and magazines often publish speculations but label them with the more imposing term *theory.*

Such a **theory of star formation** has been developed (see summary by Shu, 1996). This theory can be approached by asking: What causes some clouds to contract and not others? Why don't all interstellar clouds contract into stars and be done with it, once and for all? The answer involves the same two opposing forces we have considered before: gravity versus thermal pressure. Gravity pulls all the atoms in a cloud inward. But even at only 100 K, the atoms are dancing, striking each other, and creating an outward pressure that opposes the tendency to collapse.

Gravity increases if the density of the cloud increases, since this gets more mass into the same amount of space. Outward pressure increases if the gas heats up, since this makes the atoms and molecules move faster. We can already say, then, that the reason not all clouds contract is that not all clouds are dense enough or cool enough. Furthermore, a noncontracting cloud can be turned into a contracting cloud if it suddenly experiences some turbulence

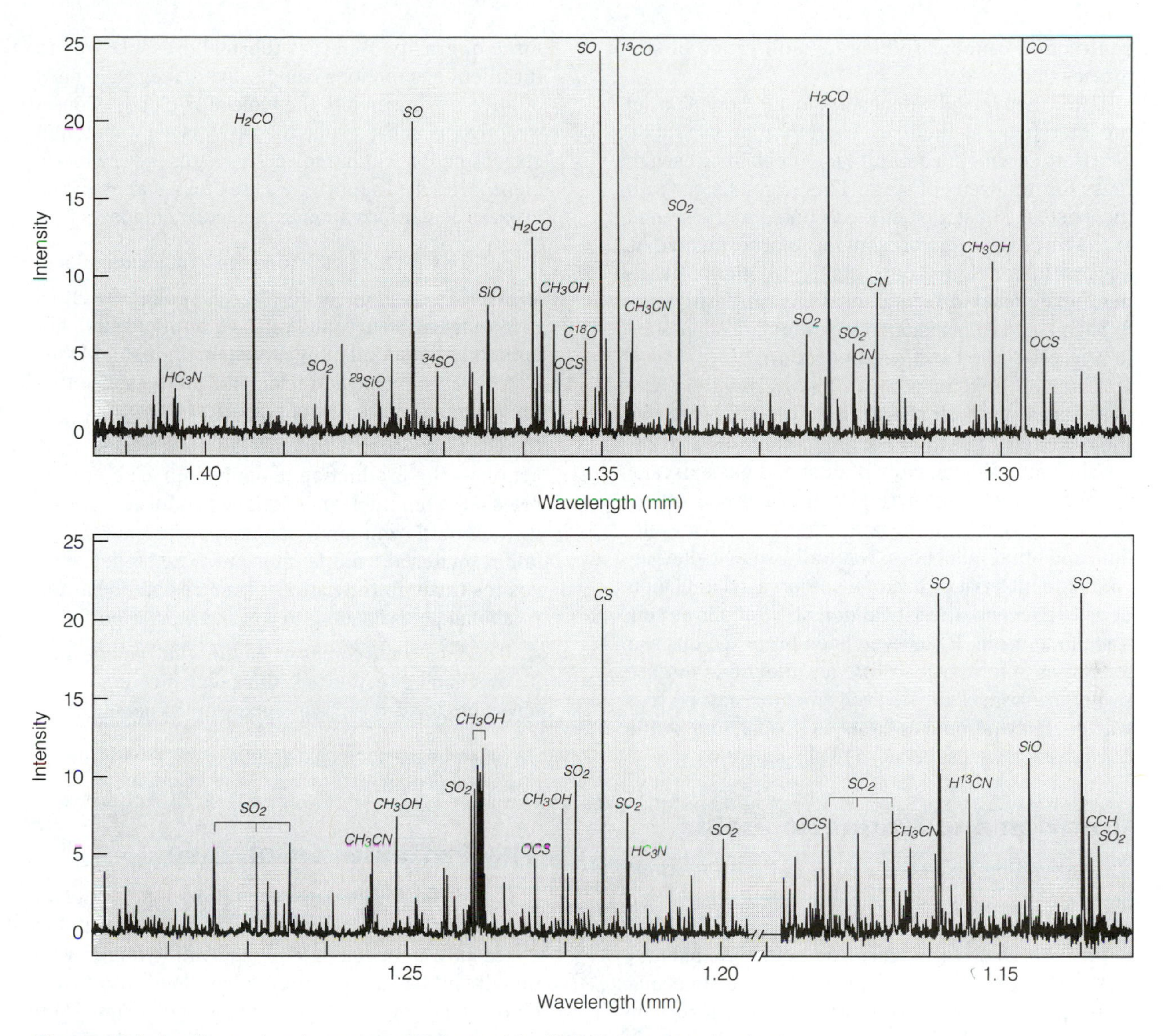

Figure 18-1 A radio wavelength spectrum of the core of the OMC-1 giant molecular cloud, made at the Owens Valley Radio Observatory. The spectrum covers a small interval in the atmospheric window at 1.3 mm. Over 800 spectral features are seen, corresponding to 29 different types of molecule. Such data are needed to understand the chemistry of cold and dense interstellar regions. (Courtesy of T. G. Phillips, California Institute of Technology)

(caused by a nearby stellar explosion, for example) that compresses it.

The Role of Gravitational Collapse

★ Astronomers want a more quantitative theory of star formation. In 1902, the English astrophysicist James Jeans calculated the combination of mass and temperature a cloud would need to have to begin its **gravitational contraction.** The calculation is idealized, because it does not consider the effects of magnetic fields or the possibility that the cloud might be rotating. For more information on how this theory can explain the collapse of matter on all sorts of scales, see our web site. Jeans' calculation showed that in the denser interstellar medium, at about 10^{-17} kg/m^3 and 10 K, masses contracting should be about 10^{33} kg, or several hundred to a thousand solar masses. This is several times larger than the most massive stars and thousands of times larger than the least massive stars. In other words, the simple process of gravitational contraction cannot produce the full range of star masses that we see.

How then do individual stars form? Conditions in the interstellar medium are such that the gas subdivides into enormous concentrations containing enough mass for hundreds of stars. This is consistent with the existence of star clusters, as noted in the second of the three proofs of present-day star formation. As a protocluster cloud contracts to still higher densities, smaller star-size condensations can form inside it. Shrinking protoclusters divide into individual stars in a process called **subfragmentation.**

The new fragments become stars, and the whole mass turns into a star cluster. It is not hard to see why the interstellar gas might begin to contract. The material is not uniform; clots of dust and gas exist and are constantly being stirred by the galaxy's rotation, material ejected from the later stages of stellar evolution, and other influences. Naturally, some clouds accumulate material or become compressed until their density exceeds the critical density that allows contraction to begin. It has even been proposed that star formation might spread like an infectious disease through a large cloud (or even an entire galaxy), as a wave of gravitational collapse is propagated by the compression with supernova explosions.

Rotation and Magnetic Fields

The preceding discussion does not tell the whole story of star formation. Gravitational collapse and subfragmentation may be relevant for the formation of globular clusters and large star clusters, but they fail to describe observed properties of molecular clouds, which account for most of the star formation in the disk of the Milky Way. For example, large-scale free-fall motions of gas have not been observed in molecular clouds. Also, star formation in molecular clouds occurs more slowly and less efficiently than the simple theory of gravitational collapse would lead us to expect.

What can keep a cloud from collapsing? Thermal pressure can oppose the force of gravity, but the cores of molecular clouds are both dense and cold, so thermal pressure is insufficient to prevent collapse. Molecular clouds could have **rotational support,** but the observed rotation is too slow to support the cloud and prevent collapse.

Current thinking is that the presence of magnetic fields prevents gravitational collapse in molecular clouds. These fields contribute a **magnetic support** in the form of a "pressure" that opposes the inward force of gravity. Whereas thermal pressure due to turbulent gas motions can dissipate, magnetic lines of force are frozen into the molecular clouds. Consequently, the collapse due to gravity is very slow, and star formation within molecular clouds is very inefficient. Modern theory envisages four stages in the process of star formation in molecular clouds:

1. A slowly rotating core forms in a molecular cloud.
2. The core becomes unstable and collapses into a protostar and a surrounding disk, both of which are embedded in an infalling envelope of dust and gas. The collapse phase is "inside-out," in the sense that the material nearest the core collapses first.
3. The third stage is the onset of the first fusion reactions—the production of deuterium. The energy released from nuclear reactions produces a stellar wind of outflowing material, in opposition to the infalling material from farther out. The stellar wind rushes through the paths of least resistance at the rotational poles, leading to jets and bipolar outflows.
4. The final stage comes when the wind widens until it flows in all directions. At this point, a young stellar object has formed, still surrounded by its nebular disk.

★ See our web site for an illustration of these early stages of star formation.

The Protostar's Collapse

The preceding discussion allows us to define a **protostar** as a cloud of interstellar dust and gas that is dense and cool enough to begin contracting gravitationally into a star. An interstellar cloud may hover on the verge of this state for millions of years. Then a nearby exploding star or some other disturbance may compress the gas cloud and thus trigger the collapse, which then proceeds very rapidly. For instance, typical protostars contract to stellar dimensions in 100,000 y—a wink of the cosmic eye. This explains why astronomers use the term *collapse.* The initial contraction is very rapid in terms of astronomical time. The collapse may not be smooth. The inner core of the cloud may collapse first, later absorbing the surrounding material; and under certain conditions, a single rotating cloud may break up into two or more stars. Of course, once the cloud begins to reach stellar dimensions (a few astronomical units, or 0.00001 pc), the atoms of gas bump into each other frequently enough to produce substantial outward pressure, so the collapse is slowed. At this point the object is still not yet on the main sequence.

THE PRE-MAIN-SEQUENCE STAGE

★ What are the forms of pre-main-sequence stars? Where do they lie on the H-R diagram? Can they actually be observed among the stars in space? Can we actually see stars being born? In fact, sites of active star formation are often shrouded in gas and dust (Cohen, 1988). The clearest view is at infrared wavelengths. Visible light is typically obscured by a factor of 10^5 to 10^6, but radiation at 2 μm in the infrared is obscured only by factors of three to five. (For an illustration of the detail visible in the infrared, see our web site, which shows optical and infrared views of M17, the Omega Nebula.)

The **pre-main-sequence star** stage covers the evolution from the end of the protostar stage to the main sequence. Astrophysicists such as the Japanese theorist C. Hayashi and his American colleague L. Henyey did pioneering calculations of the evolutionary tracks and appearance of stars contracting toward their main-sequence configurations.

The energy during most of the pre-main-sequence period does *not* come from nuclear reactions, which have not yet started inside the protostar; instead, heat is generated by gravitational contraction. After only a few thousand years of protostellar collapse, surface temperatures reach a few thousand kelvins, causing visible radiation. Hayashi's work showed that convection would transport large amounts of energy from the interiors of most newly forming stars, making them very bright for short periods known as *high-luminosity phases.* A star of solar mass, for example, contracts in less than 1000 y from a huge cloud to a size about 20 times bigger than the Sun with a luminosity about 100 times greater than the present Sun's.

These calculations have three important consequences. First, they show that stars have complicated, if short-lived, evolutionary histories even before nuclear reactions start. Second, they show that *newly forming stars have properties that place them above and to the right of the main sequence in an H-R diagram.* (This is important in identifying new stars by direct observation.) Third, the calculations indicate that stars spend only a small fraction of their lifetimes in the pre-main-sequence stage.

More massive stars evolve faster to the main sequence than less massive stars (see Figure 18-2). For instance, a 15 $M_\odot$ star reaches the main sequence in only 100,000 y. A 5 $M_\odot$ star takes about 1 million years, while a Sun-like star takes tens of millions of years; and the lowest mass stars, around 0.1 $M_\odot$, take over 100 million years. (Note: When astronomers talk about stars "moving" on the H-R diagram, they are not talking about motions or positions in space. They are describing a change in physical properties that causes a star's position to change in a plot of those properties.)

Prediction of Cocoon Nebulae and Infrared Stars

While some theorists were calculating evolving conditions in newly forming stars, others, such as the Mexican astronomer A. Poveda, hypothesized that as young stars form from collapsing clouds, remnants of the clouds might surround and obscure these stars. As in the early solar nebula, dust would form in the cooling cloud and block the outgoing starlight. After a few hundred thousand years, the nebula would eventually dissipate, revealing the star.

The term *cocoon nebula* describes the nebula that hides a star from view during its earliest formative period (perhaps a few million years for a solar-sized star). The nebula is cast off later, just as a cocoon is cast off by an emerging butterfly. The first cocoon nebula to be discovered is shown in Figure 18-3, in a sequence of photos showing real evolution in a stellar system over a period of nearly 60 y.

What would the newly forming star and cocoon nebula look like to an observer before the nebula dissipated? Could examples ever be observed? The outermost dust of the opaque cocoon would have a temperature of only a few hundred kelvins. Therefore the nebula would radiate not visible light but rather infrared light (review Wien's law), and the star-nebula complex, as seen from outside, would lie far to the right of the main sequence on an H-R diagram. Such an object could be detected as an *infrared star* at wavelengths of a few micrometers. As shown in Figure 18-2, it would evolve across the H-R diagram in less than a million years and thus be detectable as an infrared star for a relatively brief period of its life. Stars of this type should therefore represent only a small fraction of all stars.

Mass Limits, Large and Small

If a cloud is dense and cool enough to contract but has less than about 8% of a solar mass, it will contract but never develop a high enough central pressure and temperature to reach a main-sequence state (i.e., no extensive fusion of hydrogen). Protostars from about 0.01 to 0.08 $M_\odot$ may heat up temporarily due to their

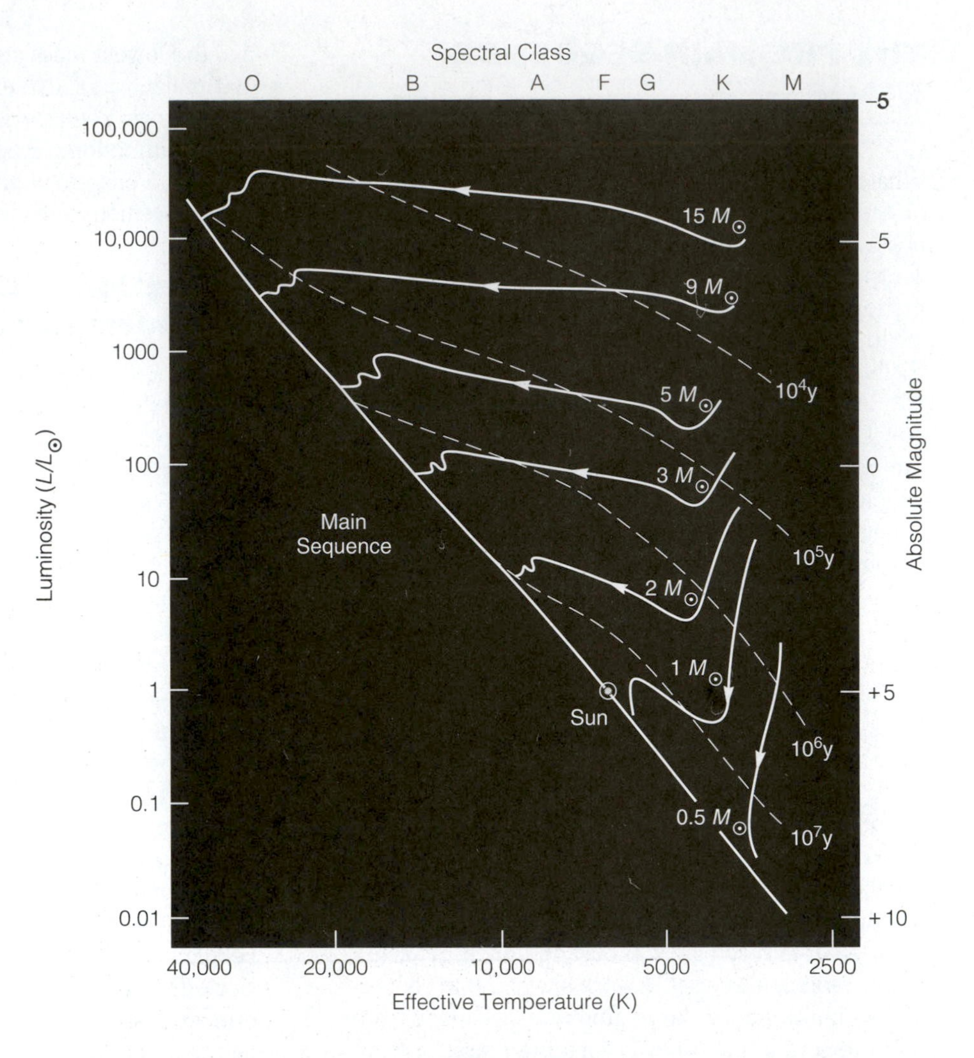

Figure 18-2 H-R diagram showing simplified pre-main-sequence tracks for stars of seven different masses. Note that the high-mass stars settle on the upper end of the main sequence, while low-mass stars are on the lower end. Dashed lines show the states reached after the indicated number of years. More detailed recent calculations—taking into account the nebula around the star and other details—show more complex "squiggles" in the evolutionary tracks. (Data from I. Iben)

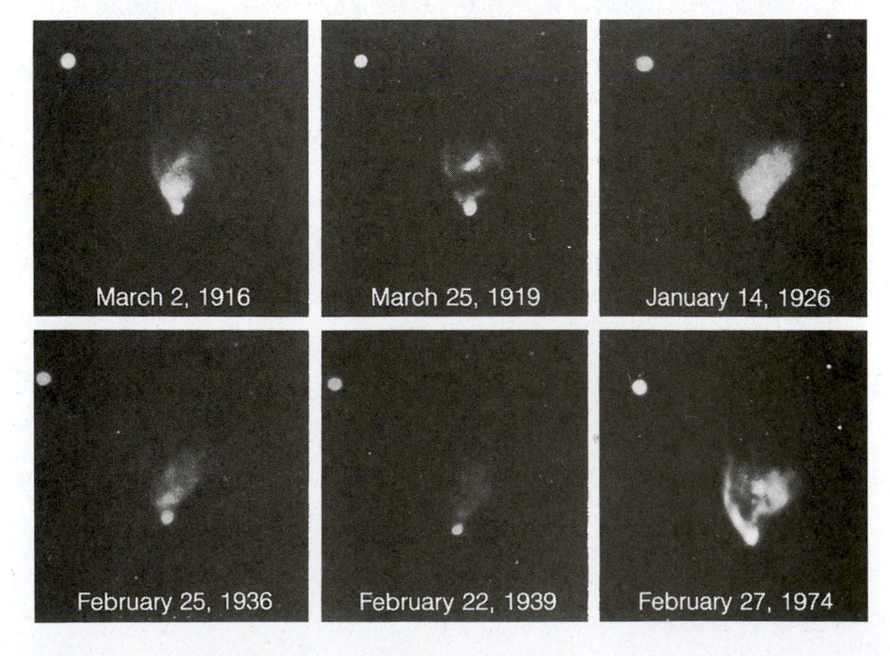

Figure 18-3 Exposures of the newly formed star R Monoceros over a 58-y period. The star is the bright object at the bottom tip of the irregular nebulosity. The nebula, called Hubble's variable nebula, changes shape from year to year, possibly due to changes in the illuminating radiation escaping from the cocoon nebula around the associated star. (Lowell Observatory; photo courtesy Alan Stockton, Mauna Kea Observatory, University of Hawaii)

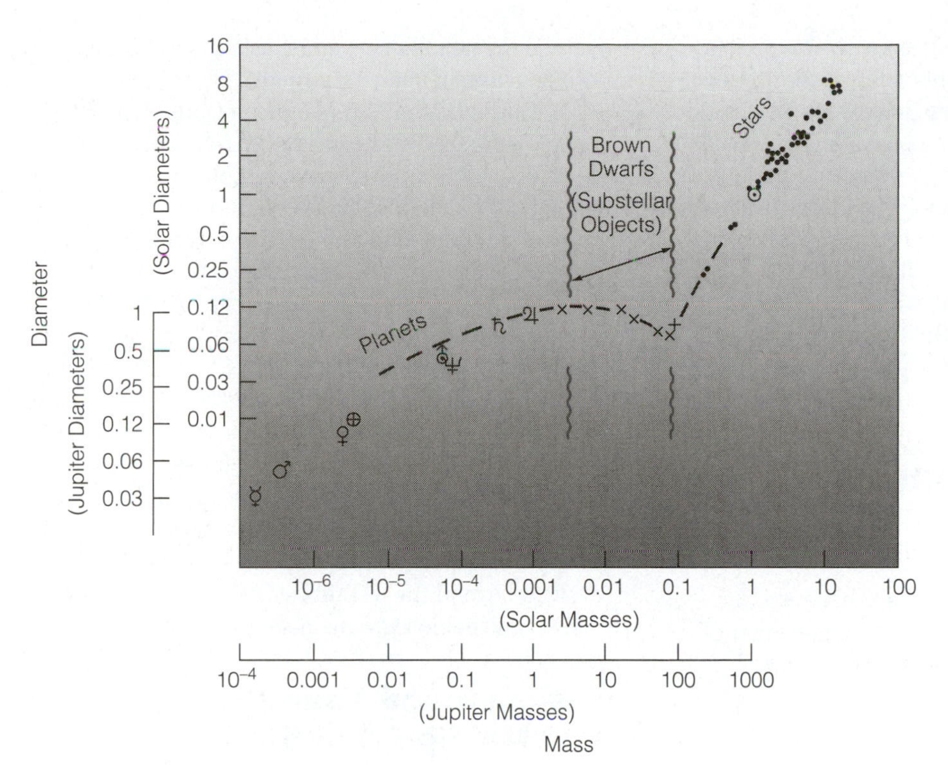

Figure 18-4 The transition from planets through brown dwarfs to stars. A combination of theoretical and observed data shows the changing size of objects as we add more mass. If we start with planetary masses (lower left) and add material of solar composition, the size levels off at about the size of Jupiter and then declines slightly due to compression by gravity. As additional mass pushes the total mass over 0.08 $M_{\odot}$, where nuclear reactions begin, the size again expands. (After data by D'Antona, Lunine, Hubbard, and others)

gravitational contraction, and may even develop a few feeble nuclear reactions, but eventually they fade without settling on the main sequence for any appreciable time. Smaller objects, including even Jupiter-size planets, also warm up temporarily due to their gravitational contraction; they may glow for a while in the infrared, but they fade before any nuclear reactions can start.

If the cloud is dense enough to contract but is more than about 100 $M_{\odot}$, the contraction is violent and produces an extremely high central temperature and pressure. Under these conditions, so much energy is generated inside the new star that the star is very luminous and may blow itself apart almost immediately without spending much time on the main sequence. This rapid destruction is more appropriate to the topic of the next chapter and will be taken up again there. We will see that explosions of massive stars explain many features of our starry surroundings.

The Transition from Planet to Star: Brown Dwarfs

Objects with less than 0.08 $M_{\odot}$ (80 times Jupiter's mass), but bigger than planets, are called **brown dwarfs,** or substellar objects, because, lacking nuclear reactions, they are not true stars. The name *brown dwarf* comes from the theory that they glow with a dim, red light. In fact, they emit nearly all of their radiation at invisible infrared wavelengths.

What would brown dwarfs really be like if we could see them up close? Suppose we could do this experiment: Start with Jupiter, keep adding solar-composition gas of hydrogen and helium, and go from planets through the brown dwarfs' size range to true stars. The result of this hypothetical experiment (or model) is shown in Figure 18-4. As we add mass to Jupiter, it gets only a little bigger until it reaches roughly 2 M_{Jupiter}, where it levels off. As we add even more mass, the object actually gets a little smaller, because gravity compresses the gas to a denser state. Then, as we approach 80 M_{Jupiter}, where nuclear reactions begin, it "turns on" as a star and again gets much bigger as we add more hydrogen fuel. Thus, conceptually, there are fairly natural dividing lines.

★ We will use the word *planet* for anything smaller than 2 M_{Jupiter} and the term *brown dwarf* or *substellar object* for objects from 2 to 80 M_{Jupiter}. They would get hot, and the larger ones would look red hot—our web site has an imaginary view. But they would lack the nuclear reactions of a true star. Only objects above 80 M_{Jupiter} have nuclear reactions and are classed as stars.

Astronomers have been searching for good examples of brown dwarfs, which would help us understand the relations between planets and stars. Brown dwarfs, of course, are hard to detect because of their faintness. A number of brown dwarfs have been detected in the past five years, found mostly using infrared imaging, which is well-suited to detecting cool objects. One has a luminosity only 0.0004 times that of the Sun—the lowest luminosity object yet found outside the solar system. It appears that about 10% of the mass density of the solar neighborhood is in the form of objects too cool to be fusing hydrogen into helium.

EXAMPLES OF PRE-MAIN-SEQUENCE OBJECTS

Several types of observed objects are believed to relate to late protostellar stages or pre-main-sequence stages of star formation (Stahler, 1991).

Infrared Stars and Cocoon Nebulae

★ Some of the best examples of infrared stars, cocoon nebulae, and other young stars are found in the *Orion star-forming region,* a large area of the sky around the constellation Orion. This region, which is about 400 to 700 pc away, is a hotbed of dense clouds and star-forming activity. When we look in this direction on a starry night, we can see many of the results of recent star formation (see a variety of images on our web site). Many newly forming objects lie hidden to our eyes inside dust clouds in interstellar space. But these are now being mapped by detectors sensitive to infrared light that passes through the clouds, even though the clouds block visible light.

These and other objects exhibit many features that tie in well not only with the developing theory of star formation, but also with evidence about conditions in the nebula that surrounded our own Sun as it formed. One of the most exciting implications of the observational work on newly formed stars is that we may be seeing dusty systems in which planets are forming now or might form in the "near" future—that is, in the next few million years (Lada, 1993).

T Tauri Stars

The most important pre-main-sequence stars are the **T Tauri stars,** named after the twentieth variable star cataloged in the constellation Taurus. Although a great many of them are found in the region of Taurus and the neighboring constellation Orion, they can also be found in many other parts of the sky.

T Tauri stars may represent a transitional stage between infrared stars surrounded by opaque nebulae and stable stars that have lost their cocoons and settled on the main sequence. There are many signs of their youth and instability. The number of T Tauri stars alone in a star-forming cubic parsec may exceed the number of *all* stars per cubic parsec near the Sun by more than a factor of 10. T Tauri stars vary irregularly in brightness. T Tauri stars lie to the right of the main sequence in the H-R diagram, where young stars are supposed to lie. Many have the infrared radiation characteristic of cocoon nebulae.

Judging from all available evidence, T Tauri stars are typically 20,000 to a million years old. This is younger than the human species! Stars that have evolved beyond the T Tauri stage would probably be indistinguishable from main-sequence stars.

Disks around Young and Middle-Aged Stars

Although there were many signs that cocoon nebulae might bear a connection to the proposed solar nebula in which the planets formed, there were no direct observations of disk-shaped nebulae around newly formed stars until the 1980s.

In 1983, however, the Infrared Astronomical Satellite (IRAS) was launched into orbit with the capability of mapping stars at the far-infrared wavelengths emitted by cool dust. IRAS promptly made a new discovery: some two dozen stars that had clouds of infrared-emitting dust extending hundreds of astronomical units from the star. The discovery led ground-based astronomers to use sophisticated techniques to make detailed images of these systems in order to discover their properties. Starting in 1984, astronomers began discovering flat, disk-shaped nebulae of dust near these stars. These discoveries seem to provide a "missing link" between the early stage of a ragged cocoon nebula and the later dusty disk required to form planets. HL Tauri, Beta Pictoris, and other main-sequence stars have been intensively studied. The edge-on disk around Beta Pictoris can be seen in Figure 18-5.

The Mystery of Mass Ejection from New Stars

What is the process by which the thick obscuring dust around a T Tauri star thins or disappears to reveal the star? Data show that as much as 0.4 $M_{\odot}$ of

Figure 18-5 The best pictorial evidence yet available of planetary dust around a nearby star, Beta Pictoris, 16 pc away. Thermal infrared radiation from the dust had been detected in 1983 by the IRAS telescope in space, leading astronomers to make this 1984 visual-light image from a ground-based telescope in Chile. The black central disk and crosshairs represent a system used to block the light of the star itself. With the star glare blocked, sensitive imaging equipment recorded bright material extending on either side of the star. This is an edge-on view of a dust disk extending to about the distance of the inner Oort swarm of comets in our own planetary system. (Photo by S. Larson and S. Tapia, taken at the Cerro Tololo International Observatory, Chile)

gas and dust is blown away from T Tauri stars as they evolve. Observations show chaotic gas motions around T Tauri stars. Doppler shifts revealed gas sometimes moving inward toward the star, but more often blowing outward as fast as 50 to 200 km/s. Astronomers in the 1970s visualized T Tauri stars as having strong *stellar winds* blowing materials out from them, analogous to the solar wind, but this did not explain occasional inward motions.

Observations in the 1980s revealed a mysterious phenomenon that we will encounter again and again, not only in T Tauri stars but in certain other kinds of stars and even in galaxies. This is the phenomenon of **bipolar jets.** Remember that gas and dust are organized in a thin disk in these systems. During bipolar outflows, the gas shoots out in opposite directions, perpendicular to the disk. Stellar winds may also blow some material away from some T Tauri stars in all directions. In many T Tauri stars, however, some gas and dust apparently spiral inward toward the star, get caught in magnetic fields around the star, and then get accelerated; they then squirt "upward" and "downward" in two diffuse jets away from the disk. The forces that accelerate these jets are probably magnetic but are poorly understood. For images showing model and actual observed bipolar jets, see our web site. Outflows are the subject of intense current research.

In 1986, a team of astronomers from Arizona and Missouri announced another "missing link" observation. They studied an infrared star named IRAS 1692A, first detected by IRAS. Bipolar jets had been found soon after the discovery of the star, but the 1986 observations revealed a gaseous disk lying perpendicular to the jets and reaching 800 AU from the star. The outer parts of the disk were found to be orbiting the star, but the inner parts were collapsing inward and feeding the jets. Astronomers have also studied small, variable nebulae called Herbig-Haro objects. In these objects, the shape and brightness can vary on a timescale of years or decades as gas is heated and ejected from the young star-forming region. These observations strongly support the theoretical model of young stellar disks and bipolar jets.

In summary, gas shedding by infrared stars and T Tauri stars, and the associated creation of bipolar jets and Herbig-Haro objects, is a challenging new area of astronomy unknown just a few years ago. It may be teaching us not only about the formation of stars, but also about what happens during the formation of planetary materials and planet systems.

Young Clusters and Associated Young Stars

Because stars form in groups, T Tauri and infrared stars are often found in groupings associated with star-forming regions. In the beginning of this chapter, we described clusters of stars proven to be very young by their dynamical properties and H-R diagram position. Because low-mass stars take the longest to

Figure 18-6 This young star cluster occurs near a site of intense star-forming activity. The star groupings have changed their appearance due to variable glows from supersonic gas masses near young stars. (George Herbig)

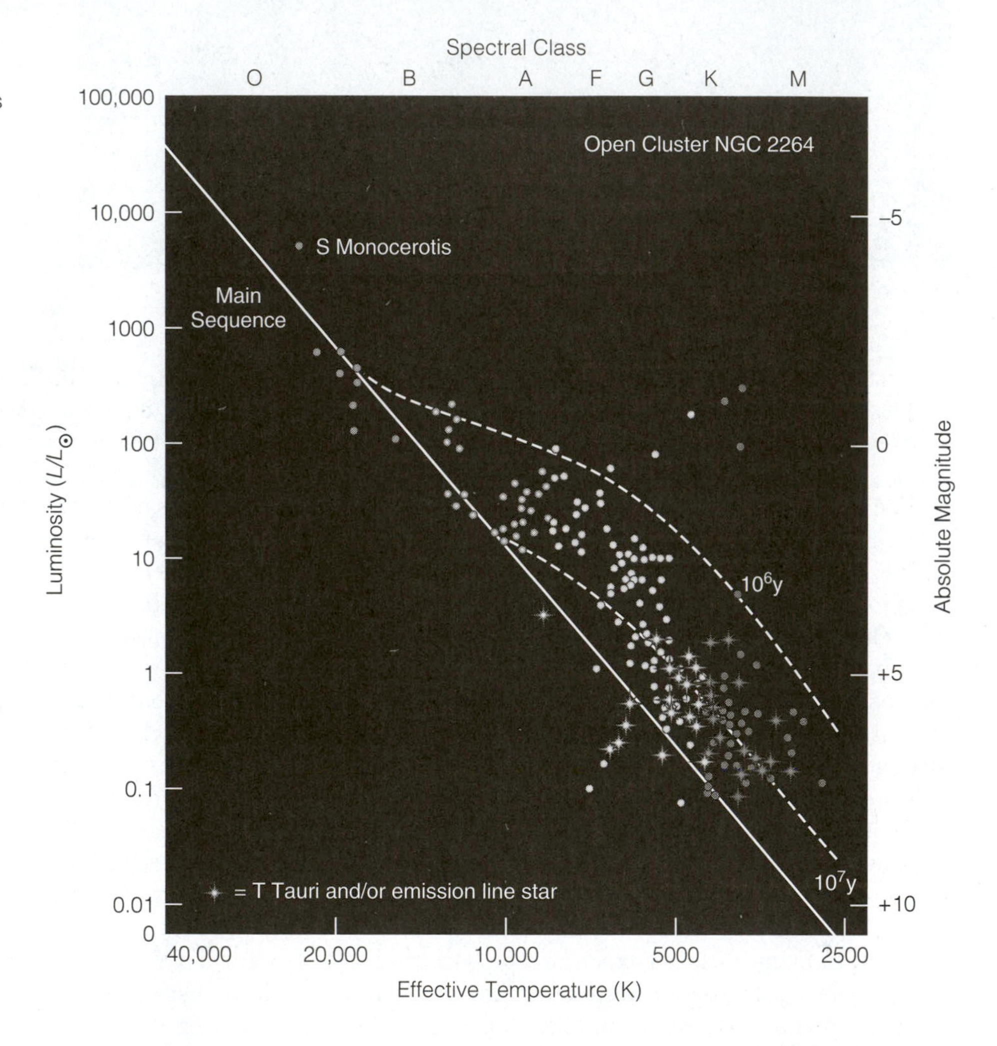

Figure 18-7 This H-R diagram of stars in the region around NGC 2264 reveals stars lying to the right of the main sequence along age lines (dashed) suggesting ages of only a few million years. (Data from M. Walker, with tracks of stellar evolution models or isochrones from I. Iben)

evolve to the main sequence, low-mass stars in these clusters ought to be found in the T Tauri stage. This has been abundantly confirmed by observations. For an evolving view of a young cluster, see Figure 18-6.

One of the best-known examples is the young star cluster NGC 2264. Most of the low-mass stars (spectral classes A to K) in this cluster lie distinctly to the right of the main sequence in an H-R diagram, as shown in Figure 18-7, and many of these are identified as T Tauri stars. The calculated lines of constant age (dashed lines in Figure 18-7) show that the T Tauri stars match the positions predicted for an age of 3 to 30 million years. This demonstrates how the H-R diagram can be used, together with theoretical data, to determine the ages of open clusters.

Recall our rule of thumb from Chapter 17 that massive stars evolve fastest. Brilliant, massive, bluish, O-type giants or supergiants last only a few million years. For this reason they are found only in star-forming regions a few million years old. By the time stars have dispersed from such regions, the O giants or supergiants have already burnt out. Thus they are rare among representative field stars, but they dominate star-forming regions.

THE MAIN-SEQUENCE STAGE

A star reaches the **main sequence** when it begins to generate energy by consuming hydrogen in nuclear reactions deep in the star's central regions. Prior to that time, the star generates energy primarily by its gravitational contraction, which raises the temperature and pressure in the central regions. As the pressure and temperature near the star's center increase, individual atomic nuclei eventually hit each other so hard that they begin to fuse together. These nuclear fusion reactions generate new heat and pressure that stop the contraction, causing the long-term stability of the star on the main sequence.

In the H-R diagram, an imaginary array of stars with different masses, which have all just reached the main sequence and just begun to consume hydrogen, is called the **zero-age main sequence.** Calculations show that it would be a very narrow band of stars on the H-R diagram. In reality, the main sequence on the H-R diagram is a bit broader because it contains stars of different ages, which have converted different fractions of their hydrogen to helium. Thus they have slightly different structures and slightly different positions on the H-R diagram.

Once the star reaches the zero-age main sequence, it begins its life as a true star and commences a long sequence of various nuclear reactions. During each such reaction, tiny amounts of matter are converted to energy, providing the heat and light of the star. The basic cause of further stellar evolution must be stressed: *The nuclear reactions convert light elements into heavier elements, changing the star's composition and its energy-generation rate; this in turn causes the star to alter its structure.* To clarify these changes, we will now review the nuclear reactions in stars in more detail.

Nuclear Reactions in Low-Mass Main-Sequence Stars

As described in Chapter 15, studies of our own main-sequence star, the Sun, reveal that its energy comes from a series of nuclear reactions called the **proton-proton chain.** We summarize it again because of its importance for stellar evolution.

$$^{1}\mathrm{H} + {}^{1}\mathrm{H} \rightarrow {}^{2}\mathrm{H} + e^{+} + \text{neutrino}$$

$$^{2}\mathrm{H} + {}^{1}\mathrm{H} \rightarrow {}^{3}\mathrm{He} + \text{photon}$$

$$^{3}\mathrm{He} + {}^{3}\mathrm{He} \rightarrow {}^{4}\mathrm{He} + {}^{1}\mathrm{H} + {}^{1}\mathrm{H} + \text{photon}$$

The proton-proton chain is the primary energy-producing process not only inside the Sun but also inside all main-sequence stars less massive than about 1.5 $M_{\odot}$. It dominates if the central temperatures are less than about 15 million kelvins.

Nuclear Reactions in High-Mass Main-Sequence Stars

In main-sequence stars more massive than about 1.5 $M_{\odot}$ where interior temperatures are higher than about 15 million kelvins, another reaction series dominates in producing energy. This is the **carbon cycle,** sometimes called the *CNO cycle* to reflect the involvement of carbon, nitrogen, and oxygen. The reactions are as follows:

$$^{12}\mathrm{C} + {}^{1}\mathrm{H} \rightarrow {}^{13}\mathrm{N} + \text{photon}$$

$$^{13}\mathrm{N} \rightarrow {}^{13}\mathrm{C} + e^{+} + \text{neutrino}$$

$$^{13}\mathrm{C} + {}^{1}\mathrm{H} \rightarrow {}^{14}\mathrm{N} + \text{photon}$$

$$^{14}\mathrm{N} + {}^{1}\mathrm{H} \rightarrow {}^{15}\mathrm{O} + \text{photon}$$

$$^{15}\mathrm{O} \rightarrow {}^{15}\mathrm{N} + e^{+} + \text{neutrino}$$

$$^{15}\mathrm{N} + {}^{1}\mathrm{H} \rightarrow {}^{12}\mathrm{C} + {}^{4}\mathrm{He}$$

Again the net result is that hydrogen atoms are used up to produce helium-4 atoms with an associated release of energy. In a sense, carbon acts as a catalyst (a stimulant of change), because carbon-12 reappears at the end of the cycle, to be used again in the first reaction of a subsequent cycle.

In main-sequence stars, one or the other of these reactions consumes hydrogen in the core. Major structural changes occur as the hydrogen is used up. In the next chapter, we will see what dramatic events ensue.

SUMMARY

Even today stars are forming within a few hundred parsecs of the Sun and in more distant regions of space. The starry sky is not a static scene but the site of continual births of new stars out of interstellar dust and gas. Many stars and star systems are less than a few million years old—much less than 1% of the age of our galaxy. Some have become visible since humanity evolved, though prominent newly formed stars have probably not appeared in our sky during recorded history.

Star formation begins with protostars, which are clouds of dust and gas that begin to contract due to their own gravity. They collapse fairly rapidly to stellar dimensions and become pre-main-sequence starlike objects. The process of star formation depends strongly on magnetic fields and rotation in the collapsing gas cloud. There has been much theoretical progress in the last two decades on understanding the basic features of star formation. Many of these features have been confirmed by observation, especially by infrared equipment, which detects the radiation from low-temperature dust in nebulae around the newly formed stars.

Among the objects revealed in this way are groups of stars evolving toward the main sequence. Many groups of stars and individual stars are surrounded and obscured by cocoon nebulae consisting of dust particles (probably silicates and ices similar to those that formed the first planetary material in our own solar system). More evolved objects, such as the T Tauri stars, appear to be shedding their cocoon nebulae and have almost reached the main sequence.

Once the main sequence is reached, energy is generated in the star as hydrogen converts into helium by means of the proton-proton cycle for smaller stars and the carbon cycle for larger stars.

CONCEPTS

theory of star formation	brown dwarf
gravitational contraction	T Tauri star
subfragmentation	bipolar jets
rotational support	main-sequence star
magnetic support	zero-age main sequence
protostar	proton-proton chain
pre-main-sequence star	carbon cycle

PROBLEMS

1. Did the sun and solar system form in the first half or last half of our galaxy's history?

2. Suppose that you magically smoothed out all inhomogeneities in the interstellar gas so that it was all uniform.

a. Would this help or hinder star formation?

b. What processes would keep the gas from staying uniform indefinitely?

c. Would star-forming conditions tend to return to normal?

3. Why do stars form in groups instead of alone?

4. How do theories of solar system formation and theories of star formation support each other? Contrast the sources of information on these two subjects.

5. Compare the time scale for significant evolution of massive pre-main-sequence stars with the time during which astronomers have recorded observations of such stars. Is it reasonable that some young, pre-main-sequence stars might show evolution-related fluctuations in their properties within the time that they have been observed?

6. Suppose a cluster of stars formed 3 million years ago. Why would you expect the H-R diagram of the cluster to show no stars on either the very high-mass end of the main sequence or its very low-mass end?

7. Suppose a new 1-$M_\odot$ star began forming about 10 pc from the Sun. What would Earth-based observers see during the next few million years? Consider infrared observers as well as naked-eye observers.

8. Why doesn't gravity immediately cause the collapse of all interstellar clouds?

9. Why does the structure of a star stabilize when it reaches the main sequence?

ADVANCED PROBLEM

10. Suppose you observe an infrared nebula whose strongest radiation comes at wavelength 10μm and which has absorption lines of solid silicates in its spectrum, as well as faint lines indicating a G-type star. Suppose the silicate lines are blueshifted by about 10^{-3} of their wavelength compared with the star's wavelength. What conclusions can you draw about this system?

PROJECT

1. On a clear night (an early evening in February or a late evening in December is ideal) scan the region of Orion with your naked eyes and compare it to other regions of the sky. Note the concentration of bright blue O-, B-, and A-type stars (such as Sirius and Rigel) and star clusters (such as Pleiades, Hyades, and the Orion Belt region) in this broad area. How do these features indicate that star formation has been going on in this general direction from the Sun in the last few percent of cosmic time?

CHAPTER

19 Stellar Evolution II: Death and Transfiguration

WHAT THE READER SHOULD WATCH FOR IN THIS CHAPTER

At every point in its life, a star is governed by the balance between gravity and internal pressure. This is called hydrostatic equilibrium. The cycle of star birth and death has distributed many of the chemical elements into space. Heavy elements are produced in massive stars and recycled into space as stars age and die. After a star has exhausted its hydrogen fuel, it must evolve to a different state. Late in their evolution, stars once again begin to shed mass, either in a steady wind or by blowing large bubbles of gas. Many stars go through a stage when their light output varies. The final state of a star is determined by the mass of its core. Most stars have low mass, and they become white dwarfs that slowly cool over billions of years. The rarer, high-mass stars die suddenly in a supernova explosion. After the violence of a supernova, the core may be turned into dense and extraordinary states of matter: neutron stars and black holes. Astronomers use Einstein's theory of relativity to understand the intense gravity that results from the final collapse of a star. ■

Stars are born and stars die. By this we mean that the nuclear reactions converting mass to energy in stars have a beginning and an end. The Sun, for example, spent some 10^7 y in its formative stages; it will spend about 10^{10} y on the main sequence, roughly 10^9 y in the giant state, and a shorter time in later unstable states consuming heavier elements until energy generation stops. This evolutionary sequence might be likened to the periods of human life: 9 mo in the womb, 65 y of normal life, 6 y of rapid aging, and perhaps a year of terminal illness. As we will soon see, stars do evolve to pathological terminal states, involving incredible forms of dense matter and processes unimagined until a few years ago.

Writer Ben Bova summed up star deaths by quoting one of Ernest Hemingway's characters, who is asked how he went bankrupt. "Two ways," he says. "Gradually and then suddenly." The gradual part is the slow expenditure of a star's hydrogen, which is scarcely noticeable to an observer. The sudden part—the subject of this chapter—comes as the star runs out of hydrogen. It goes into fits of activity, searching for new sources of fuel until, as Bova says, "gravity forecloses all the loans."

HYDROSTATIC EQUILIBRIUM

Stars are spherical balls of hot gas. Despite the complexity of the gas motions and the nuclear reactions, the internal structure of a star is governed by a simple balance between pressure and gravity. This is called hydrostatic equilibrium. A star is held together by the gravitational forces among all the gas particles. The gravity force

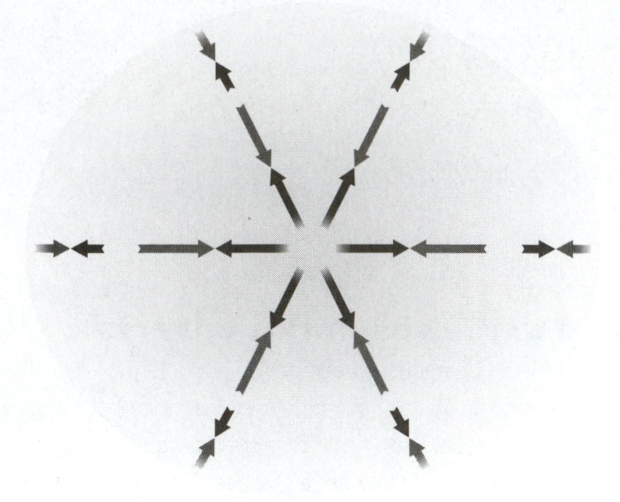

Figure 19-1 A stable star is in hydrostatic equilibrium, which means that at every point within the star the inward force of gravity is balanced by the outward pressure due to energy released from nuclear reactions.

is directed toward the center, where it compresses the gas and forces it to heat up. Greater heat means more rapid motion of the gas particles, which causes them to resist being close to each other. This internal pressure is the force that opposes gravity and stops the star from collapsing to a tiny size. Figure 19-1 shows the balance between pressure and gravity that exists at every point within a star.

The principle can be seen with a more down-to-Earth analogy. Take a bicycle pump, cover the end of the valve, and lean down hard on the plunger. Your weight will compress the air column within the pump, and you should notice two things. First, you cannot push the plunger down all the way. As you compress the air, it exerts an opposite pressure. After the air column has been compressed by a certain amount, the plunger motion stops, and an equilibrium is reached. Second, you will also notice that the base of the pump becomes warm. The gas has been compressed and heated to counter your downward force (which is analogous to gravity). If a larger person leans on the plunger, the air column will get smaller and warmer. A new equilibrium has been reached.

Almost all of a star's properties are determined by its mass and chemical composition, as discussed in Chapter 17. The mass dictates the gravitational force, which in turn determines how much pressure is needed to balance gravity. The internal pressure dictates the temperature, which in turn determines the energy that is released from the interior. Thus, there is a simple relationship between mass and luminosity for main-sequence stars. The hydrostatic equilibrium is disrupted only when there is a change in the energy supply from the nuclear reactions in the stellar core.

Main-Sequence Lifetimes

All stars on the main sequence are powered by the energy release from the fusion of hydrogen into helium. We know that 0.7% of the hydrogen mass is released as energy. (Equivalently, 0.7% of the proton mass is released in each reaction.)

We can use the relation between mass and luminosity and the energy-production rate to estimate the lifetime of a main-sequence star. The results are surprising. A hot star at the high-luminosity end of the main sequence may have a mass of about 100 $M_\odot$, but it has a luminosity of about 10^6 $L_\odot$, so it uses its energy a million times faster than the Sun. The estimated main-sequence lifetime is $100/10^6 = 10^{-4}$ that of the Sun, or only a million years (this is a slight underestimate, because high-mass stars use a larger fraction of their hydrogen than low-mass stars).

A cool star at the low-luminosity end of the main sequence may have a mass of only 0.1 $M_\odot$, but with a luminosity of only 10^{-3} $L_\odot$, it is using its energy 1000 times slower than the Sun. The estimated main-sequence lifetime in this case is $0.1/10^{-3} = 100$ times that of the Sun, or about 10^{12} y. This is much longer than the age of the universe, so all such low-mass stars are still on the main sequence. By contrast, high-mass stars evolve quickly and die. This is one of the reasons that a census of stars shows most of them to be low-mass stars. The other reason is that the star formation process tends to produce many more low-mass stars than high-mass stars.

The most massive stars are like old gas-guzzling cars. Despite large fuel tanks, such cars burn fuel at such a high rate that a full tank doesn't last very long. By contrast, low-mass stars on the main sequence are like small economy cars. The fuel tank is modest, but the rate of burning fuel is slow, and a full tank lasts a long time.

Beyond the Main Sequence

After a star leaves the main sequence, its destiny is governed by two forces. Gravity is an inward force trying to make the star collapse, and the energy-producing fusion reactions create a pressure that provides an outward force. When the core of a main-sequence star runs out of hydrogen, the balance between these forces—hydrostatic equilibrium—is disrupted. At first, the core may burn its way into outer layers that still have hydrogen. But as hydrogen fusion diminishes, so does the pressure that supports the star, and the core begins to collapse. The rapid production of energy in the collapsing core can cause the outer layers to expand so that the star temporarily balloons to a giant size. But eventually the hydrogen-burning reactions wind down. This means that less outward pressure is generated by heat released in nuclear reactions. Thus the core must continue to contract due to gravitational forces pulling inward. Although some of the upper layers may be blown outward, most of the star's mass contracts to a small size.

What happens after the hydrogen is exhausted and the core starts to shrink? Recall from the discussion of Helmholtz contraction that as a mass of gas contracts, it must grow hotter. Thus the atoms in the core move faster and collide even harder than before. They reach a condition where the helium atoms in the core collide hard enough to start fusion reactions. We say that the core has stopped burning hydrogen and is now burning helium. After the helium has burned, more contraction occurs and still other elements may become fuels. Eventually, there are no more fuels to react, and no new energy release to create a pressure that can oppose gravity. The star must therefore contract to a very dense state. This chapter discusses four main stages of stellar old age: the giant stage, variable stages, explosive stages of several types, and terminal high-density stages (Eichler, 1994).

Note that the most common stars—those about one solar mass—go through the giant phase and then contract rather smoothly to a small, high-density stage (Jastrow, 1990). But the rare, high-mass stars beyond about 8 $M_{\odot}$ go through explosive instabilities. Eventually, they reach even higher density stages than Sun-like stars, because they have more mass and a stronger gravity force. They produce some of the most exotic objects yet discovered by astronomers, such as pulsars and black holes. Table 19-1 gives examples of stars in the late stages of evolution.

THE GIANT STAGE

As described briefly in Chapter 17, a star evolves off the main sequence as hydrogen fuels run out. At this time a thin shell, in which hydrogen is still burning, surrounds the collapsing core. Outside the shell, temperatures have never risen high enough to "ignite" the hydrogen nuclei (that is, to fuse them into helium) so there is still plenty of hydrogen fuel. The collapsing core releases energy, which creates pressure, and this in turn drives the hydrogen-burning shell to layers further and further out. The increased temperatures cause great expansion of the outer layers, making the star evolve rapidly toward the giant state. The outermost atmosphere becomes huge, thin, and cool, even though the inner core is smaller, hotter, and denser than ever. From the outside, the outer atmospheric layers are seen to glow with a dull red color, and the star is perceived as an enormous **red giant.**

The largest giants are truly huge—approaching 1000 times the size of the Sun. If the Sun were replaced by one of these stars, its thin, red-glowing outer atmosphere would reach nearly to the orbit of Jupiter!

As charted on the H-R diagram (Figure 19-2), stars from all points along the main sequence display a funneling effect: As they move off the main sequence, their evolutionary tracks funnel into the red giant region. They resemble patients from all walks of life crowding into the same hospital because they are victims of the same malady—hydrogen exhaustion.

Meanwhile, the core of a star as massive as the Sun contracts until it reaches a temperature near 200 million kelvins. This is hot enough to begin to fuse helium nuclei in the central region, primarily by the **triple-alpha process** (named after the alpha particle, another name for the helium-4 nucleus):

$$^{4}\text{He} + {}^{4}\text{He} \rightarrow {}^{8}\text{Be} + \text{photon}$$

$$^{8}\text{Be} + {}^{4}\text{He} \rightarrow {}^{12}\text{C} + \text{photon}$$

In this process, three helium-4 nuclei combine to produce a carbon-12 nucleus. (Because beryllium-8 is unstable, some beryllium atoms may break up before completing the process—but this merely reduces the efficiency of the process.) Notice that we are now fusing helium to make an even heavier element, carbon.

The triple-alpha process produces prodigious energy. In low-mass stars, this energy release rapidly heats the core, creating a burst of helium burning called the **helium flash.** In some stars, the helium flash may consume the central core's helium in only a few seconds. Since the energy from the flash diffuses through the star slowly, the heating effects seen at the surface may last thousands of years. The flash occurs as the star enters the giant region. In a solar-type star, the flash may occur 300 million years after evolution off the main sequence.

Note that the core's evolution begins to be independent of the evolution of the outer atmosphere. The characteristics of the star as perceived by an astronomer on Earth are those of the outer atmosphere. These characteristics determine the star's position on the H-R diagram. When you go outside at night and look at Betelgeuse or Antares, you are seeing the cool, red, outer atmosphere of a star. Hidden inside is a very hot, dense core. Thus, although we may say

TABLE 19-1

Examples of Stars in Middle and Late Evolution

Stage of Evolution[a]	Distance from Earth (pc)	Mass ($M/M_\odot$)	Radius ($R/R_\odot$)	Luminosity ($L/L_\odot$)	Spectral Type	Average Density[b] (kg/m^3)
Main Sequence						
Sun	<1	1	1	1	G2	1420
α Centauri B	1.3	0.85	1.2	0.36	K4	700
Procyon A	3.5	1.7	2.3	6	F5	200
Algol A	31	4	3.0	100	B8	210
Giant						
Arcturus	11	~4	25	115	K1	0.36
Aldebaran A	21	~4	45	125	K5	0.07
β Pegasi	64	~5	140	400	M2	0.003
Supergiant						
Antares A	160	~12	700	9000	M1	0.00005
Betelgeuse	170	~18	~700	11,000	M2	0.00007
VV Cephei A	200–1200?	20–80?	400–1600?	350–10,000?	M2	0.00003?
Variable and Explosive Stars						
Mira A (long-period variable)	77	2	230	Up to 1100	M6	~0.0002
Polaris A (Cepheid variable)	120	~6	~25	830	F8	~0.5
HD 193576B (Wolf-Rayet)	?	12	~7	—	"WR"	~50
White Dwarf						
Sirius B	2.7	1.0	0.005	0.002	A3	130,000,000
Procyon B	3.5	0.65	0.01	0.0005	F	120,000,000
Pulsar (neutron star)						
Crab Nebula pulsar	1100	2–3	<0.00002	High in UV, X ray	—	~10^{17}

Sources: Data from Burnham (1978), Liebert (1980), Baize (1980), and Bonneau and others (1982).
[a] Listed in order of increasing age.
[b] Compare these densities of familiar materials: atmosphere at sea level, 1.2 kg/m^3; water, 1000 kg/m^3; lead, 11,300 kg/m^3.

"the star" is cooling and getting redder, it is really the star's atmosphere that is cooling. All this time, the core is shrinking and growing hotter.

As the core contracts and gets hotter, it initiates reactions involving even heavier elements. The elements synthesized depend on the mass of the star. Imagine a set of massive red giants with different masses: 4, 6, 8, 10 $M_\odot$ and so on. Those of about 4 or 6 $M_\odot$ will have helium-rich cores hot enough to ignite the helium nuclei and fuse them into carbon, in the triple-alpha process just described. Those of around 8 $M_\odot$ will have hot enough cores to ignite the carbon

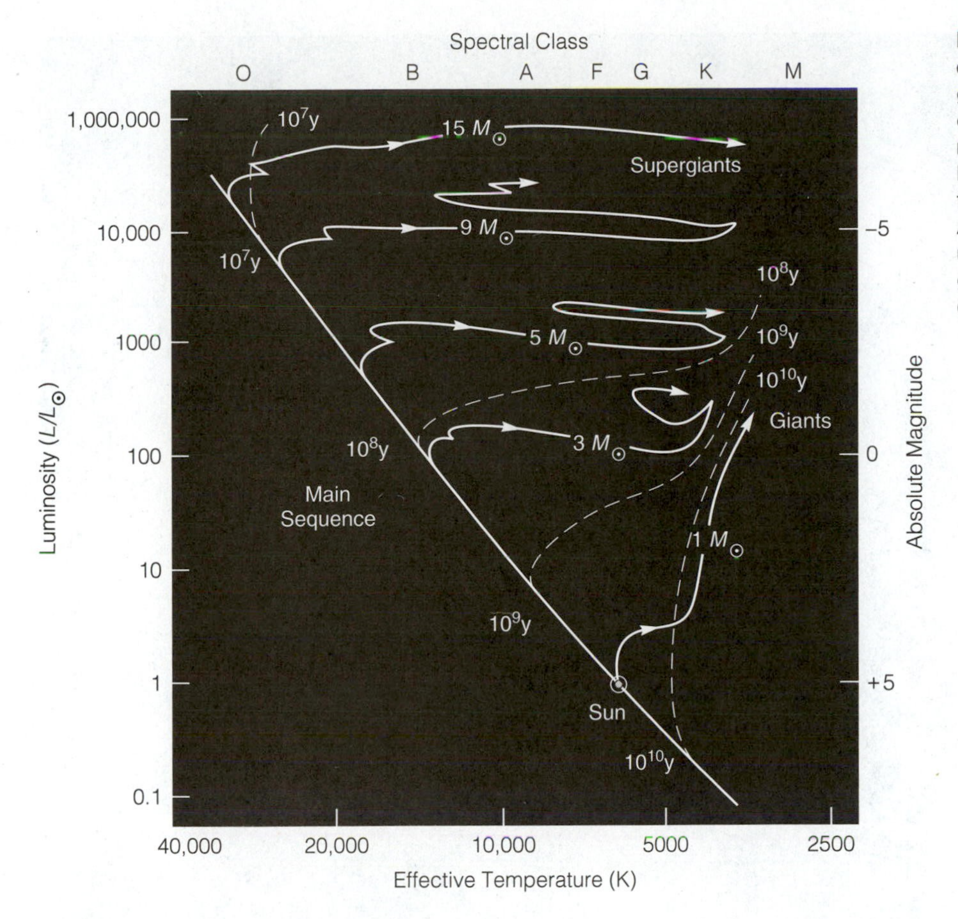

Figure 19-2 Evolutionary tracks off the main sequence toward the giant region, plotted on the H-R diagram for stars of different mass. Dashed lines give the length of time since star formation; massive stars evolve fastest. All the stars evolve toward the upper right corner of the H-R diagram, the region of giant stars. (After calculations by I. Iben)

and fuse it into heavier elements such as oxygen, neon, and magnesium. The reactions in which helium nuclei are fused to heavier nuclei include:

$$^{12}C + {}^4He \rightarrow {}^{16}O + \text{photon}$$

$$^{16}O + {}^4He \rightarrow {}^{20}Ne + \text{photon}$$

These reactions require temperatures above 500 million kelvins. Heavy nuclei can also fuse with each other, although in this case the electrical repulsion between protons in the nuclei is stronger, and a temperature above 1 billion kelvins is needed for the reactions to proceed:

$$^{12}C + {}^{12}C \rightarrow {}^{24}Mg + \text{photon}$$

$$^{16}O + {}^{16}O \rightarrow {}^{32}S + \text{photon}$$

$$^{28}Si + {}^{28}Si \rightarrow {}^{56}Ni + \text{photon}$$

The last of these reactions produces nickel-56, which is radioactive. It decays rapidly, first into cobalt-56, then into a normal iron-56 nucleus. Note that the continued addition of helium nuclei leads to the preferential formation of heavy elements with nuclear masses that are multiples of four. In stars of around 10 to 12 $M_\odot$, a long series of reactions will fuse nuclei into elements as heavy as iron. The process has an important limitation at this point. Iron has the most stable nuclear configuration of any element. This means that energy is *consumed,* not produced, as iron nuclei fuse into heavier elements. Thus the iron cores of stars do not continue to ignite if they contract and get hotter.

Therefore, if the star is massive enough, the core-building process leads to a core consisting of shells of different elements, surrounding an inner core of iron, as illustrated in Figure 19-3. The shell structure can be so complex that different fusion reactions can be happening at the surfaces of different shells, all at the same time.

Although helium capture does not make elements heavier than iron, there is another mechanism that can. Many of the reactions described so far release floods of neutrons that strike nearby nuclei. Since neutrons have no electric charge, heavy nuclei can be built up by ***neutron capture***.. In these ***s-process reactions,*** neutrons are slowly or gradually added to nuclei to build still heavier elements. This process may sound esoteric, but it is the origin of much of the copper and silver in the coins in our pockets and the gold in the jewelry on our bodies.

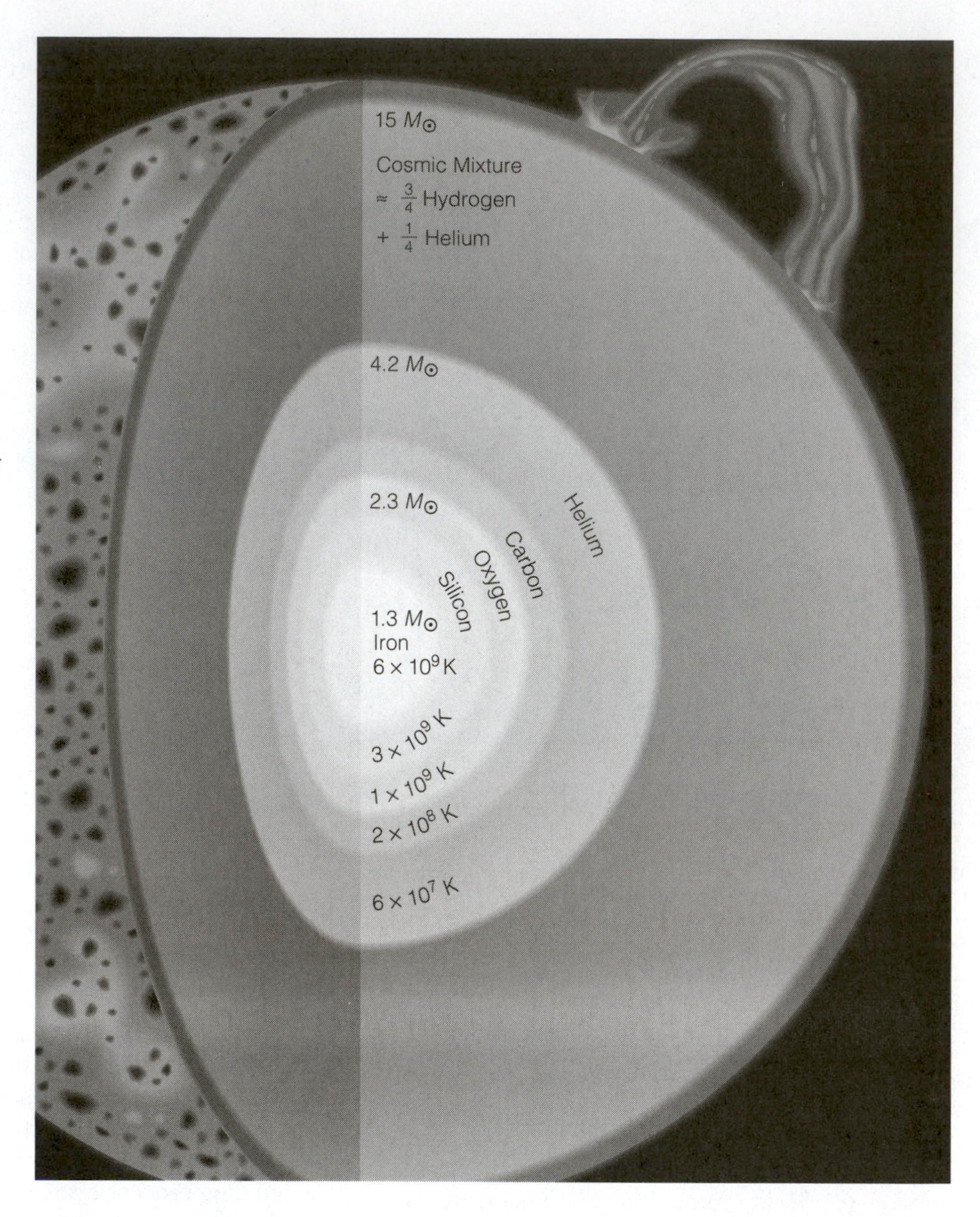

Figure 19-3 Schematic diagram of onion-skin layering of different elements as calculated for the pre-collapse core of a 15 $M_\odot$ star. The dominant element in each layer is given at the top; each layer's mean temperature is given at the bottom. Figures also give the mass of the material out to the designated layer. Nuclear reactions in successively deeper layers fuse nuclei into successively heavier elements. The core and inner layers are greatly enlarged to show their structure. (After diagram by Burrows, 1987)

THE VARIABLE STAGE

A **variable star** is a star that varies in brightness on a time scale of hours to years. Most variable stars apparently represent postgiant stages of evolution (although, as we saw in the last chapter, some variables are pre-main-sequence stars). Some pulse with a constant period; others flare up sporadically, often brightening by hundreds of percent and then fading again.

Over 25,000 variable stars have been cataloged and divided into as many as 28 types. Among these are about 2000 eruptive stars and about 400 explosive stars, whose variations are particularly violent. Most variables have regular light fluctuations. The type most important to astronomers is the **Cepheid variable,** which has regular variations in brightness, usually with periods from 5 to 30 d. (Polaris is, in fact, a Cepheid variable.) In 1784, the 19-y-old English astronomer John Goodricke[1] discovered the first Cepheid

[1]Goodricke offers an inspirational example of accomplishment during a short, difficult life. Deaf from infancy, he lived during the first European generation to recognize that deafness was not the same as idiocy. He attended the first English school for deaf children, took up astronomy, and made numerous interesting observations, but died at age 21 from pneumonia, possibly contracted from nighttime exposure during prolonged observing efforts.

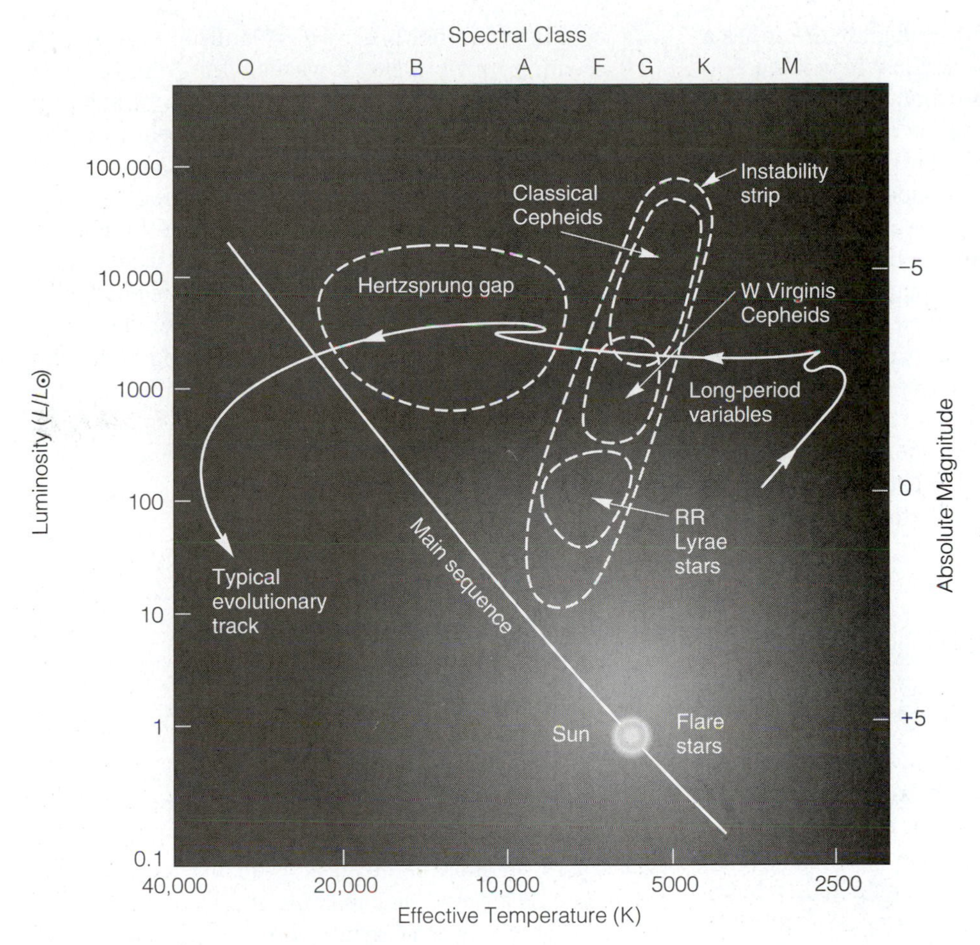

Figure 19-4 An H-R diagram showing the later stages of evolution, indicated by a typical evolutionary track (in white). The star evolves across an instability strip that contains Cepheid variable stars and RR Lyrae stars and then rapidly across the Hertzsprung gap.

variable, Delta Cephei. It varies by nearly a factor of two in apparent brightness with a period of 5.4 d; Cepheids were named after that star. Over 600 have been cataloged.

Cepheids are important for two reasons. First, because their variations are regular, they are somewhat better understood than stars whose brightness changes are unpredictable (called *irregular variables*). Second, and more important, the period of variation of each Cepheid directly correlates with its average luminosity. This relation was discovered in 1912 by one of the most famous women astronomers, Henrietta S. Leavitt, who measured hundreds of photos of Cepheid variables in the first years of the twentieth century at Harvard Observatory. Two types of Cepheids and a related type of variable called an RR Lyrae star were eventually found to have distinct period-luminosity relationships. These relationships allow astronomers to determine the luminosity of any Cepheid at any distance simply by measuring its period. This, in turn, leads to a new way to measure the distances of stars (to be discussed in Chapter 22): Find a Cepheid; measure its period and hence its luminosity, then measure its apparent brightness and derive its distance by the inverse square law.

Cause of Brightness Variations

Most variables change brightness because of changes in the way that energy flows from their centers. Variables vary in physical size and other properties as well as in brightness. Astrophysicists have long sought the precise cause of the internal physical disturbances that make Cepheid variable stars pulsate. Various researchers have established that the H-R diagram contains a nearly vertical *instability strip,* shown in Figure 19-4. Stars become unstable where their evolution gives them physical properties corresponding to this strip. Inside such stars, ionized helium absorbs outgoing radiation and thus becomes doubly ionized. In the instability strip, this absorption occurs in the outer atmosphere, not far below the star's visible surface, where it temporarily "dams up" the radiation. Again, we are seeing a phenomenon of the star's *outer* layers; the inner core may still be very hot, small, and dense.

Just like a bell (or any other object), a star has a certain frequency or time period in which it tends to vibrate in response to a disturbance. If the time required to dam up the radiation is close to the star's natural oscillation period, the star will begin to oscillate repeatedly. This is what happens in a Cepheid. As the radiation is blocked, pressure builds up, and the outer layers of the star expand. But once the expansion gets going, momentum and the star's natural tendency to oscillate at this speed carry the expansion too far. The atmosphere expands and becomes thinner, allowing radiation to escape easily. Then the star subsides, and the radiation once again begins to be dammed, restarting the cycle. In other types of stars, where the time scales of radiation damming and oscillation are not synchronized, the variation occurs irregularly.

MASS LOSS AMONG EVOLVED STARS

As the outer atmospheres of evolved stars expand, some of the gas may be blown entirely free of the star. After a red giant expands, for example, some of the gas of the outer atmosphere may continue to move outward and be blown out of the star altogether. Many red giants are losing mass. This process can be observed from blueshifted absorption lines arising in gas layers moving out of the star toward the observer.

Supergiants

The more massive a star, the more rapid and violent its evolution. Stars of more than a few solar masses are very hot and evolve rapidly. When their atmospheres expand, these stars leave the main sequence at a luminosity already brighter than most giants. They are thus called **supergiants**—at a given temperature, they must have a larger surface area to emit a larger amount of energy. Supergiants are the largest and most luminous stars on the H-R diagram. As described at the beginning of the chapter, the collapsing core generates energy, which drives the outer atmosphere of the star into space, resulting in mass loss. Because massive stars evolve quickly, the mass loss phase happens only a few million or tens of millions of years after the star's formation.

Planetary Nebulae

The gas being blown out of mass-shedding stars expands into space. It may cool enough for grains of dust, such as carbon grains, to condense in it. It may collide with other nearby gas clouds at high speed, creating glowing shock waves. Often, however, it is shed in relatively spherical bubbles of gas that surround the central star. Ultraviolet light from the star excites and ionizes the gas atoms, causing them to glow. Clouds of gas in space are called nebulae, and decades ago these particular nebulae came to be called **planetary nebulae** because the palely glowing bubbles looked like disks of planets in small telescopes (Soaker, 1992). This term is a misnomer, however, because these nebulae have nothing to do with planets. They are among the most beautiful celestial features, with wispy symmetry and delicate colors. At their cores, planetary nebulae often leave behind a white dwarf.

THE DEMISE OF SUN-LIKE STARS: WHITE DWARFS

Let us now consider the further evolution of Sun-like stars, by which we mean stars with initial mass similar to that of the Sun, say, from 0.1 to a few solar masses (Whitmire and Reynolds, 1990). The outer atmosphere is large and diffuse and may be shedding its outer layers, but most of the mass is still in a dense and hot core. As always, the fate of the star is determined by the battle between gravity and the pressure generated by nuclear reactions. Low-mass stars cannot generate nuclei heavier than carbon. Eventually, the core runs out of fuel, and without pressure support, the core collapses. Since the core cannot get hot enough to ignite any new nuclear fuel, this final collapse creates a very small, very dense star the size of Earth or other planets. Such a star is known as a **white dwarf.** Several thousand have been cataloged.

The Discovery and Nature of White Dwarfs

In 1844, the German astronomer Friedrich Bessel studied the motions of the brightest star in the sky, Sirius, and found that it was being perturbed back and forth by a faint, unseen star orbiting around it. This star was not glimpsed until 1862, when American telescope maker Alvan Clark detected it. It is almost lost in the glare of Sirius, as shown in Figure 19-5. In 1915, Mt. Wilson observer W. S. Adams discovered that the companion was hot and bluish-white, with properties that place it below the main sequence on the H-R diagram. It has about the mass of the Sun (determined from the binary orbit), but it is has such a low luminosity that the total radiating surface

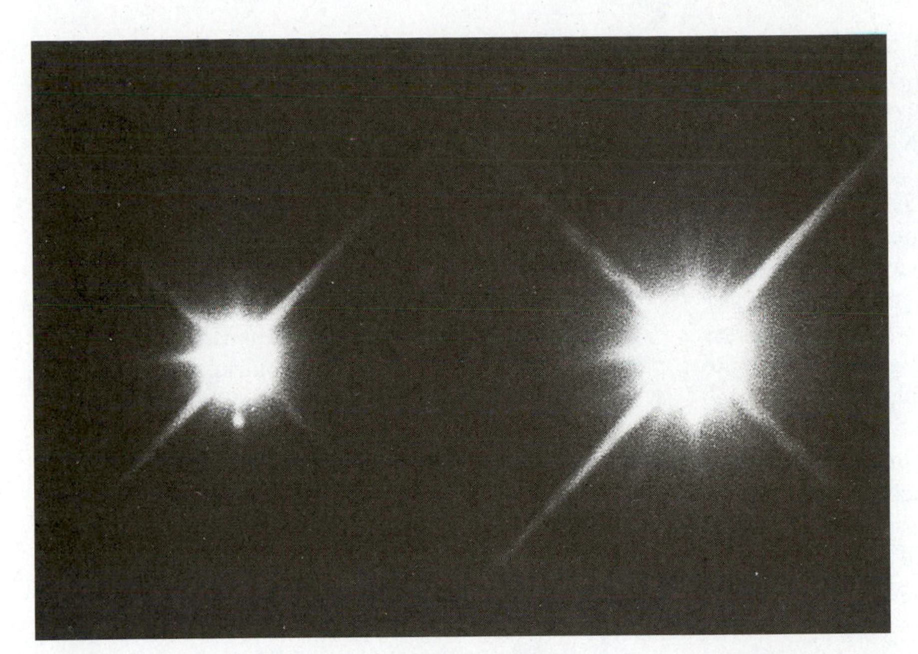

Figure 19-5 Two exposures of Sirius with the Lick 3-m telescope. Short exposure (left) reveals the faint white dwarf (below Sirius) that is the binary companion roughly 25 AU from Sirius; in the long exposure (right), the white dwarf is lost in the overexposed image. "Rays of light" are artifacts of the telescope optics. (Lick Observatory)

cannot be much more than that of the Earth (determined by the Stefan-Boltzmann law). Thus Sirius' companion turned out to be the first white dwarf to be discovered.

Theoretical astrophysicists such as William Fowler and S. Chandrasekhar explained the evolution of white dwarfs. Once no more energy is available to generate outward pressure, a star collapses until all its atoms are jammed together to make a very dense material. But what does "jammed together" really mean? At these temperatures and densities, electrons move at almost the speed of light, and matter loses its familiar properties. Stellar matter stops behaving like a perfect gas. In fact, it is no longer either gas, liquid, or solid, but a new form known as **degenerate matter.**

Degenerate matter, in which the spacing between particles is determined by quantum effects, is composed of atomic nuclei held apart by a sea of electrons at densities of about 10^8 to 10^{11} kg/m^3. Such a density is extraordinary; a teaspoon full of white dwarf matter brought to Earth would weigh as much as an elephant. White dwarfs are sometimes called *degenerate stars* (Kaler, 1991).

A physical law called the *Pauli exclusion principle* states that no two electrons can have the same set of properties (energy, position, velocity, and so on). At the microscopic level, this principle, which describes the basis for quantum effects, leads to a so-called degeneracy pressure in the dense electron gas of degenerate matter. The positively charged atomic nuclei align themselves in orderly patterns, governed by the electrical forces that act between them. The degenerate electrons move freely through this crystalline lattice, just as normal electrons move through the copper lattice of a wire when an electric current flows. This will be the fate of the Sun in about 5 billion years.

★ Most white dwarfs have temperatures of 10,000 to 20,000 K; the hottest exceed 100,000 K (see our web site for the image of one of the hottest stars known). White dwarfs have low luminosities and large amounts of stored internal energy, so they take a long time to radiate enough energy to cool significantly. White dwarfs thus last a long time. The coolest white dwarfs have temperatures under 4000 K and luminosities less than 0.0001 $L_{\odot}$, and they may have cooled for at least 10 billion years to reach these temperatures. Indeed, some white dwarfs may be among the oldest stars we can observe (Kawaler and Winget, 1987).

Although the interiors of white dwarfs may have densities of billions of kilograms per cubic meter, the outer layers of some may consist of ordinary matter—possibly hot gas at the surface and crystalline rock-like or glasslike solids in a crust 20 to 75 km deep. At the base of these crusts, densities may be as high as 3 million kilograms per cubic meter! Theoreticians predicted pulsations with periods of a few minutes in white dwarfs of certain temperatures, and these pulsations have been confirmed observationally.

The Most Massive White Dwarf: The Chandrasekhar Limit

White dwarfs cannot have more mass than 1.4 $M_{\odot}$ because the white dwarf structure becomes unstable at this point. If you tried to dump more mass on the surface of a white dwarf of 1.4 $M_{\odot}$, its gravity would become so strong that it would overcome the resistance of the electrons to denser packing. A still denser state of matter would arise.

This critical mass, 1.4 $M_{\odot}$, is called the **Chandrasekhar limit** after its Indian-American astrophysicist discoverer. The Chandrasekhar limit applies only to the *core* mass of the star and not to the *initial* mass. Current data suggest that stars with initial masses from about 0.08 to about 8 $M_{\odot}$ evolve into white dwarfs. The more massive examples do so by developing strong stellar winds (like the solar wind) or eruptive explosions that blow off mass until they are below the Chandrasekhar limit. According to calculations, the Sun will probably blow off about 40% of its mass when it goes through its red giant stage about 5 billion years from now. Thus it will collapse into a white dwarf of about 0.6 $M_{\odot}$.

THE DEMISE OF VERY MASSIVE STARS: SUPERNOVAE

At the moment of collapse, the core of a massive red giant has consumed as much fuel as possible, burning its way out into the outer parts of the star. The core collapse starts because there is no point in the core where there is both potential fuel *and* the possibility to create a hot enough temperature to ignite it. In a star with an initial mass of 6 $M_{\odot}$, the core may include the central 1.1 $M_{\odot}$. In a star with an initial mass of 8 $M_{\odot}$, the core may encompass 1.4 $M_{\odot}$. In a larger star, the eventual core mass can exceed the Chandrasekhar limit. What is the state of matter of such a massive core?

The collapse of the core to tiny size is aided by the fact that the iron "slag-heap" core will not ignite, even though it gets extremely hot and dense. The gravitational energy of the collapse is released as a prodigious burst of energy. The outburst of energy blows off all the outer layers of the star in a titanic explosion called a **supernova** (plural: *supernovae*).

The Nature of a Supernova Explosion

The advanced evolutionary stages of a massive star represent a crescendo of activity (Marschall, 1994). After millions of years spent fusing hydrogen into helium, each of the subsequent fusion stages (carbon, neon, oxygen, and silicon burning) takes less than 1000 y. The conversion of silicon and sulfur to iron takes only a few days. As pointed out earlier, iron has the most stable nuclear configuration, so no more energy can be released by rearranging the nucleus. The next stage is **core collapse.**

A supernova represents a rare astronomical event of unspeakable violence. Astrophysicists have gained insight from computer models, but the details are uncertain. Calculations suggest that core collapse takes only a few seconds! The density rises by a factor of a million as a volume the size of the Earth shrinks to a radius of about 50 km. The iron-rich core collapses at nearly a quarter of the speed of light. As the core gets compressed to the density of an atomic nucleus, forces between particles cause it to rebound, and on its way out it meets material that is still falling in. When matter meets matter traveling at supersonic speeds, the result is a *shock wave,* giving compression and a rapid temperature rise up to billions of degrees. The energy of the explosion is easily sufficient to eject the outer envelope of the star and to momentarily take the star to 10 billion times the luminosity of the Sun!

The death of a massive star takes only seconds, but it has long-term consequences. The first is **explosive nucleosynthesis.** In the high temperatures of the explosion, elements more massive than iron can be synthesized. The high flux of free neutrons leads to a rapid form of neutron capture, called the *r-process reactions.* A combination of s-process and r-process reactions creates small amounts of elements all the way up to plutonium-242. Moreover, the explosion flings all the heavy elements formed in the star's brief lifetime deep into interstellar space. This is the source of most of the rare elements and precious metals in the world (a portion comes from a slower process of neutron capture), and it is the source of much of the lighter elements such as carbon, nitrogen, and oxygen, on which life depends.

Another outcome is a *neutrino burst.* Neutrinos are the weakly interacting, massless particles that are produced in nuclear reactions of many kinds, including those in the Sun's core (see Chapter 15). During the core collapse, protons and electrons are forced to merge, creating a sea of neutrons and a huge number of neutrinos. The neutrinos flood into space at the speed of light, carrying away a huge amount of energy, about 10^{47} watts, equivalent to the mass-energy of 50 Earths. For a brief few seconds, a supernova exceeds the luminous intensity of the entire rest of the universe!

The final result comes months after the star itself fades. During the explosion, an expanding cloud of gas is launched. Initially, it is too close to the star to be seen from Earth, but years later the site will be marked by a colossal, expanding nebula, called a **supernova remnant.**

Examples of Supernovae

Many supernovae have been close enough to the solar system to produce temporarily prominent "new stars" that were recorded by ancient people. The ancient Chinese called them "guest stars." Some astronomers have suggested such an explanation for the Star of Bethlehem.

★ The most famous supernova was the explosion that produced the Crab Nebula (see our web site). It was visible in broad daylight for 23 d in July 1054 and at night for the subsequent six months. It was recorded in Chinese, Japanese, and Islamic documents, and perhaps in Native American rock art (see our web site). The nebula is the expanding, colorful gas shot out of this supernova. Other supernova remnants are scattered throughout our galaxy. Figure 19-6 shows the nearly spherical remnant of a supernova that exploded in 1680. In recorded history, there have been 14 supernovae in our galaxy, and a careful examination of the statistics leads us to expect an average of one every 40 years (although many of these will be invisible from Earth due to obscuration in the plane of the Milky Way). By this reckoning, we are long overdue, as the last supernova in the Milky Way was nearly 400 y ago. Any one of us might live to see one bright enough to be visible in daylight. In galaxies beyond the Milky Way, about 150 supernovae are discovered each year.

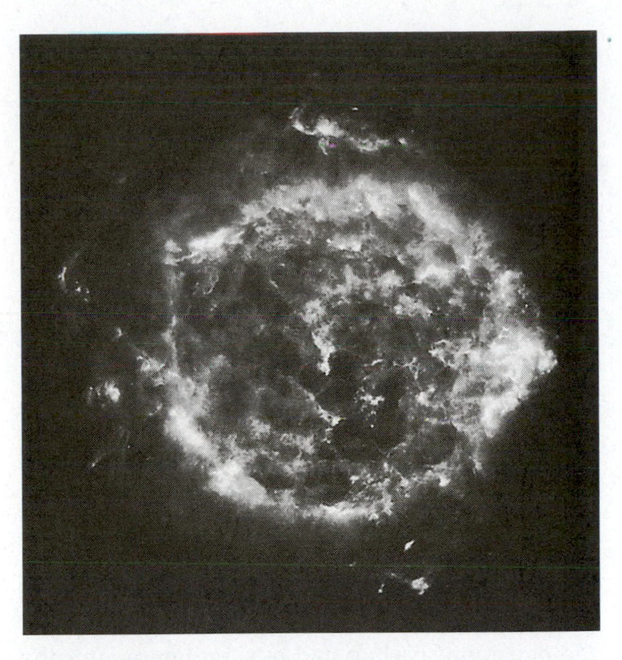

Figure 19-6 Cassiopeia A, a spherical, expanding shell of gas blown out of a supernova visible around A.D. 1680. This nebula is not prominent visually at the moment, but it has been imaged in X rays and radio waves. This image was made with 6-cm radio waves. The cloud diameter is probably a few parsecs. (National Radio Astronomy Observatory, R. Tuffs, R. Perley, M. Brown, and S. Gull)

The Supernova of 1987

At 7:35 A.M. Universal Time on the morning of February 23, 1987,[2] the Earth witnessed the core of a 20-$M_\odot$ star in the neighboring galaxy called the Large Magellanic Cloud become unstable. Within a second, the central iron plasma core, about the size of Mars, collapsed at a quarter of the speed of light down to a size of about 100 km. As temperatures reached 30 billion kelvins, iron nuclei were fragmented, and the star exploded in a blast of neutrinos and a flash of ultraviolet light brighter than any other stars in the whole galaxy. This event was some 52,000 pc away, much further than any stars we've discussed so far. Yet, it was easily detected from Earth. (The light corresponding to the event reached Earth in early 1987. Since the star's location was nearly 170,000 light-years away, the light took some 170,000 y to reach us. The explosion actually occurred around 168,000 B.C.!) We are currently overdue for a supernova much closer to Earth, which would almost certainly be bright enough to be visible during the day.

The first trace of the supernova came from the neutrino burst. The unseen wave contained so many neutrinos that about 10 billion passed through the body of every person on Earth (in the United States these neutrinos passed up through our feet, since the Large Magellanic Cloud is in the southern hemisphere). At the distance of the Large Magellanic Cloud, only 1 in 10,000 people suffered even a single (harmless) neutrino interaction. Sensitive detectors in Japan and Ohio detected a total of 19 neutrinos. The detection of neutrinos marked the first experimental confirmation of a 50-y-old theoretical prediction—and the first application of neutrino astronomy outside the solar system.

Light from the explosion began to arrive a few hours later. The extra time was required for the shock wave to climb from the center of the dying star to its surface. As brilliant as the sight is, the light from a dying star is less than 0.01% of the total energy of the event, much of which is carried off in the form of

[2]The date and universal time (clock time at the 0° longitude meridian in Greenwich, England) are the time when the light corresponding to the event reached Earth.

neutrinos. The newly brightened star was sighted by astronomers in Chile and New Zealand, and word was flashed to observatories around the world. For many days it was bright enough to see with the naked eye, but only at equatorial and southern latitudes. Telescopes on Earth, on the Russian Mir space station, and on robotic satellites were pointed at the supernova in a coordinated observational effort.

★ The supernova consumed the attention of astronomers for many months, and it offered a remarkable opportunity to test basic aspects of the theory of stars. Detective work has allowed us to recreate the history of the doomed star. The star was a B3 supergiant, which had a mass of about 20 $M_{\odot}$ when it left the main sequence. Calculations by S. Woosley and collaborators at Lick Observatory suggest that the star had an iron core of 1.5 $M_{\odot}$ at the time of the detonation and a temperature of 10 billion (10^{10}) kelvins. Our web site shows supercomputer calculations of the density of carbon and oxygen hours after the explosion.

As the first nearby supernova in nearly four centuries, Supernova 1987A has given astronomers a ringside seat at a stellar explosion that can be studied with the full array of modern astronomical techniques (Woosley and Weaver, 1989). It is all but certain that the collapsed core of the supernova is an example of one of the most bizarre objects in the universe—a neutron star.

NEUTRON STARS (PULSARS): NEW LIGHT ON OLD STARS

What is left after a supernova explosion? As early as 1934, American astronomers W. Baade and F. Zwicky speculated that one result might be a **neutron star.** The concept of a neutron star is an extension of the concept of a white dwarf. We have discussed states of matter of increasing density. In an ordinary star, electrons are stripped from atomic nuclei and collide violently in the high-temperature and high-pressure gas. The collisions are violent enough to overcome the electrical repulsion between nuclei, and so heavier elements are created by fusion. In a white dwarf, the energy supply from fusion reactions is exhausted, so gravitational forces cause the star to collapse. The new stable configuration is supported by the pressure of degenerate electron matter.

But if a burnt-out star core is still more massive than a white dwarf, the gravity is so strong that it can overcome even the pressure of degenerate electrons. Theory indicates that this could happen in objects between 1.4 $M_{\odot}$ and about 2 to 2.5 $M_{\odot}$ (the lower bound may be as low as 1.2 $M_{\odot}$, since a neutron star formed by a supernova loses up to 0.2 $M_{\odot}$ in the form of neutrinos). The force of gravity causes electrons and protons to coalesce and form neutrons. In the absence of electrical forces and electron degeneracy pressure, the core collapses to a state of dense neutron matter. Once again, the new stable configuration is supported by degeneracy pressure, this time due to neutrons. In effect, the neutrons are so tightly packed that they nearly "touch" each other.

Neutron stars are truly remarkable objects. The matter is the densest in the observable universe, a phenomenal density of 10^{17} kg/m^3. A thimbleful brought to Earth would weigh 100 million tons! A neutron star could contain all the mass of the Sun but be no larger than a small asteroid—perhaps 20 km across. Neutron stars rotate at up to 10% of the speed of light and have surface magnetic fields of 10^{12} Gauss, a million times stronger than the strongest magnetic fields that have been produced on Earth. Because the density is comparable to the density of the nucleus of an atom, some astronomers have pictured a neutron star as a giant atomic nucleus with atomic mass around 10^{57}.

The Discovery of Neutron Stars

For decades, astrophysicists talked about neutron stars, but, as with the weather, nobody did anything about them, because nobody *could* do anything. No known observational technique could detect them, and no one could prove they existed. But in November 1967, a 4.5-acre array of radio telescopes in England detected a strange new type of radio source in the sky. Analyzing the surveys (each equaling a 120-m roll of a paper chart), a sharp-eyed graduate student, Jocelyn Bell, was astonished to find that one celestial radio source (about a centimeter of data on the chart) emitted "beeps" every 1.33733 s!

At first, project scientists speculated that they might have actually discovered an artificial radio beacon placed in space by some alien civilization! But within two months, another source was found, pulsing at a different frequency, arguing against the beacon hypothesis. By careful analysis, they were also able to rule out terrestrial sources for the unusual signals. Analysis showed that a rapidly repeating pulse could only be produced by a very compact source. The estimated size was less than 4800 km across, much smaller than ordinary stars.

These pulsing radio sources came to be called **pulsars.** In February, Anthony Hewish and his col-

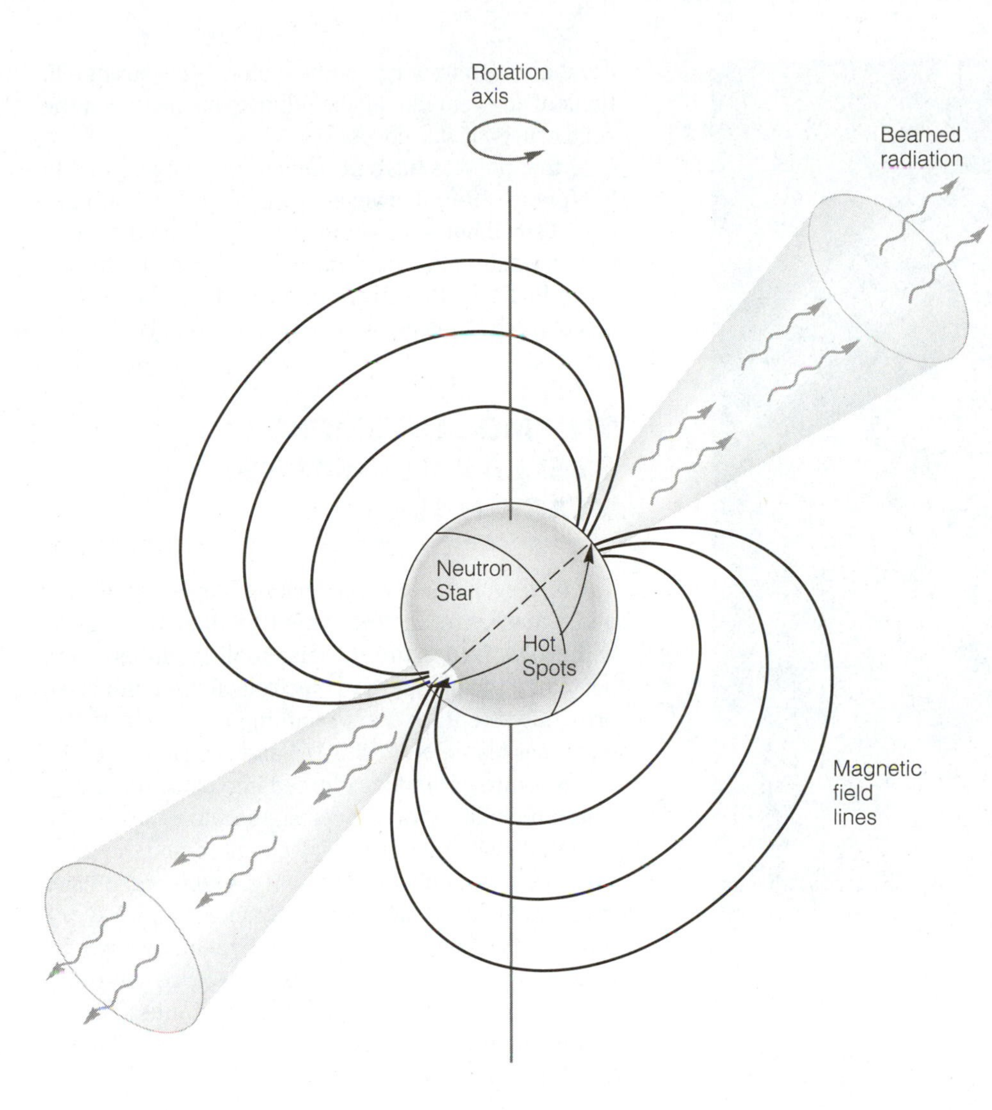

Figure 19-7 This model of neutron star emission accounts for many of the observed properties of pulsars. The neutron star is spinning rapidly and has a strong magnetic field. Charged particles are accelerated near the hot spots and produce a beam of radiation that travels out from each magnetic pole. The rotation axis is offset from the magnetic axis, so the beams sweep the sky as the pulsar rotates. If the beam sweeps across the Earth, we see a pulsar.

leagues published an analysis suggesting that the pulsars might be superdense vibrating stars that could "throw valuable light on the behavior of compact stars and also on the properties of matter at high density."

The mysterious pulsars turned out to be the long-sought neutron stars. In an exciting burst of research, the number of scientific papers on pulsars jumped from zero in 1967 to 140 in 1968. Over 800 pulsars have now been discovered. The co-directors of the original discovery project, Anthony Hewish and Martin Ryle, shared the 1974 Nobel Prize in physics.

Properties of Pulsars

Pulsars are just the subset of neutron stars that exhibit strong, pulsed radio emission. Why do neutron stars pulse? Following the core collapse, neutron stars spin very fast and have very strong magnetic fields. Recall from Chapter 14 that any spinning object—even a figure skater—spins faster as it contracts. Collapse by a factor of a million produces a corresponding increase in the spin rate. This is a consequence of the conservation of angular momentum. Similarly, the magnetic lines of force that thread a normal star will be greatly concentrated if it collapses to a neutron star. In 1968, Cornell researcher Thomas Gold showed how charged particles trapped in magnetic fields of spinning neutron stars produce strongly focused radio radiation. The pulsar acts like a lighthouse with a beam sweeping around rapidly. We see a pulsar only if the beam happens to periodically sweep across the direction to the Earth. This model of pulsar emission is illustrated in Figure 19-7.

The discovery of a pulsar in the center of the Crab Nebula (Figure 19-8) and in some other supernova remnants demonstrates that pulsars are related to supernovae. It is believed that each of the hundreds of known pulsars is the neutron star corpse of an extinct

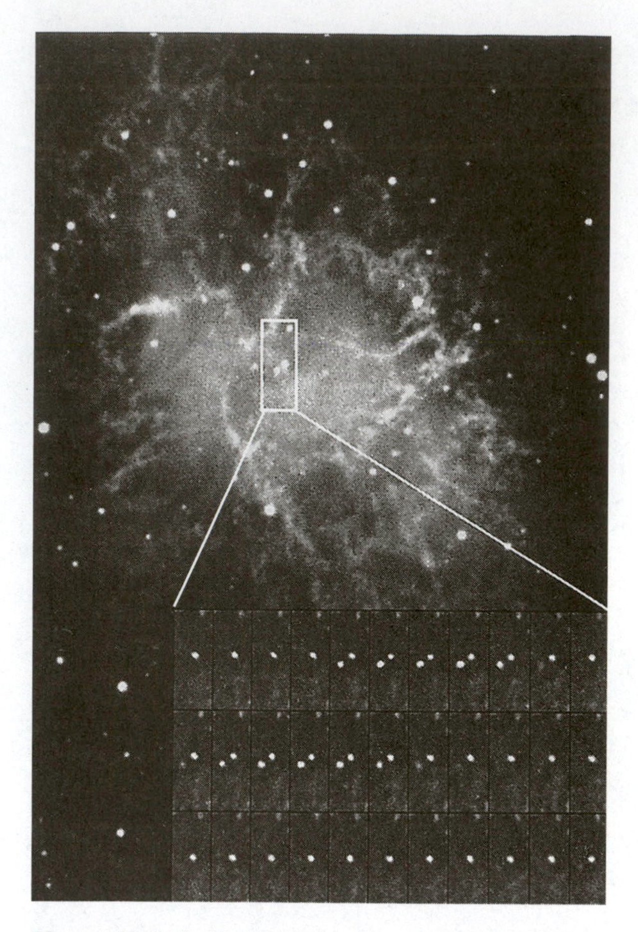

Figure 19-8 The Crab Nebula, with a sequence of 33 close-ups showing the region of the central pulsar. The entire cycle, including two flashes, lasts about $\frac{1}{30}$ s, equaling one complete rotation of the pulsar. The Crab pulsar, at a distance of about 2000 pc, is one of the very few where optical pulses have been observed. (N. Sharp/NOAO)

massive star. A few of the youngest pulsars have been shown to pulse not only in radio waves, but also in X rays and visible light. In these cases, the visible and X-ray emission is caused by intense heating of material near the pulsar and does not represent the temperature of the pulsar itself.

Pulsars make excellent clocks. However, the rotation rate is not absolutely constant. The typical pulsar is *slowing down,* with its rotation rate diminishing by a 30-millionth of a second per year. (No other timekeeping device can keep such good time!) Even this gradual slowing down corresponds to an enormous rate of energy release. Occasionally, radio astronomers have detected tiny abrupt changes in the spin rates of pulsars. These *glitches* are attributed to changes in the intense magnetic field, which change the mass distribution and so the rotation rate. The solid crust can also undergo sudden shifts. It is strange to think of an event like an earthquake happening in the solid crusts of star corpses!

Most pulsars have periods in the range of 0.2 to 2 s (Figure 19-9). However, in Chapter 21 we will discuss a small but interesting class of pulsars that complete a rotation in a tiny fraction of a second. The last entry in Table 19-1 gives informtion on the properties of the famous pulsar in the Crab Nebula.

THE MOST COMPACT STELLAR REMNANTS: BLACK HOLES

The final evolutionary state of a star depends critically on its mass. Stellar remnants less than 1.2–1.4 $M_\odot$ will evolve into white dwarfs, cooling embers supported by the pressure of electrons forced into close proximity. Neutron stars resulting from supernovae have masses below 2–2.5 $M_\odot$ and are supported by the pressure of neutrons forced into close proximity. These boundaries are uncertain because the physics of high-density matter is very complex, and the models depend not only on mass but also on rotation and magnetic fields. A stellar core of more than 2.5–3 $M_\odot$ has gravity strong enough to overwhelm neutron degeneracy pressure. Since no known force can resist the force of gravity, the collapse continues. The result is one of the strangest objects in astronomy: a **black hole.**

Einstein's Theories of Relativity

In the early part of the twentieth century, Albert Einstein revolutionized physics with new ideas on the nature of space and time. Although the mathematical results are beyond the scope of this book, some of the main results can be used to illuminate our discussion of black holes. Einstein's **special theory of relativity** starts with the observation that nothing travels faster than the speed of light. In addition, it is observed that the measured speed of light is a universal constant, *independent of the motion of the observer.* A little thought will show that this is a bizarre result. For example, if you are traveling in a car at 100 km/h and throw a ball forward at 50 km/h, you would expect that the ball would reach someone standing on the ground at 100 + 50 = 150 km/h. What if you were traveling in a spaceship at half the speed of light,

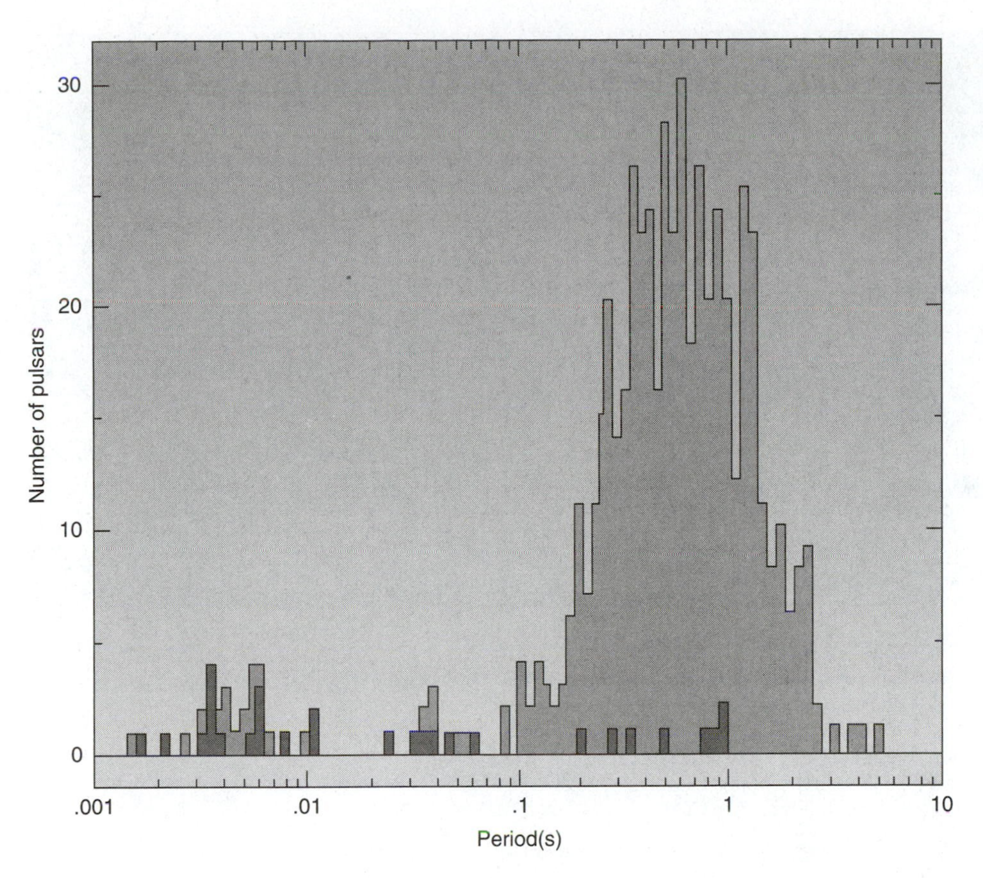

Figure 19-9 The distribution of periods of 558 pulsars. Most pulsars have periods of a fraction of a second. Binary pulsars are indicated by the dark regions; most of them have short periods.(J. H. Taylor, R. N. Manchester, and A. G. Lyne)

150,000 km/s, and you shined a beam of light forward at the speed of light, 300,000 km/s? Intuition tells you that an outside observer would measure the light arriving at 150,000 + 300,000 = 450,000 km/s. However, observations and special relativity say that the measured speed will always be the speed of light, or 300,000 km/s!

The universal constancy of the speed of light has some fascinating consequences. Newtonian concepts of absolute space and time must be abandoned. Objects moving at close to the speed of light, or *relativistic* speeds, suffer a physical contraction and actually get shortened in the direction of motion. Clocks moving at close to the speed of light slow down relative to a clock held by a stationary observer. The masses of particles increase as they approach the speed of light. None of these effects are tricks or illusions; they are real physical phenomena. Physicists have confirmed that accurate atomic clocks traveling on a jet plane keep slightly slower time than an identical clock on the Earth. The predictions of special relativity are confirmed in physics laboratories thousands of times a day. Although special relativity supersedes Newton's laws of motion, the unusual effects are significant only *close to the speed of light*. At everyday low speeds, Newtonian equations work extremely well.

After outlining the special theory of relativity, Einstein set about including gravity in its framework. His **general theory of relativity** starts with the observation that acceleration due to gravity cannot be distinguished from acceleration due to any other force. In other words, the effects of special relativity that apply when objects are accelerated to near the speed of light—space being compressed, clocks slowing down, and masses increasing—must also apply to objects moving in intense gravitational fields. According to the general theory of relativity, gravity *distorts* space and time, causing space to become curved and time to slow down. Einstein's conception of gravity is totally different from the traditional Newtonian picture. As with special relativity, the consequences of general relativity often seem nonintuitive. Whereas Newton postulated absolute and uniform time and space, general relativity predicts that any mass alters the properties of time and space around it. As we will see in Chapter 27, this result extends to the discussion of the entire universe!

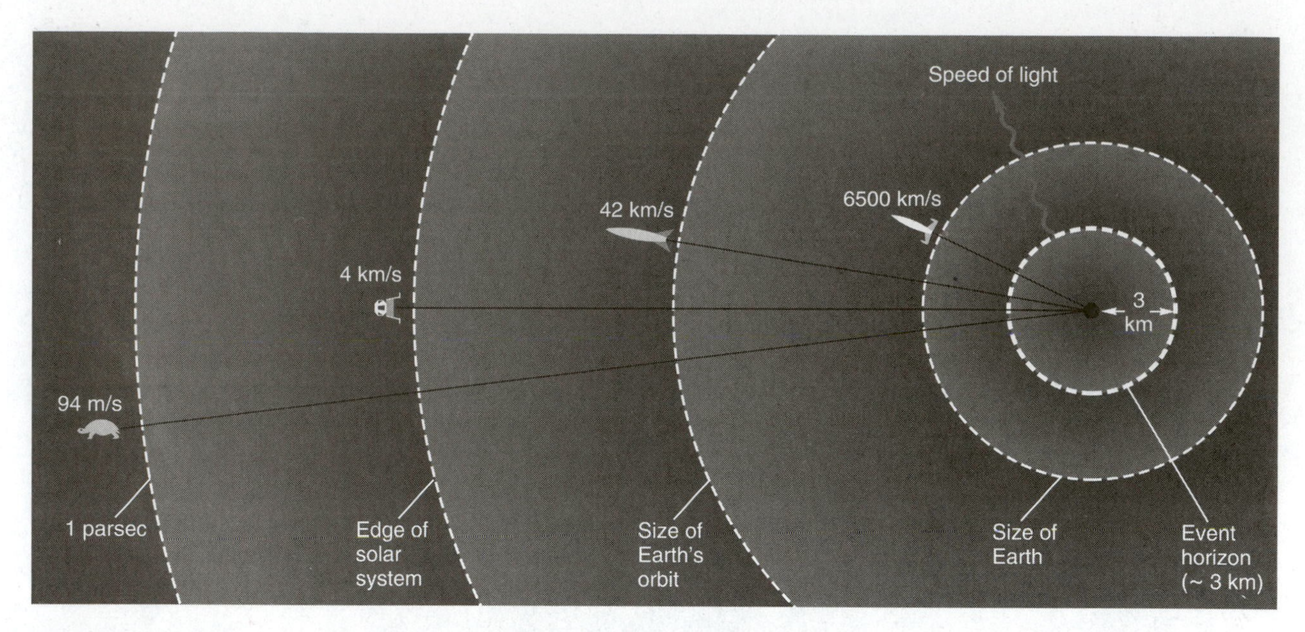

Figure 19-10 Exploring the gravitational field of a black hole of 1 $M_{\odot}$ (right). The schematic diagram (not to scale) shows the velocity needed to escape from the black hole at different distances from it. At a parsec away (left), a mere 94 m/s would suffice, but this speed increases as we get closer. At a distance of about 3 km, a body would have to be moving at the speed of light to escape; this distance is called the event horizon. No object on a ballistic trajectory could escape from inside this distance.

Einstein also demonstrated the equivalence of mass and energy, through the famous equation $E = mc^2$. This interchangeability of mass and energy is a fundamental principle of physics. We can use it to understand the nature of the speed of light as a universal limit. As a particle is accelerated closer and closer to the speed of light, it never exceeds 300,000 km/s. What happens is that the energy being put into the particle to speed it up actually goes into *increasing its mass,* according to the equation above. The particle becomes more and more massive, and it approaches but never exceeds the speed of light.

★ The equation also removes the distinction between particles and radiation, and it implies that light must respond to gravity just as particles do. Photons leaving a massive object suffer a **gravitational redshift,** caused by the photons losing energy as they climb out of the gravitational "well." Once again, the predictions of general relativity can be confirmed experimentally (see Will 1986, and material on our web site for more details). Atomic clocks do run slower on the Earth's surface than at the top of a tall building, where the gravity is slightly weaker. Photons do lose energy in leaving the gravitational field of the Sun. It has been demonstrated that light is indeed bent by gravity. Pulsars have been used as accurate clocks to confirm predictions of general relativity, resulting in the award of the 1993 Nobel Prize in physics to Joseph Taylor and Russel Hulse. Although general relativity supersedes Newton's law of gravity, the effects are only large in *strong gravitational fields.* Some of the most extreme consequences can be seen in the collapsed final states of stars.

The Physical Nature of Black Holes

We can understand the nature of a black hole in terms of the idea of *escape velocity.* Imagine that the Sun has somehow been compressed into a black hole of 1 $M_{\odot}$ (Figure 19-10). A rocket passing at a great distance would experience the same gravity field as a rocket at a great distance from the Sun. At 1 AU from the black hole, for example, the velocity needed to escape into interstellar space would be 42 km/s, the same as the speed needed to leave the Earth's orbit. But as we get much closer to the black hole, the escape velocity increases. Larger speeds are needed to escape the stronger and stronger gravity. At a distance of 3 km, the speed needed to escape would be the speed of light. The imaginary sphere with a radius of 3 km is called the **event horizon.** Inside this surface, no object, no particle, no information, not even light can escape. Any star that collapses within its event hori-

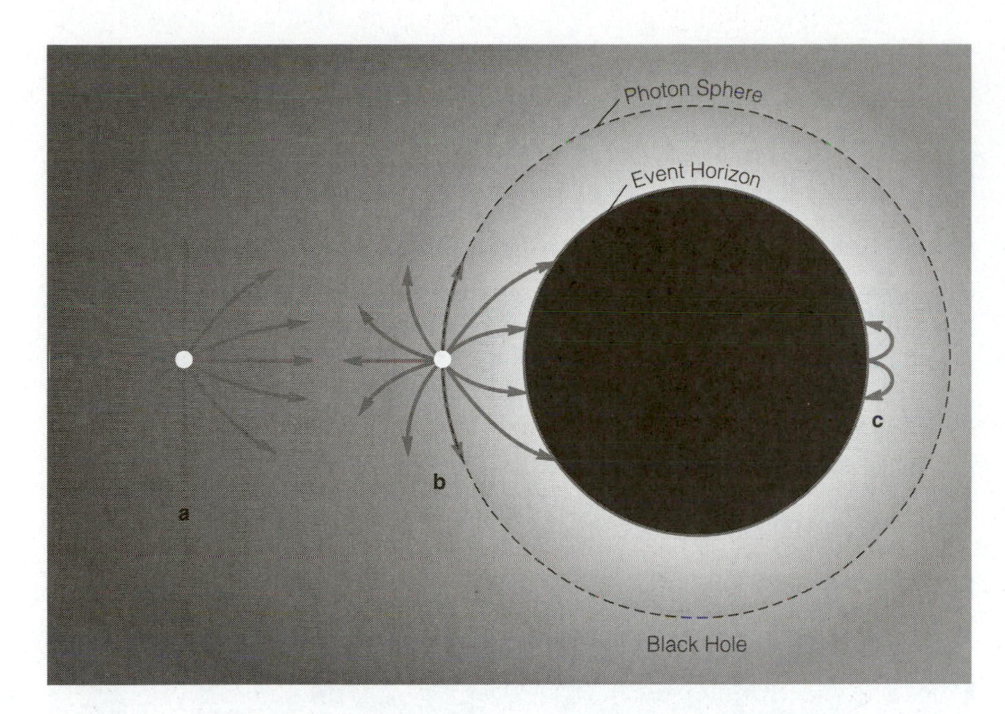

Figure 19-11 Environment of a black hole. **a** Far from a black hole, light travels mostly straight away from a source, except for light traveling near the direction to the black hole, which is deflected. **b** At a distance of 1.5 Schwarzschild radii, half of the light escapes. The surface at this radius is called the photon sphere. **c** At the Schwarzschild radius, all light is trapped by the black hole. The surface at this radius is called the event horizon.

zon disappears from the universe, betraying its presence only by its gravitational field (Thorne, 1994).

The radius corresponding to the event horizon is called the **Schwarzschild radius,** after the German astronomer who was the first to solve Einstein's equation of general relativity. *Any* object can become a black hole if it is sufficiently compressed; the Schwarzschild radius is proportional to the mass. For example, the Earth would become a black hole if it were compressed in an enormous vise down to a size of 1 cm! For the Sun, the Schwarzschild radius is 3 km. For a 3-$M_\odot$ stellar core, it is 9 km. How is a black hole produced? Any star that ends its life with a core mass of 3 $M_\odot$ or more will become a black hole, because no known force in nature can prevent its collapse within its event horizon.

Figure 19-11 illustrates the environment of a black hole and the effect of the gravity on light rays at different distances from the event horizon. At a large distance from a black hole, light travels away from a light source uniformly in all directions. As the black hole is approached, light passing near the hole will be slightly deflected. Closer to the event horizon, some light rays are deflected by the strong gravity and are captured by the black hole. At 1.5 Schwarzschild radii, half the light escapes. Photons emitted at right angles to the black hole are trapped in circular orbits. These orbits define the *photon sphere.* At the Schwarzschild radius, the deflection of light is so severe that no light can escape. This defines the event horizon.

Another analogy can be used to convey the extreme *space-time curvature* caused by black holes. General relativity predicts that *any* mass will distort the space and time around it. Figure 19-12 presents the space curvature in two dimensions, as the distortion in a thin rubber sheet. In the absence of any matter, space will be flat and have no curvature (Figure 19-12a). For a concentrated mass, the distortion is large enough to clearly deflect matter and radiation that pass near it (Figure 19-12b). In the extreme case of a black hole, the curvature is complete. We can imagine a piece of space and time being "pinched off" and permanently removed from communication with the rest of the universe (Figure 19-12c).

If you were unfortunate enough to fall into a 1-$M_\odot$ black hole, you would be killed by tidal forces long before you reached the event horizon. (Essentially, the difference between the gravity force on your head and that on your feet would rip you apart!) Assuming that somehow you could survive the descent, you would see clocks far from the black hole keeping slower and slower time, until as you neared the event horizon they appeared to stop altogether. Seen from the outside, your clock would appear to slow down as you took an infinite time to reach the event horizon! If you carried a light source with you as you fell into the black hole, a distant observer would see the photons suffer a larger and larger gravitational redshift (to you, the light would stay the same color).

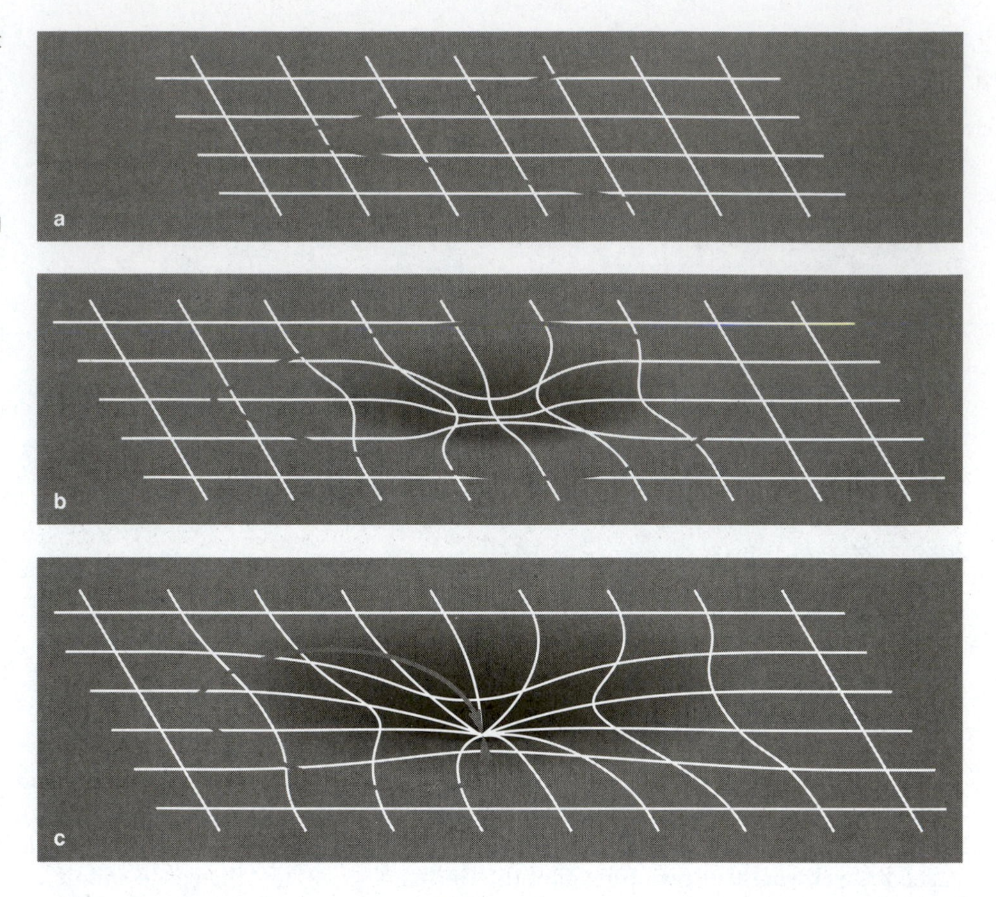

Figure 19-12 Rubber sheet analogy for the curvature of space. **a** Empty space is flat and does not cause particles or radiation to deviate as they pass through it. **b** A concentrated mass distribution will distort space in its vicinity and cause the deflection of particles and radiation. **c** A black hole causes such extreme curvature of space that particles and radiation can be completely trapped.

Seen from the outside, the photons would be infinitely redshifted to zero energy as you reached the event horizon.

What lies within the event horizon of a black hole? Nobody really knows. The event horizon is not a physical barrier, just an information barrier. Einstein's theory predicts that matter will keep collapsing gravitationally until it has shrunk to a point of zero volume and infinite density! This endpoint is called a *singularity.* However, a singularity represents the breakdown of our theory of gravity—it has been said that general relativity contains the seeds of its own downfall.

Black holes are not entirely black. In the 1970s, English physicist Stephen Hawking calculated that black holes could create subatomic particles near their event horizons and slowly radiate away their energy, or "evaporate." This so-called Hawking radiation is expected to be dramatic for microscopic black holes, but barely noticeable for solar-size black holes. Far more important is the fact that any material falling *toward* the event horizon will be subject to enormous gravitational forces. The friction and heating of infalling material are expected to be released in the form of X rays. Therefore, a black hole may be a *source* of energy due to the death spasms of matter falling into it.

Detecting a Black Hole

Can we ever hope to detect a black hole? Yes. Outside their event horizons, black holes have gravity fields indistinguishable from those of ordinary stars of the same mass. Thus they can orbit around stars just like planets or binary star companions. If we observed such a star from a distance, we would not see the black hole, but we could see the star's orbital motion and calculate the mass of the unseen companion, just as astronomers routinely do in the case of ordinary faint companions. The result would indicate an unusually high-mass companion for an X-ray source—maybe 5 or 10 $M_{\odot}$—which should tip us off that we are dealing with a black hole candidate. Furthermore, such strong gravity exists close to black holes that any matter falling into them undergoes terrific acceleration.

★ Suppose the black hole is orbiting around an evolved star that has expanded into the giant state and is shedding mass. Some of the expanding gas would fall toward the black hole at relativistic speeds.

Because this gas would, on the average, have some angular momentum around the star, rather than falling directly toward the star, it would form a disk of gas spiraling inward toward the black hole (see our web site). This disk is called an **accretion disk.** Its gas would be extremely hot, because it would be constantly hit by new gas streaming in from the other star. Because of the high temperature, the disk would radiate very short-wave radiation, such as ultraviolet, X-ray, and gamma-ray radiation (see our web site). Therefore X-ray and gamma-ray telescopes launched into orbit around the Earth have played an important role in searching for black holes.

The best evidence for a black hole would be a massive, high-temperature X-ray or gamma-ray source orbiting another normal star. There are currently 11 excellent candidates and another dozen or so plausible candidates. It is important to point out that the argument for a black hole in these systems has three steps. First, the X-ray emission is similar to more than 40 well-understood cases where the object is accreting and is at least as dense as a neutron star. Second, the orbital velocity of the companion star and the orbital period of the binary system imply that the mass of the compact object exceeds 3 $M_\odot$. Third, it is assumed that general relativity is the correct theory of gravitation and that neutron stars with masses above 3 $M_\odot$ cannot exist.

Uncertainties enter into the analysis in two ways. The parameters of the orbit, such as the period and the inclination to the observer's line of sight, have observational errors. Also, the nature and mass of the normal companion must be known to determine the mass of the compact object. The *minimum* possible mass for the compact object is the important quantity. In each case, the unknown orientation of the orbit also leads to an uncertainty in the mass of the dark companion. In all the best candidate systems it is likely that the dark companion is too massive to be a neutron star, but the evidence is necessarily indirect and circumstantial. This is unlike the evidence for the existence of neutron stars, which is direct and overwhelming. The observational case for stellar black holes continues to get stronger, but their existence is not yet proved beyond all doubt.

Black holes of different sizes have been implicated in a variety of phenomena in astrophysics. Energy released from the environment of a black hole may be responsible for the unusual source of positron emission near the center of the Milky Way (see Chapter 23). Supermassive black holes (over a billion times the mass of the Sun) may account for the enormous luminosity of quasars and for the high-velocity jets that are seen emerging from radio galaxies (see Chapter 25). As astrophysicist and mathematician Roger Penrose has noted, "There is no shortage of unexplained phenomena in astronomy today that might conceivably be relevant" to black holes.

SUMMARY

The life and death of stars of all types is governed by the irresistible force of gravity. Figure 19-13 summarizes the life histories of stars of different masses. At the low-mass end, stars are dim, red, and slow to evolve. The coolest-sequence stars have not yet evolved off the main sequence in the entire age of the universe. Every star more massive than the Sun goes through a phase of mass loss, involving either a wind or a more violent explosion. Stars below 1–2 $M_\odot$ end their lives quietly, as cooling white dwarf embers. Massive stars are rare, and they evolve quickly toward a spectacular demise. Supernovae are not only responsible for the production of neutron stars and probably black holes, but they also recycle rare and important heavy elements into the universe.

The search for the forms of aging stars has yielded some of the most fascinating objects now being studied both by physicists and by astronomers. Stellar old age leads to two basic phenomena: high-energy nuclear reactions, as ever-more-massive elements interact, and inexorable contraction, as energy sources are eventually exhausted.

During the first stages of old age, as stars evolve off the main sequence, high energy production causes expansion of stars' outer atmospheres, producing giants and supergiants. As heavier elements go through quick reaction sequences, various kinds of instability may produce variable stars and slow mass loss. After energy generation declines to a rate too low to resist contraction, low-mass stars contract to a dense state known as a white dwarf, with final mass less than 1.4 $M_\odot$. Rarer massive stars, which start out with as much as 8 $M_\odot$ or more, undergo supernova explosions and blow off much of their initial material.

If the remnant cores of stars end up between 1.4 $M_\odot$ and about 2.5 to 3 $M_\odot$, they form dense, rapidly rotating neutron stars known as pulsars. If the remnant cores have more than about 3 $M_\odot$, they may form black holes. Although virtually no radiation escapes from these strange objects, they may be detectable by orbital motions of their companion stars and by high-energy radiation from material falling into them. The detailed physics of these dense, small star forms is an area of intense current research.

Complications may arise if the evolving star has a nearby co-orbiting companion. Mass may be blown off one star and fall onto accretion disks around the other, changing the second star's mass and causing sudden instabilities. This probably accounts for some types of supernovae and some sources emitting X rays and gamma rays. Ultraviolet, X-ray, and gamma-ray astronomy con-

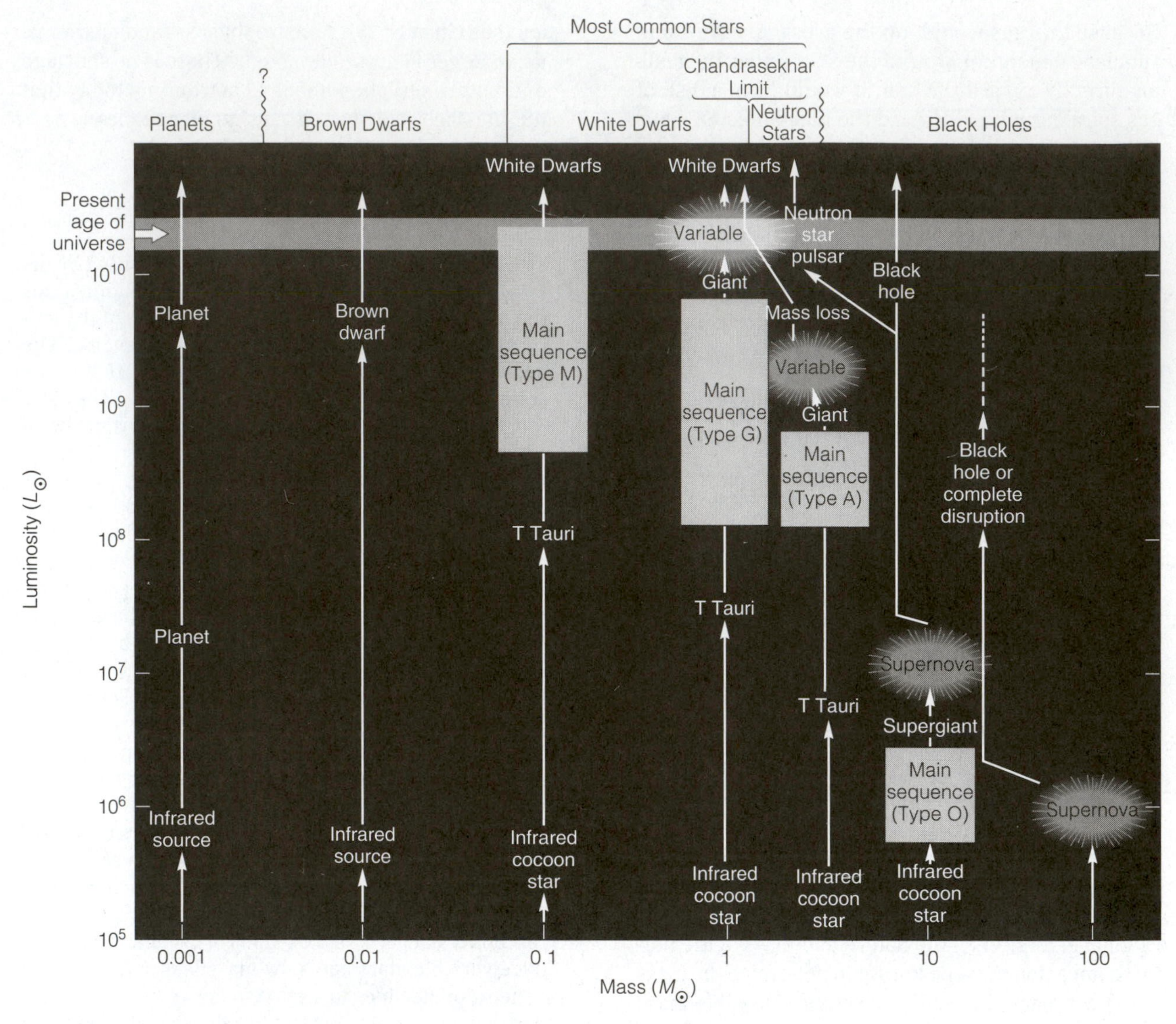

Figure 19-13 A schematic summary of stellar evolution showing the inexorable march toward high density. Evolutionary histories are shown for objects with different initial masses and the associated terms described in this and preceding chapters. Some states are hypothetical and are still being intensively researched by theoreticians and observers.

ducted on orbiting satellites is an exciting area of new research on accretion disks, neutron stars, and black holes.

CONCEPTS

red giant
triple-alpha process
helium flash
variable star
Cepheid variable
supergiant
planetary nebula
white dwarf
degenerate matter
Chandrasekhar limit
supernova
core collapse
explosive nucleosynthesis
supernova remnant
neutron star
pulsar
black hole
special theory of relativity
general theory of relativity
gravitational redshift
event horizon
Schwarzschild radius
accretion disk

PROBLEMS

1. Why do stars just moving off the main sequence expand to become giants instead of starting to contract at once? Why does contraction ultimately win out?

2. Many red giants are visible in the sky, even though the red giant phase of stellar evolution in relatively short-lived. Why are so many red giants visible?

3. List examples of evidence that certain stars can lose mass.

4. Which stars still eventually become:

a. white dwarfs?

b. neutron stars?

c. black holes?

What will be the ultimate fate of the Sun?

5. A main-sequence B3 star has about 10 times the mass of the Sun and therefore has about 10 times as much potential nuclear fuel. Why then does it have a main-sequence lifetime only $\frac{1}{200}$ as long as that of the Sun?

6. According to the law of conservation of angular momentum, a figure skater spins faster as she pulls in her arms. How does this principle help explain why neutron stars spin much faster than main-sequence stars?

7. Comment on the roles and relations of theorists and observers in the three decades of work on white dwarfs, pulsars, and black holes. Are black holes fully understood today?

ADVANCED PROBLEMS

8. Suppose a 1-$M_{\odot}$ star has reached a terminal evolutionary state where it has the same diameter as the Earth.

a. What would be the velocity of any possible material (such as captured meteoritic debris) in a circular orbit just above the star's surface?

b. What velocity would be needed to blow material off its surface?

c. Compare these values with values for the Earth.

d. What type of star would this object be?

9. Use the Stefan-Boltzmann law to prove the surface of a star such as FG Sagittae, evolving to the right on the H-R diagram (keeping constant total luminosity but decreasing in temperature), must be expanding.

10. Many white dwarfs have spectral types and surface temperatures similar to A or F main-sequence stars, but are much smaller. Use the Stefan-Boltzmann law to prove that this statement requires white dwarfs to lie below the main sequence on the H-R diagram.

11. Suppose a supernova occurred in the nearby star-forming region of Orion, 500 pc away. If the cloud expanded at an average velocity of 1000 km/s, how long would terrestrial observers have to wait before they could see details of the cloud's shape with telescopes resolving $\frac{1}{2}$ second of arc? (*Hint:* Use the small-angle equation; 1 y is about $\pi \times 10^7$ s.)

12. Prove that a gamma-ray telescope (sensitive to wavelengths less than about 0.1 nm) would be best suited to observing radiation from a 1 billion kelvins accretion disk surrounding a black hole. Could such observations be made from a ground-based observatory?

13. The discussion of supernovae noted that the nuclei in stellar cores begin to break apart into protons and neutrons and other subatomic particles at temperatures of a few billion kelvins. Laboratory experiments show that simple nuclei will fragment in this way when struck by particles moving fast enough to carry about 3.5×10^{-13} joules of energy. Prove that a gas would have to have a temperature of several billion degrees before its particles collided hard enough to begin to fragment nuclei of atoms. (*Hint:* In deriving Optional Basic Equation VI, we saw that the mean kinetic energy of a particle in an ordinary gas is $\frac{3}{2}$ kT. Gases in supernovae, accretion disks, and so on, at billions of degrees are unlikely to behave as perfect gases, but assume that they do for the purpose of this estimate. Assume also that the faster particles in the gas have four times the energy of the average particle mentioned above. Then you can estimate that the energies mentioned would be reached by the faster atoms in a gas at some 4 billion kelvins.)

14. Suppose the Sun were replaced with a black hole that had one solar mass. How would the Earth's orbital velocity change if the Earth remained in a circular orbit?

PROJECTS

1. Locate the star Mira (R.A. = 2^h14^m; Dec. = $-3°.4$) with a small telescope and determine whether it is in its faint or bright stage. If it is bright enough to see with the naked eye, record its brightness nightly by comparing it with other nearby stars of similar brightness. By checking brightnesses of these stars with star maps showing magnitudes, plot a curve of Mira's brightness over time.

2. With binoculars or a small telescope locate the star Delta Cephei (R.A. = 22^h26^m; Dec. = $+58°.1$) and compare it from night to night with other nearby stars of similar brightness. Can you detect its variations from about 4.4 to 3.7 magnitude in a period of 5.4 d?

PART

6

Environment and Groupings of Stars

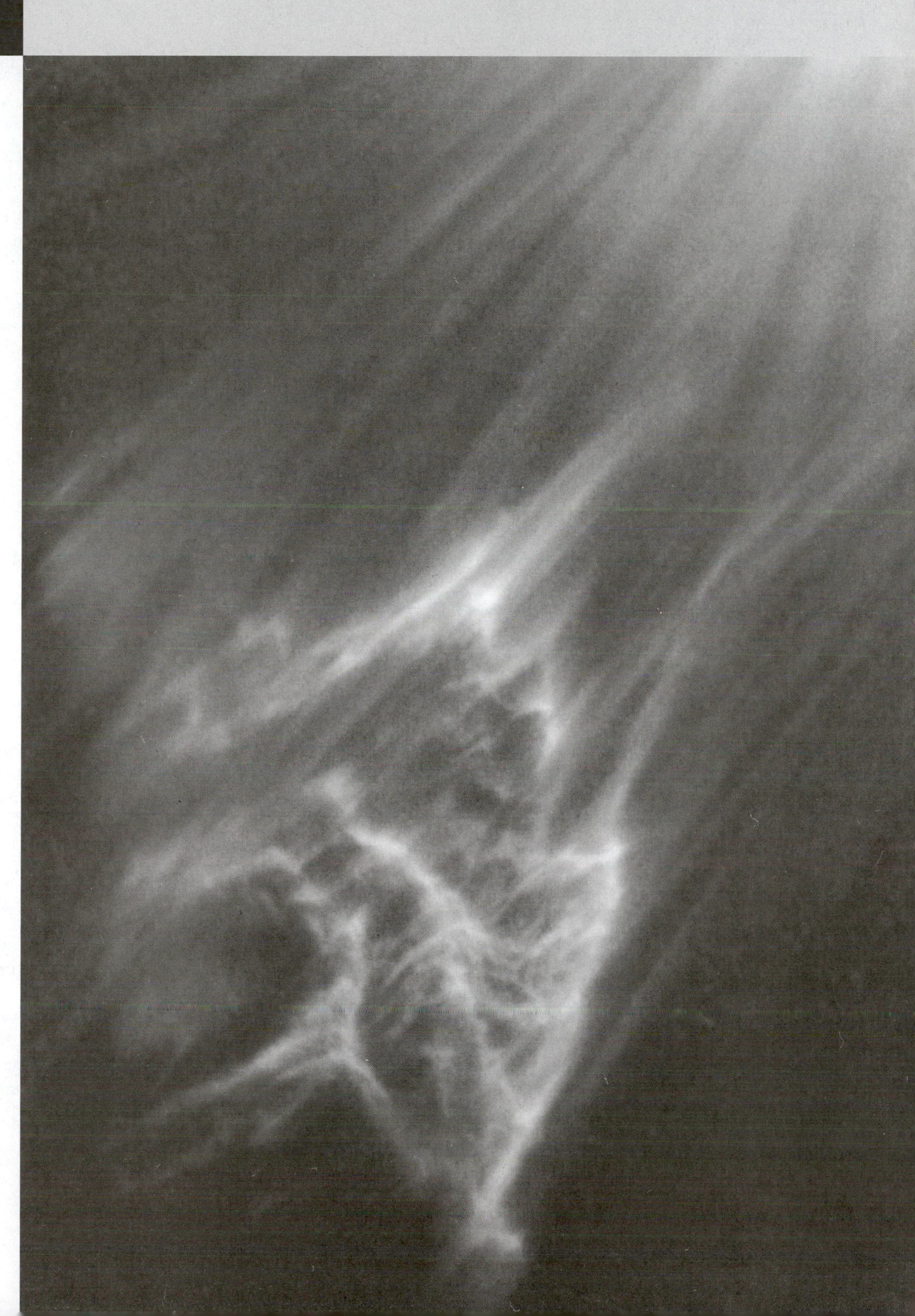

This ghostly apparition is actually an interstellar cloud caught in the process of destruction by strong radiation from a nearby hot star. Located in the Pleiades star cluster, the cloud is called IC 349 or Barnard's Merope Nebula. (NASA and The Hubble Heritage Team (STScI/AURA), acknowledgement to George Herbig and Theodore Simon (Institute for Astronomy, University of Hawaii))

CHAPTER

20

Interstellar Atoms, Dust, and Nebulae

WHAT THE READER SHOULD WATCH FOR IN THIS CHAPTER

Although the space between stars may seem empty, it is not a perfect vacuum. Interstellar space contains dust and gas that redden and extinguish starlight as it travels across vast distances. The densest regions where stars are currently forming can be opaque to visible light. The coldest regions of the interstellar medium contain dust grains and complex molecules whose properties can best be studied at infrared and millimeter wavelengths. The hotter regions near stars contain gas where atoms have been excited by ultraviolet radiation, creating the beautiful colors of the diffuse nebulae. The Orion Nebula is the most spectacular nearby example of the interaction between stars and the surrounding material. ■

The preceding chapters discussed stars as if they were isolated, individual objects. But they are not isolated. They are engulfed in a thin but chaotic medium of gas, dust, and radiation. They form from this thin material, interact with it, recycle it, and expel it to form new interstellar material.

Vivid examples of these individual clouds of gas and dust were cataloged by the French astronomer Charles Messier in 1781. They are thus known by their *Messier numbers,* or M numbers. The well-known Orion Nebula, for example, is M 42. Others are known by *NGC numbers* or *IC numbers,* based on the more recent New General Catalog and Index Catalog, respectively. Tradition also names most bright nebulae according to their appearance in small telescopes; examples include the Crab nebula (M 1), the Dumbbell Nebula, and the Ring Nebula. Some examples are listed in Table 20-1.

We sometimes casually say that "space is a vacuum," but this is not quite true. Although space is a better vacuum than can be achieved in labs (Table 20-2), its material cannot be neglected (Wynn-Williams, 1992). What significance can this thin material have for us? For one thing, it dots our sky with **nebulae,** or vast clouds of dust and gas, some twisted into beautiful wispy forms, some dark, and some glowing with different colors. For another thing, as Chapter 14 pointed out, our solar system, the Earth, and we ourselves are formed from atoms that were once part of the interstellar gas and dust. More provocatively, recent discoveries have demonstrated that interstellar material does contain complex organic molecules, and some scientists have speculated that primitive processes, perhaps related to the origin of life, may have occurred in nebulae. Perhaps the most important point is that there are universal aspects of chemistry that make organic molecules common, both in living things on Earth and in inanimate objects such as comets and interstellar clouds.

TABLE 20-1

Characteristics of Selected Nebulae

Name	Constellation	Approx. Distance from Earth (pc)	Approx. Diameter (pc)	Estimated Atoms per m^3	Mass ($M_{\odot}$)	Spectral Type of Associated Star
Nebulae Probably Associated With Young Objects						
NGC 2261 Hubble's (R Mon)	Monoceros	700	10^{-5}	10^{18}	10^{-1}	F
Kleinmann-Low IR	Orion	500	0.1	10^{12}	100	O
Dark Nebulae						
Coal Sack	Crux	170	8	2×10^7	100	None
IC 434 Horsehead	Orion	350	3	2×10^7	0.6	B
Emission Nebulae						
M 42 Orion (central)	Orion	460	5	6×10^8	300	O
Eta Carina	Carina	2400	80	2×10^8	1000	Peculiar
M 8 Lagoon	Sagittarius	1200	9	8×10^7	1000	O
M 20 Trifid	Sagittarius	1000	4	10^8	1000	O
Reflection Nebulae						
M 45 Pleiades	Taurus	126	1.5	?	?	B
Cocoon	Cygnus	1600	2	7×10^7	7	B
Planetary Nebulae						
M 57 Ring	Lyra	700	0.2	10^9	0.2	White dwarf?
M 27 Dumbbell	Vulpecula	220	0.3	2×10^8	0.2	Whitedwarf?
NGC 7293 Helix	Aquarius	140	0.5	4×10^9	0.2	White dwarf?
Supernova Remnants						
M 1 Crab	Taurus	2200	3	10^9	0.1	Pulsar
NGC 6960/2 Veil (Cygnus Loop)	Cygnus	500	22	10^8	?	?
Gum	Puppis-Vela	460	360	10^5	100,000	Pulsar?

Sources: Data from Allen (1973); Maran, Brandt, and Stecher (1973); and other sources.

TABLE 20-2

Gas Densities in Different Environments

Locale	Density[a] (kg/m^3)	Particles per m^3	Typical Distance Between Particles
Air at sea level	1.2	10^{25}	1 nm
Circumstellar cocoon nebula	10^{-5}	10^{22}	50 nm
"Hard vacuum" in terrestrial laboratory	10^{-9}	10^{18}	1 μm
Orion Nebula	10^{-18}	10^{9}	0.1 cm
Typical interplanetary space	10^{-20}	10^{7}	0.5 cm
Typical interstellar space	10^{-21}	10^{6}	1 cm
Interstellar space near edge of galaxy	10^{-25}	10^{2}	20 cm
Typical intergalactic space	10^{-28}	10^{-1}	2 m

[a]Average density of observable matter in the whole universe is estimated to be about 3×10^{-28} kg/m^3 (Shu, 1982), but many astronomers believe the average may be somewhat higher (3×10^{-27} kg/m^3?) due to nonluminous unseen matter.

THE EFFECTS OF INTERSTELLAR MATERIAL ON STARLIGHT

Dispersed particles, whether floating in space or in the atmosphere, interact with radiation. When light from a distant star passes through clouds of interstellar atoms, molecules, and dust grains, the interaction changes the light's properties, such as intensity and color (Knapp, 1995). Several complex physical laws describe these changes in some detail, but the changes can be grouped under two main principles:

> **1. When radiation (ultraviolet light, visible light, infrared, radio waves, or any other type) interacts with particles, the type of interaction depends on the types of particles and their sizes relative to the wavelength of the light.**
>
> **2. The appearance of the light and the particles may depend on the direction from which the observer looks.**

Three important types of interaction between radiation and matter involve atoms, molecules, and dust grains. In reality, the interstellar material is always a mixture of gas and dust, but it is easier to understand the effects if we imagine separate interactions of light with atoms, molecules, and dust grains.

Interaction of Light with Interstellar Atoms

As shown in Figure 20-1, several things can happen if starlight passes through a cloud of interstellar gas atoms. Suppose a star radiates light of all wavelengths, and the photons of light enter a cloud of gas. **Excitation** occurs when photons corresponding to certain wavelengths have just enough energy to knock electrons from lower to higher energy levels. **Ionization** is a more extreme process where photons knock electrons clear out of the atoms. Each time a photon excites or ionizes an atom, that photon is consumed or degraded in energy and disappears from the light beam. Thus an observer looking at the star through the cloud would see absorption lines created by the interstellar material, as shown in Figure 20-1.

But most of the energy absorbed by the cloud is reradiated rapidly. The absorbed energy that is not reradiated contributes to the heating of the gas and to the chemical activity within it. The balance between heating by absorption and cooling by reradiation governs the temperatures of the gas and the dust. This reradiation occurs as the electrons cascade back down through the energy levels of the atoms, creating emission lines. The photons in these emission lines leave the cloud in all directions, as shown in Figure 20-1, so an observer off to one side would see the cloud glowing in the various colors corresponding to the emissions.

★ Colors of nebulae are hard to see with the eye, even with large telescopes, because the light's intensity is low and the eye's color sensitivity is poor at low light levels (the reason why a moonlit scene looks less colorful than in daylight). Sensitive films and other detectors can record the extraordinary colors quite accurately, however. Photos in different colors often reveal different patterns. Since hydrogen is the most

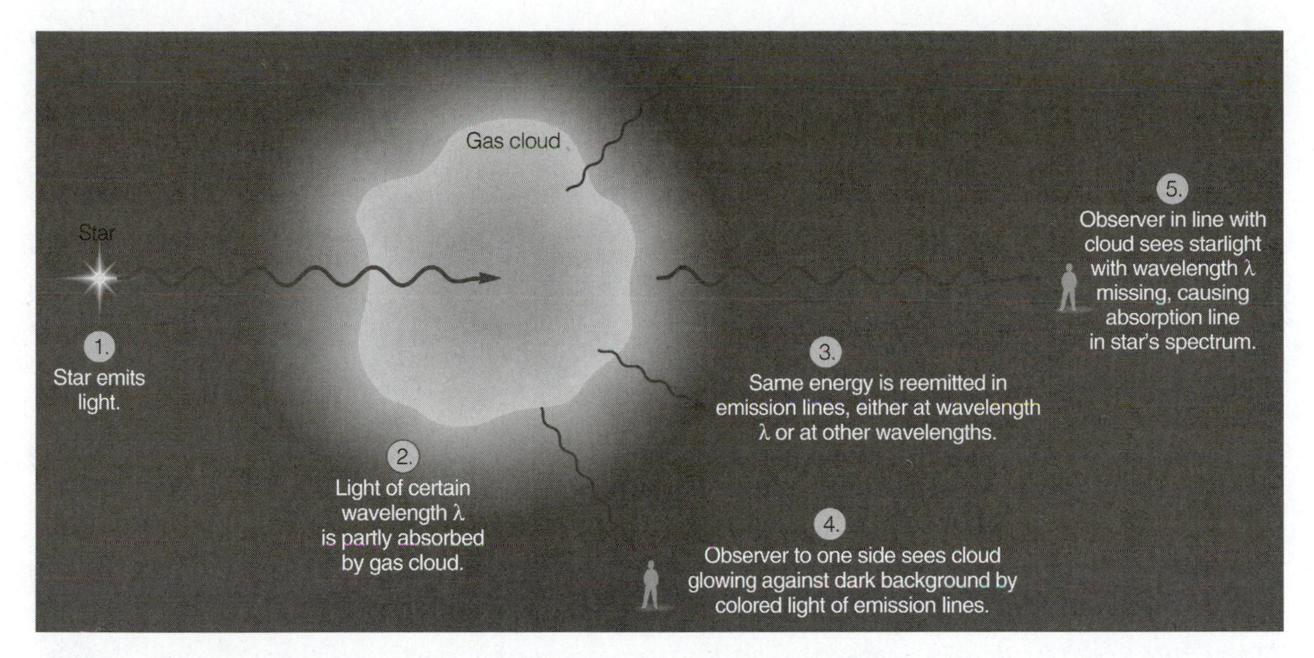

Figure 20-1 A cloud of gas (atoms and molecules) illuminated by a star and seen by an observer to one side (item 4) and an observer in line with the cloud and the star (item 5).

abundant gas, and since the red Hα emission line is one of its strongest transitions, many clouds of excited gas glow with a beautiful deep red color. Other colors also appear in nebulae—see the selection of beautiful color images of nebulae on our web site. These color images were made by combining digital CCD images from the HST in blue, green and red filters.

Interaction of Light with Interstellar Molecules

Molecules are two or more atoms bound together by weak electrical forces. Like atoms, molecules have characteristic spectra that are related to the structure of their internal energy levels. Molecular spectra are typically more complicated than those of single atoms because the rotations and vibrations of molecules add together many closely spaced energy states. When observed with low spectral resolution, this can give the appearance of an absorption or emission *band* rather than a single narrow line. Tiny amounts of energy are needed to excite vigorous rotation in molecules, which explains why many molecular absorption and emission features appear at infrared or millimeter wavelengths. A small amount of energy released from an atom or molecule appears as a long-wavelength photon, as discussed in Chapter 5. It also explains why these features are observable even in very cold, low-density interstellar environments.

Interaction of Light with Interstellar Dust Grains

Whereas atoms are as small as 0.0001 times the wavelength of visible light, and molecules are a few percent of this wavelength, interstellar dust grains range in size from 0.3 to 300 nm, and their interactions are quite different. They absorb some starlight, dimming distant stars. They also affect colors over a much broader range of wavelengths than individual spectral lines or bands. The most important effect is that redder light (longer wavelengths) passes through clouds of dust, whereas bluer light (shorter wavelengths) is scattered out to the side of the beam (see Figure 20-2).

Scattering of light is different from absorption, mentioned a few lines earlier, but we will not emphasize the difference. Think of absorption as a disappearance of a photon into an atom (or ion or molecule), making the atom more energetic. Think of scattering as a bouncing of a photon off an atom out of the light beam into a new direction. Either process reduces the amount of light in the beam.

Thus an observer who looks through a dust cloud at a distant star sees most of its red light, but not much of its blue light. In this way, interstellar dust makes distant stars look redder than they really are—an effect called **interstellar reddening.** An observer who looks at the dust cloud from the side will see the blue light scattered out of the beam, however,

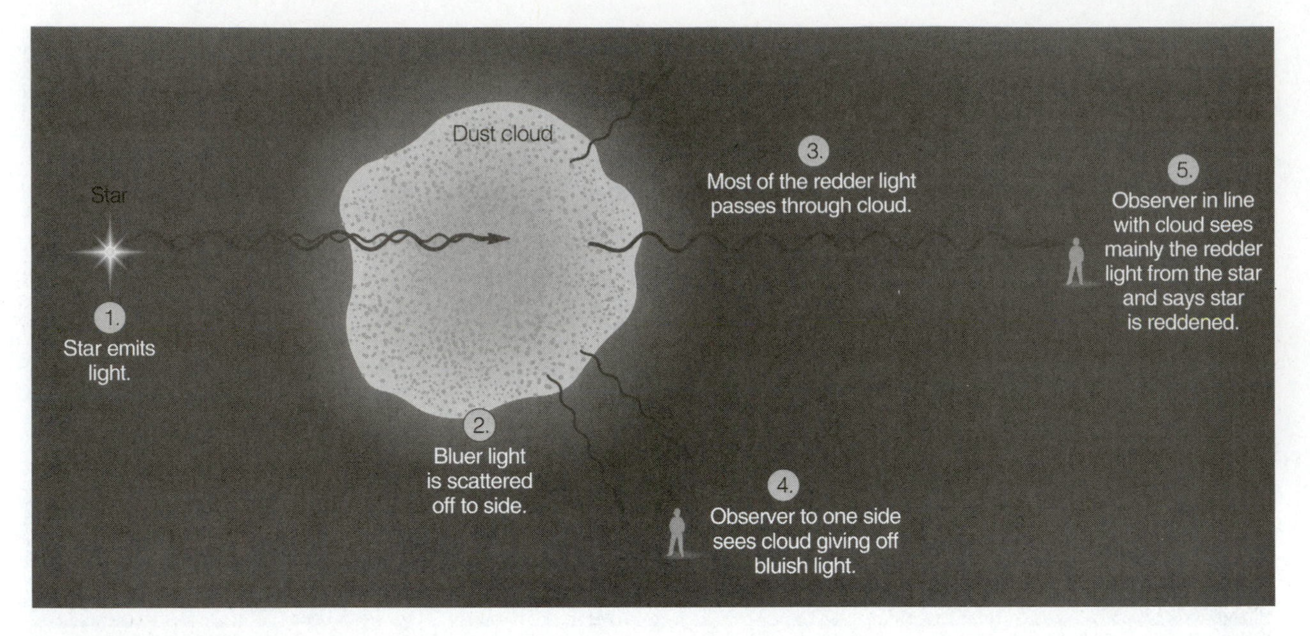

Figure 20-2 A cloud of dust grains illuminated by a star and seen by an observer to one side (item 4) and an observer in line with the cloud and the star (item 5). Compare with Figure 20-1, showing effects for a *gas* cloud.

so a nebula illuminated in this way will have a bluish color. The light from a reflecting nebula arises from scattering by dust rather than atoms or molecules. The blue color of the nebulosity is caused partly by dust scattering and partly by the fact that the light being reflected comes from a hot, blue star.

Why Is the Sunset Red and the Sky Blue?

These same rules also apply to material in our own atmosphere. The atmosphere is full of gas molecules and tiny dust particles. The molecules are much smaller than the wavelength of light, so scattering is more efficient for short wavelengths of light. When we look at the nearest star, our Sun, through these particles, the same effects can be seen. If the Sun is high in the sky, we look through the minimum amount of gas and dust, as shown in Figure 20-3a; thus the reddening is minimal, and the Sun is perceived as yellow.

At sunset, as shown in Figure 20-3b, the sunlight passes through much more gas and dust, and much of the blue light is lost from the beam. This strongly reddens the Sun and adjacent parts of the sky. At any time of day, if we look at some other part of the sky, as in Figure 20-3c, the light we see is the blue light scattered out of the beam of sunlight and then scattered by air molecules and dust back toward our eyes.

OBSERVED TYPES OF INTERSTELLAR MATERIAL

★ Interstellar material includes gas (that is, atoms and molecules), microscopic dust grains, and possibly larger objects (Verschuur, 1989). The full range of colors that can be produced by the interstellar medium is illustrated by the lustrous blues and reds of the Triffid Nebula (see the image on our web site).

Interstellar Atoms

Atoms of interstellar gas were discovered in 1904, when German astronomer Johannes Hartmann detected their absorption lines. While studying the spectra of a binary star, he accidentally discovered absorption lines caused by ionized interstellar calcium atoms. Certain other **interstellar atoms,** such as sodium, were soon found to produce additional prominent interstellar absorption lines. These lines are identified as interstellar by the fact that they are very narrow and have different Doppler shifts than the stars in whose spectra they appear.

Further studies have convinced astronomers that, although atoms such as calcium and sodium have prominent absorption lines, the most common interstellar gas is the ubiquitous hydrogen. Like the Sun and stars, interstellar gas is about three-fourths hydrogen and nearly one-fourth helium by mass.

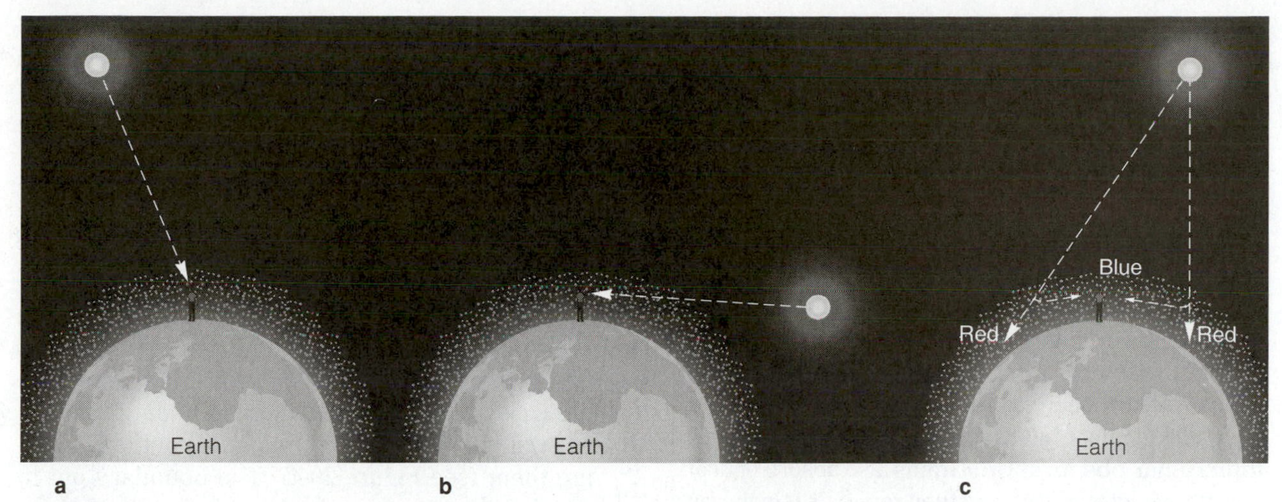

Figure 20-3 Explaining colors in the sky. **a** Light from the Sun high in the sky passes through minimal gas and dust and is minimally reddened. **b** Light from the Sun at sunset passes through the maximum amount of gas and dust and is strongly reddened. **c** Light from the sky at any time of day is the blue light scattered out of the sunlight beam.

Radio Radiation from Interstellar Atoms In 1944, the Dutch astronomer H. C. van de Hulst predicted that the most important type of interstellar radiation would be an emission line with wavelength 21 cm, caused by a change in the spin of hydrogen atoms' electrons. A low-energy radio photon is emitted when the electron spin changes from being aligned with the proton spin to being misaligned with the proton spin. Such a long wavelength is not visible light, but radio radiation. The predicted emission was confirmed in 1951 when Harvard astronomers, using radio equipment, detected this **21-cm emission line** of atomic hydrogen.

This discovery not only confirms the importance of hydrogen as a main constituent of interstellar gas, but it also gives radio astronomers a tool to detect where clouds of interstellar gas are concentrated. Because of the long wavelength of this radiation, it can penetrate much greater distances through the interstellar gas and dust than ordinary light. The 21-cm line of interstellar hydrogen is thus one of the most important emission lines in radio astronomy.

Interstellar Molecules

By 1940, astronomers at Mt. Wilson Observatory in California had built spectrographs that detected absorption due not only to interstellar atoms but also to **interstellar molecules,** such as CH and CN. Because of molecular structure and the cold, low-density environment of interstellar clouds, a great many of the molecular absorption features lie in the infrared or radio parts of the spectrum. Not until after World War II did astronomers have available the technology of infrared and radio detectors to search for interstellar molecules. After being developed in the 1960s, new detectors on large telescopes sparked an explosion of interstellar discovery in the late 1960s and 1970s (Figure 18-1 showed the variety of molecular lines that can now be observed).

More than 100 varieties of interstellar molecules have been cataloged (Hartquist and Williams, 1995). The most important in astrophysical processes are molecular hydrogen (H_2) and carbon monoxide (CO). The atoms recurring again and again in these large molecules comprise the quartet carbon, hydrogen, oxygen, and nitrogen—the "building blocks of life!" Repeatedly, these building blocks have been found in space in complex large molecules. It is intriguing that two of the detected molecules, methylamine (CH_3NH_2) and formic acid (HCOOH), can react to form glycine (NH_2CH_2COOH), one of the *amino acids.* These large molecules can join to form the huge protein molecules that occur in living cells.

Thus research on interstellar molecules has raised two exciting questions. First, does the existence of complex, carbon-rich molecules in space suggest that life could have originated elsewhere in space? The answer may be yes, and we will discuss this possibility in more detail in Chapter 28. Second, how do so many atoms come together to form these molecules? They form both in ordinary interstellar gas, where collisions between atoms are extremely rare, and in the denser regions, such as the clouds in which stars form.

Interstellar Grains

Interstellar grains are even bigger than interstellar molecules. Typical grains are about the size of smoke particles in our air. Grains account for about 1% of the total interstellar mass, although there is only 1 for every 10^{12} hydrogen atoms or molecules. Interstellar grains have several observational effects. The reddening has already been explained, and they also cause a general dimming of starlight at all wavelengths, called **interstellar obscuration.** Grains also absorb optical and ultraviolet starlight, and they reradiate it in the far infrared, with a thermal spectrum that reflects the cold temperature of the grains. Some grains are concentrated in distinct clouds, where they can dramatically extinguish the background light. Others are widely distributed throughout the interstellar gas, producing a general haze or "interstellar smog."

Nature and Origin of the Grains

To summarize, astronomical studies reveal various properties of the interstellar grains: (1) They range in size from about $\frac{1}{100}$ to twice the wavelength of visible light; (2) they are concentrated in clouds; (3) compositions may vary somewhat from cloud to cloud; (4) many grains are elongated; and (5) many elongated grains are aligned parallel to other nearby grains, possibly because they are iron-rich and aligned with magnetic fields in space.

Grain composition has long been debated. In 1967, Sri Lankan astrophysicist N. Wickramasinghe proposed that grains are a form of carbon condensed in cooling gas blown off carbon-rich giant stars. Others have suggested silicates and ices, likening grains to the dust in our primordial solar system. All these materials probably exist in interstellar grains. Carbon compounds, silicates, and ices have been identified spectroscopically, especially in star-forming regions. Reactions initiated by ultraviolet light in these materials apparently create complex organic molecules on many grains' surfaces.

FOUR TYPES OF INTERSTELLAR REGIONS

Interstellar gas and dust are far from uniform. There are thick clouds of material (nebulae) and regions of different temperature. Generally, astronomers recognize four different types of regions, defined by the condition of the gas: (1) Cold regions of dark clouds of dust and gas at about 10 K are called **molecular clouds** (Scoville and Young, 1984). Here most gas exists as molecules, including not only molecular hydrogen (H_2) but many of the more complex molecules discussed earlier. (2) Regions at around 200 K contain neutral hydrogen atoms and are called **HI regions** after the symbol for neutral hydrogen (HI). HI also exists in a warm form at 7000 to 10,000 K. (3) Hot regions at around 10,000 K surround hot O and B stars, whose ultraviolet photons excite the surrounding H atoms or even knock electrons out of most of them, ionizing them (see Figure 20-4). The boundary out to which hydrogen is ionized by the central star is called a Strömgren sphere (even though it is not usually perfectly symmetric). These are called **HII regions** after the symbol for ionized hydrogen (HII). (4) Very hot regions, at about 10^6 K, exist

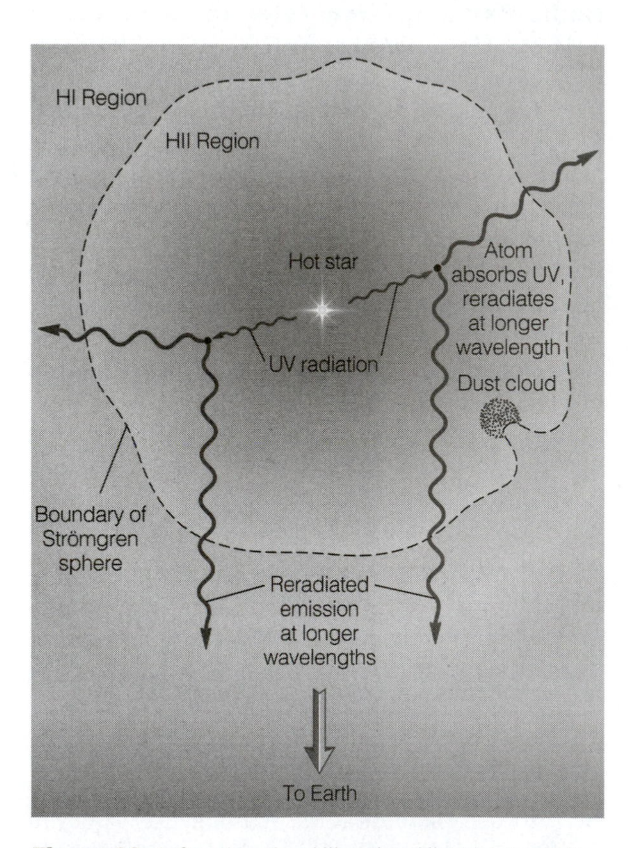

Figure 20-4 Creation of an HII region. Ultraviolet radiation from the central hot star is absorbed by hydrogen atoms, ionizing all of them out to a certain distance (dashed line). Recombination of electrons with atoms causes reradiation of various spectral emission lines in all directions, including towards Earth, causing the HII region to be visible as a glowing nebula.

where gas has been superheated by blasts from supernovae. This is the **hot interstellar medium.**

Radio telescopes and satellite X-ray telescopes have helped reveal molecular clouds and the hot interstellar medium, respectively, only in recent years. Hence it is not yet clear exactly what fractions of space are occupied by the four types of regions. Although as much as 50% of our galaxy's gas may be molecular, as much as 90% of the volume is in the form of the hot interstellar medium.

LIFE CYCLE OF A STAR-FORMING REGION

In localized regions where gas gets compressed to higher-than-average densities (perhaps by winds from neighboring expanding clouds), atoms and ions collide more often and molecules may tend to grow. Dust grains get concentrated along with the gas. They often make the cloud opaque and dark, shielding its inner parts from nearby hot stars. More important, the dust grains can radiate heat more easily than the gas, so a dense region rich in dust grains gets cooler than surrounding clouds. Recall from Chapter 18 that the denser and colder a region of gas, the easier it is for stars to form in it by gravitational collapse. Thus molecular clouds are extremely important as regions of star formation (Blitz, 1982).

★ Figure 20-5 is a schematic view of the cycle of star birth and star death. Hot O and B stars form and heat portions of the molecular cloud, creating glowing HII regions. This explains why cold molecular clouds are found in close association with hot HII regions. The most massive, short-lived stars will soon explode, blowing out million-degree gas that pushes against adjacent gas. The resulting compression starts new star formation on the outskirts of the expanding cloud. This leads to new supernovae and more expansions until the whole star-forming complex is disrupted. Beyond the regions where stars are actively forming, astronomers have observed diffuse gas and dust grains that radiate at infrared wavelengths. This wispy material, called cirrus, can be seen in a color image on our web site.

Example: The Orion Nebula, a Nearby Star-Forming Region

★ In the direction of the constellation Orion lies a region up to 500 pc away dominated by nebulae and star-forming activity. Orion itself is the constellation of the hunter—a great figure raising a club over his head. Three bright stars mark his belt, and three below it mark his sword. Orion's shoulder on the left side is the bright red giant star Betelgeuse, 180 pc away. His opposite knee is the blue B-type star Rigel, 270 pc away. Our web site contains a wide-angle view of the constellation and a close-up view of the Orion Nebula.

If you look toward Orion on a February evening, swinging your head back and forth past this part of the sky to compare it with other regions, you will see a great concentration of bright stars. Orion contains 7 of the 100 brightest stars in the sky, and except for Betelgeuse, all of them are massive, hot O- and B-type stars 140 to 500 pc away. Since massive stars are short-lived, their mere existence shows that they and other stars must have formed recently in the nebulosity around Orion.

Consistent with its fame as a star-spawning region, Orion is full of vast clouds of gas and dust. Since the dust-cloud temperatures are around 30 K, the dust radiates in the far infrared, where it has been imaged by satellites. Here we see the most intense dust clouds centered on the Orion Nebula in Orion's "sword" and at the left end of the "belt." By coincidence, a shell of dust has been blown outward from stars in the head region, making a ring-shaped infrared halo about 7 degrees in diameter around Orion's head.

Since O and B stars are extremely hot, they radiate prodigious amounts of ultraviolet (UV) radiation. For this reason, UV photos of Orion reveal it to be one of the most dazzling regions in the UV sky. The massive stars have ionized much hydrogen, and the hot gas is being blown outward from the bright stars in central Orion.

★ The **Orion Nebula** is perhaps the most famous of nebulae. It is about 460 pc away, in the heart of the Orion star-forming area. To the naked eye it appears as the middle star in the sword, but even a small telescope or good pair of binoculars reveals it as a misty luminous haze about 5 pc across. As early as two centuries ago, English astronomer William Herschel examined this region with his pioneering large reflecting telescope and prophetically described it as "an unformed fiery mist, the chaotic material of future suns." And the details are spectacular: intermingled wisps of red-glowing hydrogen, dark dust, and filaments dominated by pale greens of ionized oxygen and colors of other excited atoms. Here new stars are being born. A cross section of the inner Orion region appears on our web site.

A time-lapse movie of Orion beginning about 6 to 10 million years ago, with frames every 10,000 y, would reveal star formation followed by a cosmic explosion

Figure 20-5
Schematic drawings showing the evolution of interstellar material in a certain region. The first box shows a random concentration of gas and dust that initiates star formation. The largest of the new stars then explode as supernovae. This expands the gas and eventually creates a "superbubble" of very hot expanding gas. The surrounding gas is finally blown away, and the star formation ends.

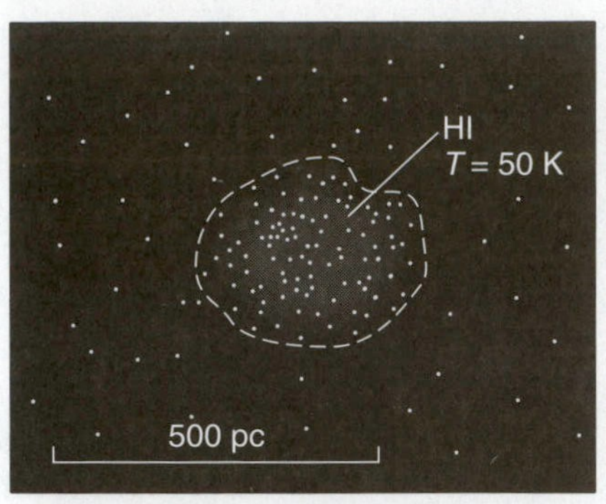

a Dense, cold molecular cloud, contracting

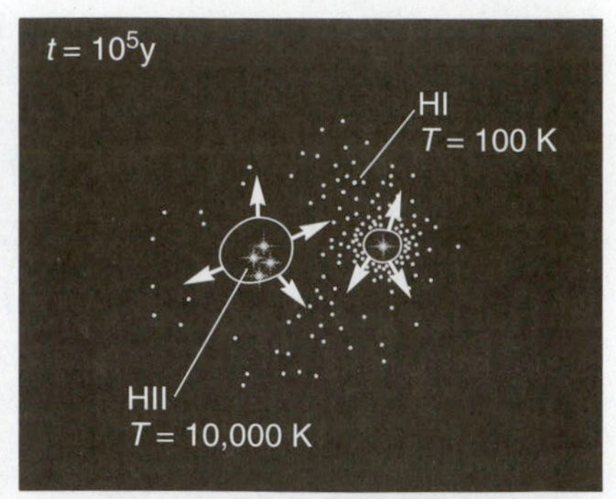

b Star formation, HII regions, molecular clouds

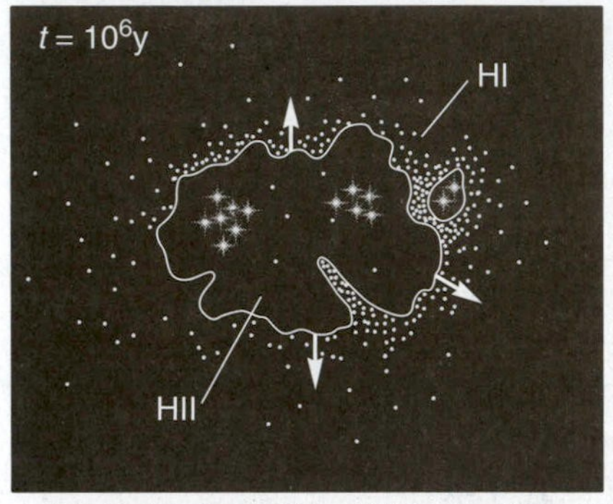

c More star formation, HII expansion by heating

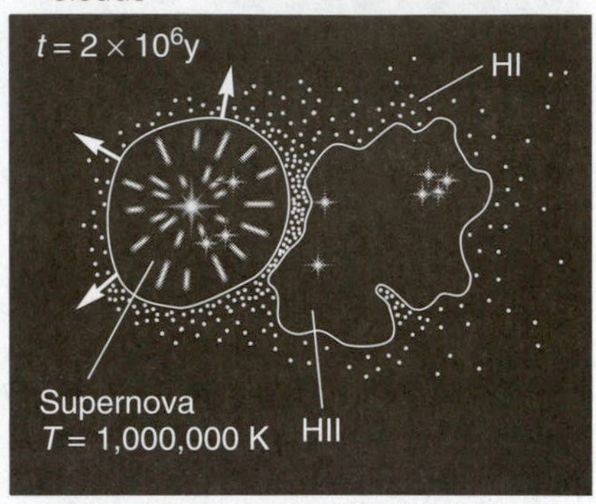

d Supernova expands, new star formation

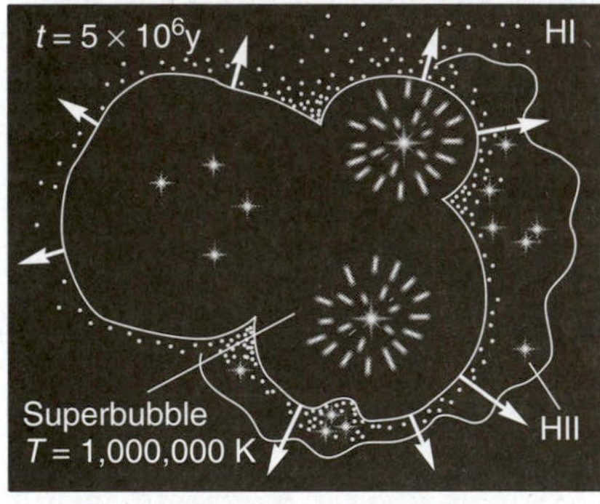

e More supernovae, superbubbles, and star formation

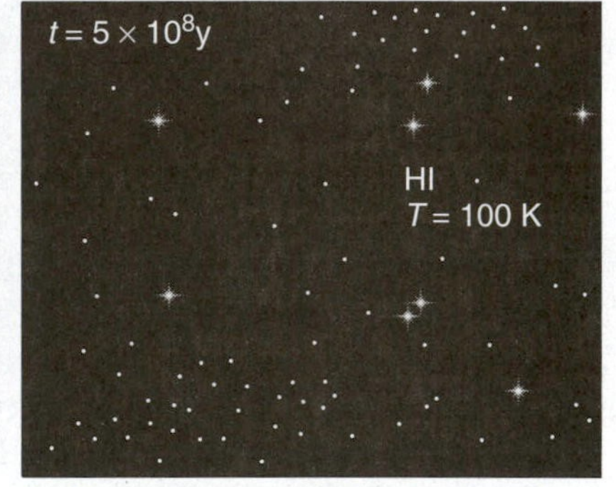

f Clouds and star clusters dispersing star formation and supernovae over

of gas in the region of Orion's sword. The movie would also show certain stars racing out from central Orion like sparks from a blast. Orion has changed dramatically in the last few million years, since humans emerged on the plains of Africa. New stars have blazed up, clouds of hydrogen have been expelled, and star formation is apparently continuing there today.

SUMMARY

Space is not empty but thinly filled with atoms and molecules of gas, grains of dust, and possibly bigger debris, which partly obscure distant stars and redden their light. Concentrations of these materials are nebulae. Some are excited to fluorescence by nearby stars, forming emission nebulae; some merely reflect the light of nearby stars; and some are dark silhouettes against distant backgrounds. Various processes of emission, scattering, and absorption give nebulae different colors. Various processes of expansion and turbulence give nebulae different shapes.

Nebulae are the materials from which stars are born and into which the larger stars blow some of their material when their fuel runs out. Their existence shows that matter in our galaxy has not dispersed uniformly and become stable, but rather is continually stirred, formed into clouds, dispersed, and disturbed by influences such as the formation of new stars, the explosions of old stars, and movements of all local material around our galaxy's center.

From nebulae and the general interstellar complex we know that all local material visible from the Earth has participated in a vast cosmic recycling. Interstellar matter forms clouds; clouds may contract to form stars; young and old stars blow out their material, replenishing and enriching the interstellar medium. Nebulae also reveal clear evidence of cosmic events that have markedly changed our celestial, and perhaps even terrestrial, environment within the last million years—a period less than 0.01% of cosmic time.

CONCEPTS

nebula	interstellar grains
excitation	interstellar obscuration
ionization	molecular cloud
interstellar reddening	HI region
interstellar atom	HII region
21-cm emission line	hot interstellar medium
interstellar molecule	Orion Nebula

PROBLEMS

1. Since sunlight is white (a mixture of all colors), why does the Sun look red at sunset? What happens to the blue light? Why does the part of the sky away from the Sun look blue?

2. Why was the sky red as photographed on Mars by Viking cameras?

3. How do massive stars help keep interstellar gas stirred up?

4. How do interstellar molecules illustrate the fact that complex organic chemistry is likely elsewhere in the universe?

5. How do masses of prominent nebulae compare with masses of single stars—are stars more likely to form singly or in groups?

6. Why are O-type supergiant stars likely to be associated with large emission nebulae, whereas solar-type stars are not?

7. Why do planetary nebulae often have simple, nearly spherical forms, whereas typical large emission nebulae, such as the Orion Nebula, are ragged, irregular masses?

ADVANCED PROBLEMS

8. If the Sun's material were redispersed into space at a density of 10 atoms/cm^3, typical of some clouds, how big a cloud would it make? (*Hint:* The Sun contains about 10^{57} atoms. 1 pc = 3×10^{16} m = 2×10^5 AU.)

a. Compare with the size of the solar system.

b. Compare with prominent nebulae.

9. Suppose spectral lines from gas on the near side of the Crab Nebula have a Doppler blueshift of 0.5%.

a. At what velocity is it expanding?

b. If the Crab Nebula is 1500 pc away (or 4.5×10^{19} m), use the small-angle equation to derive the expansion velocity of the nebula if it is observed to expand at an angular rate of 0.2 second of arc per year. (*Hint:* $1\ \text{y} \cong \pi \times 10^7$ s.)

c. Confirm that the above two methods of estimating expansion velocity in the Crab Nebula give consistent results.

10. Suppose a tight cluster of newly formed O stars totals 100 solar masses within a region a parsec wide (like the heart of the Orion Nebula). Several of them explode as supernovae, creating a superbubble. The superbubble expands to a point 10 pc from the center of the cluster and has a temperature of 1 million kelvins. Compare the thermal velocity of a hydrogen atom at the edge of the superbubble with the velocity of a hydrogen atom at the edge of the superbubble with the velocity it would need to escape from the gravitational pull of the star cluster into interstellar space. Would you expect the superbubble to continue to expand and dissipate, assuming it did not run into dense clouds of adjacent material?

PROJECTS

1. Observe a cloud of cigarette or match smoke illuminated by a single light source, preferably a shaft of sunlight in a darkened room or a strong reading lamp. Compare the color of light transmitted through the smoke (by looking into the beam) with the color of light scattered out of the smoke (by looking at right angles across the light beam). Are there any differences? What can you conclude about the size of particles in the smoke cloud, assuming that the light wavelength is mostly 400 to 800 nm? Compare forms in the drifting smoke cloud with forms of nebulae illustrated in this book and on the web site.

2. In a dark area away from city lights, by naked eye, observe and sketch the Milky Way in the region of Cygnus. Can you observe the dark "rift" that divides the

Milky Way into two bright lanes in this region? The rift is caused by clouds of obscuring interstellar dust close to the galactic plane that are between us and the more distant parts of the Milky Way galaxy.

3. Observe the Orion Nebula with a telescope. Sketch its appearance. Locate the Trapezium (four stars near the center). The dark wedge radiating from the Trapezium is a dense mass of opaque dust. If a large telescope (50–100 cm) is available, look carefully for color characteristics. Generally, the eye is unresponsive to colors of very faint light, but large telescopes gather enough light so that colors can sometimes be perceived, especially with fairly low magnifications, giving a compact, bright image.

4. Observe the Ring Nebula in Lyra or other nebulae such as the Crab (in Taurus) or the Trifid (in Sagittarius). Comment on differences in form and origin.

CHAPTER 21

Companions to Stars

WHAT THE READER SHOULD WATCH FOR IN THIS CHAPTER

Most stars are not alone but are locked in a gravitational embrace with one or more companions. Various types of binary systems can be revealed by imaging and spectroscopic techniques, and a binary system offers the best way to measure a star's mass. In close binary systems, each star can affect the evolution of the other. In fact, mass transfer between stars in a binary system can lead to some of the most spectacular phenomena in any star's life—eruption as a nova, detonation as a supernova, and intense X-ray emission from an accretion disk. Binary and multiple stars can be created in several different ways. Enormous excitement has been aroused by the discovery of alien planets. These low-mass companions to stars near the Sun are generally Jupiter-mass or greater, and they often orbit their stars on tight, elliptical paths. The discovery that our solar system is not unique is a major new step in the Copernican revolution. ■

We have been soft-pedaling a fundamental fact: Most stars are not solitary wanderers in space. Four of the first six stars we encounter beyond the solar system have known companion stars. Surveys of this kind suggest that more than two out of every three stars are involved in co-orbiting star systems.

Each pair of co-orbiting stars is called a **binary star system,** or sometimes just a binary star. Each system with more than two stars is called a **multiple star system.** How do these systems affect our understanding of the universe? For one thing, we would like to know if our own multiple system, the Sun and its planets, is related to other multiple star systems, or whether it is a different kind of phenomenon. Second, we must be sure that our theories of star formation and evolution account for binaries.

OPTICAL DOUBLES VS. PHYSICAL BINARIES

Among the star pairs that appear to be close together in the sky, some are at different distances and are merely aligned by chance (as seen from Earth)—they are not actually co-orbiting. Such star pairs are called *optical double stars.* They can be identified by the absence of any orbital motion around each other (revealed by photos or Doppler shifts) and are of little consequence in astronomy.

Pairs that are close enough to orbit around each other are called **physical binary stars.** They are the ones that clarify our knowledge of star properties and are the ones we will discuss here. Early observers thought that *all* close star pairs were merely optical doubles, but in 1767 John Mitchell pointed out that so many

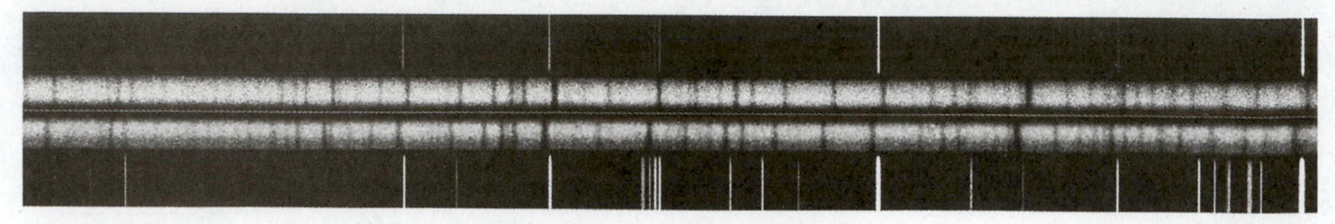

Figure 21-1 A portion of the spectrum of a single-line spectroscopic binary, Alpha Geminorum. The bright vertical lines at top and bottom are reference emission lines produced in the spectrograph; the middle two bright spectra crossed by dark, vertical absorption lines are spectra of the star on two dates. Offsets of these lines to the red and then the blue show that the star is receding and then approaching because it is orbiting around another star. (Lick Observatory)

chance alignments were unlikely. He proposed a physical association. This was confirmed in 1804 when William Herschel discovered that Castor (brightest star in the constellation Gemini) has a companion orbiting around it. This marked the *first discovery of gravitational orbital motion beyond the solar system,* an important confirmation that gravitational relations are universal.[1]

Among physical binaries, the brighter star is usually designated A and the fainter star B—for instance, Castor A and Castor B. Analysis of orbits reveals which is the more massive star, and it is usually called the *primary.* The less massive one is called the *secondary.* Normally, the primary is also the brighter one, or star A.

TYPES OF PHYSICAL BINARIES

Astronomers classify physical binaries according to how they are detected. To understand what we can learn from binaries, it helps to understand these different methods of detection.

A **visual binary** is a physical binary in which the orbiting pair can be resolved (seen separately) with a telescope. Some 65,000 have been studied.

In a **spectroscopic binary,** orbital motion is revealed by periodic Doppler shifts in the spectral lines, but the individual stars cannot be resolved. In some spectroscopic binaries, only *one* spectrum can be detected (Figure 21-1); in others, *two* sets of spectral lines are seen (Figure 21-2). The latter yield more information. Over 1000 pairs have been measured.

An **eclipsing binary** is a binary pair (generally unresolved) whose orbit is seen nearly edgewise. Because our line of sight lies in, or nearly in, the orbital plane, the stars alternately eclipse each other. Eclipses are detected by plotting **light curves,** or plots of brightness vs. time, as shown in Figure 21-3. Depending on the relative brightness and size of the stars, the eclipse of the primary may produce a marked, short-term decrease in brightness. The star then returns to its normal brightness. About 5000 eclipsing binaries have been found in surveys of stars (Kopal, 1990).

Most eclipsing binaries are really *eclipsing-spectroscopic binaries,* in which both Doppler shifts and eclipses can be detected. This is the most informative type of binary, permitting very detailed analysis of the motion, mass, and size of the stars. To take a simple example, suppose the Doppler shifts reveal that one star moves in a circular orbit at 100 km/s, and timing of the eclipse shows that the star takes 10,000 s (about 3 h) to pass in front of the other star. Then the diameter of the primary star must be about 1 million kilometers (100 km/s times 10,000 s). This type of observation is one of the most accurate checks on theories of stellar structure.

★ Imagine a binary consisting of a bright star and a companion too faint to see from Earth. The bright star is moving around the unseen companion. Extremely careful measurements of the bright star's position, relative to background stars, can reveal its motion, in turn revealing that it is a binary and has an unseen companion. Such a binary is called an **astrometric binary.** In such systems, information can be derived about the *unseen* companion, as well as the visible star. One interesting facet of binary stars is that very different types of stars can be paired:

[1]Herschel realized that this discovery was much more important than his original goal of simply measuring distances, because the orbital motions reveal many properties of the stars. This illustrates how important, unexpected scientific results often derive from mundane research on another topic. Herschel reportedly likened himself to the biblical character Saul, who went out to find his father's mules and discovered a new kingdom.

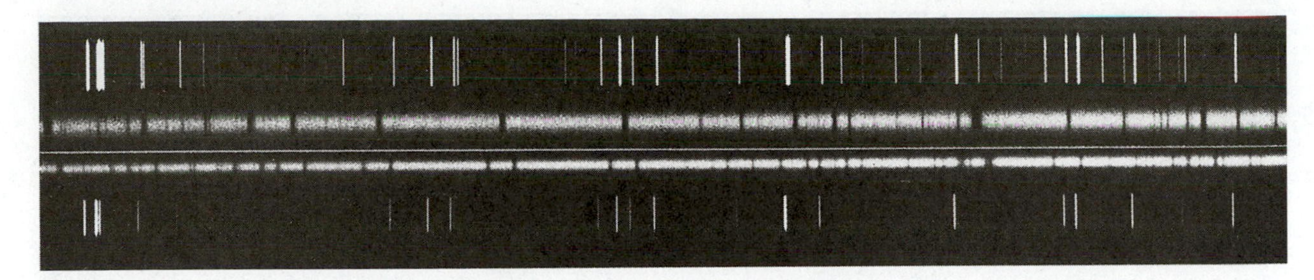

Figure 21-2 A portion of the spectrum of a double-line spectroscopic binary, Kappa Arietis. Arrangement as in Figure 21-1. Two sets of absorption lines in the top stellar spectrum reveal two stars, one receding (redshifted) and one approaching (blueshifted). The lower stellar spectrum, when orbital motions are perpendicular to the line of sight, shows lines merged, with no Doppler shift. (Lick Observatory)

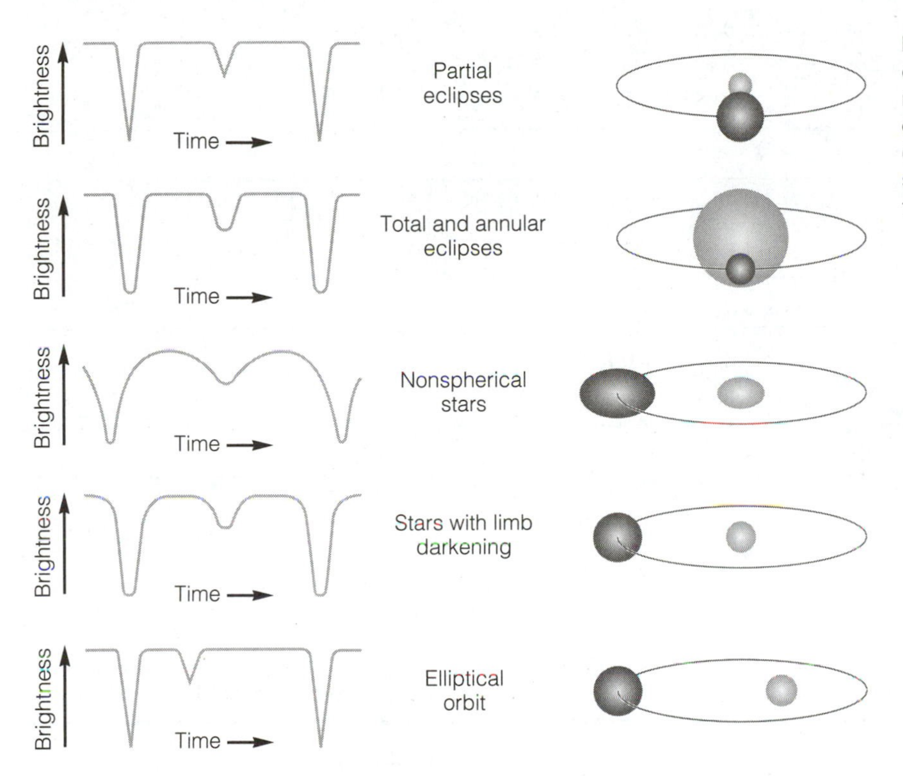

Figure 21-3 Different types of light curves (left) from eclipsing binaries reveal different geometric properties of the eclipsing stars and their orbits (right), even though the stars themselves cannot be resolved with the telescope.

massive and not so massive, giant and dwarf, red and blue (see our web site for an imagined example). The variety offers challenging opportunities for both measurement and imagination. Table 21-1 gives some selected examples of binary and multiple star systems.

WHAT CAN WE LEARN FROM BINARY STARS?

The various types of physical binaries yield much information. *The most important application of binary star studies is in determining star masses.* The mass of an isolated single star can only be determined in terms of a model of stellar structure.

According to Kepler's laws as modified by Newton, when any two cosmic objects are in orbit around each other, the period of revolution (the time it takes to complete an orbit) increases as the distance between the objects increases and as the sum of the masses of the two objects decreases. Many pairs of stars orbit around each other. For many of these binary stars, we can measure both the period of revolution and the distance between the two stars. Thus we can calculate the sum of the masses, designated $m_A + m_B$, where A and B designate the two stars. But we want to know each individual mass, not the sum of the two. Newton showed that in a system of orbiting

TABLE 21-1

Selected Binary and Multiple Stars

	Component A		Component B			Component C		
System Name	Distance (pc)	Mass ($M_\odot$)	Separation from A (AU)	Mass ($M_\odot$)	Eccentricity	Separation from A (AU)	Mass ($M_\odot$)	Eccentricity
Visual Binaries								
α Centauri	1.3	1.1	24	0.9	0.52	10,000	0.1	
Sirius	2.6	2.2	20	0.9	0.59			
Procyon	3.5	1.8	16	0.6	0.31			
Eclipsing Binaries								
α Coronae Borealis	22	2.5	0.19	0.9?				
Algol (β Persei)	27	3.7	0.73	0.8	0.04	2?	1.7	0.13
β Aurigae	27	2.4	0.08	2.3				
Eclipsing-Spectroscopic Binaries								
β Scorpii	118	13	0.19	8.3	0.27			
η Orionis	175	11	0.6	11	0.02			
Astrometric Binaries								
Krüger 60	4.0	0.28	9.5	0.16	0.41			
Barnard's star	1.8	0.14	2.7?	0.001?				
Visual and Astrometric Binaries								
L 726-8	2.6	0.11	5.3	0.11				
61 Cygni	3.5	0.6	83	0.6			0.0008?	
BD + 66°34	10	0.4	41	0.13	0.05	1.2	0.12	0.00
Solar System								
Sun, Jupiter, Saturn	0.0	1.0	5.2	0.001	0.05	9.5	0.0003	0.06

Sources: Heintz (1978); Popper (1980).
Note: One of the components of η Orionis has been shown to be a triple.

bodies, each body orbits around an imaginary point called the **center of mass** and that by measuring the distance of each star from the center of mass, we can measure the ratio of the masses.

Now we have both the sum of the masses and the ratio of the masses. From this information, we can get each individual mass. For example, take the double star Sirius, consisting of the bright star Sirius A (which has the greatest apparent brightness of any star in the night sky) and the faint star Sirius B. The preceding procedure shows that the sum of the masses of A and B is 3 $M_\odot$. The ratio indicates that A is twice as massive as B. From these facts, the only possible solution is that Sirius A has a mass of 2 $M_\odot$ and Sirius B, 1 $M_\odot$. Study of many stars by this technique reveals an important fact: Stars with nearly identical spectra usually have nearly identical masses. This fact allows the masses of many stars to be estimated roughly from their spectral properties alone.

In the case of many binaries, very accurate orbits have been measured during many years of work (Figure 21-4). Physical binaries also offer good laboratories for studying stellar evolution. Recall that stellar mass is the single most important parameter in the study of stars. The mass determines the stellar structure, the rate of evolution, and the end state of a star.

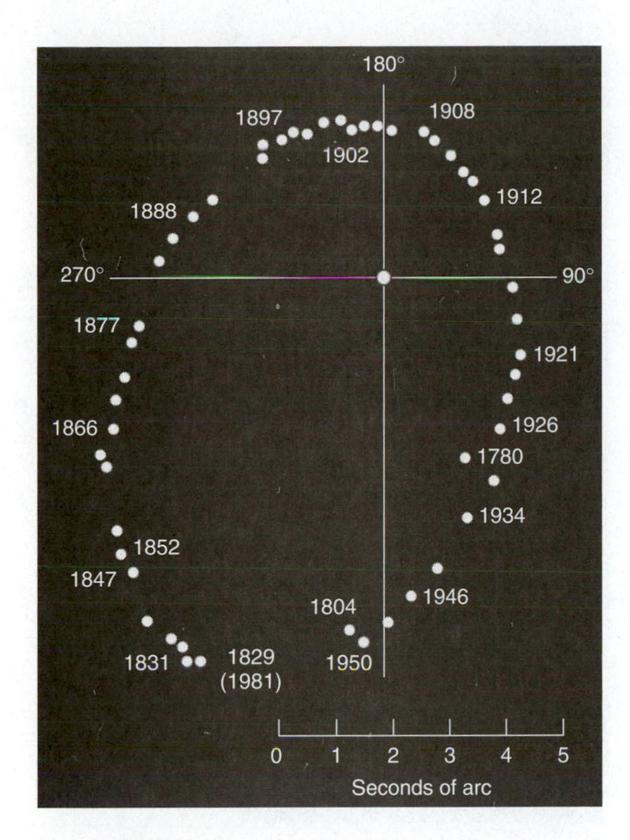

Figure 21-4 Observed positions of the fainter component of the visual binary Xi Boötes from 1780 to 1950, plotted relative to the brighter component, reveal Keplerian elliptical motion. (The apparent ellipse is the projection of the true elliptical orbit in the plane of the sky.) (After a diagram courtesy of *Sky and Telescope*)

HOW MANY STARS ARE BINARY OR MULTIPLE?

This question is a challenge to observers. The nearest stars are easiest to observe but give too small a statistical sample to be reliable. At greater distances there are more stars, but faint companions might not be detected. Spectroscopic binary statistics are biased toward pairs with small separation distances, because according to Kepler's laws these have the fastest velocities and greatest Doppler shifts; thus they are the most likely to be discovered. Visual binary statistics are biased toward wide separation distances, which make the two stars easier to resolve. All these biases, which tend to make the data nonrepresentative of the whole population, are called **selection effects.**

★ Table 21-2 lists three estimates of the *incidence of multiplicity* for systems ranging from single to sextuple. By the time we reach six-member systems, definitions of multiple systems become hazy. It is unclear whether close groupings like the Trapezium in the Orion Nebula should be counted as multiple systems. The 1982 Yale Catalog of 9096 prominent stars lists multiple systems ranging as high as one system with 17 members. Such systems would present an interesting spectacle from nearby (see an imagined view at our web site). We don't know if there is a physical relation between such large multiple systems and small clusters or if they are different phenomena. Even though Table 21-2 illustrates our uncertainty about the frequency of binaries, it does show that single stars are a minority.

TABLE 21-2

Incidence of Multiplicity Among Stars (Estimated Percentage of Systems with *n* Members)

n	25 Systems Within 4 pc	Average of 7 Estimates by Various Authors[a]	Estimate by Batten (1973)[b]
1	48%	41%	30%
2	36%	41%	53%
3	12%	14%	13%
4	4%[c]	3%	3%
5	—	1%	0.8%
6	—	—	0.2%

[a]Data from Batten (1973); Abt (1983, p. 345)
[b]Batten's estimates attempt to average over all stars, using data from various sources. Differences between the estimates are measures of our uncertainty about multiple systems.
[c]Solar system.

Kitt Peak astronomers Helmut Abt and Saul Levy found that about two-thirds of all stars have detectable companions, consistent with Table 21-2. But from statistics of companions' masses, they estimated that the other seemingly single stars probably all have companions too small to detect! Some companions might be brown dwarfs with only a few percent of a solar mass; still smaller ones may be planets. According to this logic, virtually all stars have at least one companion.

Thus although many people mistakenly assume that most stars in the night sky are single, *binary and multiple systems are more common than single stars.* Obviously, then, we must understand the origin of systems of two, three, four, and more stars in order to claim any understanding of stars in general. We will return to the problem of origin after reviewing some related observational data.

EVOLUTION OF BINARY SYSTEMS: MASS TRANSFER

The evolution of individual systems provides a natural way to classify them. The theory of dynamics of binary stars indicates that a system of two co-orbiting stars can be pictured as containing an imaginary **Roche surface** or lobe. Its cross section is a figure 8, with one lobe around each star, as shown in Figure 21-5a. Slow-moving material inside either lobe orbits around the star in that lobe. Material that moves out of either lobe is not gravitationally bound to either star individually. Mass transfer from one star to the other occurs through the point of contact between the Roche surfaces. When either star evolves toward a red giant state, it expands until it fills its lobe, as in Figure 21-5b. Its outer layers then assume the teardrop shape of the lobe, and the Roche surface becomes the real surface of the star. Therefore three classes of binaries exist, as shown in Figure 21-5:

1. Systems in which neither star fills its Roche lobe
2. Systems in which one star fills its Roche lobe
3. Systems in which both stars fill their Roche lobes; that is, stars in contact with each other (**contact binaries**)

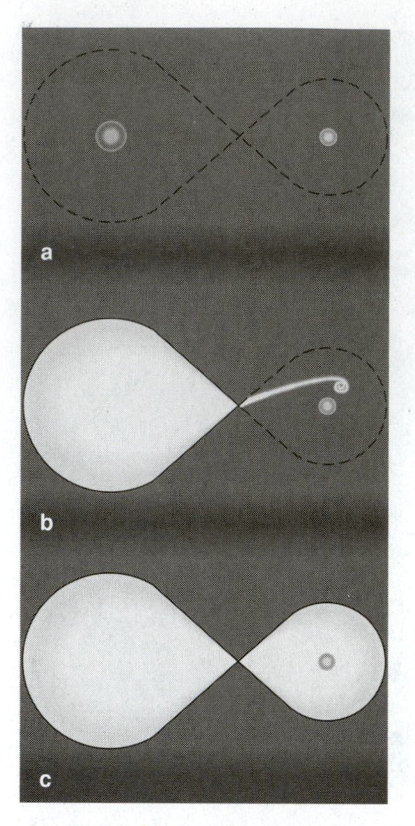

Figure 21-5 Evolution of a binary pair is related to the configuration of Roche lobes. **a** In the first class of systems, neither star fills its lobe (dashed line). **b** After the larger star expands to become a giant, it may fill its lobe, taking on a teardrop shape and perhaps ejecting some mass through the tip, which interacts with the second star. **c** In the third class of systems, both stars fill their lobes.

Evolution can take a single binary through all three types. The pair starts as class 1. The more massive star swells into a giant as it evolves. This giant may fill its Roche lobe, creating class 2. Any further tendency to expand causes matter to be shed, mostly through the point common to the two lobes, as shown in Figure 21-5b. If the giant were a single star, it would be spherical, and what little mass it did lose would stream off in all directions. But pressure within the giant and the companion's gravity force the bloated giant to lose gas through its distorted, pointed tip. Like sand in an hourglass, this gas enters the lobe of the second star. Most of this gas spirals around the small star and crashes onto it, making it gain mass as the large star loses mass.

Since stellar evolution is determined by mass, the evolution of *both* stars will be altered, compared to their destinies had they been single. Later, the second star evolves to the giant state, partly through its normal evolution, but partly because of the added mass. Both stars now fill their lobes, producing class 3, a contact binary. The two stars actually touch as they orbit each other as a single unit! If two stars start very far apart, the giant expansion may not fill either Roche lobe, and the binary may never evolve beyond class 1.

NOVAE: EXPLODING MEMBERS OF BINARY PAIRS

Consider the case of a class 2 binary when the smaller star is a white dwarf of nearly 1.4 $M_{\odot}$. If the larger star's atmosphere dumps unburned hydrogen onto the surface of a white dwarf, the hydrogen will be compressed by the intense gravity of the white

TABLE 21-3

Selected "Guest Stars" (Novae, Supernovae, and Variables That Have Become Brighter than Apparent Magnitude 2)

Star	Date Observed	Maximum Brightness (apparent magnitude)	Type
Lupus supernova	1006	–5?	Supernova
Crab Nebula explosion	1054	–2 to –6?	Supernova
Tycho's star	1572	–4?	Supernova
Kepler's star	1604	–2?	Supernova
Cassiopeia A[a]	1680?	+2 to +5?	Supernova
Eta Carina	1843	–0.8	Nova?
T Corona Borealis	1866	+1.9	Recurrent nova[b]
GK Persei	1901	+0.2	Nova
DQ Herculis	1934	+1.3	Nova
Nova Cygni	1975	+1.9	Nova
o Ceti (Mira)	—	+1.0	Long-period variable (period 331 d)

Sources: Data from Glasby (1968); Lupus data from *Sky and Telescope* (July 1976).

Note: "Guest star" is the ancient Chinese term for temporarily visible stars. Fourteen supernovae ranging back 2000 y have been tentatively identified from ancient records. Certain expanding nebulae must also mark prehistoric supernovae.

[a]Cassiopeia A is the brightest radio source, and its expanding cloud has been imaged by X-ray telescopes in space. It may be the transient "star" observed by Flamsteed in 1680 [*Nature 285* (1980): 132]. See Figure 19-6.

[b]Second recorded flare-up in 1946 reached only +3 magnitude.

dwarf. Under these conditions, the gas is heated as it is compressed, and the hydrogen may ignite in nuclear reactions that blow excess gas outward. This type of explosion is called a **nova** (plural: *novae*).

Some novae occur in cycles. Hydrogen may accumulate on a dwarf until nova explosions occur, anywhere from 100 to 10,000 y apart. Since the individual explosions blow off only a small fraction of the star's mass, leaving the rest intact, the process can start over. Among closer novae, the cloud of expanding debris can sometimes be seen in telescopes a few years after the explosion.

Novae vs. Supernovae

Historically, the term *nova* was introduced long before the term *supernova.* Nova comes from the Latin root for *new* and was the term used for all "new stars" that appeared in the sky in past centuries. In ancient Chinese records, the novae were called "guest stars." In the twentieth century, astronomers discovered from their rate of light variation and other properties that there were two types of "new stars"—the novae and a much more energetic type that came to be called *supernovae.* Novae were found to be much more common. Of the 100 billion stars in our galaxy, 30 to 50 explode as novae each year, but supernovae occur only every 40 y or so. These exploding stars were important in humanity's discovery that the heavens are not immutable, as once thought, but are actually evolving, even as we watch. See Table 21-3 for historical examples.

★ In the 1960s, observers proved that novae involve mass transfer onto white dwarfs in binary systems. Supernovae, however, came to be recognized as involving several types of explosions. As mentioned in Chapter 19, one type also involves mass transfer in binary systems (Cannizzo and Kaitchuk, 1997). Recall that above the Chandrasekhar limit, a white dwarf can't be stable. If the red giant dumps too much gas on the white dwarf too rapidly, the dwarf will exceed the limit and collapse to form a neutron star, blowing off a fraction of the excess mass in a titanic supernova explosion. A second type of supernova, emphasized in Chapter 19, is the natural outcome of evolution of very massive single stars and need not involve a binary system. For an interview with an astronomer who works on binary stars, go to our web site.

CONTACT BINARIES AND OTHER UNUSUAL PHENOMENA

★ Among the most unusual stars are the contact binaries, stars of class 3 in the evolution sequence described above (Figure 21-5c). A contact binary would make a strange "sun" in the sky of some imaginary world (see our web site). The most famous contact binaries are the W Ursae Majoris stars, named after the prototype in the constellation Ursa Major (better known as the Big Dipper). These consist of stars of rather similar mass, both filling their Roche lobes. They have total masses ranging from 0.8 to 5 $M_{\odot}$, and because they are so close together, their common revolution periods are very short, less than 1.5 d.

How did such a pair form? Some theorists believe they formed from protostars rotating so rapidly that they split, or *fissioned,* into two components. In contrast to the widely separated types of binaries, which have mass ratios expected from random pairings of field stars, close binaries show a tendency toward mass ratios of 1:1, consistent with the fission theory of origin. Theorists have also proposed mechanisms that might cause widely spaced pairs to evolve into contact binaries.

Possibly related are stars with very close orbits and extremely short orbital periods. The shortest known period for any binary, only 11 min, was discovered in 1986 by the European orbiting X-ray observatory, EXOSAT. The binary is apparently a neutron star orbiting a white dwarf, cataloged as 4U1820-30 and about 6000 pc away. Theoretically, mass loss could produce contact binaries with periods as short as 2 min! Somewhere in our galaxy there might be a planet with a giant glowing figure 8 in its sky, doing cartwheels like some bizarre advertising gimmick. Several stars once thought to be short-period variables have been recognized as probable short-period binaries in which matter is streaming from one star to another, causing irregular variations in brightness. We are discovering that many peculiar types of stars are explained by processes of evolution in binary or multiple systems.

Millisecond Pulsars

In the 1980s, a class of pulsars with a phenomenal spin rate was discovered. The fastest-spinning pulsar found so far was discovered in 1982 and flashes with a 0.0016-s period. Among human technologies, only the very fastest turbine engine can spin at this rate. Yet stellar collapse has produced a mountain-size ball of nuclear matter that spins 642 times every second! Over 50 **millisecond pulsars** are now known. Most of the slower pulsars are concentrated in the plane of the Milky Way (Figure 21-6a), but a surprising number of millisecond pulsars have been found in globular star clusters, which are distributed away from the plane of the Milky Way. This discovery is puzzling because the stars in globular clusters are perhaps 10 billion years old, yet a pulsar should spin down in only 10 million years. In other words, the rapid rotation of millisecond pulsars must have been caused long after their birth by some more recent event, such as a binary encounter (Kaspi, 1995).

Astronomers used to think that the evolution of neutron stars was very simple: They form as supernovae, cool, and spin down. Now it is clear that pulsars can evolve in binary systems. A neutron star in orbit around a normal star will draw material from the normal star onto it. As the material spirals in, it "spins up" the neutron star into a millisecond pulsar. In some cases, the binary system is disrupted by the close passage of another star, leaving an isolated millisecond pulsar.

High-Energy Emission from Binary Systems

The existence of millisecond pulsars is a confirmation that some of the most spectacular phases of stellar evolution result from binary systems. As an example, imagine the evolution of a binary system with two massive stars, one of 10 $M_{\odot}$ and the other of 20 $M_{\odot}$. The more massive star evolves faster and becomes a red supergiant. It then releases a wind that dumps large amounts of mass on its companion. The core of the originally more massive star detonates as a supernova. The companion now evolves rapidly, because it has increased its mass, and it will also become a supernova. A binary neutron star is one possible end result for this system. At several stages in the evolution of a massive binary system, gas will form a disk around one or both of the stars. The intense gravity of the compact star heats the gas up to millions of kelvins. The result is copious emission of X rays (van den Heuvel and van Paradijs, 1993).

In the mid-1970s, astronomers discovered a class of objects called **X-ray bursters.** These objects are concentrated near the plane and center of the Milky Way, and they are presumed to be due to neutron stars in binary systems. As matter is torn from a nor-

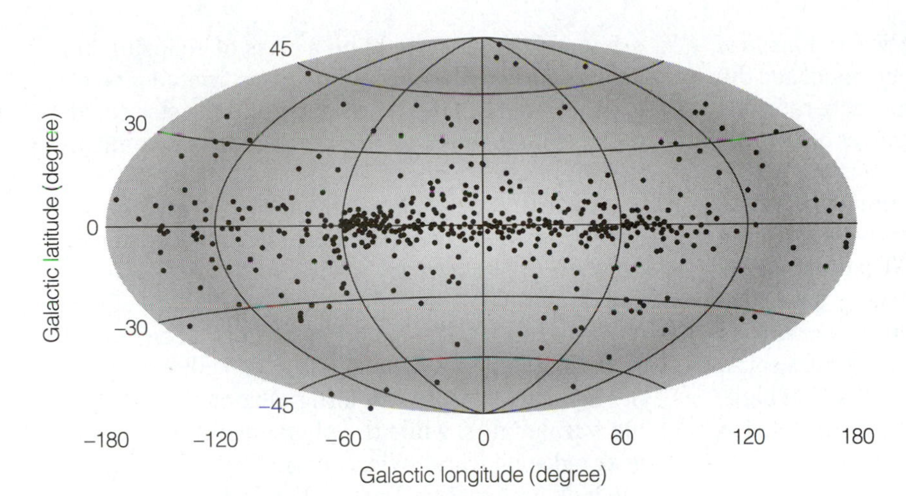

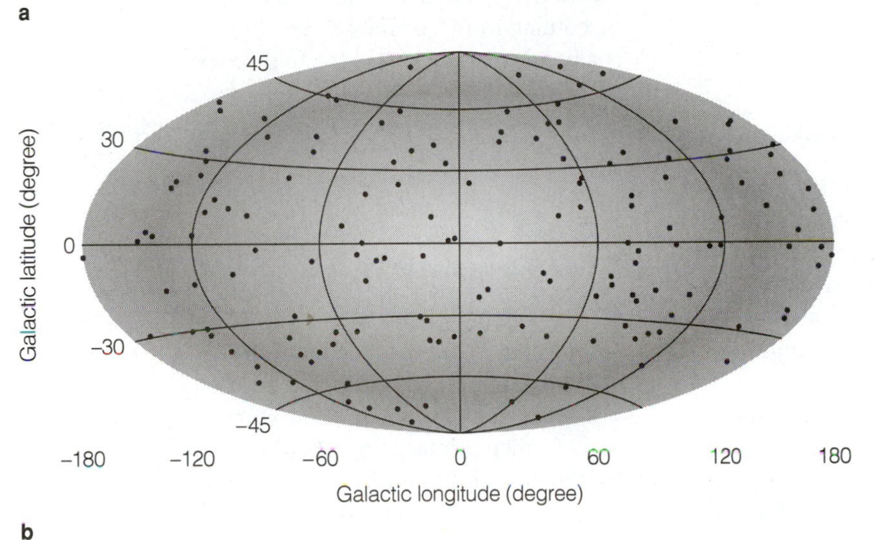

Figure 21-6 a Distribution of 558 pulsars in galactic coordinates, using an equal-area projection. The pulsars are concentrated toward the plane of the galaxy. (J. Taylor, R. Manchester, and A. Lyne) **b** Distribution of 153 gamma-ray bursts in galactic coordinates. The bursts show no correlation with the directions of the majority of stars in the Milky Way. The center of both plots is the galactic center. Over 2000 bursts have now been detected. (C. Meegan and collaborators)

mal main-sequence star onto a neutron star, the gravitational compression heats up the gas enormously. The pressure is so high that hydrogen is converted to helium by the fusion process and emits a steady level of X rays. The neutron star is then surrounded by a blanket of helium. When the helium layer reaches a certain thickness, its temperature is sufficient for helium burning. In this case, however, the thermonuclear reaction is explosive, and a strong pulse of X rays is produced. X-ray bursters can generate thousands of times more energy than the Sun in a pulse that lasts only a few seconds. An analogous process of explosive nuclear reactions on a white dwarf produces a nova (see the earlier discussion).

Even more spectacular are objects called **gamma-ray bursters.** The unknown nature of these objects represents one of the greatest mysteries in high-energy astrophysics. In the 1970s, satellites designed to monitor Soviet compliance with the Nuclear Test Ban Treaty discovered gamma-ray flashes coming from the sky! No known astronomical source has ever been reliably identified with one of these gamma-ray bursts; when these sources are quiet, they are difficult to detect in any part of the electromagnetic spectrum. It is speculated that these bursts may result from high-speed electrons moving near the surface of a magnetized neutron star or from interactions between a binary pair of neutron stars. Unlike pulsars, the gamma-ray bursters show no preference for the plane of the Milky Way (Figure 21-6b). In fact, their uniform distribution in the sky means that they may lie either close to the Sun or far beyond the Milky Way in the distant universe!

In the last couple of years, astronomers have made great progress in the study of gamma-ray bursters. Nearly 2000 are now known—NASA's Gamma Ray Observatory detected the bursts at a rate of about one per day for several years. However, the surge in gamma rays lasts only about a second, and the optical counterparts proved to be extremely faint. With great effort, the fading optical remnants of more than a dozen bursters have been found, and the Keck Observatory has been used to measure redshifts for six of them. The objects are at distances of 3–10 billion light-years, far beyond the Milky Way galaxy! This implies that they have enormous luminosities. For a brief moment, a gamma-ray burster rivals the luminous intensity of the visible universe! Theorists speculate that the only possible source of such prodigious energy is the collision of two black holes or of a black hole and a neutron star. As the compact stars spiral toward each other, they release a "fireball" of high-energy radiation. A large research effort is devoted to understanding these extraordinary events.

THE SEARCH FOR ALIEN PLANETS

Astronomers are interested not only in stars as companions to other stars, but also in brown dwarfs and planets as possible companions to other stars. One of the most provocative searches in modern astronomy is the search for such alien planets. The answer will tell us about the role of our solar system and ourselves in the cosmic scheme of things.

New techniques are being developed that should make it possible within a decade to detect such planets, if they exist. First is improvement in *astrometry,* the technique for making precise measures of a star's position in the sky. If a star has an unseen Jupiter-like planet moving around it, the star itself "wiggles" back and forth due to gravitational pulls of the planet first on one side of its orbit and then on the other. Thus, the star is revealed as a class of astrometric binary, and in some cases, the mass of the unseen companion can be estimated. A mass smaller than 2 M_{Jupiter} could be considered a planet. A slightly larger mass, from 2 to 85 M_{Jupiter}, would be classified as a brown dwarf.

A second technique is extremely precise measurement of Doppler shifts. If the same "wiggle" motions are detected in this way, the star would be a class of spectroscopic binary. Again, an estimate of the companion's mass might be possible, especially if this technique is combined with one of the other techniques. And the mass estimate might reveal the companion to be a planet.

A third technique is to seek eclipses. If a large planet crossed in front of the star, as seen from Earth, it would block a fraction of the star's light. This would cause a slight dimming of the star once during each orbit, and the star would be a class of eclipsing binary. Such repeated "micro-eclipses" would give at least a hint of a planet's presence, and the star could then be studied by the other techniques to confirm the planet.

A fourth promising technique is to detect the infrared radiation coming from the planet. You might think direct photography with giant space telescopes could reveal a planet, but generally the glare of the star is expected to be too bright for that technique to work. In the infrared part of the spectrum, however, the hot star is putting out less radiation than at visible wavelengths, while the planet is putting out its peak radiation (recall Wien's law). Thus astronomers can look for excessive infrared emission that might be coming from a planet at a few hundred kelvins. Already this technique has led to discovery of systems of dust particles, possibly from asteroids or comets, near other Sun-like stars—a promising harbinger of possible planets.

Planet Formation vs. Star Formation

By contrasting planet formation (Chapter 14) with star formation (Chapter 18), we can see the problem we face in interpreting possible future discoveries of low-mass companions around stars. Planets apparently formed primarily by a condensation of dust and ice particles followed by accumulation during their collisions; the surrounding gas medium eventually blew away. This happened inside a cocoon nebula near a star. Most stars formed when the entire contents of an interstellar region of gas and dust collapsed gravitationally, trapping all the gas, dust, and ice particles in the region.

Suppose we find an intermediate-size object of, say, 3 M_{Jupiter} orbiting a star. Did it form by dust particle aggregation or by gravitational collapse? Could there be some unusual planets and brown dwarfs that formed by collapse? Could there be some unusual companion stars in binary systems that formed by particle aggregation within the cocoon nebula? Today's astronomers are uncertain of the answers, but the search for planetary systems and brown dwarf companions may clarify the situation.

The Discovery of Extrasolar Planetary Systems

In recent years, researchers have made great progress on the detection of **extrasolar planets** (Lemonick, 1996). The conclusion: our solar system is not unique! The technique that finally worked was the indirect detection of planets by the Doppler effect. Over 70 extrasolar planets are known, and 5 to 10 more are discovered each year.

How did astronomers manage to detect extrasolar planets? A stellar spectrum has narrow spectral features that are associated with specific elements in the star. Wavelengths of these narrow spectral features can be measured with high accuracy. The researchers use a clever trick at the telescope to be sure of what is actually moving. They pass light from the star through a glass cell of iodine vapor. The vapor imprints a set of narrow spectral features on the spectrum, which act like a set of "tick marks" against which the spectral features of the star can be measured. With care, it is possible to reach an accuracy of 3 m per second for a single velocity measurement. A second requirement of the search is patience. Giant planets are expected to be far from a star. By Kepler's third law, large orbits correspond to longer orbital periods. The searches that are currently bearing fruit were begun a decade ago. Bearing in mind the uncertainties in these difficult measurements, what have we learned about extrasolar planets?

Extrasolar planets are not extremely rare, but they are not ubiquitous either. About 10% of Sun-like stars have at least one giant planet. Most of the planets detected so far have masses from about Jupiter's to about 10 times that of Jupiter. About half of the extrasolar planets have rapid orbital periods of well under a year and orbits within 1 AU of their host star. Several of the planets have orbits that are highly elliptical. Figure 21-7 shows a histogram of the masses of the first 34 extrasolar planets to be discovered. Most of the planets are within a factor of three of Jupiter's mass, but lower-mass planets might be common. Current techniques cannot detect a planet below the mass of Neptune, and Earth-like planets might be detectable within 10 years.

Now that the science of extrasolar planets has been established, we have a new set of clues to work with. Scientists are puzzled. These planets tend to have masses of 1 to 10 M_{Jupiter}, yet much larger planets of 10 to 100 M_{Jupiter} could have been detected. Since they were not, we conclude that they are rare. Apparently, some mechanism stops planets from growing much bigger than 10 times Jupiter's mass. Several of these massive planets are in highly noncircular orbits. The layout of our solar system may not be typical. Among the recent discoveries is the detection of three Jupiter-size planets around Upsilon Andromeda—the first evidence of a "system" of planets like our own. In 2000, the first extrasolar planet was detected by an occultation technique as it dimmed the light of its parent star by 2%.

Many of the extrasolar planets are moving in fast, tight orbits close to the host stars. For example, the star 51 Pegasi has a planet of roughly half the mass of Jupiter at a distance of only a sixth the distance of Mercury from the Sun. It whips around in its orbit in just over 4 d! The star 70 Virginis has a planet of 7 M_{Jupiter}, located at about Mercury's distance from the Sun. These examples go against our conventional idea of how the solar system formed. Theorists are struggling to explain how a giant planet can form or survive in the hot inner regions of a solar system. Current thinking is that giant planets form at Jupiter-like distances from their stars, and then migrate in to their current tight orbits. Unfortunately, the Doppler technique reveals the mass of a planet but not its size. Without a size estimate, we cannot say whether these giant planets are gassy, icy, or rocky. Earth-like planets are beyond the sensitivity of the current generation of imaging and Doppler experiments, but might be detected within the next decade.

Science has just taken a major new step in the Copernican revolution. Earth is a rocky cinder sheltering in the glow of a nearby star. It is natural to wonder how often the universe has created such planets and how often the history of planets involves the creation of life. For the first time we have the evidence to move beyond pure speculation. It will be a few years before we know how common planets are. It will be even longer before we know the conditions on these planets and their likelihood of harboring life. A new adventure is beginning.

THE ORIGIN OF BINARY AND MULTIPLE STARS

Clearly there are many questions about the origin of multistar systems. What determines whether a single star forms or one with companions? What determines the distribution into systems of two, three, four, or more members? (This may be a question of the angular momentum and its distribution in the original collapsing protostellar cloud.) And what determines whether a companion to a star is a planet, a brown dwarf, another star, or a mixture of these? Most astronomers believe several processes may be at work in the formation of binary and multiple systems. Different processes—fission, capture,

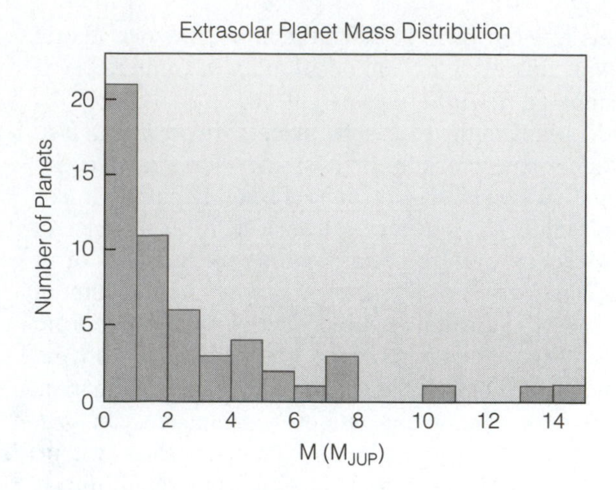

Figure 21-7 Histogram of the masses of 34 extrasolar planets, discovered by the Doppler technique. Planets with more than 10 times the mass of Jupiter are rare; planets below 1 M_{Jupiter} may be common, but current techniques cannot easily detect them. (Courtesy G. Marcy, San Francisco State University)

subfragmentation—may produce different types of systems (Boss, 1991).

Fission Theories that picture a fast-spinning protostar as splitting in two are called *fission theories.* Many researchers believe that very close pairs such as the W Ursae Majoris contact binaries are formed in this way. However, Kitt Peak astronomer Helmut Abt and co-workers believe that fission might be less common than was once thought, and that contact binaries may form as orbits evolve closer together by other mechanisms.

Capture According to *capture theories,* systems of widely separated stars that share weak gravitational attraction are chance configurations arising when one star approaches another. The chance of encounters among random field stars is far too slight to explain the observed numbers of binaries. Furthermore, randomly paired stars would be of widely different ages, but this is not observed among binaries. Where could stars of similar ages interact in a closely packed group? In a newly formed star cluster. Massachusetts astronomers T. Arny and P. Weissman showed that fully half the protostars in a cluster probably undergo collisions or close encounters. Therefore astronomers believe at least some binaries and multiples formed inside newly formed open clusters as protostars approached and, by various gravitational interactions, became bound in orbit around one another. Many wide pairs probably formed in this way.

Subfragmentation During Gravitational Collapse In a third category of theories, the *subfragmentation theories,* the prestellar gravitationally contracting cloud shrinks because of its own gravity, but instead of forming a flat disk with a central star, its mass distribution or angular momentum distribution may make it split into two or more co-orbiting clouds. These clouds then collapse independently into stars.

Each of these three theoretical processes may produce binaries of a certain type. A complication in sorting out binaries of these different types is that orbits of binary and multiple stars evolve through gravitational influences. Mass transfer in close pairs can alter orbits. Widely spaced pairs formed inside larger star clusters can evolve into closely spaced pairs as the clusters break apart. In one theoretical study of some 800 imaginary triple-star systems, about 97% were found to be gravitationally unstable, eventually kicking out one star and becoming binary systems. Sometimes the ejection velocities are quite high; 58 km/s is a typical example. This in turn helps explain the stars observed speeding out of some young star-forming areas, as described in Chapter 20. Thus the observed statistics of binaries and multiples may not reflect their original characteristics.

SUMMARY

At least half of all the seemingly single stars in the sky are binaries or multiple systems. Many of these may have formed by interactions of stars in crowded new clusters, but some may have formed by other means such as fission or common condensation.

Binary and multiple systems can be detected in different ways. Binaries were once classified by these different methods of detection, which yield different types of knowledge. Examples include spectroscopic binaries, eclipsing binaries, and astrometric binaries. Certain types of binaries give the best available data on certain properties of stars, such as mass and diameter.

A classification based on evolution includes binaries that have filled neither Roche lobe, one Roche lobe, or both Roche lobes. Once one lobe has been filled by expansion of a star to the red giant stage, gas may flow from one star to the other, causing flare-ups, X-ray emission, and nova explosions.

Companions to stars include not only other stars but also brown dwarfs and probably planetary systems. In an exciting recent development, astronomers have detected massive planets around other stars—proof that our solar system is not unique. This result is another step in the Copernican revolution, reaffirming that

Earth may not occupy a unique place in the scheme of things. If even 1% of stars have planetary systems like ours, then millions of Earth-like planets could exist among the stars of our galaxy.

CONCEPTS

binary star system
multiple star system
physical binary star
visual binary
spectroscopic binary
eclipsing binary
light curve
astrometric binary
center of mass
selection effect
Roche surface
contact binary
nova
millisecond pulsar
X-ray burster
gamma-ray burster
extrasolar planet

PROBLEMS

1. Describe verifications of Kepler's laws other than the planets' motions around the Sun. What was the first verification outside the solar system?

2. How do binary and multiple star systems generally differ from planetary systems?

3. Give evidence that at least a subclass of binary and multiple star systems might be generically related to planetary systems.

4. How are novae related to binaries? Are supernova related to binaries?

5. How will the evolution of a 1-$M_\odot$ star in orbit close to a 3-$M_\odot$ star differ from the evolution of a 1-$M_\odot$ star by itself?

6. Why are binaries more likely to have formed in star clusters than as isolated field stars?

ADVANCED PROBLEMS

7. If a star of low mass (about 0.05 $M_\odot$, for example) were orbiting in the Earth's orbit around the Sun, what would be its period of revolution?

8. Use the circular velocity equation to derive the orbital velocity of a Jupiter-size body around a 1-$M_\odot$ star if the separation distance is 5.2 AU (7.8×10^{11} m). (*Hint:* 1 $M_\odot = 2 \times 10^{30}$ kg.)

a. How does this compare with the actual orbital velocity of Jupiter?

b. What would be the orbital velocity if the central star had two solar masses?

9. Suppose you observe a 0.5-$M_\odot$ star in circular orbit around a 5-$M_\odot$ star and can tell from Doppler shifts that the orbital velocity is 47 km/s. What would you conclude is the separation distance between the stars in astronomical units?

10. A star of mass *m* is in a circular orbit around a star of mass M = 10 $M_\odot$. Star *M* explodes and rapidly blows away 0.6 of its mass.

a. What is the orbital velocity of *m* before the explosion?

b. How does the total force of attraction on *m* from *M* change during the explosion?

c. Describe the future history of *m*. Will it remain in orbit around the remnant of *M*? (*Hint:* This sort of event has been suggested as a source of stars moving very rapidly within clusters.)

PROJECTS

1. Observe Mizar and Alcor with the naked eye. They are the middle "star" (actually a close pair) in the handle of the Big Dipper. Can you see the faint star Alcor? Sketch its position. (Inability to see Alcor may be due to insufficiently keen eyesight, a hazy sky, or a sky illuminated by city light.)

2. Observe the eclipsing binary Algol with a telescope or binoculars each evening for 10 to 20 d in a row. (This can be done as a class project with rotating observers.) Using neighboring stars as brightness reference standards, estimate the brightness of Algol. Can you detect the eclipses, which occur at 2.9-d intervals?

3. The star Epsilon Lyrae is famous as the "double double." It consists of a binary pair 208 seconds of arc apart, easily seen in a small telescope. But each of these is a binary only 2 to 3 seconds apart. These pairs are a test of good optics and good atmospheric observing conditions. Does your telescope reveal the two close pairs? Sketch them.

CHAPTER

22

Star Clusters and Associations

WHAT THE READER SHOULD WATCH FOR IN THIS CHAPTER

Building upon William Herschel's pioneering observations in the nineteenth century, astronomers have mapped out various groupings of stars. Close to the Sun are open clusters and associations, along with the hot gas and debris of active star formation regions. These mostly hot and blue young stars tend to dwell in the flattened plane of the Milky Way. The most impressive groups of stars are the globular clusters, which are mighty swarms of old, red stars that orbit far from the plane of the Milky Way. By a combination of techniques, astronomers can measure the distances to faint stars and map out the structure of the Milky Way. The Sun is offset from the center of an enormous disk of stars thousands of light-years across. A spherical halo of stars and globular clusters surrounds this disk. ■

If you could roam through space to ever greater distances, you would eventually lose track of individual stars and see the galactic disk defined primarily by clusters of stars. Writing about clusters in 1930, Harvard astronomer Harlow Shapley pointed out that "their problems are intimately interwoven with the most significant questions of stellar organization and galactic evolution." (Shapley, 1930)

Yet at the same time Shapley noted that scientific study of clusters had hardly begun. Nobody knew how to measure their distances or plot their distribution in space until the 1920s. Although some clusters, such as the Pleiades (or Seven Sisters), are easy to recognize with the unaided eye, others are so far away that they require large telescopes to detect. Still others are so close that they cover much of our sky and were not even recognized until recent years.

In other words, our cosmic journey has brought us to clusters so far-flung that they were recognized as a class only in the twentieth century (Hoskin, 1986). Clusters have been of major importance because they have revealed to us the shape and age of our own galaxy, as will be clarified in this chapter.

THREE TYPES OF STAR GROUPINGS

The three basic types of clusters are *open clusters, associations,* and *globular clusters.* Examples are listed in Table 22-1.

Open Star Clusters

Open star clusters are moderately close-knit, irregularly shaped groupings of stars. They usually contain 100 to 1000 members and are usually about 4 to 20 pc in

TABLE 22-1

Selected Star Clusters and Associations

Name	Distance (pc)	*Z*[a] (pc)	Diameter[b] (pc)	Estimated Mass ($M_\odot$)	Estimated Age (y)
Open Clusters					
Ursa Major	21	18	7	300	2×10^8
Hyades	42	18	5	300	5×10^8
Pleiades	127	54	4	350	1×10^8
Praesepe	159	84	4	300	4×10^8
M 67	830	450	4	150	4×10^9
M 11	1900	99	6	250	2×10^8
h Persei	2250	156	16	1000	1×10^7
χ Persei	2400	167	14	900	1×10^7
O Associations					
I Orionis	470	150	?	3000	$<10^8$
I Persei	1900	164	?	180	$<10^8$
T Associations					
Ori T2	400	132	28	800	$<10^8$
Tau T1	180	52	?	50?	$<10^8$
Globular Clusters					
M 4	2000	550	9	150,000	$1.2–1.4 \times 10^{10}$
M 22	3500	460	9	530,000	$1.2–1.4 \times 10^{10}$
47 Tuc	4400	3100	5	1,600,000	$1.2–1.4 \times 10^{10}$
M 13	6600	4300	11	660,000	$1.2–1.4 \times 10^{10}$
M 5	7600	5500	12	850,000	$1.2–1.4 \times 10^{10}$
M 3	10,000	9900	13	1,100,000	$1.2–1.4 \times 10^{10}$

Source: Globular cluster data from C. H. Peterson (1986, private communication).
[a] *Z* = perpendicular distance north or south of Milky Way plane.
[b] Diameter of the bright core, in the case of globular clusters.

diameter. Our Sun is possibly inside or on the edge of a loose open cluster centered only about 22 pc away toward the constellation Ursa Major, many of whose stars belong to this cluster, as shown by Figure 22-1. The best-known clusters, the Hyades and the Pleiades, lie 12° apart in our winter evening sky, about 42 and 127 pc away, respectively. Both clusters were recognized long before the invention of the telescope (see Table 22-2). Figure 22-2 shows the Pleiades in more detail. About 900 open clusters are concentrated along the Milky Way band, indicating that they lie in the plane of our galaxy. (Open star clusters are sometimes called *galactic clusters* for this reason.)

As described in Chapter 20, stars form in open clusters, and most open clusters have prominent young stars or associated clouds of star-spawning gas. Then why aren't all stars in clusters? The reason is that most open clusters break apart into individual stars within only a few hundred million years because of dynamic forces acting on them. In comparison with most cosmic lifetimes, open clusters are short-lived.

TABLE 22-2

First Recognition of Selected Clusters

Cluster	Type	First Recognition
Pleiades	Open	Prehistoric[a]
Hyades	Open	Prehistoric[a]
ω Centauri	Globular	Ptolemy, c. A.D. 140[b]
Praesepe	Open	Galileo, c. 1611
M 22	Globular	Ihle, 1665
M 11	Open	Kirch, 1681
M 5	Globular	Kirch, 1702
M 13	Globular	Halley, 1714
27 globular clusters and 29 open clusters, cataloged		Messier, 1781
I Persei	Association	Ambartsumian, 1949

Source: Data from Shapley (1930).

[a] First recognition as a star cluster similar to telescopic examples was in Messier's 1781 list.

[b] Not resolved into stars, but listed as a bright, fuzzy patch.

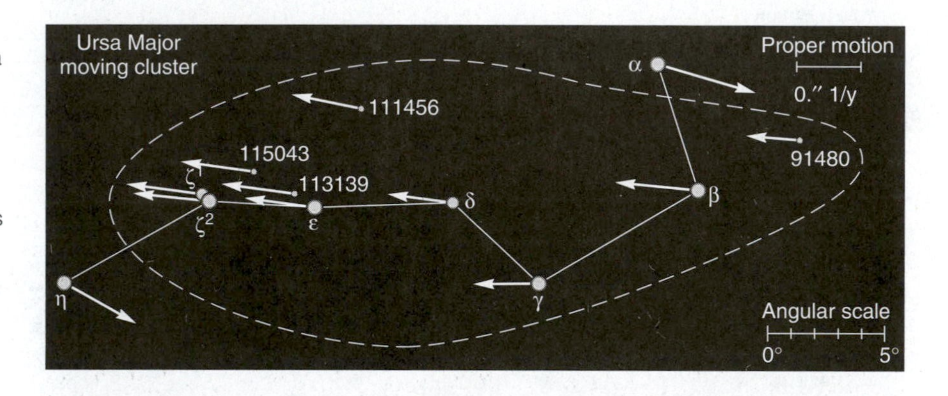

Figure 22-1 Some members of the closest open cluster, the Ursa Major cluster, form part of the familiar figure of the Big Dipper (solid lines). Arrows show proper motions of the stars, or the rates of angular motion relative to distant background stars. All but two of the stars in the figure are moving together at almost exactly the same rate, thus defining the cluster (dashed line). Stars at each end of the dipper are moving in different directions and are not true cluster members.

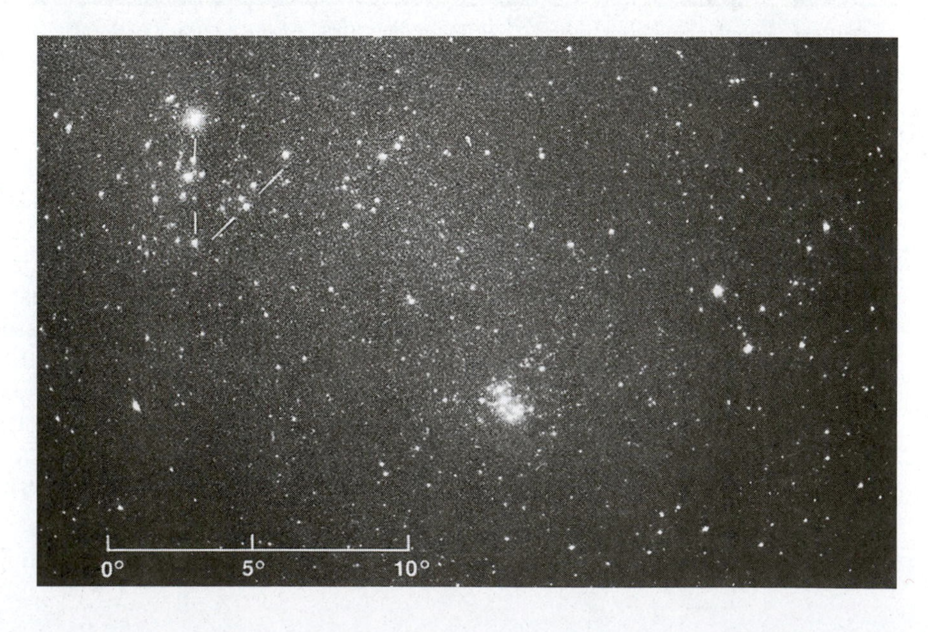

Figure 22-2 Two open clusters are prominent in this portion of the night sky. The Pleiades, also known from mythology as the "Seven Sisters," is an open cluster of young stars approximately 80 to 150 million years old. It is about 127 pc away and about 2 pc in diameter, seen just below the center of the photograph. Closer to us is the Hyades, a V-shaped group in the upper left, about 42 pc away. The brightest star in the upper left tip of the Hyades is the 1st-magnitude red giant Aldebaran. (Ten-minute exposure with 35-mm camera, f1.4, 50-mm lens, 2475 recording film)

Associations

Associations are cousins of open clusters. They often have fewer stars but are larger in size and have a looser structure. Some large associations include an open star cluster within them. They may have 10 to a few hundred members and diameters of about 10 to 100 pc. They are rich in very young stars, such as O and B stars (which burn their fuel too fast to last long), or T Tauri stars (which evolve toward the main sequence too fast to last long). Associations are classified as *OB associations* or *T associations,* depending on whether the prominent stars are O and B blue stars or T Tauri variables. Over 100 OB associations have been cataloged.

The smallest associations grade into small, multiple-star-like groups, such as the Trapezium in the Orion Nebula, which might be a link between multiple stars and small clusters. Like open clusters, associations are involved with regions of recent star formation and are short-lived.

Globular Star Clusters

Globular star clusters are quite different from the other two types. They are much more massive, more tightly packed, more symmetrical, and very old. Figure 22-3 emphasizes the remarkable symmetry and compactness of globular clusters. They typically contain 20,000 to several million stars, although many of these stars crowd too closely to be resolved by Earth-based telescopes, especially in the central regions.

Typical diameters of the central concentrations range from only 5 to 25 pc. To imagine conditions inside a globular cluster, picture 10,000 stars placed around the Sun at distances no farther than Alpha Centauri, our nearest star. If we lived in the core of a globular cluster, our night sky would blaze with starlight 10 times brighter than the light of the full moon!

Even the nearest globular clusters are thousands of parsecs away from us. It is only because they have so many and such very bright stars that we see them at all. Yet a modest backyard telescope can reveal many prominent examples. About 150 globular clusters have been found around our galaxy, and over 200 have been found around our sister galaxy, M 31 in Andromeda. In three-dimensional space, they are not confined to the galactic plane, but are distributed in a spherical *halo* surrounding our galaxy. They are among the oldest objects in our Milky Way galaxy, with oldest ages estimated in the range of 12–14 billion years.

MEASURING DISTANCES OF CLUSTERS

Because of clusters' enormous range of distances, astronomers have devised varied and sometimes ingenious methods to determine their distances (Reddy, 1983). Details of all the methods are beyond the scope of this book, but a few examples can be given.

Parallax The most basic method of measuring stellar distance is the geometric method of parallax, described in Chapter 16. The current limit for reliable distances is about 50 pc from ground-based observations, but that limit has been extended to several hundred parsecs by tens of thousands of new astrometric measurements from the Hipparcos satellite. The Hyades cluster, at about 42 pc, is pivotal in the galactic distance scale, not only because it has parallax measurements, but also because the distance can be measured by a variety of methods, which allows a more reliable determination (Hodge, 1988).

Main-Sequence Fitting A very effective method of determining distances to open clusters uses the *collective* properties of the stars in the cluster. If we know the distance to the Hyades, for example, we can compare the H-R diagram of a more distant cluster to the H-R diagram for the Hyades. The concentration of stars on the main sequence should define two lines of the same shape. The amount the two lines have to be displaced to lie on top of each other gives the apparent brightness difference between the two clusters. Using the inverse square law and the known distance to the Hyades, the distance of the fainter cluster can easily be calculated.

Cepheid Variables as Distance Indicators Luminosity is the key to distance measurement, and this brings us to the most useful technique for measuring distances of remote objects. As described in Chapter 19, Henrietta S. Leavitt discovered in 1912 that the periods and luminosities of **Cepheid variable stars** are correlated. The Cepheid variables have periods—the duration of their brightness variation—of 1 to 50 d. Once you measure the period of one of them, say, 5 d, that tells you the luminosity of the star (Figure 22-4).

Thus measuring the distance of a cluster is a process requiring several steps. A Cepheid in the cluster must be located, its period and type measured, and its luminosity read from the appropriate period/luminosity diagram and combined with the apparent brightness to calculate the distance.

Figure 22-3 A high resolution Hubble Space Telescope image of the globular cluster M 15 (NGC 7078), one of the densest clusters in the Milky Way galaxy. (Courtesy of the Hubble Heritage Team, STScI/AURA, M. Shara and D. Turek, AMNH, and F. Ferraro, ESO)

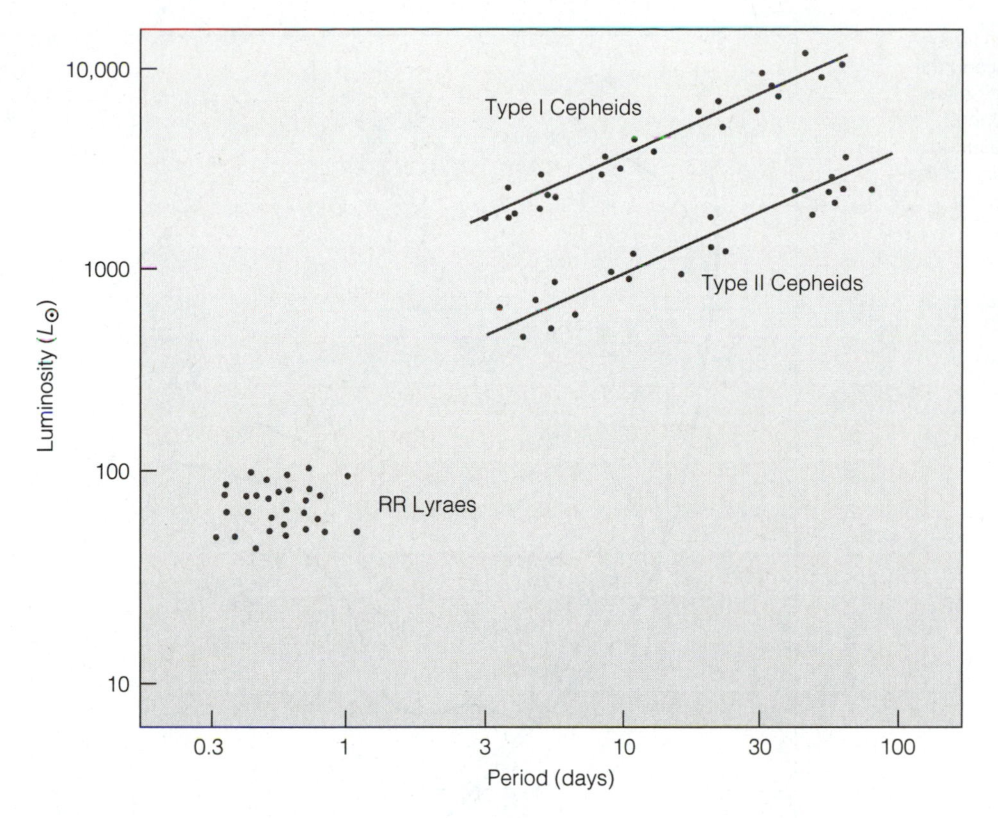

Figure 22-4 The relationship between period and luminosity for different types of variable stars. Type II Cepheids and RR Lyraes are found in older stellar populations and have few heavy elements. Type I Cepheids are found in young stellar populations and have more heavy elements. The unique relationship between period and luminosity allows Cepheids to be used as distance indicators.

Few astronomers are equipped to carry out all the necessary observations, such as measuring light variations and periods, determining spectral properties, measuring luminosities, and measuring interstellar reddening and obscuration. Thus astronomers have specialized. Some study periods of variable stars; some make photometric measures of absolute brightness; some study interstellar reddening. The simple statement that cluster A is *x* parsecs from the Earth may represent years of work by many astronomers.

The Effect of Interstellar Obscuration

Before 1930, these methods were carried out under the assumption that all intervening interstellar space is transparent. However, early estimates of distances, brightnesses, and sizes of clusters gave inconsistent results. In 1930, Robert Trumpler used the cluster results to show that the problem lay in the erroneous assumption that space is clear. He showed that diffuse interstellar dust dims stars and clusters that are more than a few dozen parsecs away. This throws off the estimates of stars' luminosities and distances. Fortunately, the total amount of dimming can be estimated by measuring the amount of interstellar reddening, or color change caused by the dust. The more reddening observed, the greater the degree of dimming. Once the obscuration is measured and taken into account, distances can be accurately measured if the luminosity of any star or class of stars in the cluster is known.

THE NATURE OF OPEN CLUSTERS AND ASSOCIATIONS

The preceding discussion mentioned that open clusters and associations are groups of stars formed from the contraction of large gas clouds, which later break apart into individual stars. We will now describe observations and theory shedding more light on the ages, evolution, and ultimate fate of these star groups.

Open Clusters: Ages and Ultimate Disruption

It is relatively straightforward but time-consuming to construct the **H-R diagram of a cluster**. The apparent brightnesses and color temperatures of hundreds of stars must be measured. If the distance to the cluster is known, the apparent brightnesses can be

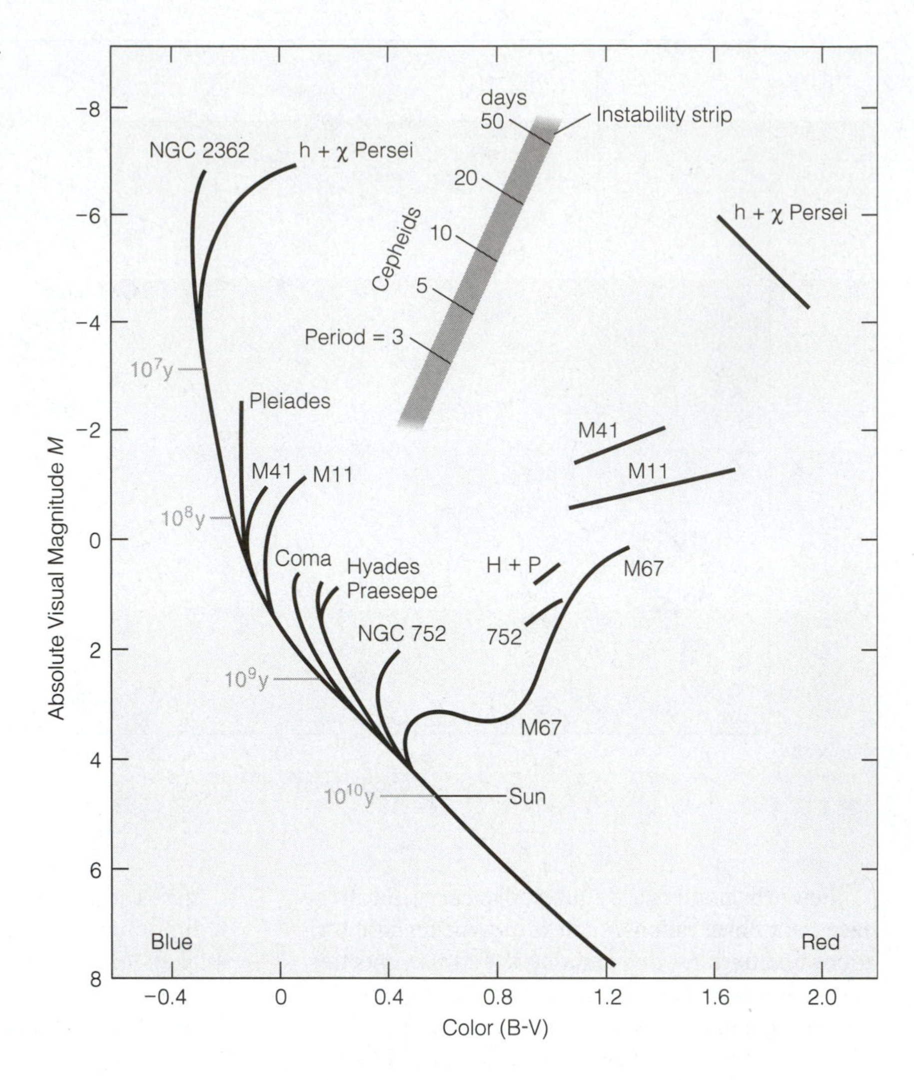

Figure 22-5 An H-R diagram for several open clusters. Numbers along the main sequence give the age for stars just turning off the main sequence. Clusters can be dated by measuring these turnoff points. The position of the Cepheid instability strip is shown. (Adapted from A. Sandage)

converted to luminosities. Astronomers have plotted the H-R diagrams of many clusters.

The importance of such work, of course, lies in the evolutionary information it yields. Figure 22-5 brings together data on several open clusters. Marked along the main sequence is the age at which various stars evolve off the main sequence. (Compare with Figure 19-2, where the dashed lines show predicted positions of stars of various ages.) Recall from Chapter 19 that the point at which stars have left the main sequence in a cluster is a measure of the cluster's age. Note in Figure 22-5 that some clusters, such as NGC 2362, are so young that hardly any stars have evolved off the main sequence. **Ages of open clusters** range from about 1 or 2 million years for NGC 2362 to as much as 7 billion years for NGC 188 and 8 or 9 billion years for a cluster called NGC 6791. Of the published ages for 27 open clusters, about half (55%) are less than 100 million years old. The solar system is nearly 50 times as old as this. It is remarkable to think that we can examine a randomly chosen open cluster and eventually state that it is only $\frac{1}{50}$ (or some other fraction) of the age of the Earth.

★ These results confirm that in cosmic terms, *open clusters are mostly young*, and that (as stated in Chapter 18) star formation is continuing in open clusters to this day. Many open clusters are associated with dense nebulae where star formation is still proceeding (see our web site for an example). The brightest stars are the hot, blue O and B stars, which formed recently, burn their fuel fast, and cannot last long. In Chapter 18, we mentioned two groupings where the least massive stars have not yet even evolved onto the main sequence, the Trapezium association and cluster NGC 2264. They are only about 1 and 3 million years old, respectively. In slightly older clusters some of the more massive stars may have evolved into red giants, adding interesting color effects. You can see these colors with the naked eye in some cases.

Velocities of cluster stars, measured by Doppler shifts, indicate that some clusters are expanding or losing members, or both. Some high-velocity stars exceed the escape velocity of their parent clusters. Thus these clusters are breaking apart as we watch. Clusters tend to disperse after a few hundred million or a billion years. This disruption of open clusters occurs by several simultaneous mechanisms. The primary means of mass reduction are stellar winds and supernovae that disperse the gas in the cluster and thereby reduce its hold on the member stars. According to the gravitation law, stars on the side of the cluster closer to the galactic center orbit around the galactic center faster than stars on the other side, thus stretching and shearing the cluster. The Hyades, for example, are barely stable now, and the outer parts of the Pleiades are already dissipating, though the central, tighter grouping may be stable.

Associations: Ages and Ultimate Disruption

Like open clusters, associations are young. Because of their loose structure, they may break up even faster than ordinary open clusters. For example, a T association of eight T Tauri variables, about 100 pc away and 25 pc across, has been estimated to be only about 10 million years old; it may soon break apart because of the tidal gravitational forces of the galaxy acting on the cluster. Most associations cannot last more than a few tens of millions of years, because of the disruptive tidal forces and the tendency for their stars to follow individual orbits around the center of the galaxy.

In many associations, there are large masses of neutral hydrogen gas, some of which exceed the mass of the stars. The hydrogen may be debris left over from the formation of incorporated stars, or it may be material ejected from the fastest-evolving stars. Often this gas is carried out of the association by stellar winds and supernovae.

THE NATURE OF GLOBULAR CLUSTERS

The technique of H-R diagram analysis is especially well suited to determining the ages of globular clusters because they have well-defined turnoff points along the main sequence. It turns out that globulars are extraordinarily old. Star formation has ceased in them. In all globulars of our galaxy, the O, B, and A stars have already evolved off the main sequence and have become red giants, as seen in Figure 22-6. For this reason, most bright stars in almost all globulars have a reddish color.

Assigning numerical ages to globulars must be done with some care, because spectra show that their stars contain fewer heavy elements than the more familiar stars of the solar neighborhood (Iben, 1970). This means that the details of their inner processes of energy generation and transport are different, according to the Russell-Vogt theorem. Since the evolutionary time scales may thus be different, different calculations are used to determine the ages corresponding to various main-sequence turnoff points.

Theoretical models of stellar evolution are required to determine the age of a globular cluster. The models must include details such as the heavy-element abundance, the amount of helium diffusion in the stellar cores (that is, the amount of helium that migrates from the cores to layers farther out), and the amount of reddening in the cluster. Systematic errors in any of these calculations lead to a systematic error in the age estimate. The largest uncertainty is introduced by the uncertain distances to most globular clusters. The model is used to predict stellar tracks in an H-R diagram as a function of time. Figure 22-6 shows these models for M 92. This method of determining the age is called **isochrone fitting.** We will return to this important relation between globulars and our galaxy in the next chapter.

The best techniques set the **age of globular clusters** at about 12–14 billion years. It now appears that globular clusters did not all form simultaneously; rather, they show an age spread of 20%, some 2–3 billion years. This is important, because it means that the halo did not form rapidly, as was thought for many years. Instead, it may have formed by the merger of a number of gas clouds, each smaller than a galaxy, over an extended period of time.

Shapes of Globular Clusters

Why are globular clusters globular in shape, as seen in Figure 22-3? Many cosmic systems form flat disks because of their rotation—for example, the rings of Saturn, the primeval solar nebula, and even the galactic disk itself. A disk is the final stable state for a contracting, rotating, self-gravitating system with substantial initial rotation. The final shape of a system is determined by how much angular momentum it had when it started to form. Globular clusters are not spherical, however, but are slightly flattened. They

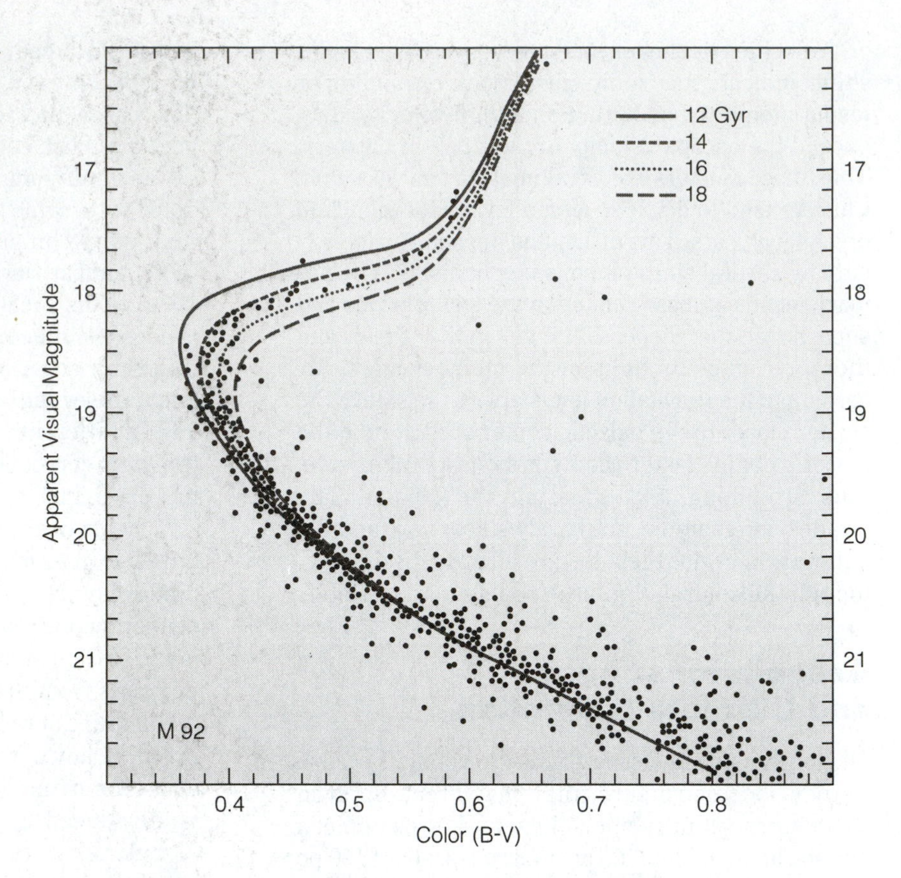

Figure 22-6 H-R diagram of the globular cluster M 92. The main-sequence turnoff is clearly seen, and the solid lines represent isochrones fitted to the main sequence. The age estimate depends on details of metal abundance and energy transport in the stars, but in this model the age is estimated to be about 14 billion years. (Adapted from Demarque and others)

rotate slowly with rotation axes distributed approximately at random with respect to the galaxy.

★ Inside a globular cluster, the orbits of individual stars must be very complex. The cluster's overall gravity field, the spatial distribution of its stars, their relative speeds, and the effects of near encounters among stars are all-important in determining how an individual star orbits in complex loops around the globular's central regions. Even in these crowded conditions, however, actual collisions between two stars are very rare or nonexistent. Furthermore, as the cluster orbits the galaxy, it passes through the galactic disk every 100 million years or so. We can imagine that such passages might allow spectacular close-up views of globular clusters from planets in the galactic disk, as shown in an artistic realization on our web site.

Globular Clusters as X-Ray Sources

Satellites above the Earth's obscuring atmosphere have mapped celestial sources of high-energy X rays. Most sources are in the galactic disk, perhaps associated with binary pairs in which one star dumps material onto another. But by the 1970s globular clusters had been found to be sites of strong X-ray radiation. Globular cluster X rays do not have the smooth periodic variations that are caused by orbital motions in binary pairs. Instead, they have irregular variations, sometimes over weeks or months, but sometimes doubling in intensity within a few minutes.

These surprising findings led to new thinking about conditions inside globular clusters. Generally, the X-ray sources are not exactly at the centers of clusters. They seem to come from binary systems. Studies in the 1980s showed that inner core regions of globulars are so crowded with stars that binary pairs may form more often than usual under these conditions. One idea is that globulars may contain many neutron stars and other stellar remnants and, because of the overcrowding, these may often capture passing stars into binary orbits. As the captured star evolves and blows off mass, gas is transferred onto the dense stars, and the resulting high-energy impact of gas onto the neutron star or its accretion disk may produce X rays (see the discussion of millisecond pulsars in Chapter 21). There is little doubt that the dense packing of stars in the central few cubic parsecs of a globular cluster makes these regions extraordinary stellar environments.

ORIGIN OF CLUSTERS AND ASSOCIATIONS

Astronomers have a reasonably good understanding of how the different types of clusters formed. The gas in halos around newly formed galaxies was extremely thin. This means that large masses around 10^{35} kg became gravitationally unstable and began to contract into discrete entities. These masses—equivalent to around 10^5 $M_\odot$—became globular clusters on the fringes of galaxies, probably around 14 billion years ago. In cool dust clouds in the disks of galaxies, gas densities were 10^5 to 10^9 times higher. This led to the contraction of clouds containing a few hundred solar masses; these became the open clusters and associations strewn through galactic disks.

The amount of heavy elements in a stellar system gives an idea of its age and formation process. The clusters and associations in the galactic disk have heavy-element fractions like that of the Sun, whereas the globular clusters in the galactic halo have heavy-element fractions near zero. This suggests a simple picture where the globular clusters form quickly in the initial collapse of the galaxy, while clusters in the disk have a more continuous history of star formation continuing to the present day (Henbest and Couper, 1994). A complicating factor is the existence of young globular clusters in the *outer* halo, implying that the galaxy has added material since its formation. Current theories invoke an initial collapse followed by the slow accretion of material from mergers with smaller nearby galaxies.

SUMMARY

Although three types of star groups have been defined, they can be grouped in two main categories. The open clusters and associations are young groups of about 10 to 1000 newly formed stars located in the galactic disk. The globular clusters, which are old groups of 20,000 to several million stars, are located within a spherical volume above and below the galactic plane, which is centered on the galaxy's center. They formed about 12–14 billion years ago.

Star clusters are important for three major reasons. First, they have played an extremely important role in mapping our galaxy and clarifying its history. Second, they clarify stellar evolution by presenting groups of stars formed at about the same time, so that their H-R diagrams clarify evolutionary tracks. Third, their stars reveal two different stellar populations in our galaxy. Stars in the disk and in open clusters and associations have solar-type compositions, with a few percent heavy elements, a certain type of Cepheid variable, and other distinctive properties. Stars in globular clusters have very few heavy elements, a different type of Cepheid, and other distinctive properties. The importance of these properties of clusters will be clarified in the next chapter as we turn our attention to our galaxy as a whole.

CONCEPTS

open star cluster	H-R diagram of a cluster
association	age of open clusters
globular star cluster	isochrone fitting
Cepheid variable star	age of globular clusters

PROBLEMS

1. Why are O and B stars the brightest in open clusters? Why are red giants the brightest stars in globulars?

2. If you saw the galaxy from a great distance, which would be brighter, open or globular clusters? Which redder? Which farther from the galactic disk?

3. Sketch the H-R diagrams of open and globular clusters and associations.

4. Describe a view of the sky near the center of a globular cluster.

ADVANCED PROBLEMS

5. If the Pleiades have 350 stars in a diameter of 4 pc, how many stars per cubic parsec are there? Roughly how far apart are the stars? Compare these numbers with those in the neighborhood of the Sun. (*Hint:* Volume of a sphere = $4/3\ \pi r^3$.)

6. Assuming globular cluster M 3 has 200,000 stars in a diameter of 13 pc, make the same comparison with the solar neighborhood as in Problem 5.

7. If a telescope could resolve 1 second of arc, what would be the smallest details it could reveal in:

a. An open cluster 1000 pc away?

b. A globular cluster 10,000 pc away?

8. A star is in a circular orbit around a globular cluster:

a. What would be its orbital velocity? Assume a cluster mass of 300,000 $M_\odot$ (6×10^{35} kg) and a radius of 5 pc (1.5×10^{17} m).

9. Use the distance and angular scale given in the caption of Figure 22-2 to get a rough confirmation that the size of the Pleiades diameter is about 2 pc.

PROJECTS

1. Observe the Pleiades with your naked eye and make a sketch. How many stars can you count in the group? Can you see all "seven sisters"? (The number of stars seen depends on keenness of vision, darkness of the observing site, and the clarity of the atmosphere.)

2. Observe the Pleiades and Hyades or open clusters h and χ Persei in a telescope. Move the telescope and compare star fields in and out of the cluster. Estimate how many times more stars are in the cluster than in the background region.

3. Locate a globular cluster with the telescope. Make a sketch. Can you resolve individual stars? Compare the view in the telescope with photos, where the central region is often everexposed and "burned out."

PART 7

Galaxies

A detailed view of the Whirlpool galaxy's spiral arms and dust clouds, which are the birth sites of massive and luminous stars. This galaxy, also called M 51 or NGC 5194, has numerous clusters of bright, young stars. (NASA and The Hubble Heritage Team)

CHAPTER

23

The Milky Way Galaxy

WHAT THE READER SHOULD WATCH FOR IN THIS CHAPTER

All the stars visible in the night sky are part of a vast collection of stars called the Milky Way galaxy. Our galaxy has three major components: a disk, a bulge, and a halo. The disk, in which the Sun is located, contains gas and mostly young stars; the disk is rotating, and we can map out the rotation using the techniques of radio astronomy. The bulge is a nearly spherical concentration of stars around the center of the galaxy. The galactic center contains a high concentration of stars, and there is growing evidence for a supermassive black hole there—a collapsed object millions of times more massive than a stellar remnant. The halo of the galaxy seems inconsequential, but it contains far more mass than the disk. The biggest surprise in the study of the Milky Way is the indirect detection of dark matter. This unseen material drives the rotation of the disk and gives the Milky Way most of its mass. ■

Our exploration of space has taken us out to distances of a few thousand parsecs. By looking at the distribution of stars and clusters throughout volumes of this size, we begin to perceive the **Milky Way galaxy.** To a remote observer, the Milky Way would be a disk with a central bulge. The disk is about 30,000 pc across, 400 pc thick, and packed with open clusters, individual stars, dust, and gas, mostly arranged in ragged spiral arms. Globular clusters surround the disk in a spherical swarm concentrated toward the center of the disk (Trimble and Parker, 1995).

These distances are difficult to comprehend. In a model of the Milky Way galaxy the size of North America, stars like the Sun would be microscopic specks less than a thousandth of a centimeter across and scattered a block apart. The solar system would fit in a saucer.

The view from the Earth is not from the outside of the galaxy, of course, but from the inside. From a point partway out in the disk, we see a band of unresolved, faint, distant stars when we look out along the plane of the disk. This is the Milky Way, shown in a wide-angle photographic view in Figure 23-1.

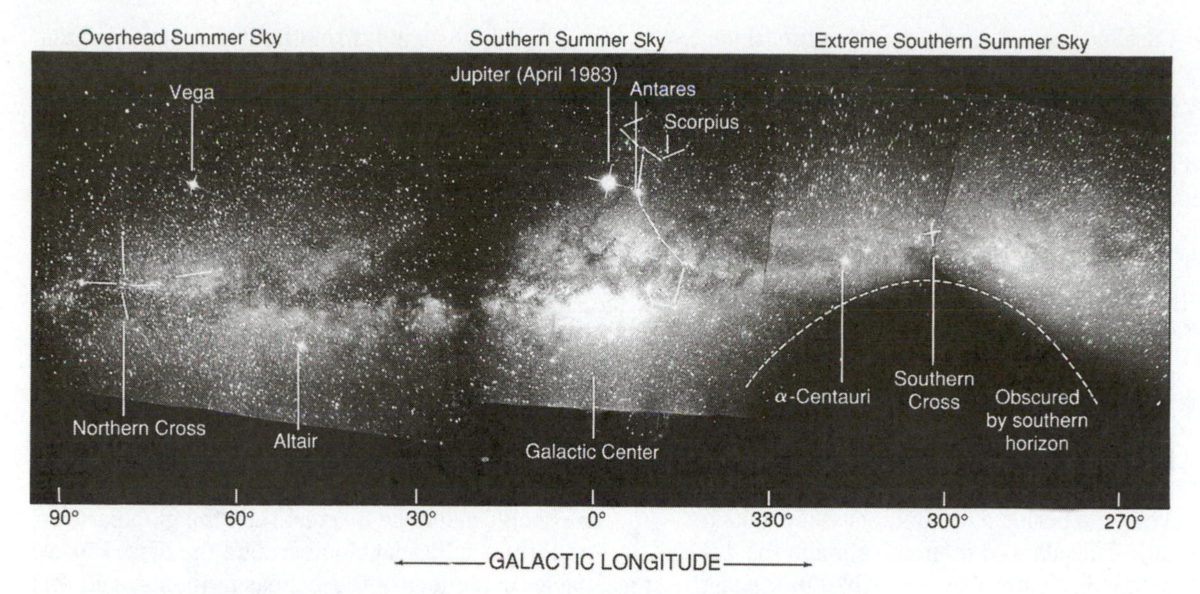

Figure 23-1 A panorama of the inner Milky Way galaxy from our position near the outer edge. This 180° view clearly shows the bright central bulge and the ragged dust clouds that lie along the plane of the galaxy between us and the center. The galaxy's nucleus is hidden behind dark dust clouds at the center of the photo. Seasons of visibility for the different regions, as seen in U.S. evening skies, are at the top. On the date of the photo, Jupiter happened to lie north of the galactic center. (Mosaic of photos from Hawaii by WKH, exposures 32 to 77 min with 24-mm wide-angle lens, f2.8, 35-mm camera with commercially available 2475 recording film)

DISCOVERING AND MAPPING THE GALACTIC DISK

Even before the invention of the telescope, people could plainly see a band of light arching across midnight skies at certain seasons.[1] Democritus (c. 400 B.C.) correctly attributed this glow to a mass of unresolved stars, which came to be called the *Via Lactea*, or Milky Way. In 1610, Galileo turned his telescope on the Milky Way and confirmed Democritus' idea. In 1750, the English theologian Thomas Wright correctly hypothesized that the galaxy must be a slablike arrangement of stars. Other theoreticians, such as Immanuel Kant, analyzed this idea more rigorously with Newton's laws in the 1750s and 1760s.

Herschel's Star Counts

Around 1773, a German-born composer and musician in England, William Herschel, bought some astronomy books and began building his own telescopes. Fascinated by the stars, he finally shifted his career from music to astronomy and within a few years built a telescope with a 1.2-m mirror (a size not surpassed until the 1840s). Backed by what we would call federal support—an annual stipend from King George III beginning in 1782 (after Herschel discovered Uranus)—Herschel used the following method to figure out what he called "the construction of the heavens" (Hoskin, 1896).

★ He swept the skies, counting stars in each direction. In all his observations, he was ably assisted by

[1]The naked-eye prominence of the Milky Way is unknown to most modern urbanites. It can't be overemphasized that faint celestial displays must be viewed away from urban lights. Coauthor William Hartmann relates, "My own astonishment at the Milky Way's clarity was greatest when I was living in a deserted region at about 3 km altitude on Mauna Kea volcano in Hawaii, prior to the establishment of the observatory at that site. One night when my eyes were already dark-adapted and I stepped outside, I thought the sky was partly cloudy, because I could see dark patches blotting out parts of the softly glowing Milky Way. Then I realized the air was crystal-clear. The dark patches were clouds, all right, but instead of being clouds of water droplets 1 km away, they were clouds of dust grains 100 million billion kilometers away! Such a view makes one believe we really do live in an immense, disk-shaped system of dust, luminous nebulae, and distant stars!"

his sister Caroline. Herschel assumed that the fainter the stars he could see in any direction, the farther away they were. He found far more faint stars in the Milky Way than in other directions (see the color all-sky view on our web site). He correctly took this as evidence that stars are scattered farther from us along the Milky Way than in other directions. He correctly mapped the Milky Way system as a disk of stars, with us inside it, but he did not know how big the **galactic disk** was.

Other Galactic Components

The next major advance came as astronomers began to apply Henrietta Leavitt's 1912 discovery of the relation between the period and luminosity of Cepheid variable stars. This allowed them to estimate the distances of globular clusters, as described in the last chapter. By 1918, Harvard astronomer Harlow Shapley showed that globular clusters are distributed in a spherical swarm extending above and below the disk. This swarm of clusters, which also includes some sparsely scattered individual stars and gas, is called the **galactic halo**. Shapley also showed that the halo is centered not on the Sun but on a distant point in the disk in the direction of the constellation Sagittarius. He correctly hypothesized that this point is the center of the galaxy.

Astronomers also realized that there was a concentration of stars toward the galactic center, spheroidal in shape but much smaller than the halo. These stars appeared to be mostly old and red. This third component of the Milky Way is called the **galactic bulge.**

Further work refined estimates of the *dimensions of our galaxy*. Trumpler discovered the effects of dust obscuration in the Milky Way about 1930. Soon afterward, astronomers were deriving approximately the correct dimensions. According to current estimates, the disk is about 30,000 pc across, and the Sun is about 8500 pc from the center. Although the International Astronomical Union (IAU) has adopted 8500 pc as the distance to the center of the galaxy, recent studies of the space distribution of globular clusters and X-ray bursters suggest that the number may be somewhat lower. We will continue to use 8500 pc in this book.

When we consider objects in distant parts of our galaxy, we are dealing with distances of thousands of parsecs. For this reason, astronomers often use a unit of distance still larger than our earlier units of astronomical units, light-years, and parsecs. This unit is a **kiloparsec (kpc)**—note the use of the same prefix, kilo-, that is used throughout the metric system to indicate 1000:

$$1 \text{ kpc} = 1000 \text{ pc}$$

Galactic astronomers say, for example, that we are 8.5 kpc from the galactic center and that our galaxy is roughly 30 kpc in diameter. Table 23-1 lists some distances to various parts of our own galaxy (Henbest and Couper, 1994).

The Shape of the Milky Way

The shape of the Milky Way is revealed by the distribution of the two types of clusters discussed in Chapter 22. The distribution of clusters is not random: The open clusters lie in a disk, and the globular clusters form a spherical cloud around the disk. Harlow Shapley, a pioneer of these measurements, said that the clusters reveal "the bony frame of our galaxy." Pioneering work on galactic structure was also carried out by the Dutch astronomer Jan Oort (Bok and Bok, 1980).

Figure 23-2 shows the distribution of open and globular clusters plotted in galactic longitude and latitude. The system of galactic coordinates defines zero longitude ($l = 0°$) toward the center of the Milky Way in the constellation Sagittarius. The direction away from the galactic center ($l = 180°$) lies toward Taurus, near the Hyades and Pleiades clusters. Zero latitude ($b = 0°$) defines the galactic equator. The direction "straight up" out of the galactic disk is the north galactic pole ($b = 90°$), and it has its southern equivalent in the opposite direction. As seen in Figure 23-2, the open clusters define the plane of the Milky Way, and the globular clusters form a symmetric cloud around the galactic center.

★ Different aspects of the Milky Way are revealed by observations at different wavelengths. These different views of our galaxy can be seen on our web site. The cold atomic gas is measured with radio telescopes using the 21-cm line of neutral hydrogen. These measurements show the thinness of the disk, but also complex structures of loops and filaments rising out of the galactic plane, outlining the ejected gas of supernova remnants. Far-infrared emission comes from cool dust mixed in with the gas and stars that form the plane of the galaxy. Near-infrared emission comes from stars in the thin disk and the central bulge. Together, these images reveal structures traced out by gas, dust, and stars. The halo is extremely diffuse and does not show up clearly at any wavelength of observation. A schematic view of the major components of the Milky Way is given in Figure 23-3.

TABLE 23-1

Distances to Selected Destinations in the Milky Way Galaxy

Destination	Distance (pc)	Travel Time of Light from Object to Earth (y)[a]
Nearest star beyond Sun (α Cen)	1.3	4.2
Sirius	2.7	8.8
Vega	8.1	26
Hyades cluster	42	134
Pleiades cluster	125	411
Central part of our spiral arm (Orion arm)	400	1300
Orion Nebula	460	1500
Edge of galactic disk in *Z* direction (perpendicular to plane)	1000	3300
Next-nearest spiral arm (Sagittarius arm)	1200	3900
47 Tucanae globular cluster	4600	15,000
Center of galaxy	8500	29,000
M 13 globular cluster[b]	11,000	36,000
Far edge of galaxy	24,000?	78,000?
Full diameter of galaxy	30,000?	98,000?

[a]These numbers, by definition, also equal the distance as expressed in light-years.
[b]Target of first beamed radio transmission from Earth directed to hypothetical extraterrestrial civilizations, November 1974. Signal will reach the cluster in A.D. 38,000.

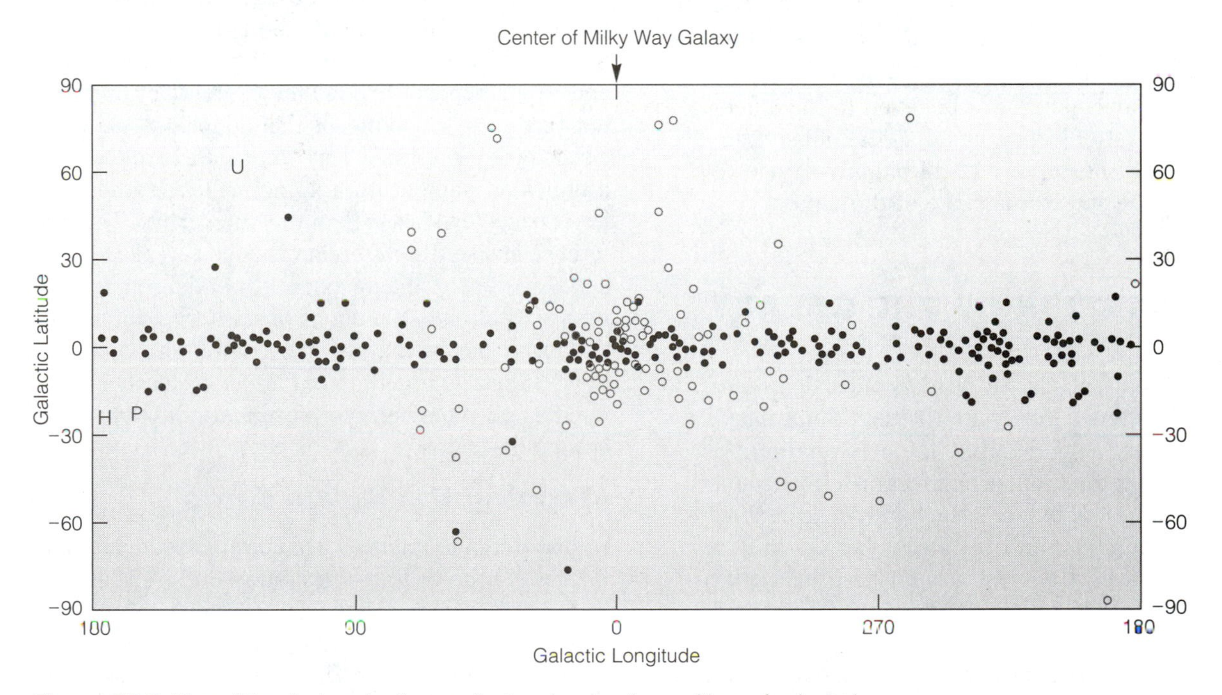

Figure 23-2 Map of the sky in galactic coordinates showing the positions of selected open clusters (filled circles) and the 93 most prominent globular clusters (open circles). The Ursa Major, Hyades, and Pleiades clusters are indicated by U, H, and P, respectively. The galactic center is the line at the center of the plot. Open clusters concentrate along the plane of the Milky Way (the line at zero latitude), and globular clusters form a spherical swarm around the galaxy's center.

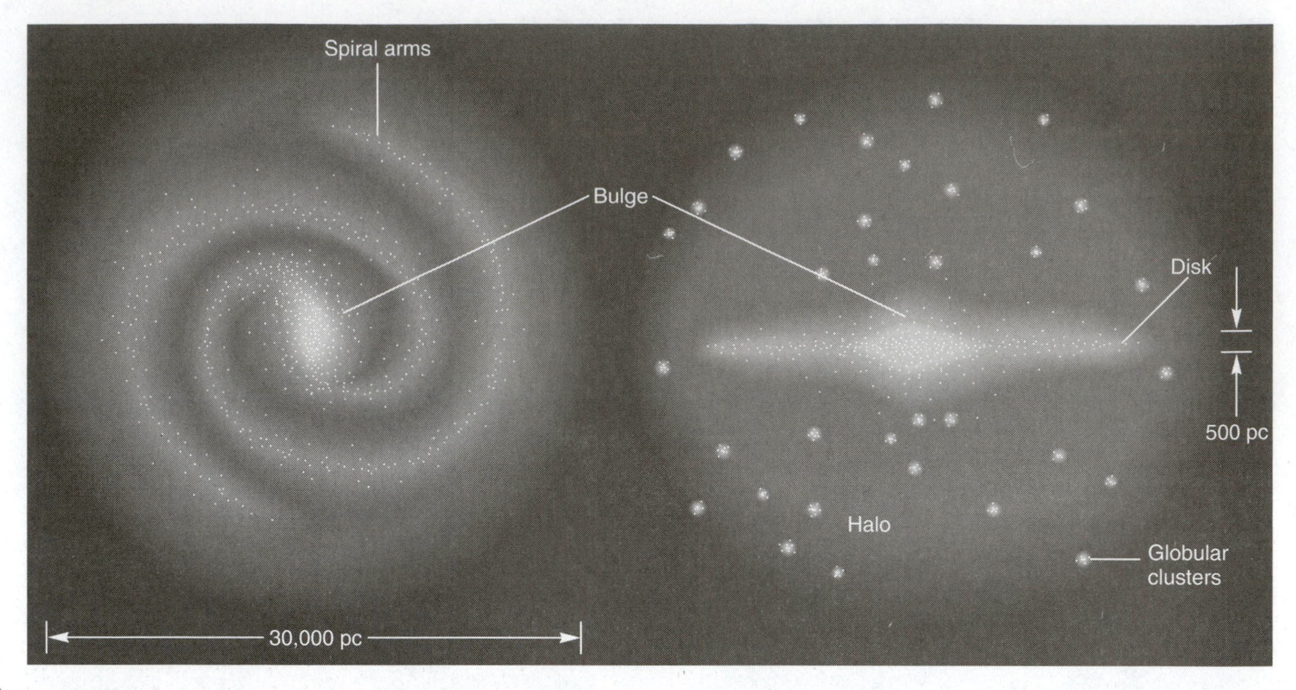

★ **Figure 23-3** Schematic view of the Milky Way from overhead and from the side, showing the dimensions and main components (for an artistic representation, see our web site).

THE AGE OF THE GALAXY

In the last chapter we saw that globular clusters have an average age of around 12–14 billion years. Because globular clusters are believed to have been among the first objects formed as the galaxy took shape, this age of roughly 12–14 billion years is believed to be the approximate **age of the galaxy.**

THE ROTATION OF THE GALAXY

We have noted that cosmic systems of particles tend to become flattened if they are rotating. The galaxy's flattened shape suggests that it, too, is rotating. All stars, including the Sun, are in fact orbiting around the massive central bulge. If the Sun travels at about 225 km/s, then the Sun's revolution period, or time required to travel all the way around its circular orbit, which has a radius of about 8500 pc, is nearly 240 million years.

Nearby stars are at nearly the same distance from the center as we are, so they are moving around the center at almost the same speed as our Sun. Orbital speeds around the center are different at different distances from the center. (The relation of orbital speed to distance departs from Kepler's third law, because the galaxy's mass is spread throughout many stars in the large central bulge, rather than being concentrated in one central object at the center, as in the case of the solar system.) Both the linear speeds and angular speeds around the center vary at different distances, which means that no galaxy rotates as a solid disk. No spiral galaxy, for instance, rotates like a pinwheel painted on a spinning CD; rather, the inner parts turn faster than the outer parts. This difference in speed at different distances is called *differential rotation* of the galaxy. Differential rotation, together with random motions of stars (typically about 20 km/s), means that the stars of the galaxy do not move smoothly together, but are ceaselessly changing their positions relative to each other as they move around the center.

Mapping the Spiral Arms

Spiral arms of galaxies are spiral-shaped patterns formed by the brightest hot stars and their associated, bright emission nebulae. Does our galaxy have spiral arms like the Andromeda galaxy and some other nearby galaxies? The objects that need to be mapped in order to seek evidence of spiral structure in our galaxy lie in the disk, not in the halo. Therefore, astronomers began mapping positions of associations of bright, young O and B stars and young clusters (which are bright, easiest to detect over

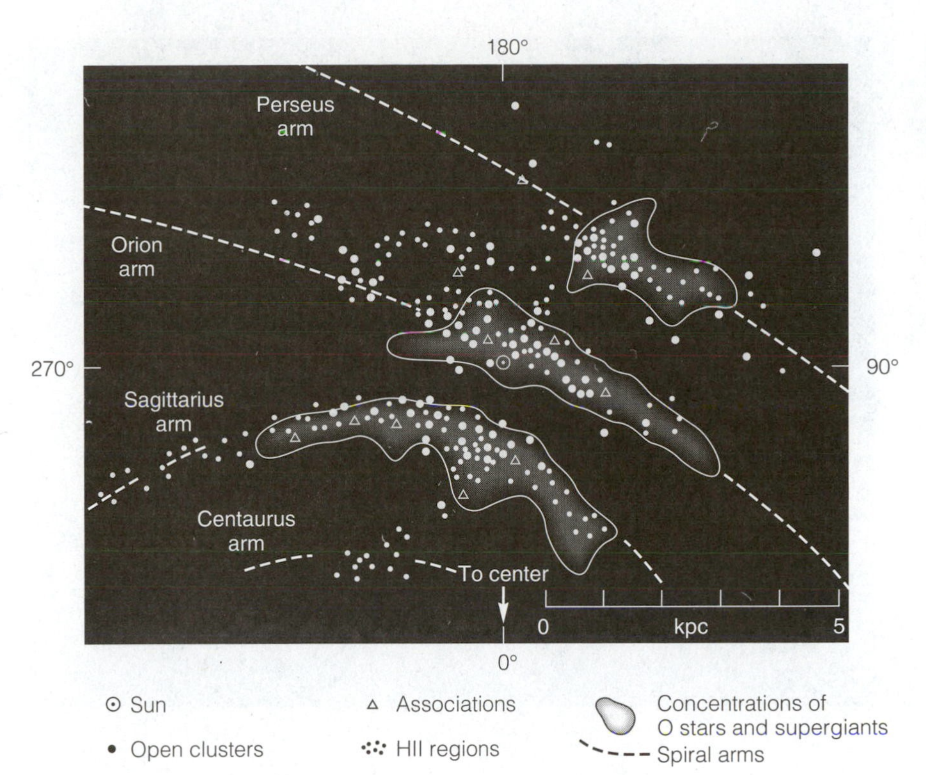

Figure 23-4 Evidence of spiral arm features in the neighborhood of the Sun. Galactic longitude is plotted at the edges. Clusters, associations, HII gas clouds, and young stars are concentrated in spiral arms around a distant center toward $l = 0°$. (Adapted from data of Klare and Neckel, Moffat and Vogt, and Walborn)

large distances, and good markers of star-forming sites). That is, the astronomers plotted the positions of the young stars and clusters on the sky and their distances in order to place them in three-dimensional space. The maps revealed that young stars and nebulae are not distributed at random, but lie in arms, coiling out at an angle from our galaxy's center.

The spiral arms are named for the constellations in the directions of prominent features in each arm. The next arm beyond us is called the Perseus arm. Our arm is the Orion arm or sometimes the Cygnus arm. The next arm in toward the center is the Sagittarius arm. Figure 23-4 shows a schematic view of the galactic environment of the Sun.

We are located on the inner edge of a spiral arm. On an evening with a clear sky, rural observers have a commanding view of the galaxy from our home in the Orion arm. If you look at the brightest parts of the Milky Way in the southern summer sky, in the middle of Figure 23-1, you are looking toward the center of the galaxy. The true center is hidden behind 8500 pc of intervening gas and dust. Higher in the sky, stretching toward Cygnus, the Milky Way is divided by a band of dust clouds that obscure the background stars, as can be seen in a detailed view in Figure 23-5. Because we live on the inner edge of a spiral arm, our view away from the galactic center directly crosses our arm. The view in this direction is spectacular on a winter evening, as seen in Figure 23-6.

Mapping the Distribution of Gas

All the mapping just described reveals only the closest parts of nearby spiral arms of our galaxy—the regions not obscured by the interstellar dust. What about the rest of the galaxy? Is there any way to map the spiral features farther around the disk, such as on the other side of the center? Visual light is useless, because we can see only about 1000 pc through the dust. But longer waves pass much farther through the interstellar dust. The **21-cm radio emission** produced by neutral hydrogen, or HI, is especially useful for galactic mapping, because it allows us to detect HI clouds, which are concentrated in the spiral arms. Similarly, it is possible to map out the galaxy in the millimeter wavelength emission of carbon monoxide (CO).

Suppose we start scanning along the Milky Way with a radio telescope tuned to the 21-cm wavelength. If we find a strong signal near $l = 30°$, as in Figure 23-7, a concentration of HI must lie in this direction. To plot it on a map, we need its distance. We don't know whether it is nearby (cloud A) or far away (cloud B). The Doppler shift provides guidance, as seen in Figure 23-7. All clouds in the disk are revolving around

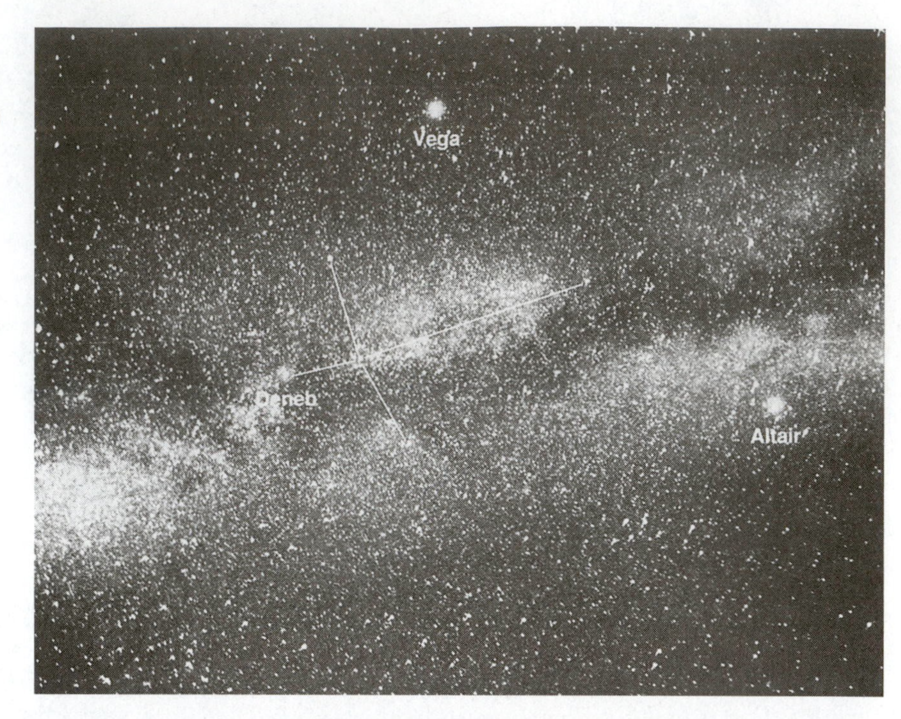

Figure 23-5 This 60° view of the summer Milky Way shows the dark rift in the region of Cygnus (Northern Cross, outlined), caused by obscuring dust clouds between us and the distant parts of the Milky Way. Names identify bright stars comprising the "summer right triangle." Compare with the left end of Figure 23-1. (35-mm camera, 24-mm lens at f2.8; 16-min exposure on 2475 recording film; photo by WKH)

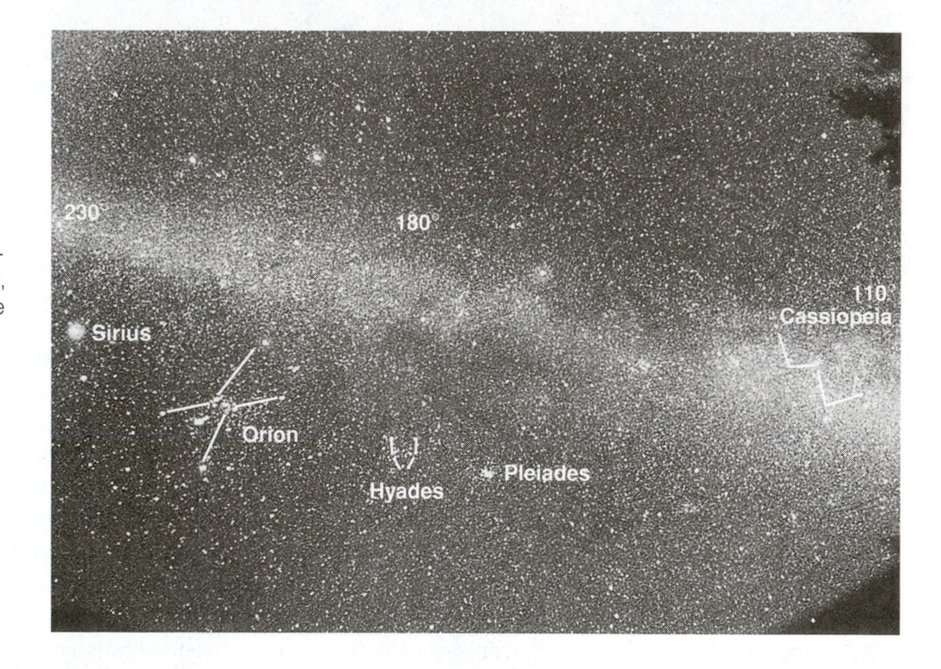

Figure 23-6 A 120° panorama of the winter Milky Way, looking away from the galactic center. Sirius and the Orion star-forming region lie in the direction looking down from our local spiral arm. The Hyades and Pleiades clusters are in our arm in a direction opposite from the center. The W-shaped constellation Cassiopeia lies to the right. (35-mm camera, 15-mm lens at f2.8; 25-min exposure on 2475 recording film; photo by Floyd Herbert and WKH)

the center, and objects nearer the center orbit somewhat faster than objects in our neighborhood. Thus the velocity of cloud B away from the Sun would be greater than that of cloud A. This effect is enhanced because V_B is directed along the line of sight, while V_A is not.

Radio astronomers have combined theoretical dynamical laws with observations of objects in our galaxy and in other galaxies to estimate the rotation rates in our galaxy as a function of distance from the center. These models give the velocity V_B of a cloud at position B, or V_A for a cloud at position A, and velocities for objects at all other positions in the galaxy. Thus, once a cloud's velocity is measured, its position can be plotted.

In this way HI clouds have been mapped over a large part of the galaxy. Their positions roughly define the spiral arms and offer our first view of our galaxy's spiral shape. Most of the galaxy's material is concentrated in a thin disk only about 200 to 400 pc

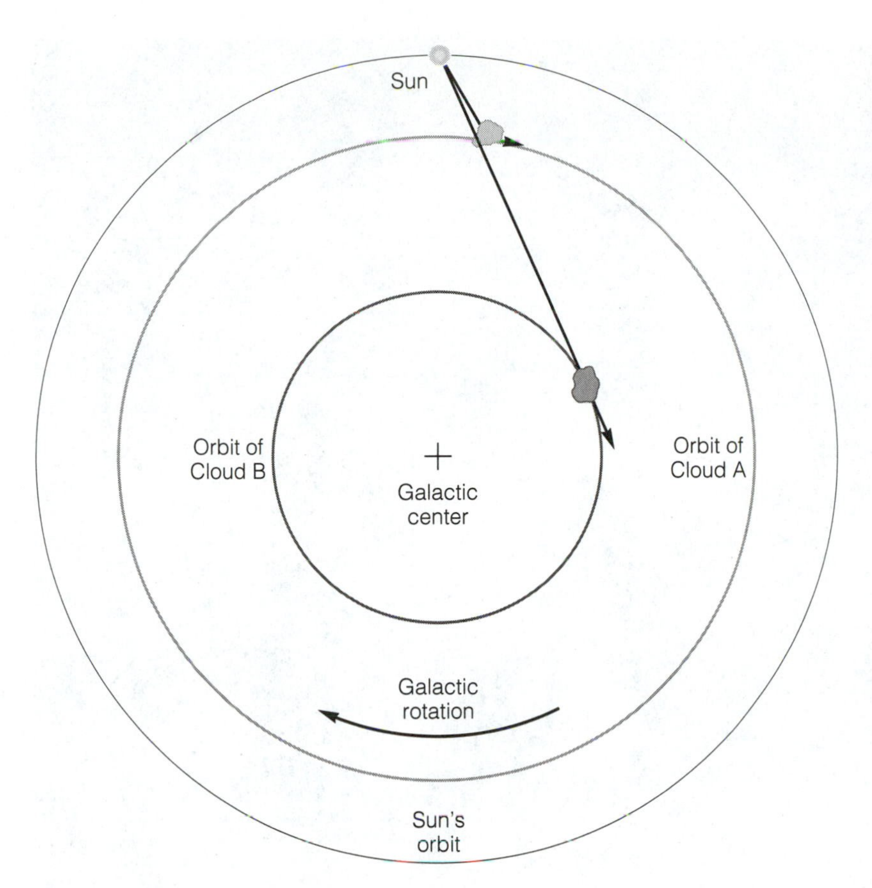

Figure 23-7 A view of the galactic plane from the north shows two hydrogen clouds, A and B, lying in nearly the same direction from the solar system. Different Doppler shifts, caused by different velocities V_A and V_B, allow radio astronomers to distinguish the positions of the clouds.

thick. Some hydrogen clouds "above" or "below" the disk are falling toward the disk, and some are moving away. They might be material dragged out of the disk by passing globular clusters or blown out of the chaotic central regions.

★ Radio, millimeter, and infrared techniques allow us to probe to large distances in the galactic disk. We can map structures that lie some 15,000 pc away, even farther than the galactic center. However, there are regions directly on the far side of the center that we cannot map. Nonetheless, radio and far-infrared maps have revealed the structure of our galaxy and have given us a good estimate of the gas and dust content. Figure 23-8 is a guide to what our galaxy might look like if we could see it from "above." A visualization of our galaxy can be found on our web site. We will see in the next chapters that many other galaxies have a similar appearance.

Why Does the Galaxy Have Spiral Arms?

This question has puzzled astronomers for many years. Some plausible "commonsense" explanations do not fit well with observations. For example, a commonsense analogy might be a rotating garden sprinkler, where the water jets spray out in spiral arm forms. The trouble with this model is that most material in the galactic spiral arms is not moving radially outward from the central region, as the water droplets do.

A better analogy is a stirred cup of coffee with a few drops of cream. Because the coffee surface rotates faster at the center than at the rim, which slows it, the cream is sheared into long spiral streamers that look like galactic arms. The trouble with this model is that if arms are primordial features of the galaxy, twisted by rotation, they should be very old. Since the age of the galaxy is about 14 billion years and the rotation time is nearly 240 million years, there has been time for spiral arms to be twisted into about 60 complete windings. In contrast, observations show that spiral arms of various galaxies (including ours) consist of the youngest stars and clusters. They rarely show more than one complete winding.

Though the final answer is still uncertain, two modern theories help explain some features of galaxies. The **density-wave theory** emphasizes that spiral arms may not be fixed features of specific star groups, but rather waves in the galactic material. The crest of an ocean wave consists of certain molecules

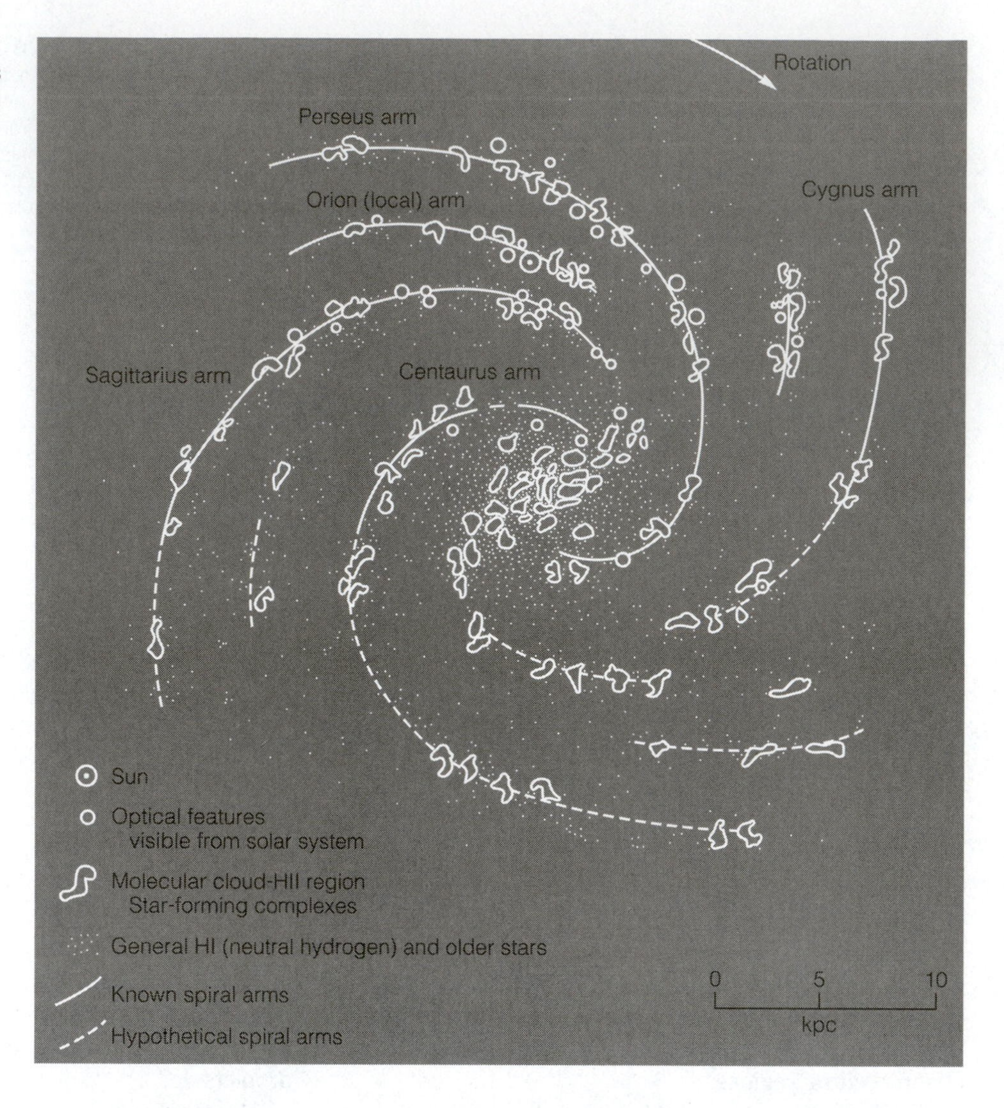

Figure 23-8 Currently known and estimated features of the Milky Way galaxy viewed from the north. Features nearest the Sun are the most certain; features on the far side are unmapped.

at one moment and certain other molecules the next, yet an outside observer sees a single wave that seems to have a history of its own. Just so, spiral arms may be persistent concentrations of material, with individual stars entering an arm, passing through, and finally emerging on the other side. The galactic gas tends to pile up in the spiral arms, reaching high densities in the giant molecular clouds. Higher densities trigger more gravitational collapse, thus explaining why star formation occurs mainly in the arms. This in turn explains the prominence of the arms, since newly formed massive stars are a galaxy's brightest stars. According to this view, the pattern of spiral arms does not rotate at the orbital speed of its constituent stars, but more slowly, with different stars defining the arms at different times. Modern density-wave theory is primarily due to University of California astronomers D. Lin and F. Shu.

A second theory is a modified coffee cup model, called the **stochastic star-formation theory**. It is based on the fact that star formation has not happened smoothly and continuously since the galaxy formed, but rather has occurred in chain-reaction bursts. As discussed in the last chapter, star formation occurs in open clusters, and the expanding gas from massive supernovae in one cluster compresses neighboring clouds of gas and dust, initiating formation of new, adjacent clusters. Therefore, during a period of up to 100 million years, a large region of new clusters containing brilliant, hot, massive, short-lived stars may be produced in one part of a galaxy. During the galaxy's rotation in 200 million years or so, the inner edge of this region pulls ahead of the outer edge because of differential rotation (compare the rates of clouds A and B in Figure 23-7), and the region of bright, new stars is sheared into a spiral segment. Following a few hundred million years, this arm segment runs out of young stars and fades, which explains why the spiral pattern rarely achieves more than one winding.

In summary, both theories suggest that spiral arms are shifting features associated with star formation; their individual member stars come and go, but the pattern persists.

MEASURING THE GALAXY'S MASS

Knowing the Sun's velocity around the galactic center (about 225 km/s) and its distance from the center (about 8500 pc, or 2.5×10^{17} km), we can calculate the amount of mass in the central bulge around which we are orbiting. This is essentially an application of Kepler's laws, giving a result of 4×10^{41} kg, or 200 billion $M_{\odot}$, in the central region, within our orbit. This calculation is an approximation since not all the stars are on circular orbits.

Until the 1970s, this was thought to represent most of the mass of our galaxy. Since then, there has been increasing evidence that much more mass lies in a dark halo surrounding the galaxy. The first indication came from the fact that the circular velocities of stars in the disk do not drop off with increasing radius, as would be expected from an application of Kepler's law. A large mass is also indicated by the rapid motions of satellite companions to the Milky Way. Recent estimates of the **mass of the galaxy** amount to 1000 billion (10^{12}) $M_{\odot}$ or even more. Much of this unseen halo mass is not composed of stars and gas and dust, but must be a new form of *dark matter.* Dark matter has important implications for cosmology, and it will be discussed in more detail in the next few chapters.

THE TWO POPULATIONS OF STARS

One of the most startling discoveries about our galaxy is that it contains a range of star types of different composition, age, distribution, and orbital geometry. These are commonly divided into two major groups called Populations I and II. **Population I stars** have compositions similar to the Sun's, are relatively young, and are distributed in nearly circular orbits in the galactic disk. A few percent of the mass of Population I stars consist of "heavy elements"—the elements heavier than helium, including carbon, oxygen, silicon, and iron. **Population II stars** are nearly pure hydrogen and helium with only a fraction of a percent of heavy elements. They are old and are associated with the bulge of stars in the center of our galaxy and with globular clusters that have orbits taking them far above and below the galactic plane. You can recall the two types more easily by remembering that Population I stars were the *first* group astronomers became familiar with, since they are the type located near the Sun: Population II stars were discovered *second.* Key properties of the two populations are listed in Table 23-2.

The evidence for chemical differences between the two populations is seen in Figure 23-9, which compares spectra of a Population I G-type star (the Sun) and a Population II G-type star. They have hydrogen lines of equal strength, but the latter has very weak lines of metals and other heavy elements, showing that Population II stars have very low abundances of heavy elements.

Ages of Populations I and II

Ages of populations are especially important for understanding how the galaxy formed and evolved. Population I stars, like the Sun and other stars in the galactic disk, have varied ages, from billions of years to zero age in regions where stars are still forming. But studies of H-R diagrams of Population II star clusters show that these stars are all around 12–14 billion years old—an indication that the globular clusters and other halo stars are the oldest objects in our galaxy.

Motions and Orbits of Populations I and II

Consider stars in our local region of the galactic disk. These Population I stars can be thought of as a swarm having nearly random velocities of about 20–50 km/s, but all moving together around the distant galactic center at a speed of about 225 km/s. We might represent this by analogy with a swarm of gnats randomly darting among each other at 0.2 m/s, while the whole swarm is moving through the air at 1 m/s. These stars can be represented by the nearly concentric orbits shown in the upper part of Figure 23-10. Disk stars have a slight up and down undulation in their circular orbits caused by the slight amount of motion they have perpendicular to the plane of the Milky Way.

Measurement of stellar velocities led to the discovery of Population II stars. Since Population I stars move in circular orbits, they have low velocities with respect to the Sun—random speeds of 20–50 km/s. But as shown in the lower part of Figure 23-10, Population II stars do not share these motions; they have highly inclined, elongated orbits.

TABLE 23-2

Stellar Populations and Their Properties

Property	Extreme Population I	Intermediate Populations[a]	Halo Population II
Orbits	Circular	Elongated, perturbed by galaxy	Elliptical
Distribution	Patchy, spiral arms	Somewhat patchy	Smooth
Concentration toward galactic center	None	Slight	Strong
Typical Z range (pc)[b]	120	400	2000
Heavy elements (%)[c]	2–4	0.4–2	0.1
Total mass ($M_{\odot}$)	2×10^9	5×10^{10}	2×10^{10}
Typical ages (y)	10^8	10^9	10^{10}
Typical peculiar velocities (km/s)	10–20	20–100	120–200
Typical objects	Open clusters, associations, gas and dust, HII regions, O and B stars	Sun, RR Lyrae stars ($P < 0.4$ d,[d] A stars, planetary nebulae, giant stars, novae, long-period variables	Globular clusters, RR Lyrae stars ($P > 0.4$ d),[d] Population II Cepheids

Source: Data based on tabulations by D. O'Connell, A. Blaauw, J. Oort, C. Allen, and others.
[a]Includes older Population I and intermediate Population II stars.
[b]Z = distance above or below galactic plane.
[c]Elements heavier than helium, sometimes loosely referred to as metals.
[d]P = period of light variation.

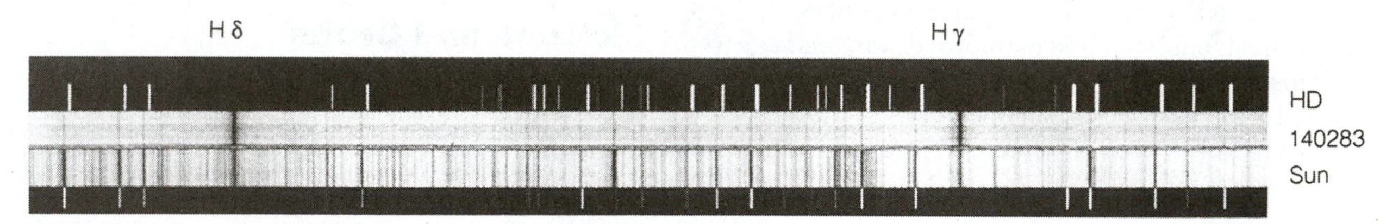

Figure 23-9 Comparison of spectra of a Population II star (upper, HD 140283) and a Population I star (lower, the Sun) of similar spectral type. (Bright lines at top and bottom are matching comparison spectra produced in the laboratory.) Both stars have prominent absorption lines of hydrogen (Hδ and Hγ), but the Sun has many additional absorption lines caused by various heavy elements. In the Population II star, these lines are very weak or absent, indicating that the heavy elements are virtually absent from its gases.

Astronomers in the early 1900s began detecting a few stars speeding through the solar neighborhood with velocities sometimes exceeding 120 km/s with respect to the Sun. The Sun and stars near us in the disk move around the galactic center at 225 km/s (Figure 23-10, top). A Population I star moving with us seems to have low velocity, relative to us (like a neighboring car on a racetrack). But a Population II star moving at 225 km/s on an inclined or retrograde orbit (Figure 23-10, bottom) seems to zip by at high speed (like a car moving the wrong way around the track).

Origin of Populations I and II

The composition of Population II stars includes only *a fraction of a percent of elements heavier than hydrogen and helium.* Yet such elements comprise about 2% of the Sun and other Population I stars. These elements include such important planet-forming materials as silicon, oxygen, nickel, and iron. Astrono-

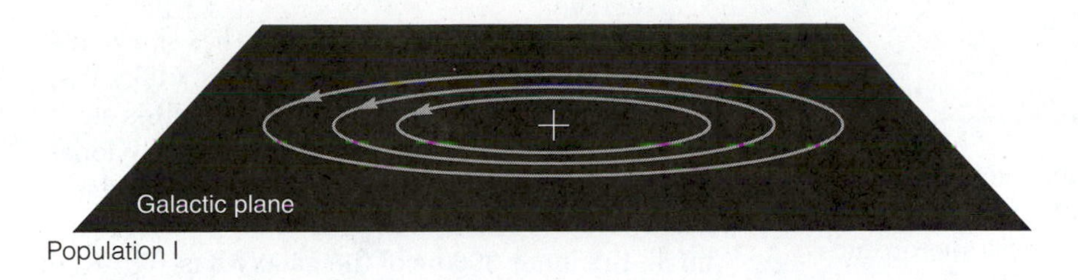

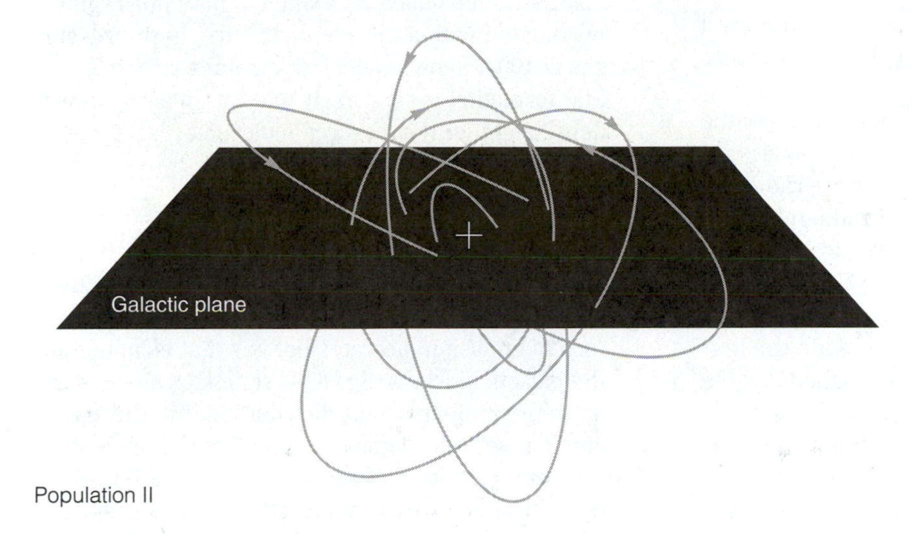

Figure 23-10 Comparison of orbits of Population I (disk) and Population II (halo) stars. Stars in circular orbits in the galactic plane have a slight component of vertical motion that gives them undulating trajectories. Stars in the galactic halo follow elliptical orbits and spend most of their time at large distances from the galactic center (by Kepler's law).

mers believe that Population II stars formed from gases with few heavy elements. Since Population II objects are older than Population I objects, the chemical composition of stellar populations must be telling us something about the history of the Milky Way.

The Formation of the Milky Way

Our understanding of the formation and evolution of the Milky Way has changed in the last few years. The traditional picture, dating back 30 y was of a protogalactic gas cloud composed of 75% hydrogen (by mass), 25% helium, and virtually no heavy elements. The volume of this system must have matched the volume now occupied by the most remote globular clusters in the outer halo. In the traditional picture, this gas cloud collapsed quickly in just the time it takes the particles to reach the center of the system by gravitational attraction. The globular cluster stars form quickly out of gas distributed over the whole volume, and the rest of the gas collapses to the center to form a disk. For the Milky Way, this time is a few hundred million years.

The implication of this picture is that the halo formed in the first few percent of the galaxy's lifetime. Population II stars are therefore very old and contain the original "primordial" abundance of helium and heavy elements (as we will see in Chapter 26, this composition was set by primordial nucleosynthesis during the first few minutes of the universe!). The disk formed later and more slowly, and it has since been recycling its gas into generation after generation of Population I stars. As described in Chapter 19, nuclear reactions inside stars fuse light elements into heavier ones through successive stages of stellar evolution. Heavy elements are *created* inside stars and ejected into interstellar space by disruptive phases of stellar evolution such as red giants, planetary nebulae, and supernovae. Later generations of stars form from this interstellar gas that has been enriched with heavy elements, and more heavy elements are steadily added to the gas (Binney, 1995).

This simple picture must be modified for several reasons. First, it is now clear that the globular clusters did not all form quickly and at the same time. There is an age spread of 2–3 billion years between the youngest and oldest globular clusters. Also, some stars in the galactic bulge have a relatively large fraction of heavy elements and so must have formed later than most of the Population II stars. Astronomers now believe that the galaxy may have been formed from smaller units, which merged and formed stars over a period of several billion years.

PROBING THE GALACTIC CENTER

★ In many galaxies, most of the light comes from a brilliant central core, or **galactic nucleus,** buried in the heart of a large central bulge. What mysteries are hidden in a galactic nucleus? Does our galaxy have such a nucleus? A large optical telescope does not help us to find out, because from our vantage point in a spiral arm, our galaxy's center is hidden behind kiloparsecs of dust (as shown on our web site).

In the 1930s, Bell Telephone Laboratories put the young researcher Karl Jansky to work on sources of radio static interfering with long-distance radio signals. Jansky built, in effect, the first radio telescope and discovered that one major source was a steady hiss from the Milky Way galaxy. In the 1940s, an amateur astronomer and radio buff, Grote Reber, built a radio telescope in his backyard and made the first radio maps of the Milky Way. He established that the strongest radio emission comes from the galactic center. This type of mapping reveals the nucleus because radio waves (along with infrared, X-ray, and gamma-ray wavelengths) can penetrate all the way from the galactic center, even though visible light is blocked by dust. In fact, the extinction (the amount by which visual radiation is extinguished) due to dust toward the galactic center is a factor of 10^{12}. We might as well be trying to look through a closed door!

HOMING IN ON THE GALACTIC NUCLEUS

★ From the time of Jansky's first observations in 1932, it has been clear that something extraordinary is happening in the nucleus of the Milky Way. At the long wavelengths he used, the galactic center is the brightest object in the sky, brighter even than the Sun. The unusual activity near the center has spawned suggestions of a massive black hole or a prodigious burst of star formation (Townes and Genzel, 1990). To understand this complex region, we will describe the different structures revealed by radio, infrared, and gamma-ray observations as we "home in" on the galactic nucleus. Color pictures corresponding to these various views can be found on our web site.

The Stellar Bar

Observations by the Cosmic Background Explorer (COBE) satellite have clearly revealed the bulge of the Milky Way. Photometric evidence has shown that the bulge is elongated in the plane of the Milky Way; our galaxy has a *stellar bar.* The effect of this structure is to drive gas clouds in the inner few kiloparsecs away from their circular orbits and cause gas to fall toward the center of the galaxy. As a result of this infall, the inner 300 pc of the galaxy has 10^8 $M_\odot$ of molecular gas, 100 times the surface density of molecular gas of the galaxy as a whole. This same region is permeated with a high-temperature, high-pressure gas at 100 million kelvins, which emits strong X rays. Star formation is very likely to occur in a region with dense and highly pressured molecules.

Sagittarius A

In the central few parsecs of the Milky Way is an intense region of radio emission called Sagittarius A. The brightest part is called Sgr A West. A spiral pattern of emission indicates hot gas that is falling into the galactic nucleus. Sgr A West, in fact, sits in a cavity in the molecular gas distribution, but there is an ample reservoir of gas to funnel into the nucleus. At the center of Sgr A West is a compact radio source called Sgr A*, which many astronomers suspect is the actual nucleus of the Milky Way. Very close to Sgr A* is the center of one of the densest star clusters ever discovered.

The Central Star Cluster

The central parsec of our galaxy contains over a million stars; its density of stars is 10 million times that of the solar neighborhood. If we lived on a planet around a star near the galactic center, our night sky would be ablaze with stars as bright as the Moon! Near-infrared cameras resolve the center of the star cluster into discrete sources. A number of these individual stars have been shown to be young, blue, and massive supergiants. The key issue is the nature of the compact radio source Sgr A*. Sgr A* is known to be smaller than 10 AU in size, which means it must be a stellar source. One possible explanation is a black hole. However, Sgr A* is just a single stellar source. Stellar motions over a much larger region of space indicate the gravity of a much larger compact object.

A Massive Black Hole?

There is considerable circumstantial evidence for an extremely massive black hole of a few million $M_\odot$ at the galactic center. The evidence comes from modeling the observed velocity distribution of the stars in the dense stellar core. The existence of Sgr A* already demonstrates that there is some kind of compact object near the center. Infrared astronomers have found a faint stellar counterpart to Sgr A*, with a plausible explanation in terms of a massive black hole.

★ Questions remain about the black hole interpretation. A black hole is certainly not required to account for the total luminosity in the central parsecs. A dense cluster of young, luminous stars will suffice. Also, Srg A* does not mark the exact center of the dense star cluster, yet a massive black hole would be expected to sit at the center of such a region of intense gravity. However, in recent years the evidence for a black hole has become stronger, with the measurement of proper motions of stars near the very center of the galaxy. Their directly measured speeds point unmistakably to a massive black hole. (For a hypothetical view, see our web site.)

SUMMARY

Our discussion in the last few chapters can now be fitted into a compelling theory of the origin and evolution of the galaxy. The range of globular cluster ages gives us an average estimate of the age of the galaxy—about 12–14 billion years.

What was happening 12–14 billion years ago? Evidently, the galactic mass was a single cloud of hydrogen-rich gas, with probably 25% helium (by mass) and virtually no heavier elements. The mass of this protogalaxy equaled the present galactic mass, roughly $10^{12}\ M_{\odot}$. The protogalaxy must have been rotating and contracting. When it reached a rather spherical shape perhaps 30 to 40 kpc across, it became dense enough that separate, large, self-gravitating masses formed within it. These became globular clusters with certain characteristics we observe today:

Age: about 12–16 billion years

Composition: few heavy elements

Distribution: spheroidal halo 30–40 kpc across; elongated orbits

Mass: 10^5 to $10^6\ M_{\odot}$ each

Because of rotation and collisions among the atoms and clouds of gas, the inner part of the protogalaxy, still containing some $10^{11}\ M_{\odot}$, flattened into a disk shape. The globular clusters, too far apart to interact, were left behind and did not form a disk. Thus two populations (with some intermediate objects) arose from the earlier- and later-formed systems. The stars and gas in the disk took on the properties we see today in nearby space:

Age: youthful; a few billion years for stars; only tens of millions of years for nebulae

Composition: few percent heavy elements

Distribution: flat disk; 30 kpc across, 400 pc thick

Mass: stars formed in clusters of a few hundred $M_{\odot}$.

In the center, an extraordinary, massive black hole may have formed, and violent explosions may have taken place. In the disk, nebulae traveling on elliptical orbits quickly collided with other nebulae until the gas, dust, and nebulae all moved together in relatively circular orbits. The spiral arm pattern probably emerged after a number of galactic rotations, perhaps within a billion years. In the spiral arms, the densest clouds contracted and spawned associations and open star clusters. Each group broke apart into scattered stars a few hundred million years after its formation, but new star groups continued to form, so the galaxy kept its present general appearance. Supernovae blew out gas laced with heavy elements created inside stars, so later generations of stars had more heavy elements than the earlier stars. Perhaps 7 or 8 billion years after the galaxy's formation, in one of the spiral arms, an obscure star formed—our Sun—and in its surrounding dusty nebula, the Earth was born.

CONCEPTS

Milky Way galaxy
galactic disk
galactic halo
galactic bulge
kiloparsec (kpc)
age of the galaxy
spiral arms
21-cm radio emission
density-wave theory
stochastic star-formation theory
mass of the galaxy
Population I star
Population II star
galactic nucleus

PROBLEMS

1. During what percent of human recorded history (define as you think appropriate) have people *not* known that we live in an isolated galaxy of stars similar to other remote galaxies?

2. From the appearance of the Milky Way, how do we know that the solar system is in the galactic disk and not far above or below it? How might the appearance of the central region differ in the latter case? (*Hint:* See Figure 23-1.)

3. Compare the shapes of the volumes occupied by the swarm of open clusters and by globular clusters. Relate the difference to differences in stellar populations.

4. How do we know the size of the galaxy?

5. How do we know the location and distance of the galactic center?

6. Describe evidence for spiral structure in the Milky Way. Which of the following types of objects reveal spiral structure when their positions are mapped on the Milky Way plane?

a. O stars
b. M stars
c. HII clouds
d. Open clusters
e. Globular clusters
f. Star-forming regions
g. Supernovae

7. Will the present constellations be recognizable in the Earth's sky 100 million years from now? Why or why not?

8. Why is the term *high-velocity star* a misnomer?

9. Why are no O- or B-type stars found in the galactic halo?

10. Summarize evidence for violent, energetic activity in our galaxy's central region.

ADVANCED PROBLEMS

11. What is the linear size, in parsecs, of a feature subtending an angle of 1 second of arc, located at the galactic center?

a. Could it be seen with a telescope resolving 1 second of arc?

b. Could it be seen with a radio telescope resolving 1 second of arc?

12. If the Sun moves in a circular orbit at 225 km/s and is 8500 pc from the orbit's center, calculate the time required to complete one circuit. (*Hint:* Circumference of a circle = $2\pi r$; 1 pc = 3×10^{16} m; 1 y $\approx \pi \times 10^{7}$ s.)

13. Using the relations in Problem 12, confirm the calculation of the galaxy's mass given in the text. Why would it be incorrect (or at least not meaningful) to quote the galaxy's mass to three significant figures (such as 2.34×10^{41} kg)?

14. What percentage of the galactic diameter could be crossed in one lifetime (say, 70 y) if we could travel at half the speed of light?

15. If an asteroid could be hollowed out and converted to a spaceship on which many generations of people could live (as in some science fiction stories), how many generations would live (at 20 y each) on the way to the Orion Nebula at the speed of light?

16. Suppose interstellar hydrogen atoms could fall freely onto the surface of an accretion disk 10 AU from a $10^{5}\ M_{\odot}$ black hole at the galactic core.

a. Neglecting any relativistic effect and recalling that material falling from far away falls at about escape velocity, estimate the velocity at which the gas would hit the accretion disk.

b. If much gas could flow onto the accretion disk at this speed, so that particles were maintained at such velocities, estimate the approximate temperature of the gas in the impact regions.

c. Show that the peak thermal radiation escaping from such regions would be extremely short-wave gamma radiation. (Other radiation from nonthermal processes would probably occur also.)

d. Comment on how detection of such processes in galactic cores requires nontraditional astronomical instruments, preferably above the atmosphere (see Figure 5-15).

PROJECTS

1. Compare views of the Milky Way with the naked eye, binoculars, and a small telescope. Scan along the Milky Way with each instrument and record the number of stars per square degree in different constellations (or at different galactic longitudes). Relate these densities to the actual structure of the galaxy. Note that if observations of the summer evening Milky Way can be obtained in the regions of Scorpio and Sagittarius, the direction toward the center can be studied. Compare star counts with each instrument in the Milky Way plane with counts near a point 90° from the plane, where we are looking directly "up" out of the disk. Can you account for the differences?

2. Compare the Milky Way as seen with the naked eye or binoculars:

a. In the heart of your city

b. On the edge of town

c. As far from lights as you can possibly get

City lights have little effect on the telescopic views of individual bright stars. But when it comes to broad areas of faint nebulosity or unresolved star clouds, even a single street light can illuminate local smog or fog and cause the iris of the eye to contract, thus destroying faint contrast and the ability to see faint glows.

CHAPTER

24 The Local Galaxies

WHAT THE READER SHOULD WATCH FOR IN THIS CHAPTER

Early in the twentieth century, Edwin Hubble discovered that many of the faint and fuzzy "nebulae" in the night sky are also galaxies. They are at distances so large that their light takes millions of years to reach us. Hubble provided a scheme for classifying galaxies according to their shape or morphology. There are three major categories: spiral galaxies like the Milky Way, elliptical galaxies, and irregular galaxies. A survey of our neighbor galaxies in the Local Group revealed the Andromeda spiral, the irregular Magellanic Clouds, and a number of dwarf galaxies. Galaxies come in an enormous range of masses and sizes and give us further evidence of dark matter: The mass of virtually every galaxy that we can study in detail is dominated by this material, which makes its presence felt only by the force of gravity. Galaxies are separated by vast gulfs of space, but they interact by means of gravity in ways that can alter their evolution. Although astronomers know much about galaxies, the process of galaxy formation is still poorly understood. ■

What lies in the vastness of space beyond our own galactic disk and its surrounding swarm of globular star clusters? Early in the twentieth century, astronomers discovered that many of the "nebulae" (diffuse patches of light scattered across the night sky) were galaxies of stars like our own Milky Way. At a stroke, the size of the known universe became larger by a factor of nearly 100. It is a bold and austere vision: the idea of our own galaxy as one of many flung across empty voids of space. It completes a progression that began with Copernicus, in which the Earth has systematically been displaced from its position of central importance in the cosmos (Ferris, 1988).

In this chapter, we will begin to explore the universe beyond the Milky Way. This journey starts with the realization that our galaxy is one of many, separated by vast distances. We will discuss the ways in which astronomers measure the distances to galaxies and the reasons why those distances are often uncertain. Then we will do an inventory of the nearest galaxies to the Milky Way. This is followed by a discussion of the classification and properties of galaxies. We will pay particular attention to the mass, since most of the mass in galaxies appears to be invisible to telescopes! Finally, we will outline some of the current ideas about the origin and evolution of galaxies.

DISTANCES TO GALAXIES

The measurement of distance is absolutely fundamental to astronomy. Without knowing an object's distance, astronomers cannot derive basic parameters such as size, mass, and luminosity. For example, stars can differ in their intrinsic brightness by factors of millions. So two stars with the same apparent brightness could differ in their distances by factors of thousands. The goal in distance determination is to find objects whose intrinsic brightness or size is well known; these objects are in general called **standard candles.** Knowing the absolute brightness (from the well-understood physics of the source) and the apparent brightness (measured through a telescope), we can easily calculate the distance. It sounds simple, but the story of the nature of the nebulae shows that accurate distances are exceedingly hard to measure.

A crucial assumption underlies all extragalactic distance measurements: that the laws of physics are constant across the universe. This has been called the assumption of the *uniformity of nature.* Recall that the laws of physics are confirmed only in terrestrial laboratories and, to a more limited degree, from our study of Moon rocks and particles in interstellar space. Beyond our galaxy, their application requires this powerful and unifying assumption. Edwin Hubble was relieved to find Cepheids and other familiar stars in distant galaxies, because their presence indicated to him that "the principle of the uniformity of nature thus seems to rule undisturbed in this remote region of space."

Discovering the Distances to Galaxies

At the turn of the twentieth century, there were two schools of thought on the nature of the nebulae. The *local hypothesis* held that they were clouds or whirlpools of gas condensing to form stars. Spectroscopy provided some support for the local hypothesis. The London amateur William Huggins showed that the spectra of a number of planetary nebulae and unresolved nebulae like that in Orion were due to hot gas and not unresolved stars (recall the emission lines from a hot, glowing gas discussed in Chapter 5). Theoretical work also argued for the local hypothesis. Sir James Jeans demonstrated that a collapsing cloud of gas would tend to form a disk shape and might even show spiral structure.

The alternative, and much more radical, hypothesis held that the nebulae were *distant systems of stars.* Spectroscopy of the Andromeda Nebula showed the features expected of a composite of stars, demonstrating that not all nebulae were purely gaseous. Also, Jeans's theory had a fundamental flaw, because it predicted far too much angular momentum in the center of the collapsed object. Some observers were guilty of wishful thinking, claiming to have detected both parallax and rotation in several of the spiral nebulae. Both effects do exist, but at levels thousands of times smaller than could have been seen by telescopes of the day. Resolving the controversy required a new and reliable indicator of distance. Henrietta Leavitt's discovery of the relationship between luminosity and period in Cepheid variables provided the indicator. Given this relationship and the apparent brightness of a Cepheid, it can be assigned a distance (see Figure 22-4).

The question of the distance of the spiral nebulae was settled by Edwin Hubble, a talented astronomer who was working with the new 2.5-m reflector at the Mt. Wilson Observatory. With this large telescope, Hubble unambiguously resolved the Andromeda Nebula into a myriad of individual stars. His crucial observation in 1924 was to identify and measure the periods of Cepheids in Andromeda. Using the already known relation between period and luminosity, he concluded that the Andromeda Nebula was a galaxy of stars about 300 kpc away, far beyond the periphery of the Milky Way. Since then, Hubble's estimate has been revised upward by a factor of two, due to yet another error in the Cepheid distance calibration. Figure 24-1 shows the chain of reasoning involved in Hubble's momentous discovery.

Hubble cemented his discovery using other distance indicators, such as novae and bright giants, and he also measured the distances to nearby galaxies M 33 and NGC 6822. Hubble made audacious use of the scientific method, using the properties of known types of stars to measure distances and extend the size of the known universe by a factor of 100. The dramatic conclusion was that the Milky Way was just one in a universe of galaxies, each containing billions of stars and scattered over millions of light-years of space.

Distances within and beyond the Milky Way

Distances within the solar system are measured in the most direct way possible by timing radar signals bounced off nearby planets. The distances to stars in the solar neighborhood are measured by their parallax. Parallax is so direct and fundamental that it gives

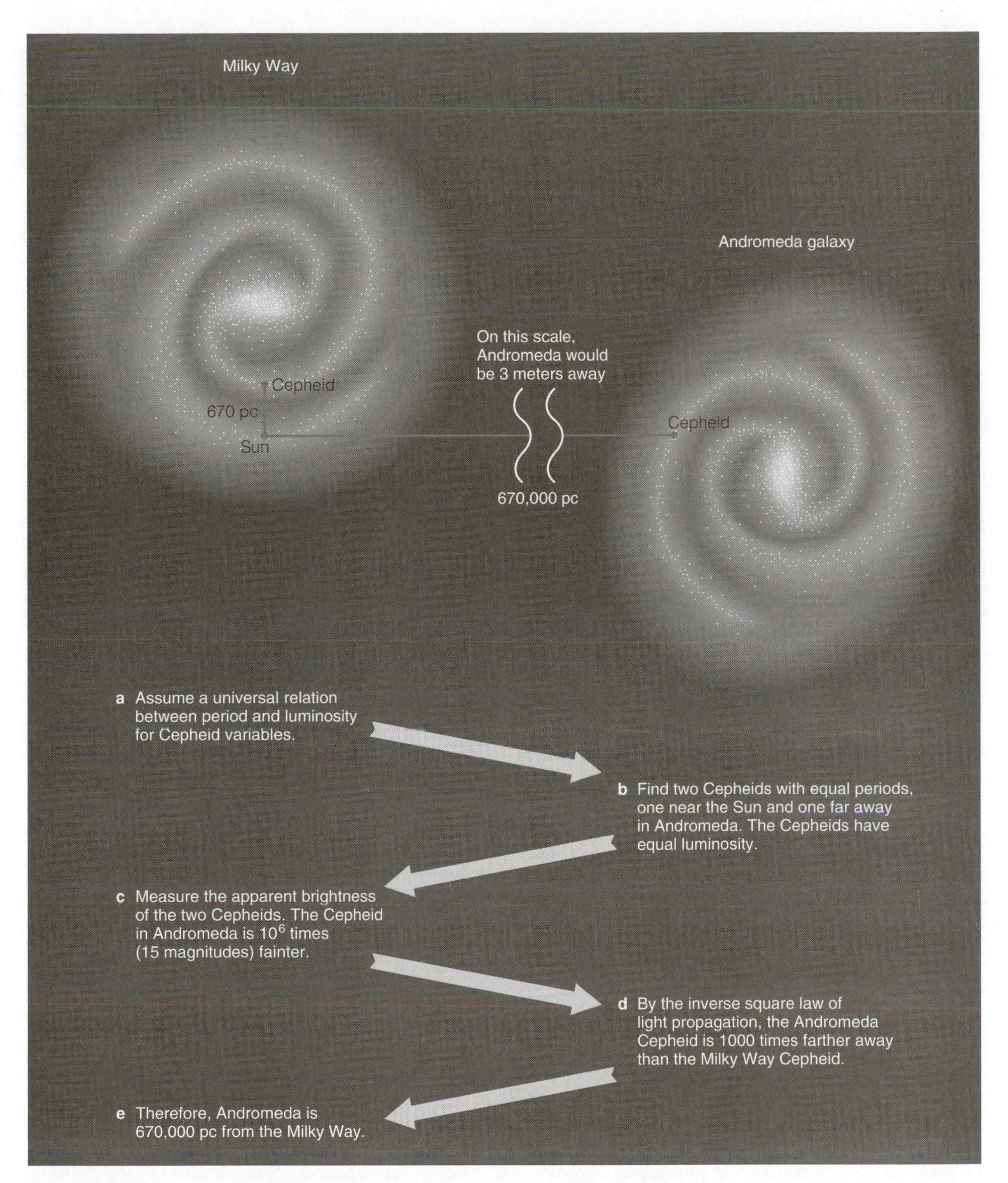

Figure 24-1 The chain of reasoning followed by Hubble in determining the distance to the Andromeda Nebula. A key assumption is that the laws of physics apply uniformly across large distances in the cosmos.

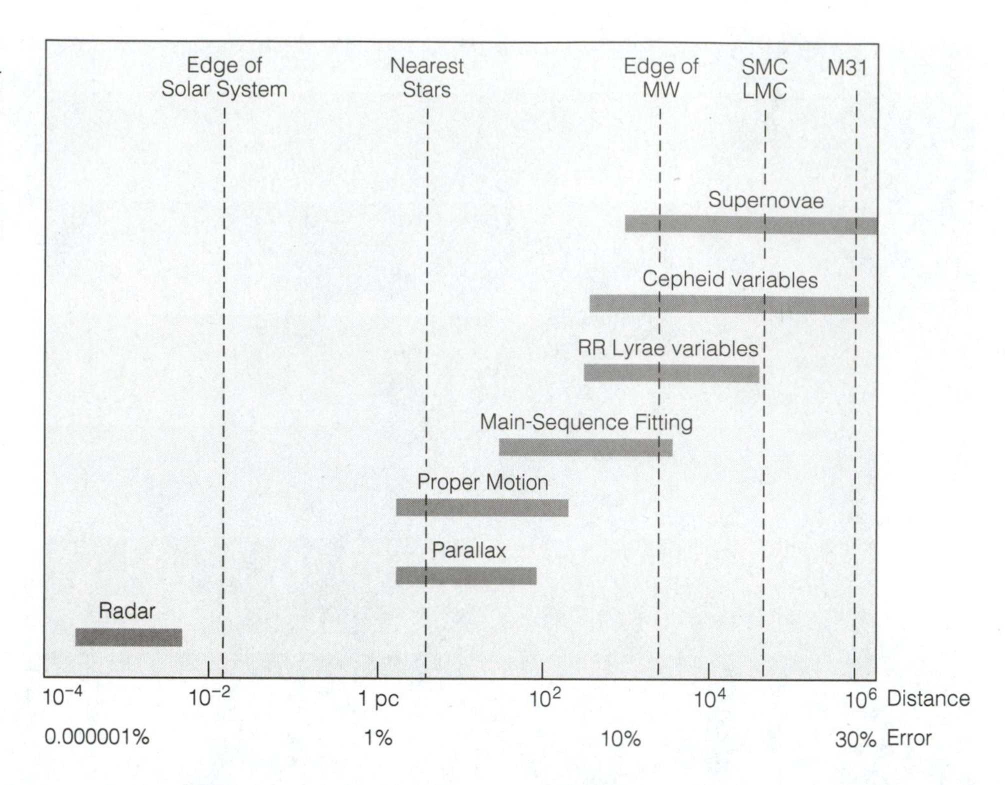

Figure 24-2 The chain of overlapping distance indicators plotted on a logarithmic scale, with a horizontal bar showing the range of applicability for each method. Important thresholds in distance are marked as vertical lines. Supernovae can be used to very large distances, but their uniformity as standard candles is not secure, and there are no calibrators in our own galaxy.

its name to the measurement unit used all the way out into the distant universe, the parsec. The parallax technique provides distances for stars out to 100 pc. Very nearby stars can also show observable proper motions.

The Hyades cluster is a crucial landmark in the determination of distances. Various methods are used to try and get a reliable distance to the Hyades. Since stars in the cluster are all nearly at the same distance, it is easy to calculate intrinsic brightness from apparent brightness, allowing the calibration of the technique of main-sequence fitting on an H-R diagram. Now we can plot H-R diagrams for other open clusters, and the amount by which the main sequence has to be shifted in brightness to overlay the Hyades is a measure of the relative distance of the two clusters (see the discussion in Chapter 22). This technique takes us with some reliability out to the edge of the Milky Way.

To calibrate distances to nearby galaxies, a more luminous standard candle is needed than a typical solar-type star. At this point, we have distances for enough clusters to include RR Lyrae and Cepheid variables. The well-established relationship between the period of variability and the luminosity for these rare stars allows a distance to be calculated, given the apparent brightness. Cepheids, in particular, are up to 20,000 times more luminous than the Sun and so can be seen with large telescopes out to a distance of 5000–10,000 kpc. Finally, it is possible to use supernovae (see Chapter 19), which can outshine the normal stars of an entire galaxy and can be seen out to distances of hundreds of thousands of kiloparsecs. Unfortunately, a supernova is such a fleeting stage of stellar evolution that one is observed only every 40-50 y in a galaxy. Therefore, there is no guarantee that we can use this measure in a specific galaxy.

Conceptually, the methods of the **distance scale** form a *pyramid.* Nearby methods are more direct and accurate. Moving farther from the Sun, we must employ different techniques. Each technique depends on the reliability of those that work at smaller distances. A range of overlap in distance is required to calibrate each new method. Errors continue to accumulate and grow as we reach toward the galaxies (Rowan-Robinson, 1985). The overlapping ranges and increasing errors of distance estimators to the nearby galaxies are illustrated in Figure 24-2.

Why use so many different methods? Why is it impossible to find one reliable technique to measure all distances in the universe? Part of the reason is the *sheer immensity of empty space.* The type of standard candle found near the Sun must also be found 10,000 times farther away to map out the Milky Way. By the inverse square law, this increase in distance corresponds to a dimming by a factor of 100 million. The type of standard candle found across the Milky Way must also be discovered a thousand times farther away to be useful in measuring the distances to other galaxies, a dimming by a factor of 1 million. Nearby distance indicators lose their usefulness as the standard candles become too faint to observe. Distant ones are ineffective in the local universe because luminous standard candles are very rare. Cepheids can be seen in galaxies up to 20 million parsecs away, but the nearest one to the Sun is about 200 pc, too far for a reliable parallax measurement. Also, it is sheer bad luck that although we might expect a supernova every 40 y or so in our galaxy, none has been seen since Kepler's supernova of 1604.

SURVEYING AND CLASSIFYING GALAXIES

Having established the extragalactic nature of the nebulae, Hubble took photographs of many galaxies and tried to classify them according to their appearance or **galactic morphology.** The standard scientific approach in a new area of research is to classify the different objects by type and then to arrange them in a system that shows smooth transitions from one type to another. The act of classification does not lead directly to a physical theory, but the hope is that the relationship between galaxy types will have a physical underpinning. For example, the galaxy types might be related by age, or gas content, or mass differences, or rotational differences. Hubble found three major types of galaxies in his study of the Local Group and regions of space beyond: spirals, ellipticals, and irregulars (Sandage, 1961).

A Classification System

Spirals Our own galaxy and the Andromeda galaxy are examples of large **spiral galaxies.** The components of a spiral galaxy are the disk, the bulge, and the halo (some spirals also may have a nucleus or a bar). The disk has spiral arms with bright emission nebulae; large amounts of dust are often mixed in. The dust forms an obscuring band when the disk is viewed edge-on. When the disk is viewed face-on, the spiral arms appear clearly outlined by luminous, young, blue stars. Roughly one-third of spiral galaxies have bright barlike features in their central regions. In such a *barred spiral galaxy,* the spiral arms originate at the ends of the bar, rather than originating in the nucleus itself.

★ In both normal and barred spirals, Hubble noticed a gradual transition of morphological types (Lake, 1992). Among normal spirals, the sequence from *Sa* to *Sb* to *Sc* corresponds to less and less tightly wound spiral arms and less and less prominent central bulges. The corresponding barred spirals are classified from *SBa* to *SBb* to *SBc.* Galaxies are flung at random orientations in space, so some disks will be seen face-on and some edge-on. Examples are given on our web site. The size of the central bulge and the degree of winding of the spiral arms go hand in hand, so we can classify spirals according to Hubble's scheme even if they are viewed edge-on. It is also observed that spirals with more loosely wound arms have much more prominent luminous stars, star clusters, and emission nebulae that outline the arms. All spirals rotate in the sense that the arms trail, as observed for the Milky Way.

Another type of galaxy with prominent bulges has disks that lack spiral arms. These are classified as *S0 galaxies* and are sometimes called *lenticular galaxies* due to their lenslike shape.

What about the third component—the halo? It turns out that the halo is too diffuse to be visible, even on deep images of galaxies. This is important since, as we shall see, the halo contains most of the mass of a spiral galaxy!

Ellipticals **Elliptical galaxies** come in a wide range of sizes, ranging from tiny dwarfs to giant elliptical galaxies that are three to four times larger than the Milky Way. In general, ellipticals have smooth spherical or ellipsoidal shapes and no spiral arms. They are classified according to how round or flat they look, and according to their size. The numerical scheme from E0 to E7 runs from circular to highly elongated galaxies. Since galaxies are three-dimensional objects distributed in space, our perspective from the Earth may not give us a true indication of the shape. An E7 elliptical *must* be a relatively flat galaxy seen edge-on. However, an E0 elliptical *need not* be a spherical galaxy. Both a flattened distribution of stars viewed face-on and a cigar-shaped distribution of stars seen end-on would be classified E0. Ellipticals have smooth distributions of reddish stars, with little indication of gas, dust, or young, luminous stars.

Irregulars **Irregular galaxies** come in many shapes, and they are usually small. Some irregulars show a degree of spiral structure, but without the high symmetry of spirals. Others have chaotic morphologies without any obvious symmetry. Many of these chaotic irregulars appear to have undergone collisions or to be in the process of merging. Irregulars have regions of intense star formation, with conspicuous O and B stars and HII regions. The Magellanic clouds are the only two examples visible to the naked eye. Finally, it is important to point out that a number of galaxies have *peculiar* morphologies and

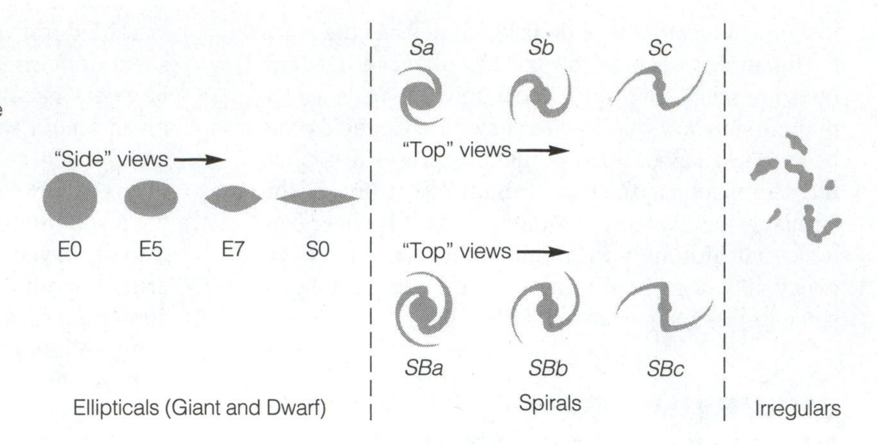

Figure 24-3 The simplified classification scheme for galaxies known as "Hubble's tuning-fork diagram" after the originator of the system. We now know that this scheme does not correspond to any kind of evolutionary sequence.

resist being shoehorned into Hubble's classification scheme. Peculiars may have loops or tails or other extended structures not seen in irregulars.

Classification is an important first step in organizing the richness of the extragalactic universe, but it does not automatically lead to physical understanding. Hubble organized galaxies into a tuning-fork diagram, in which the normal and barred spirals form parallel sequences (Figure 24-3), and the S0 galaxies are a transition type between the spirals and ellipticals. This system correlated well with certain galaxy properties. For example, compared to spirals, elliptical galaxies have older stars and smaller amounts of gas and dust. For some time it was believed that the Hubble classification implied an *evolutionary* sequence, in which spirals gradually used up their gas, the stars aged and faded, and the final result was an elliptical. We now know that this cannot be true, since spirals contain old halo populations, and many must be as old as ellipticals. Morphology alone does not explain the differences in galaxy type.

THE NEARBY GALAXIES

To understand the nature of the neighborhood around the Milky Way, we will take a reconnaissance journey, describing the nearby galaxies in order of distance. At about 50 kpc are the Large and Small Magellanic Clouds, which are close to the south celestial pole and so are not visible from most northern latitudes. The Magellanic clouds are companions to the Milky Way, in close orbit around it. Also scattered around the Milky Way, in the range of 50 to 250 kpc, are at least eight other dwarf companions. The Andromeda galaxy, M 31, lies much farther away at about 670 kpc. M 31 is similar in size and structure to the Milky Way, and it has seven known dwarf satellites, as well as the more substantial spiral companion M 33. These galaxies plus a few others that make up the whole population of galaxies out to about 1000 kpc are called the **Local Group.**

★ A three-dimensional sketch of the Local Group can be found on our web site. Our cosmic neighborhood teaches us a very important feature of the universe: Galaxies are not randomly distributed through space; they tend to cluster in different-size groups. Most of the galaxies in the Local Group are clumped into two subgroups, those around the Milky Way and those around the Andromeda galaxy. See Table 24-1 for a list. Every year or so, a new member of the Local Group is found, but astronomers are confident that they have identified all members except for the tiniest dwarf galaxies (Hodge, 1994).

To visualize the distances, imagine the Milky Way as a dinner plate. The Magellanic clouds would be crumpled balls of cotton 8 or 10 cm across, within a meter of the plate. A dozen or so galaxies of various shapes, 1 to 15 cm across, would be scattered across the room. Andromeda, the nearest galaxy resembling the Milky Way, would be another dinner plate 7 m away.

The Magellanic Clouds

Medieval Arab astronomers recorded a glowing patch in the southern sky, just visible from southern Saudi Arabia. Since there is no bright south polar star, the clouds helped navigators to mark the pole. Well-described during Magellan's around-the-world expedition of 1518 through 1520, they came to be called the **Magellanic clouds.** The clouds are in fact small galaxies, moving in orbits around the Milky Way. The Large Magellanic Cloud is 50 kpc away, and the Small Magellanic Cloud 63 kpc, less than three times the distance to the far edge of our own galaxy.

What kind of galaxies are these companions of ours? The large cloud is only 19 kpc across, and the small cloud only 8 kpc, compared with about 30 kpc

TABLE 24-1

Local Group Galaxies out to 100 kpc

Galaxy	Type	d (kpc)	L_V ($10^7\ L_\odot$)	Radial Velocity (kms^{-1})	Gas Mass ($10^6\ M_\odot$)
Milky Way	Sbc	8	1500	–10	4000
Sagittarius	dSph	25	1	170	none
Large MC	SBm	49	170	274	700
Small MC	Irr	58	34	148	650
Ursa Minor	dSph	64	0.02	–247	none
Draco	dSph	72	0.02	–293	none
Sculptor	dSph	72	0.14	107	≤0.1
Sextans	dSph	83	0.04	225	none
Carina	dSph	100	0.03	223	none
Fornax	dSph	120	1.4	53	none
Leo II (DDO 93)	dSph	207	0.06	76	none
Leo I (DDO 74)	dSph	270	0.5	285	none
Phoenix	dIrr/dSph	420	0.08	56	0.2
NGC 6822	Irr	490	30	–56	—
And II	dSph	590	0.3	–188	—
NGC 185	dE	600	10	–202	0.1
Leo A (DDO 69)	dIrr	690	2	20	20
IC 1613 (DDO 8)	dIrr	715	10	–231	60
M32 (NGC 221)	E2	750	30	–203	none
NGC 147	dE	760	12	–193	none
Pegasus (DDO 216)	dIrr	760	8	–183	3
And VII/Cas dSph	dSph	760	0.5	–307	—
M31 (NGC 224)	Sb	770	2700	–299	5700
And I	dSph	770	0.5	–370	none
And III	dSph	770	0.1	–352	<0.1
Cetus	dSph	775	0.08	—	—
Sagittarius DIG	dIrr	800	0.1	–78	4
LGS3 (Pisces)	dIrr/dSph	810	0.06	–281	0.2
And V	dSph	810	0.03	–387	—
IC 10	Irr	820	20	–344	150
And VI/Peg dSph	dSph	830	0.3	–341	—
M33 (NGC 598)	Sc	850	550	–183	1500
NGC 205	dE	850	40	–241	0.4
Tucana	dSph	870	0.05	—	none
Aquarius (DDO 210)	dIrr	950	0.2	–137	3
WLM (DDO 221)	dIrr	970	4	–120	80

Sources: Tully (1988), Nearby Galaxy Catalog, and Sparke and Gallagher (2000), Galaxies in the Universe.

for the Milky Way. Star counts and measures of neutral hydrogen indicate that these clouds are only a few percent as massive as the Milky Way. In addition, the clouds do not show the Milky Way's beautiful spiral structure; they are irregular galaxies. Each contains a softly glowing, barlike structure composed of stars. Somewhat off the end of the bar in the Large Magellanic Cloud is the spectacular Tarantula Nebula, also known as 30 Doradus. This luminous nebula can be seen with the naked eye. In fact, if it were moved to the distance of the Orion Nebula, it would fill the whole constellation of Orion and be bright enough to cast shadows on Earth! In its center is a cluster 60 pc in diameter containing thousands of massive, bluish supergiant stars.

We can best understand other galaxies and their evolution by considering what populations of stars they contain. The Large Magellanic Cloud has the red HII regions and luminous blue stars that are a hallmark of Population I, as found in our own spiral arm of the Milky Way. Faint star photometry has also identified red giants and main-sequence stars. The bulk of the star formation in both clouds occurred between 3 and 1 billion years ago, but it continues to the present day. There is little dust in the two clouds, except in the prominent young nebulae. Overall, the stellar populations are younger and more deficient in heavy elements than the solar neighborhood.

The two clouds are connected by a bridge of diffuse hydrogen gas called the *Magellanic Stream.* Australian radio astronomers showed that this long filament of HI also extends from the small cloud in an arc beyond the south galactic pole, and in the other direction it reaches into the plane of the Milky Way. This filament resembles the bridge between the two clouds themselves. The Magellanic clouds are satellites of our own galaxy, gravitationally bound to the Milky Way. Their orbits are likely to take them through the Milky Way disk, and it is speculated that the Magellanic stream is a tail of gas drawn out during such an encounter an estimated half billion years ago.

The Magellanic clouds are important to modern astronomers because they provide a cornucopia of stellar types at essentially identical distances. Use of the Magellanic clouds as a stellar laboratory was given a boost by the explosion of Supernova 1987a in the Large Magellanic Cloud. The *Hubble Space Telescope (HST)* has been used to study the ring of hot gas that was thrown off by the death of this star, deriving an accurate distance of 50 kpc using simple geometric arguments. The door is now open to calibrate many distance indicators, using the rich stellar nursery of the Large Magellanic Cloud.

Dwarf Galaxies

The Local Group contains a number of small stellar systems. Most of these are dwarf elliptical galaxies, and a few are irregulars. The elliptical or spheroidal galaxies resemble giant globular clusters. They are more symmetrical than the Magellanic clouds, but lack the disk and spiral arms of the Milky Way.

Dwarf ellipticals are dominated by old Population II stars and have little gas or dust. In most respects, they are less impressive than our own giant spiral disk, with its chaotic clouds of gas and dust and regions of continuing star formation. However, they appear to be more active than globular clusters, whose stars are all around 12 to 14 billion years old. Analysis of H-R diagrams of the individual stars in dwarf ellipticals indicates that some of their stars are relatively young, in the range of 3 to 9 billion years old. Dwarfs are the most common type of galaxy, but their diffuseness makes them difficult to detect. The satellite companions of the Milky Way are 5 to 30 times smaller and 1000 to 100,000 times less luminous than large spirals like M 31 and the Milky Way. Companions to the Milky Way are still being discovered; the most recent was first identified in 1998.

The Andromeda Spiral Galaxy

★ At 670 kpc, we encounter the first spiral galaxy truly comparable to the Milky Way, along with several of its smaller satellite galaxies. This galaxy must have been known since prehistoric times, since it is visible to the naked eye as a hazy patch on a clear, dark night. It was first recorded in a star catalog by Arab astronomer al-Sufe in A.D. 964. Edwin Hubble used Cepheid variables in the **Andromeda galaxy** to finally settle the debate over the nature of the so-called spiral nebulae. The Andromeda galaxy is slightly larger than the Milky Way and similar in stellar content. The naked eye sees it as a faint patch (Figure 24-4), but this is really only the brightest, innermost region, a few kiloparsecs across. For a color view of how to find Andromeda in the night sky, go to our web site. Images made with large telescopes show that the spiral arms form a disk at least 20 kpc across (Figure 24-5). Like the Milky Way, Andromeda has globular clusters and a halo of HI gas reaching perhaps 100 kpc in diameter.

★ The Andromeda galaxy played an important role in the discovery of the two main stellar populations, and studies showed that the pattern of populations in Andromeda matches that in our own galaxy (Hodge, 1993). The Andromeda galaxy gives us a chance to look directly at the nucleus of a galaxy like ours, rather than trying to study it through 8.5 kpc of dust. HST images of M 31 reveal a surprise. The galaxy has a double nucleus, with the dimmer component actually corresponding to the true center of the galaxy (see our web site). As with the nucleus of our galaxy, there are two interpretations: The nucleus is either a dense cluster of massive blue stars and supernova remnants, or we can believe the dynamical evidence,

Figure 24-4 The region of the Andromeda galaxy, showing an angular extent approaching 3°. Vertical height of the photo is about 8°. The bright star at the upper right is Beta Andromedae. The box shows the outline of the view in Figure 24-5. (35-mm camera; 135-mm lens at f2.8; 10-min exposure on 2475 recording film)

which points to the existence of a massive black hole. The explanation for the intriguing double nucleus is not yet known. Other bright galaxies in the nearby universe are listed in Table 24-2.

Properties of Galaxies

Distance is the most basic property of a galaxy. Knowing the distance, we can calculate many other intrinsic properties (Ferris, 1985). Size and luminosity can be calculated directly from the apparent diameter and apparent brightness. As we have already seen, however, it is difficult to measure distance with accuracy better than 30–50%, and our knowledge of luminosity and size is no better than the error on the distance estimate. Colors can be measured by observing a galaxy through filters to determine the brightness at different wavelengths. Measures of colors and galaxy rotation generally do not depend on distance. The mass deserves special attention, since it is difficult to measure directly. The general properties of galaxies are summarized in Table 24-3.

Size Although it sounds simple, measuring the size of a galaxy is not a trivial matter, because galaxies do not have sharp edges! As can be seen from the photographs in this chapter, the light from a galaxy smoothly fades away until it becomes indistinguishable from the background sky level. The conventional way to measure a diameter involves summing up the total light from the galaxy through a very large aperture, so the total light measured does not depend very much on the particular size of the large aperture used. The diameter is then defined as the size enclosing half of the total light (or a quarter, or 90%; the fraction doesn't matter as long as it is applied consistently from galaxy to galaxy).

Measured in this way, galaxies vary widely in size. Dwarf ellipticals and irregulars can be as small as a few kiloparsecs; several companions to the Milky Way are this size. The disks of spiral galaxies range in size from 10 to 50 kpc. The largest galaxies in the universe are giant ellipticals with diameters up to 200 kpc, much larger than the distance from the Milky Way to the Magellanic clouds.

Luminosity The luminosity of a galaxy is derived directly from its distance and apparent brightness. As with the size, measuring the apparent brightness of an extended object is not simple, because the brightness of the galaxy falls off toward the edge. Two strategies are used. The first is to measure the flux through a large enough aperture that essentially no light is missed. The other is to use the fact that the light from most galaxies falls off with distance from the center in a simple and predictable way. Using this strategy, we can make a *model* of the light distribution and use it to *calculate* the total brightness and hence the luminosity. Galaxies have a wide range of luminosity, from $10^{11}\,L_\odot$ for giant ellipticals to $10^5\,L_\odot$ for the satellites of the Milky Way.

★ **Color** The colors of galaxies provide useful information on the types of stars that dominate the light output. Recall that Population I stars are young, blue, and rich in heavy elements—these are the stars found in spiral disks. Population II stars are old, red, and dominated by only hydrogen and helium—these are the stars found in ellipticals, the bulges of spirals, and the globular clusters that orbit the halos of many galaxies. Some of this information can be seen in the images of the giant elliptical M 87 (Figure 24-6) and

Figure 24-5 Wide-field Schmidt telescope photograph of M 31, the Andromeda galaxy. M 31 is a type *Sb* spiral with a diameter of 30 kpc. The small galaxy below M 31 is a dwarf elliptical companion, M 32. (National Optical Astronomy Observatory)

the spiral M 51 (Figure 24-7). For a color version of M 51, see our web site.

★ A connection can be made between the Hubble type of a galaxy, its dominant stellar populations, and its colors. Irregular galaxies have predominantly young Population I stars and are correspondingly blue and gas-rich. Spiral galaxies contain both old and young populations, and they follow a sequence from Sc to Sa of increasing amounts of light in the old bulge population (note the yellowish and reddish tones in the central bulge of Andromeda in the figure on our web site). Elliptical galaxies have predominantly old Population II stars and are correspondingly red and gas-poor. Therefore, the colors of a galaxy offer some insight into its age and history.

Rotation Additional information is provided by the rotation of galaxies. Since individual stars can be resolved in only a few nearby cases, the observation usually involves the *average* motion of stars in some large chunk of the galaxy. The disks of spirals and S0 galaxies rotate, and this rotation can be mapped using either the Doppler shift of stellar absorption lines, or the Doppler shift of emission lines from HII regions, or the Doppler shift of the 21-cm line of neutral hydrogen. Note that this rotation cannot be observed when a disk is face-on, because in this case the motion is transverse to the line of sight and there is no Doppler shift.

It was once suspected that rotation also caused some elliptical galaxies to become flattened. Ellipticals can rotate, but the amount is too slight to cause flattening. In fact, spectroscopy shows that the motions of stars in ellipticals are quite random. The fundamental physical quantity that accounts for the shape of galaxies is angular momentum, which is related to the amount of mass and the velocity at which it is rotating. Spiral disks have high angular momentum, bulges have somewhat less, and halos have the least of all. Spiral and elliptical galaxies therefore have very different amounts of angular momentum. Since

TABLE 24-2

Luminous Galaxies out to 15,000 kpc

Catalog Number	Distance (kpc)	Diameter (kpc)	Mass ($M_{\odot}$)	Absolute Magnitude (M_v)	Type[a]	Radial Velocity (km/s)
NGC 55	2300	12	3×10^{10}	–20	Sc	+ 190
NGC 253	2400	13	10^{11}	–20	Sc	– 70
M 82 (NGC 3034)	3000	7	3×10^{10}	–20	Irr	+ 400
M 81 (NGC 3031)	3200	16	2×10^{11}	–21	Sb	+ 80
M 83 (NGC 5236)	3200	12	10^{11}	–21	SBc	+ 320
NGC 5128 (Centaurus A)	4400	15	2×10^{11}	–20	E0p[b]	+ 260
M 101 Pinwheel (NGC 5457)	7200	40	2×10^{11}	–21	Sc	+ 402
M 51 Whirlpool (NGC 5194)	7600	9	8×10^{10}	–20	Sc	+ 550
M 104 Sombrero (NGC 4594)	12,000	24	5×10^{11}	–22	Sa	+1050
M 87 (NGC 4486)	13,000	20	4×10^{12}	–22	E1	+1220

Source: Data from Allen (1973).
[a]S = spiral (subtypes 0, a, b, c—see discussion in chapter)
Irr = irregular
SB = barred spiral (subtypes 0, a, b, c—see discussion in chapter)
E = elliptical (subtypes 0 through 7)
P = peculiar
[b]Centaurus A is a strong radio source, appearing as an elliptical galaxy with a peculiar dense dust lane across its face.

TABLE 24-3

General Characteristics of Galaxies

	Spirals	Ellipticals	Irregulars
Mass ($M_{\odot}$)	10^9–10^{12}	10^6–10^{13}	10^8–10^{11}
Luminosity ($L_{\odot}$)	10^8–10^{11}	10^6–10^{11}	10^8–10^{11}
Mass-to-light ratio	2–10	5–30	1–3
Diameter (kpc)	5–50	1–200	1–10
Stellar populations	Old halo and bulge (II) Young disk (I)	Old (II)	Young and intermediate ages (I and II)
Composite spectral type	A (Sc) to K (Sa)	G to K	A to F
Interstellar material	Gas and dust in the disk	Small amounts of gas and dust	Copious amounts of gas, some dust
Large-scale environment	Small groups, low-density regions	Rich clusters	Low-density regions

Figure 24-6 The giant elliptical galaxy M 87, located in the Virgo cluster. The smooth brightness distribution represents the summed light of hundred of billions of individual stars. (Courtesy of NOAO/AURA/NSF)

angular momentum is always conserved, a single galaxy cannot evolve from one type to another.

Mass One of the most difficult properties of a galaxy to measure is the mass. The H-R diagram reveals that a main-sequence star of a certain mass has a unique luminosity and temperature. A useful way to characterize a galaxy is in terms of the ratio of its mass, in $M_{\odot}$, to its luminosity, in $L_{\odot}$. Clearly, if a galaxy consisted entirely of stars like the Sun, it would have a **mass/luminosity ratio** of 1. However, a galaxy is a *composite* of many millions of stars of dif-

Figure 24-7 Central region of M 51, the Whirlpool galaxy, as imaged by the Hubble Space Telescope. Complex details of star formation regions in the spiral arms can clearly be seen. (Courtesy of the Hubble Heritage Team, StScI/AURA, N. Scoville, Caltech, and T. Rector, NOAO)

fering ages and masses. The global mass/luminosity ratio of a galaxy depends on the relative numbers of stars of different types.

It turns out that the visible light from galaxies of all types has a mass/luminosity ratio in the range of 3 to 20. Spiral galaxies (with more young stars) tend to be at the bottom of that range, and elliptical galaxies (with more old stars) tend to be at the top. All of these calculations refer only to the visible stellar populations in galaxies. All mass estimates must be reconsidered in light of one of the most profound discoveries in astronomy—the discovery of large amounts of dark matter.

DARK MATTER

The mass of a galaxy can be measured by various methods using motions of stars or gas. These are called dynamical methods. The results are surprising. Regardless of the technique or the Hubble type of the galaxy, about 90% of the mass of the galaxy is found to be in the form of **dark matter,** mass that makes its presence felt by gravitational forces but does not emit light. This is startling, because it means that all the visible light from galaxies represents only a small fraction of the universe!

As an analogy, consider an accounting problem. On one side of the ledger, we derive the mass of a galaxy from the motions of stars in the galaxy. On the other side of the ledger, we use the mass/luminosity ratios discussed above to infer the mass of the galaxy from the visible light produced by populations of stars. The mass based on motions is always about 10 times larger than the mass associated with the observed starlight. The typical mass/luminosity ratios of galaxies based on dynamical arguments are 50–100 (Crosswell, 1996). The only alternative to the dark matter hypothesis is to jettison Newton's law of gravity. However, all available evidence indicates that Newton's law provides a good description of the motions of stars and galaxies.

Motions within Galaxies

The mass distribution of spiral galaxies can be mapped out using a **rotation curve.** A rotation curve is a map of the velocity of different parts of a galaxy with respect to the nucleus. The velocity is measured from the Doppler shift of emission lines produced by gas or absorption lines produced by stars. Figure 24-8 shows the rotation curves for three spiral galaxies. The striking feature of all these rotation curves is that they are flat; the velocities stay high out to the limits of the visible material in the spiral disk.

The interpretation of a flat rotation curve can be understood in terms of Newton's law of gravitation. We have seen that the velocity of a small body orbiting a planet or a star gets smaller with increasing distance from the planet or star (the orbital velocities of the planets in the solar system provide a good example). In the case of a galaxy, we would therefore expect the circular velocity of stars in the galaxy to begin to fall at a point where the orbit enclosed most of the mass of the galaxy. This does not happen. In many spirals, the gas disk extends further than the stellar disk; in that case, we can use the 21-cm line of neutral hydrogen to map the rotation curve out to very large distances. Since the velocity stays constant out to the largest radii studied, we must conclude that even these outer orbits do not enclose most of the mass of the galaxy. The bulk of the mass must be invisible material in the halo.

Motions of Galaxies Themselves

Another way to probe the mass of a large galaxy is to measure the motions of any dwarf galaxy companions that are in orbit around it. These orbits cannot, of course, be followed in time because they typically take hundreds of millions of years. We have to make do with a snapshot of the instantaneous velocity of each companion. The companions are then used as test particles to measure the mass of the large primary galaxy. Dwarf ellipticals in orbit around the Milky Way give a total mass for the Milky Way of about $10^{12}\ M_{\odot}$, distributed up to 100 kpc from the center of the galaxy. This work has been extended to other spirals and to ellipticals (where a rotation curve cannot be measured because the orbits are not circular). The results are always the same: About 90% of the mass of all bright galaxies is in the form of a dark halo. Dark matter is a ubiquitous feature of galaxies.

Direct Detection of Dark Matter

Various attempts are being made to detect dark matter directly. The stakes are very high: The nature of dark matter is probably the largest unanswered question in astronomy. The most likely dark matter candidates are massive collapsed objects like black holes, or stars that are not massive enough to liberate energy by hydrogen fusion, the so-called brown dwarfs. Cold, free-floating planets are also a possibility. More exotic suggestions include massive neutrinos or a variety of exotic particles.

Recently, advances in CCD detectors have permitted a search for dark matter using the technique of **gravitational microlensing.** The idea is based on a simple prediction of Einstein's general theory of relativity. If a dark, compact object passes between us and a more distant star, the dark object can act as a gravitational "lens," amplifying the light of the star. In this way, the dark object reveals its presence. The problem is that stars are very small and space is very empty, so the probability of a crossing is tiny. The microlensing phenomenon is not like a conventional eclipse; the amplification is caused by the bending of light by an intervening mass (see the discussion in Chapter 25). If the dark halo of the Milky Way is composed entirely of dark compact objects, only one in a million background stars will be lensed at any

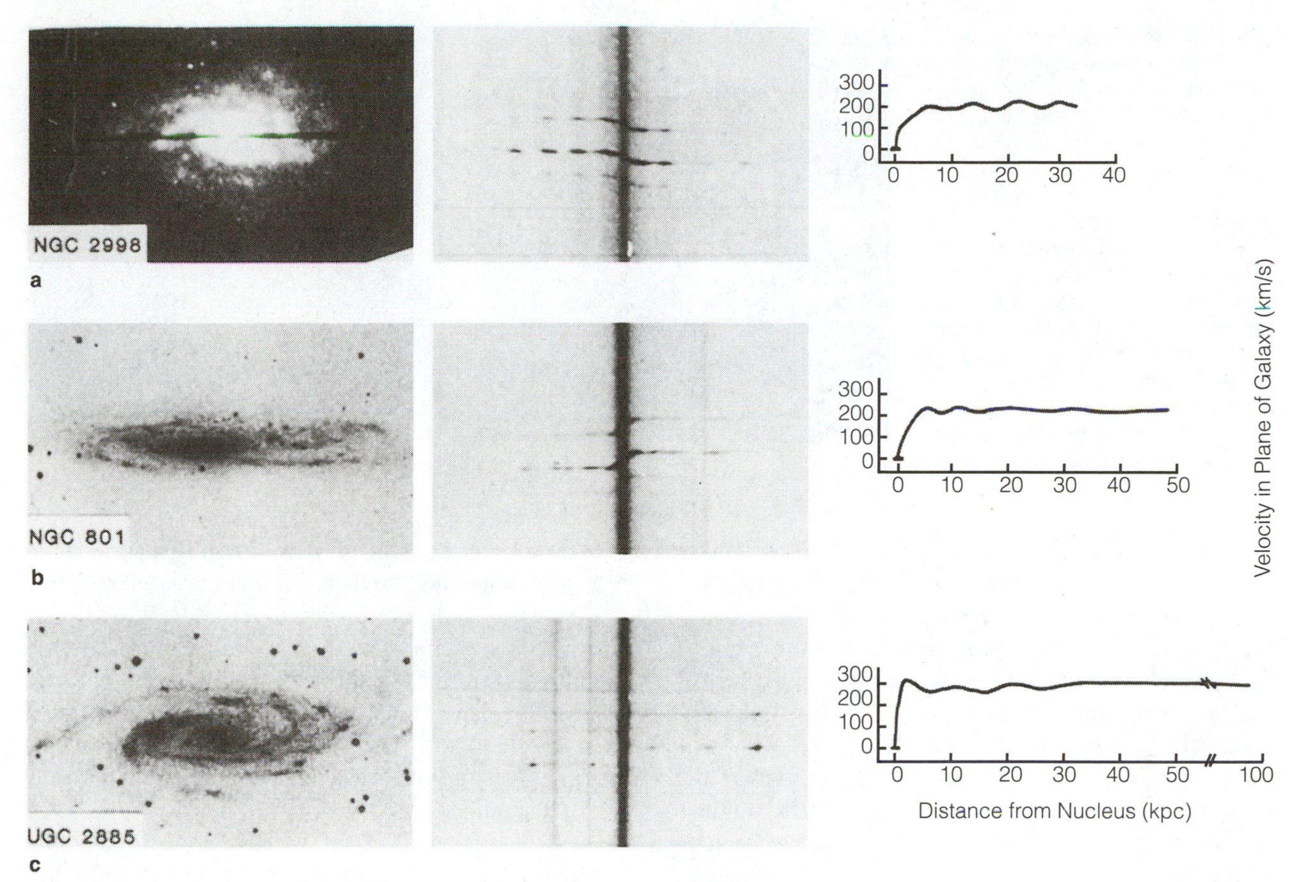

Figure 24-8 Rotation curves for three spiral galaxies: **a** NGC 2998, **b** NGC 801, and **c** UGC 2885. The slit of a spectrograph is placed along the major axis of the galaxy (as shown on the left). In the spectrum for each galaxy (center), the wavelength runs vertically, and the position across the galaxy runs horizontally. Each of the resulting rotation curves (right) shows constant amplitude of rotation out to the edge of the galaxy. The jump in wavelength near the center of each galaxy corresponds to the steeply rising part of the rotation curve. (Courtesy Vera Rubin, Carnegie Institution of Washington)

given time. To optimize the chances of detecting such an event, astronomers are searching fields toward the galactic bulge or the Magellanic clouds, which have the largest number of background stars.

The search for gravitational microlensing is the ultimate "needle in the haystack experiment." Millions of stellar images per night must be examined in the hope of finding a few events. Variable stars are a source of potential confusion, but the lensing signature is very specific: The signal should rise and fall symmetrically, with the same amplitude at all wavelengths, and it should not repreat. This exciting technique is sensitive to everything from sub-Earth-size planets to solar mass black holes. Since 1993, nearly a hundred events have been reported by three international groups of astronomers. The lensing masses are in the range 0.05–0.40 $M_{\odot}$. However, the lensing rate is too low for the dark matter halo of the Milky Way to be made entirely of brown dwarfs or white dwarfs. Some other more exotic form of matter is required.

Black Holes in Galaxy Nuclei

Most of the dark matter in galaxies is distributed at distances of 100–200 kpc in the form of a halo. However, careful study of the nuclei of nearby galaxies reveals dark matter within the central few parsecs. The favored explanation is a supermassive black hole, an object first mentioned in the context of our own galactic center. Two types of evidence support this conjecture.

★ The first involves high-resolution imaging of the cores of bright galaxies. This can be done at a level of 0.1 second of arc using the HST or at a level of 0.3–0.4 second of arc using ground-based telescopes on excellent sites. M 32 is a small elliptical galaxy in

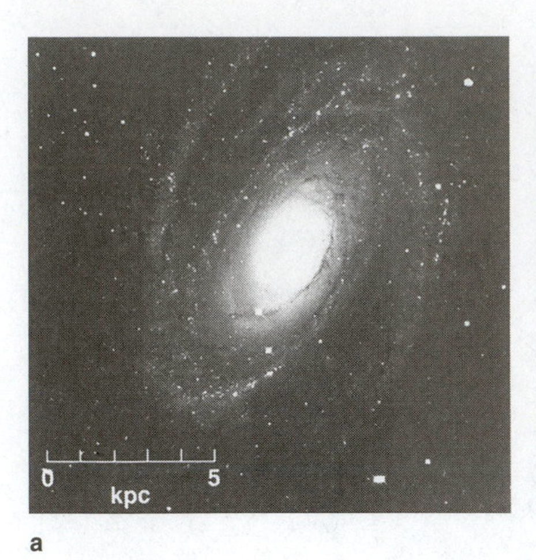

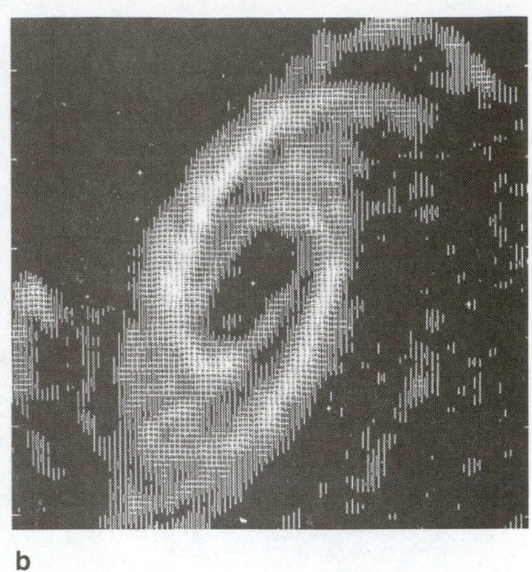

Figure 24-9 Optical and radio images of the spiral galaxy M 81 (NGC 3031), about 3200 kpc away. **a** The optical image shows stars and star clusters. The scale bar applies in the unforeshortened direction. **b** The radio image shows radiation from neutral hydrogen gas. The scale is about the same in both images. Important results are that (1) neutral hydrogen gas coincides with the position of Population I stars in the spiral arms, (2) there is virtually no neutral hydrogen in the center, and (3) the hydrogen spiral pattern is much larger than the visible star pattern. (Leiden Observatory)

the Local Group. The density of stars in the center of M 32 is over 100 million times that in the Sun's neighborhood (for an HST image of the core of M 32, see our web site). Models of this light distribution indicate a black hole of 3 million $M_{\odot}$. However, a central light peak does not point uniquely to a black hole, as opposed to a very dense star cluster.

The second type of evidence uses spectroscopy to measure the mean velocities of stars near the center of the galaxy. If the motions are too rapid to be accounted for by the stellar populations known to inhabit the nuclear regions, a case can be made for a compact object like a black hole. The case for supermassive black holes can be made with some confidence in six cases. A recent survey of nearby galaxies by J. Kormendy and others working in Hawaii finds evidence for black holes in about 25% of them. Unseen dark forces are at work in the hearts of many galaxies.

THE ENVIRONMENT AND EVOLUTION OF GALAXIES

Galaxies are so widely separated in space that it is natural to think of them as isolated entities. Yet galaxy research over the past two decades has shown that galaxy morphology and evolution are strongly influenced by the large-scale environment. Our snapshot of galaxies frozen in time is misleading: Galaxies can interact and merge, they can accrete and eject gas, and they can collide to trigger bouts of star formation.

Stellar Populations

A simple way to think about the **evolution of galaxies** is in terms of stellar populations. Population I stars are young, mostly blue, luminous, and rich in heavy elements. Population II stars are old, red, less luminous, and poor in heavy elements. Irregular galaxies contain almost pure Population I. Spiral galaxies contain a mixture of the two populations, with Population I star-forming regions concentrated in the spiral arms and Population II stars in the central bulges. Elliptical galaxies contain almost purely Population II.

★ This information can be seen clearly in the color photographs of the elliptical M 87 and the spiral M 83 (see the figures on our web site). The nucleus of M 83 has the pale yellowish-orange color of old Population II giants. But the spiral arms are made out of bluish clusters of young O and B stars, along with red-glowing HII regions scattered like rubies in a jeweled brooch.

The galaxies we see in the local universe reflect the consequences of stellar evolution. Imagine a population of stars that formed billions of years ago. The overall light would initially be dominated by young, hot stars. As the galaxy aged, the most massive stars would evolve off the main sequence, and the galaxy would become dimmer and redder. Astronomers construct theoretical models of stellar populations to match their observations. In principle, this approach can be used to measure the age of a galaxy.

Ellipticals produced almost pure Population II and show little signs of recent star formation. Spirals also produced a generation of Population II stars, which we see in the bulges and halos. Most of the gas in the central bulges was used up in the first generation of stars. This can be seen in Figures 24-9 and 24-10, where comparisons of optical (stars) and radio (gas) images show most gas in the spiral arms and little in the center. Disk stars, by contrast, show a wide range of ages. It is likely that the disks of galax-

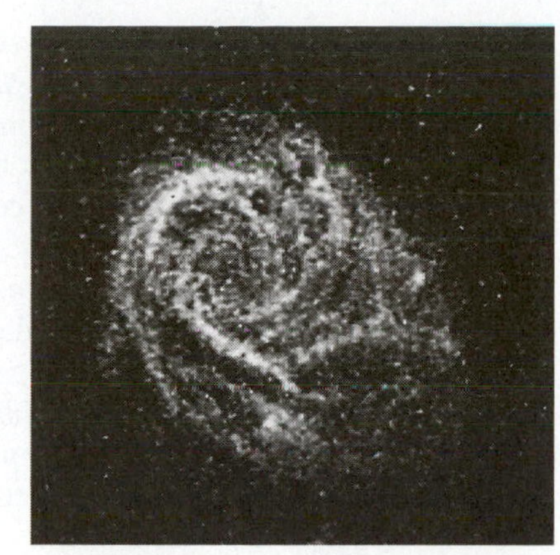

Figure 24-10 **a** Optical and **b** radio images of the spiral galaxy M 101 (NGC 5457), the Pinwheel Galaxy, about 7200 kpc away. Comparison shows the concentration of gas in the arms and the lack of neutral gas in the center. (Image *a*, Kitt Peak National Observatory; image *b*, Westerbork Synthesis Radio Telescope, courtesy R. Allen and E. Jenkins)

ies have been assembled over a period of several billion years.

The current star-formation rates in ellipticals and spirals have some interesting implications. Although the initial formation process of ellipticals should not have consumed all the gas and dust, these galaxies show little current star-forming activity. Any galaxy with significant amounts of gas and dust will inevitably show star formation. In addition, the old stars out of which ellipticals are made will lose mass by stellar winds, replenishing the interstellar medium. Thus gas and dust are somehow being *removed* from elliptical galaxies. The puzzle for spirals like the Milky Way is the converse. They are actively forming stars from gas and dust in the disk. However, the star-formation rate is so rapid that it could not have been going on for the *entire lifetime* of the galaxy (as measured by the oldest stars in the halo). One possibility is that star formation in the disk started relatively recently and so has not had time to exhaust the gas supply, but there is no good evidence for this. The other possibility is that gas has been *added* to spirals over their lifetimes, either by gradual infall or by swallowing small gas-rich companions. Either way, environment plays an important role in the evolution of galaxies.

Collisions, Interactions, and Mergers

The Milky Way galaxy and the Andromeda galaxy are separated by a distance of only 15 to 20 of their own diameters. Galaxies in rich clusters have even smaller separations. As galaxies drift through space, they can collide. Colliding galaxies present an extraordinary situation, very different from that of stars. Stars in galaxies are separated by millions of their own diameters. Therefore, despite their motions, few if any stars in a galaxy have collided with each other during the whole history of the universe.

If the random motions of two galaxies cause them to penetrate each other, individual stars are unlikely to collide. The dusty gas in each galaxy, on the other hand, is more like a continuous medium, so when galaxies collide at speeds of hundreds of kilometers per second, the gas interacts dramatically. You might think of the water in two buckets being sloshed toward each other—the two masses of water hit and make a splash. When gas atoms in a galaxy collide, the gas and its entrained dust get heated. The gas atoms are excited or even ionized. Hydrogen is the most abundant element, so colliding galaxies are often intense sources of radio radiation from excited hydrogen and infrared radiation from hot dust (Barnes, Hernquist and Schweitzer, 1991).

Galaxies do not actually have to collide for their properties to be changed. Gravity has a long reach, so a close encounter will produce a **tidal interaction.** Just as the Moon raises a tidal bulge in the Earth's oceans, so a nearby companion may distort the shape of a galaxy. Stars and gas will be pulled out on either side, and rotation will sling the material out into graceful, curving arcs (Figure 24-11).

★ In a merger of two nearly equal-size galaxies, gas clouds collide and are compressed (Miller, 1992). They become hotter and denser, making star formation much more likely. Star formation in turn heats the surrounding dust, causing intense far-infrared emission. In one statistical study, 14% of all galaxies showed distorted shapes from past collisions or close encounters. This means either that about one in six galaxies has close encounters, or that every galaxy suffers encounters that affect it for $\frac{1}{6}$ of its lifetime. A sequence of images on our web site shows a simulation of the collision between two disk galaxies of similar size, performed using a supercomputer. The transient

Figure 24-11 The interacting galaxies NGC 4038 and NGC 4039. A deep photograph shows curved streams of material flung off during the interaction. (Hale Observatory/California Institute of Technology)

streamers that can be thrown off in such a merger are clearly seen.

★ If two galaxies of very different size meet, the larger galaxy will gobble up the smaller one. This is called **galactic cannibalism.** As a small galaxy approaches a much larger one, two things will happen. First, the gravity of the small galaxy will be too weak to retain its own outer stars, so they will be ripped away onto the larger galaxy. Second, as the core of the smaller galaxy plows through the outer regions of the larger galaxy, it will lose energy and decelerate, while the stars in the large galaxy heat up and accelerate. This process is called *dynamical friction.* As a result, the core of the small galaxy spirals into the center of the larger galaxy. A second sequence of images on our web site shows a supercomputer simulation of a small galaxy being devoured by a much larger disk galaxy.

At the center of a rich cluster of galaxies, galactic cannibalism has a fascinating effect. As galaxies pass through the dense cluster core on their high-velocity orbits, they are continually swept free of gas and dust. Galaxies will gradually merge over time in the cluster core. The result of a large galaxy devouring a number of smaller galaxies could be a giant elliptical. This may be the reason that the largest galaxies known are at the centers of rich clusters.

Implications for Galaxy Formation

Any theory of the formation and evolution of galaxies must explain the presence of two very different stellar components: (1) the thin, rotating disks of gas and young stars, and (2) the bulges and halos, with older stars and much less rotation. Disks appear to form only in a relatively narrow range of masses, from 10^9 to 10^{11} $M_\odot$. By contrast, galaxies can have spheroidal components (bulges and halos) covering a wide range of masses, from 10^5 to 10^{12} $M_\odot$. In addition, galaxies of all types have dark matter. Another important distinction is that ellipticals are found primarily in regions that are dense with galaxies, whereas spirals are often quite isolated in space. Galaxy formation is analogous to star formation—only on a much larger scale—in that it involves the conversion of gas into stars.

We do not yet have a complete **theory of galaxy formation.** Current evidence suggests that galaxies formed billions of years ago by gravitational collapse from primordial gas clouds called protogalaxies. The causes and details of these events are still poorly understood (Silk, 1987). Theorists speculate that if star formation occurs early in the collapse, the galaxy will retain its nearly spherical shape because the stars are too widely separated in space to collide. This description resembles an elliptical galaxy. If star formation proceeds slowly, the galaxy remains gaseous. The gas will radiate, lose energy, and collapse into a disk. The conservation of angular momentum ensures that the slow rotation of the initial configuration speeds up in the eventual gas disk. This description resembles a spiral galaxy. As mentioned previously, since both major galaxy types contain old stellar components (halos), we know that ellipticals do not evolve *into* spirals.

Why does star formation progress quickly in some cases and slowly in others? The *initial* rate of star formation probably depends on the average density of matter. In a dense region, the stars form early, and the shape of the galaxy remains "frozen" in a spheroidal form. This idea is consistent with the fact that ellipticals are found in high-density regions of the universe like rich clusters of galaxies. In a low-density region, the protogalactic cloud collapses into a disk before widespread star formation occurs. This is consistent with the fact that spiral galaxies tend to be found in low-density regions of the universe (the Local Group is one such region).

This picture of galaxy formation raises other questions. For example, why did gas in the early universe form clumps of just the right size and mass to turn into galaxies? Speculative answers are deferred until we consider the formation of large-scale structure in the universe in Chapter 25. The largest uncertainty of all is the role of the ubiquitous dark matter. The classification of galaxies is based on the roughly

5–10% of the mass of the galaxy that emits light. We are only beginning to map the large dark halos, and we still have no idea what they are made of! No theory of galaxy formation and evolution will be persuasive without a better understanding of dark matter (Bartusiak, 1997).

SUMMARY

The Milky Way is adrift in a vast volume of space that is loosely sprinkled with other star systems, or galaxies (Tully, 1988). Large new telescopes built early in the twentieth century demonstrated that many of the so-called nebulae were in fact distant galaxies. Measuring the distances to galaxies is still one of the greatest challenges in astronomy. Many types of errors combine to give us distance estimates that are only accurate to 20–30%.

The Local Group of galaxies, out to about 1000 kpc, contains a variety of shapes and sizes. Combining observations of our neighbors with data on more distant galaxies, we find that most galaxies can be classified as elliptical, spiral, or irregular. A more elaborate classification scheme accounts for the many morphological details. Most galaxies lie in groups or clusters. For every giant galaxy, there are many more small ones, and the most common type of galaxy in the universe is a dwarf. The observations discussed here combine with theories of stellar evolution (Chapters 17 to 19) to explain how galaxies evolve and use up their gas and dust, thus producing different stellar populations.

The largest galaxies are giant ellipticals with Population II stars and little gas and dust. Spirals are intermediate in size, with young Population I stars in a rotating disk mixed with gas and dust, and Population II stars in a central bulge and an extended halo traced out by the globular cluster system. Other galaxies of intermediate and small size have elliptical and irregular shapes. All considerations of galaxy properties are overshadowed by the presence of dark matter, which forms about 90–95% of the mass of all galaxies, large and small. Dark matter reveals itself in terms of rapid motion of stars and gas within a galaxy or of galaxies themselves in groups. The physical nature of dark matter is unknown.

Some galaxies do not fit into the simple classification scheme—they have peculiar morphologies or appear to be interacting with a companion. In an astronomical version of the nature vs. nurture debate, there is clear evidence that the life story of a galaxy is affected by its environment. A galaxy's morphology and star-formation history can be influenced by tidal interactions, collisions, mergers, and galactic cannibalism. This stately violence has played out over billions of years of the history of the universe.

CONCEPTS

standard candle	mass/luminosity ratio
distance scale	dark matter
galactic morphology	rotation curve
spiral galaxy	gravitational microlensing
elliptical galaxy	evolution of galaxies
irregular galaxy	tidal interaction
Local Group	galactic cannibalism
Magellanic clouds	theory of galaxy formation
Andromeda galaxy	

PROBLEMS

1. Which type of galaxy tends to be biggest? Brightest? To contain fewest young stars? (See Tables 24-1 and 24-2.)

2. Of giant ellipticals, ordinary ellipticals, spirals, barred spirals, and irregulars, which type is most common? (See Tables 24-1 and 24-2.)

3. How many years would a radio signal take to reach the Andromeda galaxy?

4. Two very faint and distant galaxies were detected on photos but were too distant to allow identification of spiral arms and other typological features. If spectra showed that one was reddish with a spectrum of K-type stars, while the other was bluish with a spectrum of B and A stars, what types would you expect the two galaxies to be?

5. If you were to represent the Milky Way and Andromeda galaxies in a model by two cardboard disks, how many disk diameters apart should they be to represent the true spacing of the two galaxies?

6. In the Earth's geography, we usually designate the "up" direction as the northern rotation axis. When discussing the solar system, we use the northern ecliptic pole defined by orbital revolution. When discussing the Milky Way, we use the northern galactic pole defined by galactic rotation. Has any asymmetry or special direction appeared in this chapter that would define a preferred orientation for discussion of the distant galaxies outside the Milky Way? If so, describe it.

7. Why are small ellipticals not found in catalogs of the more distant galaxies, such as Table 24-2?

8. Irregular galaxies are dominated by stellar associations, open clusters, and gas and dust clouds, all of which indicate stellar youthfulness. Does this prove that the galaxies themselves have only recently formed?

9. Using principles of star formation and stellar evolution from Chapters 17 and 18, explain why the prominent light from star-forming regions in galaxies comes from massive, hot, blue stars.

a. Why are these stars not seen in regions where star formation has ended?

b. What population is indicated by hot, blue stars?

ADVANCED PROBLEMS

10. Use the small-angle equation to solve the following problems:

a. What is the angular diameter of the main stellar part of the Andromeda galaxy if it is 40 kpc across and 670 kpc away? How does this compare with the angular size of the Moon?

b. What is the angular diameter of the 3-pc-diameter bright nucleus at the center of the Andromeda galaxy?

11. If spectral lines of stars observed on the right side of a galaxy were redshifted 0.165 nm relative to those at the galaxy's center, while those on the left side were blueshifted by the same amount, what would you conclude to be the rotational velocity of these stars? Assume that the spectral lines normally occur at a wavelength of 500 nm.

12. If the stars in Problem 11 were measured to be on the outer edge of the galaxy, at a distance of 4 kpc from its center, what would be the mass of the galaxy? (*Hint:* Use the circular velocity equation.)

PROJECTS

1. Locate the Andromeda galaxy by naked eye. Compare its visual appearance with its appearance in binoculars and telescopes of different sizes. Across what diameter can you detect the galaxy? ($1° = 12$ kpc at the Andromeda galaxy's distance.) Why do most photographs show a large central region of constant brightness, while visual inspection reveals a sharp concentration of light in the center? Can dust lanes or spiral arms be observed? (Check especially with low magnification on telescopes with apertures of 0.5 to 1 m [about 20 to 40 in.], if available.)

2. If photographic equipment is available, take a series of exposures with different times, such as 1 min, 10 min, and 100 min. Describe some of the differences in appearance. What physical relations are revealed among different parts of the galaxy?

3. Make similar observations of other nearby galactic neighbors of the Milky Way.

CHAPTER

25

Galaxies and the Expanding Universe

WHAT THE READER SHOULD WATCH FOR IN THIS CHAPTER

The Milky Way is a single galaxy in a vast universe of galaxies. Astronomer Edwin Hubble launched the modern study of the universe by showing that galaxies are systems of stars remote from the Milky Way. Hubble's second major discovery was the expansion of the universe. Hubble observed that the light from galaxies is redshifted by an amount that increases with increasing distance from the Milky Way. Astronomers interpret the redshift as a sign that the universe is expanding. If we project the expansion back in time, we conclude that galaxies used to be much closer together and that the universe was much hotter and denser. As you will see, galaxies are not randomly distributed in space. Astronomers have traced enormous structures that involve thousands of galaxies and span millions of light-years. Gravity has sculpted these structures over billions of years. Certain galaxies have extraordinary events occurring in their nuclei. Active galactic nuclei are characterized by rapid motions of stars and gas, strong emission across the electromagnetic spectrum from radio waves to gamma rays, and evidence for a central compact object. The most distant and luminous examples of active galaxies are quasars. Astronomers believe that the extreme emission from active galaxies involves a supermassive black hole that is devouring gas and stars near the center of the galaxy. ■

The universe contains a rich variety of galaxies. As described in Chapter 24, we know a fair amount about the structure and morphology of nearby galaxies. But as we move out into the universe, galaxies become so distant that we lose sight of their detailed properties. Instead, we can use them as test particles or probes to learn more about the **large-scale structure** of the universe—in other words, the spatial distribution of clusters and superclusters of galaxies. (Astronomers speak of galaxies as "test particles," imagining them as part of a grand physics experiment. In terms of gravity, the entire effect of millions of stars can be reduced to that of a single particle.) Because of the vast distances involved, we introduce a new unit of distance, the **megaparsec** (abbreviated as Mpc), which is 1 million parsecs or 3×10^{19} km!

Exploring large distances in the universe has two profound consequences. First, we will find that distant galaxies are all rushing away from us. This observation leads to the concept of an **expanding universe.** Second, the light from distant galaxies has taken a very long time to reach us. It takes tens of thousands of years to cross the Milky Way, millions of years to span the distance between neighboring galaxies, and billions of years to reach us from the most distant galaxies. Thus we see those galaxies as they were when light left them. This idea is called **look-back time.**

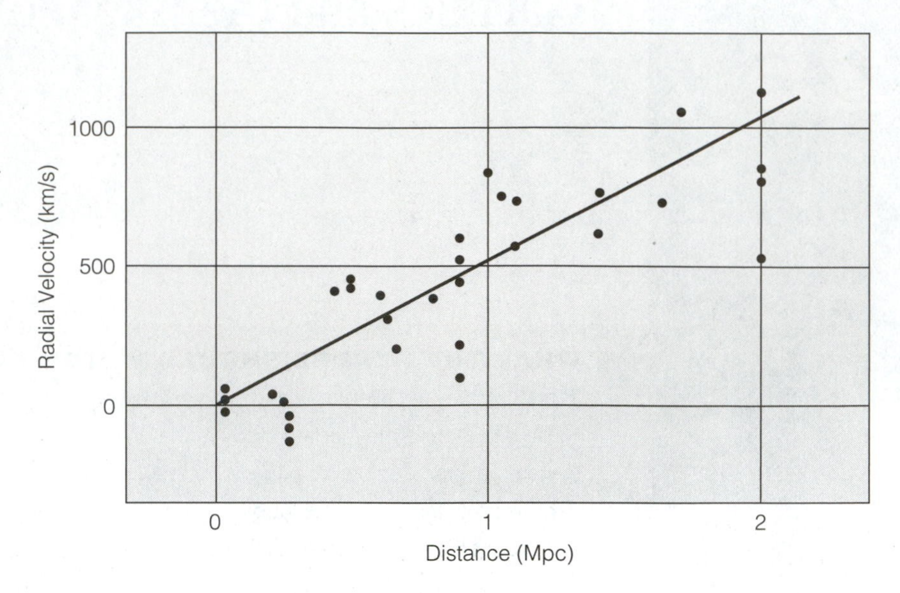

Figure 25-1 The linear relationship between radial velocity and distance of galaxies, discovered by Edwin Hubble in 1929. The slope of Hubble's original plot is steeper than the current value because of an error in the Cepheid calibration of the distance scale.

Modern physics has shown that a journey out into space is a journey back in time. Astronomers make this explicit by measuring distances in terms of the *time* it takes for light to travel across space. This is the same as saying that it is a 10-min drive to the mall or that a neighboring town is two hours away. Large telescopes can be used as armchair time machines (Eichler, 1995). We see distant galaxies and quasars as they were soon after their formation. Therefore, the study of the most distant objects brings us face to face with questions about the origin and large-scale structure of the universe itself.

INTERPRETING THE REDSHIFT

The Hubble Relation

After Hubble's discovery of galaxies outside the Milky Way, he began to measure radial velocities for the galaxies. Recall from the discussion of star motions in Chapter 16 that radial velocity is just the part of a star's motion that is along the line of sight. Spectroscopy can also be used to measure a radial velocity for an entire galaxy. By 1925, Hubble had accumulated radial velocities for 40 galaxies, and they were almost all positive. The velocities were also surprisingly large, thousands of kilometers per second. In other words, the galaxies are almost all speeding away from us, a phenomenon called the **redshift of galaxies.**

When Hubble compared the radial velocities for galaxies with his distance estimates, he found a clear correlation between the radial velocity—inferred from the redshifted spectral lines—and the distance. The relation is linear. The redshift, usually symbolized as z, is directly proportional to the distance (Figure 25-1). This fundamental result of extragalactic astronomy is called the **Hubble relation.**

Figure 25-2 shows the distances and redshifts of five galaxies. The most distant, more than 2 billion light-years away, is receding at over 20% of the velocity of light. These data extend the Hubble relation hundreds of times farther into space than Hubble's original work. Note, too, that the angular size of the galaxies decreases with increasing redshift, consistent with a larger estimated distance.

It is natural to interpret the redshifts as Doppler shifts. However, we should make a distinction between a *Doppler redshift,* which results from motion through a medium, and a *cosmological redshift,* which results from the expansion of the medium itself. This concept has been known as the expanding universe since 1933, when English astrophysicist Arthur Eddington gave his book on the subject that title.

How can we interpret the observation of galaxy redshifts? Figure 25-3 shows two different possibilities, to guide our thinking. In a static universe (Figure 25-3a), the galaxies would mill around randomly in space, separated by large distances. They would change position over time, but their typical separations would be constant. If this were true, on average we would observe equal numbers of redshifts and blueshifts, and a plot similar to the one made by Hubble would have points scattered around the line of zero velocity, as shown in Figure 25-3b. By contrast, in an expanding universe (Figure 25-3c), galaxies would recede with a velocity that increases with distance. This is the situation actually observed and plotted as the Hubble relation (Figure 25-3d).

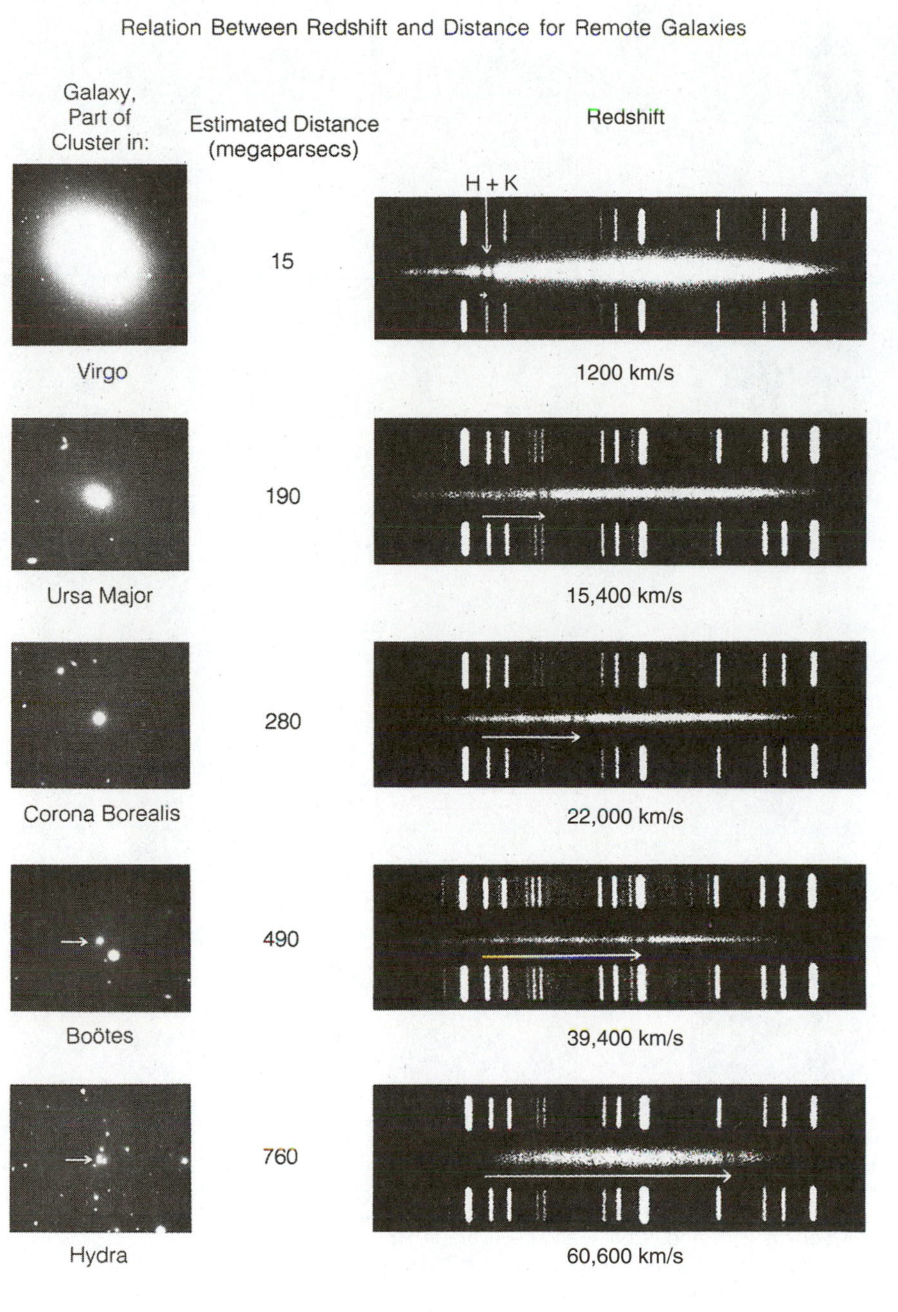

Figure 25-2 Photographic evidence of a relation between redshift and distance for remote galaxies. The left column shows galaxies. The right column shows spectra: the white lines at the top and bottom of each spectrum are emission lines produced in the instrument for comparison, being similar in each spectrum. A pair of dark absorption lines (the *H* and *K* lines of gaseous calcium) can be detected in each galaxy's spectrum, above the head of the white arrow. This pair is farther right (red) in each succeeding galaxy. The center column shows the inferred distance. (Hale Observatories)

This leads to an obvious question. If we see all galaxies moving away from us, does this mean we are at the center of the universe? Figure 25-4 shows a section of the expanding universe at three successive times. The distance between a galaxy and *every other galaxy* increases with time. In other words, there is no central position in this universe, because all galaxies are moving away from each other. An observer on a planet around a star in a distant galaxy would measure exactly the same Hubble relation that we do.

In our universe, the Hubble relation indicates the mutual recession of galaxies. If we "play the film backward" to imagine how our universe might have looked in the distant past, we see that the galaxies must have been much closer together. The implication is that there was a time in the distant past when all the mass in the universe was concentrated in a state of extremely high density. We appear to be riding out the aftermath of an ancient and vast explosion.

Are Redshifts Cosmological?

It is relatively simple to measure a redshift for a galaxy; only a spectrum is required. However, as

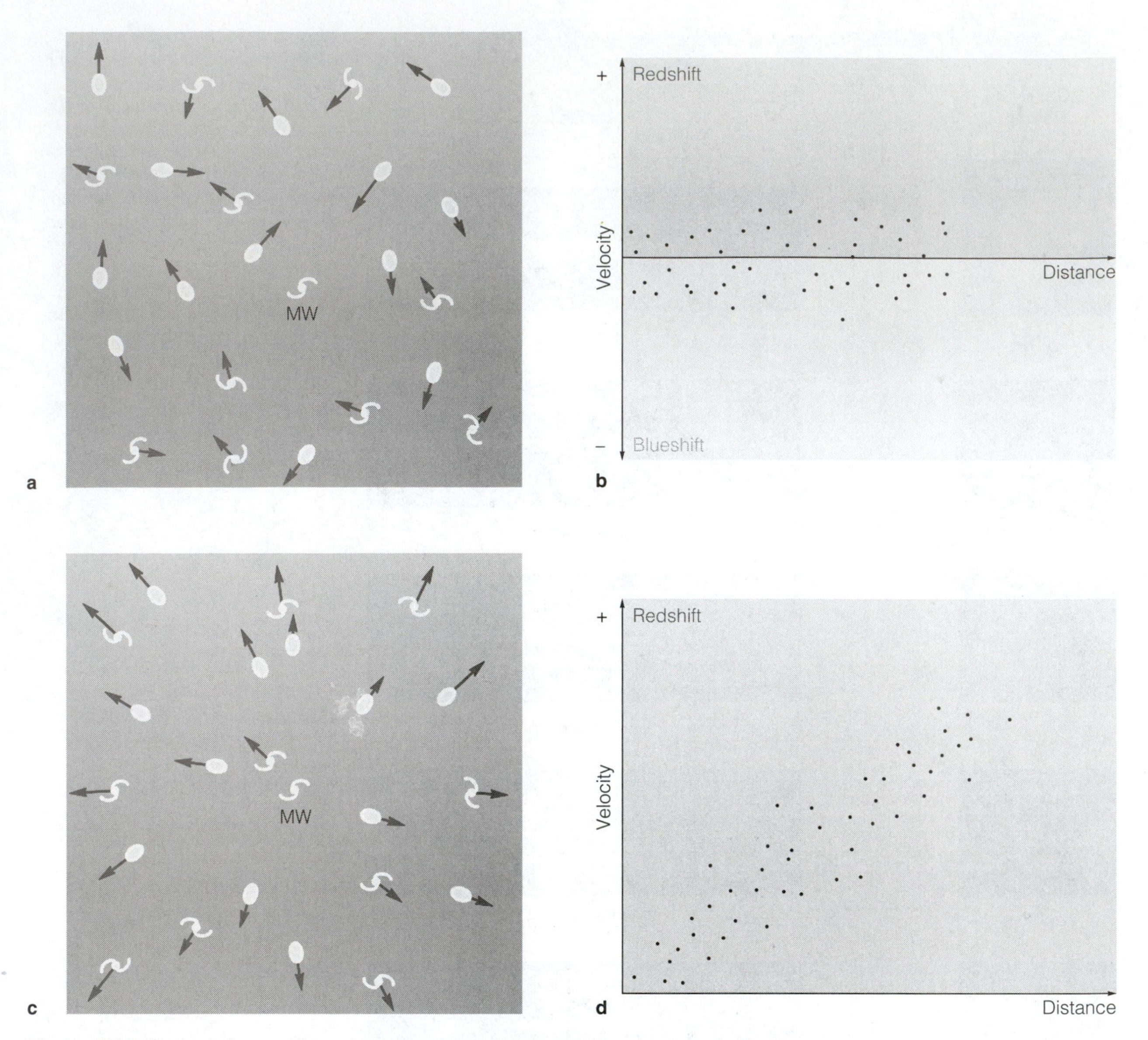

Figure 25-3 Static and expanding universes compared. **a** In a static universe, galaxies move randomly with respect to the Milky Way; the length of each arrow indicates the size of the velocity. **b** There is no correlation between distance and velocity, and the average velocity is zero. **c** In an expanding universe, galaxies move away from each other. **d** The recession velocity increases with increasing distance, the Hubble relation.

Chapter 24 showed, establishing a reliable distance can be very difficult. According to the Hubble relation, *redshift can be used as a distance indicator.* In other words, the Hubble relation allows us to measure the redshift for a galaxy and assign it a distance, without measuring the distance directly. So far we have assumed that the observed redshift should truly be interpreted as a cosmological redshift, caused by the expansion of space. The Doppler effect has certainly been confirmed with planets and stars. But the idea that the expansion of space explains *all* galaxy redshifts is an assumption, called the theory of **cosmological redshifts.** There are other possibilities.

Light escaping an intense gravitational field, such as the environs of a black hole, will suffer a gravitational redshift (see Chapter 19). Given what we know about galaxy masses, however, this effect is far too small to cause galaxy redshifts. Some researchers have proposed *tired light theories,* in which photons traveling toward us lose energy and get redshifted in traversing the vast distances of interstellar space. Such theories make few specific predictions and so are difficult to rule out.

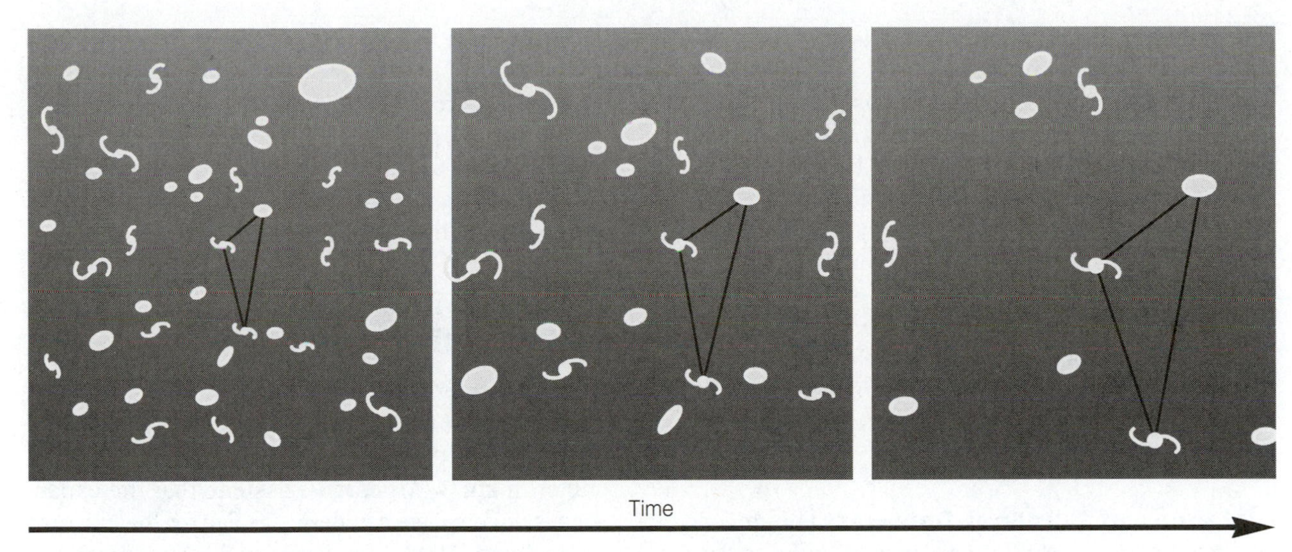

Figure 25-4 An expanding universe (in two dimensions) viewed at three successive times. The distance between any two galaxies increases with time. An observer on any galaxy would measure the same Hubble relation, and the expansion has no detectable center. Note that each galaxy remains the same size because it is held together by its own internal gravity.

★ The argument that some galaxies have noncosmological redshifts is based on certain curious situations where a galaxy with a large redshift appears to be close on the sky to a galaxy at a much lower redshift (see our web site). Under the cosmological assumption, they should be at very different distances. Many of these unusual associations have been found by Halton Arp at the European Southern Observatory. Apparent associations are intriguing, but there are no reliable statistics about their occurrence. Some objects with very different redshifts will be projected close together on the sky purely by chance. Redshift is correlated with the apparent brightness and angular size of galaxies, both of which are indicators of distance. The great bulk of evidence favors the cosmological interpretation.

The Hubble Parameter

Hubble's relation expresses how the recession velocity increases with distance. The slope of the plot, the ratio of velocity to distance, is known as the **Hubble parameter,** *H*. The Hubble parameter sets the rate of the expansion and thus the size and age of the universe. *It is the single most important number in extragalactic astronomy.* In most cosmological models, the rate of the expansion changes with time, so astronomers use a separate symbol, H_0, to represent the current rate of expansion.

How do astronomers estimate the Hubble parameter? First, they take the spectra of many galaxies to measure the redshifts. Then, they derive the distance for *each* galaxy. For galaxies, a redshift is generally much easier to measure than a distance. Finally, the redshift and distance of each galaxy can be plotted. The slope of the line that gives the best fit to the data is H_0. The individual stellar types used as distance indicators within the Local Group (see Chapter 24) cannot be used for the enormous distances to the most remote galaxies, for two reasons. First, they are not luminous enough to be seen at large distances. Second, the individual stars of a distant galaxy cannot be spatially resolved. Astronomers therefore use the *properties of entire galaxies* as distance indicators.

Which properties should we use? We are looking for a property that will indicate either a well-defined luminosity, often called a *standard candle,* or a well-defined diameter, often called a **standard measuring rod.** Because galaxies cover an enormous range in luminosity (Table 24-2), the apparent brightness of a galaxy gives very little clue to its distance. Similarly, galaxies range in size from 1-kpc dwarfs to 100-kpc giants, so the apparent diameter of a galaxy gives very little clue to its distance. As it turns out, astronomers have developed different techniques for measuring the distances to spiral and elliptical galaxies.

Distances of Spiral Galaxies The *Tully-Fisher relation* is based on the discovery that more massive spiral galaxies are more luminous. The mass of the galaxy is not measured directly, but is inferred from the rotational speed of the gas disk. From its mass comes an estimate of the galaxy's absolute luminosity. The apparent brightness is then observed, and the distance is calculated using the inverse square law. The Tully-Fisher relation is an excellent distance indicator, with a scatter of only 10–15%. However, it works best when used on galaxies with intermediate inclinations to the line of sight. Face-on spirals show *no* Doppler motions due to rotation, because the disk motion is on the plane of the sky, and the total brightnesses of edge-on spirals are difficult to measure due to the obscuring effects of dust in the disk.

Distances of Elliptical Galaxies The *Faber-Jackson relation* for ellipticals is analogous to the Tully-Fisher relation for spirals. It relates the range of stellar velocities, or the *velocity dispersion,* to the size of the galaxy. In this case, the velocity dispersion gives an estimate of the galaxy's size at a certain brightness level. This size is correlated with luminosity. The apparent brightness is observed, and once again the distance is calculated using the inverse square law.

Best Estimate of the Hubble Parameter The key in all methods of distance determination is to have an accurate calibration at small distances. The calibration sets the luminosity of each new standard candle—all of the uncertainty in H_0 comes from this procedure. Once the calibration is known, *relative distances* are easy to calculate accurately using the inverse square law. For example, a standard candle that is four times fainter is twice as far away.

Figure 25-5 shows the overlapping ladders of the distance scale, extending the indicators beyond the Local Group (compare with Figure 24-2). At the distance of the Virgo cluster and beyond, the accuracy of H_0 is about 20–30%. Remember, though, that redshift is a measure of velocity, not distance. It is useful as a distance indicator only in the context of a cosmological model. There are several crucial benchmarks in the extragalactic distance scale. One is the Large Magellanic Cloud, which has an accurate geometric distance from measurements of the dust shell of Supernova 1987A and contains a "stellar zoo" for the calibration of Cepheid and RR Lyrae variables. Another is the Virgo cluster, about 15 Mpc distant. The *Hubble Space Telescope (HST)* has been used to measure Cepheids in Virgo cluster galaxies, allowing the calibration of the Tully-Fisher method for even greater distances.

Astronomers have begun to reach a consensus as to the value of the Hubble parameter. After a vast observational effort over the past 25 y, various distance indicators all point to a Hubble parameter in the range of H_0 = 60 km/s/Mpc to H_0 = 85 km/s/Mpc. We will use an intermediate value of 75 km/s/Mpc for the rest of this book.

The Size and Age of the Universe

The Hubble relation states that $v = H_0 d$, where v is the velocity of the galaxy in km/s, d is the distance in Mpc, H_0 is the current value of the Hubble parameter in km/s/Mpc. If we assume that the expansion is uniform, we can derive an age for the universe (Freedman, 1992). Essentially, this measures the time in the past when all galaxies were together in space. Since the expansion is actually slowing down, this estimate is the ***maximum*** age of the universe. At a constant velocity, distance traveled is equal to the velocity multiplied by the time. Equivalently, the travel time (T) is the distance divided by the velocity. In terms of the Hubble relation, this means that $T = 1/H_0$.

We can now understand the implications of low and high estimates of the Hubble parameter. Assume H_0 = 50 km/s/Mpc. This means that for each megaparsec of distance from the observer, the recession velocity increases by 50 km/s. This value represents a slow expansion, a long distance scale, and a large universe. Tracing the recession backward in time, the inferred age of the universe is 20 billion years. At the other extreme, assume that H_0 = 100 km/s/Mpc, in which case the recession velocity increases by 100 km/s for each megaparsec of distance. This value represents a rapid expansion, a short distance scale, and a smaller universe with an inferred age of 10 billion years. These age estimates are crude because, as we shall see in Chapter 26, the expansion was more rapid at early epochs (Kinney, 1996).

The difference between these two cases gives a good idea of the level of consensus among researchers on the distance scale and the age of the universe. An uncertainty in the Hubble parameter of 60–85 km/s/Mpc corresponds to an age range of 12–17 billion years—the low Hubble parameter gives the high age, and the high Hubble parameter gives the low age. This range is an uncertainty of about 30% in the expansion rate or age of the universe. A massive observational effort will be required to bring the uncertainty below 10%. For more on the Hubble relation and the age of the universe, see Optional Basic Equation IX.

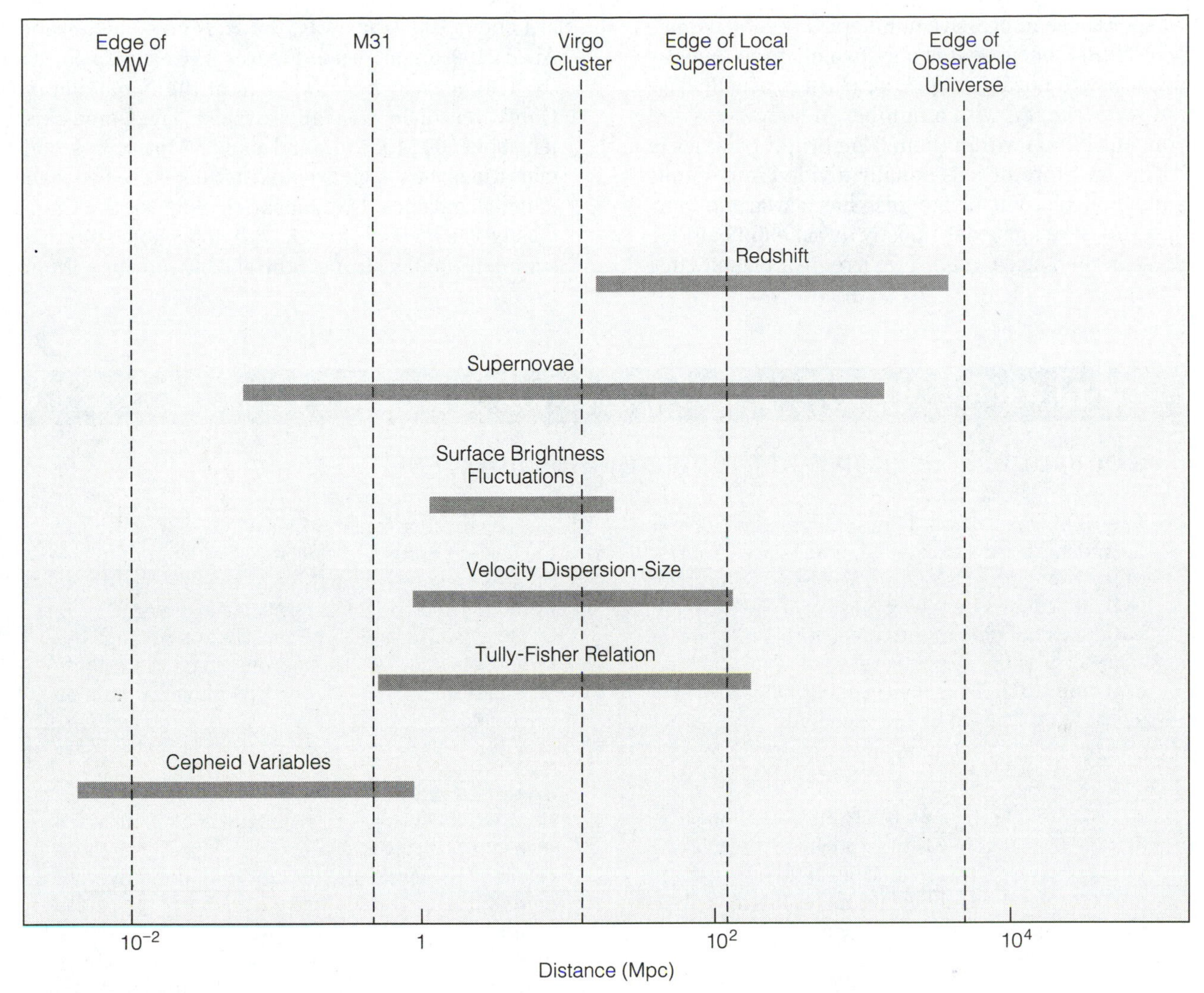

Figure 25-5 The ladder of distance indicators used to measure extragalactic distances, with the horizontal bar showing the range of applicability. Many of these indicators use the properties of an entire galaxy as a standard candle. Redshift is a useful distance indicator only if we assume a particular model for the expanding universe. Moreover, at distances less than 20 Mpc, redshift is a poor indicator of distance because of the gravitational interactions of galaxies in the Local Supercluster.

LARGE-SCALE STRUCTURE

Galaxies are not scattered randomly throughout the universe, but instead concentrate into **clusters of galaxies.** As Chapter 24 describes, we live in a modest cluster called the *Local Group,* which includes three moderate-size spiral galaxies, each with a few small companions, mostly dwarf ellipticals. These galaxies are loosely sprinkled over a region about 1 Mpc across. Beyond the Local Group, we encounter the Sculptor and M 81 groups at 2–3 Mpc. Each is like the Local Group, with two or three large galaxies and a number of smaller ones. At a distance of 5–6 Mpc, we find a loose cluster centered on the giant spiral galaxy M 101, the Pinwheel. Finally, at the edge of our cosmic neighborhood about 15 Mpc away, we encounter the Virgo cluster, which is the nearest large cluster of galaxies (West, 1997).

Clusters of Galaxies

The Virgo cluster is a sprawling mass of hundreds of galaxies, with three giant ellipticals near its center—M 84, M 86, and M 87. It is about 15 Mpc away and 1.5 Mpc across (half again as large as the Local Group). The dense part of the Virgo cluster covers 6° on the sky, or 12 times the diameter of the Moon.

Despite these impressive numbers, it is considered a *poor cluster,* because it is only two or three times as dense as the Local Group. Poor clusters usually have *irregular shapes,* with a number of subgroups and concentrations within them. The bright galaxies in Virgo are more or less equally divided into spirals and ellipticals. The cluster also has a swarm of hundreds of dwarf galaxies, mostly dwarf ellipticals like those in the Local Group. Two irregular clusters that are about 100 Mpc away, the Hercules cluster and Abell 2199, are shown in Figures 25-6a and 25-6b.

The nearest *rich cluster* is in the constellation Coma Berenices (Berenice's Hair). The Coma cluster, about 100 Mpc away and about 8 Mpc across, contains thousands of galaxies. At its center are two giant elliptical galaxies. Like most rich clusters, the Coma cluster is a *regular* cluster, with noticeable spherical symmetry and a strong central concentration. Most

OPTIONAL BASIC EQUATION IX

The Hubble Relation and the Age of the Universe

The Hubble relation is a linear correlation between the redshifts of galaxies and their distances from the Milky Way. The redshifts of galaxies are interpreted as being caused by the expansion of the universe, but the calculation of the wavelength shift has the same form as the Doppler effect (see Optional Basic Equation VIII). Thus the redshift of a galaxy, symbolized as z, is given by the equation

$$z = \frac{\Delta\gamma}{\gamma} = \frac{v}{c}$$

As described in the text, the Hubble relation states that $v = H_0 d$, where v is the velocity of the galaxy in km/s, d is the distance in Mpc, and H_0 is the current value of the Hubble parameter in km/s/Mpc. Combining these two results relates the distance of a galaxy to its redshift:

$$d = \frac{zc}{H_0}$$

This equation gives a reliable measure of distance only in regions of the universe governed by smooth Hubble flow (Hubble flow is the component of a galaxy's motion that is caused by the smooth expansion of the universe). In the vicinity of mass concentrations like clusters, galaxies will have peculiar velocities caused by gravitational interactions (peculiar velocity refers to the component of a galaxy's motion that is caused by gravitational interaction, and not by the smooth expansion of the universe). When the peculiar velocity is a significant fraction of the total velocity, the Hubble relation does not give a reliable measure of distance. In the vicinity of the Milky way, redshifts in excess of 5000 km/s give good approximations to Hubble flow.

As we saw in the text, if we assume that the expansion is uniform, we can derive an age for the universe, using the following equation:

$$T = \frac{1}{H_0}$$

The maximum age comes out in years if we use the correct units. This means we must multiply by 3×10^{19} to convert from Mpc to km and then divide by 3×10^{7} to convert from seconds to years.

The two sample problems below present different sets of observations from which the Hubble parameter and the age of the universe can be calculated.

Sample Problem 1 A bright galaxy has a recession velocity of 7000 km/s. An independent measure of the distance gives a value of 115 Mpc. Given these values, what is the maximum age of the universe? *Solution:* The Hubble parameter based on these numbers is v/d = 60 km/s/Mpc. In other words, there is 60 km/s of cosmological redshift for every megaparsec in distance. The maximum age of the universe in this case is

$$T = \frac{1}{H_0} = \frac{1}{(60\ \text{km/s/Mpc})} \times \frac{(3 \times 10^{19}\ \text{km/Mpc})}{(3 \times 10^{7}\ \text{s/y})}$$

$$= 16.7\ \text{billion y}$$

Sample Problem 2 A cluster of galaxies has a mean redshift of 12,500 km/s. Distance estimates from Fisher-Tully observations of the brightest spirals in the cluster give a distance of 139 Mpc. Given these values, what is the maximum age of the universe? *Solution:* The Hubble parameter based on these numbers is v/d = 90 km/s/Mpc, or 90 km/s of cosmological redshift for every megaparsec in distance. The maximum age of the universe in this case is

$$T = \frac{1}{H_0} = \frac{1}{(90\ \text{km/s/Mpc})} \times \frac{(3 \times 10^{19}\ \text{km/Mpc})}{(3 \times 10^{7}\ \text{s/y})}$$

$$= 11.1\ \text{billion y}$$

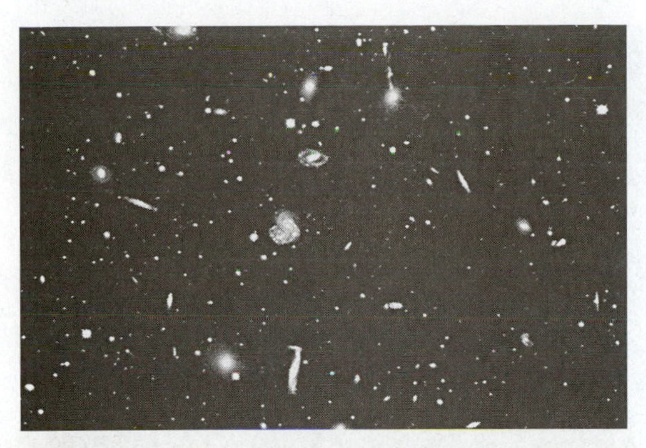

a

b

Figure 25-6 Two galaxy clusters. **a** An image of the irregular Hercules cluster, which has a core membership of a few hundred galaxies and no dominant central galaxy. (N. Sharp/NOAO) **b** An image through a red filter of the irregular cluster Abell 2199. The dominant giant elliptical galaxy has multiple nuclei. (W. Harris and N. Sharp/NOAO)

of its galaxies are ellipticals, but it has a moderate number of S0 galaxies, which are like spirals but without the spiral arms and interstellar matter. Only the most luminous galaxies in a cluster appear in most of the images made by astronomers. We are seeing the "tip of the iceberg" of a much larger population of (mostly dwarf) galaxies concentrated in space.

Astronomers have discovered that the relative number of galaxies of different Hubble types in a region depends on the *density* of the environment. This **morphology-density relation** is a key to understanding how galaxies interact and evolve. Let's consider the relative fraction of spiral, S0, and elliptical galaxies in regions that cover a range of 1 million in terms of galaxy space density. Below a density of about five galaxies per cubic megaparsec, the spiral fraction is 60–70% and independent of density. Above this density, the spiral fraction drops steadily, until in the dense cores of the richest clusters (with thousands of galaxies per cubic megaparsec), virtually all the galaxies are S0s or ellipticals.

The morphology-density relation quantifies the observation in Chapter 24 that spirals inhabit low-density regions and ellipticals inhabit high-density regions. It can be understood in terms of the *time* it takes for galaxies to interact in different environments. Astronomers define an interaction as an encounter or close passage where the force of gravity leaves measurable effects on one or both of the galaxies. Below a density of a few galaxies per cubic megaparsec, this time is longer than the age of the universe. The mostly spiral population in low-density regions has never had much interaction. At the density of a rich cluster, however, the time scale for interaction is "only" a few hundred million years. There has been time for dozens of interactions in the age of the universe. Presumably, the low fraction of spiral galaxies in a cluster results from collisions and mergers that strip away the gas from gas-rich galaxies.

Superclusters and Voids

Gravity has a long reach and can cause galaxies to cluster on an enormous scale. Nearly 40 y ago, George Abell cataloged over 2700 of the richest clusters of galaxies in the northern sky. Astronomers naturally speculated that even larger structures might exist. Then, Gerard de Vaucouleurs suggested that the Milky Way is in fact part of an enormous flattened **supercluster,** a cluster of clusters of galaxies. Much earlier, Clyde Tombaugh, the discoverer of Pluto, had made the first map of a supercluster. He showed it to Edwin Hubble, who refused to believe the observation—no doubt because no one at that time expected to see such large structures!

Placing galaxies accurately in space requires a measure of distance. But even without such distance

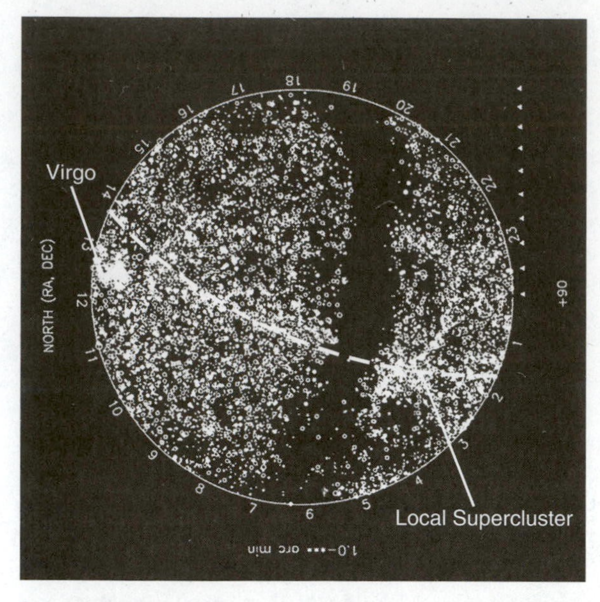

a

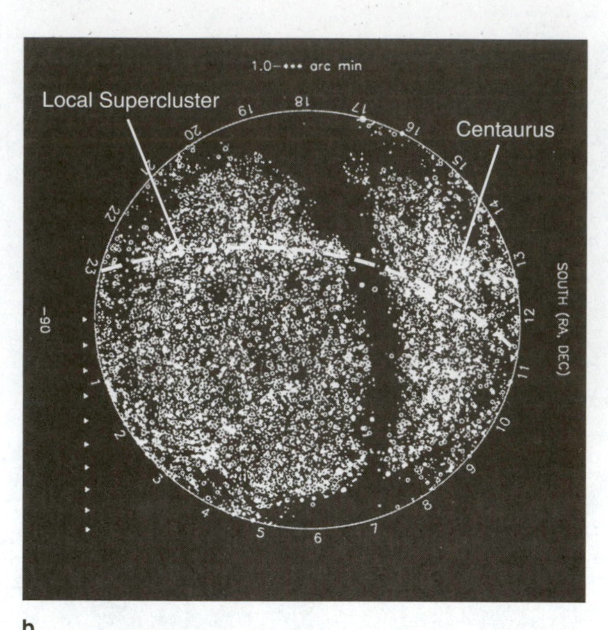

b

Figure 25-7 (See the essay on astronomical coordinates at the end of the book.) Equal-area projections of the northern and southern sky, in which each point is a galaxy from large photographic surveys. Right ascension (RA) is marked around the circumference. Declination (DEC) is the radial coordinate with the pole at the center and 10° spacings given by the triangular tick marks. The blank bands mark the zone of obscuration by the Milky Way. **a** The northern sky. **b** The southern sky. (O. Lahav and D. Lynden-Bell, Institute for Astronomy, Cambridge)

measures, we can see some of the *large-scale structure* of the universe. Equal-area projections of the brightest galaxies in the northern and southern sky out to a distance of about 50 Mpc are shown in Figure 25-7. The blank region is the zone of obscuration by the Milky Way; distant galaxies in this plane cannot be seen due to the veil of gas and dust in our own galaxy. Various bright clusters show up as denser regions of galaxies. There is also a high concentration of galaxies running in a strip across the sky, nearly at right angles to the Milky Way. This flattened structure of galaxies, analogous to the flattened disk of stars in the Milky Way, is called the **Local Supercluster.**

★ The Local Supercluster contains the Local Group, the Virgo and Coma clusters, and about 100 other clusters. It measures about 20 Mpc across by 2 Mpc thick and contains a total of $10^{16}\ M_{\odot}$. The gravitational center of the Local Supercluster is near the Virgo cluster; the Milky Way is near the outskirts. The most remarkable aspect of the Local Supercluster is that so much of it is completely empty of luminous matter. As many as 98% of the galaxies occupy only 5% of the volume. Brent Tully at the University of Hawaii has found possibly the largest structure in the universe, called the Pisces-Cetus supercluster (see our web site). Over 100 rich clusters of galaxies are grouped in a huge flattened structure that is 400 Mpc long and 60 Mpc across!

The map in Figure 25-8 shows galaxies 10 times as faint as the map in Figure 25-7. It reaches to a typical distance of 150 Mpc (nearly 500 million light-years), but it covers only 10% of the sky. It is obvious that the galaxies are not uniformly distributed. There are knots and filaments of galaxies and regions with very few galaxies. However, maps like these cannot be used to give a true sense of galaxies in space. Galaxies that appear projected close to each other on the plane of the sky may not be close in space. All the structures at different distances are collapsed into one plane. To fully understand the complex patterns, galaxies must be positioned in three dimensions. This requires distance estimates.

Slices of the Universe The most detailed information on large-scale structure has come from painstaking redshift surveys of galaxies in narrow "slices" of the universe. The first astronomers to make three-dimensional maps of the positions of large numbers of galaxies were Margaret Geller, John Huchra, and their collaborators at the Harvard-Smithsonian Center for Astrophysics. They placed

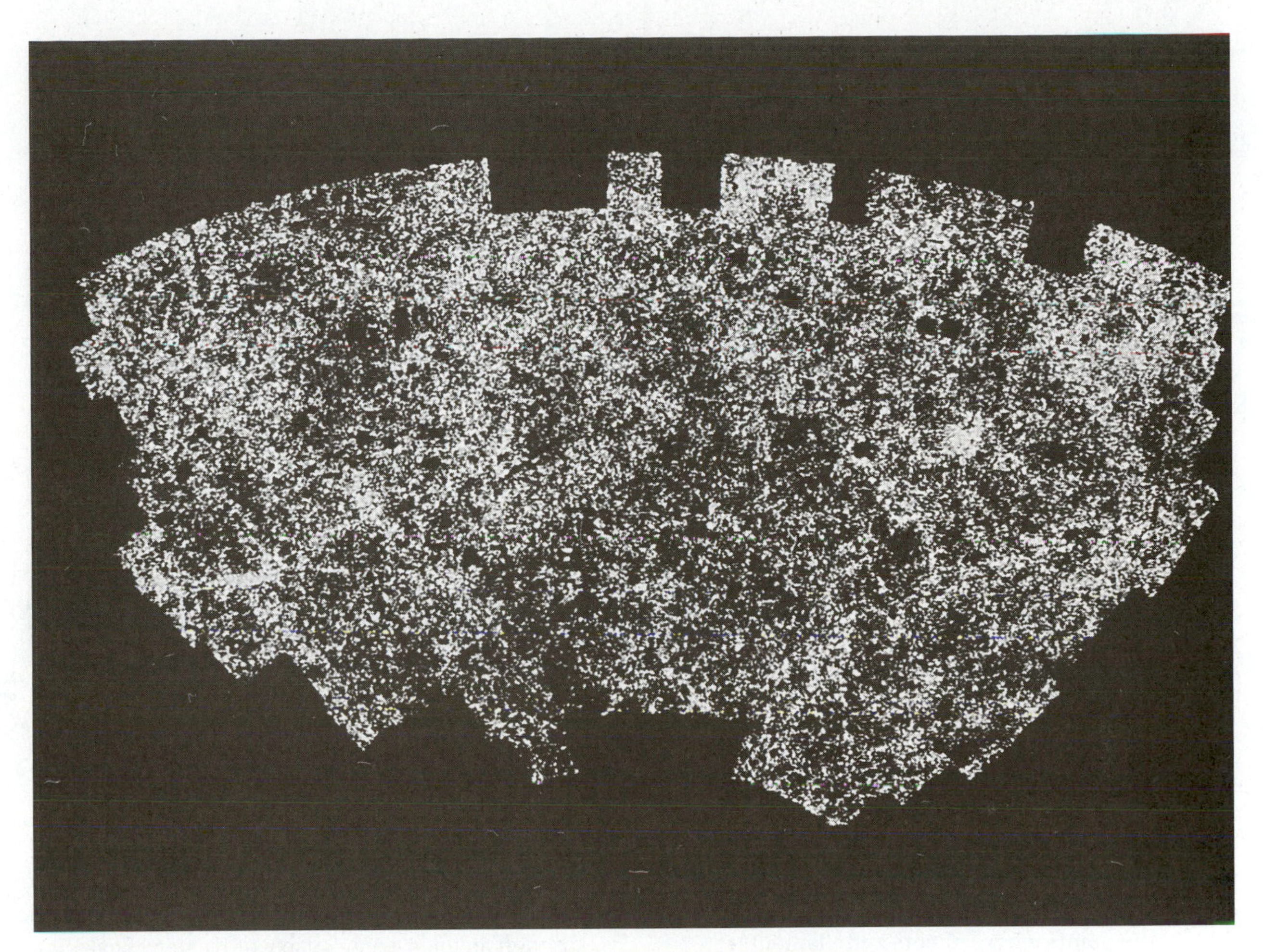

Figure 25-8 A map of the distribution of 2 million galaxies covering about 10% of the sky. The brightness of each dot represents the number of galaxies in each patch of sky. Black is empty, gray is a number between 1 and 19, and white is 20 or more. A complex pattern of clusters, filaments, and voids can be seen. A statistical analysis shows that on scales larger than 100 Mpc, the galaxy distribution is essentially smooth. (S. Maddox, W. Sutherland, G. Efstathiou, and J. Loveday, Oxford Astrophysics)

galaxies in space according to their redshift. In other words, rather than try the almost impossible task of measuring distances for thousands of galaxies, they used *recession velocity as a distance indicator.* Recall that the Hubble relation relates velocity (or redshift) and distance.

Figure 25-9 shows the results from the first slice of the Harvard survey (Geller and Huchra, 1991). A series of separated slices can be used to reconstruct the distribution of galaxies over a three-dimensional volume. In three dimensions, it can be seen that the **voids** are nearly circular in cross section and are truly empty of bright galaxies. The Coma cluster, the bright concentration near the center of the wedge, is elongated vertically due to the large velocity dispersion of the cluster environment. All clusters show these elongations in redshift space toward and away from the observer. These early surveys measured only a few thousand galaxies. More recent surveys have measured redshifts for over 200,000 galaxies.

Geometry of Large-Scale Structure Surveys of galaxies are revealing the architecture of the cosmos, patterns on the largest scales that have been sculpted by gravity over billions of years. To avoid being buried in the details, astronomers have resorted to colorful language to describe the structures they see. Historically, it was thought that the rich clusters were set in a uniform sea of galaxies, like "meatballs." The discovery of strings and filaments of galaxies added a component of "spaghetti." The mapping of more slices has shown that the

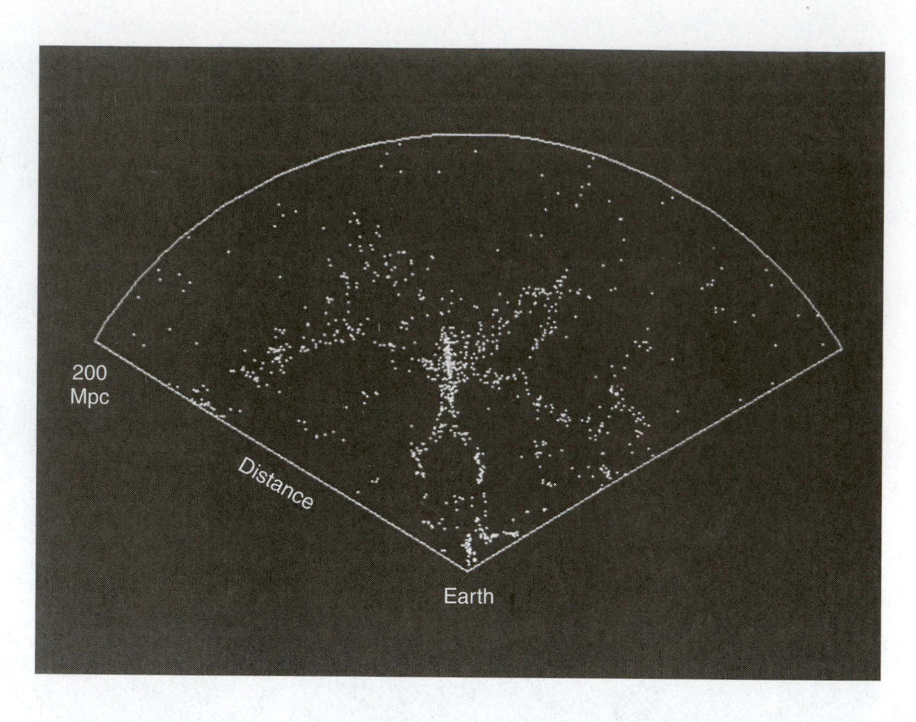

Figure 25-9 A three-dimensional "slice of the universe." Redshifts were measured for about 1100 galaxies in a thin wedge stretching 120° across the sky. The Earth is at the apex of the wedge, and its far edge is 200 Mpc away. The galaxies form strikingly thin sheets; regions between the sheets are nearly devoid of luminous matter. Surveys of adjacent wedges have shown that these structures are genuine "sheets" rather than "strings." The pattern is reminiscent of a cross section through a foam of soap bubbles. (M. Geller, J. Huchra, M. Kurtz, and V. de Lapparent, Harvard-Smithsonian Center for Astrophysics)

galaxies form large connecting structures around empty voids 20–100 Mpc across. This has naturally been referred to as "Swiss cheese."

Perhaps the most accurate way to think of the geometry traced out by galaxies in space is as a froth of soap bubbles. Galaxies lie in sheets and filaments, with large regions in between nearly devoid of galaxies. At the points where sheets and filaments intersect, we find the great clusters. These frivolous terms should not distract us from the important issues. The study of large-scale structure has implications for how galaxies form, for how the dark matter is distributed, and for the physical conditions in the early universe.

Dark Matter in the Universe

In Chapter 24, we saw that only about 10% of a galaxy's mass emits radiation at any wavelength. The part of a galaxy that emits visible light is embedded in a dark matter halo. On larger scales, we find additional evidence for the existence of dark matter, particularly in rich clusters with a high central concentration of galaxies. There are two types of evidence.

Back in the 1930s, Caltech astronomer Fritz Zwicky observed that the *velocity dispersion* (spread or scatter in velocities) of galaxies in rich clusters was too high to be accounted for by the visible matter. The number is typically 1000 km/s. If the only mass in the cluster was that represented by the galaxies, the galaxies would fly apart—the cluster could not survive! The implied mass/luminosity ratio is in the range of 300–500, which is five or six times larger than the mass/luminosity ratio of individual galaxies determined dynamically (see Chapter 24). The dark matter in clusters is *in addition to* the dark halos surrounding the galaxies themselves.

★ A second probe of dark matter on a large scale involves **gravitational lensing.** One of the basic predictions of general relativity is that mass can bend light. When light from a distant galaxy passes through an intervening galaxy or cluster of galaxies, it is deflected by curved space in the vicinity of the cluster. Just like an optical lens, a gravitational lens can also magnify or demagnify an image (corresponding to the amplification and de-amplification of the signal). Go to our web site to see the geometry that leads to lensing and a computer simulation of gravitational lensing using an actual CCD image of the galaxy NGC 3992. The simulation shows what effect variations in alignment and in the symmetry of the intervening cluster will have.

★ In addition to distorting the light of normal galaxies, gravitational lensing can also distort the light of active galaxies and quasars; these distant objects will be discussed in the next section. If the alignment of the distant object and the foreground galaxy is perfect, the light from the distant object is turned into a ring. If the alignment is not perfect, the ring becomes a segment of a circle, or an arc. If the lensing object is an elliptical galaxy, the ring is split into two or four distinct images. For examples of these "mirages," go to our web site. The existence of gravitational lensing is a wonderful confirmation of Einstein's theory of relativity and the idea that mass can bend light.

Note that gravitational lensing of light can be caused by an intervening object of *any mass.* Also, galaxies and clusters do not have to be used as probes for dark matter. The lensing effect can be observed even when the intervening mass is *completely dark.* When viewed on the largest scales, the universe is overwhelmingly composed of dark matter. On a galactic scale, about 90% of the mass is dark; on the scale of clusters of galaxies, the fraction reaches 95–99% (Trefil, 1988). Some of the dark matter may be brown dwarfs or free-floating planets, but at least part of it is likely to be exotic particles of a variety not yet discovered in laboratories and accelerators on Earth. Even if we do not know what the dark matter is, the large-scale structure of the universe gives us an idea of some of its properties.

Active Galaxies and Quasars

Almost everything we know about galaxies can be understood in terms of *stellar processes* (ignoring for a moment the issue of dark matter). Galaxies are made of stars, and the gas and dust from which stars form. Galaxy morphology can be understood in terms of stellar orbits and the gross features of stellar birth and death. Galaxy spectra are just the sums of the spectra of billions of individual stars. However, the placid exterior of some galaxies can conceal events of great violence. A fraction of all galaxies harbor *nonstellar processes* in their nuclear regions. These processes can take several forms: torrents of radio or X-ray emission, high gas velocities, the implied presence of a massive black hole. Galaxies with any of these characteristics are referred to as **active galaxies.**

Our own galaxy has an active nucleus. In Chapter 23, we saw that the central parsec contains a compact radio source, a region of intense star formation, and a large quantity of fast-moving ionized gas. There is strong evidence for a massive black hole at the galactic center. It turns out that the activity in our galaxy is quite modest. The nuclei of some distant active galaxies dwarf the starlight in the surrounding galaxies so that they appear to be stellar objects. These extraordinary beacons, called **quasars,** can be seen across the universe. Quasars are the outer sentinels of the observable universe, and their discovery has transformed extragalactic astronomy.

Nonthermal Radiation

One of the key indicators of activity in galaxies is **nonthermal radiation.** Most astronomical objects emit *thermal radiation,* but recall from Chapter 5 that thermal radiation is caused by constant collisions among the particles of a gas (or other state of matter). The *average* amount of energy for a particle is set by the *temperature* of the gas, and the wavelength of the peak emission is governed by the temperature of the gas as well. Thermal radiation is the hallmark of a system that is in equilibrium, where the energy gained and lost by particles is always in balance.

For nonthermal radiation, the system is not in equilibrium because there is a source of energy for the gas. Synchrotron radiation, an important type of nonthermal radiation, depends on the presence of a magnetic field. We know that a magnetic field makes charged particles move in curved trajectories, which accelerates them. The accelerated charged particles emit electromagnetic radiation. This is how a radio transmitter works; electrons racing up and down a wire generate a radio wave. In a hot gas of hydrogen threaded by a magnetic field, as electrons spiral around the magnetic lines of force at nearly the speed of light, they emit nonthermal radiation. The spectrum of synchrotron radiation (or any other nonthermal radiation) is quite different from the spectrum of thermal radiation. Since the electrons are not in equilibrium, they have no *average* energy, so there is no corresponding peak in the spectrum. Instead, the spectrum is generally smooth and flat, extending over a broad range of wavelengths. When the electrons are very energetic, the spectrum can extend to X rays or even gamma rays.

Active Galaxies

In 1908, Edward Fath discovered intense emission lines coming from the central regions of the bright galaxy NGC 1068. V. M. Slipher and Edwin Hubble discovered other galaxies with similar lines. By the 1940s, Carl Seyfert had studied active galaxies in detail and noted their common features. Active galaxies are characterized by *one or more* of the following features: a bright compact nucleus, strong and very broad emission lines, intense radio emission, and a peculiar morphology.

Seyfert Galaxies Spectroscopy of a normal spiral galaxy will show emission lines coming from hot gas in the spiral disk. Each line will be Doppler-broadened by an amount corresponding to the amount of rotation in the disk. As can be seen from the rotation curves in Figure 24-8, the total width is a few hundred kilometers per second. Essentially, emission from gas at each different velocity has a slightly different redshift, and when the contributions are added together, the result is a smeared-out emission line.

In **Seyfert galaxies,** which are often gas-rich spirals with bright bluish nuclei, the emission lines are much broader, indicating a velocity of thousands of kilometers per second. Gas at such a high velocity could not be gravitationally bound; it would fly away from the center. There are two possibilities. Either the gas is actually being ejected from the nucleus of the galaxy, or a dark massive object in the nucleus is keeping the gas bound at that high velocity. In either case, something unusual is going on in the nuclear regions. Spectroscopy is often required to reveal the activity; although many Seyfert galaxies have bright starlike nuclei, the images of others appear quite normal.

Radio Galaxies About 1% of all galaxies (and 10% of all active galaxies) show extraordinary levels of radio emission. The first radio telescope was built in 1936 by Grote Reber, an amateur astronomer (see Chapter 23). For nearly 10 years, Reber was the only radio astronomer in the world! By 1944, he had detected strong radio sources in the constellations of Sagittarius, Cassiopeia, and Cygnus. The Sagittarius radio source corresponded to the Galactic Center, and the Cassiopeia source to a supernova remnant, but the position of the Cygnus source could not be specified accurately until 1951, when Walter Baade and Rudolf Minkowski located a faint, distorted-looking galaxy at the radio position. Cygnus A, the brightest radio source in the constellation of Cygnus, was the first known **radio galaxy**. Its radio luminosity is 10 million times that of the Milky Way, and it has a flat nonthermal radio spectrum. Radio galaxies are generally ellipticals.

A radio image of Cygnus A shows the main features of a typical radio galaxy (Figure 25-10). The bright radio *nucleus* at the center corresponds exactly to the position of the optical galaxy. A pair of **radio jets,** only dimly visible on one side, join the nucleus to two enormous radio *lobes.* The lobes have a complex structure, with large regions of wispy emission and intense emission at the outer edges. Baade and Minkowski measured the redshift of Cygnus A to be 17,000 km/s, nearly 6% of the velocity of light. The galaxy is therefore 230 Mpc or 750 million light-years away. Yet it emits radio waves strong enough to be detected by amateur astronomers with backyard equipment!

★ The elliptical galaxy M 87 in Virgo is another example of a strong nearby radio galaxy; it has a core and a one-sided jet but no extended lobes. There is also good dynamical evidence for a central mass concentration of $3 \times 10^9\ M_\odot$ in the nucleus of M 87. Recent HST observations have detected a gas disk near the nucleus and have made it almost certain that the galaxy contains a supermassive black hole. Images of M 87 can be found on our web site.

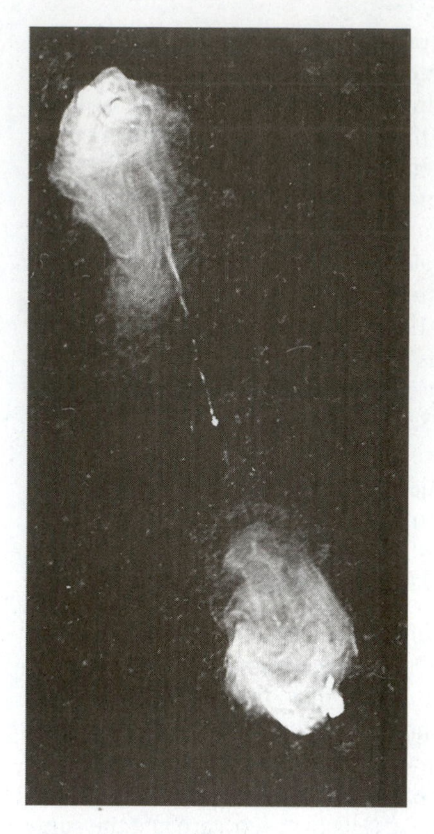

Figure 25-10 High-resolution radio image of the giant elliptical galaxy 3C 405, also known as Cygnus A. The jets of hot, fast-moving particles shooting out from the bright central nucleus expand into large, wispy radio lobes extending tens of kiloparsecs from the galaxy. (National Radio Astronomy Observatory, courtesy R. Perley, J. Dreher, and J. Cohan)

Peculiar Galaxies Besides broad emission lines and nonthermal radio emission, active galaxies often have a *peculiar morphology.* The classic example is the irregular galaxy M 82, which harbors large amounts of nuclear gas and dust and has chaotic filaments of excited gas streaming out from the nucleus (Veilleux, Cecil, and Bland-Hawthorne, 1996).

★ Studies of galaxy morphology have been advanced by the capabilities of the HST. For example, its high-resolution imaging shows that the Seyfert galaxy NGC 1275, a peculiarly shaped elliptical in the Perseus cluster, has young and blue globular clusters. Their formation is apparently connected with the nuclear activity. A close-up look at the peculiar galaxy Arp 220 reveals gigantic young star clusters near the nucleus. The furious rate of star formation is probably triggered by the collision of two spiral galaxies. Images of these two peculiar galaxies are on our web site.

It is important not to associate peculiar morphology with nuclear activity too strongly, however.

Some peculiar morphologies can be the result of normal stellar processes (see the "antennae" in Figure 24-11), and some apparently normal galaxies have active nuclei.

Quasars

Throughout the 1950s, radio astronomers used more and more sensitive telescopes to compile lists of radio sources. Many were identified with distant galaxies, and some had no visible optical counterpart. However, a few appeared to be coincident with bluish stars; they were called quasars (quasi-stellar radio sources). Spectra of the objects showed emission lines that could not be identified with the lines of any known element. At that time, no normal stars had been found to emit radio waves, with the exception of the Sun, whose radio emission is very weak and can be detected only because the Sun is so close. Astronomers were mystified.

In the early 1960s, astronomers at Palomar Observatory used the 200-in. telescope to take spectra of two of these quasars, 3C 48 and 3C 273 (the 48th and 273rd objects in the Third Cambridge Catalog of radio sources). Maarten Schmidt noticed that four prominent lines in the spectrum of 3C 273 had the same relative spacings as the Balmer series of hydrogen (Preston, 1988). However, the lines were shifted to the red by about 16%, symbolized as $z = 0.16$. Then he recognized the same lines in the spectrum of 3C 48, redshifted by 37% (or $z = 0.37$). This is an astonishing result. The apparent velocities are so high that according to the Hubble relation, quasars must be at distances of thousands of megaparsecs, and their light must have taken billions of years to reach us. The exact distances and look-back times for quasars depend on the cosmological model assumed.

The Most Distant Active Galaxies? What are the implications of such large redshifts? If we assume that the redshift is caused by the cosmological expansion of the universe, then quasars must be very distant, and we must be observing light emitted when the universe was much younger than it is now. Also, if quasars are very distant, their luminosity or intrinsic brightness must be enormous. Assuming a cosmological redshift, 3C 273 is 620 Mpc or 2 billion light-years distant and shines with the light of 10^{14} Suns, an almost unimaginable amount of energy. It is no wonder that the interpretation of quasar properties is controversial!

There is a crucial distinction between galaxy redshifts and quasar redshifts. Galaxies follow the Hubble relation, in which estimated distance is correlated with radial velocity. This correlation has a natural interpretation in terms of the expansion of the universe. In the nearby universe, it is a relatively reliable procedure to measure a galaxy's redshift and assign it a distance. But *quasars do not have direct distance estimates,* and their redshifts are far higher than the range over which the Hubble relation has been tested. The distance to a quasar is meaningful *only in the context of a cosmological model.* The enormous distances are the result of the assumption of cosmological redshifts.

The Nature of Quasar Redshifts If quasars were much closer than their cosmological redshifts would indicate, then quasar luminosities are correspondingly lower, and finding a physical explanation is not as challenging. The evidence for and against *noncosmological redshifts* must be considered. A number of astronomers have pointed out cases where quasars with high redshifts appear to be associated on the sky with nearby galaxies that have low redshifts. A lot is at stake. If noncosmological redshifts are confirmed, the entire edifice of observational cosmology would have to be revised. It all comes down to statistics. The sky is filled with thousands of quasars and hundreds of thousands of galaxies, so some alignments are bound to occur *by chance.* The statistical evidence for noncosmological redshifts is not persuasive.

In fact, a large body of evidence supports the idea that quasars *are* at the distances indicated by their cosmological redshifts. As the name implies, quasars are often not completely stellar in appearance. Images of quasars at redshifts $z = 0.3$ to 1 reveal fuzz or nebulosity surrounding the bright core. The nebulosity has the size and brightness expected of a luminous galaxy at that redshift. In addition, quasars often lie in clusters of galaxies at the same redshift; this association is strong circumstantial evidence that the quasars are at cosmological distances. Gravitational lensing provides another line of argument that quasars are distant objects. General relativity predicts the splitting and magnification of quasar images; multiple quasar are extremely difficult to explain any other way. For example, the probability of three or four quasars being found close together with identical redshifts and spectra is infinitesimally small.

Another strong argument that quasars are at cosmological distances is the *continuity* in observed structure among normal galaxies, active galaxies, and quasars. Nonthermal radiation is rare in the cores of normal galaxies, but a few nearby examples can be found. Searching larger volumes of space turns up even rarer but more luminous active nuclei, but at such large distances that the nebulosity from

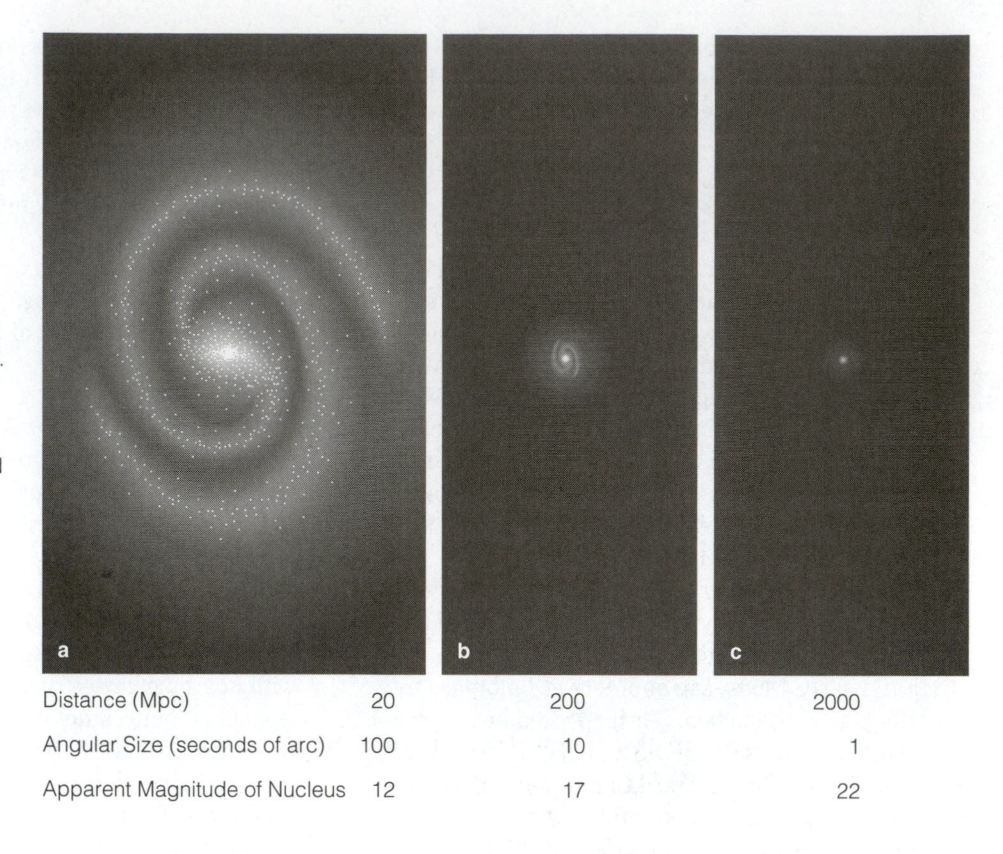

Figure 25-11 The appearance of an active galaxy as a function of distance. **a** At a distance of 20 Mpc, the active nucleus is a bright core in a well-resolved galaxy. **b** At 200 Mpc, only the basic features of the host galaxy are visible. **c** At 2000 Mpc, a redshift close to 0.5, the galaxy—now classified as a quasar—is so faint and small that it can barely be resolved. If we reversed this reasoning and placed a quasar in the nucleus of our own galaxy (about 10 kpc away), it would rival the Sun's brightness in the daytime sky!

the surrounding galaxy is difficult to discern. At the greatest distances, only highly luminous objects can be seen, and the light from the surrounding galaxy is correspondingly weak. This progression is indicated in Figure 25-11. For an active galaxy with a redshift of $z = 0.5$, the host galaxy is barely visible. At this point, it is called a quasar.

Properties of Quasars

The first quasars were discovered using radio surveys, and by the end of the 1960s, a few hundred were known. Radio searches for quasars are very efficient, because strong radio emission is a good indication of nonthermal activity. However, it turns out that relatively few quasars are strong radio sources, so most must be found by their optical emission. Optical searches are more difficult because the sky is crowded with faint stars and galaxies. Luckily, the extremely blue colors and broad redshifted emission lines of quasars distinguish them from stars. Optical techniques are now the most efficient way of finding quasars. Currently, more than 20,000 quasars are known, and surveys now under way will increase this number to over 100,000.

Redshift How far away is the most distant quasar? The record for the highest redshift has crept steadily upward as astronomers have sifted through thousands of faint quasar candidates. There is even a good-natured but intense rivalry among several research groups to find the most distant quasar. The record is currently held by the object at a redshift of $z = 6.3$ (quasars at such high redshift are extremely rare; hundreds of times rarer than at $z = 2$). Depending on the exact cosmological model used to calculate look-back time, the light we are seeing left this quasar when the universe was only 3–5% of its current age!

Size Quasars must be very luminous if their redshifts are cosmological, but it would also be useful to have some idea of their sizes. The apparent size of a direct quasar image is not a very good indicator of its physical size because the Earth's atmosphere blurs up the incoming light from *all* astronomical objects. Variability of the light gives a much more interesting limit on size.

Most quasars vary in brightness, usually over a span of weeks or months. Because of the finite velocity of light, these data translate into an upper limit on an object's size. Imagine the object is 1 light-year in diameter. If the light output from the entire object varies, the signal from the far side of the object will take one year longer to vary than the signal from the

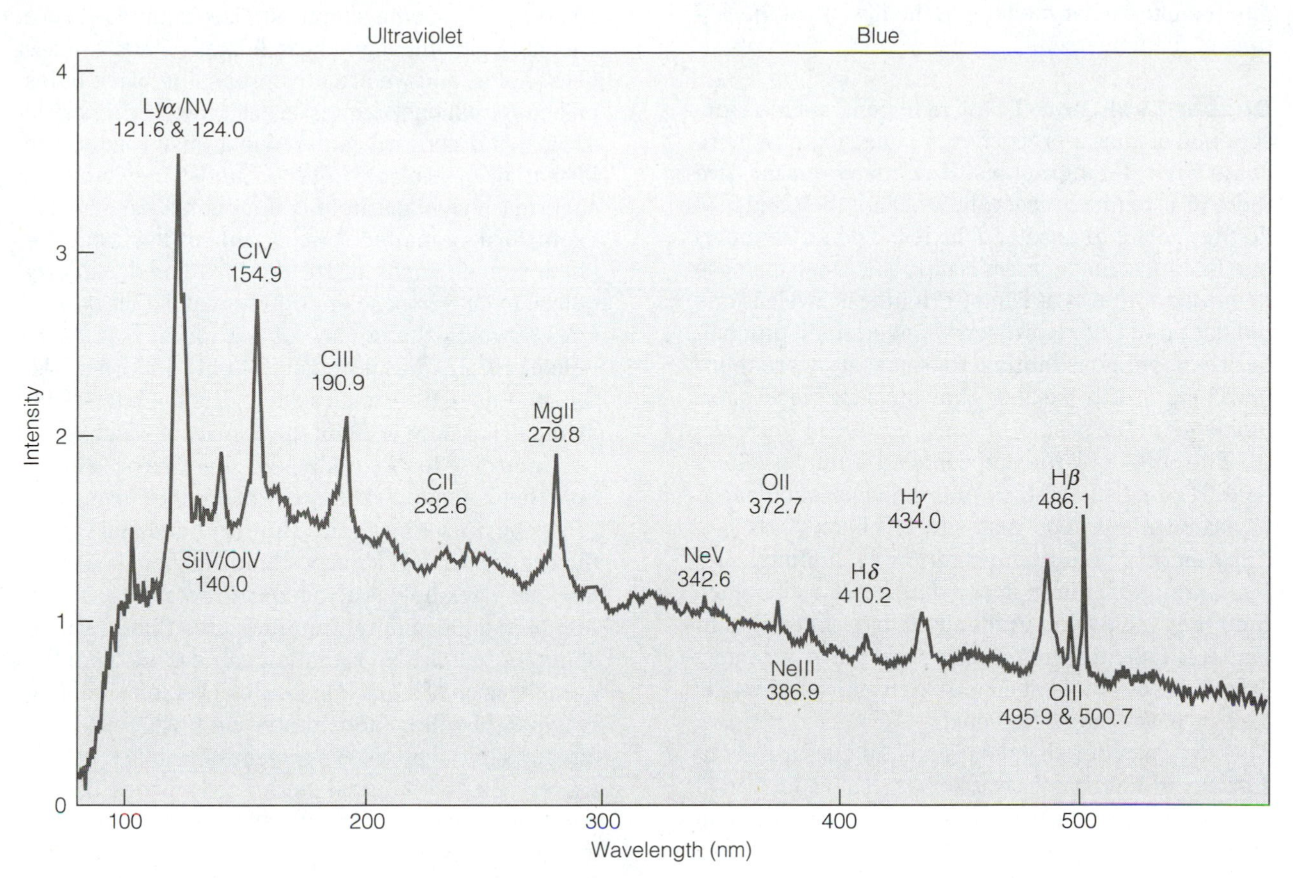

Figure 25-12 A composite quasar spectrum compiled from the largest published survey of bright quasars. The composite includes data from 740 quasars, all with different redshifts. Each spectrum was blueshifted to zero velocity, and then the spectra were added together to give a very sensitive measure of the spectrum of an "average" quasar. The numbers above each emission line give the wavelength of each line in nanometers, and the Roman numerals following the element designation give the level of ionization: II means one electron removed, III means two electrons removed, and so on. (P. Francis and others, reproduced from the *Astrophysical Journal* by permission of the authors)

near side. An object's brightness cannot vary any faster than the time it takes light to travel from one side of the object to the other. If a quasar varies over one week, its diameter must be less than 1000 AU, only about 10 times that of the solar system.

The fantastic energy output of quasars comes from a tiny volume at the center of the host galaxy. As an analogy for this remarkable concentration of energy, imagine flying above a large city at night, with the lights of the city below you. There are probably about a 100 million house, street, and car lights, spread over a 50-km region. Think of each light as a star. A quasar is then a light source 1 cm across, with an intensity equal to 1000 times the sum of all the lights in the city! Packing so much energy into such a small volume is a real challenge for theories of the quasar power source.

Spectrum The spectra of quasars show strong, broad emission lines. Figure 25-12, a composite spectrum of 740 different quasars, illustrates the typical features of a quasar. If the width of the lines is due to Doppler shifts, then the hot gas emitting the lines must be moving between 10,000 and 20,000 km/s, or roughly 5% of the velocity of light. The strongest emission lines are due to ionized hydrogen, carbon, magnesium, neon, oxygen, and nitrogen. Heavy elements have been observed in the spectra of even the most distant quasars, with look-back times of about 12 billion years. Two important conclusions can be drawn. First, the laws of physics seem to be unaltered over large distances—the elements we see in the far reaches of the cosmos are the same elements we find in the Sun. Second, heavy elements had been produced by stars and dispersed into the interstellar medium in the first 5% of the lifetime of the universe.

Quasar Evolution To move beyond a simple description of quasar properties, we need to know more about their demographics. How many quasars are there in a volume of space, how old are they, and how do they relate to normal galaxies? The most important finding is that quasars *evolve.* The steep increase in quasar numbers at fainter brightness levels is one indication of this. However, we need to distinguish between two possibilities: that quasars were more *luminous* in the past, or that quasars were more *numerous* in the past.

The study of large and complete samples allows us to recount the history of the quasar population. Quasars were first born at least 10 billion years ago. They increased in number for 2 or 3 billion years, reaching a maximum space density at $z = 2$. Since then they have been gradually fading, like brilliant embers. The most luminous quasars are extremely rare now; you would need to search a box about 4600 Mpc on a side just to find one!

If we consider all galaxies and all quasars, about 1 galaxy in 1000 has quasar activity. This information can be interpreted two ways. Perhaps 1 galaxy in 1000 is special in some way that makes it develop a quasar core. Or perhaps *every* galaxy goes through a phase of quasar activity lasting 1/1000 of the age of the universe or about 10 million years. We are limited to a snapshot of the universe, an inevitable consequence of our own short lives and the universe's great age. At any given time, only 1 in 1000 galaxies will be switched on and show quasar activity. The two results look the same, and we cannot tell the difference with a simple census.

A Model for Quasars

Understanding the quasar power source is one of the most challenging tasks in astronomy. A luminosity equal to 1000 times the light from an entire galaxy is contained in a volume not much bigger than the solar system. Since normal stellar processes are not efficient enough, astronomers have proposed an exotic alternative: **supermassive black holes.** Figure 25-13 is a simplified cartoon of the central region of a quasar showing the main components. The black hole model for quasars is speculative and does not have the secure status of, for example, the model of stellar evolution.

The Power Source To power a quasar, a black hole would have to be very massive—in the range of 10^6 to 10^9 $M_\odot$. Some skepticism is warranted. There are only a handful of good candidates for stellar-mass black holes, and we are now proposing black holes millions or billions of times larger! However, theorists argue that if stars are gathered in a small volume, the formation of a black hole is almost inevitable. A supermassive black hole is not even a particularly exotic form of matter. A black hole of 1 $M_\odot$ has the phenomenal density of 10^{19} kg/m^3. The density required to curve space enough to make a black hole goes down as the square of the mass, so a black hole of 10^9 $M_\odot$ has a density of only 10 kg/m^3. Although this is 10^{20} times as dense as the interstellar medium, it is only 1/100 of the density of water.

According to the model, the supermassive black hole that is the quasar's "central engine" forms and grows by accreting (gathering) gas and stars from the central parts of a galaxy. Since most galaxies rotate, the black hole is also likely to rotate; conservation of angular momentum indicates that it will be spinning rapidly. No radiation can escape from the event horizon of a black hole. However, material that is falling in will be accelerated and will emit huge amounts of radiation. A supermassive black hole is therefore a gravitational engine, converting the potential energy of infalling material into radiation and kinetic energy of outflowing particles. This process is very efficient. In principle, a rotating black hole can convert 20–30% of infalling mass into pure energy. This is 30 to 40 times the efficiency of stellar fusion.

The Accretion Disk The second major ingredient of the quasar model is an **accretion disk,** the reservoir of hot gas and dust that feeds the black hole. Rotation of the infalling material ensures that it forms a disk rather than a spherical cloud. The rapid motion and the proximity of an intense radiation source ensure that the disk will be hot. Thermal spectra with temperatures of 10,000 to 20,000 K have been observed in a number of quasars, consistent with the expected radiation from an accretion disk. Such a hot disk would "puff up" into a torus (doughnut), rather than flatten like a pancake. This geometry would obscure our view of the black hole in the equatorial plane, making it visible only along the poles of the spin axis.

Emission-Line Clouds Outside the accretion disk are the **emission-line clouds,** small dense clouds of gas orbiting the supermassive black hole. Clouds at distances of under a parsec are partially ionized by strong ultraviolet radiation from the central engine, and they move at tens of thousands of kilometers per second in the deep gravitational well created by the

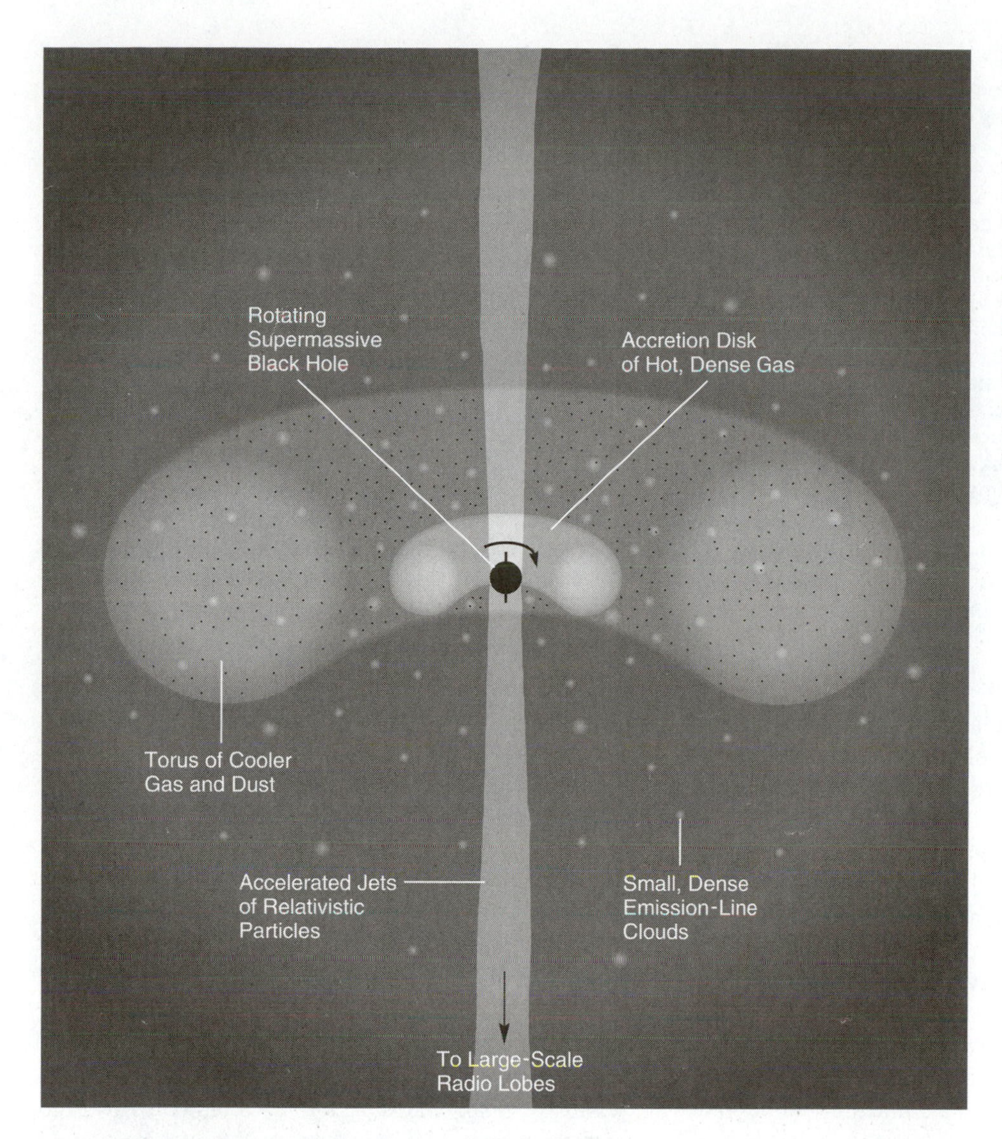

Figure 25-13 Schematic cross section of the central parsec of a galaxy with an active nucleus and jets. The rotating black hole is fueled by infalling gas from the hot accretion disk. Along the poles of rotation, particles can be accelerated into two oppositely directed relativistic jets. Outside the accretion disk is a large dusty torus of obscuring material. Small, dense emission-line clouds are distributed with a density that increases toward the center.

black hole. This region gives rise to the broad lines seen in quasar spectra. All quasars show broad emission lines, but many lower-luminosity active galaxies in addition show narrower emission lines, with widths of hundreds of kilometers per second. The region that gives rise to the narrow lines is much farther from the central engine; the clouds are distributed over kiloparsec scales.

★ **Radio Jets** Some quasars have the most spectacular signature of nuclear activity: radio jets (Blandford, Begelman, and Rees, 1982). Most quasars are weak radio sources; jets are observed only in quasars that are strong radio sources. The region around a supermassive black hole can act as a powerful particle accelerator. The accretion disk blocks most of the outflow, but material is free to be accelerated in the two directions perpendicular to the disk, along the spin axis. The gravitational energy of infalling material is converted into the kinetic energy of a directed particle beam. The region in which the jet forms is extraordinarily complex, with high-energy particles, radiation, shock waves, and magnetic fields. See our web site for a supercomputer simulation of the structure and evolution of a jet; also shown is a visualization of how the night sky might look in a galaxy with a luminous optical jet. Radio observations reveal a rich variety of astrophysical jets; examples are shown in Figure 25-14.

★ The black hole model for quasars has a some striking implications. Ten billion years ago, black holes grew quickly in the centers of massive galaxies, fueled by the frequent interactions between galaxies in the early, dense universe. Since their peak about 5–8 billion years ago, quasars have been gradually fading. Supermassive black holes cannot disappear (!), so they must grow dim from a lack of fuel. As the universe expands, there are fewer interactions among

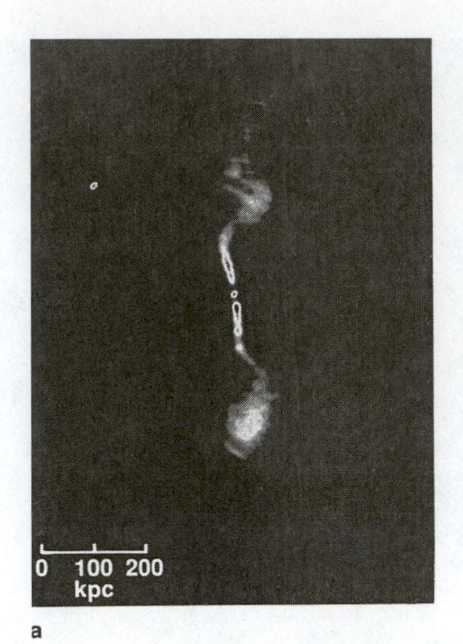

a

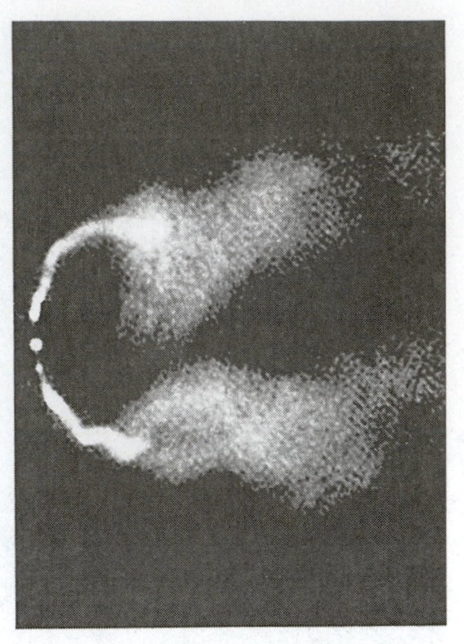

b

Figure 25-14 High-resolution radio images of jets being emitted from radio galaxies (bright central objects). **a** Linear jets moving out from elliptical galaxy 3C 449, at an estimated distance of 66 Mpc. Approximate scale is indicated. (Computer processing yielded dark cores in images of the galaxy and jets; they are actually bright.) **b** The jets from elliptical galaxy NGC 1265 are believed to be bent to the right due to leftward motion of the galaxy at about 2000 km/s through intergalactic gas in the Perseus cluster of galaxies. (National Radio Astronomy Observatory, operated by Associated Universities, Inc. under contract with the National Science Foundation; image *b*, courtesy C. O'Dea)

galaxies, and as stellar populations evolve, the amount of mass loss available to fuel nuclear activity diminishes. By now, most quasars have been starved into silence!

If we add up all the quasar luminosity at a redshift of 2 and assume that 3% of the infalling mass is converted into radiating energy (according to the equation $E = mc^2$), we can deduce that *every bright galaxy* must contain at least $5 \times 10^6\, M_\odot$ of black hole. In practice, this dark mass would be distributed unevenly, so most galaxies will have $10^6\, M_\odot$ black holes, 10% will have $10^7\, M_\odot$ black holes, 1% will have $10^8\, M_\odot$ black holes, and so on. Amazingly enough, searches for dark matter in nearby galaxies have found the predicted number of black holes (see Chapter 24). The glory of quasars occurred billions of years ago, but we still see their dark shadows, lurking in the hearts of nearby galaxies.

Quasars as Probes

More than 30 y after their discovery, quasars continue to perplex astronomers. However, even though astronomers do not understand the power source, they can use quasars as cosmological *probes* of intervening material. One technique is to use **quasar absorption lines** to measure the properties of galaxy halos and cool gas clouds.

Light from a quasar travels for billions of years before reaching the Earth. If that light intercepts a galaxy along the way, the wavelength at which it will be absorbed depends on the redshift of the intervening galaxy. There are two types of absorption lines. One type is generated by heavy elements like carbon, magnesium, and silicon, presumed to originate in the disks and halos of normal galaxies. The other type consists of hydrogen lines and is believed to represent unprocessed gas clouds of primordial abundance.

The positions of absorption lines in a quasar spectrum represent a *map* of the absorbers in redshift. One quasar might have a limited number of absorbers, but many quasars with different lines of sight can be combined to get a map of the absorbers' positions in space. By measuring the properties of absorbers using lines of several different elements, it may soon become possible to trace the *chemical evolution* of galaxy halos back to $z = 5$ (since an average galaxy at that redshift is extremely faint, direct study would be impossible).

Another application of quasar absorbers is to study the *clustering* properties of the intervening material. There is even evidence that the heavy-element absorbers associated with galaxy halos are clustered on enormous scales of 10–20 Mpc, providing unique data on the large-scale structure of the universe at high redshift. Unlike the heavy-element absorbers, the hydrogen-gas-cloud absorbers are not clustered in space and do not appear to be associated with normal galaxies. Astronomers speculate that they might be the detritus left over from galaxy formation. Cold hydrogen can be detected in emission, using the 21-cm line of neutral hydrogen, but absorption techniques using quasar light can detect clouds 1 million times smaller and 100 times farther away. Quasars can be used to probe for the tiniest amounts of cold gas at enormous distances.

SUMMARY

The underpinning of modern cosmology is Hubble's discovery of a linear relationship between redshift and distance. All distant galaxies are receding from the Milky Way; in fact, according to the expanding universe model, every galaxy is rushing away from every other galaxy, carried by the expansion of space itself. The Milky Way is not at the center of this expansion; observers on any other galaxy would measure the same Hubble relation. Moreover, the enormous size of the universe, coupled with the finite speed of light, means that we do not see the distant universe as it is now. Rather, the farther out in space we look, the further back in time we can see.

If this model of an expanding universe is correct, astronomers can use redshift as a measure of distance. This is called the assumption of cosmological redshifts. Some astronomers have questioned the assumption, pointing to apparent associations between galaxies and quasars that should be widely separated in space. However, the vast weight of evidence indicates that galaxy redshifts are indeed cosmological.

The Hubble parameter measures the rate of expansion of the universe, which determines its size and age. Astronomers do not agree on the value of the Hubble parameter, despite major observational efforts with the world's large telescopes. One complication is the difficulty of finding standard candles and extending the accurate calibration of distances in the Milky Way to nearby galaxies. The best methods use bright stellar components of galaxies, such as Cepheid variables and supernovae. At larger distances, the properties of entire galaxies are used as distance indicators.

Galaxy surveys have mapped out the large-scale structure of the universe in rich detail. Galaxies are gregarious. Most live in small groups or large clusters, and the clusters are often bound by gravity into enormous superclusters. These sheetlike structures are separated by voids around 100 Mpc across. Yet, these impressive structures describe only 10% of the matter in the universe! The other 90%—the dark matter—does not emit radiation and makes its presence known only through gravity. On much larger scales, the universe appears smooth or homogeneous.

A small fraction of galaxies have violent events occurring in their nuclei. This activity can be manifested as strong radio emission or high-velocity gas near the nucleus. There is a strong relation between nuclear activity and galaxies with peculiar morphologies or strong bursts of star formation. A quasar is an active galaxy in which the active nucleus outshines the light from the entire galaxy. The most luminous quasars cram 1000 times the luminosity of an entire galaxy into a region not much larger than the solar system. Quasars are distant beacons in the universe; thousands have been discovered. Light from the most distant examples was emitted when the universe was only 5% of its current age.

Theorists face a stiff challenge in explaining quasars. The best model for the power source involves a supermassive black hole feeding on gas and stars from the surrounding galaxy. The central regions of quasars and other active galaxies can act as enormous particle accelerators, spitting out relativistic jets of glowing material. Radio jets have been discovered in a number of quasars. Even without understanding the power source, astronomers can use quasars to probe cold material across vast distances in the universe.

The simple fact of the Hubble expansion does not prepare us for the exotic phenomena of the distant universe. Gravity has sculpted structures of great complexity and delicacy, a loose filigree of sheets and knots and voids covering hundreds of megaparsecs. And concealed among it all is the dark matter. A similar darkness lurks in the hearts of some galaxies, in the form of supermassive black holes. Although the peak of the quasar era is billions of years past, the dead remnants of quasar activity can still be detected in nearby galaxies. This inventory prepares us for cosmology, the study of the universe as a whole.

CONCEPTS

large-scale structure
megaparsec (Mpc)
expanding universe
look-back time
redshift of galaxies
Hubble relation
cosmological redshifts
Hubble parameter
standard measuring rod
cluster of galaxies
morphology-density relation
supercluster of galaxies
Local Supercluster
voids
gravitational lensing
active galaxy
quasar
nonthermal radiation
Seyfert galaxy
radio galaxy
radio jet
supermassive black hole
accretion disk
emission-line clouds
quasar absorption lines

PROBLEMS

1. How many miles is a megaparsec?

2. Suppose observers located in the Coma cluster of galaxies observe Doppler shifts in the spectra of our Local Group of galaxies, including the Milky Way.

a. Would they see a redshift of a blueshift?

b. What sizes of shift would they observe? (*Hint:* The observed recession velocity of the Coma cluster is 7000 km/s.)

c. What would they conclude about our Local Group's velocity if they believed the theory of cosmological redshift?

3. In what ways do Seyfert galaxies bridge the gap between ordinary galaxies and quasars?

4. Explain how the existence of groups of galaxies having members with widely discrepant redshifts, such as Stephan's Quintet (which can be seen on our web site), implies that quasars might not be at the extreme distance usually assumed.

5. Since our galaxy's nucleus is a radio source, why is the Milky Way not considered to be a typical radio galaxy?

6. Why is the study of the most distant galaxies we can see related to the study of conditions around the time that our galaxy (and perhaps others) was forming?

7. Under the assumption of cosmological redshifts, galaxy velocities are telling us something about the state of the universe.

a. What is the implication if galaxies showed equal numbers of redshifts and blueshifts?

b. What would be implied if *all* galaxies showed blueshifts, with a size of blueshift that increased in proportion to the distance?

c. What could you deduce if all galaxies on one half of the sky showed redshifts, while all galaxies on the other half of the sky showed blueshifts?

ADVANCED PROBLEMS

8. Explain why Doppler shifts don't cause significant drifting or detuning of car radios as one drives toward or away from radio station transmitters.

9. Suppose a galaxy of stars emitted most of its light at wavelength 500 nm, where the eye is sensitive. Suppose it was as far away as the quasar OH 471, which has a redshift of 3.4 (*shift* in wavelength = 3.4 × original wavelength).

a. Assuming that redshifts are cosmological, find the wavelength at which most of its light would appear.

b. If it was an ordinary galaxy, would it look brighter or fainter than quasar OH 471?

c. Comment on how the results of parts *a* and *b* would affect attempts to detect the galaxy.

10. A certain faint galaxy is found to have a recession velocity of 5700 km/s. How far away is it if its redshift is cosmological?

11. Using the small-angle equation, estimate the angular size of the galaxy in Problem 10 as seen from the Earth if it has a diameter of 30 kpc, like the Milky Way.

PROJECT

1. Use a small telescope (with an aperture of at least 30 cm) to locate the Seyfert galaxy NGC 1068 (R.A. = 2^h 42.7^m; Dec. = $-0° 01'$) in the constellation of Cetus. What is distinctive about the nucleus of the galaxy? How does it compare with the nuclei of other, normal galaxies, such as Andromeda (M 31)?

PART

8

Frontiers

A special place in the universe. The evolution of life as we know it apparently depends on water and a changing but relatively stable planetary climate. Modern astronomy leads us to the question of whether planets exist near other stars and whether life has evolved on them. (Photo by WKH)

CHAPTER

26

Size and Structure of the Universe

WHAT THE READER SHOULD WATCH FOR IN THIS CHAPTER

Perhaps the greatest intellectual adventure humans have undertaken is the quest to understand the universe. Cosmology is the study of the size, shape, and evolution of the universe. Modern cosmology began with the discovery that galaxies are distant systems of stars moving away from the Milky Way. The modern mathematical basis of cosmology—Einstein's general theory of relativity—governs the physical properties of the universe. This theory holds that space can be curved due to the sum of all the matter in the universe. You will also learn about the scientific story of creation—the big bang model. According to the big bang model, the universe was once in a state of extremely high temperature and density. Since then, galaxies have been carried apart by the expansion of space for billions of years. The fate of the expansion is governed by two measurements that can be carried out with modern telescopes: the current expansion rate and the mean density of matter and radiation. ■

In the opening pages of this book, we compared humanity to explorers on a strange island in an unknown sea. To extend the analogy, at this point in our explorations we know that the island is made of rocks and grains of sand, we know how big it is, we know that the sea is large, and we know that there are many other islands out there in the distance. Now we ask: Do the islands go on forever? Does the sea go on forever? Does our world have an edge, or is it infinite? These are questions of **cosmology.**

Cosmology is the study of the size and structure of the universe—in other words, the "geography" of the universe as a single, orderly system. This takes us to a new level of discussion, because our subject is the **universe,** defined as all matter and energy in existence anywhere, observable or not. The method of science is stretched thin in our speculation about the universe. Our universe is unique, so we cannot learn about it by comparisons, as we can with planets and stars and galaxies. Cosmologists approach their subject by making simple and testable assumptions and by developing theories that describe the present state, origin, and fate of the universe. Speculation should be constrained by observation, and theories must agree with known physical laws.

Cosmology begins with some fundamental questions about the nature of time and space, mass and energy. By answering these questions, cosmologists have been led to some startling ideas about the universe as a whole—ideas that lie far outside everyday experience.

EARLY COSMOLOGIES

Cosmology is not the domain of scientists alone. For thousands of years, poets and priests and philosophers have romanced the universe and tried to understand its nature (Ferris, 1988). Innate in the character of curious, restless humanity is the desire to understand our surroundings and to know where we came from (Brush, 1992).

Mythology and Philosophy

Cosmology is as old as the first ancients who looked at the stars set against the velvet backdrop of night. The universe is described in the earliest surviving writings of the Babylonian, Egyptian, Greek, Chinese, and Indian civilizations. This early phase of *mythological cosmology* links celestial phenomena to the spiritual life of people; creation myths are found in almost all cultures. These cosmologies provide us with an enduring link with our early ancestors.

The next phase of cosmology dates back to the birth of scientific inquiry in the sixth century B.C. on the shore of Asia Minor (as described in Chapter 2). The *philosophical cosmology* of the ancient Greeks marked the application of the power of reason to the universe. Observations were not central to Greek cosmology; the telescope would not be invented for another 2000 y. Progress was made by bold hypothesis, logic, and abstract reasoning. Thinkers like Anaximander believed that the universe evolved from a state of primordial chaos to the order and structure that we see today. Others, like Aristotle, thought that the universe was perfectly ordered and unchanging. The twin themes of chaos and order find a strong echo in modern cosmology.

By the third century B.C., mathematical reasoning had become part of astronomy. Primitive common sense held that the stars lay on a two-dimensional backdrop, nestled snugly around the Earth. Euclid pulled together the theorems of geometry and laid the foundation for the idea of infinite space. Imaginary Euclidean triangles could be extended into space; distances to celestial bodies could be measured. Aristarchus used geometry to anticipate the Sun centered cosmology of Copernicus by nearly 1800 y. The Greeks also wrestled with the uncomfortable implications of an infinite universe.

Physics and Mathematics

With Isaac Newton, we enter the stage of *physical cosmology.* Newton had described gravity as a force by which every particle in the universe attracts every other particle. He realized that this principle might allow a simple description of the structure of the whole universe. Newtonian cosmology pictured the universe as infinite in extent and filled with randomly moving particles. Newton's theory of gravity was brilliant and audacious. For the first time in human history, the mundane motions of objects on Earth had been unified with the stately orbit of heavenly bodies. However, there were serious conceptual problems with Newtonian cosmology. Gravitational force falls off as the inverse square of the distance between any two masses. The inverse square of an enormous number is a tiny number, but it is never zero. In other words, gravity has an *infinite* range. In mathematics, infinities tend to spawn more infinities. When gravitational forces acting on an infinite number of bodies spread over infinite space are added up, the *amount of gravity* is infinite too!

Perhaps the universe was not infinite? Newton argued that no other assumption would make sense. If the universe were not infinite, or if the particles were all in one part of the universe, then gravity would eventually cause all matter to clump together in one place. Only in an infinite universe does each particle feel balanced gravitational forces from other particles in all directions in the sky. There is motion in a Newtonian cosmology, but the universe is *static*—its appearance is unchanging in time.

Sometimes the simplest questions provoke the most profound thoughts. The largest objection to Newtonian cosmology results from asking: "Why is the sky dark at night?" The concern was first raised by Edmund Halley in 1720, but it came to be associated with the German astronomer Wilhelm Olbers one hundred years later. **Olbers' paradox** can be described as follows. In an infinite universe, filled with stars, every line of sight must *eventually* intercept a star. Moving out from Earth, the brightness of a star reduces as the inverse square of the distance. However, the number of stars in any spherical shell increases with the square of the distance. The net effect is that the contribution of light from distant stars continues to pile up. In an infinite universe, the sum of all light from distant stars is infinite: The night sky should be ablaze with light!

The modern response to Olbers' paradox is subtle, invoking the age and recession of distant galaxies (Harrison, 1974). First, there is a distance beyond which we can see no galaxies or stars. This distance does not represent an edge to the universe, but a distance corresponding to a light travel time of 12 to 14 billion years, beyond which we see no galaxies because none had formed that long ago. In other words, the total number of photons emitted by the galaxies in their finite lifetimes is too low to create the kind of pervasive bright glow described by Olbers. A secondary effect that helps to explain Olbers' paradox is the expansion of the universe. The redshifts of receding galaxies cause the apparent energies of the photons we receive from them to be reduced from high energies (short wavelengths) to low energies (long wavelengths), because the photons are "stretched" in the expanding space. Photons received from galaxies with redshifts close to the speed of light are strongly reduced in energy.

Newton's theory of gravity left another very basic question unanswered. Just what was this force that operated across vast distances through the vacuum of space? Newton was acutely aware of this issue and went so far as to call the idea of gravity acting at a distance "an absurdity." He absolved himself of his ignorance about the cause of gravity by saying, "I have not been able to discover the cause of these properties of gravity from phenomena, so I frame no hypothesis." A more profound understanding of the force of gravity had to wait until early in the twentieth century.

MODERN COSMOLOGY

Assumptions

To make a model of the universe as a whole, cosmologists must make certain assumptions. These assumptions may be difficult to prove or verify in practice, but they form an essential starting point for cosmology. The first is the principle of the *uniformity of nature,* the assumption that the laws of physics can be applied to the universe as a whole. As noted in Chapter 24, Hubble made this assumption to demonstrate that many of the "nebulae" were distant galaxies. It is a very bold assumption, because our laws of physics may not apply exactly over all time and space. Nevertheless, we will assume the uniformity of nature in everything that follows. We can sense an echo of the Greek idea that a rational order governs the universe.

A second assumption is that the universe is **homogeneous,** that its structure is more or less the same at all points. Clearly, from the discussion of galaxy clustering in Chapter 25, this is not strictly true! Homogeneous does not mean that all regions of space should appear identical, only that the same types of structures—stars, galaxies, clusters, and superclusters—are seen everywhere. Viewed up close, a beach consists of grains of sand and shells and pebbles of many different sizes. From afar, all we see is a beach. Figure 26-1 illustrates the distinction between a homogeneous and an inhomogeneous universe.

A third assumption is that the universe is **isotropic**—that it looks the same from all vantage points. In other words, no observation can be made that will identify an edge or a center. Figure 26-2 illustrates isotropic and anisotropic universes. The fact that the same Hubble relation is measured in different directions in the sky is good support for the concept of isotropy. The assumptions that the universe is homogeneous and isotropic are together called the **cosmological principle.**

It is difficult to test the cosmological principle. The isotropy of the universe is reasonably well confirmed, because observers looking in different directions from the Earth see essentially the same patterns and structures. We cannot test homogeneity, however, because we cannot travel to different locations to see if things look any different. When we do look to great distances, we are also looking back to a time when the universe was hotter and denser, so the situations may not be comparable. All available evidence supports this principle, but our degree of certainty is not very high.

Cosmologists are like designers crafting a toy model of the universe. If the model is too realistic and tries to contain everything, it will be hopelessly complicated. If it is too simplified, it will not represent essential features of the universe. The assumption of homogeneity is particularly optimistic, because it relegates all the rich structures of matter, the stars, galaxies, and clusters to details of the model! In effect, galaxies are simply markers in space. Another complication is the fact that the *observable* universe is a subset of the *physical* universe, since there are distant regions whose light has not yet reached us. Despite these limitations, modern cosmology is successful in explaining the basic features of the universe. Our bold assumptions have been rewarded with increased understanding.

General Relativity

Modern cosmology began with Albert Einstein. First, his special theory of relativity showed that time and space are supple, and not the rigid, linear measures proposed by Newton (see Chapter 19). Starting with the experimentally verified statement that the speed of light is a constant number, regardless of the observer's state of motion, he showed that fast-moving clocks slow down and fast-moving rulers shrink in the direction of motion. Then, Einstein tackled

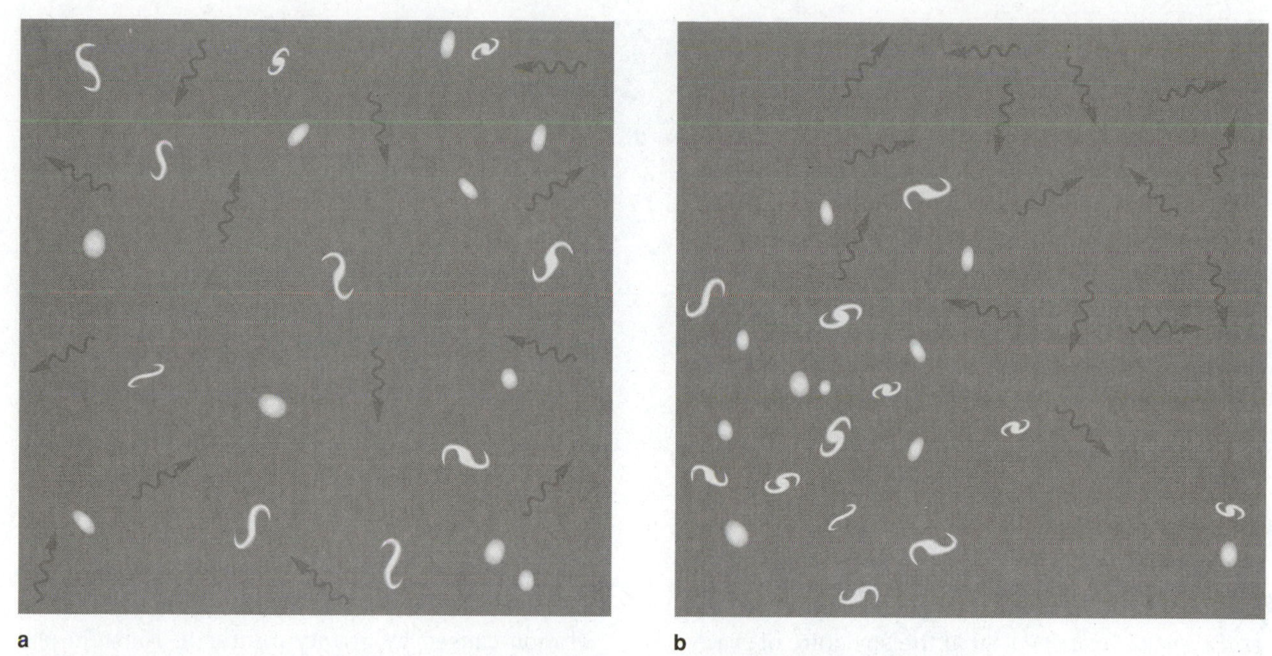

Figure 26-1 Homogeneous and inhomogeneous universes compared. **a** In a homogeneous universe, matter and radiation are distributed uniformly over large scales. The contents of any two nearby volumes of the universe are the same. **b** In an inhomogeneous universe, matter and radiation are not distributed uniformly. Different volumes can have quite different contents.

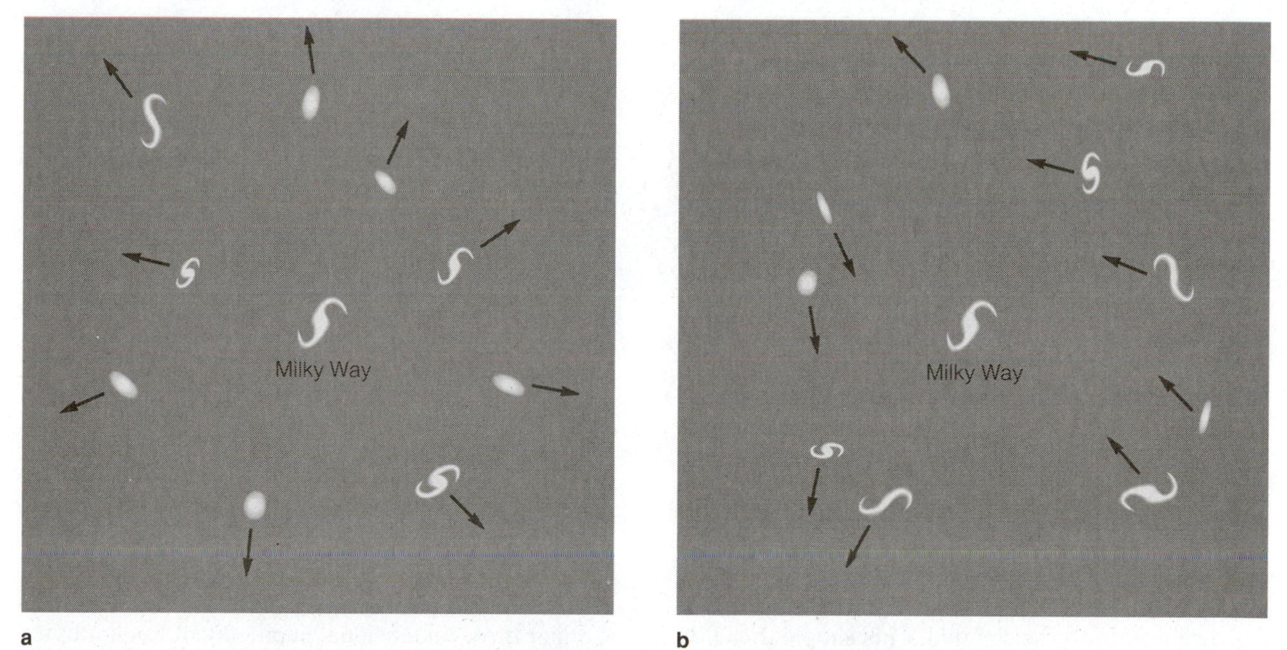

Figure 26-2 Isotropic and anisotropic universes compared. **a** In an isotropic universe, space has the same properties in any direction. The expansion is smooth and uniform, and the same Hubble relation is measured in any direction. **b** In an anisotropic universe, motions of galaxies may be systematic, but they are not the same in different places. A Hubble relation is not observed.

Figure 26-3 Einstein's thought experiment showing the equivalence of acceleration due to gravity and acceleration due to any other force. **a** In an elevator at rest on the Earth's surface, an apple (or any mass) will fall with an acceleration of 9.8 m/s/s. **b** A second elevator is in deep space far from any gravitational influence. If it is being accelerated at 9.8 m/s/s, the apple (or any mass) will fall exactly the same way. There is no experiment that can distinguish the two situations.

Newton's notion of gravity as a force acting remotely across linear time and space. He was led to the *general theory of relativity*—that the structure of spacetime is affected by the mass-energy of the universe—by pondering what seemed to be a coincidence in physics (Einstein, 1923).

The **gravitational mass** of an object is its response to a gravitational field or, in other words, is proportional to its weight. The inertial mass of an object is proportional to the resistance it presents to a change in its motion. At first glance these appear to be very different forms of mass. Imagine a large, smooth piece of iron at rest on a slick, icy surface. Gravitational mass dictates the force with which the iron presses down on the ice. **Inertial mass** is the resistance of the iron to a change in speed, either trying to speed it up or slow it down. Inertial mass has nothing to do with gravity, since the motion on the ice is horizontal, but gravity acts vertically. Nevertheless, inertial and gravitational masses are measured to be equal to an exquisite degree of precision: Modern experiments find the difference to be less than 1 part in 10^{15}. To Einstein, this was more than a coincidence; he believed that it held the key to understanding gravity.

Acceleration and Gravity In 1907, Einstein was working as a lowly clerk in the Swiss patent office in Bern. The job was easy, so he had plenty of time to think. He recalled later: "All of a sudden a thought occurred to me: If a person falls freely he will not feel his own weight. I was startled. This simple thought made a deep impression on me. It impelled me toward a theory of gravitation." He imagined two experiments featuring sealed and windowless elevators. In the first (Figure 26-3), one elevator is at rest on the surface of the Earth, and the other is in free space, but accelerating at 9.8 m/s/s, the same acceleration caused by gravity on Earth. Einstein proposed that there was no way people could tell which elevator they were in! An apple dropped in the elevator on the Earth would fall downward in the familiar way, pulled by gravity. But the apple in the second elevator would behave exactly the same way because the acceleration of the elevator would cause the floor to rush up to meet it.

In the second experiment (Figure 26-4), one elevator is plunging toward the Earth in free fall. A second is floating in free space. Far from the effects of gravity, a person would be weightless, and a released apple would also float freely. One situation is clearly more ominous than the other, but again Einstein realized that no experiment could reveal which elevator an occupant was in!

Einstein interpreted the equivalence between inertial and gravitational mass in the following way. There is no way to distinguish between acceleration caused by gravity and acceleration from any other force. Gravity is just a convenient way to describe how the presence of mass causes an object to change its motion. With this concept of gravity, Einstein proceeded to "generalize" his special theory, that is, to show how gravity could *distort space and time.* This approach led him to a theory of gravity based on *non-Euclidean geometry.*

Curved Space Newton's gravity relied on the familiar three-dimensional geometry of Euclid. In the late 1800s, mathematicians in Germany, Italy, and Russia became fascinated with geometries that were quite different from those of everyday experience. None of those mathematicians dreamed that their

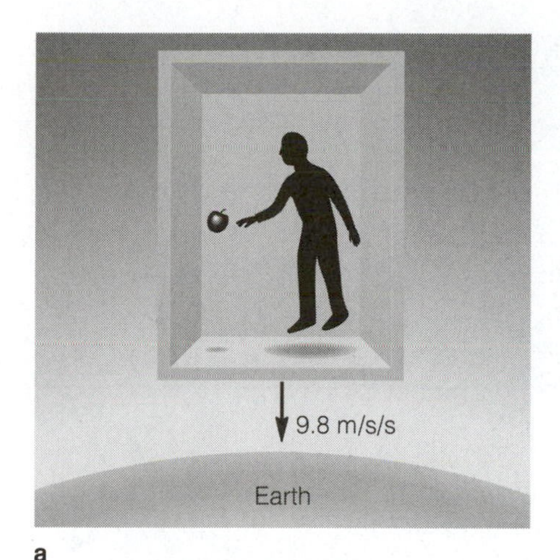

Figure 26-4 Einstein's thought experiment showing the equivalence of inertial and gravitational mass. **a** In an elevator falling toward the Earth's surface at 9.8 m/s/s, local gravity is eliminated, and the observer and the apple float freely. **b** A second elevator is in deep space far from any gravitational influence. If this elevator is at rest or has a constant velocity, the apple and the observer will float freely. No experiment can distinguish the two situations.

esoteric work would be applied to the entire universe. Analogies must be used in cosmology because the actual mathematics is beyond the scope of this book. We use a two-dimensional analogy for curved space because experience and intuition help in understanding the two-dimensional situation, whereas the three-dimensional situation is counterintuitive.

The work of Euclid describes a *flat geometry* (Figure 26-5a). The sum of the angles in a triangle is 180°, and parallel lines will never meet no matter how far they are projected. Euclidean space has no curvature. Two classes of non-Euclidean space exist. A *positively curved geometry* in our analogy is like the surface of a sphere (Figure 26-5b). Slightly less familiar is a *negatively curved geometry,* which in our analogy is shaped like a saddle or a hyperbola in two dimensions (Figure 26-5c). Features of the three types of geometry are summarized in Table 26-1.

The surface of an open space is infinite. In three dimensions, this corresponds to an infinite volume. The surface of a closed space is finite, and so is its volume. The Earth's two-dimensional surface, for example, is finite but unbounded: It has a definite area but you can travel in one direction forever without coming to an edge. By analogy this suggests that a finite universe can exist, in which the galaxies stretch into space in every direction, but where there is nevertheless no edge.

Everyday experience gives us no clue as to whether or not space is curved. Similarly, out on the desert or on the ocean, the planet appears flat. No local surveying technique would show any departure from Euclidean geometry. Observations over a large distance are needed to measure the curvature. We are comfortable with the notion of a curved two-dimensional *surface,* but the notion of a curved three-dimensional *volume* is harder to comprehend. Nevertheless, the idea is supported by observations of our universe.

If space is actually curved, how can we measure it? The best way is through the deflection of light. Einstein showed that all energy has an equivalent mass, which is expressed as $E = mc^2$; this predicts that light should respond to space curvature just as particles do. Thus a light beam sent across an elevator accelerating through space will be deflected by a tiny amount (Figure 26-6a), because during the time it takes to cross the elevator, the elevator has moved. For an elevator at rest on the Earth's surface, the same amount of deflection is predicted due to the gravity of the Earth (Figure 26-6b). Einstein used the deflection of light by the Sun to confirm the general theory of relativity. He predicted that starlight should be deflected around the limb of the Sun by 1.8 seconds of arc (only 0.1% of the Sun's angular diameter!). In dramatic fashion, this prediction was apparently confirmed by the solar eclipse expedition of 1919. The first measurement turned out to have a large error, but other tests have followed, and the theory has passed each with flying colors.

General relativity is a towering achievement. It replaces the force of gravity with the geometry of space-time itself. The familiar Newtonian idea of masses placed in smooth and uniform space is replaced with the counterintuitive idea of space that is distorted by the masses it contains. Matter curves space, and light and particles follow the undulating paths dictated by the curvature. The geometry of the universe can be determined, provided we can make sufficiently accurate measurements over large enough distances. The stage is set for observational cosmology.

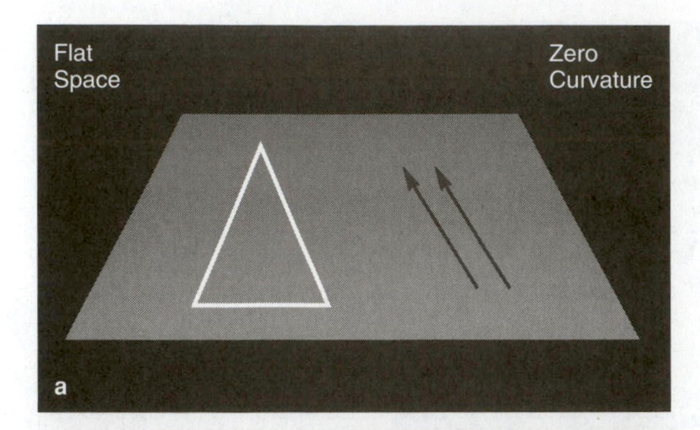

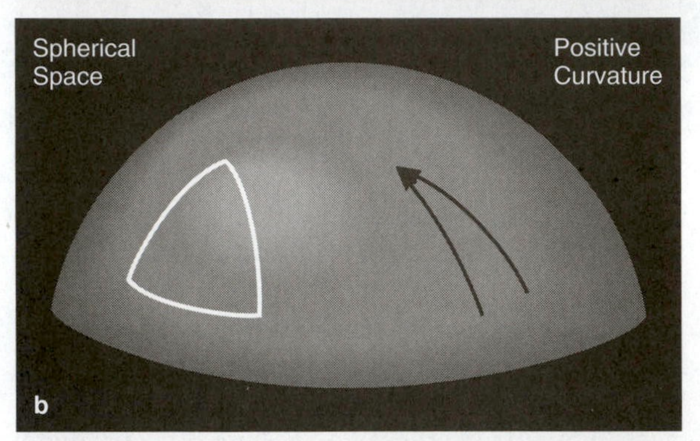

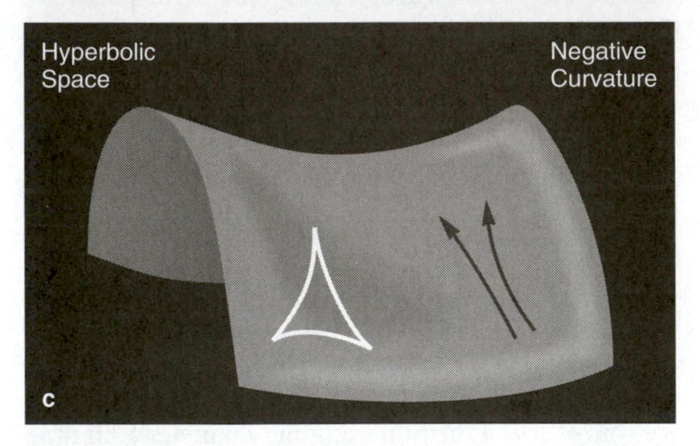

Figure 26-5 Two-dimensional analogies for different curvatures of space. **a** Euclidean geometry describes flat space, in which the angles in a triangle add up to 180°, and parallel lines never meet. **b** Spherical space has positive curvature. Angles in a triangle add up to more than 180°, and parallel lines converge. **c** Hyperbolic space has negative curvature. Angles in a triangle add up to less than 180°, and parallel lines diverge.

The Expanding Universe

In 1917, Einstein applied the equations of general relativity to the universe as a whole, making the usual assumptions about homogeneity and isotropy. However, he also assumed that the universe was *static,* an assumption that seemed reasonable to most astronomers at that time. (Ironically, the astronomer Vesto Slipher was already gathering spectra that would reveal the recession of the galaxies, but he was unaware of general relativity!)

TABLE 26-1

Euclidean and Non-Euclidean Geometries

Type of Space	Curvature	Sum of Angles in a Triangle	Parallel Lines
Euclidean	Zero	180°	Stay parallel
Spherical	Positive	>180°	Converge
Hyperbolic	Negative	<180°	Diverge

No matter how Einstein solved the equations, they stubbornly indicated a *dynamic* universe, one that was either expanding or contracting. To force a static solution to the equations, he added an arbitrary **cosmological constant** (the Greek symbol lambda, Λ). The constant represented a pressure meant to balance the force of gravity in a static universe and stop the galaxies from falling toward each other. Einstein later admitted that the cosmological constant was "the greatest blunder of my life." Because of it, he missed the chance to predict the expansion of the universe ten years before Hubble observed it.

With Hubble's discovery, what had originally seemed to be a purely abstract, theoretical model was supported by observations. The equations of general relativity do seem to describe the observable universe. Once again we can fall back on an analogy—the surface of a balloon with small beads glued on to represent galaxies (Figure 26-7). Keep in mind that this is a two-dimensional representation of a positively curved space that exists in three dimensions. As the balloon is being blown up, its expansion reveals several relevant features of the expanding universe.

The analogy is accurate in the way that the beads are carried farther apart by the stretching rubber of the balloon. When we hear that galaxies are moving away from the Milky Way, we tend to think of galaxies moving through space. But it is the *fabric of space itself* that is expanding, carrying the galaxies with it. The beads follow a Hubble relation, with recession velocity proportional to distance. No bead is at the center of the balloon, and no bead is at the edge.

Although the space is expanding, the beads remain the same size. Similarly, although galaxies and clusters are moving farther apart, their internal gravity keeps them from expanding in size. These regions of nonexpanding space are well described by Newton's laws. The solar system is not expanding, nor is your house.

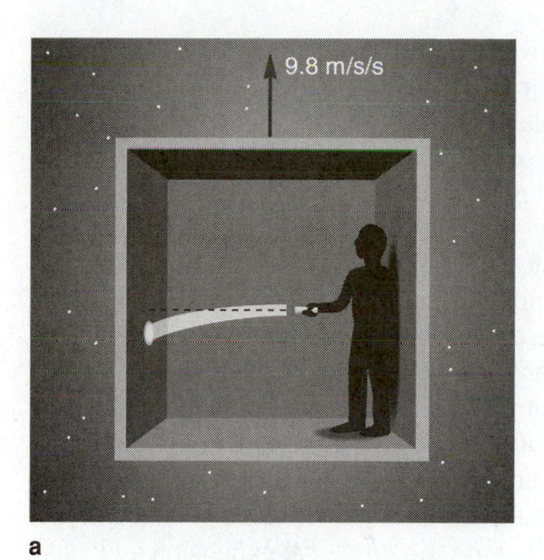

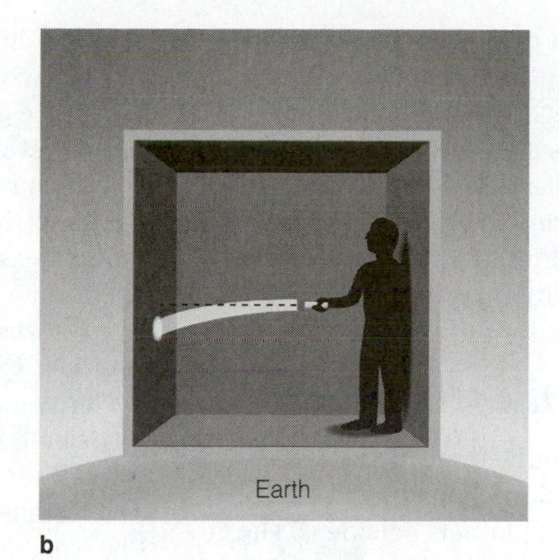

Figure 26-6 Gravitational deflection of light. **a** In an elevator accelerating through space at 9.8 m/s/s, a light beam will be deflected due to the motion of the elevator. **b** The same amount of deflection will occur in an elevator at rest on the Earth's surface. The amount of deflection is exaggerated here; in practice, it is a tiny fraction of a millimeter.

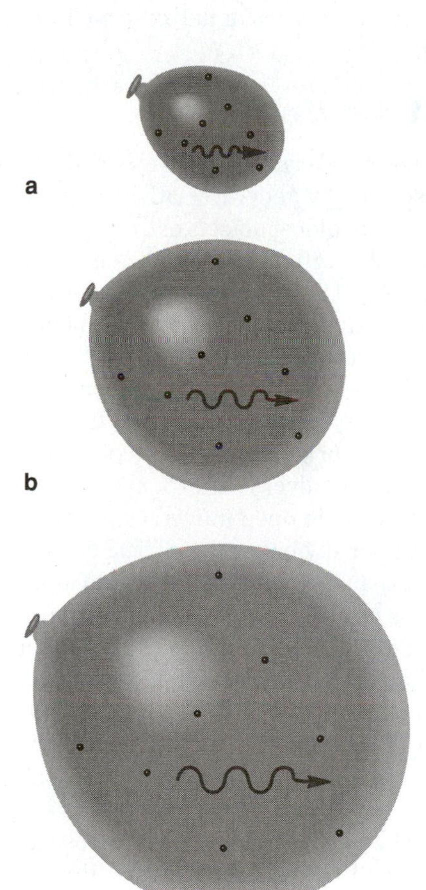

Figure 26-7 An expanding balloon as an analogy for the expanding universe. Beads glued to the surface of the balloon are galaxies. Radiation is represented as a wave drawn on the surface of the balloon. As the balloon expands, each galaxy is carried farther and farther away from every other galaxy. Radiation in the universe redshifts to longer wavelengths and lower energies as the universe expands.

As the balloon expands, the curvature of the space decreases. Imagine the difference between a balloon the size of your fist and one the size of a house; the degree of curvature of any particular patch of the larger balloon is much smaller than the curvature of the smaller one. There is even an analogy for the cosmological redshift. Imagine a wave of light drawn on the balloon while it is small (Figure 26-7). As the balloon expands, the wavelength becomes stretched. In the real universe, light travels through expanding space, and we see a redshift that increases with the distance (time) that light has traveled.

★ The discovery of the universal recession of the galaxies leads to an emphatic conclusion: The universe is evolving, and it has not always been in the same state. Our observations of galaxies represent a single frame in a movie that has been playing for billions of years. Play the movie backward, and what would we see? All the galaxies move closer together. As the universe gets smaller, its volume contracts, until the galaxies are all cheek by jowl in a tiny universe. At some stage, all structure is lost. Galaxies and stars break down into a seething hot gas. The universe tends toward a state of infinite temperature and density, which astronomers call the "big bang." We will return to this topic in the next chapter. Although the big bang model has strong observational support, a discussion of alternative theories is presented on our web site.

AGE AND STRUCTURE

The evolution of the universe can be understood in terms of the way its scale changes with time. Astronomers use the symbol R to represent the scale of the universe. R can be thought of as the size of the universe, but more accurately it represents the distance between any two points. As the cosmological principle implies, any two points are moving apart at the same rate, except for peculiar velocities. Departures from a smooth Hubble expansion are discussed later in the chapter. Thus the entire dynamical history of the universe is described by the way that R varies with time.

The Structure of the Universe

In all possible expanding universes, the rate of expansion decelerates (slows down) with time, because galaxies are pulling on all other galaxies. The strength of the deceleration depends on the mean density of matter. The deceleration is also related to the curvature of space, which determines the fate of the universe. It is a fundamental consequence of general relativity that the structure of space-time (which we measure as the curvature of space) is related to the amount of matter in the universe (which we measure as the density). In an almost empty (that is, low-density) universe, the deceleration is small, and R increases almost linearly with time. The curvature of space is negative, and such an **open universe** expands forever.

At a certain **critical density,** the universe has zero curvature. It continues to expand to some maximum size at an ever-decelerating rate, taking an infinite amount of time to come to a halt. This special case, called a **flat universe,** is the boundary between open and closed universes.

A universe in which the mean density is above the critical density has a positive curvature and is a **closed universe.** The mutual attraction of matter in such a universe is eventually enough to overcome the Hubble expansion. After R reaches a maximum value, the universe collapses, and all the galaxies show blueshifts. The most useful quantity to define is the **density parameter** (the Greek symbol omega, Ω), which is the ratio of the observed density to the density that is *just* sufficient to halt the expansion. The current value is written Ω_0. If $\Omega_0 < 1$, the universe is open and will expand forever; if $\Omega_0 > 1$, the universe is closed and will recollapse.

The best analogy for the relationship involves a rocket launched from the surface of the Earth. We know that the escape velocity is about 11 km/s. A rocket launched with an initial velocity of less than 11 km/s will decelerate as it rises. The Earth's gravitational attraction will eventually overcome the upward velocity and force the rocket back to the planet's surface. On the other hand, a rocket with an initial velocity above 11 km/s will escape from the Earth forever. Of course, the rocket will continue to slow down since the planet's gravity has a long reach, but it will never reverse its direction and fall back to Earth.

In our universe, the Hubble parameter is analogous to the launch velocity of the rocket and the mean density to the mass of the planet. For each possible launch velocity of a rocket, there is a mass of planet where that velocity will just equal the escape velocity. Conversely, for every planet there is a single velocity that will allow the rocket to just escape gravity's pull. Larger planets have larger escape velocities. By analogy, for every possible value of the Hubble parameter, there is a mean density that will just be enough to stop the universal expansion and close the universe. A larger Hubble parameter requires a larger mean density to close the universe. Both quantities can in principle be measured. Table 26-2 shows the features of the three types of universe we have been discussing.

The Age of the Universe

There is one obvious and crucial test of the model of the universe described by general relativity. We can measure the age of the oldest objects in the universe and compare that value with the age predicted by the model (Kinney, 1996). Clearly, no objects in the universe should be older than the age given by the model. The age of the universe is normally written as T_0.

Figure 26-8 shows how the scale of the universe changes with time, depending on whether the universe is open, closed, or at the critical density. Remember that the open universe shown as case A is just one of many possible open universes; any universe with a mean density below the critical density will be open and expand forever. Similarly, the closed universe shown as case C is just one of many possible closed universes, and any universe with a mean density above the critical density will be closed and eventually collapse. There can be only one value for case B, however. A universe with exactly the critical density is a very special transition case between the broad range of open and closed universes.

The point at which the curves all meet represents the present. In Chapter 25, we described how the age of the universe could be approximated by reversing the Hubble expansion. The time in the past when all the galaxies were in one place was the origin. For a linear Hubble expansion, this age is $1/H_0$, called the **Hubble time.** We can see now that any expansion will have a deceleration, so the age estimate of $1/H_0$ is correct only in the artificial case of an empty universe (represented by the dashed line in Figure 26-8).

TABLE 26-2

Geometry of the Universe

Type of Universe	Curvature	Density Parameter (Ω_0)	Deceleration Parameter (q_0)	Fate of Universe
Flat	Zero	1	$\frac{1}{2}$	Barely expands forever
Closed	Positive	>1	$>\frac{1}{2}$	Eventual recollapse
Open	Negative	<1	$<\frac{1}{2}$	Endless, decelerating expansion
Empty	Negative	0	0	Endless and constant expansion

That is, the Hubble time is an *upper limit* to the age of the universe.

Calculations using the currently favored value of H_0 = 75 km/s/Mpc show that open universes will have ages in the range of 9 to 13 billion years, and closed universes will have ages less than 9 billion years. How does this compare with direct age measurements? The Earth's age is well determined, but the Sun contains heavy elements produced in prior generations of stars. A number that will get us closer to the age of the universe is the age of the Milky Way. One approach to measuring the age of the galaxy uses radioactive isotopes—a technique called *nucleocosmochronology.* Radioactive isotopes of the heavy elements thorium and uranium are believed to be made during supernova explosions. Theory predicts the initial ratio of atoms of the different isotopes, and their decay rates are well known. Thus their observed ratio in terrestrial rocks gives the number of years they must have been decaying. Allowing for the (considerable) uncertainties, this method puts the age of the galaxy in the range of 12 to 14 billion years.

Another approach is to measure the age of the oldest components of the galaxy—globular clusters' stars. The full theory of stellar evolution is folded into models of their changing color and brightness. These models yield stellar tracks that can be matched against H-R diagrams to derive an age estimate; the technique is called *isochrone fitting.* The main constraint is the position at which the stars evolve off the main sequence (see the discussion in Chapter 22 and Figure 22-6). The estimated ages can be affected by systematic errors. For example, if the material in the stellar interior is more efficiently mixed than assumed in the model, evolution would be accelerated, lowering the age estimates. Estimates for the ages of the oldest globular clusters

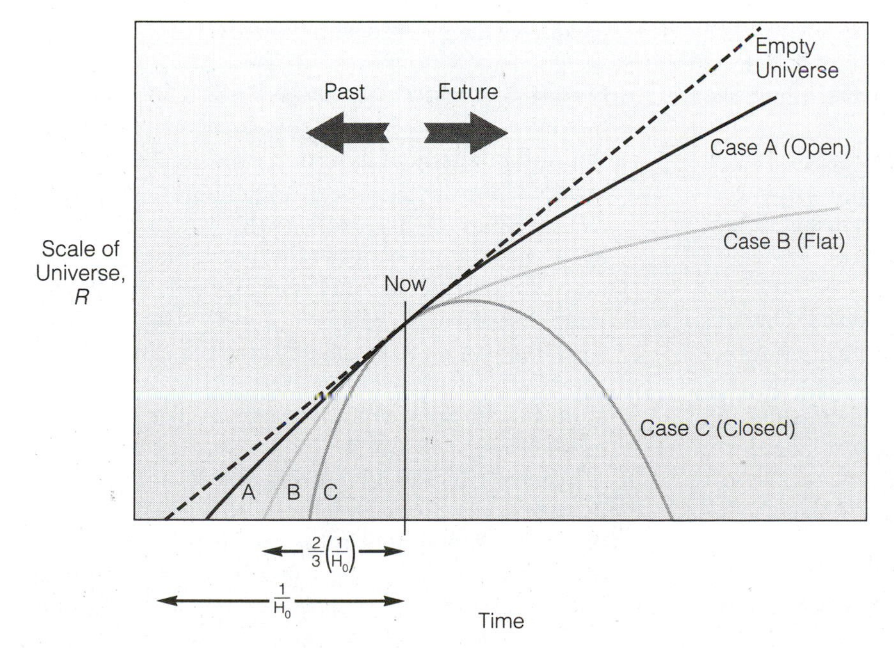

Figure 26-8 The change in the scale of the universe with time. *R* represents the distance between any two points in an expanding universe. Case A is an open universe with less than the critical density, which will expand forever although it will continue to decelerate. Case B is a critical density universe, which will continue to expand at a decreasing rate toward a maximum size far in the future. Case C shows a closed universe with a density greater than the critical density. This universe will eventually recollapse. The dashed line represents an empty universe, in which the expansion is not slowed by gravity. The age of an open universe is between $\frac{2}{3}(1/H_0)$ and $1/H_0$, and the age of a closed universe is less than $\frac{2}{3}(1/H_0)$.

using the most recent stellar evolution models fall in the range of 12 to 14 billion years.

Estimates for the age of the Milky Way are not the same thing as estimates for the age of the universe, though. We have to add an increment for the time between the big bang and the formation of the Milky Way. It is not clear how to estimate this time scale, since our knowledge of galaxy formation is so rudimentary. However, it is likely to be approximately 1 billion years. Adding this increment, we find that the three independent estimates are in plausible agreement at around 14 billion years.

In astronomy, ages are as vexing to determine as distances. Finding a standard clock is as difficult as finding a standard candle. Systematic errors due to unknown physics are likely to be significant. However, bearing in mind the uncertainties, an age of 14 billion years has very interesting implications for the big bang model: It implies that the universe has insufficient matter in it to be closed. A closed universe would be too young to have produced the objects we see around us. A closed universe is possible in the big bang model, but only if the Hubble parameter is lower than most current estimates.

Measuring the Curvature of Space

General relativity predicts that mass will cause the curvature of space. This prediction was verified by observing the gravitational deflection of light rays near the Sun. However, general relativity also predicts that the universe will have a *global* curvature due to the *total* density of matter in it. This curvature is subtle enough that it can be detected only with observations over a significant fraction of the observable universe. This means using very distant objects as probes.

To test the curvature of our universe, we cannot measure the angles of gigantic triangles in space. What we can do is measure the properties of distant objects compared with nearby objects.

The Distribution of Objects in Space One test of the geometry of the universe involves the way that objects are distributed through space. Using the two-dimensional analogies in Figure 26-9, consider the surfaces with galaxies randomly scattered on them. Now imagine flattening the curved surfaces onto a plane so that we can measure the linear distance between any two points. The map of the flat surface gives a result expected from Euclidean geometry: The number of galaxies with a distance R of a given point increases as R^2. To flatten the positively curved surface onto a plane, the edge must be stretched. This means that the number of galaxies increases *more slowly* than R^2. To distort the negatively curved surface onto a flat plane, the edge must be scrunched up. This leads to a number of galaxies that increases *faster* than R^2.

In three dimensions, the analogy holds. In other words, the curvature of the universe is revealed by whether the number of galaxies per cubic megaparsec increases more slowly or more quickly than R^3. The idea is promising, but this approach is susceptible to large systematic errors, because is presumes that we can select similar populations of galaxies at different redshifts. In fact, in any brightness-limited sample, we are always more likely to select luminous galaxies at high redshift and less luminous galaxies at low redshift.

Using Standard Measuring Rods A second test relies on the way that angles change in curved space. In flat, Euclidean space, more distant objects have smaller angular sizes on the sky, with the angular diameter inversely proportional to distance. The curvature of space can distort the images of distant galaxies, and since it is mass that bends light, the distortion increases with the density of the universe. For an extremely distant galaxy, distortion can actually *magnify* the image. Intuitively, we can understand this effect: At large redshifts, a galaxy of a certain size would subtend a larger angle in what was then a smaller universe. Unfortunately, such measurements are useless for making cosmological tests unless we can be sure that the *intrinsic* properties of the objects used have not evolved with cosmic time.

Using Standard Candles We are familiar with the inverse square law of light propagation—the fact that the intensity of light decreases as the inverse square of the distance. Over cosmological distances, the curvature of space affects the way in which the apparent brightness of an object decreases with distance. In this case, redshift is being used as an indicator of distance. The goal is to identify a set of objects with fixed luminosity, called standard candles, and see how their apparent brightness varies with redshift. In principle, this trend can be used to distinguish between open and closed cosmological models.

Unfortunately, high-redshift objects are necessarily seen at large look-back times, so the application of this test is complicated by the need to understand cosmic evolution. In the past few years, astronomers have come to prefer supernovae to galaxies as stan-

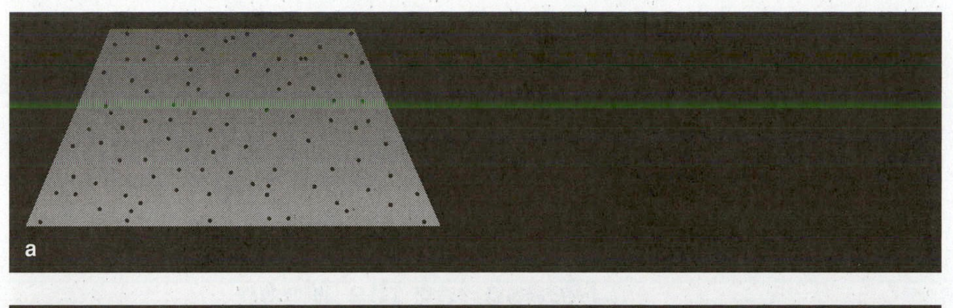

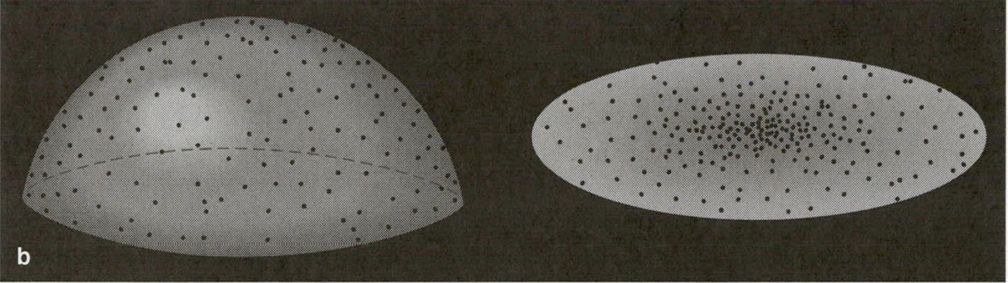

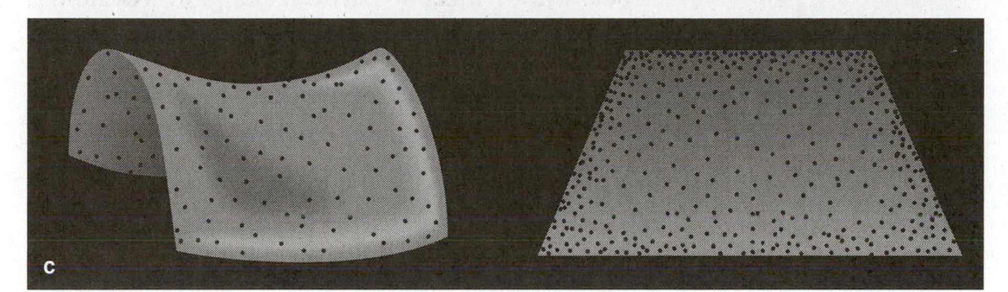

Figure 26-9 Two-dimensional analogies demonstrating that the distribution of objects depends on the curvature of space. **a** A flat or zero-curvature universe is represented by a flat sheet marked with a random distribution of points. The number of points increases as the square of the distance from the center of the sheet. **b** A closed or positive-curvature universe is represented by a spherical shape randomly marked with points. When the surface is stretched flat, the number of points increases *more slowly* than the square of the distance from the center. **c** An open or negatively curved universe is represented by a saddle shape randomly marked with points. When the surface is flattened out, the number of points increases *more quickly* than the square of the distance from the center.

dard candles. In a binary star system, mass can flow from one star onto the second star until it just exceeds the mass needed to detonate as a supernova. In this case, the supernova is a "standard bomb." Since a supernova at peak intensity rivals the brightness of an entire galaxy, supernovae can be seen out to large enough redshifts to perform cosmological tests.

Recently, this technique has provided evidence that the expansion of the universe is *accelerating.* This surprising result can be explained in terms of the cosmological constant, which provides a repulsive force to counter the attraction of gravity. Although Einstein disliked the cosmological constant, it is part of the mathematical apparatus of general relativity. If other astronomical techniques confirm the results from supernovae, then we will have to accept that the universe might not be as simple as we had imagined. The addition of a cosmological constant increases the variety of big bang models of the universe. For example, the universe might have spent a long time at virtually the same size, meaning that the age might be much greater than the current expansion indicates. Also, the universe might be accelerating rather than decelerating, as indicated by observations of distant supernovae. Unfortunately, there is no known law of physics that explains why the vacuum of space should contain enough energy to alter the expansion of the entire universe.

The Issue of Evolution Distant galaxies are, of course, young galaxies (Eichler, 1995). Even if we can be sure our standard candles have not undergone mergers or other catastrophic events, they do consist of billions of stars, and stars evolve.

The simplest stellar system is an elliptical galaxy. Assuming that most of its stars formed at the same time, an elliptical galaxy will fade and become redder. This will happen quickly at first, as the early light is dominated by luminous and short-lived blue stars. Then it will continue more slowly, as stars of lower and lower mass move off the main sequence. Figure 26-10 shows an evolutionary model of a galaxy with a single burst of star formation. The change in energy distribution, particularly the loss of blue and ultraviolet light, is dramatic in the first billion years.

The evolution of stellar populations complicates cosmological tests for two reasons. First, more distant galaxies are likely to be brighter and bluer, simply because they are younger. An unavoidable consequence of look-back time in cosmology is that we are

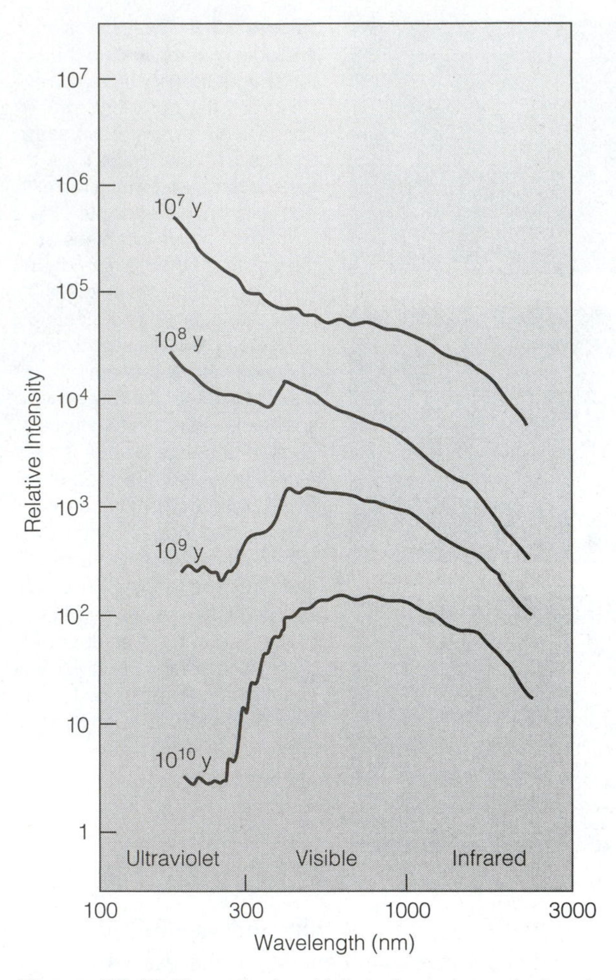

Figure 26-10 The optical and infrared spectrum of a burst of star formation is shown at four different times after the star formation ceases. The energy distribution fades and reddens with time. Ten billion years after the burst, the ultraviolet radiation has faded by a factor of a million, 3000 times more than the fading of the infrared radiation. (Adapted from S. Charlot and G. Bruzual)

forced to compare nearby objects with distant ones that are *younger.* Second, galaxies are measured over a fixed range of optical wavelengths, usually defined by a filter. Distant galaxies are redshifted, so we observe successively bluer parts of the energy distributions of more distant galaxies. Both of these effects mean that galaxies are *expected* to appear brighter at high redshift for reasons that have nothing to do with cosmology.

To conclude, a vast observational effort using large telescopes around the world cannot determine with any certainty whether we live in an open or a closed universe. The problem is not a failing in the data, but the difficulty of finding standard candles and the difficulty in understanding the evolution of galaxies. However, the geometry has been well enough measured to say that it is *close* to the boundary between open and closed universes. As we will see in the next chapter, astronomers have begun to use observations of the very early universe to measure the geometry of space. Those exciting observations indicate that the universe is flat.

Measuring the Mean Density of Matter

General relativity predicts that the geometry or curvature of the universe is determined by the distribution of matter. This suggests a second way of measuring whether the universe is open or closed. By measuring the mean density of matter in the local universe, we can determine whether it is above or below the critical density needed to overcome the universal expansion. Matter in the universe is lumpy: Atoms coalesce into stars, and stars are bound by gravity into galaxies. The key to the concept of mean density is to imagine all the matter in the universe broken up into atoms and *smoothly* distributed through space.

The value of the critical density depends on the Hubble parameter, since a faster expansion requires a larger density to overcome it. For our assumed value of H_0 = 75 km/s/Mpc, the critical density is 2×10^{-29} g/cm^{-3}. This tiny number is equivalent to 4 hydrogen atoms in a cubic meter of space or the density of a grain of sand distributed over the volume of the Earth.

For mathematical convenience, astronomers work with the density parameter, Ω_0, which is the ratio of the measured mean density to the critical density. Since the mean density of the universe is decreasing, the density parameter also decreases with time (as does the Hubble parameter). At the present epoch, an open universe has Ω_0 between 0 and 1, a closed universe has Ω_0 greater than 1, and a flat universe has $\Omega_0 = 1$. Calculating the density parameter sounds easy. Just count the galaxies in some large volume of space, add up their masses, and divide by the volume containing them to get the mean density. Then calculate the critical density using the best estimate of the Hubble parameter. Compute the ratio and see whether it is greater or less than 1.

Visible and Dark Matter The simple calculation described above has been carried out for the local universe. Even though the typical galaxy is a dwarf galaxy, most of the total light is contributed by luminous galaxies. Then we convert from total light to

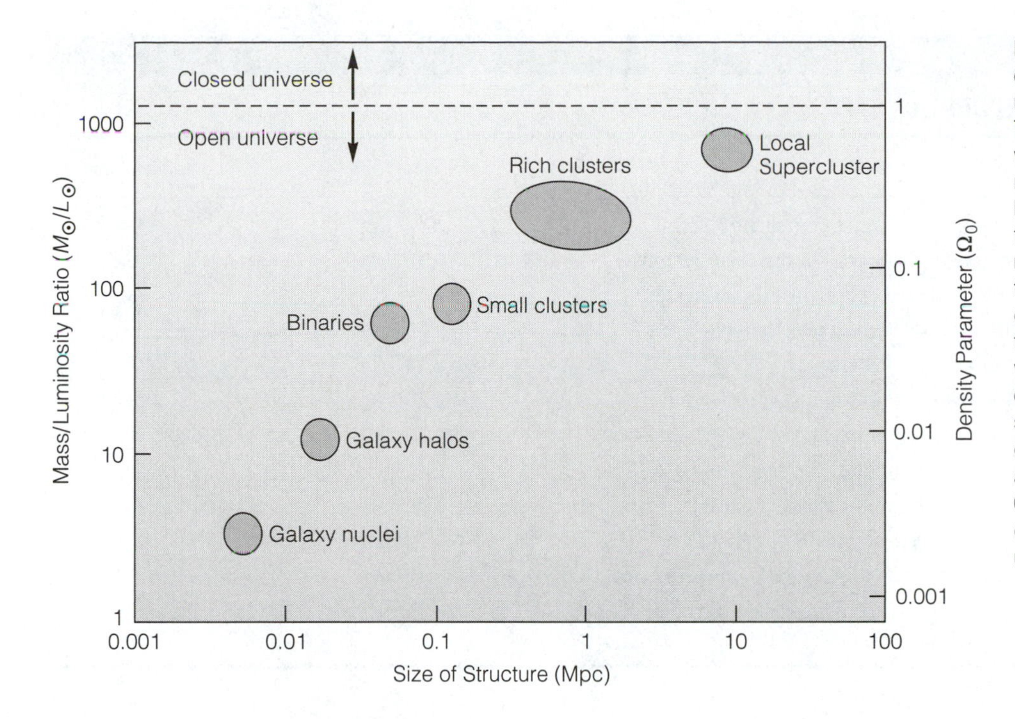

Figure 26-11 The amount of dark matter found in the universe on various scales. The mass is shown in two ways: as the mass/luminosity ratio in units where the Sun has $M/L = 1$, and as the contribution to the density parameter Ω_0, or the fraction of the mass density needed to close the universe. Most of the dark matter in the universe is beyond the scale of individual galaxies. Measured on the scales of superclusters of galaxies, Ω_0 approaches a value of 0.2–0.3, below the critical density required for a closed universe.

total mass using the mass/luminosity (M/L) ratios for the stellar populations in galaxies of each type. The result is $\Omega_0 = 0.002$–0.003. This value shows that the visible mass of galaxies contributes less than 1% of the mass needed to close the universe. The value of Ω_0 depends on the Hubble parameter, but not strongly. Over the full plausible range for H_0, Ω_0 only varies by a factor of 2.

We cannot emphatically conclude that the universe is open, however, because we have not yet taken account of dark matter. As we saw in Chapter 24, about 90% of the mass of most large galaxies is dark. Including this mass in our calculations raises the mean density by a factor of 10, implying that $\Omega_0 = 0.02$–0.03. This is still only 2–3% of the mass needed to close the universe. However, substantial amounts of dark matter are distributed in the extended halos of galaxies. The census of dark matter is not complete until we can make measurements on even larger scales.

Mapping the Motions of Galaxies Measuring the density parameter on the largest scales is one of the most exciting and difficult tasks in observational cosmology. It requires the concerted efforts of large groups of observers and theorists. Although dark matter does not shine, it betrays its presence by the force of gravity. Therefore, the *motions* of galaxies can be used to measure the amount of dark matter.

Every galaxy has two components to its motion. The Hubble flow is a smooth and predictable variation of velocity with distance. The residual velocity (the excess or deficit with respect to Hubble flow) is called the ***peculiar velocity.*** The peculiar velocities show how galaxies are tugged around in response to the clumpy mass distribution of the universe. It is as if the galaxies are shiny marbles rolling on an undulating surface of black velvet. The undulations represent the high- and low-density regions of dark matter. The marbles will roll toward concentrations of dark matter and away from regions where dark matter is sparse. A larger excess of density will generate a larger motion. The map of peculiar velocities leads to a map of the distribution of dark matter.

The best calculations of the density parameter using this technique give values in the range $\Omega_0 = 0.2$–0.3, representing 100 times more matter than is seen in the starlight of galaxies. In other words, the motions of galaxies in their departures from Hubble flow are too small to indicate a closed universe. Figure 26-11 summarizes the mass measurements on different scales, showing that the vast majority of mass in the universe is dark and seen on very large scales (Schramm, 1994). In fact, the density measures are really lower limits since there could be a smoothly distributed component of dark matter that would be very difficult to detect.

The Most Distant Galaxies

★ What do the deepest surveys of the optical universe reveal? Large telescopes and electronic detectors

TABLE 26-3

History of the Size of the Universe

Date	Description	Size	Look-back Time
Ancient	Sky as a canopy suspended above the Earth	$\sim 1 \times 10^3$ km	
A.D. 150	Ptolemy's geocentric model of the solar system	8×10^7 km	
1550	Copernicus proposed heliocentric solar system	2×10^8 km	
1672	First crude triangulation of planet distances	1.5×10^9 km	
1781	Herschel discovers Uranus	3×10^9 km	
1838	Bessel measures parallax of a star, 61 Cygni	1×10^{14} km (3.4 pc)	11 y
1924	Hubble measures distance to M 31	3×10^5 pc	1×10^6 y
1936	Hubble and Humason infer distances to clusters	5×10^8 pc	1.6×10^9 y
1963	The quasar 3C 273 discovered by Schmidt	8×10^8 pc	2.9×10^9 y
1965	Highest-redshift quasar is 3C 9 at $z = 2.015$	2.9×10^9 pc	9.6×10^9 y
1973	OQ 172 holds mark as highest-redshift quasar ($z = 3.53$)	3.3×10^9 pc	10.9×10^9 y
2000	Sloan Digital Sky Survey quasar discovered at over 95% of the lookback time of the universe ($z = 6.3$)	3.5×10^9 pc	11.4×10^9 y

make a powerful combination for searching to very faint levels. On our web site, you can see an image of the Hubble Deep Field. The ultra-deep image of the sky represents several weeks of the *Hubble Space Telescope* staring at a small patch of sky. The limiting apparent brightness is about a billion times fainter than can be seen with the naked eye. The frame is crowded with images, only a couple of which are stars in the Milky Way; virtually all of the 3500 images are distant galaxies.

If we assume that the Hubble Deep Field "core sample" of the universe is representative of the view in all directions (in other words, the cosmological principle), it is a simple step to estimate the total stellar content of the universe. We merely divide the number of galaxies in the frame by the fraction of the sky covered by one frame (or, how many nail heads would it take to cover the sky?). The result is about 5×10^{10} galaxies or 10^{20} stars in the observable universe!

There are enormous numbers of faint galaxies; each square degree (or five times the area of the Moon or Sun on the sky) has 500,000 galaxies down to the brightness limit surveyed so far (Phillipps, 1993). Imaging beyond this limit is difficult even with the largest telescopes, because the images of faint galaxies almost overlap, making measurements difficult. This crowding is a result of projection onto the image plane; in three dimensions, the galaxies are separated by vast distances. Spectroscopy is required to place galaxies in three-dimensional space, but spectroscopy is challenging because the feeble light signal must be dispersed to look for spectral features and measure a redshift. The highest-redshift galaxies have $z > 5$, which places them in the first 5% of the age of the universe.

Table 26-3 gives a chronology of human estimates of the size of the universe. You can see how modern cosmology has given us a sense of a truly vast cosmos. Astronomers routinely measure the properties of galaxies at distances of 5–10 billion light-years (1500–3000 Mpc). It is likely that we will soon observe the first epoch of galaxy formation (Dressler, 1993). Astronomers hope that their intense research efforts will clarify the nature of the faintest galaxies. These pioneers are like the cartographers and explorers who mapped the Earth hundreds of years ago.

The Fate of the Universe

What does the future hold? Even though the brief flicker of human existence seems inconsequential in the eons of the universe, it is irresistible to ask the question. The outcome depends on whether the universe is open or closed. At the moment, the favored evidence points to a flat universe, but a closed universe cannot be ruled out. A flat universe, like an open universe, will expand forever, while a closed universe will eventually collapse. The fates could not be more different.

If the universe is closed, its rate of expansion will continue to decrease for another 5 to 10 billion years.

The universe will pause momentarily at its maximum size and then begin to recollapse, like a cosmic sigh. As all galaxies begin to approach each other, universal redshifts will be replaced by universal blueshifts. At first this will happen only to the nearest galaxies, because those are the ones we see most recently. Eventually, however, distant clusters will also be blueshifted. As space contracts, the photons that fill space will be blueshifted to higher energies. The end state of the contracting universe is often called the "big crunch."

As the universe accelerates toward its second appointment with a singularity, galaxies will crowd cheek by jowl. Humans or their descendants will not be around to care, because the Sun will have bloated into a red giant billions of years previously. The background radiation will be energetic enough to strip atoms of their electrons. Stars and planets will be obliterated. Black holes will coalesce in a crescendo of gravity. According to the work of Stephen Hawking, the big crunch would be the final victory of entropy or disorder. It is a prospect of crushing finality.

If, as seems likely, the universe is open, its galaxies will retreat farther and farther from each other. Since a universe with $\Omega = 1$ never contracts, this is also the fate of a flat universe. Over many billions of years, stars will use up the store of hydrogen and helium for thermonuclear fusion. More and more mass will become locked up in stellar remnants; less and less will go into massive stars that can return gas into the interstellar medium. The cycle of star birth and death will be broken. After about 10^{13} y, all nuclear fuel will be exhausted. All matter will be in the form of cold embers like white dwarfs or collapsed objects like black holes.

Over even longer periods of time, the dead stars within galaxies will interact. Some of the stars will gain energy from the encounters and leave the galaxy, "evaporating" into deep space. Others will lose energy and spiral into the center of the galaxy, becoming fuel for the massive black hole typically found there. What happens next is highly speculative. Particle physics theories predict that protons are unstable, with a lifetime in excess of 10^{32} y. As the protons gradually decay, they release electrons, positrons, and neutrinos, and a small amount of energy as photons. Since the rules of quantum mechanics allow black holes to emit a small amount of radiation, they will eventually evaporate. The large black holes in the centers of galaxies might last as long as 10^{100} y, but the final state of an open universe would be a thin and structureless gas of electrons, positrons, neutrinos, and photons. It is a bleak, if distant, prospect.

SUMMARY

Cosmology is as old as the first civilizations. The human mind has always been drawn beyond the pressures of simple survival to wonder about the size and nature of the cosmos. Cosmology has matured in three stages through the application of mathematics. First, the Greeks applied the geometry of Euclid and were able to measure the sizes and distances of celestial bodies. In the second stage, Newton interpreted the motions of planets and stars in terms of the physical force of gravity and showed that the same laws govern the motions of terrestrial and celestial bodies. Early in the twentieth century, Einstein developed a new theory of gravity that replaced the idea of force with the geometry of space itself. Mass and energy both create and respond to the curvature of space.

Modern observational cosmology began with Hubble. He demonstrated, first, that galaxies are distant systems of stars and, second, that galaxies are all being carried away from one another by the expansion of the universe. Solving the equations of general relativity leads to a cosmological model of impressive power and simplicity. The overriding assumption is that the universe is homogeneous and isotropic, or appears the same to all observers at all points in space. This is called the cosmological principle.

Two numbers govern the past, present, and future of a universe described by the big bang model. One is the present rate of expansion, measured as the Hubble constant, H_0. The other is the mean density of matter, divided by the density required to just overcome the expansion, Ω_0. The effect of gravity means that the expansion has decelerated since the origin event. The deceleration, the curvature of space, and the mean density are all related through the theory of general relativity. Recently, astronomers have found evidence that a third number, the cosmological constant, Λ, might be needed in our model of the universe. The cosmological constant represents a form of vacuum energy whose fundamental nature is still mysterious.

Tests of the type of universe we live in are enormously difficult. The rate of expansion is known to an accuracy of only 20–30%. Measurements of the curvature and the deceleration (or acceleration) both require observations over a significant fraction of the observable universe. A comparison of nearby and distant objects amounts to a comparison between old and young objects. The poorly understood evolution of most standard candles has thwarted the application of both of these tests. Measurement of the density parameter is complicated by the presence of large amounts of dark matter, which dwarfs the mass contribution of visible stars that make up galaxies. The best evidence suggests that the universe is close to flat. This would imply a geometry with little or no space curvature and an expansion that continues forever.

There is a certain grandeur about the quest to understand the size and shape of the universe. We have applied our theory of gravity across the voids of space and back through the eons of time. We have used our telescopes to gather light from ancient galaxies and to detect the shimmering afterglow of the creation event. We should be mindful of Thomas Carlyle's warning: "I don't pretend to understand the universe—it's a great deal bigger than I am." One of the most surprising things about cosmology may be that we *can* understand the universe despite our microscopic part in it. Yet many important questions remain to be answered, and each new answer leads to many more questions. As Einstein said, "The more the universe seems understandable, the more it is also incomprehensible."

CONCEPTS

cosmology	cosmological constant
universe	open universe
Olbers' paradox	critical density
homogeneous	flat universe
isotropic	closed universe
cosmological principle	density parameter
gravitational mass	Hubble time
inertial mass	dark matter

PROBLEMS

1. Progress in many scientific fields, such as studies of stars, plants, or animals, has come by classification of different types of specimens, followed by comparisons of the different classes. How does a cosmologist suffer a disadvantage in this regard?

2. Telescopes much larger than present-day designs, perhaps located in space, would have much more resolving power and light-gathering ability and could reveal much fainter objects. Give examples of how this development would clarify current cosmological problems.

3. How justified is the cosmological assumption that the universe is homogeneous and isotropic?

4. Why is the sky dark at night?

5. Many cosmologies, such as the big bang model, assume the cosmological principle that the universe is homogeneous at any given time, given large enough scale. Yet quasars seem to be more common per unit volume at very great distances than near our galaxy. Does this refute the big bang theory? Why or why not?

6. Give examples of observations that seem to refute:

a. The Newtonian static model of the universe.

b. The steady-state model (discussed on the web site).

7. What is a standard candle in cosmology? What is a standard measuring rod?

8. The measurement of cosmological parameters requires distant astronomical targets whose properties are well understood.

a. Why is it difficult in practice to use galaxies for cosmological tests?

b. Quasars are the most distant known objects. Why don't astronomers use them to measure the size and shape of the universe?

9. How does the existence of dark matter affect the measurement of the density parameter Ω_0? What type of observations have been successful in mapping out dark matter on the scales of clusters and superclusters of galaxies?

ADVANCED PROBLEMS

10. The Hubble parameter H can be thought of as specifying the speed at which galaxies have traveled to reach any specified distance from the (arbitrarily chosen) central site of the big bang.

a. Show by logical deduction that $1/H$ ought to equal the age of the universe (the time since the big bang).

b. Confirm that $1/H$ has the dimensions of time.

c. Confirm numerically that $1/H$ equals 13 billion years, if $H = 78$ km/s per Mpc. (*Hint:* The problem really deals with conversion of units. Note that $1\ \mathrm{y} \approx \pi \times 10^7$ s.)

11. Quasars are not found in our region of the universe. Why is this not necessarily a violation of the cosmological assumption of homogeneity?

12. What is the maximum age of the universe for a Hubble parameter of 50 km/s/Mpc, 100 km/s/Mpc? Show that the ages of galactic globular clusters can be used to put a limit on the maximum value of H_0. How is this limit changed if Ω_0 is very low (much less than one)?

13. If the average distance between galaxies is 3 Mpc and the average mass of a galaxy is $2 \times 10^{11}\ M_\odot$, what is the mean density of the universe ($1\ M_\odot = 2 \times 1^{33}$ g, $1\ \mathrm{Mpc} = 3 \times 10^{24}$ cm)? How does this compare to the critical density for $H_0 = 75$ km/s/Mpc? Does this imply the universe is open or closed?

CHAPTER

27

Origin and Evolution of the Universe

WHAT THE READER SHOULD WATCH FOR IN THIS CHAPTER

Our confidence in the big bang model rests on three pieces of evidence. The first is the cosmological redshift of galaxies: the more distant a galaxy is, the faster it is moving away from us. The second is the abundance of helium, created a few minutes after the big bang when the entire universe had the temperature of the interior of a star. Finally, we observe the cold, featureless radiation left over from the big bang, redshifted by the expansion to microwaves. Astronomers have explored the big bang model from the first second after creation. They have come up with tentative explanations for some basic aspects of the universe, including the four fundamental forces of nature, the flatness of space-time, and the excess of matter over antimatter. Most important, humans can actually understand the universe of which they are such a small part. ■

The big bang model has been successful in describing the main features of the evolving universe. It explains many known observations, it makes new predictions, and it has the possibility of being refuted. However, the big bang model need not be unique; there may be other theories that also describe the physical universe. Nevertheless, it gives us a good framework for further discussion (Hogan, 1998).

Cosmologists look back toward the big bang and wonder how far they can push the model. Can it describe the first few minutes of the universe, the first fractions of a second? They are motivated by a desire to understand the early phases of the expanding universe. Moreover, they realize that the early universe is an unmatched laboratory for studying the structure of matter. The first instants of the big bang reached extraordinarily high temperatures and densities as all the matter in the universe was squeezed in a gravitational vise. These conditions may yield clues to the fundamental forces of nature.

The conventional wisdom is that science proceeds by describing *how* things happen. The nuts and bolts of astronomy are indeed devoted to understanding how a star makes heavy elements, or how planets form, or how galaxies evolve. But many of the questions on the frontiers of a subject involve asking *why?* "Why" questions can eventually lead to a more profound understanding, as we have seen from the answer to the simple question, "Why is the sky dark at night?" Later in this chapter we will ask (but not always satisfactorily answer!) other such questions. Why is the universe nearly flat? Why is the universe made of matter and not antimatter? Why are there so many more photons than particles in the universe? Why are there four forces of nature?

There are also questions that science cannot answer. As we probe back toward the creation event, it is tempting to ask: Why did the big bang take place? Such a question is more in the domain of philosophy and religion than science. Many people feel uncomfortable at the point where science and religion brush against each other. Yet faith and reason have coexisted in scientists as notable as Newton and Einstein. As long as we do not overreach the limits of science, we can continue and tell the fantastical story of the early universe. In this chapter, we will explore the big bang model in detail and use it to understand the physical conditions in the early universe.

THE BIRTH OF THE UNIVERSE

In 1929, an obscure Belgian priest and mathematician, Georges Lemaître, was the first to hypothesize a time when the universe might have been as small as an atomic nucleus. He proposed that the origin was a *cosmic singularity,* "a day without a yesterday," when the universe was infinitely small and infinitely curved, and all matter and energy were concentrated in a single mathematical point. Needless to say, many astrophysicists found the idea distasteful, and the idea was occasionally disparaged with the name "big bang." The label stuck, and scientists continue to call the creation of the universe the **big bang model.**

Lemaître described the big bang model lyrically in one popular account: "The evolution of the world could be compared to a display of fireworks just ended—some few red wisps, ashes, and smoke. Standing on a well-cooled cinder we see the slow fading of the suns and we try to recall the vanished brilliance of the origin of the worlds." Thinking of the big bang as an explosion is tempting, but it is also misleading. In an explosion on Earth, debris flies *through space.* In the big bang, the initial singularity *contains all space and matter.* Time itself begins with the big bang. The evolution of the universe is the unfolding of time and space from a condition of incredible heat and density to a cold and enormous state billions of years later.

The impetus for the big bang concept was the discovery of the recession of the galaxies. However, its status as the leading model for the origin of the universe was cemented by two other discoveries—cosmic nucleosynthesis and cosmic background radiation.

Cosmic Nucleosynthesis

The fusion of light elements in the hot universe soon after the big bang is called **cosmic nucleosynthesis.** About 25% of the mass of the universe is observed to be in the form of helium. Yet stellar fusion in stars like the Sun converts only about 10% of their hydrogen mass into helium in about 10 billion years. Low-mass dwarfs, which form most of the mass of the Milky Way, produce helium at an even lower rate, about 2% of the hydrogen mass in 10 billion years. In addition, helium shows a uniform distribution throughout the Milky Way and in many other galaxies. This contrasts with the abundance of heavier elements, which decreases with distance from the center of the Milky Way and is correlated with the number of supernovae.

These two points provide strong circumstantial evidence that helium has a pregalactic origin. It turns out that the density and temperature predicted for the early phases of the big bang are just right to produce helium nuclei by the fusion process. By contrast, heavier elements cannot be made in this way because the expanding universe cools too quickly; heavier elements are created by the late stages of stellar evolution.

Cosmic Background Radiation

In the 1940s, Russian émigré physicist George Gamow realized that the early hot universe would expand and cool but would never become completely cold. Short-wavelength, high-energy photons of the early universe would be stretched out and lowered in frequency by the cosmic expansion. In effect, they would be redshifted to much longer wavelengths or lower energy. He predicted that the universe should be bathed in the relic radiation of the big bang, redshifted to microwaves and with a temperature only a few degrees above absolute zero. Unfortunately, radio astronomy was in its infancy, and no one had a telescope sensitive to microwave radiation, so Gamow's prediction fell into obscurity.

Now flash forward to the early 1960s. Arno Penzias and Robert Wilson, two engineers at Bell Telephone Laboratories, were testing a sensitive microwave horn antenna designed to relay telephone calls to Earth-orbiting communication satellites. As part of their tests, they were mapping the faint microwave radiation emitted by the Milky Way. An unknown source of noise affected their measurements even after they had carefully checked their equipment. The excess noise did not appear to change intensity with direction in the sky, time of day, or season, and it was not associated with any known astronomical source. At one point, they noticed that pigeons were roosting inside the antenna. Concerned that a residue left by the birds might be affecting their measurements (Penzias and Wilson tastefully called it a "thin, white, dielectric film"), they cleaned the horn and started over. The noise persisted.

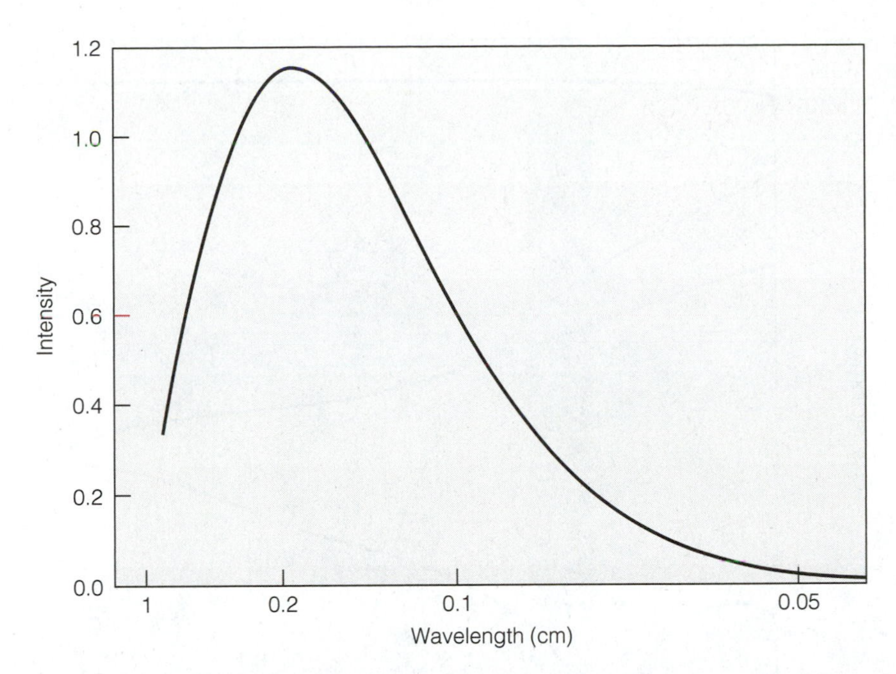

Figure 27-1 Spectrum of the microwave background radiation, as measured by the COBE mission. The radiation matches almost perfectly a thermal spectrum with a temperature of 2.726 K. Deviations from a thermal spectrum are less than 0.3%. (Courtesy COBE Science Working Group/NASA)

Through a colleague, Penzias and Wilson learned that a group of physicists at Princeton University led by Robert Dicke was building a receiver to look for Gamow's proposed **cosmic background radiation.** But Penzias and Wilson found it first and were eventually awarded the Nobel Prize for their discovery. Observations of the cosmic background radiation have continued, with the most spectacular results coming from the *Cosmic Background Explorer* satellite (COBE), launched in 1989, and from recent measurements by high-altitude balloons and Antarctic observatories.

★ The cosmic background radiation has two striking features. First, the spectrum is thermal, falling with great accuracy along a curve that defines a temperature of 2.726 K (Figure 27-1). Second, the radiation is highly *isotropic,* meaning that it has the same intensity in all directions. There are two slight modifications to this isotropy. The radiation does reflect the Sun's motion around the center of the galaxy, the galaxy's motion in the Local Group, and the Local Group's motion toward the Virgo cluster. These combined motions cause us to have a slight Doppler shift with respect to the cosmic background radiation. The background is slightly warmer, or blueshifted, in the direction we are moving (toward the constellation Leo) and slightly cooler, or redshifted, in the opposite direction (toward the constellation Aquarius). The other departure from isotropy is fluctuations in the radiation, which were revealed by COBE. These fluctuations of less than 0.001% represent the seeds of galaxy formation. Maps of the microwave background and its departures from isotropy can be found at our web site (see also Talcott, 1992).

The discovery of the cosmic background radiation is a striking confirmation of the big bang model. The temperature matches that expected for redshifted radiation emitted by a hot gas soon after the universe began. The uniformity is a direct verification of the cosmological principle, an indication that at least the early universe was homogeneous and isotropic. Most important of all, the cosmic background radiation is evidence that the universe has evolved. It is a fossil that tells of a hot, dense, and structureless universe out of which galaxies, stars, planets, and people were forged.

EXPLORING THE BIG BANG

Despite its frivolous title, the big bang is a very sophisticated model. If we accept its basic assumptions, we can use known physical laws to calculate how the density and temperature of the universe have changed with time (Peebles and others, 1994). These calculations lead to predictions that can be tested with astronomical observations. Cosmologists have considerable confidence in the big bang model going back to the first seconds of the universe (Davies, 1992). The next few sections describe the evolution of the universe from the first few seconds until the galaxies formed. Speculations about the first tiny fractions of a second will be deferred until later in the chapter.

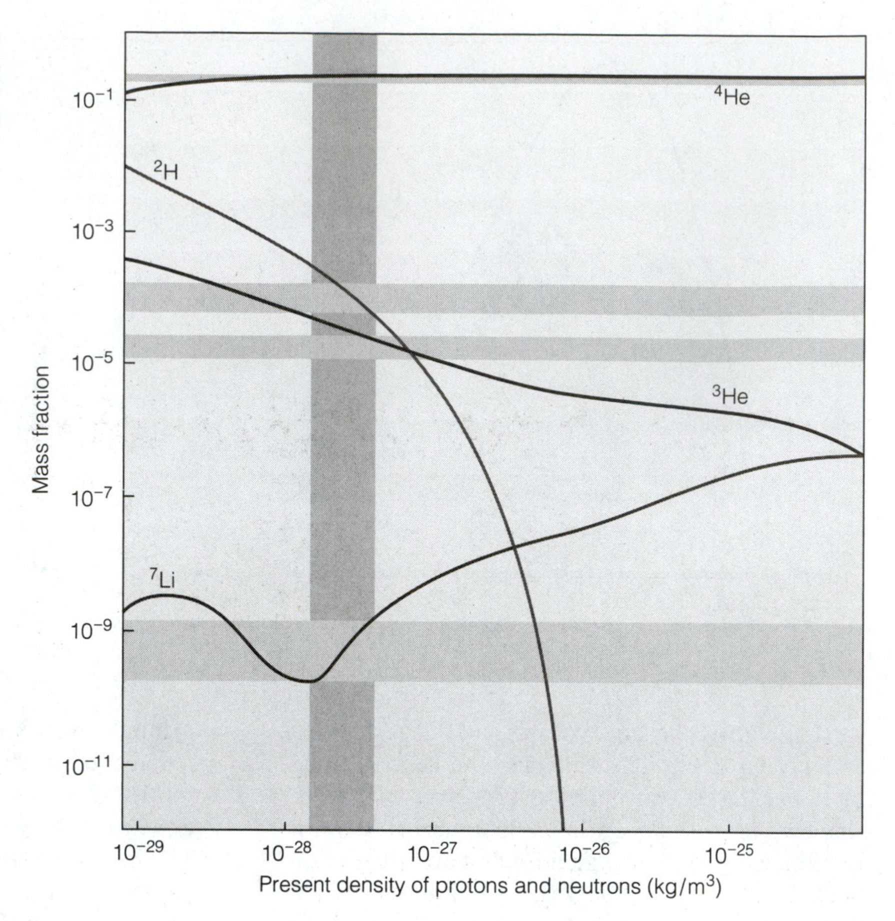

Figure 27-2 The predicted abundances of the light elements from nucleosynthesis calculations (solid curves) compared with the observational bounds on the abundances (light horizontal bars). There is a relatively narrow range of density (dark vertical bar) for which the observations of all four elements agree with the model. (Adapted from Walker and others)

★ Due to the finite speed of light and the large size of the universe, we see objects as they were at some time in the past. Distant light is ancient light. In Einstein's general theory of relativity, space and time are formally linked, so cosmologists refer to *space-time.* This concept is explored in more detail in material on our web site (see also Odenwald, 1996).

The Epoch of Cosmic Nucleosynthesis

In the first seconds of the universe, the temperature was billions of degrees. This was too hot for nuclei to stick together, so the universe was a dense hot broth of radiation (photons) and particles (such as protons, neutrons, electrons, and neutrinos). Photons outnumbered particles by about a billion to one, and most of the interactions were driven by the intense radiation; so at this time we say that the universe was **radiation-dominated.** The entire universe was compressed into a volume the size of the Sun. Every particle in your body was once a part of this cauldron.

After about a minute, when the temperature had fallen to a billion degrees, nuclear reactions began to take place. Neutrons and protons combined to form deuterium nuclei, symbolized as ^{2}H and sometimes called ***heavy hydrogen.*** Deuterium can capture another neutron to make tritium (^{3}H, one proton and two neutrons) or another proton to make helium-3 (^{3}He, two protons and one neutron). After one more stage, helium-4 nuclei (^{4}He) were created, using up almost all the available neutrons. This process was complete 4 min after the big bang, by which time about 25% of the mass of the universe had turned into helium. During the next half-hour, tiny amounts of lithium-7 (^{7}Li) and beryllium-7 (^{7}Be) were created. The creation of light elements in the big bang is called cosmic nucleosynthesis (Parker, 1988). After this, the reactions stopped. The universe became too cool and diffuse to synthesize heavier elements. In the simple big bang model, all heavier elements are produced much later, in the interiors of stars.

Figure 27-2 shows the comparison between theory and observation for four light elements—helium-4, deuterium, helium-3, and lithium-7. These calculations were carried out for a range of densities in the early universe; in the plot these are converted into densities at the present time. There is a range of density for which *all four* calculated curves agree with the corresponding observations, even though the abundances cover an enormous range.

Cosmic nucleosynthesis also leads to a constraint on the type of universe we live in. The narrow range of density in which the big bang predicts the correct abundances can be compared with the critical density of the expanding universe model. Cosmic nucleosynthesis implies that Ω_0 is in the range 0.04 to 0.06. At first sight, this looks like another argument for an open universe, but remember that big bang calculations include only conventional or *baryonic particles*—such as protons, neutrons, electrons, and neutrinos. The hypothetical candidates for the dark matter are *nonbaryonic particles,* which interact very weakly and so do not enter into the nucleosynthesis calculations.

A more detailed analysis leads to two statements. The *lower bound* on the matter density from nucleosynthesis is *higher* than that contributed by the visible stars in galaxies (recall that Ω_0 in the form of stars in galaxies is only 0.002–0.003). Therefore, some of the dark matter is likely to be baryonic. Also, the *upper bound* on the matter density from nucleosynthesis arguments is *lower* than that inferred from large-scale motions (recall that Ω_0 = 0.2–0.3 from dynamical studies). Therefore, some of the dark matter must be nonbaryonic. This is the astronomical argument for a totally new form of matter, as yet unobserved in the Earth's laboratories!

Radiation and Matter

After the feverish activity of nucleosynthesis, the radiation-dominated universe expanded uneventfully for thousands of years in thermal equilibrium (see Chapter 5 to review this concept). The universe was so hot that it was a *plasma;* electrons had too much energy to remain bound to atomic nuclei. As the universe expanded, its density and temperature diminished. According to Wien's law, the radiation shifted to longer wavelengths as the temperature went down. After one year, when the temperature was several million kelvins, the thermal radiation of the universe consisted of X rays. After several thousand years, at a temperature of about 100,000 K, the radiation had cooled down enough to emit ultraviolet waves.

An alternative way to think of the changing radiation of the big bang is in terms of redshift. As the volume of the universe expanded, the wavelength of the radiation was stretched by a corresponding amount. The photons therefore lost energy due to the cosmological redshifting of space. High-energy photons with a short wavelength lose energy due to the expansion of space and shift to longer and longer wavelengths.

Matter Domination The *energy density of radiation* in a given volume is proportional to the average energy of a single photon times the number of photons in the volume. We can also calculate the *energy density of matter,* using Einstein's famous equivalence, $E = mc^2$. The energy density of matter is therefore proportional to the number of particles in the volume. As the volume increases, the number of photons and particles stays the same, but the photons lose *extra* energy due to the effect of redshift. Therefore, the energy density of radiation goes down *faster* than the energy density of matter as the universe expands. Some 10,000 y after the big bang, an important transition occurred; the universe became **matter-dominated.** This is illustrated in Figure 27-3.

Decoupling After 300,000 y, the temperature had fallen to about 3000 K, and the thermal radiation from the big bang had been redshifted to visible light. The events of these first few hundred thousand years are completely hidden by an obscuring fog of photons and particles in constant interaction. Even a hypothetical observer at that time would not have seen much, because the early universe was opaque. As the universe expanded, particles became more thinly dispersed, and radiation was redshifted. At some point, the radiation no longer had enough energy to keep negatively charged electrons and positively charged protons from combining to form stable hydrogen atoms. The sudden formation of hydrogen atoms is called *recombination.* By the time the universe reached an age of 1 million years, all but one out of every 100,000 protons and electrons had been mated into hydrogen atoms.

The time when hydrogen atoms formed is often called the era of **decoupling,** because radiation and matter could no longer interact with absolute freedom. Free particles can have any energy, so they are very effective at interacting with photons. Hydrogen atoms, however, can absorb and emit light only at a few specific wavelengths, corresponding to the transitions between energy levels of the single electron (see Chapter 5). Very few photons in the broad spectrum of thermal radiation couple with atoms, so radiation travels through the expanding universe unimpeded.

The most dramatic effect of decoupling was that the universe became *transparent.* Consider a cloud.

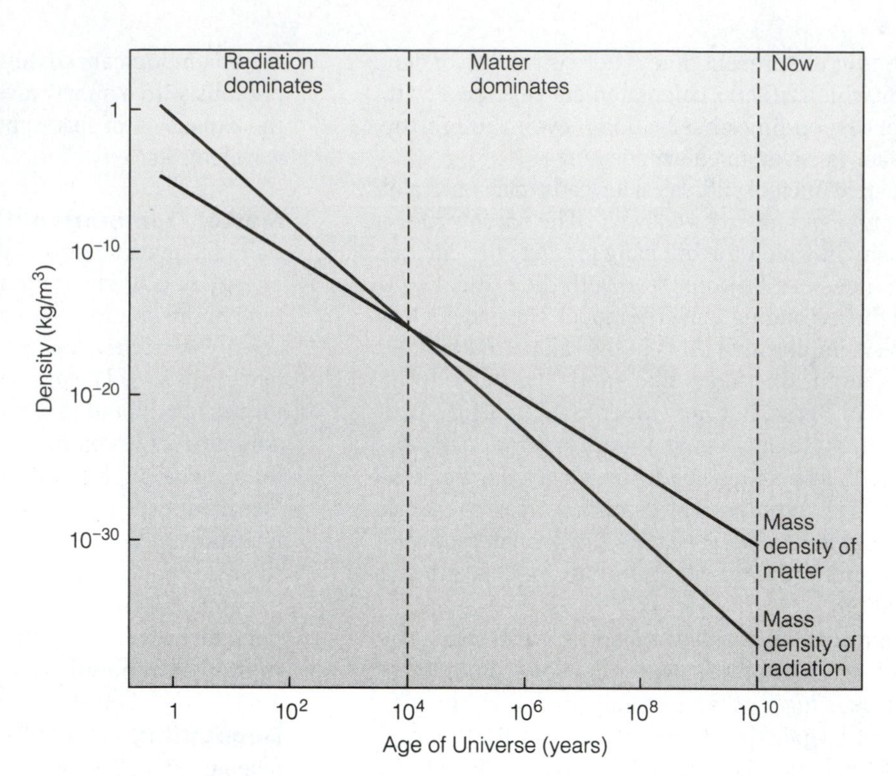

Figure 27-3 The time evolution of the density of the universe. Until 10,000 y after the big bang, the mass density of radiation exceeded the mass density of particles, making the universe radiation-dominated. The effects of cosmological expansion caused radiation to lose energy faster than matter. From 10,000 y onward, the universe was matter-dominated, and gravity could cause structures to form.

A cloud is opaque because light within it is bouncing around off many tiny water droplets. The cloud appears to have a sharp edge, but in fact its temperature and density change very gradually from inside to outside. At a certain point, the density of water droplets falls off to the point where light no longer interacts with them. From this point, the light streams to us directly, and we see an edge. The apparent "edge" of a star's photosphere is also produced by this effect. Similarly, as the universe evolves, there comes a time in the steadily decreasing temperature and density when photons can travel freely, and the fog lifts.

Since decoupling, radiation from the big bang has continued to diffuse and cool in the ever-growing volume of space. Figure 27-4 shows how temperature has changed from the era of nucleosynthesis to the present day. The universe has grown in size by a factor of 1000 since decoupling, and so the temperature of radiation from the cosmic "fireball" has been reduced from 3000 K to just under 3 K. Today, it is stars and galaxies that catch our eye, but the faded glory of the early universe has not disappeared. There are still vastly more photons than particles in the universe, and every cubic meter of space contains a billion microwave photons. (This radiation does not pose a health hazard; the power is only 10^{-5} watts, or a billionth of the power of a typical microwave oven!)

To see the microwave background radiation for yourself, tune your TV between stations. About 1% of the noise specks on the screen is due to interactions with the cosmic background radiation. The big bang is all around us.

A Cosmic Reference Frame

Although the microwave background dates from 300,000 y after the big bang, the spectrum also gives us information about much earlier times. The most recent time any physical processes could have distorted it was a few months after the big bang, when the temperature was 10 million kelvins. The thermal spectrum therefore tells us that the universe has been expanding smoothly and in thermal equilibrium since it was only a few months old.

★ The largest departure from isotropy is a broad *asymmetry* from one side of the sky to the other; the microwave background signal was $\frac{7}{1000}$ of a degree hotter in one direction than in the opposite direction (see our web site for the COBE map that confirmed this result). This asymmetry occurs because the Earth is not stationary. In terms of temperature, the radiation is a little hotter in the direction of motion and a little cooler in the opposite direction. This asymmetry leads to a fundamental question: What's moving and what's not? We define the universe as a

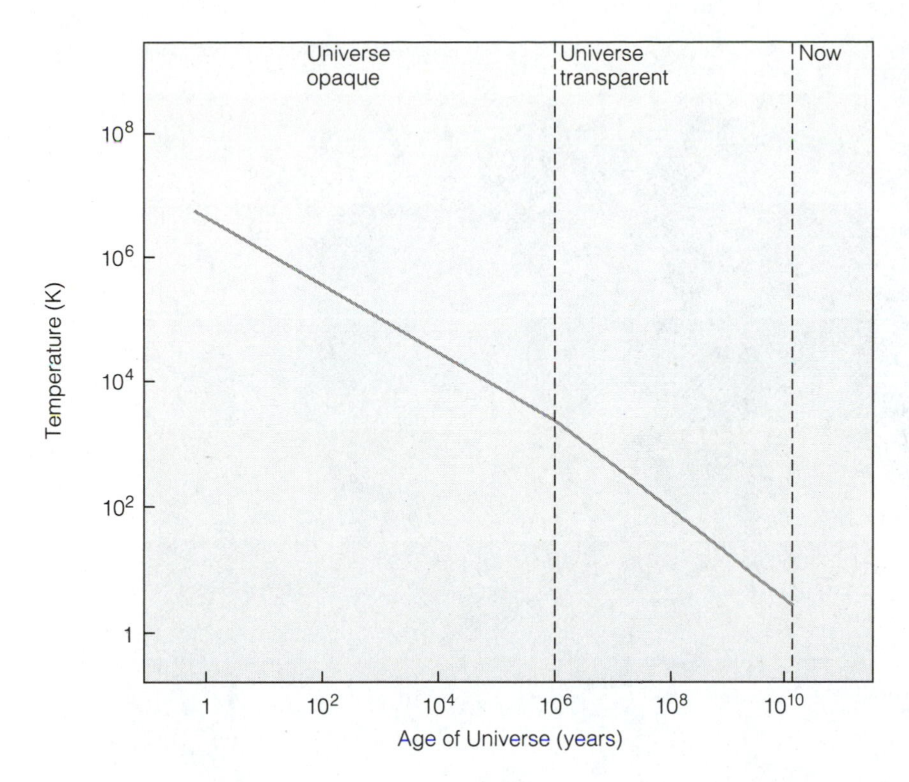

Figure 27-4 The time evolution of the temperature of the universe. Until about 300,000 y after the big bang, the radiation was intense enough to keep atoms ionized, and since photons interact frequently with electrons, the universe was opaque. After about 1 million years, the protons and electrons had all combined into hydrogen atoms, leaving photons to travel freely, and the universe became transparent.

whole as being stationary. Astronomers assume that the microwave background represents an absolute reference frame for motion, because it comes from the *entire observable universe.* The Doppler asymmetry means that the Earth's motion with respect to the microwave background is 390 km/s. In theory, all of the gravitational motions on larger scales should combine to produce this number.

Figure 27-5 shows the motions that must be understood and combined if we expect to recover a set of galaxies that is not moving with respect to the microwave background radiation. On scales of 100 Mpc and larger, astronomers can find a set of distant galaxies that approximates Hubble flow and so is at rest with respect to the microwave background. The basic assumption of homogeneity requires that the universe is smooth when observed on large enough scales. With about 10% surveyed, that smoothness has been seen (see also Figure 25-8).

The Seeds of Galaxy Formation

The contrast between the microwave background and the present-day universe could not be more striking. The radiation 1 million years after the big bang is almost perfectly smooth. Yet matter today is clumped into stars and galaxies, with virtually nothing in between. The early universe must have contained the "seeds" for today's large-scale structure (Cowan, 1994). So it was with great excitement, and more than a little relief, that scientists associated with the COBE satellite announced the discovery in 1992 of **microwave background fluctuations.** These temperature variations—70 millionths of a degree between different parts of the sky—are even smaller than those due to the motion of the Milky Way.

These tiny ripples were the starting points for galaxy formation, and decoupling was a vital step in this process. Before decoupling, collisions between the large number of photons and the much smaller number of particles acted as a kind of pressure that prevented matter from clumping. After decoupling, gravity could begin to act. How much time did it take to form the first galaxies? A few galaxies have been found at redshifts above $z = 5$. In addition, astronomers working with the Sloan Digital Sky Survey have discovered quasars with redshifts above $z = 6$. The galaxies that host these quasars must have formed at even higher redshifts. The time between $z = 6$ and $z = 1000$ (the era of decoupling) depends on the cosmological model, but is typically 300–500 million years. This is the amount of time available for the transition from shimmers in the primeval fireball to the stark beauty of spiral and elliptical galaxies (Macchetto and Dickinson, 1997).

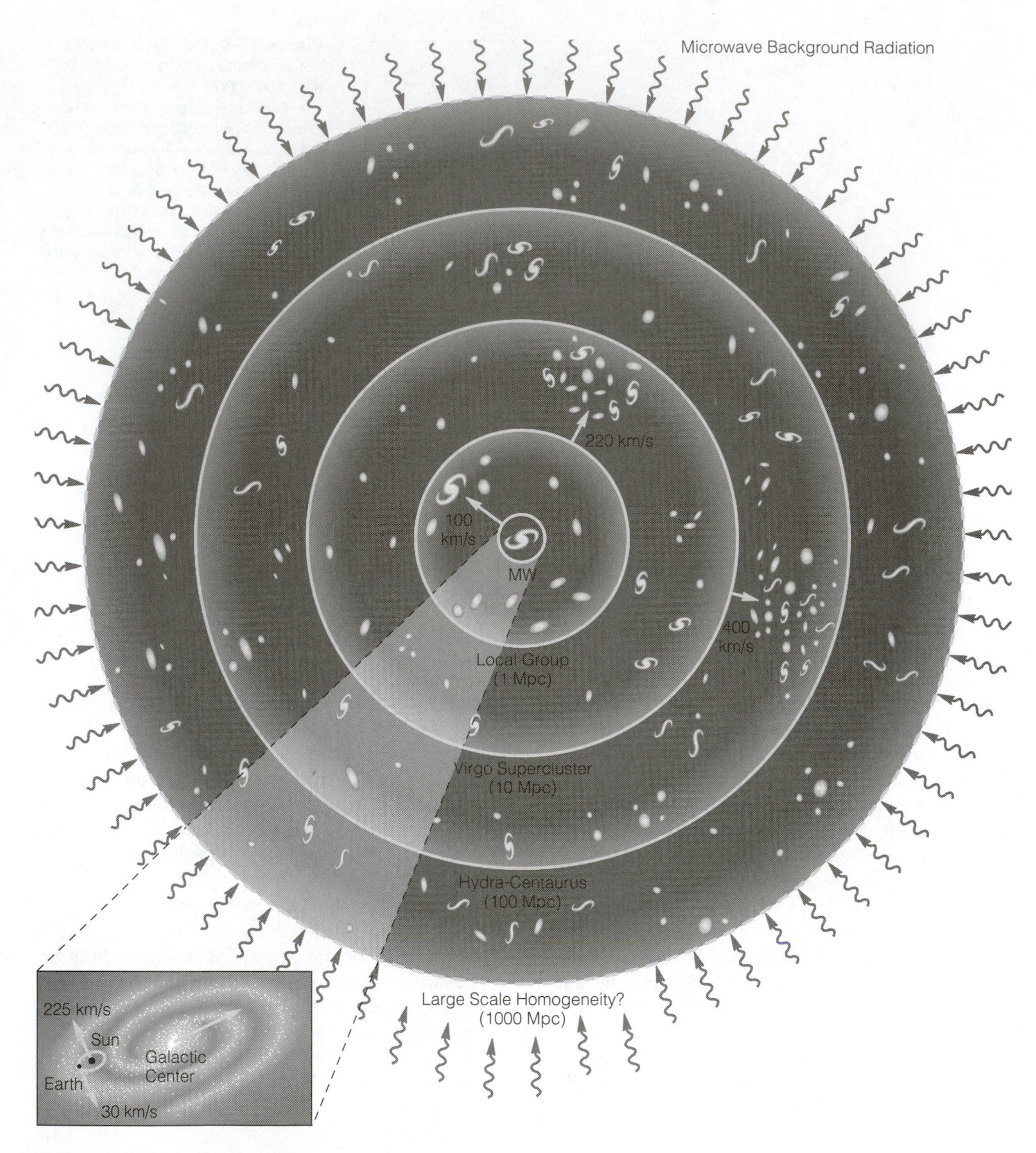

Figure 27-5 The cosmological frame of reference for measuring motions. The Earth moves around the Sun at 30 km/s, and the Sun moves around the center of the Milky Way galaxy at 225 km/s (the inset shows an expanded view of the motions within the Milky Way). The Milky Way is falling toward the Andromeda galaxy (M 31) at 100 km/s. The entire Local Group, including the Milky Way, M 31, and a few dozen dwarf galaxies, is falling toward the Virgo cluster at about 220 km/s. The Local Supercluster is being tugged toward the region of the Hydra and Centaurus clusters at about 400 km/s. The galaxy distribution appears to be homogeneous on scales much larger than 100 Mpc. (Adapted from a diagram by J. Silk)

Galaxy formation is one of the most hotly debated topics in modern cosmology. Why is a galaxy the basic mass unit of the universe? Why isn't the universe dominated by planets, or grains of dust, or objects larger than galaxies? Luckily, physics provides some guidance. *Density fluctuations* mean that regions of slightly higher and lower density were interspersed. Regions of enhanced density have a higher gravity and so attract nearby material. As the collapse begins, it is resisted by higher pressure within the compressed region. This is the same tussle between gravity and pressure that governs the life history of a star (see the discussion of star formation in Chapter 18). An object can form only if gravity wins the battle.

At the time of decoupling, the size of the inhomogeneity where gravity and pressure were just balanced was about $10^5\ M_\odot$. Lumps smaller than this would not form. Nor would lumps larger than about $10^{12}\ M_\odot$ form, because a very large cloud cannot cool *fast enough* to collapse. It may be a coincidence, but this range for the mass of collapsing lumps at a redshift of $z = 1000$ is very close to the mass range of present-day galaxies (see Table 24-3).

A new generation of experiments is taking observations of the microwave background to a new level. Since the Earth's atmosphere absorbs microwaves, such observations are best done from space, and the future *MAP* and *Planck* missions will view the radiation with an unprecedented level of detail. Meanwhile, useful observations can be made from high-altitude balloons or from an observatory at a dry site like the Antarctic plateau. In 1999, the *Boomerang* balloon experiment measured the spectrum of microwave fluctuations and found a peak at a position that indicates flat space. This is a powerful and direct test of space geometry using radiation that dates back to the first 2–3% of the age of the universe. An indirect consequence of this result is to strengthen the evidence for a cosmological constant, since there is not enough matter in the universe to make space flat. It appears that some kind of mysterious energy from the vacuum of space stretches space out to be geometrically flat. This vacuum energy counters the force of gravity and accounts for the apparent acceleration of the universe as traced by supernovae, a result described in the last chapter.

The Role of Dark Matter We know that there are larger structures in the universe, however. Clusters and superclusters range up to $10^{17}\ M_\odot$. Which formed first, galaxies or these larger structures? Not surprisingly, the answer depends on the nature of the dark matter. In the absence of any experimental information, cosmologists discuss two scenarios—top-down and bottom-up structure formation.

The scenario of *top-down structure formation* is based on *hot dark matter,* particles that were moving at relativistic velocities at the time of decoupling. The first structures to form were superclusters. Galaxies formed later, by fragmentation within the supercluster. The alternative scenario, *bottom-up structure formation,* is based on *cold dark matter,* particles that were moving slowly at the time of decoupling. With cold dark matter, the first structures had masses in the range 10^6 to $10^{11}\ M_\odot$. The larger the scale, the later the growth of fluctuations, so structures form in a hierarchical fashion, from small groups of galaxies up to superclusters.

These two scenarios make quite different predictions. A key observational test is whether galaxies or clusters formed first. Indirect evidence suggests that clusters are considerably younger than the galaxies they contain. As a cluster evolves, it becomes more and more symmetrical; the irregular shapes of some clusters suggest that they are very young. On the largest scales, superclusters may only just now be forming. All this argues for bottom-up clustering guided by cold dark matter.

★ No set of equations can describe the rich structure of the universe. The most powerful way to test the models is to create a "universe in a computer." In recent years, astrophysicists have used powerful supercomputers to make *numerical simulations* of the evolution of large-scale structure (for examples, see our web site). In the simulations, the motion of individual particles is calculated according to Newton's law of gravity, while the volume in the calculation is steadily increased to represent the expanding universe. The enormous power of supercomputers allows tens of millions of calculations to be carried out each second. In general, models with cold dark matter give a better match to the clustering of the observed universe than models with hot dark matter.

Limitations of the Big Bang Model

The big bang model is very successful in describing the evolution of the universe from the age of 1 second to 1 million years. After 1 million years, our understanding of galaxy formation and the evolution of large-scale structure is limited by ignorance of the nature of dark matter. In the first second, the limitation is uncertain particle physics and the lack of a unified theory of the forces of nature. However, there are several key aspects of the universe that the simple big bang model does not explain.

The first relates to the mass density of the universe. A variety of observational tests indicate that Ω_0 is in the range 0.2–0.3. Although less than 1, the mean density is *quite close* to the critical density. To see why this is remarkable, consider that in an expanding universe, the behavior of Ω is unstable. If it differs from 1 in the early universe, then Ω will rapidly become either very large or very small. Only if it is exactly 1 in the early universe will Ω still be close to 1 after billions of years. Now reverse the argument. If we measure Ω to be close to 1 now, billions of years after the big bang, then it must have been exquisitely close to 1 in the early universe. The apparent flatness of space has no explanation in the standard big bang model.

A second mystery is the smoothness or uniformity of the early universe. The universe in many ways resembles a gas. If heat is introduced to one region of a volume of gas, collisions between molecules will gradually distribute the heat, until the entire volume reaches the same temperature. The microwave background radiation—with a temperature constant to within 0.007%—certainly indicates that the early universe was in nearly perfect thermal equilibrium. The problem is that in an expanding universe, there is a limit to how fast thermal equilibrium can be established—a limit given by the speed of light. If two regions of space are moving away from each other very rapidly, no signal can pass between them, and they cannot equalize their temperatures. About 300,000 y after the big bang, the expansion was so fast that there is no possible way that different patches of sky could have come into thermal contact and equalized their temperature. Yet opposite directions in the sky have virtually identical microwave background temperatures.

Finally, the microwave background shows tiny fluctuations out of which galaxies eventually grew. The big bang gives no explanation for either the origin or the form of these fluctuations. To gain a deeper understanding of the big bang, we must look at the exotic conditions of the very early universe.

THE VERY EARLY UNIVERSE

We have followed the story of the big bang from the first few seconds to the present day. This saga includes major events such as the explosive nucleosynthesis of light elements, the separation of radiation and matter, the formation and clustering of stars and galaxies, and the evolution of humans to ponder the meaning of it all.

As we probe back toward the origin, the universe becomes microscopic and approaches a state of infinite temperature and density. Is the universe comprehensible in the first fraction of a second? The material that follows is speculative. The ideas are abstract and remote from common experience. But the speculation is supported by known laws of physics and is fueled by a desire to understand the first instants of creation. The early universe was an extraordinary laboratory for the creation and destruction of the fundamental constituents of matter. The temperature and density were so extreme that no laboratory on Earth can duplicate the conditions. Nevertheless, fossil relics and fragments of information from the early universe may lead to an understanding of the forces of nature. There is also the hope of deeper insights into the meaning of matter, space, and time.

★ There is a limit to our knowledge of the early universe. The general theory of relativity treats space and time as a smooth continuum. In the current large, low-density universe, this is a good assumption, but the very early universe was dominated by quantum fluctuations of matter. General relativity fails when the scale of space curvature equals the quantum uncertainty in the position of a particle. This means that general relativity does not explain the very early universe—the first 10^{-43} s when the universe was only 10^{-35} m in diameter. This incredibly small scale, 10^{-35} m, is called the *Planck distance* after one of the pioneers of quantum mechanics. The fantastically small iota of time it takes light to travel this distance, 10^{-43} s, is called the *Planck time.* We have no quantized theory of gravity that can properly describe the universe before the Planck time. For more details on the physics of the very early universe, see the additional material on our web site.

Energy, Matter, and Antimatter

To understand the early universe, we must enter the world of the quantum. Early in the twentieth century, physicists discovered that the smooth and continuous surfaces of objects concealed a microscopic discreteness. Matter is made of atoms. Even more surprising, they found that energy, too, is subject to this discreteness. There is a fundamental unit of energy that cannot be further subdivided—the quantum.

As we saw in Chapter 5, the electrons around the nucleus of an atom can have only certain fixed energy levels. When electrons change orbits and emit or absorb photons, those photons are also constrained to have certain fixed energies or quanta. The quantum of energy is a tiny number; the graininess of the quantum is not apparent in the everyday world. The quantization of energy applies to *any* particle or photon. In the classical view, particles are smooth and clearly delineated, like billiard balls or grains of sand. In the quantum view, matter has a fundamental fuzziness; uncertainty is always involved in defining the position (or velocity, or any other property) of a particle. This is expressed as the *Heisenberg uncertainty principle.* This uncertainty has nothing to do with limitations in the measuring apparatus. It is a basic consequence of *making* a measurement. To know a particle's position, you must measure it, and in the quantum world, the act of measurement changes the position. This represents a fundamental limit to our knowledge of the physical world.

Mass and Energy We are used to thinking about the essence of radiation as *energy* and the essence of matter as *mass.* In the special theory of relativity, however, Einstein related these two quantities by the expression $E = mc^2$. This essential attribute of the quantum world is called **mass-energy equivalence.** This equivalence has several implications. For example, since energy of all types has an equivalent mass, it must be responsive to gravity. Another implication is that matter is not as permanent as it seems. Matter can decay into pure energy, and matter can be fabricated from pure energy. Making a tiny bit of mass takes a prodigious amount of energy—but there was enough radiant energy in the early universe to create particles in copious quantities.

Matter and Antimatter In the microscopic world of the quantum, matter and antimatter are on an equal footing. **Antimatter** is material that has properties equivalent to matter but that just happens to be rare in the universe. Every particle has an equivalent antiparticle, with an opposite set of quantum properties. Charged particles have antiparticles with the opposite charge. For example, the antiparticle of the electron is the positively charged positron, and the antiparticle of the proton is the negatively charged antiproton. Neutrons, which are neutral particles, have neutral antineutrons, and neutrinos have antineutrinos. The photon is its own antiparticle.

Particles and antiparticles can be created out of pure energy, as specified by Einstein's mass-energy equivalence. The mechanism is called *pair creation.* This is the way that physicists (and nature) create antimatter on Earth. Antimatter is created momentarily in the upper atmosphere by cosmic rays, or expensively and in minute quantities by enormous particle accelerators. The annihilation process is defined as the disappearance of a particle/antiparticle pair, releasing high-energy radiation or gamma rays. The annihilation of particles and antiparticles, called *pair annihilation,* is a perfectly efficient process; 100% of the mass-energy is liberated.

The ordinary world is composed overwhelmingly of matter. When an antiparticle is created, it cannot survive long before encountering a particle. The result is annihilation in a dazzling display of gamma rays. Nevertheless, since the organization of matter and antimatter is so symmetrical, many have wondered whether there are large concentrations of antimatter somewhere in the universe. It is hard to hide antimatter. There are no antimatter stars anywhere near the Sun; they would react with the interstellar medium to produce far more gamma rays than are observed. Perhaps there are distant antigalaxies composed of antistars, shielded from annihilation by the depths of intergalactic space? Yet in the early universe, all galaxies must have been much closer together, and the result would still be a large annihilation signal, redshifted by the expansion to lower energies. We conclude that the universe is made of matter.

The First Few Seconds Now we can apply these ideas from particle physics to the context of the early universe. For most of the age of the universe, matter has been dominant. Even though there are 1 billion photons for every particle, photons have low energy and interact too weakly with matter to be significant. The universe is governed by the force that molds matter—gravity. At one-millionth of a second (10^{-6} s) after the big bang, the situation was entirely different. The temperature was a prodigious 10^{14} K, and photons had enough energy to create particle/antiparticle pairs of many kinds. The processes of particle creation and destruction occurred at equal rates, so there were roughly equal numbers of particles and photons.

About a millisecond (10^{-3} s) after the big bang, cosmological expansion had redshifted the radiation such that gamma rays no longer had enough energy to create protons and antiprotons. Now we face an interesting situation. If particles were being made only by pair creation, there would be *equal* numbers of protons and antiprotons. When the universe cooled, these would all annihilate, leaving only radiation and no matter! Thus, there must have been a tiny excess of matter over antimatter in the very early universe. For every billion antiprotons, there must have been 1 billion plus 1 protons. When the protons and antiprotons annihilated and created a flood of gamma rays, a slight residue of matter remained. Slightly later, one second after the big bang, the same story was played out with electrons and positrons.

Now, billions of years later, the original gamma rays of the big bang have been redshifted to feeble microwaves. The photons of the microwave background have a temperature of just under 3 K. Although they are feeble, there are enormous numbers of them: Each cubic meter of space contains about 10^8. Most of the particles in the universe long ago annihilated with their corresponding antiparticles, and the residue of neutral matter has formed all the gravitational structure in the universe. In other words, our existence depends on a tiny asymmetry between matter and antimatter in the very early universe!

The Forces of Nature and the Early Universe

For a deeper knowledge of the very early universe, we must consider the fundamental forces of nature. Like the Greeks before them, modern scientists hope to discover the simplest possible description of the physical world. They start with several assumptions, the most basic being that the universe is understandable. Humans are very attracted to this viewpoint, but it *need not be true!* A second assumption is that the universe is governed by physical laws that can be described mathematically. A third assumption is that theories that most closely represent the physical world will be simple. This does not mean that the mathematics is necessarily simple; it can often be fiendishly difficult. Rather, it means that successful theories will have as few assumptions and arbitrary parameters as possible (Hogan, 1998).

The success of science has strongly affirmed these three assumptions. Think of the rich diversity of structures in the cosmos, from the swirl of spiral galaxies to massive black holes to the filigree of arcs and voids in the distribution of galaxies. The single force of gravity has shaped all of these structures. Think also of the variety of the material world, from lustrous precious metals to inert and colorless gases to the radioactive fizzing of heavy elements. Atomic forces yield this variety.

The final ingredient in modern theories is the notion of **symmetry,** the idea that diverse physical phenomena have a simple underlying basis. The conservation laws of physics (mass, energy, charge, and so on) all express an underlying symmetry. Though the symmetry may not always be obvious in the everyday world, it is contained in its most pure form by the theories that represent nature.

The symmetries of nature become apparent at the high energies of the early universe. Mass and energy are interchangeable, and an equality is established between the creation and annihilation of particles. The symmetry between matter and antimatter is also restored. The present-day universe is massively skewed toward matter, and antimatter can be produced only in tiny, ephemeral amounts. The early universe created and destroyed particles and antiparticles in equal numbers. The most profound asymmetry in our everyday world is the arrow of time. The sense of time as forward motion is too deeply rooted for us to question. Yet at the subatomic level, time-reversed particle interactions make equal sense and are equally likely to occur.

The Four Fundamental Forces The quest for symmetry continues with the forces of nature. There are four forces in the physical world. The two with which we are most familiar—**gravity** and **electromagnetism**—operate over an infinite range and contribute to the structure of the macroscopic world. The electromagnetic force is 10^{38} times stronger than gravity. If you doubt this, take an air-filled balloon and notice how the slight static charge of the balloon rubbed against your head will hold it on the ceiling against the full attractive force of the entire Earth. Similarly, a modest magnet holds a nail against the Earth's gravity. In normal matter, electromagnetic forces are literally neutralized by the equal numbers of negatively charged electrons and positively charged protons.

The other two forces of nature operate over very short distances within an atom. The **weak nuclear force** is actually far stronger than gravity, but its range is 100 times smaller than an atomic nucleus. It is responsible for the radioactive decay of massive nuclei and for the decay of a neutron into a proton, an electron, and an antineutrino. The weak force is also responsible for the interaction of neutrinos with other particles. The **strong nuclear force** has a range about the size of an atomic nucleus, and it acts to bind neutrons and protons inside a nucleus. To do so, it must overpower the electromagnetic force, which would otherwise cause the positively charged protons in a nucleus to repel each other. Table 27-1 summarizes the properties of the four forces.

The Idea of Unification

The ultimate expression of physicists' belief in symmetry is the search for **unification,** a theory demonstrating that the four forces of nature are manifestations of a single unified force. Our low-energy world may represent a situation of broken symmetry, in which the underlying similarities between forces and particles are concealed. Imagine you spin a roulette wheel. At first, the ball has high energy and moves in

TABLE 27-1

The Fundamental Forces of Nature

Force	Relative Strength	Range	Boson Carrying the Force	Interacting Particles	Role
Strong nuclear	1	10^{-15} m	Gluon	Quarks	Holds atomic nuclei together.
Electromagnetic	10^{-12}	Infinite	Photon	All charged	Holds atoms together, governs propagation of electromagnetic waves.
Weak nuclear	10^{-14}	10^{-17} m	Vector boson	Quarks, electrons, neutrinos	Involved in radioactive decay.
Gravitation	10^{-40}	Infinite	Graviton	All	Holds planets, stars, and galaxies together and governs their motions.

circles. At the end, when the wheel has stopped, the ball rests in one of 37 numbered slots. We might watch the results of many spins and conclude that the ball has 37 different states. However, at high energy the ball always has the same motion; the symmetry is broken only at low energy.

As described so far, the four forces of nature could not appear more different. Some have infinite range, some have short range; they are carried by particles with different spins and different masses, or no mass at all; and they differ by a factor of nearly 10^{40} in strength! It is apparent that the elegance and economy of a single theory of nature can be tested only at exceptionally high energies—for this reason, particle physicists have become extremely interested in the very early universe.

The expectation is that the four forces of nature were on an equal footing in the first instant of the universe, the Planck time. At 10^{-43} s after the big bang, the density of matter was 10^{64} times that of lead. The entire visible universe was contained in a region 0.1 mm across! Imagine it: Hundreds of millions of galaxies on the head of a pin!

As we trace history back toward the big bang, general relativity contains the seeds of its own downfall. At the point where the scale of quantum fluctuations rivals the scale of space curvature due to gravity, the theory breaks down. A quantized theory of gravity is required. The universe was a seething cauldron of matter and energy appearing and disappearing out of a vacuum, and space and time twisted and fractured like foam. Figure 27-6 shows the theoretical expectation that the interaction strengths of the four fundamental forces are equal at the extremely high energies or temperatures seen in the very early universe.

Unified Theories The first steps toward unification were made in the 1970s with an *electroweak theory* that united the electromagnetic and weak nuclear forces at temperatures above 10^{15} K. This energy occurred in the early universe some 10^{-12} s after the big bang. At lower energies, the symmetry is broken, and the weak force operates only over short distances. The electroweak theory was convincingly confirmed when the particles that carry the weak force were observed in 1983.

Emboldened by this success, theorists have attempted to bring the strong nuclear force under the umbrella of unification in **grand unified theories (GUTs).** None of these theories has yet been completely successful, and none incorporates the force of gravity. The proposed epoch of GUTs occurs after only 10^{-35} s, when the universe had a phenomenal temperature of 10^{28} K. The basic idea behind GUTs is simple. At a high enough energy, the weak, strong, and electromagnetic forces should be equal. The only possible laboratory for GUTs is the very early universe.

GUTs predict that ordinary matter should be unstable! We think of protons as stable particles, but the uncertainty principle allows that very occasionally a proton can decay. The probability of a proton decay is so low that you would have to wait over 10^{32} y to see it happen (much, much longer than the age of the universe)! However, the alternative is to gather 10^{32} protons and wait only one year. Particle physicists have set up experiments in deep mines (to shield their detectors from contaminating cosmic rays), hoping to register proton decay, but so far they have not done so. The observed limit on the proton lifetime is already good enough to rule out some of the simpler GUTs.

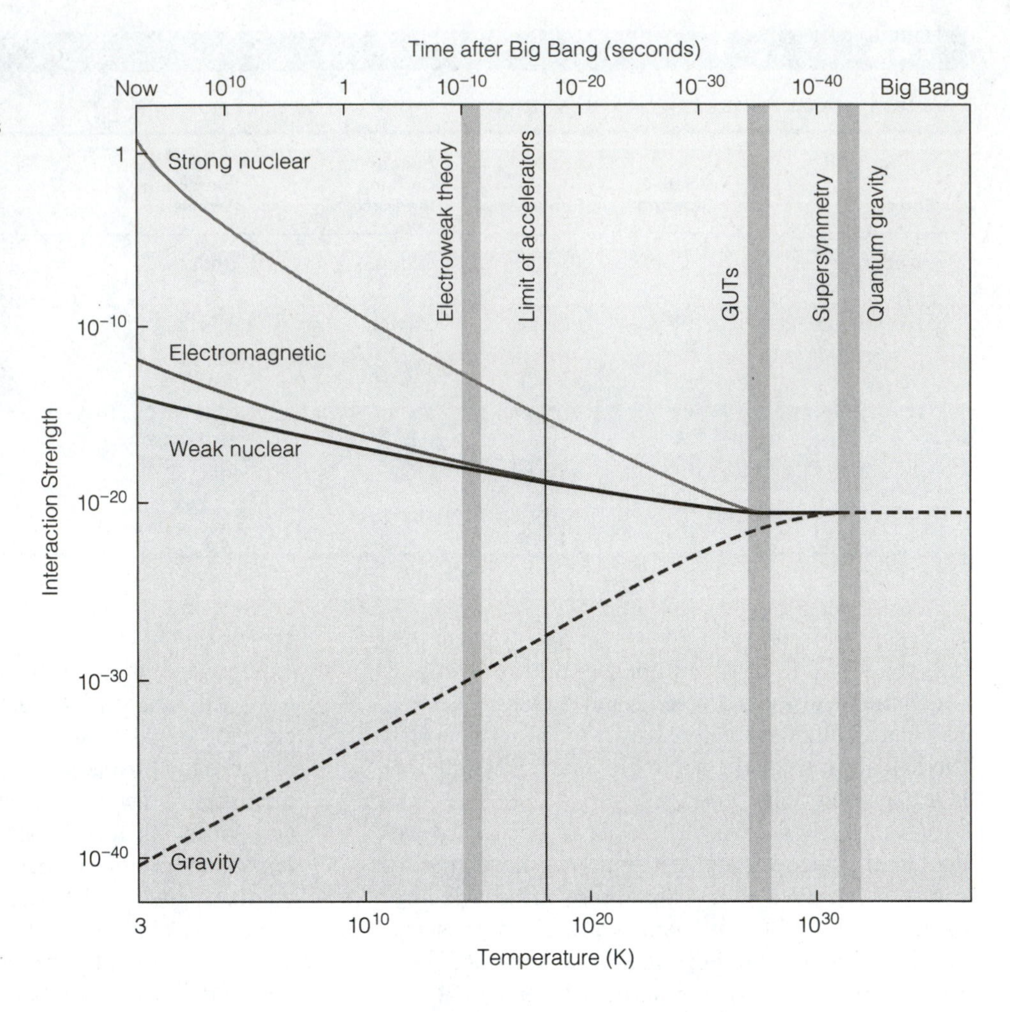

Figure 27-6 A plot showing the theoretical expectation that the interaction strengths of the four fundamental forces will become equal at a high enough energy or temperature. The top scale shows the time after the big bang corresponding to each temperature. The vertical bands show the unification temperatures of the forces, and the vertical line shows the maximum temperature that will be realized by planned accelerators. Gravity is shown as a dotted line because unification theories that incorporate gravity are very speculative.

GUTs might also explain one of the mysteries of the big bang: the excess of matter over antimatter. It turns out that a slight asymmetry between the behavior of matter and antimatter was discovered experimentally in 1964. GUTs can incorporate this effect to naturally produce an excess of matter over antimatter at temperatures near 10^{28} K. Had the excess gone the other way, we would just have called antimatter matter, and matter antimatter! Unfortunately, GUTs give no concrete prediction of the *amount* of the excess. Unfortunately, most GUTs require that 30 or more numbers be "fed in" to the theory or experimentally determined before the theory can be used to make predictions. This is not the elegance and simplicity that physicists had in mind. For this reason, and because GUTs do not incorporate gravity, scientists have recently focused their search on a theory that unifies *all four* forces of nature.

Current theoretical work on unification involves the idea of treating particles not as points but as *superstrings.* This conceptual shift make the calculation of particle interactions much more tractable. Superstrings are a Planck-length long, or 10^{-35} m, and the properties of the known particles are generated by *vibrations* of the superstrings. A host of new particles are predicted, too. Superstrings can be open, or they can be closed in a loop; they can interact, merge, join, and split with great complexity due to their vibrations. There is a twist, however: The mathematics that preserves the symmetry of superstring interactions works only in 10-dimensional space-time! According to the theory, the extra dimensions remain "balled up" in a region the size of the Planck length. This is the ultimate broken symmetry, with the familiar four dimensions of space and time masking the loss of six dimensions. Superstrings are at the speculative edge of theoretical physics. The search for the ultimate theory of nature is far from over.

Inflation

The most important modification to the standard big bang model was an idea proposed by Alan Guth in 1981. The **inflationary universe** model proposes

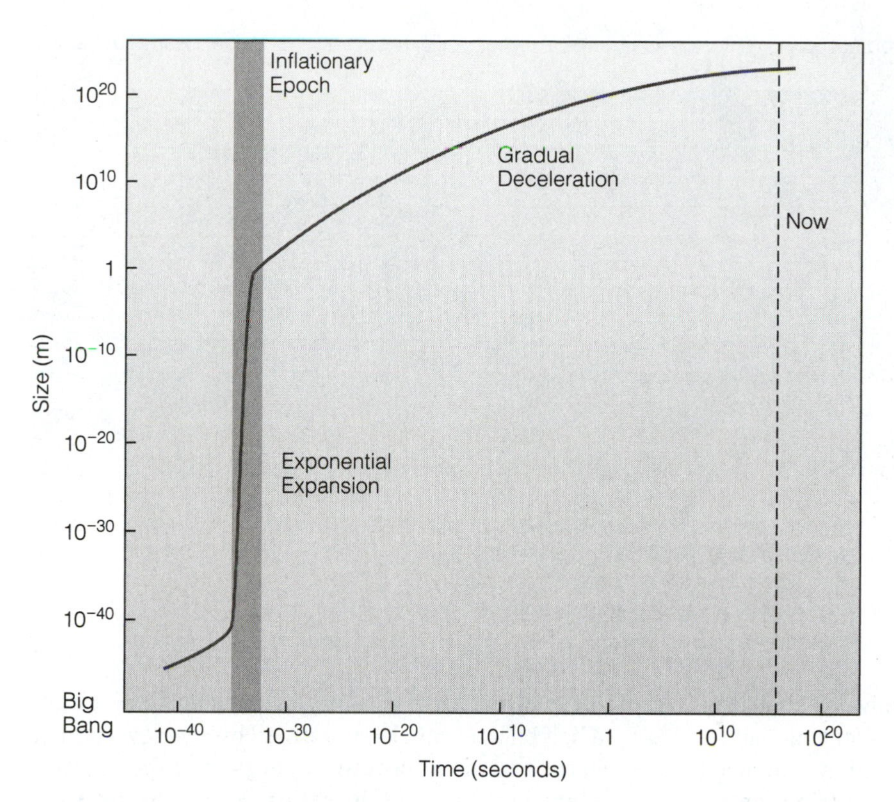

Figure 27-7 Graph showing the enormous and rapid expansion of the universe predicted by the inflationary model. After the inflationary epoch, the expansion decelerates slowly due to the gravitational attraction of matter in the universe.

that the infant universe went through a period of extremely rapid expansion just after the Planck time. The term *inflation* indicates that the universe expanded at an increasing rate rather than a decreasing rate, as it does today. During the inflationary era, the universe expanded in size by many billions of times between 10^{-35} and 10^{-33} s after the big bang (Figure 27-7). The inflation moved matter that originally was near us (within a few meters) to a position far outside today's observable universe (much more than 10 billion light-years). The observable universe is therefore expanding into space containing matter and radiation that was once in close contact with our location. It is a truly audacious theory that seeks to explain the entire universe in terms of such a microscopic and brief event.

The inflationary universe model explains two of the mysteries of the big bang model—the horizon problem and the flatness problem. The smoothness or isotropy of the cosmic background radiation is puzzling, because different regions of the early universe were not in causal contact. Causal contact means that radiation had time to travel between the different regions. That is the horizon problem. During inflation, a region about 1 billion times smaller than a proton expanded to a size of many billions of light-years. The very different regions of the sky from which we now detect the cosmic microwave background were originally in very close contact. That is why they have the same temperature.

Whatever curvature the universe might have had before the inflationary era was stretched out by the enormous inflation. Imagine a tiny balloon that is rapidly inflated to many times its original size (Figure 27-8) that any section we choose to explore is almost perfectly flat. The inflationary model predicts that space should be extremely close to flat. A flat universe is no longer a coincidence; it is an inevitable consequence of the inflationary era.

Inflation appears to violate the idea that nothing can travel faster than the speed of light. Remember, however, that the constraints of relativity apply only to particles or radiation moving *through* space. Inflation was the rapid and prodigious *expansion of space itself*. The inflationary model presents us with a cosmos that has stretched far beyond our horizon. Therefore, the observable universe is probably a tiny fraction of the total physical universe. Whatever the power of our largest telescopes, we are consigned to a humble corner of an immense cosmos.

The inflationary model is still being developed. Many of its consequences are speculative due to our uncertain knowledge of particle physics at the energy of the GUTs. All this would be no more than esoteric speculation if inflationary models did not make *predictions* that can be tested by observation. Inflation

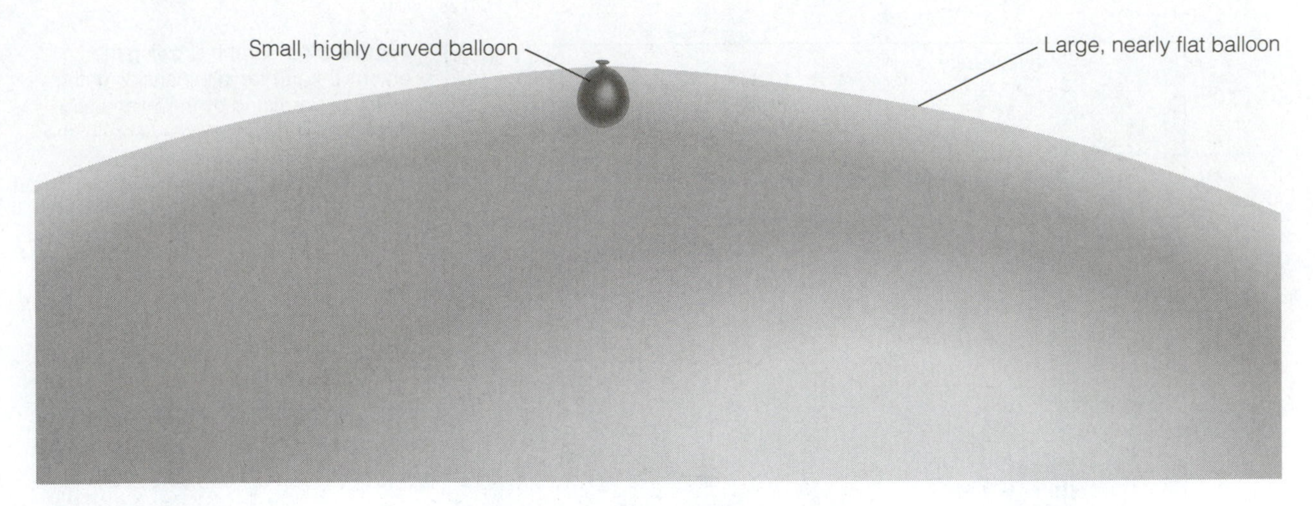

Figure 27-8 The inflationary model's solution to the flatness problem. By the end of the inflationary era, the universe had expanded by such a large factor that, whatever the original curvature, it was indistinguishable from being flat. In a similar way, measurements made over a small region of the Earth cannot detect the curvature of the surface.

makes two clear predictions. The universe should be exactly flat, that is, $\Omega_0 = 1$. The jury is still out, but recent studies of large-scale structure have deduced that $\Omega_0 = 0.2$–0.3. The remaining contribution required to make space flat may be vacuum energy in the form of a cosmological constant. Also, inflation smooths out the universe so effectively that the only ripples remaining are *quantum fluctuations.* These ripples are the seeds from which galaxies must eventually grow. The model predicts that quantum effects should produce equal numbers of large-scale and small-scale ripples (technically, this is called a flat power spectrum). Remarkably, the tiny fluctuations in the microwave background observed by the COBE satellite and the recent *Boomerang* experiment have exactly this property.

The universe began with the origin of space, time, matter, and energy. After an inflationary epoch that flattened space, particles combined to leave radiation dominant. Once the three other forces merged, gravity was free to exert its attractive pull. Gradually, it sculpted the ripples in the early universe into structures on all scales. This process was influenced by the large amount of unseen dark matter. One billion years after the big bang, enormous gas clouds collapsed into galaxies with great gaps between them. Within them, legions of stars switched on for the first time. For 6 or 7 billion years, stars followed a pattern of birth and death. Most stars died a quiet death, as slowly fading embers. The massive ones exploded, seeding the space between stars with heavy elements. Many stars ended up surrounded by the debris of their own creation. On one such rocky fragment, life evolved out of a watery, organic broth. Four and a half billion years later, warmed by a yellow star and bathed in the microwave afterglow of the creation event, humans point their telescopes at the skies and ponder the meaning of it all.

SUMMARY

The attempt to understand the early universe is one of the loftiest goals of the human intellect. Modern cosmology sees humans as insignificant observers of a vast universe, riding out the aftermath of an enormous explosion billions of years ago. The big bang is only a model, but it is a very successful model. One consequence of the rapid expansion is the limited amount of the universe that can be observed at any time. The finite speed of light and the finite age of the universe limit our view to an observable fraction of a larger and possibly infinite physical universe.

The big bang accounts for the observed abundance of the light elements, where helium in particular was created by nucleosynthesis in the early hot phase. The other unequivocal reminder of the big bang is the microwave background radiation. This has been redshifted and diluted by the universal expansion to just the level predicted by the model. The microwave radiation is amazingly smooth, with just two deviations from total isotropy. The first reflects the motion of the Milky Way with respect to this universal background. On scales of 100 Mpc or larger, galaxy velocities approximate Hubble flow, and galaxies mark out a reference frame that is at rest with respect to the expanding universe.

Recently, a second, even tinier anisotropy has been discovered; it represents the fluctuations from which galaxies eventually evolved. Despite its successes, the big bang leaves some basic questions unanswered. The search for a more complete model leads to the physics of the very early universe, when particles and antiparticles were created from the intense radiation field. Cosmologists routinely speculate about the first fractions of a second of a universe that is currently more than 10 billion years old. The inflationary universe model proposes that the infant universe underwent a prodigious and rapid expansion—this supposition explains the isotropy of the microwave background and leads to the prediction that the geometry of the universe should be exactly flat.

Cosmologists suppose that four forces of nature were unified and indistinguishable very early in the evolution of the universe. As the universe cooled, a slight asymmetry among the forces led to the universe's consisting overwhelmingly of matter rather than antimatter. The early universe reflects the words of Roger Bacon, a medieval philosopher of science: "There is no excellent beauty that has not some strangeness in the proportion." At the dawn of the scientific age, Plato was well aware of the limitations of our view of reality. He likened us to cave dwellers trying to discern the truth by interpreting the flickering shadows on the cave walls. For all its successes, cosmology has acquainted us with a shadowy world that we do not perfectly understand.

CONCEPTS

big bang model
cosmic nucleosynthesis
cosmic background radiation
radiation-dominated universe
matter-dominated universe
decoupling
microwave background fluctuations
mass-energy equivalence
antimatter
symmetry
gravity
electromagnetism
weak nuclear force
strong nuclear force
unification
grand unified theories (GUTs)
inflationary universe

PROBLEMS

1. Which phrase describes the big bang theory best?

a. An assumption with logical consequences that turn out to be verified by observation

b. A fact proved by repeated observation

c. A revelation

2. Why was the discovery of 3-K radiation from all over the sky heralded as strong evidence in favor of the big bang theory? Does comparing the radiation from different parts of the sky give any evidence of asymmetry or inhomogeneity in the universe?

3. Why would planets such as Earth be unlikely to exist in globular clusters? (*Hint:* Consider the Earth's composition.)

4. How would spectroscopic observations help reveal that a cluster of very distant galaxies, which looks like a mere grouping of fuzzy stars on a photograph, is not a group of stars inside our own galaxy?

5. Why does the radiation from the microwave background as observed today have such long wavelength (radio waves) instead of the very short wavelengths (ultraviolet) that might be expected from the high temperatures theorized for the early universe?

6. Do you find any fundamental disagreement between the description of the universe's origin and structure, as described in this and the previous chapter, and any philosophical or religious beliefs you may hold? If so, do you believe such a disagreement might be clarified by further observations, or do you believe further observations are superfluous?

7. Where did the big bang occur? Explain your answer. Where did the photons in the microwave background come from?

8. The inflationary big bang model proposes that the universe underwent a brief period of rapid expansion only 10^{-35} s after the big bang.

a. How does this model solve the horizon problem?

b. How does this model solve the flatness problem?

c. Is there any observational evidence in favor of this model?

ADVANCED PROBLEMS

9. What defines the edge of the observable universe? How far away is this horizon? Assume the universe is spherical, and calculate the amount of volume by which the observable universe gets bigger each day. (*Hint:* The volume of a spherical shell of thickness Δr and radius r is $4\pi r^2 \, \Delta r$.)

10. If Planck's constant is such a tiny number, why are quantum processes so important in the early universe?

CHAPTER

28

Life in the Universe

WHAT THE READER SHOULD WATCH FOR IN THIS CHAPTER

The possibility of life beyond the Earth is one of the most profound issues in science. The topic is interdisciplinary and involves astronomy, geology, chemistry, and biology. The first issue is the definition of life and the conditions that are necessary for life to exist on planets. This initial speculation leads to the question of whether there are planets that could support a similar evolution of life elsewhere in the universe. A long and not fully understood process led to the formation and development of life on Earth. With considerably less certainty, scientists can speculate about the factors involved in the evolution of intelligence and the eventual appearance of species that have the technology to explore their universe. All of the arguments combine to estimate a probability about whether intelligent life actually exists on planets beyond the solar system. We then discuss the optimal search strategy and summarize previous and planned attempts at communication. The idea of life in the universe has been the focus of human dreams and speculations for hundreds of years. Science is now beginning to provide the answer to the question: Are we alone? ■

Are we alone in the universe? Or is our planet only one of many on which chemical processes and a benign climate have led to the origin of life? Are a million civilizations scattered across our galaxy? These questions form a natural "bottom line" to astronomy. They constitute a new and somewhat controversial field, called **exobiology,** which is defined as the study of life and biochemistry outside the Earth's biosphere. The field is somewhat controversial because, after all, we have only one example of an inhabited planet in the universe, and unkind critics have called exobiology a field without any subject. On the other hand, we now know of a wide range of complex chemical processes occurring on other planetary bodies and even in interstellar clouds, and we know of several potentially habitable worlds. Thus scientific studies of such subjects may take us closer to understanding the origins of life and its distribution in the universe.

Exobiology takes us full circle back to the first chapters of this book, which described how our ancestors looked up at the star-strewn sky and wondered about the relationship of human beings to the rest of the universe. The question of whether extraterrestrial life exists or does not exist is a perfect scientific question, because either answer is profound. The nineteenth-century Scottish writer Thomas Carlyle commented about these two possibilities with a touch of sardonic humor: Other worlds, he said, offer "a sad spectacle. If they be inhabited, what a scope for misery and folly. If they be not inhabited, what a waste of space."

What does modern astronomy say about these two possibilities? Extraterrestrial amino acids found in carbon-rich meteorites prove that **organic chemistry** (chemistry based on complex molecules with carbon-hydrogen bonds) occurs in other parts of our solar system. Claims that fossil microbes have been found in Martian rock samples remain controversial, however. The discovery of **organic molecules** (complex molecules based on carbon-hydrogen bonds), such as formaldehyde and sugars, in interstellar gas proves that organic chemistry occurs even between the stars. Still we have no proof of the existence of aliens, even though Hollywood treats them as known commodities.

In this chapter, we will consider the issue of alien life in several steps. First, we will discuss what we mean by *life* and what conditions would be required for life to exist on planets. Second, we will review the long process that led to the evolution of life on Earth, as best we understand it. Third, we will ask whether there are planets that could support a similar evolution of life elsewhere in the universe. Finally, we will try to estimate the probability of intelligent life actually existing elsewhere and whether we might contact it, or it, us.

THE NATURE OF LIFE

What do we mean by "life"? Our culture tends to make us conceive of life as a quality or entity, such as the "animate spirit of life" that inhabits living beings and departs at death. A deeper understanding may come from viewing **life** not as a status but as a process—a series of chemical reactions using carbon-based molecules, by which matter is taken into a system and used to assist the system's growth and reproduction, with waste products being expelled. A traditional analog is a flame; it is "alive" not because of the presence of an unseen essence, but because a chemical process of oxidation continues until the necessary fuels are used up. In the same way, however else we may view life, we must recognize that it is a chemical process based on the ability of complex carbon-based molecules (especially DNA molecules) to reproduce themselves (Dawkins, 1995).

The system in which life's processes occur is the cell. All known living things are composed of one or more cells.[1] A **cell** is in essence a container filled with an intricate array of organic and inorganic molecules (protoplasm). An important step in the evolution of life was the chemical evolution of the cell membrane, which walls off the contents of the cell from the surrounding fluid (Maddox, 1994). Codes for cellular processes are contained in very complex molecules, such as the famous DNA and RNA molecules, which are important in reproducing the cell. In a primitive type of cell called a prokaryote, the DNA molecules are dispersed, but in a type of cell called a eukaryote (regarded as a later, more evolved form), they are concentrated in a central body called the nucleus. Early life on Earth was probably composed of prokaryote cells, microbes such as bacteria. You can remember the difference because "prokaryote" comes from root words meaning "before the nucleus." Bacteria have this type of primitive cell with no nucleus. All cells in our bodies are eukaryotes, with nuclei where the DNA and genetic machinery are housed.

As mentioned above, organic molecules depend on bonds between carbon and hydrogen and often involve the ability of the carbon atom to form big molecules by linking with other carbon atoms in long chains, rings, and other complex forms. It is these large molecules that encourage the complicated chemistry of genetics and reproduction. The elements most prominent in organic molecules are carbon, hydrogen, oxygen, and nitrogen—all common in planetary matter. This is why the discovery of so-called CHON particles—rich in carbon (C) , hydrogen (H), oxygen (O), and nitrogen (N)—in comets, was regarded as important. Even if planets had no organic compounds (which is unlikely), impacts of comets would have "seeded" them with abundant organic molecules. Phosphorus, also important to life (in small amounts), is also widely available. Note carefully that the term *organic chemistry* (as well as *organic molecules*) refers not specifically to life-forms but more generally to all chemistry (and molecules) involving carbon. All life (as we define it) involves organic chemistry, but not all organic chemistry involves life.

The process that defines life involves the ability of complex organic molecules to create copies of themselves by linking up with other atoms. Suppose a long molecule was made up of the atoms A, B, C, D, E, etc., so that it had the structure

A B C D E F G H I J K L M N O P Q R S T U

Now suppose that this molecule was floating in a sea of chemicals and that, chemically, the element A could link to another B, and that the B could link to

[1]Viruses might be an exception. They are simpler than cells, yet can reproduce themselves using materials from host cells. Biologists disagree on whether viruses should be considered a form of life. They might better be viewed as complex molecules that duplicate themselves when exposed to the organic materials in cells.

another C, and the C to a D, and so on. Then, eventually, a new "double molecule" might form, which would have a structure like this:

A B C D E F G H I J K L M N O P Q R S T U V

B C D E F G H I J K L M N O P Q R S T U V A

Now if this molecule split into two pieces along the horizontal space between the two lines, each of the two pieces would be essentially a copy of the original, or long pieces of the original. This is roughly a description of the molecular basis of life, in which complex molecules make copies of themselves. Cells and bigger organisms—regardless of whatever else they may be—can be described as entities that take in material (so-called nourishment or food), break it up to molecular levels, and use those molecules to make copies of their existing molecules.

We often make the mistake of thinking of ourselves as static beings instead of dynamic systems. We casually assume that we are constant entities, as if our identities depended solely on the form of our bodies. But our bodies today are not the same ones we had seven years ago. Hardly a cell is still alive that was part of that body. This dynamic conception of life is a far cry from the conception only a few generations back, when bodies were thought of as semipermanent machines whose parts gradually wore out. Even our seemingly inert skeletons are living and changing, always replacing their cells. We *must* keep changing—the cells must keep processing new materials to stay alive. When the processing stops, we call the condition death.

The nature of living beings is illustrated in an analogy made by the Russian biochemist A. I. Oparin (1962). Consider a bucket that has water pouring in at the top from a tap and flowing out at the same rate through a tap in the bottom. The water level in the bucket stays constant, and a casual observer would call it a "bucket of water." It seems to be a "thing," deserving of a name. But it is not like an ordinary bucket standing full of water. The water at one instant is not the same water as at any other instant, even though the outward appearance is constant. We are like buckets with water and nutrients and air flowing through us, but with other, much more complex attributes, such as the ability to reproduce and to be affected by genetic changes that let us evolve from generation to generation.

This analogy gives us some clue to the kinds of processes involved in the origin of life. We are looking for a process in which complex carbon-based molecules can enter cell-like systems that (1) draw material from the incoming molecules to create new molecules, (2) incorporate the new molecules into new structures, (3) eject unused material, and (4) reproduce themselves.

Scientists therefore usually choose to define life by these specific carbon-based processes. Often at this point people ask, "What about some unknown form of consciousness?[2] Or what about some unknown chemistry based on other elements, such as silicon, that can form big molecules?" The answer is that we have never observed or experimented with such life-forms, so we can say nothing substantive about them. If we admit they are plausible, we simply increase the probability of life or consciousness in the universe. But researchers usually restrict their discussion to forms on which chemical and behavioral data are available.

Whatever other concepts we invent—civilization, religion, technology, art, war, love, communication—it is the chemical processes of life that define us, just as they define the spiders, sea urchins, elephants, moths, amoebas, redwoods, and all the other incredibly varied living creatures around us. To judge whether life may exist on other planets—whether other planets are already "taken"—we must find out how those processes got started on Earth.

THE ORIGIN OF LIFE ON EARTH

In addition to the life-forming elements—especially C, H, O, and N—water was crucial to the development of life on Earth. Several facts indicate this:

1. Most of our body weight is made up of water. The percentage is even higher for plants.

2. The oceans are our heritage. The oldest fossils are of simple organisms that lived in the sea. The whole history of life on Earth involves a quest to move the sea environment itself onto land. Eggshells encapsulate a fluid environment long enough for embryos to develop. Amphibians can lay eggs on the land, but then return to the sea. Reptiles lay eggs on land and can remain on land. Mammals solve the problem by having the embryo evolve in a fluid environment inside the body of the female. Even we humans, as em-

[2]For example, in the novel *The Black Cloud,* astronomer Fred Hoyle imagined an interstellar nebular cloud with matter and electromagnetic fields organized in such a way that it developed a consciousness or will of its own.

bryos, are initially immersed in fluid and develop bodies more like fish than like mammals—for example, we have gills. In an adult, body fluids such as tears have a saltiness similar to that of the oceans.

3. Organisms deprived of liquid water quickly die. And dead organisms are shriveled and dried.

4. Taking a more theoretical view, water provides a medium in which organic molecules can be suspended and can interact—a likely place for life to begin.

Probably within 50 million years after Earth formed, heat from the initial planetary accretion and from internal radioactivity melted parts of the Earth's interior, unleashing volcanic activity. Because 1 million years is like one day in the life of the Earth, we will use that unit frequently in this chapter and abbreviate it as My. Volcanic activity released gases, especially water vapor (H_2O). The water vapor condensed into oceans, leaving carbon dioxide (CO_2) as the dominant atmospheric gas (as is still found today on Venus and Mars). Thus, even if there were no oceans at the beginning, by about 4500 to 4000 My ago there were probably bodies of surface water under an atmosphere that was rich in CO_2. Other minor constituents of the early environment included compounds of the cosmically common elements, C, H, O, and N—compounds such as methane (CH_4) and ammonia (NH_3).

Chemical reactions occurring naturally in such an environment could have produced building blocks of life, as has been proved by laboratory experiments. In the 1950s, chemist S. L. Miller conducted the now-famous experiment in which he put a gaseous mixture of hydrogen, ammonia, methane, and water vapor (to represent the primitive atmosphere) over a pool of liquid water (primitive seas) and passed electric sparks through it (simulating energy sources such as lightning). After several days, the pool began to darken. The water now contained a solution of **amino acids,** the class of molecules that join to form **proteins,** the huge molecules in cells. This so-called **Miller experiment** proved that the complex organic chemicals necessary for life could be built in a natural environment (Miller, 1955). An example of a reaction in the Miller experiment is the production of the amino acid glycine:

$$NH_3 + 2CH_4 + 2H_2O \xrightarrow{\text{energy}} C_2H_5O_2N + 5H_2$$

ammonia + methane + water → glycine + hydrogen

Early researchers, at the time of Miller's experiment, thought that Earth's early atmosphere was a hydrogen-rich concentration of the interplanetary gas. Further geochemical studies indicated that it was more rich in volcanic gases, such as CO_2, carbon monoxide (CO), and N_2, and water vapor. Solar ultraviolet light striking such an atmosphere tends to produce water and hydrogen cyanide (HCN). Further experiments showed that a Miller-like reaction produces amino acids even under these conditions (Goldsmith and Owen, 1979):

$$2HCN + 2H_2O \rightarrow C_2H_5O_2N + CN_2H_2$$

hydrogen cyanide + water → glycine + cyanamide

Other experiments showed that many kinds of energy sources—including ultraviolet light from the Sun, volcanic activity, and even meteorite impacts—could produce similar reactions. Thus we can conclude that building blocks of life *would* have formed on early Earth and in similar environments elsewhere. This conclusion has been confirmed by the discovery of extraterrestrial amino acids (but not fossils or life-forms) in several carbonaceous chondrite meteorites (Kvenvolden and others, 1970; Cronin, Pizzarello, and Moore, 1979; Kerridge, 1991, 1995) and in interplanetary dust (Zeman, 1994). Amino acids can form in either a "right-handed" or a "left-handed" symmetric molecular structure, but on Earth, some biological or prebiological process led to a strong preponderance of only one of the two structural options (called a **nonracemic mixture**). The meteoritic amino acids were proved extraterrestrial because their molecular structures had equal proportions of the two structures (called a **racemic mixture**).

From such results, researchers have concluded that molecular organic materials existed in Earth's primitive oceans, probably less than 500 My after the planet formed. Such material may have accumulated in isolated tidewater ponds (Figure 28-1). As water molecules evaporate from such a pool, they leave behind the heavy molecules, which thus become more concentrated and can interact to form complex organic substances.

★ Since many scientists now believe ancient Mars had lakes or oceans that later evaporated, samples from such sites might show whether complex organic molecules or even life-forms evolved on the red planet. As discussed on our web site, the search for organic molecules on present-day Mars has been negative, apparently because ultraviolet light from the Sun breaks up such molecules and sterilizes the soil.

Figure 28-1 Broths rich in amino acids and complex organic molecules formed in long-lived tidewater pools on ancient Earth. Evaporation of water could have concentrated the remaining organic materials, allowing complex reactions. The resulting products could have "fertilized" oceans with living organisms or protoliving materials. (Photo by WKH)

The next step is less certain. Florida biologist S. W. Fox has shown that simple heating of dry amino acids (as might happen on a dry planet such as present-day Mars) can create protein molecules. When water is added, these proteins assume the shape of round, cell-like objects called **proteinoids,** which take in material from the surrounding liquid, grow by attaching to each other, and divide. Though they are not considered living, proteinoids resemble bacteria so much that experts have trouble telling the difference in microscope views.

Engel and Nagy (1982) reported finding non-racemic mixtures of amino acids, like those in terrestrial proteins, inside an uncontaminated piece of the Murchison carbonaceous chondrite. This raises the possibility that some protobiological processes went further than once thought, producing protein-related amino acids inside moist carbonaceous chondrite parent bodies. This would be an exciting discovery about the origins of life, but the findings are still very controversial.

Possibly related to proteinoids are objects discovered in the 1930s by Dutch chemist H. G. Bungenberg de Jong. When proteins are mixed in solution with other complex molecules, both sets of substances spontaneously accumulate into cell-size clusters called **coacervates.** The remaining fluid is almost entirely free of complex organic molecules (Oparin, 1962).

★ The next step toward recognizable life-forms is still more uncertain, but many biologists theorize that cell-like proteinoids and coacervates in primeval pools of "organic broth" began reacting with fluids in the pools and with each other, accumulating more molecules and growing more complex, as suggested by Figure 28-2. Perhaps these eventually evolved into biochemical systems capable of reproducing.

Whatever the processes, microscopic cellular life must have arisen between about 4200 and 3500 My ago, because life's fossil remains have been discovered in rocks after that period. Chemical evidence of biological processes has been reported in rocks from Greenland that are 3700 and 3850 My old (Rosing, 1999; Holland, 1997), and many species of fossil microbes have been described in 3465-My-old rocks from Australia (Schopf, 1993.)

Among the best evidence of early life are fossil remains of **stromatolites,** which are colonies of blue-green algae matted together in football-size cabbagelike structures; they are found even today in certain shallow tidal ponds or bays on the seacoast. A good example dated at 3500 My was reported from Australia (Walter, Buick, and Dunlop, 1980). Nisbet (1980) lists other 2700- to 3500-My-old stromatolites from Canada, Zimbabwe, and Australia, with possible evidence of 3500-My-old rocks with biological carbon isotope chemistry from western Greenland. Life must have spread rapidly after about 3700 My ago.

An interesting factor is that life probably could not have evolved very long before about 4200 My ago because of the intense early meteorite bombardment, though biochemical evolution must have begun in that period. Thus life-forms such as bacteria and blue-green algae apparently evolved rapidly around 3400 to 3700 My ago. Some researchers have thus concluded that life arose "as soon as it could;

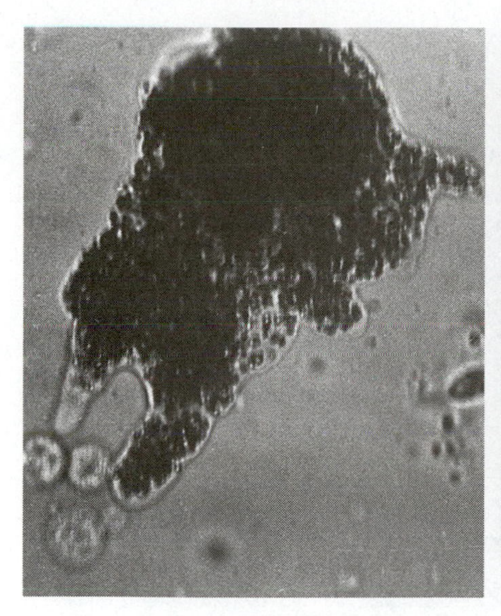

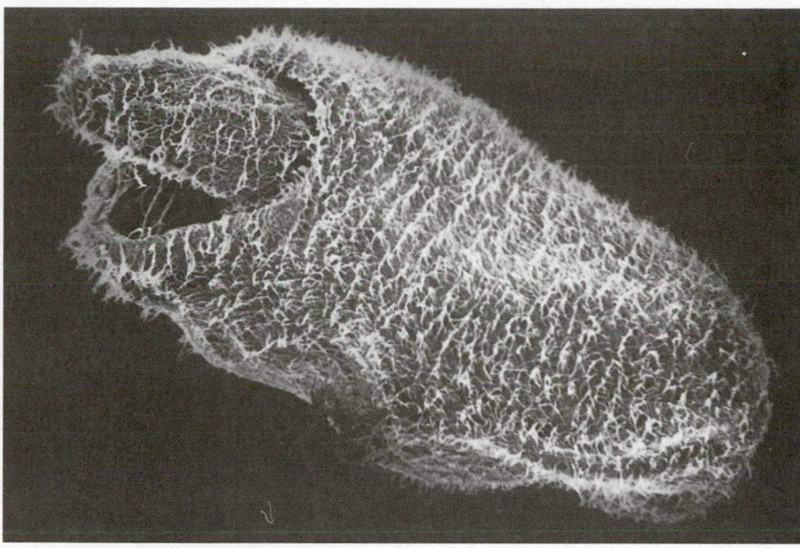

a b

Figure 28-2 The importance of a fluid medium for primitive evolution is suggested by these photos of single-celled organisms engulfing food from surrounding fluid. **a** An amoeba flows to surround a nearby food particle. (Optical microscope photo; S. L. Wolfe) **b** The protozoan *Woodruffia* ingests a paramecium. (Electron microscope photo; T. K. Golder)

perhaps it was as inevitable as quartz or feldspar" (Gould, 1978).

Earth was not a passive backdrop. Meteorite impact explosions formed craters, and sedimentation filled them. Ocean and land configurations changed. The composition of the atmosphere evolved. The oxygen content in the initial atmosphere was small, probably only a few percent. However, the first plant organisms, such as plankton in the sea, produced molecular oxygen (O_2), and the oxygen content began to rise. Empirical evidence for this is that rocks formed and buried more than about 2500 My ago have been found to be suboxidized, or formed under more oxygen-poor conditions than exist today (Goldsmith and Owen, 1979). Oxidized red beds are rare before 2000 My ago and common afterward (Walker, 1977).

As soon as some oxygen became available, solar ultraviolet light broke some O_2 molecules, and the free pair of O's joined with other O_2 molecules to make **ozone** (O_3). This led to the production of an **ozone layer** high in Earth's atmosphere, which absorbs nearly all the incoming ultraviolet. This was an important step, because too much ultraviolet breaks up complex molecules and thus could prevent life—the apparent fate suffered by biochemistry on Mars. Thus, paradoxically, the early presence of solar ultraviolet may have helped life start by providing energy to break up old molecules and form new ones, but its absence later allowed life to evolve safely on the surface.

Living things could hardly be unaffected by all these changes. Whereas the earliest life-forms developed and existed without much oxygen, life now had to adjust to oxygen. In fact, life-forms could use the energy produced by oxidation to power their cellular machinery. Such environmental changes favored evolution—when a mutation trait appeared that happened to be well adapted to the new environment, that organism was likely to have more offspring, promoting retention of the new trait and thus evolution of new species.

Two thousand My ago, we still would have scarcely recognized our Earth. The land was still barren. Some areas must have looked like today's deserts or like Mars. Some areas were moist and washed by rains, but instead of luxurious forests we would have seen only bare dirt, eroded gullies, and grand canyons. Brown vistas stretched to the sea.

Most life was still in the oceans; these life-forms had soft bodies and rarely produced fossils. Stromatolites increased in abundance 2500 to 2000 My ago, living at the boundary of water and rock along seacoasts. Today they are found only in rare, oxygen-poor salt marshes on some seacoasts. Stromatolites produce oxygen by photosynthesis, and their abundance 2500 to 2000 My ago is more evidence for the upswing in O_2 production after that time (Walker,

Figure 28-3 The first signs of life encroaching on land did not occur until about halfway through Earth's history. This scene about 2000 million years ago shows two such signs: a colony of stromatolites (left shoreline) and some lichens or moss on the rocks (right foreground). The rest of the land (left distance) was probably barren and eroded. Earth had few surface signs of life during the first 90% of its history. (Painting by Ron Miller)

1977; Goldsmith and Owen, 1979). It is strange to realize, as Figure 28-3 emphasizes, that during most of Earth's history, life would have been hard to find on representative *land*scapes on Earth. The land was empty and highly eroded.

One sign of the adaptability of life-forms, once they evolved, is the rapid proliferation of advanced species, as indicated in the geological time scale shown in Table 6-1. Whereas it took about half the available time to go from complex molecules to algae and bacteria, it took only the last 12% of Earth's history to go from the first hard-bodied sea creatures to humans. Biologists attribute evolution in general and this rapid proliferation in particular to **natural selection,** the greater production of offspring by those individuals best adapted to the ever-changing environment. Details of the natural selection process are uncertain. Prototypes of new species may have evolved by mutation and lived in small numbers in obscure ecological niches, only to emerge rapidly when sudden environmental changes caused the extinction of earlier species better suited for earlier conditions. A perfect example is the appearance of mammals. During the age of dinosaurs, mammals were relatively obscure, small creatures who lived in burrows. But when an asteroid impact created a temporary climatic catastrophe that caused the giant reptiles to die out 65 My ago, the mammals found an Earth with fewer predators, and they expanded rapidly.

Whatever the specific mechanism, we can say from the fossil record that Earth experienced evolution from nonliving organic chemicals to small organisms in a few billion years, and that these organisms evolved in less than a billion years to species with self-intelligence.

HAS LIFE EVOLVED ELSEWHERE?

If habitable planets exist and if life evolves readily under habitable conditions, is life abundant throughout the universe? There are additional factors to consider. For one thing, planetary environments change with time, so today's habitable planet may not be habitable tomorrow. Can life survive such changes?

Effects of Planetary and Astronomical Processes on Biological Evolution

Basic planetary or stellar processes may be involved in encouraging or hindering biological evolution. It is not clear that the average planet will maintain benign conditions long enough for advanced species, or intelligence, to evolve. Even on Earth, we barely made it. For example, as we have seen, Earth's fossil record indicates an episode called the "great dying" at the end of the Paleozoic Era about 250 My ago. As many as 96% of all land and sea species became extinct in a few million years or less (Raup, 1979). This cleared the way for the evolution of the giant reptiles that dominated the Mesozoic Era from 250 My to 65 My ago.

The "great dying" was only the biggest of many mass extinction events. Raup and Sepkoski (1986) studied the evolution of 11,800 genera (a broader class than species) of marine animals during the last 270 My. They found that 61% of existing genera disappeared during the "great dying" 250 My ago and 49% of existing marine genera, including dinosaurs, became extinct at the end of the Mesozoic Era 65 My ago. In addition, many smaller mass extinctions marked the divisions between the geologic periods into which the eras are divided. Thus 43% of marine genera died out at the end of the Triassic Period 210 My ago; 30% died out at the end of the Jurassic about 140 My ago; and many smaller extinctions occurred, grading down to still smaller events that are too small to divide geologic periods.

★ Thus, even on our fertile planet, relatively short-lived catastrophes drastically affected life-forms every 100 My or so. How can these sudden mass extinctions be explained? We've already seen that the second-largest catastrophe, at the end of the Mesozoic Era 65 My ago, was probably caused by a random impact of a modest-size asteroid. Naturally, scientists have sought evidence of an asteroid impact that might explain the even greater extinction at the end of the Paleozoic, such as once-molten spherules or element anomalies at the boundary layer between the Paleozoic and the Mesozoic Eras. Strong evidence of an impact has been lacking. In 2000, however, New York biologist Michael Rampino studied drill cores from the Austrian Alps and found evidence that the great dying could have unfolded in less than 8000 y, and Chinese scientists (Jin and others, 2000) studied strata at the boundary in China and reported a sudden extinction and an increase in once-molten microorganisms 251.4 My ago. These scientists believed the spherules were mostly volcanic and suggested that the extinctions may have come from side effects of massive lava flows in Siberia, but due to the suddenness and the presence of spherules, they could not rule out impact. Our web site details a number of possible natural causes of catastrophes and mass extinctions that might make it difficult for species to evolve toward intelligence on other planets. Examples include frequent climate changes caused by volcanism, axial tilt, and other processes; changes in the radiation output of the central star; explosion of a nearby supernova; and asteroid impacts. If Earth (or any other planet) suffers catastrophic impacts or other disasters that wipe out species every few hundred million years, this would mark a radical change from classic Darwinism, where evolution is driven only by competition among species *within environments that change slowly due to terrestrial forces.*

Do such disasters necessarily hinder evolution of "advanced" life-forms? The answer might seem to be an obvious yes. But as long as such a catastrophe does not wipe out *all life,* it probably promotes the emergence of new species. We have seen how mammals emerged after an asteroid impact wiped out large reptilian predators. To take another example, if some aspects of intelligence first appeared in a benign, constant environment where food was plentiful, these traits would have had little value. But if ice ages destroyed the mild environment, then evolution would favor the cleverer groups who could figure out ways to cope with climate change.

One of our problems is that we just don't know whether Earth has experienced rare special effects that have promoted life and intelligence. Goldsmith and Owen (1979) suggest that the mere presence of our unusually large Moon is such an effect: Its tides helped produce tidal pools, where organic molecules could concentrate, and its tidal forces have helped hold Earth's axis in a fixed position, preventing axis excursions (such as probably happened on Mars) that could cause drastic climatological oscillations.

Perhaps it is no coincidence, then, that the time scale of biological change is comparable to the time scale of geological change. Life's response to changing environments is to change itself, producing new species on time scales of a few million years. Old species die and new species prosper under the new conditions. A threat to life arises when the time scale of global environmental change is too short to allow life to adapt, as when a large asteroid impact occurs or human technology produces global changes, such as ozone destruction, over a few generations.

Adaptability and Diversity of Life

The great variety of ancient and modern species on Earth and the variety of environments in which they have thrived suggest that, given time, life might have evolved to fit a wide range of conditions on other planets. Even humans have a remarkable adaptability. We thrive from equatorial wet jungles to dry deserts to Arctic plains to Andean summits, where air pressure is barely half that at sea level. During the last ice age, we survived by migrating.

But the limits for survival of simpler life-forms are even wider. Microflora are known in supercooled Antarctic ponds that remain liquid at 228 K (–49°F) because of dissolved calcium carbonate and other salts. And bacteria are known in Yellowstone hot springs at a temperature of 363 K (194°F)—a 59% variation in temperature (Sagan, 1970). In 1997, researchers discovered the highest-temperature microbes yet known, thriving in water as hot as 386 K (235°F), 3650 m below the ocean surface near geothermal "smoker" vents (Holden, 1997). This temperature is above the boiling point of water at sea level, and normally we think of boiling as destroying microbes, since the cell involves a liquid water medium. So how do these microbes survive? It turns out that under the high pressures at those depths, the temperature is below the boiling point. In laboratory experiments, common bacteria have survived at least 24 h in liquid cultures in CO_2 atmospheres at 433 K (320°F), extending the range to a 90% variation in temperature.

Habitable pressure regimes show an even greater range. Bacteria exist at altitudes where the atmospheric pressure is only 0.2 bar, and more advanced organisms live in ocean depths with pressures of hundreds of bars (Sagan, 1970). This pressure range exceeds a factor of 1000.

The harsh "dry valleys" of Antarctica were once thought to be uninhabited by life, because the dry soils exposed there tested sterile. In 1978, however, Florida State University biologist E. I. Friedmann reported an interesting discovery: In apparently lifeless, dry valleys in Antarctica, microorganisms such as bacteria, algae, and fungi thrive in pore spaces *inside* rocks. Though the rock may be in a frozen area, sunlight heats the rock itself and creates a microscopic niche suitable for life. Life, seemingly inexhaustibly accommodating, fills the niche.

Beginning in the 1920s, biologists discovered bacterial life existing at kilometer-scale depths in the rocks of Earth's crust (Frederickson and Onstott, 1996). Thriving microbes have been detected as far as 2.8 km below the continental surface, and it is believed that they may exist down to about 4 km, where the temperatures rise well above the boiling point. One study of volcanic lavas below the seafloor south of Panama (Furnes and others, 1999) revealed evidence of microbes living by chemical alteration of glassy lavas about 650 m below the seafloor itself, at temperatures around 100°C (373 K).

The discovery of animal populations around geothermal vents at a depth of 2550 m (at a pressure of 260 bars) in the Pacific is intriguing. The bacteria that form the base of the local food chain draw their energy not from sunlight but from chemical reactions involving hydrogen sulfide from the vent. This discovery overturned the notion that sunlight is always the main energy source for larger species (Karl, Wirsen, and Jannasch, 1980). Feeding on such bacteria are larger animals, clustered around the vents in the darkness of the seafloor; they include tube worms as long as 1 m. These life-forms appear almost independent of the life-forms at Earth's surface, except that in order to oxidize the hydrogen sulfide, they appear to depend on the oxygen generated by plants (Tunnicliffe, 1992). Reports in 1995 found bacteria living in groundwater aquifers within basaltic lavas more than 330 m below the surface in Washington; they concluded that these microbes conduct their life cycle without depending on either light or oxygen, gaining all their sustenance from water and rock minerals (Kaiser, 1995). The existence of such creatures suggests that organisms might evolve on remote planets where light levels are too low to depend on starlight, and where life might depend instead on geothermal energy sources or simply rock chemistry.

Experiments with terrestrial life-forms in nonterrestrial environments have shown grass seeds germinating in atmospheres of C, H, O, and N compounds, insects exhibiting normal behavior at pressures as low as 100 to 160 millibars (0.1 to 0.16 bar), insects surviving brief exposure to Martian surface pressures, and bacteria surviving in CO_2 at conditions between those of Earth and Venus (Siegel, 1970).

Conversely, alien organisms might survive terrestrial conditions and have devastating effects on Earth. There are historical examples of similar events. Plague caused by Asian bacteria that were introduced into Europe in the 1300s killed about a quarter of all Europeans; as many as 75% of the inhabitants died in some areas. Diseases introduced into Hawaii after the first European contact in 1778 killed about half of all Hawaiians within 50 y. Some 95% of the natives of Guam were wiped out by disease within a century of continued European contact (Underwood, 1975). For these reasons, early Apollo astronauts were quarantined until it was clear that they carried no lunar organisms. And future samples from Mars may be analyzed only in space labs, well above Earth's atmosphere.

Thus change and evolution in life populations are caused not only by life's adaptations to new environments but also by the invasion and destruction of some populations by others. With these facts as well as cultural competition in mind, anthropologist D. K. Stern remarked as early as 1975, "It is likely that the meeting of two alien civilizations will lead to the subordination of one by the other."

Appearance of Alien Life

If alien life did evolve somewhere else in the universe, could we expect it to evolve through any stages similar to our own? Opinion is divided on this point. MIT physicist Philip Morrison (1973) emphasized the **convergence** effect, whereby species with similar capabilities in similar habitats evolve to look alike. For example, three different species of large sea creatures evolved a similar design for fast swimming in the ocean: the extinct reptile Ichthyosaur;

the shark, which is a fish; and the dolphin, which is a mammal that returned to the sea. Similarly, aliens living on planetary surfaces in gaseous atmospheres and using "intelligence" to manipulate their surroundings with tools might well have bilateral symmetry, appendages used as hands, a pair of eyes designed (like ours but unlike most animals') for stereo vision, and so on.

On the other hand, the famous paleontologist George Gaylord Simpson (1964, 1973) argued that although life is likely to start, the long chain of environmental changes and evolutionary steps that produced humans is unlikely to be approximated elsewhere, so there is likely to be a "nonprevalence of humanoids." Simpson labels the whole attempt to estimate probabilities of alien life as nearly meaningless because of our lack of experiments or examples.

Physicist W. G. Pollard (1979) countered that we do have examples. He notes that about 180 My ago, Australia broke off from Gondwanaland and thus can be thought of as a separate Earth-like planet, "A," where evolution continued independently from a primarily reptilian stock. Similarly, South America broke off from Africa 130 My ago and can be viewed as an independent planet "S." Independent evolution also continued on planet "E" (Earth, made up of Africa and the adjoining Eurasian landmasses). In the same period of 130 to 180 My ago, independent evolution diverged on these "planets," rather than converging on humans. Humans appeared only on E, primates developed on S, and marsupials on A. Humanlike creatures, appearing on E only about 4 My ago, have existed for only 0.1% of Earth's history, but spread around the globe in that time.

In summary, natural selection seems to produce species capable of occupying any reasonably habitable environment. Thus we should not be surprised if life has evolved on other planets. But this life may look very strange to us, even if it displays recognizable intelligence. After all, if mushrooms and corals and woolly mammoths and Venus flytraps all evolved on one planet, how much greater are the differences between life-forms on two different planets likely to be? Feathers and fur and seeds and symphonies may be products of Earth only.

Effects of Technological Evolution on Biological Evolution

If life will evolve when the right conditions exist, and if those conditions probably do exist elsewhere, then should we predict that intelligent species and civilizations are common? What do these terms mean, anyway? Should we assume that other civilizations will achieve space flight or might visit us some day? This question raises the issue of technology and its role in the evolution of life. Many exobiologists define intelligence as involving, among other things, the use of tools to modify the environment—hence, technology.

★ Does the development of technology increase the chance that a species will survive? Our own technology is beginning to affect our planetary environment on continental and global scales. As we have seen, although limited environmental change can be helpful, environmental change that is too much or too fast can alter a planet so fast as to be fatal to the inhabitants! As Pulitzer Prize-winning naturalist René Dubos points out, we are umbilical to Earth. If a natural or human-caused catastrophe modifies our planet too much before we acquire an ability to leave it, we are finished. For this reason, consideration of our own case leads to the conclusion that if a species develops technology capable of altering its planet, without developing a robust capability for space flight, that species might actually destroy itself before finding any alternative places to live. This is hardly wild speculation, since we see a few nominally moral, intelligent technologists on our own world spending entire careers devising weapons solely to deal death to our own species. Normally, we assume that technology has been a good thing, but as we discuss further on our web site, technology can have possible negative consequences for a species, such as planetary damage from warfare, nuclear technology, atmospheric changes, and other rapid environmental impacts.

If we take a millennia-long point of view, we realize that even in the very first century of worldwide high-tech "progress"—the twentieth century—humanity came perilously close to worldwide nuclear damage and ozone layer depletion and launched global climate change effects, forest destruction, and species extinctions that are still proceeding. Thus, if humanity is any example, long-term survival of a planetary culture over thousands of centuries is not assured. Although the human species has been around less than a hundredth of a percent of the age of the universe, we are already having close brushes with global disaster. Thus we can speculate that if evolution produces intelligent societies that remain tied to one planet, many of them may last only a fraction of a percent of the age of the universe—in which case, there is little chance that a given culture will be around at the same time we are.

More optimistically, we have succeeded in identifying the **cultural hurdle** that we (and perhaps intelligent species on any planet) must surmount: the transition from scattered, competing nation-states with the capability to damage the planet to a stable planetary or interplanetary technological society of intelligence and imagination. Perhaps we and some other cultures will cross this hurdle. After all, we have identified some perils in time. Perhaps some cultures have completed the transition from being planetary cultures to being interplanetary, thus ensuring their survival against ecological disaster on any one planet. Such cultures might last and be detectable for billions of years instead of thousands.

Our conclusions so far are that a fraction of the stars have planets, a fraction of the planets ought to be inhabited, and a fraction of the inhabited planets ought to have either civilizations or relics of destroyed civilizations. The next question is whether any actual evidence for extraterrestrial carbon-based life as we know it exists today.

Alien Life in the Solar System?

According to our ground rules, a habitable world has (or had) liquid water and an atmosphere in which complex carbon-based organic chemistry can proceed; furthermore, these conditions must have lasted long enough for biochemical complexity to have evolved. Few worlds in the solar system fit the bill. Nevertheless, many of them may give us clues about the origins of life. As noted earlier, meteorites indicate that many ice-rich asteroids or comet nuclei have been exposed to trickles of liquid water, and that amino acids have been synthesized in those environments. Have amino acids been synthesized throughout the solar system and evolved into life in more than one place?

Planetary scientists have wavered from optimistic to pessimistic in their estimation of the chances for life elsewhere in our solar system. The estimate reached a nadir in the late 1960s: during those years the first Mars probes returned pictures of a barren, cratered planet, and all the worlds of the outer solar system were believed to be dead ice balls. Today, the prospects seem much more interesting, focusing on Mars, Europa, and Titan. Even if these worlds yield no traces of present-day life, they may reveal evidence of how organic chemistry and water combined to create at least the precursors of life.

Mars Mars offers by far the best prospects for either (1) present-day microbial life below the surface, (2) fossil evidence of extinct microbial life, or (3) evidence of early chemical conditions that might help us understand why life does or does not begin on a water-rich planet.

As we remarked in Chapter 10, present-day life on the surface of Mars is unlikely because Mars lacks an ozone layer, which in turn means that solar ultraviolet (UV) light passes through the atmosphere. Remember that UV photons are the most energetic of the solar spectrum; thus they smash and destroy organic molecules. Consistent with this, the Viking 1 and 2 landers in 1976 showed that the Martian surface soil in two different places is sterile.

But what about underground Mars? As we also saw in Chapter 10, there is abundant evidence of underground ice, at least at high latitudes. Moreover, in 2000, scientists working on the Mars Global Surveyor project announced evidence of hillside seeps or springs, suggesting that liquid water has gushed out of underground layers in very recent geologic time, perhaps within the last 100,000 y. Consistent with this finding are Martian lava flows for which counts of impact craters suggest ages as low as 10 My—not to mention meteoritic rock samples from Mars that crystallized from molten magma as little as 170 My ago. These prove that geothermal heat sources, which could melt ice and produce liquid water, have been active in the last few percent of the planet's history. What's more, many of the Martian rocks have deposits of salts and other chemicals left behind when salty water, like seawater, soaked into the rocks and then evaporated. Dating of such deposits in one 1300 My-old Martian rock shows that it was exposed to water only 670 My ago—in the last 15% of Martian history. Many researchers believe that ancient Mars had enough water—and perhaps a milder climate—to produce lakes, seas, and/or glaciers, and that modern Mars may see occasional water release by sporadic melting of ice.

According to the Miller experiment and other evidence, ancient Martian oceans or underground aquifers might provide the perfect setting for producing life! Perhaps the real issue is whether liquid water (seas? underground aquifers?) ever existed long enough on Mars to allow life to gain a toehold (or tentacle-hold).

As we reviewed in Chapter 10, one research group in the 1990s announced it had found fossil microbes in the oldest of the Martian rocks (a rock formed 4500 My ago on Mars, possibly exposed to the putative Martian seas). This announcement set off a furious controversy as to whether the supposed fossil microbes were really fossils or just rock crystal structures. This controversy, in the best tradition of science, generated much research, which proved at least that microbes on Earth can live in Mars-like lava rocks and in deep underground layers, a kilometer or more below the surface. It has been said that microbial life below the ground has more biomass than all the creatures that live on Earth's surface! If life exists elsewhere in the universe, its most common form may be simple, single-celled organisms, such as existed through most of Earth's history; we "advanced" creatures may be the anomaly. Thus the best bet is that if life ever did exist on Mars, it was in the form of microbial bacteria that evolved to live within moist soil and rock layers of the red planet.

To solve the mysteries of whether life ever existed on Mars, we may need to establish a long-term base there and drill into the crust. To sum up, the major questions are: Did life ever evolve on Mars? If so, what was it like? Was its DNA like ours? If not, why not?

Europa To the great surprise of scientists, the second most likely site for extraterrestrial life in the solar system is Europa, the icy moon of Jupiter. This became evident during the "golden age" of solar exploration in 1979, when the Voyager probe photographed a thin, young (nearly uncratered) crust of H_2O ice on Europa. The science fiction writer Arthur C. Clarke, in his novel *2010,* was probably the first to speculate in print that the liquid water ocean below this ice crust might harbor living organisms. Again, the Miller experiment might suggest that life could have formed here.

The stakes were raised in the 1990s when the Galileo probe showed that Europa's ice crust was indeed thin; the crust is constantly breaking up into giant, iceberglike pieces that drift apart, while new ice forms in between. NASA engineers have discussed the possibility of seeking the thinnest spots in the ice where torpedolike "penetration" probes might be fired through the ice to see if there are organisms in the water underneath. If organisms are there, what are they like? If not, why not?

Titan The third most promising locale for extraterrestrial life is Saturn's giant moon, Titan. The second largest moon in the solar system, Titan is larger than the planets Mercury and Pluto. The exciting factor here is the atmosphere with its methane (CH_4), discovered by the Dutch-American astronomer Gerard Kuiper in 1944, and its dominant constituent of nitrogen (N_2—the same as Earth's dominant constituent), found by the Voyager space probes in the 1980s.

The presence of a nitrogen-methane gaseous chemistry would seem to assure a relatively fertile complex of organic chemicals on Titan. Although the temperature on Titan may be too cold for life to have evolved, geothermal hot springs may possibly have created habitable locales. The Cassini probe, due to parachute below the clouds of Titan in 2004, may tell us what strange environments exist on the surface.

The questions of whether life exists or did exist on Mars, Europa, and Titan are perfect scientific questions, because either answer has such profound philosophic implications.

Our generation may be able to answer these questions. If life is found in one or more of these places, it proves we are not alone in the universe. Conversely, if we can prove that life *never* formed in any of these places, that may prove that we are more alone than we might have thought.

Habitable Planets Beyond the Solar System?

From all we have just said, we conclude that if planetary surfaces with the necessary conditions—liquid water and C-H-O-N chemicals—exist anywhere, life is likely to evolve on them. And advanced species will probably appear eventually. But are there any such planets?

Astronomers have found that the planet-forming process is likely to be a normal evolutionary development in a cocoon nebula around a single star, and planets have actually been detected around other stars. Questions remain, however, about whether planetary systems like ours, with circular stable orbits, are common. Planets are less likely to form in a double or multiple star system, and an informed guess would be that between 1% and 30% of all stars have planetary companions.

Even if planets exist near some other stars, however, there is no guarantee that they are habitable. Astronomers have proposed several conditions for a habitable planet:

1. The central star should not be larger than about 1.5 $M_{\odot}$, so that it will last long enough for substantial life to evolve (at least 2000 My), but will not kill evolving life with too much ultraviolet radiation.

2. The central star should be at least 0.3 $M_{\odot}$ to be warm enough to create a reasonably large orbital zone in which a planet could retain liquid water.

3. The planet must orbit at the right distance from the star, so that liquid water will neither evaporate nor permanently freeze.

TABLE 28-1

Estimated Fraction of Stars with Planets That Have Intelligent Life

Criterion	Fraction	
	Lower limit (?)	Upper limit (?)
1. Stars having planets	10^{-2}	0.3
2. Criterion 1: stars ever having habitable conditions on at least one planet	10^{-1}	0.7
3. Criterion 2: planets on which habitable conditions last long enough for life to evolve	10^{-1}	1
4. Criterion 3: planets on which life evolves	10^{-1}	1
5. Criterion 4: planets on which habitable conditions last long enough for intelligence to evolve	10^{-3}	0.9
6. Criterion 5: planets on which intelligence evolves	10^{-1}	1
7. Criterion 6: planets on which intelligent life endures	10^{-7}	10^{-1}
8. Fraction of duration of intelligent life during which it retains an interest in contact with Earth-like civilization	10^{-3}	1
Fraction of all stars with planets that bear intelligent life	10^{-19}	2×10^{-2}
Implication: Distance to nearest civilization	3×10^{-8} light years	15 light years

4. The planet's orbit must be almost circular to keep it at the proper distance and prevent overly drastic seasonal changes.

5. The planet's gravity must be strong enough to hold a substantial atmosphere.

In a pioneering study, Dole (1964) reviewed these criteria and concluded that roughly 6% of the stars, mostly from 0.9 to 1.0 $M_{\odot}$, have habitable planets. In spite of the discovery of some extrasolar planetary systems, surprisingly little progress has been made in refining that figure.

Alien Life Among the Stars?

The next step is to consider alien life on planets near other stars. There is no direct evidence, but groups of international scientists have put together a method for considering the possibilities described as long ago as 1973 by Carl Sagan (1973). The logic is to try to estimate the various fractions of stars having planets, of those planets that are habitable, of those habitable planets where conditions remain favorable long enough for life to evolve, of those planets where life does evolve, of those where intelligence evolves, and of the planet's life during which intelligence lasts. The formalism is sometimes called the **Drake equation,** after the American astronomer Frank Drake, who developed the scheme.

Table 28-1 shows optimistic (upper limit) and pessimistic (lower limit) estimates of the various fractions and the consequent estimates of the upper and lower limits on the fraction of stars that might have civilizations today. The optimistic figure is a few percent; the pessimistic case is only one star in 10^{19}! Note the 17 order of magnitude uncertainty! In other words, we really don't know!

Given these figures, we can also ask how far away would the civilization be? Figure 28-4 shows the answer by plotting the radial distance required to include a given number of stars. In the first case, the nearest civilization might be only 15 light-years away—amazingly close. At the speed of light, that civilization might be reached in only one generation. In the pessimistic case, the closest civilization would probably not be in our own galaxy, but roughly 300 million light-years away in a distant galaxy.

In spite of all the uncertainties, given the innumerable galaxies, it is hard to avoid the conclusion—even with the most pessimistic view—that millions or billions of technological civilizations exist outside the solar system. If our reasoning is right, then at this moment intelligent creatures may be pursuing their own lives, joys, and disasters in unknown places under unknown suns.

Where Are They?

If there really are other civilizations (whatever that means) out in space, why haven't they contacted us? Radio astronomers have undertaken several programs to listen for radio messages from alien civiliza-

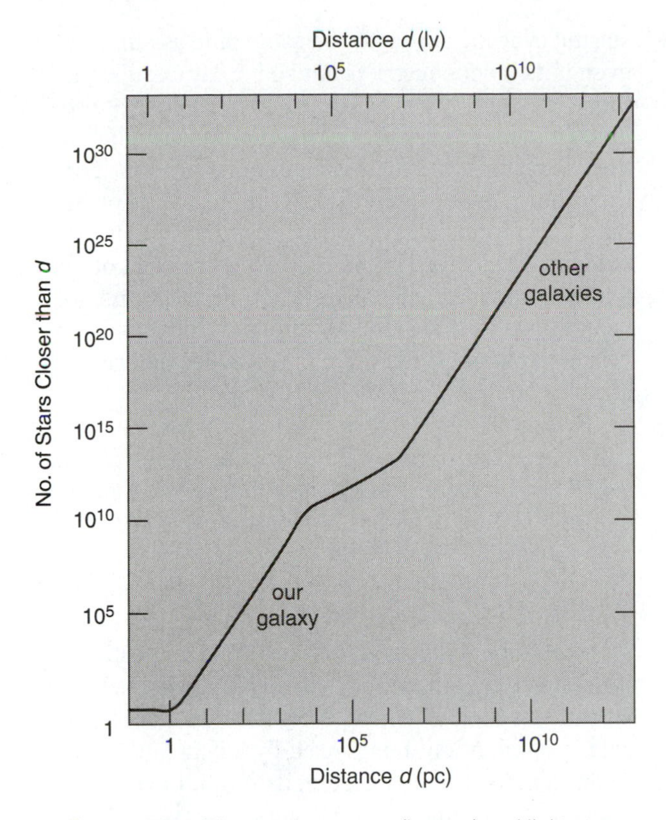

Figure 28-4 Distance in parsecs (bottom) and light-years (top) required to encounter the number of stars indicated at the left. If one star in 10^9 has life-bearing planets, the nearest one might be within a few thousand parsecs.

tions and have heard none. As early as the 1970s, for example, radio astronomers listened for broadcasts near 1420 MHz (the 21-cm wavelength that marks astronomically important radiation from neutral hydrogen atoms in space), targeting all solartype stars not known to be members of multiple star systems, within about 82 light-years (see Table 17.1); the search included 185 stars. No artificial broadcasts were identified (Horowitz, 1978). Goldsmith and Owen (1979) summarize other searches involving more than 600 stars. As of the year 2000, other more exhaustive searches have continued to be negative.

Note that a message from 1000 light-years away would come from a civilization 1000 y in the past, and no answers to our questions could come back for 2000 y. Such communication would not be dialogue, but, as physicist Philip Morrison has pointed out, more like the "messages" (books, letters, plays, and artworks) we receive from ancient civilizations such as Greece.

The negative results are not a scientific failure. The radio search for life is a perfect scientific effort, because any result is philosophically interesting. If we find aliens, that is sensational, and if we hear no broadcasts, that tells us something about our degree of solitude in the universe.[3]

Meanwhile, we have sent a few messages of our own. The Pioneer 10 spacecraft, which flew by Jupiter and left the solar system in 1973, and several later probes carried plaques designed to convey our appearance and location to possible alien discoverers of the derelict spacecraft. The first radio message was a test message sent from the large radio telescope at Arecibo, Puerto Rico, in 1974; it was beamed toward globular star cluster M 13, 27,000 light-years away.[4]

Nor do our skies seem to be overrun with alien visitors trying to contact us. All of which raises the question: Why haven't we heard from them?

★ One answer might be that we *are* being visited—witness the flying saucer reports. But although some UFO reports are intriguing, verifiable evidence for alien spaceships is abysmally poor, as discussed further on our web site.

Another answer to the question may be that the pessimistic figures are right, and that the nearest civilizations are in distant galaxies. Even their radio messages, if any, could be 10 My old by the time we receive them, and their spaceships would be unlikely to reach Earth if limited to speeds less than that of light, as current physics requires.

Most investigators of the problem, however, place the probability well above our lower limit. At a Russian-American conference as early as 1971, the favored estimate was one civilization per 100,000 stars. This would put a million civilizations in our galaxy and the nearest civilizations only a few hundred light-years away. Why, then, are there no frequent visits to Earth?

If we accept those numbers, they may suggest another answer: Biological evolution need not produce creatures who have a desire to build civilizations or travel through space. Not even all human societies evolve toward technology. Are humans fated to be explorers, bridgebuilders, and businesspeople rather

---

[3]NASA's search for extraterrestrial intelligence by radio listening was halted in late 1981 as the result of an amendment attached to the 1982 federal budget by the otherwise-respected Senator William Proxmire. The amendment explicitly forbade expenditures for a radio astronomy search for aliens on the grounds that it is a "ridiculous waste."

[4]Astronomers involved in these projects have been deluged with letters ranging from support to complaints about the nudity of the human figures on the Pioneer 10 plaque. The radio astronomers promptly received a telegram: "Message received. Help is on the way—M 13." Its authenticity might be questioned, because the round-trip radio signal travel time to M 13 is 54,000 y.

than artists, athletes, or daydreamers? Human society existed for tens of thousands of years in Australia without ever adopting technology beyond the stone age, though the aborigines there developed an intricate mythology. Chinese fleets almost rounded Africa and discovered Europe during the late middle ages, but were recalled in favor of a policy of nonexploration and isolation. Is the stereotypical aggressive Westerner more representative of the essence of humanity than the stereotypical contemplative Easterner? Isn't it possible that our aggressive technocracy is just one type of *cultural activity* rather than a universally achieved stage of *biological* evolution? Historically, many patterns we once assumed to be biologically imposed have turned out to be merely culturally imposed.

If humanity is not predestined to develop a technological civilization, how much less certain is the course of social development on other worlds? It is absurdly anthropocentric to suppose that beings on other planets would resemble us psychologically or socially. Consider again the variety of highly evolved life-forms on our planet alone. Ants live in ordered societies that do not appear to regard individual survival as important (a model for the "Borg" of Star Trek Voyager?). Dolphins communicate and have brains that seem almost comparable to ours, but they have no manipulative organs and hence no technology.

Perhaps we cannot expect aliens to be motivated by emotions that mean much to us. We certainly cannot expect, as always happens in grade-B science fiction movies, that humanlike aliens will walk out of saucers and invite us to join their democratically constituted United Planets, a galactic organization structured by documents that are curiously reminiscent of the U.S. Constitution. So why assume that other civilizations might even care to try to visit or contact us?[5]

Evolutionary Clocks and the Explorative Interval

This brings us to the fourth and perhaps most significant answer to why we have no evidence of aliens: We may be farther from aliens in evolutionary time than in physical space. Biological evolution is so persistently experimental that even if another planet started evolving at exactly the same time as ours, and even if its biochemistry produced creatures like us, those creatures are not likely to be in a phase of evolution similar to ours. If the evolutionary "clocks" on the two planets got only 0.02% out of synchronization, the other planet would be 1 My ahead of us or behind us, in terms of technology, biology, and/or psychology. Thus, even in the unlikely event that other planets produce civilizations recognizable to us, we would have to contact one of those civilizations in a very narrow time interval in order to see any recognizable common interests.

Evolution may pass through only a brief **explorative interval**, during which societies on one planet would care to reach other planets; beyond that stage, communication or space exploration might be no more attractive than a national program on our part to communicate with chimpanzees, ants, or dolphins. To be sure, a few of our scholars try this, but they "contact" an infinitesimal fraction of these creatures. What fraction of the anthills or dolphin schools have we humans tried to "contact"? By the same token, our solar system might be ignored by advanced aliens. Aliens a million years ahead of us might be no more interested in us than we are in ants.

How long might an explorative interval last? We have used tools for about 1 or 2 My, and it appears safe to assume that we will have progressed far beyond current technology in another million years, if we survive. Our explorative interval might be a few million years, then, or less than 0.1% of the history of the planet. This is the basis for the lower factor listed in criterion 8 in Table 28-1.

If civilizations are 500 light-years apart, then interstellar voyages and messages would take a millennium or more. There might be little incentive for the effort. Large spaceships in which many generations could live and die during interstellar voyages—self-contained "planets" that are a staple of science fiction stories—have also been hypothesized in published scientific models of interstellar colonization by other species (Kuiper and Morris, 1977). In any case, several authors have concluded that, given the plausible number of ships in transit at any one time from all civilizations, a visit to a specific random planet, such as Earth, would be rare.

This brings us to a fifth possible answer to the question of alien visits: They may have happened in the remote past. This is the "ancient astronaut" hypothesis, popularized in several pseudoscience books. Figure 28-5 illustrates a problem with this idea. For example, even if 100 visits occurred at random intervals in only the last 1% of Earth's history, they would still average 455,000 y apart! Given the

[5] Woody Allen (1980) summed up much of this more succinctly: "If saucers come from outer space, why have their pilots not attempted to make contact with us, instead of hovering mysteriously over deserted areas? My own theory is that for creatures from another solar system, 'hovering' may be a socially acceptable mode of relating. . . . It should also be recalled that when we talk of 'life' on other planets we are frequently referring to amino acids, which are never very gregarious, even at parties."

Figure 28-5 Even if Earth has been visited on 10,000 occasions by alien expeditions, the last visit would probably have occurred too long ago for any record likely to be identified by historical or archaeological records to remain, unless the visitors deliberately left large, prominent artifacts. (Painting by Jim Nichols)

Figure 28-6 Lines as long as a few miles in the Peruvian desert were claimed as alien landing sites requiring stupendous technologies beyond our means. Ground studies show that these lines were made by Indians around 1800 y ago and required technology no more stupendous than a broom to sweep aside dark stones from desert pavement. The lines are currently threatened by off-road vehicles, whose tracks show scale. (Photo by WKH)

rate of terrestrial erosion by water and glaciation, the prospects of finding evidence of such ancient visits are poor! As for historical visits, there is no good evidence. The Russian and American collaborators I. S. Shklovskii and Carl Sagan (1966) surveyed archaeological and mythological literature even before the hypothesis was a popular fad and found no compelling evidence for ancient astronauts. Neither Earth, the Moon, Mars, nor Venus is littered with ancient alien artifacts, and not a single mysterious artifact has been credibly advanced as physical evidence of ancient astronauts. As illustrated by Figure 28-6, popular pseudoscience books and sleaze TV have greatly exaggerated the mystery surrounding many archaeological structures and artifacts, which is too bad because the reality of the situation is fascinating enough!

Clearly, the search for life in the universe is an exciting, if highly uncertain adventure. Scientific interest in the subject is demonstrated by the continuing flow of review articles about the prospects for alien biology (Crawford and LePage, 2000; Bernstein and others, 1999; Krauss and Starkman, 2000).

SUMMARY

Life started on Earth about 3900 My ago in the form of simple microbes. Microbial life thrives in many environments on Earth, including near-boiling hot springs, deep seafloor geothermal vents, and underground rocks. Similar bacteria have been suggested inside rocks from Mars, and precursor amino acids have been found inside fragments of water-bearing asteroids. These data suggest that life may evolve readily on other planets, though no alien life-forms have been proved. The discovery of firm evidence of alien civilizations, or even alien life-forms, could be a pivotal development in human history, as suggested in Figure 28-7.

Although the limited evidence indicates that other life-forms should exist, there is no evidence that they do or that they have tried to communicate with us. We can only speculate about the reasons. Perhaps they are too far away. Perhaps most civilizations destroy themselves before successfully exploring the universe. Perhaps evolution carries them beyond a stage where they would care to communicate with us. Perhaps they are unrecognizable.

Arthur C. Clarke has remarked that any technology much advanced beyond your own looks like magic to you. Perhaps we are too limited by our own concept of civilization. After all, one creature's civilization may be another's chaos, as shown by Mohandas Gandhi's remark when asked what he thought of Western civilization: He said it wouldn't be such a bad idea. It seems likely that our first contact with aliens (if they exist) might be as incomprehensible as the dramatized contact in Clarke's novel and the Kubrick-Clarke film, *2001: A Space Odyssey.*

Clearly, we have been reduced to speculation by a lack of facts. Indeed, the whole field of exobiology has been criticized as a science without any subject matter. Exobiology recalls Mark Twain's comment: "There is something fascinating about science. One gets such wholesale returns of conjecture from such a trifling investment of fact." In the same vein, exobiology spokesman Philip Morrison (as quoted by Simpson, 1973) has admitted of exobiology: "Here is a body of literature whose ratio of results/papers is lower than any other." The only way to reduce the conjecture and increase the proportion of fact is to pursue research in many related fields—physics, chemistry, geology, meteorology, biology—and listen to the skies with radio telescopes. There may be surprises waiting out there.

CONCEPTS

exobiology
organic chemistry
organic molecules
life
cell
amino acid
protein
Miller experiment
nonracemic mixture
racemic mixture
proteinoid
coacervate
stromatolite
ozone
ozone layer
natural selection
convergence
cultural hurdle
Drake equation
explorative interval

PROBLEMS

1. Compare Alpha Centauri, Barnard's star, and a random red giant in terms of the probability of detecting radio broadcasts from intelligent creatures (See Table 17.1).

2. Describe several ways in which Earth's internal evolution has affected the evolution of life on Earth.

3. Describe several ways in which extraterrestrial events, such as solar or stellar evolution, might have affected the evolution of life on Earth or other planets.

4. Describe ways in which technology could affect the survival of intelligent life on Earth or on other planets. Construct scenarios of each of the following:

a. The possible destruction of life.

b. Guaranteeing the survival of life.

(*Hint:* In part [b] consider the impact of space travel.)

5. In your own opinion, what would be the long-range consequences of the following:

a. The arrival of an alien spacecraft and visitors in a prominent place, such as the United Nations plaza.

b. The discovery of radio signals arriving from a planet about 10 light-years distant and asking for two-way communication.

c. Proof (by some unspecified means) that life exists *nowhere* else in the observable universe.

6. Given the assumption that our technology has the potential for creating planetwide changes in environment, defend the proposition that *if* life on other worlds produces technologies like ours, then that life is likely

a

a

Figure 28-7 The discovery of positive evidence for extraterrestrial civilizations could be a dramatic and pivotal event in human history. This could occur by **a** us going out and finding such evidence (such as the discovery of an alien obelisk on the Moon in the MGM release *2001: A Space Odyssey*), or **b** such evidence arriving on Earth (such as the arrival of the "mother ship" in the film *Close Encounters of the Third Kind*).

either to have become extinct or to be widely dispersed among many planets by means of space travel.

7. In view of the devastation wrought on many terrestrial cultures by contact with more technologically advanced cultures, which would you say is the safest course: (1) aggressive broadcasting of radio signals to show where we are in hopes of attracting friendly contacts, (2) careful listening with large radio receivers to see if there are any signs of intelligent life in space, and (3) neither broadcasting nor listening, but just waiting to see what happens?

a. How would it affect the results of our listening program if other intelligent species had reached the first, second, or third conclusion?

b. If we listen at many frequencies and pick up no artificial signals, does that prove that life has not evolved elsewhere in the universe?

ADVANCED PROBLEMS

8. If an Earth-size planet (diameter 12,000 km) was circling a star 1 pc away (3×10^{16} m):

a. What would be the planet's angular diameter when seen from Earth?

b. Could this be resolved by existing telescopes?

c. If the planet orbited 1 AU (1.5×10^{11} cm) from its star, what would be its maximum angular separation from the star as seen from Earth?

d. Could this angle be resolved?

9. Derive expressions for the inner and outer radii of a toroidal "zone of habitability" around any star of luminosity L (in units of solar luminosity) if an inhabited planet orbiting the star has an effective albedo (i.e. % of light reflected) A and emissivity e, a circular orbit, and liquid water somewhere on its surface to support life.

Appendix 1: Powers of 10

It is no accident that the word *astronomical* has become a synonym for "enormous, almost beyond conception." Astronomy is full of extraordinary numbers designating great ages and distances. In astronomy (and other sciences), therefore, a convenient shorthand system of writing numbers is favored. In this system, an exponent, or superscript, designates the number of factors of 10 that have to be multiplied together to give the desired quantity—that is, the number of zeros in the quantity. For example:

$$1 = 10^0$$
$$10 = 10^1$$
$$100 = 10^2$$
$$1000 = 10^3$$
$$10{,}000 = 10^4 \text{ and so on}$$

The most often used of these large numbers are

$$\text{one thousand} = 1000 = 10^3$$
$$\text{one million} = 1{,}000{,}000 = 10^6$$
$$\text{one billion}^1 = 1{,}000{,}000{,}000 = 10^9$$
$$\text{one trillion} = 1{,}000{,}000{,}000{,}000 = 10^{12}$$

The usefulness of this system is illustrated by expressing the age of the Earth

$$4{,}600{,}000{,}000 \text{ y} = 4.6 \times 10^9 \text{ y}$$

or the distance that light travels in a year

$$6{,}000{,}000{,}000{,}000 \text{ mi} = 6 \times 10^{12} \text{ mi}$$

A similar system is used for very small numbers. Here the exponent is negative and refers to the number of decimal places. Thus:

$$0.0001 = 10^{-4}$$
$$0.001 = 10^{-3}$$
$$0.01 = 10^{-2}$$
$$0.1 = 10^{-1}$$
$$1.0 = 10^0$$

[1]Beware—the British use the term *billion* to refer not to 10^9 but to 10^{12}. This book uses *billion* to mean 10^9.

The density of gas in interstellar space, which is

$$0.000000000000000000000002 \text{ g/cm}^3$$

can be more conveniently written as

$$2 \times 10^{-24} \text{ g/cm}^3$$

One of the beauties of the metric system (discussed further in Appendix 2) is that the units are related by powers of 10. For instance, a kilometer is 10^3 m (as opposed to our own confusing English system, where 1 mi is 5280 ft). To make the system easier to use, the International Systems of units (known as the SI metric system) has a standardized set of prefixes, such as ***kilo-***, that represent multiples of 10^{-3}, 10^3, 10^6, and so forth. These prefixes are shown in Table A1-1.

TABLE A1-1

Prefixes Used in the SI Metric System of Units

Prefix	Symbol	Multiple
Tera- (TER-a)	T	10^{12}
Giga- (JIG-a)	G	10^9
Mega- (MEG-a)*	M	10^6
Kilo- (KILL-low)*	k	10^3
Hecto- (HECK-toe)	h	10^2
Deka-(DECK-a)	da	10^1
Deci- (DESS-ee)	d	10^{-1}
Centi- (SEN-tee)*	c	10^{-2}
Milli- (MILL-ee)*	m	10^{-3}
Micro- (MY-crow)*	µ	10^{-6}
Nano- (NAN-oh)	n	10^{-9}
Pico- (PEE-koh)	p	10^{-12}
Femto- (FEM-toe)	f	10^{-15}
Atto- (AT-oh)	a	10^{-18}

Note: Starred items are the most important in ordinary applications.

Appendix 2: Units of Measurement

Most scientists express distances and other measurements in metric units, and the world is in the process of converting to this system, though an official U.S. Government conversion effort, begun in the 1970s, has fizzled. This conversion will save time and money, since publications, quantities of goods, and machine parts will be measured in one uniform system. The metric system is also easier to learn and use, since it expresses units in multiples of 10 (like our money system), instead of in unpredictable multiples such as 12 inches per foot, 5280 feet per mile, and 16 ounces per pound.

Table A2-1 gives some common units in both systems. This book has emphasized the International System of metric units (abbreviated "SI" for *Système Internationale)*, which uses the meter, kilogram, and second as the basic measures of length, mass, and time, respectively. When using the SI system to solve problems involving complex quantities or constants, such as energy or the gravitational constant *G*, one must make sure that these quantities are expressed in terms of the basic units kg, m, and s. Equivalents in the centimeter-gram-second (cgs) system and in larger units are also given in Table A2-1.

TABLE A2-1

Units of Measurement

English System	SI (m-kg-s)	cgs System	Astronomical Measurements
1 inch	0.025m	2.54 cm	—
1 foot	0.305 m	30.5 cm	—
1 yard	0.914 m	91.4 cm	—
1 mile	1609 m	1.609×10^5 cm	—
1 pound	0.454 kg	454 g	—
	1.50×10^{11} m	1.50×10^{13} cm	1 astronomical unit (AU)
	9.46×10^{15} m	9.46×10^{17} cm	1 light-year (ly)
	3.08×10^{16} m	3.08×10^{18} cm	1 parsec (pc) = 3.26 ly
	3.08×10^{19} m	3.08×10^{21} cm	1 kiloparsec (kpc)
	3.08×10^{22} m	3.08×10^{24} cm	1 megaparsec (Mpc)
	1.99×10^{30} kg	1.99×10^{33} g	1 solar mass ($1\ M_{\odot}$)

While solving certain problems, such as the advanced problems in this book, we need to use certain physical constants, such as Newton's universal constant of gravitation *G*, described in Optional Basic Equation II on pages 70–71. It is important to remember that in solving any such numerical problems, all measurement units must be converted into a single system of units, and any physical constants must also be expressed in that system. The SI metric system is recommended for this purpose, since it is the international standard. Table A2-2 gives some of the more useful physical constants expressed in the SI system.

TABLE A2-2

Some Useful Physical Constants (SI system of metric units)

Constant	Numerical Value	Optional Basic Equation
Seconds of arc in one radian	206,265	I
G (Newton's universal gravitational constant)	6.67×10^{-11} N·m²/kg²	II, III
Wien's constant	2.90×10^{-3} m·K	V
k (Boltzmann constant)	1.38×10^{-23} J/K	VI, XI
c (velocity of light)	3.00×10^{8} m/s	VII, IX, X, XI
σ (Stefan-Boltzmann constant)	5.67×10^{-8} W/m²·K⁴	VIII
Mass of a proton	1.67×10^{-27} kg	XI
Mass of an electron	9.11×10^{-31} kg	XI
Solar constant	1390 W/m²	
$L_{\odot}$ (solar luminosity)	8×10^{26} W = 8×10^{26} J/s	
$M_{\odot}$ (solar mass)	2×10^{30} kg	

Glossary

AU: (See *astronomical unit.*)

absolute: Intrinsic; not dependent on the position or distance of the observer.

absolute brightness: Any measure of the intrinsic brightness or luminosity of a celestial object.

absolute magnitude: The absolute brightness (luminosity) of a star expressed in the magnitude system. The Sun's absolute magnitude is +5.

absolute reference frame: The microwave background radiation, or any group of objects whose overall motion is at rest with respect to the microwave background radiation. Neither the Milky Way nor the Local Group define an absolute reference frame.

absolute temperature scale: The temperature scale in which 0 = absolute zero, 273 = the freezing point of water, and 373 = the boiling point of water. The units are called Kelvins. One Kelvin = 1°C.

absorption: The loss of photons as light passes through a medium. A photon is lost when it strikes an electron, and the photon's energy is consumed in knocking the electron to a higher energy level.

absorption band: A dark or dim region of the spectrum, caused by absorption of light over a moderate range of wavelengths, typically about 0.1 nm, usually by molecules or crystals.

absorption line: In a spectrum, a reduction in intensity in a narrow interval of wavelength, caused by absorption of the light by atoms between the source and the observer.

accretion disk: A disk of hot gas and dust surrounding a star, usually used to denote material that has been thrown off one star onto a companion. There is weaker observational evidence for very large accretion disks in the central regions of active galaxies and quasars.

achondrite: A type of stony meteorite in which chondrules have been destroyed, probably by heating or melting.

active galaxy: Galaxy whose nucleus emits more energy than other, normal galaxies. Typical signatures of active galaxies are variable brightness, broad emission lines, and strong radio emission.

age of the Earth: The period since the Earth's formation from planetesimals, measured to be 4.6 billion years.

age of globular clusters: About 12–16 billion years.

age of the (Milky Way) galaxy: Estimated to be roughly 12–16 billion years.

age of open clusters: Time since formation of open clusters, judged by their H-R diagrams. Most are less than 100 million years. Ages from 1 million to a few billion years have been reported.

age of stars: Time since star formation, typically billions of years for smaller stars, but less than a million years for some massive stars in recently formed clusters. Age is difficult to measure for individual stars, but possible to measure for clusters of stars.

airglow: Visible and infrared glow from the atmosphere produced when air molecules are excited by solar radiation.

Airy disk: In the telescopic image of a star, a small disk caused by optical effects.

Alexandrian library: The research institution and collection of ancient works preserved after the fall of Rome at Alexandria, Egypt. Alexandrian knowledge passed to the Arabs with the Arab conquest of Alexandria, and eventually back into Europe around A.D. 100–1500.

Alpha Centauri: (1) The nearest star system, composed of three members; (2) the brightest of these three.

amino acid: A complex organic molecule important in composing protein and called a "building block of life."

Andromeda galaxy: The nearest spiral galaxy comparable to our own, about 670 kpc away.

angular measure: Any measure of the size or separation of two objects as seen from a specified point, expressed in angular units (degrees, minutes of arc, or seconds of arc), but not linear units (such as kilometers, miles, or parsecs).

angular size: The angle subtended by an object at a given distance.

anisotropic: Something that is not the same in every direction.

annular solar eclipse: An eclipse in which the light source is almost, but not quite, covered, leaving a thin ring of light at mid-eclipse.

antimatter: Material, with equivalent properties to matter, but with subatomic particles' quantum properties reversed; for example, particles' charges are opposite. Antimatter happens to be rare in our universe. Matter and antimatter annihilate on contact to produce gamma rays.

aperture: The diameter of the light-gathering objective in a telescope.

apogee: The point in an orbit around the Earth that is farthest from the Earth.

Apollo asteroids: Asteroids that cross the Earth's orbit.

Apollo program: The U.S. program to land humans on the Moon, 1961–72; first landing July 20, 1969.

apparent: Not intrinsic, but dependent on the position or distance of an observer.

apparent brightness: The brightness of an object as perceived by an observer at a specified location (but not measuring the object's intrinsic, or absolute, brightness).

apparent magnitude: Apparent brightness of one star relative to another as expressed in the magnitude system.

apparent solar time: Time of day determined by the Sun's actual position in the sky. Apparent solar noon occurs as the Sun crosses the meridian. Apparent solar time is different at each different longitude.

apparition: The period of a few weeks during which a planet is most prominent or best placed for observation from Earth.

arc-rings: Concentrations of particles along only a portion (less than 30°) of a circular ring around a planet, probably caused by gravitational forces associated with small nearby moonlets.

association: A loosely connected grouping of young stars.

asteroid: A rocky or metallic interplanetary body (usually larger than 100 m in diameter).

asteroid belt: The grouping of asteroids orbiting between Mars and Jupiter.

asthenosphere: In a planetary body, a subsurface layer that is more plastic than adjacent layers because the combination of pressure and temperature places it near (or slightly above) the melting point. Asthenospheric movements may disrupt the planet's surface.

astrology: The superstitious belief that human lives are influenced or controlled by the positions of planets and stars; this belief is rejected by modern astronomers and other scientists.

astrometric binary: A binary star system detectable from the orbital motion of a single visible component.

astrometry: The study of positions and motions of the stars.

astronomical unit (AU): The mean distance from the Earth to the Sun, about 150 million kilometers.

astronomy: The study of all matter and energy in the universe.

atom: A particle of matter composed of a nucleus surrounded by orbiting electrons.

aurora: Glowing, often moving colored light forms seen near the north and south magnetic poles of the Earth; caused by radiation from high-altitude air molecules excited by particles from the Sun and Van Allen belts.

B-type shell star: A star of spectral type B that occasionally blows off a cloud of gas, forming a gaseous shell around the star.

Balmer alpha line: A brilliant red spectral line at wavelength 656.3 nm, caused by transition of the electron in the hydrogen atom from the third-level orbit to the second-level orbit.

barred spiral galaxy: A spiral galaxy whose spiral arms attach to a barlike feature containing the nucleus.

baryonic particles: Particles such as protons, neutrons, electrons, and neutrinos out of which normal matter is made.

basalt: A type of igneous rock, often formed in lava flows, common on the Moon and terrestrial planets.

basaltic rock: Igneous rock (including basalt) with a composition resembling basalt and a relatively low content of quartz (SiO_2).

basin: Large impact crater on a planet, usually several hundred kilometers across, flooded with basaltic lava and surrounded by concentric rings of faulted cliffs.

belts: Dark cloud bands on giant planets.

big bang model: The theory that the universe started from an extremely hot and dense state, about 12–16 billion years ago. Supported by observations of the universal recession of galaxies, the abundance of light elements, and the cosmic background radiation.

big crunch: The idea that if the mean density of the universe exceeds the critical density, the universe will eventually recollapse into a final state of infinite density and temperature.

binaries, classes of: Three categories of binary stars, depending on whether neither, one, or both fill their Lagrangian lobes.

binary star system: A pair of coorbiting stars.

bipolar jets: A phenomenon in which narrow streams of gas are ejected at very high speed in opposite directions from the centers of some disks of gas, perpendicular to the disk; bipolar jetting is seen both in accretion disks around individual stars and in galactic disks. The mechanism is uncertain.

black hole: An object whose surface gravity is so great that no radiation or matter can escape from it. Some black holes discussed in astronomy are collapsed stars, but much smaller ones are theoretically possible.

blueshift: A doppler shift of spectral features toward shorter wavelengths, indicating approach of the source.

Bode's rule: A convenient memory aid for listing the planets' distances from the Sun.

body tide: A tidal bulge raised in the solid body of a planet.

Bok globule: A relatively small, dense, dark cloud of interstellar gas and dust, usually silhouetted against bright clouds.

bolometric correction: Correction that must be made to the brightness or luminosity measured through any filter to yield the total emission of a star at all wavelengths. For visual magnitudes, the bolometric correction is largest for very hot or very cool stars.

bolometric luminosity: The total energy radiated by an object at all wavelengths, usually given in joules per second (identical to watts).

Tycho Brahe: (1546–1601): Danish astronomer who recorded planetary positions, ultimately enabling Kepler to deduce laws of planetary orbits.

breccia: A rock made from angular fragments cemented together.

brecciated meteorite: A meteorite formed from cemented fragments of one or more meteorite types.

broken symmetry: A situation in which the clear relationship between different forces that is apparent at high energies is masked at low energies.

brown dwarf: A starlike object too small to achieve nuclear reactions in its center; any stellar object smaller than about 0.08 $M_{\odot}$.

bubble: A roughly spheroidal shell of interstellar gas blown outward from a star by a stellar explosion or strong stellar wind.

bulk flow: Motion of a large group of galaxies that displays a systematic departure from the smooth Hubble flow.

Callisto: Outermost of the four large Moons of Jupiter, and most heavily cratered of the four.

Cambrian period: A period from 570 to 500 million years ago during which fossil-producing species of plants and animals first proliferated.

canals: Alleged straight-line markings on Mars found not to exist by spacecraft visits to Mars.

captured moons: Satellites that did not originate in orbit around a planet but were captured into orbit from interplanetary space.

capture theory: A theory of origin of a planet–satellite or binary star system in which one body captures another body by gravity.

carbonaceous chondrite: A type of carbon-rich and volatile-rich meteorite, believed to be a nearly unaltered example of some of the earliest-formed matter in the solar system.

carbonaceous material: Black material rich in carbon and carbon compounds, found in carbonaceous chondrites and believed to color many black comets and asteroids of the outer solar system.

carbon cycle: A series of nuclear reactions in which hydrogen is converted to helium, releasing energy in stars more massive than about 1.5 $M_{\odot}$. Carbon is used as a catalyst.

Cassini's division: The most prominent gap in Saturn's rings.

catastrophic theory: A theory invoking sudden or very short (cosmically or geologically speaking) energetic events to explain observed phenomena.

catastrophism: An early scientific school which held that most features of nature formed in sudden events, or catastrophes, instead of by slow processes.

cause of eclipses: The falling of a shadow of one body onto another body.

cause of the seasons: The tilt (obliquity) of the Earth's axis to its orbit plane causes first the North Pole and later the South Pole to be tipped toward the Sun during the course of a year.

CCD: A charge-coupled device, which is an electronic instrument for detecting light or recording an image—and is much more sensitive than film.

celestial equator: The projection of the Earth's equator onto the sky.

celestial poles: The projections of the two poles of the Earth's rotation onto the sky.

cell: The unit of structure in living matter.

center of mass: The imaginary point of any system or body at which all the mass could be concentrated without affecting the motion of the system as a whole; the balance point.

Cepheid variable: Any of a group of luminous variable stars with periods of 5 to 30 d (depending on their population). The periods are correlated with luminosity, allowing distance estimates out to about 3 Mpc.

cgs system: Metric system of measurement using centimeters, grams, and seconds as the fundamental units.

Chandrasekhar limit: A mass of about 1.4 $M_{\odot}$, the maximum for white dwarfs; stars of greater mass have too great a central pressure, causing formation of a star type denser than a white dwarf.

channel: One of the riverbed-like valleys on Mars, which are possible sites of ancient Martian rivers.

chaotic rotation: A form of rotation in which dynamical forces cause the rotation period to change irregularly from one rotation to the next. Applies to Saturn's moon Hyperion.

Charon: Pluto's satellite.

chemical reaction: Reaction between elements or compounds in which electron structures are altered; atoms may be moved from one molecule to another, but nuclei are not changed and thus no element is changed to another.

chondrite: Stony meteorite containing chondrules, believed to be little altered since their formation 4.6 billion years ago.

chondrule: BB-sized spherule in certain stony meteorites, believed among the earliest-formed solid materials in the solar system.

chromosphere: A reddish-colored layer in the solar atmosphere, just above the photosphere.

circular velocity: Velocity of an object in circular orbit:

$$V = \sqrt{\frac{GM}{R}}$$

where G = Newton's gravitational constant. In SI units it is $6.67 \times 10^{-11}\ \mathrm{N \cdot m^2/kg^2}$

M = mass of central body

R = distance of orbiter from center of central body

circumpolar zone: The zone of stars, centered on a celestial pole, that never sets, as seen from a given latitude.

circumstellar nebula: Gas and dust surrounding a star.

cirrus: Diffuse clouds of dust grains in the disk of the Milky Way, which are cold and so emit thermal radiation at far infrared wavelengths. (Derived from Earth's wispy, high-altitude cirrus clouds.)

classes of binaries: See *binaries, classes of.*

closed universe: A universe with positive space curvature and more than the critical mass density. It will eventually recollapse.

cluster of galaxies: A relatively close grouping of galaxies, often with some members coorbiting or interacting with each other.

coacervate: Cell sized, nonliving globule of proteins and complex organic molecules that forms spontaneously in a water solution.

Coal Sack Nebula: A prominent dark nebula about 170 kpc away silhouetted against the Milky Way.

cocoon nebula: A dust-rich nebula enclosing and obscuring a star during its formation, but later shed.

cold dark matter: Unseen material in the universe; candidates include weakly interacting particles that were traveling slowly when the first structures formed in the universe. They may include exotic particles not yet seen in the laboratory.

collapse: Rapid contraction, especially of a cloud of gas and dust during star formation.

colliding galaxies: Galaxies undergoing interpenetration or close enough to cause major gravitational distortion of each other.

coma: (1) The diffuse part of the head of a comet surrounding the nucleus; (2) a type of distortion in some telescopes and optical systems.

comet: An ice-rich interplanetary body that, when heated by the Sun in the inner solar system, releases gases that form a bright head and diffuse tail. (See also *coma.*)

comet head: The coma and nucleus regions of a comet.

comet nucleus: The brightest starlike object near the center of a comet's head; the physical body (believed to be icy and a few kilometers across) within a comet.

comet tail: Diffuse streamers of gas and dust released from a comet and blown in the direction away from the Sun by the solar wind.

comparative planetology: An interdisciplinary field of astronomy and geology attempting to discover and explain differences between planets in properties such as climate and interior structure.

composition of Population II stars: About $\frac{3}{4}$ hydrogen and $\frac{1}{4}$ helium, by mass, with virtually no heavier elements.

compound telescope: A telescope that combines lenses and mirrors in the light-gathering system.

condensation sequence: The sequence in which chemical compounds condense to form solid grains in a cooling, dense nebula.

conduction: One of three processes that transfers heat from hot to cold regions; occurs as fast-moving molecules in the hot region agitate adjacent molecules.

conjunction: The period when a planet lies at zero or minimum angular distance from the Sun, as seen from the Earth.

conservation of angular momentum: A useful physical rule which states that the total angular momentum in an isolated system remains constant.

constellation: Imaginary pattern found among the stars, resembling animals, mythical heroes, and the like; different cultures map different constellations.

contact binary: A coorbiting pair of stars whose inner atmospheres or surfaces touch.

continental drift: The motion of continents due to (convective?) motion of underlying material in the Earth's mantle.

continental shield: A stable, ancient region, usually flat and oval-shaped, in a continent.

contingency: The idea that evolution consists of many branching points and that an organism's future evolution depends on previous branching points. This theory would predict that evolution is unlikely to produce similar species on different planets.

continuous spectrum: A spectrum made up of all wavelengths, without emission or absorption lines or bands.

continuum: In a spectrum with absorption or emission lines, the background continuous spectrum.

convection: One of three modes of transmission of heat (energy) from hot regions to cold regions; involves motions of masses of material.

Copernican revolution: The intellectual revolution associated with adopting Copernicus' model of the solar system, which displaced the Earth from the center of the universe.

core: The densest inner region of the Earth, probably of nickel-iron composition; in other planets, similar high-density central regions; in the Sun or stars, a dense central region where nuclear reactions occur; in galaxies, the densest, brightest central regions.

core collapse: Occurs when a star loses its pressure support because no more energy can be released by nuclear reactions. The core collapse drives the explosion and heavy-element creation of a supernova.

Coriolis drift: A departure from a straight-line trajectory, perceived by an observer in a rotating system; Coriolis effects in clouds were early evidence of the Earth's rotation.

corona: The outermost atmosphere of the Sun, having a temperature of about 1 to 2 million kelvins.

coronograph: An instrument permitting direct observation of the solar corona without an eclipse.

cosmic background radiation: The nearly uniform thermal radiation believed to be a relic of the hot big bang. It is observed in microwaves with a temperature of 2.7 K.

cosmic fuels: Nonfossil energy sources provided by cosmic process; for example, solar and geothermal energy.

cosmic nucleosynthesis: The fusion of light elements during the early hot phase of the big bang, to produce heavier elements. It resulted in nearly a quarter of the mass of the universe being turned from hydrogen into helium.

cosmic rays: High-energy atomic particles (85% protons) that enter the Earth's atmosphere from space. Many may originate in supernovae and pulsars.

cosmological constant: Term that can be added to the equations of general relativity to give a static solution.

cosmological principle: The assumption that the universe is homogeneous and isotropic. This assumption is crucial to modern cosmology, and it has been shown to be a good approximation to the observed state of the universe.

cosmological redshift: Any redward Doppler shift attributed to the mutual recession of galaxies or the expanding universe.

cosmology: The study of the structure of the universe. The term is often broadened to include the origin of the universe as well.

Cretaceous-Tertiary impact: The impact of one or more asteroids or comets, up to about 10 km in size, about 65 million years ago, which apparently led to extinction of most species of plants and animals living at that time, including dinosaurs.

critical density: The mass density needed to just halt the universal expansion.

crust: The outermost, solid layer of a planet, with composition distinct from the mantle and differentiated by a seismic discontinuity.

cultural hurdle: The hypothetical survival requirement for a planetary culture between the time it achieves technology capable of quickly altering its planetary environment and the time it can establish viable bases off its planet; the uncertainty of the probability of survival affects our estimates of the probability of intelligent life elsewhere in space.

dark matter: Mass that makes its presence felt by gravitational forces but does not emit light. It is usually detected by anomalously high orbital velocities in gravitational systems. Dark matter makes up about 90% of the mass of the universe, and its nature is still unknown.

daughter isotope: An isotope resulting from radioactive disintegration of a parent isotope.

deceleration parameter: The rate of deceleration of the universal expansion, measured at the current epoch it is given the symbol q_0.

declination: Angular distance north or south of the celestial equator. (Abbreviation: Dec.)

decoupling: The point in the history of the universe when the big bang radiation is cool enough for stable atoms to form. At this point, radiation and matter decouple, and photons travel freely through the universe.

degenerate matter: Matter in a very high-density state in which electrons are freed from atoms and pressure is a function of density but not temperature.

degree: An angle equaling $\frac{1}{360}$ of a circle.

density fluctuations: The seeds for galaxy formation.

density parameter: The ratio of the mean mass density of the local universe to the density required to just halt the universal expansion, given the symbol Ω_0.

density-wave theory: The leading theory for explaining the formation of spiral arms in galaxies, which posits periodicities in star, dust, and gas motions.

deposition: The accumulation of eroded materials in one place.

desert: Any of the brighter regions of Mars.

differential rotation: The differences in speed for stars at different distances from the center of the galaxy. Orbital velocities are actually slower at 5000 pc from the center than at the Sun's distance, which is 8000 to 10,000 pc from the center.

differentiation: Any process that tends to separate different chemicals from their original mixed state and concentrate them in different regions.

diffraction: The slight bending of light rays as they pass edges, producing spurious rays and rings in telescopic images of stars.

dimensions of the galaxy: As early as 1935 astronomers agreed that the Sun is about 8000 to 10,000 pc from the center, and the overall diameter of our galaxy is about 30,000 pc.

Dione: (See *Tethys, Dione, and Rhea.*)

dirty iceberg model: A theoretical description of a comet nucleus as a large icy body with bits of silicate "dirt" embedded in it.

disruption of open clusters: Gradual dispersion of stars from an open cluster as the cluster is sheared by differential galactic rotation, and as high-speed stars escape. Disruption time is usually a few hundred million years.

distance limit for reliable parallaxes: 100 parsecs.

distance scale: An overlapping set of techniques that are used to measure distances in the universe, starting with direct geometric methods such as parallax for nearby stars, and ending with global measures of galaxies. The errors in the distance scale increase with increasing distance from the Earth.

DNA (deoxyribonucleic acid): A long, replicating molecule, shaped like a twisted ladder, that is the basis of the genetic code. Information for all of life's functions is coded in the sequence of base pairs that join the two strands. DNA is found in all living organisms.

Doppler effect: The shift in wavelength of light or sound as perceived by the observer of an approaching or receding body. For speeds well below that of light, the shift is given by the equation

$$\text{Original wavelength} \times \frac{\text{radical velocity}}{\text{velocity of light}}$$

Drake equation: The statement that the fraction of stars harboring intelligent life equals the number of all stars times a sequence of fractions, such as the fraction of all stars having planets, the fraction of planets that are habitable, and so on. Named after radio astronomer Frank Drake.

dust scattering: Scattering of light by particles much bigger than the light's wavelength. Causes extinction and reddening of starlight.

dust trails: Toroidal lanes of dust stretching along elliptical orbits around the solar system; probably due to asteroid collisions or comet dust ejection.

dwarf elliptical galaxy: An ellipsoidal galaxy resembling a globular cluster but usually at least a few times larger.

early dense atmosphere: The atmosphere of a planet (if any) just after planet formation. (See also *secondary atmosphere.*)

early intense bombardment: The intense bombardment of planets and moons by interplanetary debris during their first 500 million years, from 4.5 to 4.0 billion years ago.

Earth: The third planet from the Sun.

earthquake: Vibration or rolling motion of the Earth's surface accompanying the fracture of underground rock.

eclipse: An event in which the shadow of one body falls on another body.

eclipsing binary: A binary star system seen virtually edge-on so that the stars eclipse each other during each revolution.

eclipsing-spectroscopic binary: An eclipsing binary whose motions are measurable from spectral Doppler shifts. The most informative type of binary star.

ecliptic: (1) The plane of the Earth's orbit and its projection in the sky as seen from Earth; (2) approximately, the plane of the solar system.

effective temperature: Temperature of an object as calculated from the properties of the radiation it emits.

ejecta blanket: A layer of debris thrown out of a crater onto a planet's surface.

electromagnetic radiation: Light, radio waves, X rays, and other forms of radiation that propagate as disturbances in electric and magnetic fields, travel at the speed of light, and combine to make up the electromagnetic spectrum.

electromagnetism: The force of nature that governs the structure of atoms and molecules and accounts for electromagnetic radiation.

electron: Negatively charged particle orbiting around the atomic nucleus, with mass 9.1×10^{-31} kg.

electroweak theory: Well-established theory stating that the electromagnetic and weak nuclear forces are unified at temperatures around 10^{15} K.

element: A chemical material with a specified number of protons in the nucleus of each atom. Atoms with one proton are hydrogen; with two protons, helium; and so on.

ellipse: A closed, oval-shaped curve (generated by passing a plane through a cone) describing the shape of the orbit of one body around another.

emission: Release of electromagnetic radiation from matter.

emission band: Narrow wavelength interval in which molecules emit light.

emission line: Very narrow wavelength intervals in which atoms emit light.

emission-line clouds: Small, dense clouds of gas that orbit the center of an active galaxy. Clouds near the nucleus orbit faster and so produce broader lines than clouds far from the nucleus.

Enceladus: One of the inner moons of Saturn, notable for its very bright, fissured surface of water ice, with several young, sparsely cratered areas.

energy: In physics, a specific quality equal to work or the ability to do work. Energy may appear in many forms, including electromagnetic radiation, heat, motion, and even mass (according to the theory of relativity).

energy level: The orbit of an electron around the nucleus of an atom.

English system: A nondecimal system of units using pounds, inches, and seconds. In scientific use, and in most countries, it is replaced by the more convenient metric system.

ephemeris: A table of predicted positions of a planet, asteroid, or other celestial body.

ephemeris time: A timekeeping system based on the motions of planets; more regular than conventional systems based on the Earth's rotation.

epicycle: A small circular motion superimposed on a larger circular motion.

epicycle theory: An early theory by Ptolemy that the planets move around the Earth in epicycles.

equatorial zone: On Jupiter, Saturn, and possibly other great planets, a bright cloud zone near the equator.

equinox: The date when the sun passes through the Earth's equatorial plane (occurs twice annually).

erg: The unit of energy in the cgs metric system.

erosion: Removal of rock and soil by any natural process.

escape velocity: The minimum speed needed to allow a projectile to move away from a planet and never return to its point of launch. It equals $\sqrt{2} \times$ circular velocity. (See *circular velocity.*)

eukaryote: Cell with a nucleus, that is, with DNA contained by an interior membrane; a multicelled organism. Eukaryotes first appeared about 1.4 billion years ago.

Europa: One of Jupiter's four large moons, notable for its smooth, bright, icy surface.

event horizon: Imaginary surface at the distance from a black hole where the escape velocity is the speed of light. Matter and energy cannot escape from inside the event horizon.

evidence for planets near other stars: Although planets the size of Jupiter or smaller near other stars are beyond our current detection capability, new techniques of imaging and astrometry will allow their detection within a few years, if they exist. Already some objects larger than Jupiter but smaller than stars have been found.

evidence for present-day star formation: The fact that the solar system is much younger than the galaxy; existence of young open clusters; existence of short-lived massive stars.

evidence of spiral structure (in Milky Way): Spiral distribution of hydrogen gas mapped by 21-cm radio line; spiral distribution of open clusters; spiral patterns observed in other galaxies.

evolution of galaxies: Changes in form and stellar populations of galaxies as a result of consuming gas and dust during star formation and production of heavy elements during star evolution.

evolutionary correction: A correction that must be made to the data associated with many cosmological tests, because distant objects are younger than nearby objects.

evolutionary theory: A theory in which changes occur by relatively slow processes or processes commonly growing out of the initial conditions, rather than by sudden or unusual processes.

evolutionary track: The sequence of points on the H-R diagram occupied by a star as it evolves.

excitation: The process of causing an atom or molecule to go into an excited state, that is, having some electrons in elevated energy levels; the state of being excited.

excited atoms and molecules: Atoms or molecules in which electrons are not all in the lowest possible energy levels.

excited state: The state of an atom or molecule when not all electrons are in the lowest possible energy levels.

exobiology: Study of life beyond the Earth.

expanding universe: A term popularized by Eddington to describe the mutual recession of galaxies.

explorative interval: The hypothetical interval of time during which a species actively engages in the exploration of other planets.

explosive nucleosynthesis: In the explosive expanding shell of a supernova, the formation of nuclei of heavy elements up to plutonium by means of rapid neutron capture.

extragalactic standard of rest: An assumed stationary frame of reference defined by using the nearby galaxies as reference objects.

fault: A fracture along which displacement has occurred on the solid surface of a planet or other celestial body.

fields: Entities dispersed in space but having a measurable value or magnitude that can be measured at any point in space. Examples are gravitational, electric, and magnetic fields.

fireball: An unusually bright meteor, which may yield meteorites.

first quarter: The phase of the Moon when it is one-fourth of the way around its orbit from new moon; the first quarter moon is seen in the evening sky with a straight terminator and half the disk illuminated.

flare: (1) On the Sun, a sudden, short-lived, localized outburst of energy, often ejecting gas at speeds exceeding 1000 km/s; (2) an outburst from certain types of variable stars, sometimes called *flare stars.*

flat geometry: A geometry in which parallel lines never meet, as described by Euclidean rules.

flat universe: A universe with exactly the critical mass density and zero space curvature. It will continue to expand ever more slowly.

focal length: The distance from a lens or mirror to the point where it focuses the image of a very distant object, such as the Moon.

focus: One of the two interior points around which planets or stars move in an elliptical orbit.

forbidden line: Spectral line arising from a metastable state in atoms.

force: In physics, a specific phenomenon producing acceleration of mass. Forces can be generated in many ways, such as by gravity, pressure, and radiation.

formation of galaxies: Processes that led to subdivision of the universe's gas after the big bang and its collapse into individual galaxies.

free-fall: Motion under the influence of gravity only, without any other force or acceleration, such as rocket firing.

free-fall contraction: Contraction of a cloud or system of particles by gravity only, unresisted by any other force.

frequency: Number of electromagnetic oscillations per second corresponding to electromagnetic radiation of any given wavelength.

full moon: The phase of the Moon when it is closest to 180° from the Sun and therefore fully illuminated.

galactic bulge: The spheroidal distribution of stars toward the center of the Milky Way that are intermediate in age between the disk stars and halo stars.

galactic cannibalism: The absorption of one galaxy by another during a collision, forming a new, larger galaxy.

galactic disk: The thin formation of gas and dust on circular orbits where most of the young stars in the Milky Way are found.

galactic equator: The plane of the Milky Way galaxy projected on the sky.

galactic halo: A spherical swarm of globular clusters above and below the galactic disk, centered on a point in the direction of the constellation Sagittarius.

galactic latitude: Angular distance around the galactic equator from the galaxy's center.

galactic longitude: Angular distance from the galactic equator.

galactic morphology: The appearance of a galaxy in terms of its different components: disk, bulge, halo, bar, nucleus.

galactic nucleus: The center of a galaxy.

galaxy: Any of the largest groupings of stars, usually of mass 10^8 to 10^{13} $M_\odot$.

Galilean satellites: The four large satellites of Jupiter, discovered by Galileo.

Galileo Galilei (1564–1642): Italian scientist who first applied the telescope to observe other planets, discovering lunar craters, Jupiter's moons, and other celestial phenomena.

Galileo program: A NASA mission to put an unmanned probe in orbit around Jupiter in the 1990s.

gamma-ray burster: The theoretical interpretation is unclear, but these sources display erratic bursts of high energy gamma-rays and no preference for the disk of the galaxy.

Ganymede: Largest of the four Galilean satellites of Jupiter, with a fractured and cratered ice surface.

general theory of relativity: Einstein's theory that deals with accelerations caused by gravity and other forces. It begins to differ substantially from Newton's theory only when gravitational fields begin to get very strong. It predicts that the structure of space-time is affected by the mass-energy of the universe.

geological time scale: The sequence of events in the history of the Earth.

giant elliptical galaxy: Massive galaxies with diffuse elliptical form, somewhat resembling globular clusters but much larger. They are often found in the middle of rich clusters and may be formed by the cannibalism of smaller galaxies.

giant planet: (1) Jupiter, Saturn, Uranus, or Neptune; (2) any planet much more massive than the Earth.

giant star: Highly luminous star larger than the Sun. O and B main-sequence stars are sometimes called *blue giants*; evolved stars of extremely large radius are called *red giants.*

gibbous: A phase between half-illuminated and fully illuminated, with a convex terminator (pronounced with hard *g*, as in "give").

globular cluster X radiation: X-ray radiation from globular clusters, discovered unexpectedly in the 1970s; it indicates energetic environments somewhere within them.

globular star cluster: A dense spheroidal cluster of stars, usually old, with mass of 10^4 to $10^6\ M_{\odot}$.

grain: Small (usually microscopic) solid particle in space.

grand unified theories (GUTs): Speculative theories that seek to unify electromagnetism and the strong and weak nuclear forces, probably at a temperature around 10^{28} K.

granite: A rock type of modest density and high silica content, formed in association with differentiation processes and, therefore, found primarily on the Earth.

granitic rock: A silica-rich, light-colored rock type common in the Earth's continents. Granites, being low in density, tend to float to the surfaces of planets that have had extensive melting in the outer layers.

granules: Convection cells 1000 to 2000 km across, rising from the subphotospheric layers of the Sun. Each granule rises at a speed of 2 to 3 km/s and lasts for a few minutes.

gravitational contraction: Slow contraction of a cloud, star, or planet due to gravity, causing heat and radiation.

gravitational lensing: The creation of a distorted image of a distant quasar or galaxy when its light is focused by the gravity of a galaxy between it and us.

gravitational mass: The mass of an object that reflects the way it moves in a gravitational field.

gravitational microlensing: The amplification of the light of a background star due to the presence of an object between us and the star.

gravitational redshift: Radiation leaving an intense gravitational field suffers a loss in energy, corresponding to a shift to longer or redder wavelengths.

gravitational waves: One of the predictions of general relativity, gravitational waves result when compact objects change their state. These "ripples" in the structure of space travel at the speed of light, and have been indirectly detected in binary pulsar systems.

gravity: The force by which all masses attract all other masses.

great circle: Any circle on the surface of a sphere (especially the Earth or sky) generated by a plane passing through the center of the sphere; the shortest distance between two points on a sphere.

Great Red Spot: A large, reddish, oval, semipermanent cloud formation on Jupiter.

greenhouse effect: Heating of an atmosphere by absorption of outgoing infrared radiation.

Gregorian calendar: Essentially the modern calendar system, introduced around A.D. 1600 under Pope Gregory XIII, and containing the modern system of reckoning leap years.

ground state: The lowest energy state of an atom, in which all electrons are in the lowest possible energy levels.

Gum Nebula: A large, relatively nearby nebula in the Southern Hemisphere sky, detected in hydrogen alpha light and formed by a supernova explosion estimated to have occurred around 9000 B.C.

HI region: Interstellar region in which hydrogen is predominantly neutral.

HII region: Interstellar region in which hydrogen is predominantly ionized.

half-life: In any phenomenon, the time during which the main variable changes by half its original value; often used loosely to indicate the characteristic time scale of a phenomenon. In radioactive decay, the time for half the atoms in a system to disintegrate.

Halley's comet: The most famous comet, which visits the inner solar system every 76 years, most recently in 1986 when close-up photos were made by space probes.

Hayashi track: A sharply descending evolutionary track in the H-R diagram covering the early period of stellar evolution from the high-luminosity phase to the main sequence.

Heisenberg uncertainty principle: A fundamental limitation to the precision of physical measurements. It states that we cannot know with arbitrary accuracy both the position and momentum of a system, or the energy of the system at every instant of time.

heliacal rising: A star's first visible rising during the yearly cycle.

heliacal setting: A star's last visible setting during the yearly cycle.

helium flash: Runaway helium "burning" inside a star as it evolves off the main sequence and into the giant phase of evolution. It occurs when degenerate gas at the star's center reaches a temperature of about 10^8 K.

Helmholtz contraction: Slow contraction of a cloud or system of particles by the force of gravity, which is retarded by outward gas pressure and the limited rate at which radiation can escape.

high-luminosity phase: A star's short-lived stage of maximum brightness during pre–main-sequence evolution.

high-velocity star: A star with a high velocity relative to the Sun; generally associated with the galactic halo.

homogeneous: Uniform in composition throughout the volume considered.

horizon: The boundary of the observable universe, defined by the distance that light can travel in the age of the universe.

host galaxy: The galaxy surrounding an active quasar core.

hot dark matter: Weakly interacting particles that were traveling at relativistic speeds when the first structures formed. Candidates include the neutrino.

hour angle: The number of hours since a star (or other body) last crossed the local meridian.

H-R diagram: A technique for representing stellar data by plotting spectral type (or color or temperature) against luminosity (or absolute magnitude), named after its early proponents, Hertzsprung and Russell.

Hubble flow: The component of a galaxy's motion that is caused by the smooth expansion of the universe. See also *peculiar velocity* of a galaxy.

Hubble law: The observed linear relationship between the distance of a galaxy and its velocity of recession. In modern cosmology, this is interpreted as evidence of an expanding universe.

Hubble parameter: The ratio of the recession velocity of a galaxy to its distance. In modern cosmology, this is important because it determines the size and age of the expanding universe, but it currently has a large uncertainty of at least 30%.

Hubble Space Telescope: A large orbiting telescope with 2.4-m (94-in.) mirror. Design flaws made it less efficient than intended, but it is currently producing results beyond those possible from the ground.

Hubble time: The inverse of the Hubble parameter, this gives an upper bound to the age of the universe. Deceleration due to matter means that the age of the universe will be less than the Hubble time.

hydrogen alpha line: The designation of hydrogen's red spectral line at 656.3 nm, more properly called *hydrogen Balmer alpha.*

hydrogen Balmer series: The series of all hydrogen spectral lines from 364.6 to 656.3 nm, caused by electron transitions between the second and higher energy levels.

hydrostatic equilibrium: The balance that exists at every point in a stable star between the inward force of gravity and the outward pressure due to energy released from nuclear reactions.

hyperbola: The orbital curve followed by any free-falling body moving faster than escape velocity.

Hyperion: An irregularly shaped outer moon of Saturn, noted for its chaotic rotation.

hypothesis: A proposed explanation of an observed phenomenon or a proposal that a certain observable phenomenon occurs.

Iapetus: An outer moon of Saturn noteworthy because one hemisphere has a bright, icy surface, and the other, a black carbonaceous surface.

ice ages: Intervals of geologic history during which much larger fractions of Earth's surface were covered by glaciers than is true today.

IC number: The catalog number of a cluster, nebula, or galaxy in the *Index Catalog.*

igneous rock: Rock crystallized from molten material.

impact crater: A roughly circular depression of any size (known examples range from microscopic size to diameters greater than 1000 km) caused by a meteorite impact.

impacts of interplanetary debris: Collisions of interplanetary rocky or icy bodies with Earth (or with other worlds). Small impacts are much more common than large impacts.

impact-trigger theory: The leading theory of the moon's origin, in which material was blasted off Earth's mantle and then reaccumulated to form the Moon.

incidence of multiplicity: Among stars, the fraction of systems that contain more than one star. Probably 50 to 70% of systems have companion stars, many of these having more than one companion.

inertial mass: The mass of an object that reflects its resistance to a change in its motion.

inferior planet: Mercury or Venus.

inflationary universe: A modification of the standard big bang model; it says that the universe went through a brief early period of unusually rapid expansion just after the big bang itself. Inflation helps to explain the smoothness of the microwave background radiation and the fact that the geometry of the universe is close to flat.

infrared: Radiation of wavelength too long to see, usually about 1 to 100 μm.

infrared star: A star detected primarily by infrared light.

intense early bombardment: The very intensive bombardment of planetary bodies by meteorites, from 4.6 to about 4 billion years ago, following planet formation.

interferometry: A system for obtaining high-resolution astronomical observations by linking several physically separated telescopes electronically, in effect creating a single, much larger telescope.

interstellar atom: Atoms of gas in interstellar space.

interstellar grain: Microscopic solid grain in interstellar space; interstellar dust.

interstellar molecule: Molecule of gas in interstellar space.

interstellar obscuration: Absorption of starlight by interstellar dust, causing distant objects to appear fainter.

interstellar reddening: Loss of blue starlight due to interstellar dust, causing distant objects to appear redder and fainter.

interstellar snowball: Hypothetical interstellar particle larger than an interstellar grain.

inverse square law: The relation describing any entity, such as radiation or gravity, that varies as $1/r^2$, where r is the distance of the entity from the source.

Io: The innermost Galilean moon of Jupiter, famous for its active volcanism, which is unique among moons.

ion: Charged atom or molecule.

ionization: The process of knocking one or more electrons off a neutral atom or molecule.

ionized gas: A gas in which many of the atoms have lost at least one electron, thus becoming charged particles, or ions.

iron meteorite: Meteorite composed of a nearly pure nickel-iron alloy.

irregular galaxy: A galaxy of amorphous shape. Most have relatively low mass (10^8–$10^{10}\ M_\odot$).

irregular variable: A star that fluctuates in brightness irregularly.

isochrone fitting: The method of making theoretical models of stellar evolution and fitting them to observed H-R diagrams, as a way of measuring the ages of populations of stars.

isotope: A form of an element with a specified number of neutrons in the nucleus. Each element may have many possible isotopic forms, but only a few are stable.

isotropic: Appearing uniform no matter what the direction of view.

joule: The unit of energy in the SI metric system of units.

Julian date: The date based on a running tabulation of days, starting January 1, 4713 B.C.

Jupiter's atmospheric composition: Mostly hydrogen, with additional helium, hydrogen-based compounds, and other gases.

Jupiter's infrared thermal radiation: Infrared radiation from Jupiter due to slow contraction of its interior and exceeding the incoming solar radiation.

Jupiter's interior: Beneath Jupiter's atmosphere, a high-pressure region of liquid hydrogen, liquid metallic hydrogen, and a small central core of rocky material.

Jupiter's temperature: Temperatures around –200°F are measured at and above the cloud tops in Jupiter's atmosphere, but the air temperature increases below the clouds to values around room temperature and even warmer.

Kelvin scale: The absolute temperature scale, with 0 K = absolute zero. A kelvin is the same size (some temperature difference) as a centigrade degree.

Johannes Kepler (1571–1630): Astronomer who first deduced the shapes and relations of planets' elliptical orbits.

Kepler's laws: The three laws of planetary motion that describe how the planets move, show that the Sun is the central body, and allow accurate prediction of planetary positions.

kiloparsec (kpc): 1000 parsecs.

Kirchhoff's laws: Laws describing conditions that produce emission, absorption, and continuous spectra.

Kuiper belt: The reservoir of comet nuclei in the Pluto region, from about 30 to 100 AU.

$L_{\odot}$: The luminosity of the Sun, 4×10^{26} watts.

LSR: (See *local standard of rest.*)

Lagrangian points: In an orbiting system with one large and one small body, an array of five points where a still smaller body would retain a fixed position with respect to the other two.

Lagrangian surface: An imaginary surface with a figure-8 cross section surrounding two coorbiting bodies in circular orbits and constraining motions of particles within the system.

large-scale structure: The distribution of matter in the universe on the largest scales, describing the space distribution of clusters and superclusters of galaxies.

lava: Molten rock on the surface of a planet.

life: A process in which complex, carbon-based materials organized in cells take in additional material from their environment, replicate molecules, reproduce, and do other weird things like writing books.

life in the solar system: Hypothetical biological activity on other planetary bodies, such as on Mars, under Europa's ice, or in Jupiter's atmosphere. Now regarded as unlikely.

life on Mars: Although long sought and believed possible by many scientists, biological processes on Mars were apparently ruled out in 1976 when Viking landers found no organic material there.

life outside the solar system: Hypothetical biological activity beyond the outskirts of our solar system, as on hypothetical planets around other stars. Such life is regarded as plausible, even if sparsely scattered, by most astronomers.

light curve: A plot of brightness of a star (or other object) versus time.

light-gathering power: The ability of a telescope or binoculars to gather light. It is proportional to the area of the objective, that is, the square of the aperture.

light-year: The distance light travels in one year, 9.46×10^{12} km.

limb: The apparent edge of a celestial object.

line of nodes: A line formed by the intersection of an orbit and some other reference plane, such as the plane of the solar system.

linear measure: Measurement involving linear distances, as opposed to angles or angular distances.

lithosphere: The solid rocky layer in a partially molten planet.

Local Group: The cluster of galaxies to which the Milky Way and 26 nearby galaxies belong.

local standard of rest (LSR): A frame of reference moving with the average velocity of the nearby stars (out to about 50 pc from the Sun).

Local Supercluster: A large flattened structure of about 10^{15} $M_{\odot}$ centered on the Virgo cluster, of which the Milky Way is a member.

long-period comet: A comet moving on a nearly parabolic orbit and thus having an orbital period of hundreds of thousands of years.

look-back time: The idea in cosmology that light takes a significant time to reach us from distant objects, so we inevitably see them as they were in the past.

luminosity: The total energy radiated by source per second. The luminosity of the Sun ($L_{\odot}$) is 4×20^{26} watts.

luminosity function: General term for the distribution of brightness of astronomical objects, as a function of their density in space. Can be derived for stars, galaxies, quasars, or any astronomical source.

lunar eclipse: Dimming of the Moon as it passes into the Earth's shadow.

$M_{\odot}$: The mass of the Sun, 2×10^{30} kg.

M giant: A giant star of spectral class M.

Magellanic clouds: The two galaxies nearest the Milky Way, irregular in form and visible to the naked eye in the Southern Hemisphere.

Magellanic stream: Gas filaments connecting the magellanic clouds to the Milky Way.

magma: Underground molten rock.

magma ocean: Primordial layer of molten lava on the initial surface of the Moon and (by inference) planets.

magnetic braking: The slowing of rotation of a star or planet by interaction of its magnetic field with surrounding ionized material.

magnetic field: Region of space in which a compass (or other detector) would respond to magnetism of some body.

magnetic support: The support provided when an interstellar gas cloud is laced with a magnetic field. The magnetic field resists the gravitational collapse of the cloud and so affects the rate and efficiency of star formation.

magnification: Apparent angular size of a telescope image divided by the angular size of the object seen by the naked eye.

main-sequence star: One of the group of stars defined on the H-R diagram that have a relatively stable interior configuration and are consuming hydrogen in nuclear reactions; a star on the main sequence.

mantle: A region of intermediate density surrounding the core of planets.

mare (pl. *maria*): A dark-colored region on a planet or satellite; a region of basaltic lava flow on the Moon.

Mars-crossing asteroids: Asteroids whose orbits cross that of Mars.

Mars' rotation period: The day on Mars—$24^h\,37^m$, only slightly longer than the Earth's.

Martian air pressure: The pressure exerted by the very thin Martian air. On Mars' surface, the air pressure is only about 0.7% of that at the Earth's surface.

Martian air temperature: Soil temperatures approach or exceed freezing in the day; the air is colder. Night temperatures around –123°F are recorded.

Martian atmosphere: The thin gases around Mars, composed almost entirely of carbon dioxide (CO_2).

Martian meteorites: A handful of meteorites about 1.3 billion years old, believed to have been blasted off Mars about 0.2 billion years ago.

maser (*m*icrowave *a*mplification by *s*imulated *e*mission of *r*adiation): (1) A device that amplifies microwave radio waves through special electronic transitions in atoms; (2) an interstellar cloud that acts in this way.

maser emission: Particularly intense emission from molecules in dense molecular clouds, which occurs when a large number of molecules are excited to energy levels above the ground state.

mass: (1) Material; (2) the amount of material.

mass-energy equivalence: Einstein's statement of the relationship between mass and energy, given by the famous equation $E = mc^2$.

mass extinctions: Extinctions of many species 65 million years ago, probably triggered by the impact of an asteroid.

mass/luminosity ratio: The mass per unit of light or total radiation emitted from an object such as a galaxy.

mass-luminosity relations: The relation between the mass of a main-sequence star and its total radiation rate; the more massive, the greater the luminosity.

mass of the galaxy: About 4×10^{41} kg, or $2 \times 10^{11}\ M_\odot$, as estimated from the circular velocity equation.

matter-dominated universe: The universe more than 10,000 years after the big bang, in which the mass-energy of particles exceeds the energy density of photons and gravitational structures can begin to form.

Maunder minimum: The interval from 1645 to 1715, when solar activity was minimal.

mean density: Mass of an object divided by its volume.

mean solar time: Time shown by conventional clocks, determined by the Sun's mean rate averaged over the year.

megaparsec: One million parsecs.

meridian: (1) A north-south line on a planet, moon, or star; (2) a great circle through the celestial pole and the zenith.

Messier number: The catalog number of a nebula, star cluster, or galaxy in *Messier's Catalog.*

metallic hydrogen: A high-pressure form of hydrogen with free electrons.

metastable state: In an atom, a configuration of electrons that is relatively long-lived, but is rarely found on the Earth because it is disrupted by collisions with other atoms; it may be found in interstellar atoms, creating forbidden spectral lines.

meteor: A rapidly moving luminous object visible for a few seconds in the night sky (a "shooting star").

meteorite: An interplanetary rock or metal object that strikes the ground.

meteorite impact crater: Circular depression in planetary surfaces, caused by explosions as meteorites crash into the surfaces at high speeds.

meteoroid: A particle in space, generally smaller than a few meters across.

meteor shower: A concentrated group of meteors, seen when the Earth's orbit intersects debris from a comet.

meter: 39.4 in.

microwave background fluctuations: Tiny spatial variations in the temperature of the microwave background radiation; they form the seeds for eventual galaxy formation.

Microwave Observing Project (MOP): The most ambitious SETI experiment yet attempted. This NASA-sponsored project uses radio techniques to monitor 1000 solar type stars for artificial, intelligent signals and to scan the entire sky.

Milky Way galaxy: The spiral galaxy in which we live.

Miller-Urey experiments: A series of experiments in which amino acids and other organic molecules were created in laboratory conditions that simulate the conditions of the early Earth.

millisecond pulsar: A small class of pulsars that rotate at speeds of up to 700 times per second, thought to be old neutron stars "spun up" by the infall of material from a binary companion.

Mimas: The innermost large moon of Saturn, icy and heavily cratered.

minerals: Chemical compounds, usually in the form of crystals, that constitute rocks.

minute of arc: An angle equaling $\frac{1}{60}$ of a degree.

Miranda: The innermost of five large moons of Uranus, noted for puzzling fractured and grooved terrain.

mks system: A metric system of units expressing length in meters, mass in kilograms, and time in seconds. (See also *cgs system.*)

molecular cloud: An interstellar cloud of gas and dust, with greater than average density, dust content, and high concentration of molecules.

molecules in space: More than 80 varieties of molecules have been discovered in space, some containing as many as 13 atoms. The sites for molecules are interstellar gas clouds; some have been trapped in comets and meteorites.

Moon: (1) The Earth's natural satellite; (2) any satellite.

morphology-density relation: The observation that the relative number of spiral and elliptical galaxies depends on the density of the region. High-density regions or clusters have a larger fraction of ellipticals than low-density regions.

multicelled organisms: Life forms that represented a significant increase in complexity and originated on Earth about 600 million years ago.

multiple scattering: Redirection of electromagnetic radiation (such as light waves) by repeated interaction with atoms, molecules, or dust grains in space or in an atmosphere.

multiple star system: A system of three or more stars orbiting around each other.

mutation: The fundamental mechanism for generating change in genetic material. Mutations alter the molecular structure of the DNA molecule in ways that can be helpful, harmful, or neutral to the organism.

mutual recession of galaxies: The phenomenon that all distant galaxies are moving away from us, and the farther away they are, the faster they are receding.

natural selection: The theory that states that those individuals best adapted to the ever-changing environment produce a greater number of offspring.

nebula: a cloud of dense gas and/or dust in interstellar space or surrounding a star.

negative hydrogen ions: Hydrogen atoms that have temporarily captured an extra electron, responsible for opacity in the photosphere of the Sun and many stars.

negatively curved geometry: A geometry in which parallel lines diverge; sometimes called a *hyperbolic geometry.*

Neptune: The outermost gas giant planet in the outer solar system.

neutrino: A subatomic particle created in certain nuclear reactions inside stars. It can pass through most matter. Its mass is uncertain, being either zero or a tiny fraction of an electron's mass.

neutrino burst: During the explosive death of a massive star, 0.1–0.2 solar masses of material and a vast amount of energy is carried away in the form of neutrinos. The neutrino burst was seen for the first time in supernovae 1987A.

neutron: One of the two major particles constituting the atomic nucleus; it has zero charge and mass 1.6749×10^{-27} kg.

neutron star: A star with a core composed mostly of neutrons, with density 10^{16} to 10^{18} kg/m^3. Many or most neutron stars are pulsars.

new moon: The phase of the moon when it is nearest the Earth-Sun line, hence invisible from Earth because of the sun's glare.

Isaac Newton (1642–1727): English physicist who discovered the spectrum and laws of gravitation and motion; he also developed calculus and made other discoveries. Possibly the greatest physicist in history.

Newtonian cosmology: A model of the universe with infinite volume, no expansion, and Euclidean geometry.

Newton's laws of motion: Three rules describing motion and forces. Briefly, (1) a body remains in its state of motion unless a force acts on it; (2) force equals mass times acceleration; (3) for every action there is an equal and opposite reaction.

NGC number: The catalog number of a nebula, cluster, or galaxy in the *New General Catalog.*

node: One of two points where an orbit crosses a reference plane.

noncosmological redshift: A hypothetical redshift of distant galaxies *not* caused by the Doppler effect.

nonthermal radiation: Radiation *not* due to the heat of the source; for example, synchrotron radiation.

North Star: (1) Polaris; (2) any bright star that happens to be within a few degrees of the north celestial pole during a given era.

nova: A type of suddenly brightening star (from the Latin for "new") resulting from explosive brightening when gas is dumped from one member of a binary star pair onto the other.

nuclear reaction: Reaction involving the nuclei of atoms in which a nucleus changes mass.

nucleocosmochronology: A method of measuring the age of the Milky Way by using the decay of radioactive isotopes within it.

nucleus: (1) The matter at the center of an atom, composed of protons and neutrons; (2) the bright central core (or solid body) of a comet; (3) the bright central core of a galaxy.

O association: An association of O-type stars.

objective: The major light-gathering element of a telescope; the mirror in a reflector and the lens in a refractor.

oblate spheroid: The shape assumed by a sphere deformed by rotation.

obliquity: The angle by which a planet's rotation axis is tipped to its orbit.

Occam's razor: The principle that the simplest hypothesis, with the fewest assumptions, is most likely to be correct. Named after its use to cut away false hypotheses.

ocean tide: The Moon's tidal stretching of the earth, as observed in the ocean surface (as opposed to body tide).

Olbers' paradox: The problem of why the sky is dark at night if the universe is filled with stars.

Oort cloud: The swarm of comets surrounding the solar system.

opacity: The extent to which gaseous (or other) material absorbs light.

open star cluster: A grouping of relatively young Population I stars (usually 10^2–10^3 $M_\odot$), sometimes called a *galactic cluster.*

open universe: A universe with negative space curvature and less than the critical mass density. If the universe is open, it will expand forever.

opposition: The period when a superior planet lies in an opposite direction from the Sun as seen from the Earth. At opposition, a planet appears in the midnight sky, well placed for observation.

optical double star: A pair of stars that have small angular separation but are not coorbiting.

orbital precession: A slow, cyclical change in the orientation of the plane of an orbit.

organic chemistry: Chemistry involving organic molecules.

organic molecule: Molecule based on the carbon atom, usually large and complex, but not necessarily part of living organisms.

origin of the heavy elements: The process by which light elements fused to form nuclei of heavy elements, primarily inside massive stars.

Orion Nebula: Several hundred solar masses of gas and dust composing the core of a star-forming region about 460 parsecs away. One of the most prominent nebulae in our sky, it forms the central "star" in Orion's sword.

Orion star-forming region: The larger region in which stars are forming around the Orion Nebula.

ozone layer: An atmospheric layer rich in ozone (O_3), created by the interaction between oxygen molecules (O_2) and solar radiation. On Earth, its altitude is about 20–60 km.

parabola: (1) The curved trajectory followed by a particle moving at escape velocity; (2) the curve of a Newtonian telescope's primary mirror.

parallax: An angular shift in apparent position due to an observer's motion; more specifically, a small angular shift in a star's apparent position due to the Earth's motion around the Sun. *Stellar parallax,* used to measure stellar distance, is defined as the angle subtended by the radius of the Earth's orbit as seen from the star.

parent body: A body from which a meteorite formed and later broke off as a fragment.

parent isotope: A radioactive isotope that disintegrates and forms a daughter isotope.

parsec: A distance of 206,265 AU, 3.26 ly, or 3.09×10^{16} km; defined as the distance corresponding to a parallax of 1 second of arc.

partial solar eclipse: An eclipse in which the light source is not totally obscured from an observer.

particlelike properties of light: Characteristics of light, such as concentrating of energy and momentum in discrete microscopic packets (photons) that mimic the properties of particles.

Pauli exclusion principle: A principle of subatomic physics specifying that no two electrons in a very small volume have exactly the same properties of energy, motion, and so on.

peculiar galaxy: Galaxy that does not fit into the standard Hubble sequence; its morphology can be caused by a tidal interaction with another galaxy.

peculiar velocity (of a galaxy): The component of a galaxy's motion that is not caused by the smooth expansion of the universe. Since galaxies interact gravitationally, they usually have a measurable peculiar velocity. See also *Hubble flow.*

peculiar velocity (of a star): A star's velocity with respect to the local standard of rest.

penumbra: (1) The outer, brighter part of a shadow, from which the light source is not totally obscured; (2) the outer, lighter part of a sunspot.

perigee: The point in an orbit around the Earth that is closest to the Earth.

permafrost: Semipermanent underground ice.

phase: The apparent shape of an illuminated body, varying with the "phase angle" from observer to body to illumination source.

Phoebe: The outermost moon of Saturn, a small dark moon believed to be captured.

photometry: The measurement of the amount of light, either total or in different specified colors, coming from an object.

photon: The quantum unit of light, having some properties of a wave. For each wavelength of radiation, the photon has a different energy.

photosphere: The light-emitting surface layer of the Sun.

photosynthesis: Process that converts sunlight into stored chemical energy, essential for the proliferation of advanced forms of life.

physical binary stars: Two stars orbiting around a common center of mass.

Planck's constant: The fundamental constant that relates the energy of an electromagnetic wave to its frequency.

Planck's law: A formula that describes the energy associated with each wavelength (color) in the spectrum. It shows that photons of blue light are more energetic than photons of red light.

planet: A solid (or partially liquid) body orbiting around a star but too small to generate energy by nuclear reactions.

"Planet X": A term sometimes used for a hypothetical tenth planet in the solar system, beyond Pluto.

planetary nebula: A type of circumstellar gas cloud that has spheroidal shape and often appears as a faint disk in telescopes; it has nothing to do with planets except for the rough resemblance to a planet's shape when seen telescopically.

planetesimal: One of the small bodies from which planets formed, usually ranging from micrometers to kilometers in diameter.

planetology: The study of the planets' origins, evolution, and conditions.

plasma: A high-temperature gas consisting entirely of ions, instead of neutral atoms or molecules. Because of the high temperature, the atoms strike each other hard enough to keep at least the outer electrons knocked off.

plate: Moving unit of the Earth's lithosphere, typically of continental scale.

plate tectonics: Motions of a planet's lithosphere, causing fracturing of the surface into plates. Primary example occurs on the Earth.

Pluto: Cataloged as the outermost and smallest planet in the solar system, but possibly one of many small worldlets on the fringe of the solar system.

Polaris: The North Star.

Population I: Stars with a few percent heavy elements (heavier than helium), found in the disks of spiral galaxies and in irregular galaxies.

Population II: Stars composed of nearly pure hydrogen and helium, found in the halo and center of spiral galaxies, in elliptical galaxies, and to a limited extent in irregulars.

positively curved geometry: A geometry where parallel lines converge, sometimes called a *spherical geometry.*

powers of 10: The number of times 10s must be multiplied together to give a specific number; the exponent of 10. (Example: $10^2 = 100$; the power, or exponent, is 2; see also Appendix 1.)

precession: The wobble in the position of a planet's rotation axis caused by external forces. Also, the change in a coordinate system (tied to any planet) caused by such a wobble.

pre–main-sequence star: Evolutionary state of stars prior to arrival on the main sequence, especially just before the main sequence is reached.

principle of relativity: The principle that observers can measure only relative motions, since there is no absolute frame of reference in the universe by which to specify absolute motions.

prograde and retrograde satellite orbits: Satellite orbits in which motion is in the same or opposite direction, respectively, as the planet's rotation.

prograde rotation: Spinning on an axis from west to east, or counterclockwise as seen from the North Pole (as in the case of the Earth).

prokaryote: Cell that contains a single long strand of DNA but no nucleus. Prokaryotes were the first and simplest life forms on the Earth, dating back 3.8 billion years.

prominence: A radiating gas cloud extending from the solar surface into the thinner corona.

prominent star: One of the brightest stars in our sky, but not necessarily one of the nearest.

proper motion: The angular rate of motion of a star or other object across the sky. (Most stars have proper motions less than a few seconds of arc per year.)

protein: Any of several types of complex organic molecules made from amino acids inside plants and animals, which are essential in living organisms.

proteinoid: Cell-like, nonliving spheroid of protein molecules created in the laboratory by heating amino acids and adding water; a possible step in the evolution of life.

protogalaxy: Enormous gas cloud that will collapse by gravity to form a galaxy.

proton: One of the two basic particles in an atomic nucleus, with positive charge and mass $1.6726 = 10^{-27}$ kg.

proton-proton chain: A series of thermonuclear reactions that convert hydrogen nuclei to helium nuclei, converting a tiny amount of mass into energy.

protoplanet: A planet shortly before its final formation. Sometimes hypothesized to have a massive atmosphere and greater mass than in its present state.

protostar: A gravitationally stable cloud of stellar mass contracting in an early pre–main-sequence evolutionary state.

pseudoscience: Research that has the trappings of science but does not follow the scientific method, usually lacking review and repetition of observations by independent researchers.

Ptolemaic model: The ancient Earth-centered model of the solar system, with the Sun, Moon, and other planets moving in epicycles.

pulsar: A rapidly rotating neutron star with a strong magnetic field, observed to emit pulses of radiation.

quantum: Fundamental unit of energy that governs all particle interactions and dictates the discreteness of the physical world.

quasar: An active galaxy seen at such a large distance that the bright nucleus overwhelms the surrounding galaxy, and the image appears stellar. Most quasars have very large redshifts.

quasar absorption lines: When the light from a distant quasar intercepts an intervening galaxy, an absorption line is created in the quasar spectrum at the red shift of the galaxy. Such lines are called quasar absorption lines.

quasar luminosity function: The density of quasars in space (number per Mpc^3) as a function of luminosity. The most luminous quasars are the rarest.

r-process reactions: Rapid reactions, probably occurring inside supernovae, in which heavy elements are formed as atomic nuclei capture neutrons. (See also *s-process reactions.*)

radial velocity: The velocity component along the line of sight toward or away from an observer. Recession is positive; approach is negative.

radiation: (1) Any electromagnetic waves or atomic particles that transmit energy across space; (2) one of three modes of heat (energy) transmission through stars or planets from warm regions to cool regions.

radiation-dominated universe: The universe less than 10,000 years after the big bang, in which the energy per unit volume in the form of photons is greater than the energy per unit volume represented by particles of mass.

radiation pressure: An outward pressure on small particles exerted by electromagnetic radiation in a direction away from the light source.

radioactive atom: Any atom whose nucleus spontaneously disintegrates.

radio communication: Communications sent by electromagnetic waves at radio wavelengths. This is the favored form of communication for SETI experiments, because of the high speed of transmission and the low amount of energy required to send a message.

radio galaxy: A galaxy that emits unusually large amounts of radio radiation.

radioisotopic dating: Dating of rock or other material by measuring amounts of parent and daughter isotopes.

radio jets: One-sided or two-sided jets of glowing material that are seen in some active galaxies.

radio telescope: A device that gathers and concentrates radio waves.

random error: Error that can be reduced when separate observations are combined. For example, four times the data will reduce the error by a factor of two.

ray: A bright streak of material ejected from a crater on the Moon or other planet.

Rayleigh scattering: Scattering of light by particles smaller than the light's wavelength. This process favors scattering of blue light.

recombination: The process in the early universe, completed a million years after the big bang, by which protons and electrons combined to form hydrogen atoms.

red giant: A post–main-sequence star whose surface layers have expanded to many solar radii and have relatively low temperatures.

redshift: A Doppler shift of spectral features toward longer wavelengths, indicating recession of the source.

redshift of galaxies: The shift toward longer wavelengths in light of distant galaxies, due to their recession from the solar system. It increases with galaxies' distances.

reflected radiation: Radiation that has arrived from outside a body and bounced off, as opposed to thermal radiation.

reflector: A type of telescope using a mirror as the light collector.

refractor: A type of telescope using a lens as the light collector.

refractory element: An element least likely to be driven out of a material by heating. These elements are usually concentrated in the last components to melt when a material such as rock is heated.

regolith: A powdery soil layer on the Moon and some other bodies caused by meteorite bombardment.

regression of nodes: A shifting of the R.A. = Dec. coordinate system, relative to the stars, as a result of the 26,000-y wobble of the Earth's rotation axis.

relativistic: Moving at speeds near that of light.

relativity: (See *principle of relativity.*)

replication: The basic process of reproduction for life, and the means by which genetic information is propagated. It involves the ability of some carbon-based molecules to split into two halves that can assemble copies of themselves.

representative stars: A sample of stars randomly drawn from the total population of stars in space.

resolution: The smallest angle that can be discerned with an optical system; for example, the eye can resolve about 2 minutes of arc.

resonance: A close, or simple-number, relationship between periodicities in two phenomena. For example, if one body has half the orbital period of another, they are said to be in *orbital resonance.*

retrograde motion: Revolution or rotation from east to west contrary to the usual motion in the solar system.

retrograde rotation: Spinning on an axis from east to west, or clockwise as seen from the North Pole (opposite to the spin direction of Earth).

Rhea: (See *Tethys, Dione, and Rhea.*)

rift: A major split in a planet's lithosphere due to active or incipient plate tectonic stresses.

right ascension: Longitude lines projected onto the celestial sphere. (Abbreviation: R.A.)

rille: A type of lunar valley.

RNA (ribonucleic acid): The molecule that assembles proteins from DNA instructions.

Roche's limit: The distance from a large body within which tidal forces would disrupt a satellite.

rock: Solid aggregation of minerals.

rotation curve: Orbital velocity as a function of distance from the center of a galaxy. The flat rotation curves observed for many spiral galaxies provide good evidence for dark matter.

rotational line broadening: Broadening of spectral lines due to rotation of the source.

rotational support: An outward force provided by a rotating gas cloud to resist the inward force of gravity.

RR Lyrae star: A type of variable star similar to the Cepheids that has been found associated with Population II and not Population I.

runaway star: A star rapidly moving away from a region of recent star formation.

Russell-Vogt theorem: The theorem stating that the equilibrium structure of a star is determined by its mass and chemical composition.

s-process reactions: Slow reactions in giant stars in which heavy elements are built up as atomic nuclei capture neutrons. (See also *r-process reactions.*)

Saha equation: An equation derived in 1920, by the Indian physicist Saha, that tells the percent of atoms in each different excited state, given the conditions in a gas. This in turn controls what spectral lines are emitted or absorbed by that gas.

saros cycle: An interval of 6585 d (about 18 y) separating cycles of similar eclipses, used by ancient people to predict eclipses.

satellite: Any small body orbiting a larger body.

Saturn: The sixth planet out from the Sun, famous for its prominent rings.

Saturn's atmosphere: The thick, cloudy gases around Saturn, composed mostly of hydrogen.

Saturn's ring system: A system of innumerable icy particles orbiting Saturn.

Saturn's satellite system: A family of at last 18 moons orbiting Saturn, ranging from 20 km diameter up to a size slightly exceeding that of the planet Mercury.

scale error: In distance measurement, an error that affects all measurements of larger distances.

Schwarzschild radius: The radius corresponding to the event horizon of a black hole, proportional to the mass of the black hole.

science: Study of nature using the scientific method. (See *scientific method.*)

scientific method: The method of learning about nature from making observations, formulating hypotheses, and constructing observational or experimental tests to see if the hypotheses are accurate.

search for extraterrestrial intelligence (SETI): A scientific field devoted to calculating the probability of intelligent life in the universe and designing the optimum strategies for making contact.

seasonal changes on Mars: Changes in shape and darkness of the dusky patches on Mars from summer to winter and year to year. Once thought to indicate Martian vegetation, the changes are now known to result from blowing dust deposits.

secondary atmosphere: A planet's atmosphere after modifications by outgassing and other processes. (See also *early dense atmosphere.*)

second of arc: An angle equaling $\frac{1}{3600}$ of a degree.

sedimentary rock: Rock formed from sediments.

seeing: The quality of stillness or lack of shimmer in a telescopic image, associated with atmospheric conditions. If the atmosphere is very still and the image is sharp, the seeing is said to be good.

seismic waves: Waves passing through the interior or surface layers of a planet due to a seismic disturbance, such as an earthquake or large meteorite impact.

seismology: Study of vibrational waves passing through planets, revealing internal structure.

selection effect: Any effect that systematically biases observations or statistics away from a correct understanding.

Seyfert galaxy: A type of galaxy with a bright, bluish nucleus, possibly marking a transition between ordinary galaxies and quasars.

shepherd satellites: Satellites that move near planetary rings and act to confine the ring particles onto certain orbits.

short period comet: A comet with a revolution period less than 100 y.

SI metric system: The internationally standardized scientific system of units, in which length is given as meters, mass as kilograms, and time as seconds.

sidereal: Referring to stars.

sidereal period: A period of rotation or revolution where the movement is measured relative to the stars.

sidereal time: Time measured by the apparent motion of the stars (instead of the Sun), used by astronomers to point telescopes toward celestial targets; it is the right ascension that is on the meridian at any given location.

significant figures: The number of digits known for certain in a quantity.

small-angle equation: The equation giving the relation between the distance D of an object, its diameter d, and its angular size α (expressed in seconds of arc):

$$\frac{\alpha}{206{,}265} = \frac{d}{D}$$

S0 galaxy: Also called lenticular galaxies, they have mostly old stars but also some gas and dust.

solar apex: The direction toward which the Sun is moving relative to nearby stars.

solar constant: The amount of energy reaching a Sun-facing square meter at a given planet's (usually the Earth's) orbit per unit time; for the Earth, it is 1390 W/m^2.

solar core: The Sun's central region of high-pressure gases, where nuclear energy is produced.

solar cycle: 22-y cycle of solar activity.

solar eclipse: Partial or total blocking of the Sun's light by an astronomical body (in most usages, by the Moon).

solar nebula: The cloud of gas around the Sun during the formation of the solar system.

solar rotation: Turning of the Sun on its axis in 25.4 d.

solar seismology: The study of natural vibrations and oscillations in the Sun as a way to probe the structure of the solar interior.

solar system: The Sun and all bodies orbiting around it.

solar wind: An outrush of gas past the Earth and beyond the outer planets. Near the Earth, the solar wind travels at velocities near 600 km/s, sometimes reaching 1000 km/s.

solstice: The date when the Sun reaches maximum distance from the celestial equator (occurs twice annually).

solstice principle: According to this principle, the sunrise and sunset positions of the Sun on the eastern and western horizons (respectively) shift positions according to season and the observer's latitude. The principle can be used to optimize passive solar energy input into a home or other building.

space astronomy: Astronomy that uses the unique advantages of the space environment, including sensitivity to infrared and high-energy electromagnetic waves, and freedom from the blurring effects of the Earth's atmosphere. The term usually applies to observations made with telescopes in orbit around Earth.

space-time diagram: A plot of time as the **y** coordinate and space as the **x** coordinate, with the three spatial dimensions collapsed to one dimension. It reflects the fact that events must be specified in time and space.

space velocity: A star's velocity with respect to the Sun.

special theory of relativity: Einstein's theory that deals with relative motions, and takes as its starting point the fact that the speed of light is a universal constant.

spectral class: A class to which a star belongs because of its spectrum, which in turn is determined by its temperature. The spectral classes are O, B, A, F, G, K, and M, from hottest to coolest.

spectral line strength: Measure of the total energy absorbed or emitted in a spectral line.

spectrograph: An instrument for recording a photographic image of a spectrum.

spectroheliograph: An instrument for observing the Sun in certain specified wavelengths.

spectrometer: An instrument for tracing the intensity of a spectrum at different wavelengths; the result is a graph.

spectrophotometry: The study of the amount of radiation at each wavelength in the spectrum.

spectroscopic binary: A binary star revealed by varying Doppler shifts in spectral lines.

spectroscopy: Study of spectra, especially as revealing the properties of the light source.

spectrum: Light from an object arranged in order of wavelength; specifically, the colors of visible light, arranged in this order.

spectrum binary: A binary revealed by mixture of two spectral classes in the spectrum.

speed of light: Designated as *c*, the speed of light is about 300,000 km/s and is constant as perceived by all observers.

sphere of gravitational influence: The region in which the gravitational influence of a body is the dominant influence on a passing small body's motions.

spicule: Narrow jet of gas extending out of the solar chromosphere, with a lifetime of about 5 min.

spiral arm: In spiral galaxies, one of the arms lying at an angle to the Sun-center line. The arms contain open clusters, O and B stars, and nebulae.

spiral galaxy: A disk-shaped galaxy with a spiral pattern, typically containing 10^{10}–10^{12} $M_\odot$ of stars, dust, and gas.

standard candle: An idealized astronomical source (star or galaxy) with well-understood physics and a well-determined luminosity. The known luminosity and the apparent brightness can be combined to give the distance.

standard measuring rod: An idealized astronomical source with well-understood physics and a well-determined physical size. The real size and the apparent size can be combined to give the distance.

standard time: Solar time appropriate to the given local time zone.

star: A mass of material, usually wholly gaseous, massive enough to initiate (or to have once initiated) nuclear reactions in its central region.

starburst: A relatively sudden and rapid episode of star formation in a galaxy, probably triggered in some cases by collision with another galaxy.

static universe: A theoretical universe where the typical separation between galaxies does not change with time.

Stefan-Boltzmann law: A law giving the total energy E radiated from a surface of area A and temperature t per second: $E = \sigma T^4 A$. Sigma (σ), the Stefan-Boltzmann constant, equals 5.67×10^{-8} W/m$^2 \cdot$ K^4 in SI units.

stellar evolution: Evolution of every star from one form to another forced by changes in composition as nuclear reactions proceed.

stellar populations in galaxies: Groupings of stars with different composition and age. Population I consists of young stars with a few percent of heavy elements; stars near the Sun are Population I. Population II includes older stars with virtually no heavy elements.

stochastic star formation theory: A theory of the cause of spiral arms in spiral galaxies. Star formation in a region leads to a chain reaction of adjacent star formation, and the region of young stars gets sheared into a spiral arm.

Stonehenge: A prehistoric English ruin with built-in astronomical alignments.

stony-iron meteorite: Stony meteorite that probably comes from deep within the parent body, where melted stony and metallic material coexisted.

stromatolite: A primitive life form, colonies of blue-green algae that were among the first to appear along shorelines. They are among the most abundant fossils from 3.5 to 2 billion years ago.

strong nuclear force: The force of nature that binds quarks into neutrons and protons in the atomic nucleus.

subfragmentation: Breakup of a contracting cloud into smaller condensations.

subfragmentation theory: A theory of formation of binary and multiple stars by breakup of the protostellar cloud into two or more components during its collapse from nebular to stellar dimensions.

subtend: To have an angular size equal to a specified angle. For example, the Moon subtends to $\frac{1}{2}°$.

Sun: The star orbited by the Earth.

Sun's composition: 78% hydrogen, 20% helium, 2% other gases.

sunspot: A magnetic disturbance on the Sun's surface that is cooler than the surrounding area.

Sun's revolution period: The 240-million-year period taken by the Sun to complete its orbit around the Milky Way galaxy.

superbubble: A large volume of hot gas in interstellar space, formed by coalescence of bubbles blown around supernovae.

supercluster of galaxies: Cluster of clusters of galaxies.

supergiant star: An extremely luminous star in the uppermost part of the H-R diagram.

supergranulation: Large-scale (15,000–30,000 km in diameter) convective cell patterns in the solar photosphere.

superior planet: Any planet with an orbit outside the Earth's orbit.

supermassive black hole: The hypothesized power source of a quasar or active galaxy, formed by the gradual accretion of material in the center of a galaxy.

supernova: A very energetic stellar explosion expending about 10^{42} to 10^{44} joules and blowing off most of the star's mass, leaving a dense core.

supernova remnant: The expanding and cooling shell of gas and dust that is visible for thousands of years after a supernova.

superstring theories: Highly speculative theories that attempt to unify all four fundamental forces of nature.

symbiotic stars: A pair of stars whose evolutions are affecting each other, especially by mass transfer.

symmetry: The idea that diverse physical phenomena have a simple underlying basis; one of the basic assumptions of modern science.

synchronous rotation: Any rotation such that a body keeps the same face toward a coorbiting body.

synchrotron radiation: Radiation emitted when electrons move at nearly the speed of light in a magnetic field.

synodical month: One complete cycle of lunar phases, 29.53 d.

systematic error: Error that cannot be reduced by simply increasing the number of observations. In astronomy, systematic errors are usually caused by an incomplete understanding of the physics of an astronomical object.

tangential velocity: The velocity component perpendicular to the line of sight.

Tarantula Nebula (30 Doradus): A huge HII emission nebula in the Large Magellanic Cloud.

T association: An association of T Tauri stars.

tectonics: Disruption of planetary or satellite surfaces by large-scale mass movements, such as faulting.

telescope: An instrument for collecting electromagnetic radiation and producing magnified images of distant objects.

temperature: A measure of the average energy of a molecule of a material.

terminator: The dawn or dusk line separating night from day on a planet or satellite.

terrestrial planet: (1) Mercury, Venus, Earth, or Mars; (2) a planet composed primarily of rocky material.

Tethys, Dione, and Rhea: Intermediate-sized icy moons of Saturn.

theory: A body of hypotheses, often with mathematical backing and having passed some observational tests; often implying more validity than the term *hypothesis.*

theory of cosmological redshifts: The theory that galaxies' redshifts are all due to recessional motion, increase with distance, and thus given an indicator of distance.

theory of galaxy formation: The (speculative) theory that describes how galaxies form from the gravitational collapse of enormous clouds of gas in the early universe.

theory of natural selection: Darwin's theory that evolution of species is driven, at least in large part, by competition among species and "survival of the fittest."

theory of noncosmological redshifts: The theory that at least some galaxies' redshifts are not Doppler shifts due to recession, but are due to some other cause.

theory of star formation: The theory that describes how stars form by the gravitational collapse of interstellar clouds of dust and gas.

thermal escape: Escape of the fastest-moving gas atoms or molecules from the top of a planet's atmosphere by means of their thermal motion.

thermal infrared radiation: Radiation from bodies at room temperature or planetary temperatures, primarily at infrared wavelengths. See also *thermal radiation.*

thermal motion: Movement of atoms and molecules associated with the temperature of the material; they grow faster as the temperature increases.

thermal radiation: Electromagnetic radiation emitted by a body and associated with an object's temperature; it grows greater and bluer in color as the temperature increases.

third quarter: The phase of the Moon when it is three-fourths of the way around its orbit from new moon; the third quarter moon is seen in the dawn sky with a straight terminator and half the disk illuminated.

thrust: The force generated by a high-speed discharge, as from a rocket or airplane.

tidal interaction: Close encounter between galaxies that leads to observable effects on the morphology.

tidal recession: Recession of the Moon (or other satellite) from the Earth (or other planet) caused by tidal forces.

tide: A bulge raised in a body by the gravitational force of a nearby body.

Titan: Saturn's largest moon, famous for its thick, smoggy-orange nitrogen atmosphere.

total solar eclipse: (1) An eclipse in which the light source is totally obscured from a specified observer; (2) an eclipse in which a body is entirely immersed in another's shadow. (See also *eclipse.*)

transit: (1) Passage of a planet across the Sun's disk; (2) any passage of a body with a small angular size across the face of a body with a large angular size.

triple-alpha process: A nuclear reaction in which helium is transformed into carbon in red giant stars.

Triton: Neptune's largest moon.

Trojan asteroids: Asteroids caught near the Lagrangian points in Jupiter's orbit, 60° ahead of and 60° behind the planet.

tsunami: A large ocean wave generated by earthquake or volcanic activity (the correct name for a tidal wave).

T Tauri star: A type of variable star, often shedding mass, believed to be still forming and contracting onto the main sequence.

21-cm emission line: The important radio radiation at 21-cm wavelength from interstellar neutral atomic hydrogen.

21-cm radio waves: Produced by neutral hydrogen, these waves are especially useful for galactic mapping because they allow us to detect HI clouds, which are concentrated in the spiral arms.

ultrabasic rock: A rock of high density, low silica content, and high iron content, often derives from the upper mantle of a planet or satellite.

ultraviolet radiation: Radiation of wavelength too short to see, but longer than that of X rays.

umbra: (1) The dark inner part of a shadow, in which the light source is totally obscured; (2) the dark inner part of a sunspot.

unification: The idea that the four forces of nature are just different manifestations of one basic superforce; the properties of the superforce can be realized only at phenomenally high temperatures or energies.

uniformity of nature: The fundamental assumption in astronomy that the laws of physics that are derived in terrestrial laboratories also apply throughout the universe.

universe: Everything that exists. Astronomers distinguish between the *observable* universe, the region from which light has had time to reach us in the age of the universe, and the *physical* universe, which may be much larger.

Uranus: The seventh planet outward from the Sun.

Uranus' rotation axis: Notable for its almost right-angle tilt (obliquity) of 97° to Uranus' orbital plane.

Uranus' satellites: A system of five large moons discovered from Earth, and another ten discovered by Voyager 2 in 1986.

Urey reaction: Reaction by which the Earth's carbon dioxide was concentrated in carbonate rocks after dissolving in seawater.

Van Allen belts: Doughnut-shaped zones around the Earth (or another planet with a strong magnetic field) that traps energetic ions from the Sun.

variable star: A star that varies in brightness.

velocity dispersion: The range of velocities in a self-contained dynamical system, like a globular cluster or a galaxy.

Venera 7: First spacecraft to land successfully on another planet; it transmitted data from the surface of Venus in 1970.

Viking 1: The first successful probe to land on Mars (July 20, 1976). It made the first surface photos and measures of soil composition.

Viking 2: The second successful probe to land on Mars (September 3, 1976).

virtual pairs: Particle/antiparticle pairs that appear and disappear for an instant, as allowed by Heisenberg's uncertainty principle.

visible light: Electromagnetic radiation at wavelengths that can be perceived by the eye.

visual apparent magnitude: An apparent magnitude estimate based only on visual radiation from an object (excluding infrared, ultraviolet, X rays, and so on).

visual binary: A binary in which both components can be seen.

voids: Low-density regions in the large-scale distribution of galaxies.

volatile element: Element easily driven out of a material by heating.

volcanic crater: A circular depression caused by volcanic processes such as explosion or collapse.

volcanism: Eruption of molten materials at the surface of a planet or satellite.

volcanoes: Sites where molten materials erupt from inside a planet or satellite.

wavelength: (1) The length of the wavelike characteristic of electromagnetic radiation; (2) in any wave, the distance from one maximum to the next.

wavelike properties of light: Characteristics of light, such as frequency and diffraction, that mimic properties of waves.

wave-particle duality: The concept that particles can show wavelike properties and that radiation can show particlelike properties.

weak nuclear force: The force of nature that converts neutrons into protons and is responsible for radioactive decay.

white dwarf star: A planet-sized star of roughly solar mass and very high density (10^8 to 10^{11} kg/m^3) produced as a terminal state after nuclear fuels have been consumed.

white light: A mixture of light of all colors in proportions as found in the solar spectrum.

Wien's law: A formula giving the wavelength W at which the maximum amount of radiation comes from a body of temperature T. The formula is $W = 0.00290/T$.

Wolf-Rayet star: A type of very hot star ejecting mass.

X ray: Electromagnetic radiation of wavelength about 0.01 to 10 nm.

X-ray burster: A binary system composed of a neutron star and a main-sequence star, in which the thermonuclear detonation of a layer of helium on the neutron star leads to an intense burst of X rays.

X-ray source: Celestial object emitting X rays; many are probably binary systems where mass is transferred.

zenith: The point directly overhead.

zero-age main sequence: The main sequence defined by a population of stars all of which have just evolved onto the main sequence. (Further evolution modifies the main sequence shape on the H-R diagram slightly.)

zodiac: A band around the sky about 18° wide, centered on the ecliptic, in which the planets move.

zodiacal light: A glow, barely visible to the eye, caused by dust particles spread along the ecliptic plane.

zone: Light cloud band on a giant planet.

zone of avoidance: A band around the sky, centered on the Milky Way, in which galaxies are obscured by the Milky Way's dust.

References

These references are less technical, more readily available, and recommended for general reading. They are good general references for preparation of term papers. More technical references are listed in the Instructor's Manual.

Chapter 1

Ashbrook, J. 1973. "Astronomical Scrapbook." *Sky and Telescope 46*: 300.

Bok, B. 1975. "A Critical Look at Astrology." *The Humanist,* September/October, p. 5.

Carlson, J. B. 1975. "Lodestone Compass: Chinese or Olmec Primacy?" *Science 189*: 753

Doig, P. 1950. *A Concise History of Astronomy.* London: Chapman and Hall.

Gingerich, O. 1967. "Musings on Antique Astronomy." *American Scientist 55*: 88.

______. 1984. "The Origin of the Zodiac." *Sky and Telescope 67*: 218.

Harber, H. E. 1969. "Five Mayan Eclipses in Thirteen Years." *Sky and Telescope 37*: 72.

Hawkins, G., and J. B. White. 1965. *Stonehenge Decoded.* New York: Doubleday.

Hoyle, F. 1972. *From Stonehenge to Modern Cosmology.* San Francisco: W. H. Freeman.

Jerome, L. E. 1975. "Astrology: Magic of Science?" *The Humanist,* September/October, p. 10.

Neugebauer, O. 1957. *The Exact Sciences in Antiquity.* Providence, R.I.: Brown University Press.

Pannekoek, A. 1961. *A History of Astronomy.* London: Allen and Unwin.

Stephenson, F. R. 1982. "Historical Eclipses." *Scientific American,* October, p. 170.

Stephenson, F. R., and D. H. Clark. 1977. "Ancient Astronomical Records from the Orient." *Sky and Telescope 53*: 84.

Chapter 2

Gingerich, O. 1967. "Musings on Antique Astronomy." *American Scientist 55*: 88.

______. 1986. "Islamic Astronomy." *Scientific American 254,* April, p. 74.

Hetherington, N. S. 1987. *Ancient Astronomy and Civilization.* Tucson: Pachart Publishing House.

Krisciunas, K. 1988. *The Alexandrian Museum: In Astronomical Centers of the World.* Cambridge: Cambridge University Press.

Lewis, D. 1973. *We, the Navigators.* Honolulu: University of Hawaii Press.

North, J. D. 1974. "The Astrolabe." *Scientific American,* January, p. 96.

Pannekoek, A. 1961. *A History of Astronomy.* London: Allen and Unwin.

Thomsen, D. E. 1984. "Calendric Reform in Yucatan." *Science News 126*: 282.

Wilson, E. O. 1998. *Consilience.* New York: Knopf.

Chapter 3

de Santillana, G. 1962. *The Crime of Galileo.* New York: Time.

Dreyer, J. L. E. 1953. *A History of Astronomy from Thales to Kepler.* New York: Dover. (Reprint of 1906 edition.)

Gingerich, O. 1973a. "Copernicus and Tycho." *Scientific American,* December, p. 86.

______. 1973b. *Crisis Versus Asthetic in the Copernican Revolution.* Cambridge, Mass.: Smithsonian Astrophysical Observatory.

Hartmann, W. K. 1999. *Moons and Planets.* Belmont, Calif.: Wadsworth.

Lerner, L. S., and E. A. Gosselin. 1973. "Ciordano Bruno." *Scientific American,* April, p. 86.

______. 1986. "Galileo and the Specter of Bruno." *Scientific American 255.* November, p. 126.

Pannekoek, A. 1961. *A History of Astronomy.* New York: Interscience.

Taylor, S. R. 1982. *Planetary Science: A Lunar Perspective.* Houston, Tex.: Lunar and Planetary Institute.

Chapter 4

Arnold, J. R. 1980. "The Frontier in Space." *American Scientist 68*: 299.

Banks, P., and D. Black. 1987. "The Future of Science in Space." *Science 236*: 244.

Barnsley, T. 1996. "Science in the Sky." *Scientific American, 27*: 64.

Burbidge, E. Margaret. 1983. "Adventure into Space." *Science 221*: 421.

Chaisson, Eric J. 1992. "Early Results from the Hubble Space Telescope." *Scientific American,* June, p. 44.

Clarke, A. C. 1951. *The Exploration of Space.* New York: Harper & Bros.

Dyson, F. 1969. "Human Consequences of the Exploration of Space." *Bulletin of Atomic Scientists,* September, p. 8.

Hartmann, W. K., R. Miller, and P. Lee. 1984. *Out of the Cradle.* New York: Workman Publishing.

Heinlein, R. 1950. *The Man Who Sold the Moon.* New York: New American Library.

Hetherington, N. S. 1987. *Ancient Astronomy and Civilization.* Tucson: Pachart Publishing House.

Lewis, R. S. 1969. *Appointment on the Moon.* New York: Viking Press.

Logsdon, J. 1970. *The Decision to Go to the Moon.* Cambridge, Mass.: M.I.T. Press.

Nicholson, M. 1949. *Voyages to the Moon.* New York: Macmillan.

Verne, J. 1949. *From the Earth to the Moon.* New York: Didear. (Originally published 1865.)

Chapter 5

Hjellming, R., and R. Bignell. 1982. "Radio Astronomy with the Very Large Array." *Science 216*: 1279.

Chapter 6

Davies, G. I. 1969. *The Earth in Decay.* New York: American Elsevier.

Glanz, J. 1999. In Science vs. Bible Wrangle, Debate Moves to the Cosmos. *International Herald Tribune,* October 11, p. 2.

Kasting, J. F. 1993. "Earth's Early Atmosphere." *Science 259*: 920.

Kellogg, L. B., B. Hager, and R. van der Hilst. 1999. Compositional Stratification in the Deep Mantle, *Science, 283*: 1881.

Maxwell, J. C. 1985. "What Is the Lithosphere?" *Physics Today 38*: 32.

Morris, S. C. 1987. "The Search for the Precambrian-Cambrian Boundary." *American Scientist 75*: 157.

Nelkin, D. 1976. "The Science-Textbook Controversies." *Scientific American,* April, p. 33.

Turco, R. P., and others. 1984. "The Climatic Effects of Nuclear War." *Scientific American 251,* August, p. 33.

van der Hilst, R., and H. Karason. 1999. Compositional Heterogeneity in the Bottom 1000 Kilometers of Earth's Mantle: Toward a Hybrid Convection Model. *Science, 283*: 1885.

Wuethrich, B. 1999. Lack of Icebergs Another Sign of Global Warming? *Science, 285*: 37.

York, D. 1993. "The Earliest History of the Earth." *Scientific American 268,* No. 1, p. 90.

Chapter 7

Arnold, J. R. 1979. *Journal of Geophysical Research 84*: 5659.

Feldman, W. and others. 1998. Fluxes of Fast and Epithermal Neutrons from Lunar Prospector: Evidence for Water Ice at the Lunar Poles. *Science, 281*: 1496.

Goldreich, P. 1972. "Tides and the Earth-Moon System." *Scientific American,* April, p. 42.

Lewis, J., M. Matthews, and M. Guerrieri, eds. 1993. *Resources of Near-Earth Space.* Tucson: University of Arizona Press.

Lewis, R. S. 1977. "Space Prospect: Factories and Electric Power." *Smithsonian 8* (9): 94.

Thomsen, D. E. 1986. "Man in the Moon." *Science News 129*: 154.

Watson, K., B. C. Murray, and H. Brown. 1961. *Journal of Geophysical Research 66*: 3033.

Chapter 8

Cruikshank, D. P., and C. R. Chapman. 1967. "Mercury's Rotation and Visual Observations." *Sky and Telescope 34*: 24.

Moore, P. 1954. *A Guide to the Planets.* New York: W. W. Norton.

Murray, B. C. 1975. "Mercury." *Scientific American,* September, p. 58.

Chapter 9

Adams, W. W. and T. Dunham. 1932. Absorption Bands in the Spectrum of Venus. *Publications of the Astronomical Society of the Pacific 44*: 243.

Bullock, M. A. and D. H. Grinspoon. 1999. Global Climate Change on Venus. *Scientific American, 280,* March, p. 50.

Cowen, R. 1993. "New Evidence of Ancient Seas on Venus." *Science News 143*: 212.

Kaula, W. K. 1995. Venus Reconsidered. *Science 270*: 1460.

Kerr, R. 1999. Craters Suggest How Venus Lost Her Youth. *Science 284*: 889.

Moore, P. 1954. *A Guide to the Planets.* New York: W. W. Norton.

Phillips, R. J., and V. L. Hansen. 1998. Geological Evolution of Venus: Rises, Plains, Plumes, and Plateaus. *Science, 279*: 1492.

Saunders, R. S., and others. 1991. "An Overview of Venus Geology." *Science 252*: 249.

Schubert, G., and C. Covey. 1981. "The Atmosphere of Venus." *Scientific American,* June, p. 66.

Smrekar, S. E. and E. R. Stofan. 1997. Corona Formation and Heat Loss on Venus by Coupled Upwelling and Delamination. *Science 277*: 1289.

Chapter 10

Borg, L. E. and others. 1999. The Age of the Carbonates in Martian Meteorite ALH84001. *Science 286*: 90.

Bradbury, R. 1950. *The Martian Chronicles.* New York: Doubleday.

Forget, F., and R. T. Pierrehumbert. 1997. Warming Early Mars with Carbon Dioxide Clouds That Scatter Infrared Radiation. *Science 278*: 1273.

Gibson, E. K., and others. 1997. The Case for Relic Life on Mars. *Scientific American 277,* No. 6, p. 58.

Golombek, M. P. 1998. The Mars Pathfinder Mission. *Scientific American 279,* July, p. 40.

Haberle, R. M. 1986. "The Climate of Mars." *Scientific American,* May, p. 54.

Hartmann, W. K. 1999. Martian Cratering VI. Crater Count Isochrons and Evidence for Recent Volcanism from Mars Global Surveyor. *Meteorites and Planetary Science 34*: 167.

Hartmann, W. K. and O. Raper. 1974. *The New Mars.* Washington, D.C.: National Aeronautics and Space Administration.

Head, J. W., and others. 1999. Possible Ancient Oceans on Mars: Evidence from Mars Orbiter Laser Altimeter Data. *Science 286*: 2134.

Kaplan, L. D., G. Munch, and H. Spinrad. 1964. An Analysis of the Spectrum of Mars. *Astrophysical Journal 139*: 1.

Karl, D. M. and others. 1999. Microorganisms in the Accreted Ice of Lake Vostok, Antarctica. *Science 286*: 2144.

Kerr, R. A. 1993. "An Outrageous Hypothesis for Mars: Episodic Oceans." *Science 259*: 910.

Kuiper, G. P. 1947. Infrared Spectra of Planets. *Astrophysical Journal 106*: 252.

Lowell, P. 1906. *Mars and Its Canals.* 2nd ed. New York: Macmillan.

McKay, D. S., and others. 1996. Search for Past Life on Mars: Possible Relic Biogenic Activity in Martian Meteorite ALH84001. *Science, 273*: 924.

Owen, T. C. and G. P. Kuiper. 1964. A Determination of the Composition and Surface Pressure of the Martian Atmosphere. *Communications Lunar and Planetary Laboratory 2*: 113.

Priscu, J. C. and others. 1999. Geomicrobiology of Subglacial Ice above Lake Vostok, Antarctica. *Science 286*: 2141.

Swindle, T. D. and others. 2000. Noble Gases in Iddinsite from the Lafayette Meteorite: Evidence for Liquid Water on Mars in the Last Few Hundred Million Years. *Meteorites and Planetary Science* 35: 107–115.

Wells, H. G. 1898. *The War of the Worlds.* London: W. Heinemann.

Zuber, M. T., and others. 1998. Observations of the North Polar Region of Mars from the Mars Orbiter Laser Altimeter. *Science 282*: 2053.

Chapter 11

Carlson, R. and others. 1996. Near-Infrared Spectroscopy and Spectral Mapping of Jupiter and the Galilean Satellites: Results from Galileo's Initial Orbit. *Science 274*: 385.

Carlson, R. W., R. Johnson, and M. Anderson. 1999. Sulfuric Acid on Europa and the Radiolytic Sulfur Cycle. *Science 286*: 97.

Carr, M. H. and others. 1998. Evidence for a Subsurface Ocean on Europa. *Nature 391*: 363 (and other papers in the same issue).

Chapman, C. R. 1968. "The Discovery of Jupiter's Red Spot." *Sky and Telescope 35*: 276.

Guillot, T. 1999. Interiors of Giant Planets Inside and Outside the Solar System. *Science 286*: 72.

Hartmann, W. K. 1999. *Moons and Planets,* Fourth Edition, Belmont, Calif.: Wadsworth.

Hoppa, G. V., and others. 1999. Formation of Cycloidal Features on Europa. *Science 285*: 1899.

McEwen, A. S., and others. 1998. High-Temperature Silicate Volcanism on Jupiter's Moon Io. *Science 281*: 87.

Pappalardo, R. T., J. Head, and R. Greeley 1999. The Hidden Ocean of Europa. *Scientific American 281*: 54.

Peale, S. J. 1999. Origin and Evolution of the Natural Satellites. *Annual Review of Astronomy and Astrophysics 37*: 533.

Showman, A. P. and R. Malhotra. 1999. The Galilean Satellites. *Science 286*: 77.

Stevenson, D. J. 1996. When Galileo Met Ganymede. *Nature 384*: 511 (and other papers in the same issue).

Stone, E., and others. 1979a. Special issue on Voyager I Jupiter results. *Science 204*: 945 ff.

_____. 1979b. Special issue on Voyager 2 Jupiter results. *Science 206*: 925 ff.

_____. 1981. Special issue on Voyager 1 Saturn results. *Science 212*: 159 ff.

_____. 1982. Special issue on Voyager 2 Saturn results. *Science 212*: 499 ff.

Wolfe, J. H. 1975. "Jupiter." *Scientific American,* September, p. 118.

Chapter 12

Grosser, M. 1962. *The Discovery of Neptune.* Cambridge, Mass.: Harvard University Press.

Hartmann, W. K. 1999. *Moons and Planets,* Fourth Edition. Belmont, Calif.: Wadsworth.

Chapter 13

Asphaug, E. 1997. Impact Origin of the Vesta Family, *Icarus 32*: 965.

Belton, M. J., and others. 1992. "Galileo Encounter with 951 Gaspra: First Pictures of an Asteroid." *Science 257*: 1647.

Binzel, R., M. A. Barucci, and M. Fulchignoni. 1991. "The Origins of the Asteroids." *Scientific American 265*: 88.

Brandt, J. C., and R. Chapman. 1981. *Introduction to Comets.* Cambridge: Cambridge University Press.

Brown, R. H., and others. 1997. Surface Composition of Kuiper Belt Object 1993SC. *Science 276*: 937.

_____. 1998. Identification of Water Ice on the Centaur 1997 CU 26. *Science 280*: 1430.

Chapman, C. R. 1975. "The Nature of Asteroids." *Scientific American,* January, p. 24.

Connolly, H. C., and S. G. Love. 1998. The Formation of Chondrules: Petrologic Tests of the Shock Wave Model. *Science 280*: 62.

Davis, D. R., A. Friedlander, and T. Jones. 1993. Role of Near-Earth Asteroids in the Space Exploration Initiative. In *Resources of Near-Earth Space,* ed. J. Lewis, M. Matthews, and M. Guerrieri. Tucson: University of Arizona Press.

Durda, D. D. 1996. The Formation of Asteroidal Satellites in Catastrophic Collisions. *Icarus 120*: 212.

Gehrels, T., ed. 1979. *Asteroids.* Tucson: University of Arizona Press.

Gibbs, W. W. 1998. The Search for Greenland's Mysterious Meteor. *Scientific American 279,* November, p. 73.

Glanz, J. 1997. How the Hectic Young Sun Cooked Up Stony Meteorites. *Science 276*: 1789.

Hartmann, W. K. 1975. "The Smaller Bodies of the Solar System." *Scientific American,* September, p. 143.

_____. 1982. "Mines in the Sky Are Not So Wild a Dream." *Smithsonian,* September, 70.

_____. 1999. *Moons and Planets,* Fourth Edition. Belmont, Calif.: Wadsworth.

Hartmann, W. K., R. Miller, and P. Lee. 1984. *Out of the Cradle.* New York: Workman.

Lewis, J., M. Matthews, and M. Guerrieri. 1993. *Resources of Near-Earth Space.* Tucson: University of Arizona Press.

Luu, J. X., and D. C. Jewitt 1996. The Kuiper Belt. *Scientific American 274,* May, p. 46.

McKeegan, K. D., and others. 1998. Oxygen Isotopic Abundances in Ca-Al-rich Inclusions from Ordinary Chondrites: Implications for Nebular Heterogeneity. *Science 280*: 414.

Sagan, C. 1975. "Kalliope and the Kaa'ba: The Origin of Meteorites." *Natural History 84*: 8.

Trefil, James. 1989. "Stop to Consider that Stones Fall from the Sky." *Smithsonian 20* (No. 6): 81.

Weidenschilling, S. J. 1997. The Origin of Comets in the Solar Nebula: A Unified Model. *Icarus 127*: 290.

Weissman, P. R. 1993. "Comets at the Solar System's Edge." *Sky and Telescope 85*: 26.

_____. 1998. The Oort Cloud. *Scientific American 279,* September, p. 84.

Wilkening, L., ed. 1982. *Comets.* Tucson: University of Arizona Press.

Chapter 14

Cameron, A. G. W. 1975. "The Origin and Evolution of the Solar System." *Scientific American,* September, p. 32

Cowen, R. 1999. Asteroids Formed Early on in Solar System. *Science News, 155*: 325.

Fernandez, J. A., and W.-H. Ip. 1984. *Icarus 58*: 109.

______. 1996. *Planetary Space Science 44*: 431.

Grossman, L. 1975. "The Most Primitive Objects in the Solar System: Carbonaceous Chondrites." *Scientific American,* February, p. 30.

Herbst, W., and G. Assousa. 1979. "Supernovas and Star Formation." *Scientific American,* August, p. 138.

Lee, D.-C., and A. Halliday. 1996. Hf-W Isotopic Evidence for Rapid Accretion and Differentiation in the Early Solar System. *Science 274*: 1876.

Lewis, J. S. 1974. "The Chemistry of the Solar System." *Scientific American,* March, p. 50.

Liou, J.-C. and R. Malhotra. 1997. Depletion of the Outer Asteroid Belt. *Science 275*: 375.

Lissauer, J. J. 1999. Chaotic Motion in the Solar System. *Reviews of Modern Physics 71*: 835.

Marcy, G. and P. Butler. 2000. Hunting Planets Beyond. *Astronomy 28,* March, p. 43.

Mason, B., and W. G. Melson. 1970. *The Lunar Rocks.* New York: John Wiley.

Sincell, M. 2000. Switched at Birth. *Astronomy 28,* March, p. 48

Stone, E., and others. 1979. "The Voyager 2 Encounter with the Jupiter System." *Science 206*: 925.

______. 1982. "The Voyager 2 Encounter with the Saturn System." *Science 212*: 499.

______. 1986. "The Voyager 2 Encounter with the Uranus System." *Science 233*: 39.

______. 1989. "The Voyager 2 Encounter with the Neptunian System." *Science 246*: 1417.

Weidenschilling, S. J. 1997. The Origin of Comets in the Solar Nebula: A Unified Model. *Icarus 127*: 290.

Weidenschilling, S. J., and D. R. Davis. 2000. After Oligarchy Comes Chaos: From the Middle to Late State of Planetary Accretion. Lunar Planetary Science Conference (abstract).

Wetherill, G. W. 1985. "Occurrence of Giant Impacts During the Growth of the Terrestrial Planets." *Science 228*: 877.

Chapter 15

Arnold, J. R. 1980. "The Frontier in Space." *American Scientist 68*: 299.

Bahcall, J. 2001. "How the Sun Shines." *Mercury Magazine 30*: 5, pp. 30–37.

Burch, J. 2001. "The Fury of Space Storms." *Scientific American,* April, 2001.

Hubbert, M. 1971. "The Energy Resources of the Earth." *Scientific American,* September, p. 61.

Kerr, R. A. 1986. "The Sun Is Fading." *Science 231*: 339.

Kreith, F., and R. T. Meyer. 1983. "Large-Scale Use of Solar Energy with Central Receivers." *American Scientist 71*: 598.

Lewis, R. S. 1977. "The Space Prospect: Factories and Electric Power." *Smithsonian,* December, p. 94.

Meadows, J. 1984. "The Origins of Astrophysics." *American Scientist 72*: 269.

Parker, E. N. 1983. "Magnetic Fields in the Cosmos." *Scientific American,* August, p. 44.

Pasachoff, J. M. 1973. "The Solar Corona." *Scientific American,* October, p. 68.

______. 1980. "Our Sun." *Astronomy Selected Readings,* ed. M. A. Seeds. Menlo Park, Calif.: Benjamin/Cummings.

Snell, J. E., P. Achenbach, and S. Peterson. 1976. "Energy Conservation in New Housing Design." *Science 192*: 1305.

Wilcox, J. M. 1976. "Solar Structure and Terrestrial Weather." *Science 192*: 745.

Williams, G. E. 1986. "The Solar Cycle in Precambrian Time." *Scientific American,* August, p. 88.

Wolfson, R. 1983. "The Active Solar Corona." *Scientific American,* February, p. 104.

Chapter 16

Bahcall, J., and L. Spitzer. 1982. "The Space Telescope." *Scientific American,* July, p. 40.

Shapley, H., ed. 1960. *Source Book in Astronomy.* New York: McGraw-Hill.

Shapley, H., and H. Howarth, eds. 1929. A *Source Book in Astronomy.* New York: McGraw-Hill.

Chapter 17

Aller, L. H. 1971. *Atoms, Stars, and Nebulae.* Cambridge, Mass.: Harvard University Press.

Burnham, R. 1978. *Burnham's Celestial Handbook.* New York: Dover Publications.

Humphreys, R., and K. Davidson. 1984. "The Most Luminous Stars." *Science 223*: 243

Nesme-Ribes, E., and others. 1996 "The Stellar Dynamo. *Scientific American,* August, p. 47.

Chapter 18

Boss, A. P. 1985. "Collapse and Formation of Stars." *Scientific American,* January, p. 40.

Lada, C. J. 1982. "Energetic Outflow from Young Stars." *Scientific American,* June, p. 82.

Neugebauer, G., and E. Becklin. 1973. "The Brightest Infrared Sources." *Scientific American,* April, p. 28.

Scoville, N., and J. Young. 1984. "Molecular Clouds, Star Formation and Galactic Structure." *Scientific American,* April, p. 42.

Strom, S., and K. Strom. 1973. "The Early Evolution of Stars." *Sky and Telescope 45*: 279, 359.

Welch, W., and others. 1985. "Gas Jets Associated with Star Formation." *Science 228*: 1289.

Chapter 19

Arnett, D. A., and others. 1989. "Supernova 1987A." *Annual Reviews of Astronomy and Astrophysics 27*: 629.

Bova, B. 1973. "Obituary of Stars: A Tale of Red Giants, White Dwarfs, and Black Holes." *Smithsonian 4*: 54.

Burnham, R. 1978. *Burnham's Celestial Handbook,* New York: Dover Publications.

Burrows, A. 1987. "The Birth of Neutron Stars and Black Holes." *Physics Today,* September, p. 28.

Hawking, S. W. 1977. "The Quantum Mechanics of Black Holes." *Scientific American,* January, p. 34.

Helfand, D. 1983. "Theory Points to Pulsating White Dwarfs." *Physics Today,* January, p. 21.

______. 1987. "Bang: The Supernova of 1987." *Physics Today,* August, p. 25.

Humphreys, R., and K. Davidson. 1984. "The Most Luminous Stars." *Science 223*: 243.

Kaler, J. B. 1986. "Planetary Nebulae and the Death of Stars." *American Scientist 74*: 244.

Liebert, J. 1980. "White Dwarf Stars." *Annual Review of Astronomy and Astrophysics 18*: 363.

Penrose, R. 1972. "Black Holes." *Scientific American,* May, p. 38.

Ruderman, M. 1971. "Solid Stars." *Scientific American,* February, p. 24.

______. 1972. "Pulsars: Structure and Dynamics. *Annual Review of Astronomy and Astrophysics 10*: 427.

Schorn, R. A. 1982. "The Gamma-Ray Burster Puzzle." *Sky and Telescope 63*: 560.

Shaham, J. 1987. "The Oldest Pulsars in the Universe." *Scientific American,* February, p. 50.

Wade, N. 1975. "Discovery of Pulsars: A Graduate Student's Story." *Science 189*: 359.

Chapter 20

Blitz, L. 1982. "Giant Molecular-Cloud Complexes in the Galaxy." *Scientific American,* April, p. 84.

Cash, W. and P. Charles. 1980. "Stalking the Cygnus Superbubble." *Sky and Telescope 59*: 455.

Greenberg, J. M. 2000. "The Secrets of Stardust." *Scientific American,* December, p. 70.

Herbig, G. H. 1974. "Interstellar Smog." *American Scientist 62*: 200.

Lewis, R., and E. Anders. 1983. "Interstellar Matter in Meteorites." *Scientific American,* August, p. 66.

Miller, J. S. 1974. "The Structure of Emission Nebulae." *Scientific American,* October, p. 34.

Scoville, N., and J. Young. 1984. "Molecular Clouds, Star Formation and Galactic Structure." *Scientific American,* April, p. 42.

Turner, B. E. 1973. "Interstellar Molecules." *Scientific American,* March, p. 50.

Chapter 21

Boss, A. 1996. "Extrasolar Planets." *Physics Today,* September, p. 32.

Lissauer, J. 1999. "How Common Are Habitable Planets?" *Nature,* December, p. C11.

Marcy, G., and R. P. Butler. 2000. "Planets Orbiting Other Suns." *Publications of the Astronomical Society of the Pacific. 112*: 137–140.

Margon, B. 1980. "The Bizarre Spectrum of SS 433." *Scientific American,* October, p. 54.

______. 1982. "Relativistic Jets in SS 433." *Science 215*: 247."

Warner, B. 1972. "Six Ultra-Short Period Binary Stars." *Sky and Telescope 44*: 358.

Chapter 22

Iben, I. 1970. "Globular-Cluster Stars." *Scientific American,* July, p. 26.

Jones, K. G. 1969. *Messier's Nebulae and Star Clusters.* New York: American Elsevier.

Shapley, H. 1930. *Star Clusters.* Cambridge, Mass.: Harvard University Press.

Chapter 23

Bok, B. J., and P. Bok. 1981. *The Milky Way.* 5th ed. Cambridge. Mass.: Harvard University Press.

Kraft, R. P. 1959. "Pulsating Stars and Cosmic Distances." *Scientific American,* July, p. 48.

Saunders, J. 1963. "The Globular Cluster Omega Centauri." *Sky and Telescope 26*: 133.

Shapley, H. 1930. *Star Clusters.* Cambridge, Mass.: Harvard University Press.

Struve, O. 1960. "A Historic Debate About the Universe." *Sky and Telescope 19*: 398.

Whitney, C. A. 1971. *The Discovery of Our Galaxy.* New York: Alfred A. Knopf.

Chapter 24

Bok, B. J. 1966. "Magellanic Clouds." *Annual Review of Astronomy and Astrophysics 7*: 95.

Ferris, T. 1988. *Coming of Age in the Milky Way.* New York: Morrow.

Hirshfeld, A. 1980. "Inside Dwarf Galaxies." *Sky and Telescope 59*: 287.

Rowan-Robinson, M. 1985. *The Cosmological Distance Ladder.* New York: Freeman.

Rubin, V. 1983. "Dark Matter in Spiral Galaxies." *Scientific American,* June, p. 96.

Sandage, A. 1961. *The Hubble Atlas of Galaxies.* Washington, D.C.: The Carnegie Institution.

Schweizer, F. 1986. "Colliding and Merging Galaxies." *Science 231*: 227.

Shapley, H. 1957. *The Inner Metagalaxy.* New Haven, Conn.: Yale University Press.

Silk, J. 1987. "The Formation of Galaxies." *Physics Today 40,* no. 4, p. 28.

Strom, K., and S. Strom. 1982. "Galactic Evolution: A Survey of Recent Progress." *Science 216*: 571.

Toomre, A., and J. Toomre. 1973. "Violent Tides Between Galaxies." *Scientific American,* December. p. 38.

Tully, R. B. 1988. *Atlas of Nearby Galaxies.* Cambridge: Cambridge University Press.

Chapter 25

Arp, H. C. 1987. *"Quasars, Redshifts, and Controversies."* Interstellar Media. Berkeley, Calif.

Blandford, R. D., and others. 1982. "Cosmic Jets." *Scientific American,* May, p. 124.

De Lapparent, V. M., M. Geller, and J. Huchra. 1986. "A Slice of the Universe." *Astrophysical Journal 302*: L1.

Eddington, A. 1933. *The Expanding Universe.* Cambridge: Cambridge University Press.

Geller, M., and J. Huchra. 1989. "Mapping the Universe." *Science 246*: 897.

Gregory, S. A., and L. Thompson. 1982. "Superclusters and Voids in the Distribution of Galaxies." *Scientific American,* March, p. 106.

Hodge, P. W. 1984. "The Cosmic Distance Scale." *American Scientist 72*: 474.

Osmer, P. S. 1982. "Quasars as Probes of the Distant and Early Universe." *Scientific American,* February, p. 126.

Rees, M. J. 1990. "'Dead Quasars' in Nearby Galaxies?" *Science 247*: 817.

Silk, J. 1989. *The Big Bang.* New York: Freeman.

Turner, E. 1984. "Quasars and Gravitational Lenses." *Science 223*: 1255.

Chapter 26

Einstein, A. 1923. *The Meaning of Relativity.* Princeton, N.J.: Princeton University Press.

Ferris, T. 1988. *Coming of Age in the Milky Way.* New York: Morrow.

Harrison, E. R. 1974. "Why the Sky Is Dark at Night." *Physics Today 27*: 5.

North, J. D. 1965. *The Measure of the Universe.* Oxford: Oxford University Press.

Rowan-Robinson, M. 1985. *The Cosmological Distance Ladder.* New York: Freeman.

Silk, J. 1989. *The Big Bang.* New York: Freeman.

Wesson, P., K. Valle, and R. Stabelle. 1987. "The Extragalactic Background Light and a Definitive Resolution of Olbers' Paradox." *Astrophysical Journal 317*: 601.

Chapter 27

Cowen, R. 1992. "A River Runs Through It? Mapping the Flow of the Universe." *Science News 142*: 408.

Davies, P. C. W., and J. Brown. 1988. *Superstrings: A Theory of Everything?* Cambridge: Cambridge University Press.

Dicus, D., and others. 1983. "The Future of the Universe." *Scientific American,* March, p. 90.

Hawking, S. 1977. "The Quantum Mechanics of Black Holes." *Scientific American,* January, p. 34.

Lightman, A. P. 1991. *Ancient Light: Our Changing View of the Universe.* Cambridge, Mass.: Harvard University Press.

Penrose, R. 1989. *The Emperor's New Mind.* Oxford: Oxford University Press.

Silk, J. 1989. *The Big Bang.* New York: Freeman.

Weinberg, S. 1977. *The First Three Minutes.* New York: Basic Books.

Wilczek, F., and B. Devine. 1988. *Longing for the Harmonies.* New York: Norton.

Chapter 28

Bernstein, M. P., S. Sandford, and L. Allamandola. 1999. Life's Far-Flung Raw Materials. *Scientific American 281,* July, p. 42.

Blake, D., and P. Jenniskens. 2001. "The Ice of Life." *Scientific American,* August, p. 45.

Courtillot, V. 1999. *Evolutionary Catastrophes.* Cambridge: Cambridge University Press.

Crawford, I. 1999. Where Are They? *Scientific American 283,* July, p. 38.

Dawkins, R. 1995. God's Utility Function. *Scientific American 273,* November, p. 80.

Frederickson, J. K., and T. Onstott. 1996. Microbes Deep inside the Earth. *Scientific American 275,* October, p. 68.

Furnes, H., and others. 1999. Depth of Active Bio-alteration in the Ocean Crust: Costa Rica Rift (Hole 504B). *Terra Nova 11*: 228.

Gonzalez, G., D. Brownlee, and P. Ward. 2001. "Refugees for Life in a Hostile Universe." *Scientific American,* October, p. 60.

Holden, C. 1997. Ocean Yields Hottest Life Yet. *Science 275*: 933.

Holland, H. D. 1997. Evidence for Life on Earth More Than 3850 Million Years Ago. *Science 275*: 38.

Ingber, D. E. 1998. The Architecture of Life. *Scientific American 278,* January, p. 48.

Jin, Y. G. 2000. Pattern of Marine Mass Extinction Near the Permian-Triassic Boundary in South China. *Science 289*: 432.

Kaiser, J. 1995. Can Deep Bacteria Live on Nothing but Rocks and Water? *Science 270*: 377.

Kerridge, J. F. 1991. A Note on the Prebiotic Synthesis of Organic Acids in Carbonaceous Meteorites. *Origins of Life and Evolutionary Biosphere 21*: 19.

______. 1995. Origin of Amino Acids in the Early Solar System. *Advanced Space Research 15*: 107.

Krauss, L. K., and G. Starkman. 1999. The Fate of Life in the Universe. *Scientific American 281,* November, p. 58.

Index

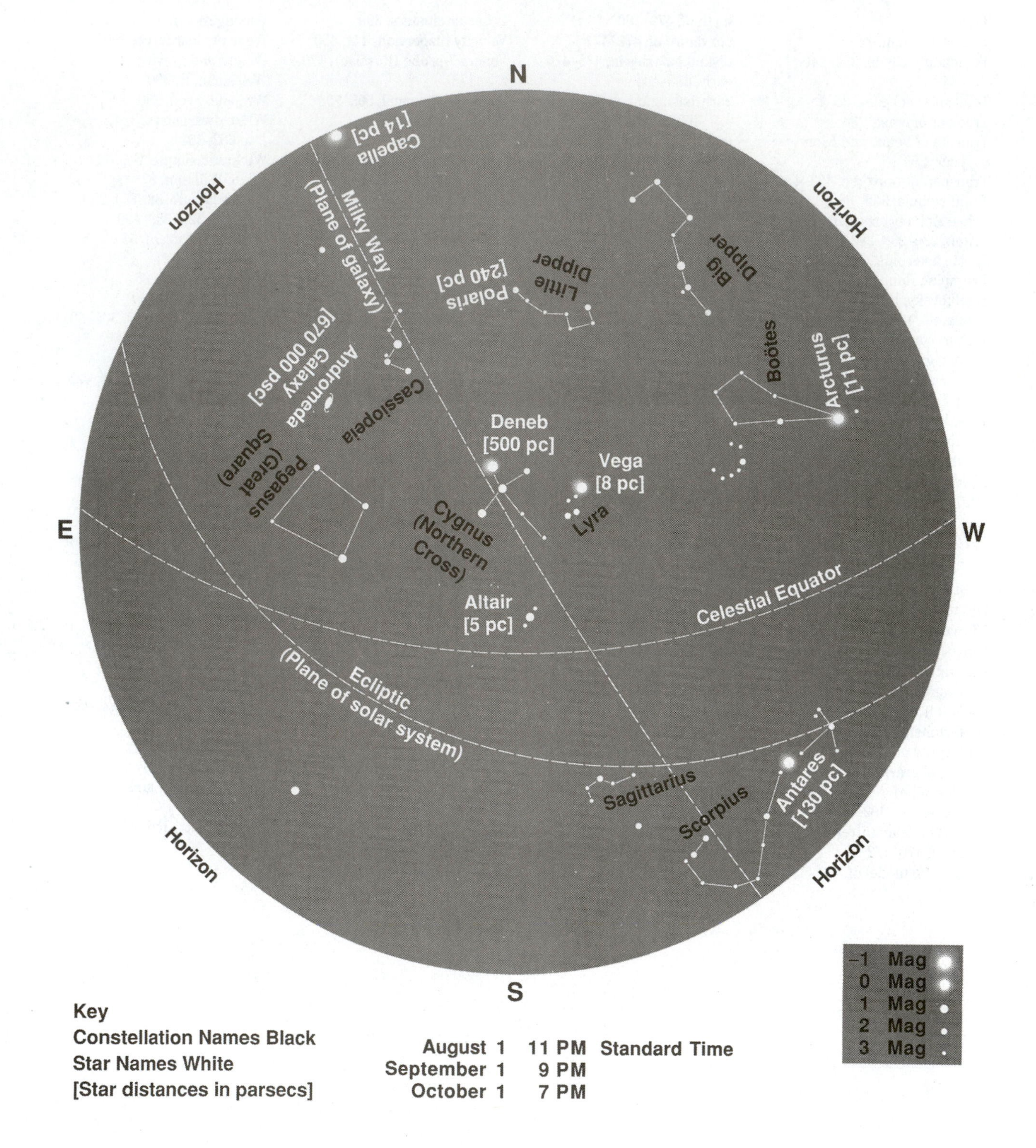
N
E
W
S
Horizon
Horizon
Horizon
Horizon
Capella [14 pc]
Milky Way (Plane of galaxy)
Polaris [240 pc]
Little Dipper
Big Dipper
Andromeda Galaxy [670 000 psc]
Cassiopeia
Boötes
Arcturus [11 pc]
Deneb [500 pc]
Vega [8 pc]
Lyra
Pegasus (Great Square)
Cygnus (Northern Cross)
Altair [5 pc]
Celestial Equator
Ecliptic (Plane of solar system)
Sagittarius
Scorpius
Antares [130 pc]
–1 Mag
0 Mag
1 Mag
2 Mag
3 Mag
Key
Constellation Names Black
Star Names White
[Star distances in parsecs]
August 1 11 PM Standard Time
September 1 9 PM
October 1 7 PM

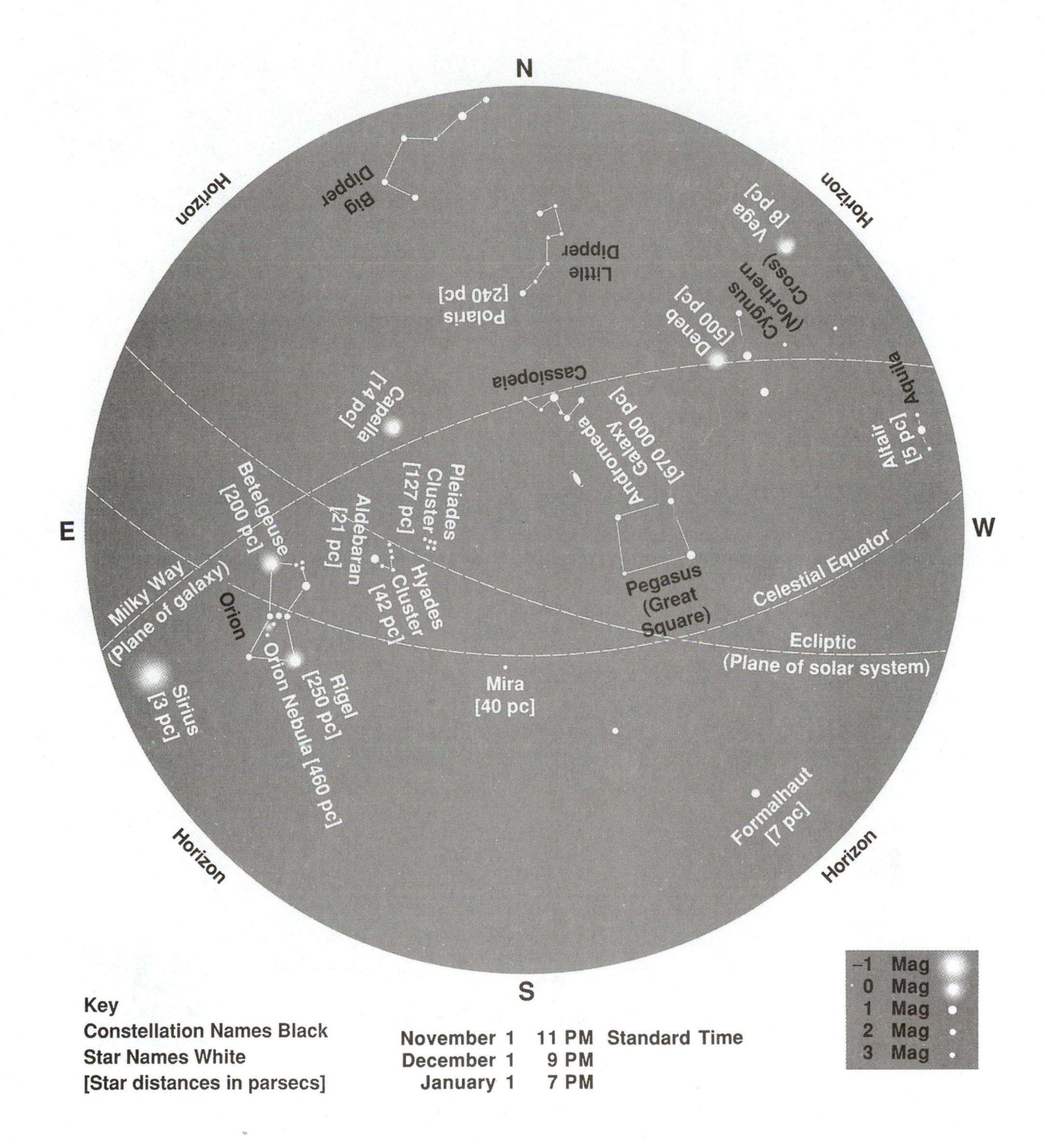
N
E
W
S
Horizon
Horizon
Horizon
Horizon
Big Dipper
Little Dipper
Polaris [240 pc]
Vega [8 pc]
Cygnus (Northern Cross)
Deneb [500 pc]
Cassiopeia
Aquila
Altair [5 pc]
Capella [14 pc]
Andromeda Galaxy [670 000 pc]
Pleiades Cluster [127 pc]
Aldebaran [21 pc]
Hyades Cluster [42 pc]
Betelgeuse [200 pc]
Orion
Milky Way (Plane of galaxy)
Pegasus (Great Square)
Celestial Equator
Ecliptic (Plane of solar system)
Sirius [3 pc]
Rigel [250 pc]
Orion Nebula [460 pc]
Mira [40 pc]
Formalhaut [7 pc]
–1 Mag
0 Mag
1 Mag
2 Mag
3 Mag
Key
Constellation Names Black
Star Names White
[Star distances in parsecs]
November 1 11 PM Standard Time
December 1 9 PM
January 1 7 PM

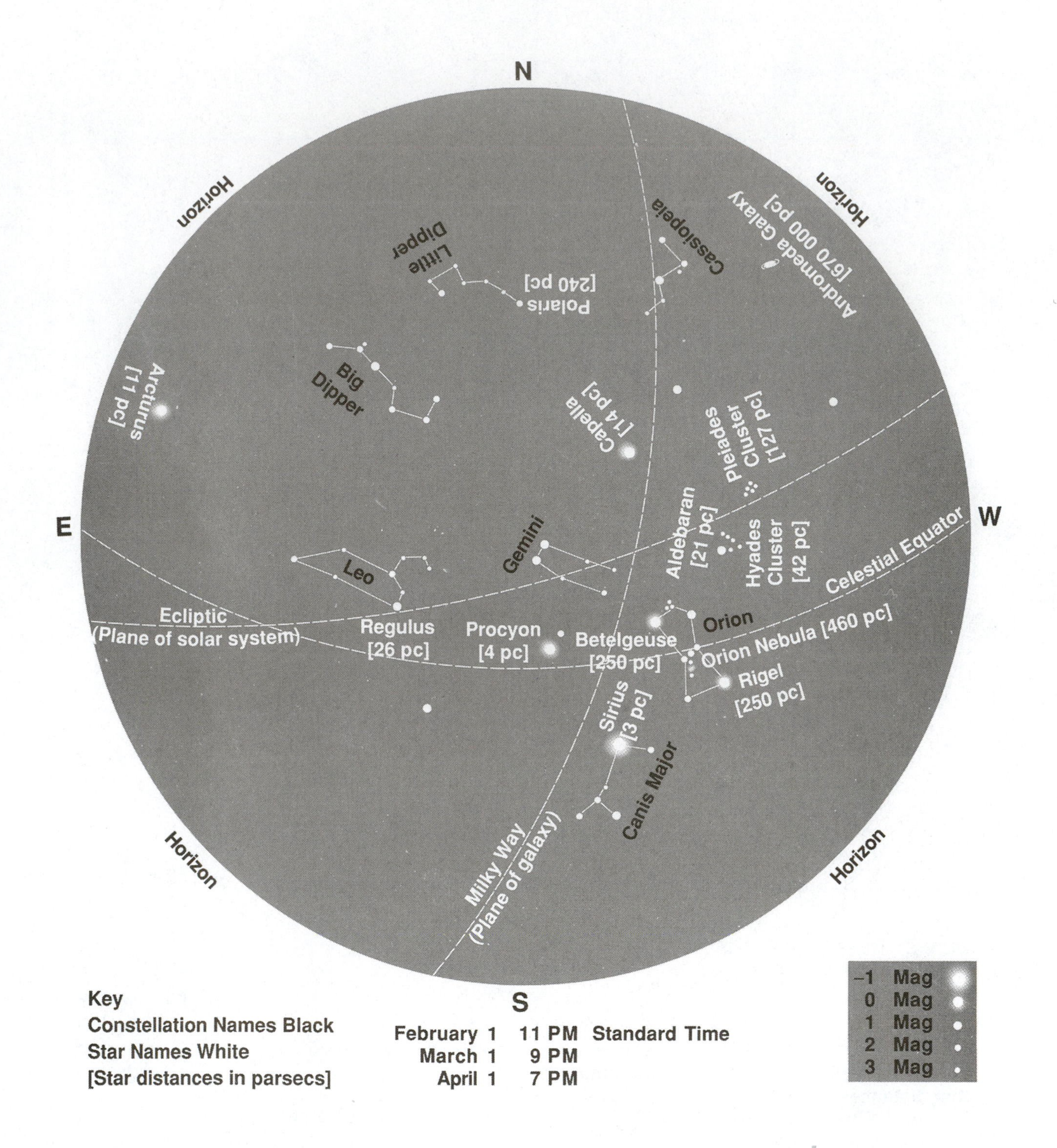

N
S
E
W
Horizon
Horizon
Horizon
Horizon
Little Dipper
Polaris [240 pc]
Cassiopeia
Andromeda Galaxy [670 000 pc]
Arcturus [11 pc]
Big Dipper
Capella [14 pc]
Pleiades Cluster [127 pc]
Aldebaran [21 pc]
Hyades Cluster [42 pc]
Celestial Equator
Gemini
Leo
Ecliptic (Plane of solar system)
Regulus [26 pc]
Procyon [4 pc]
Betelgeuse [250 pc]
Orion
Orion Nebula [460 pc]
Rigel [250 pc]
Sirius [3 pc]
Canis Major
Milky Way (Plane of galaxy)
Key
Constellation Names Black
Star Names White
[Star distances in parsecs]
February 1 11 PM Standard Time
March 1 9 PM
April 1 7 PM
–1 Mag
0 Mag
1 Mag
2 Mag
3 Mag

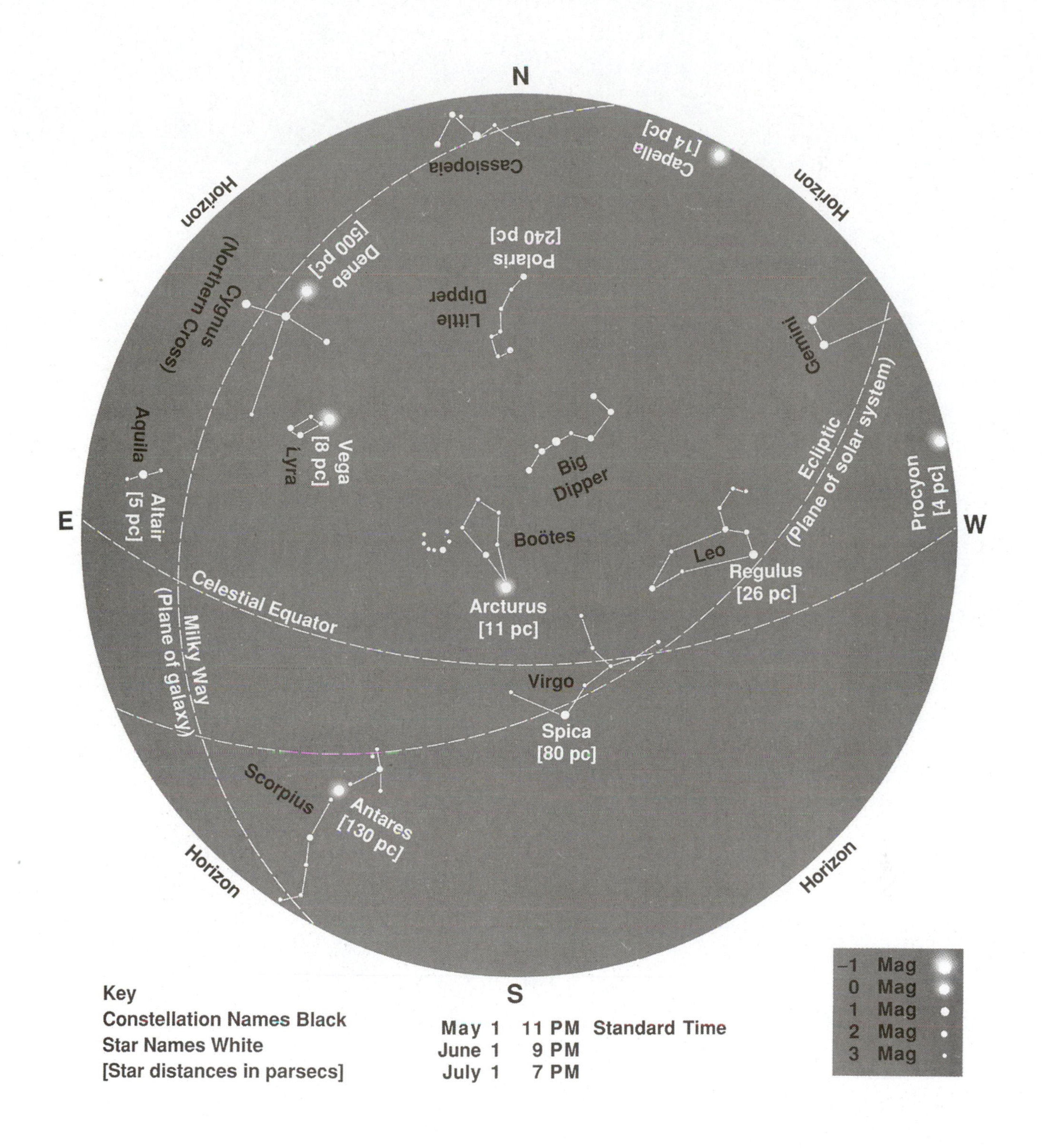
N
Cassiopeia
Capella
[14 pc]
Horizon
Horizon
Deneb
[500 pc]
Polaris
[240 pc]
Cygnus
(Northern Cross)
Little
Dipper
Gemini
Aquila
Vega
[8 pc]
Lyra
Big
Dipper
Ecliptic
(Plane of solar system)
Procyon
[4 pc]
E
Altair
[5 pc]
W
Boötes
Leo
Regulus
[26 pc]
Celestial Equator
Arcturus
[11 pc]
Milky Way
(Plane of galaxy)
Virgo
Spica
[80 pc]
Scorpius
Antares
[130 pc]
Horizon
Horizon
S
Key
Constellation Names Black
Star Names White
[Star distances in parsecs]
May 1 11 PM Standard Time
June 1 9 PM
July 1 7 PM
–1 Mag
0 Mag
1 Mag
2 Mag
3 Mag